INFORMATION SYSTEMS RESEARCH
Relevant Theory and Informed Practice

IFIP – The International Federation for Information Processing

IFIP was founded in 1960 under the auspices of UNESCO, following the First World Computer Congress held in Paris the previous year. An umbrella organization for societies working in information processing, IFIP's aim is two-fold: to support information processing within its member countries and to encourage technology transfer to developing nations. As its mission statement clearly states,

> *IFIP's mission is to be the leading, truly international, apolitical organization which encourages and assists in the development, exploitation and application of information technology for the benefit of all people.*

IFIP is a non-profitmaking organization, run almost solely by 2500 volunteers. It operates through a number of technical committees, which organize events and publications. IFIP's events range from an international congress to local seminars, but the most important are:

- The IFIP World Computer Congress, held every second year;
- Open conferences;
- Working conferences.

The flagship event is the IFIP World Computer Congress, at which both invited and contributed papers are presented. Contributed papers are rigorously refereed and the rejection rate is high.

As with the Congress, participation in the open conferences is open to all and papers may be invited or submitted. Again, submitted papers are stringently refereed.

The working conferences are structured differently. They are usually run by a working group and attendance is small and by invitation only. Their purpose is to create an atmosphere conducive to innovation and development. Refereeing is less rigorous and papers are subjected to extensive group discussion.

Publications arising from IFIP events vary. The papers presented at the IFIP World Computer Congress and at open conferences are published as conference proceedings, while the results of the working conferences are often published as collections of selected and edited papers.

Any national society whose primary activity is in information may apply to become a full member of IFIP, although full membership is restricted to one society per country. Full members are entitled to vote at the annual General Assembly, National societies preferring a less committed involvement may apply for associate or corresponding membership. Associate members enjoy the same benefits as full members, but without voting rights. Corresponding members are not represented in IFIP bodies. Affiliated membership is open to non-national societies, and individual and honorary membership schemes are also offered.

INFORMATION SYSTEMS RESEARCH

Relevant Theory and Informed Practice

IFIP TC8 / WG8.2 20TH Year Retrospective: Relevant Theory and Informed Practice—Looking Forward from a 20-Year Perspective on IS Research July 15–17, 2004, Manchester, United Kingdom

Edited by

Bonnie Kaplan
Yale University, USA

Duane P. Truex III
Florida International University, USA
Georgia State University, USA

David Wastell
University of Manchester, United Kingdom

A. Trevor Wood-Harper
University of Manchester, United Kingdom
University of South Australia, Australia

Janice I. DeGross
University of Minnesota, USA

KLUWER ACADEMIC PUBLISHERS
BOSTON / DORDRECHT / LONDON

Distributors for North, Central and South America:
Kluwer Academic Publishers
101 Philip Drive
Assinippi Park
Norwell, Massachusetts 02061 USA
Telephone (781) 871-6600
Fax (781) 681-9045
E-Mail <kluwer@wkap.com>

Distributors for all other countries:
Kluwer Academic Publishers Group
Post Office Box 322
3300 AH Dordrecht, THE NETHERLANDS
Telephone 31 78 6576 000
Fax 31 78 6576 254
E-Mail <services@wkap.nl>

Electronic Services <http://www.wkap.nl>

Library of Congress Cataloging-in-Publication Data

A C.I.P. Catalogue record for this book is available from the Library of Congress.

Information Systems Research: Relevant Theory and Informed Practice
Edited by Bonnie Kaplan, Duane P. Truex III, David Wastell, A. Trevor Wood-Harper and Janice I. DeGross
ISBN 978-1-4419-5474-9 e-ISBN 978-1-4020-8095-1
ISBN 1-4020-8095-6 (eBook)

Printed on acid-free paper.
Printed in United Kingdom by Biddles / IBT Global

CONTENTS

Part 4: Action Research

Part 5: Theoretical Perspectives in IS Research

Part 6: Systems Development: Methods, Politics, and Users

Part 7: Panels and Position Papers

FOREWORD

We are grateful for the support of the sponsoring and host organizations. Without their involvement, endorsement and financial support, this conference would not have been feasible. We would therefore like to extend our sincere thanks to the following organizations: the International Federation of Information Processing (IFIP), Technical Committee 8 of IFIP and WG 8.2 in particular, the School of Informatics at the University of Manchester, and Salford City Council. We provide a brief introduction to each of these organizations by way of providing historical context and information.

1 ABOUT IFIP

The International Federation for Information Processing (IFIP) was established in 1960. It is a multinational federation of professional and technical *organizations* (or national groupings of such organizations) concerned with information processing. In any one country, generally only one such organization—which must be representative of the national activities in the field of information processing—is admitted as a Full Member. On March 25, 2004, 47 countries were represented by Full Member organizations.

The Federation is governed by a General Assembly which meets once every year and consists of one representative from each Member organization. The Federation is organized into the IFIP Council, the Executive Board, and the Technical Assembly. The Technical Assembly is divided into 11 Technical Committees and two Specialist Groups. These committees and groups are in turn divided into Working Groups, of which IFIP WG 8.2 is one (under Technical Committee 8).

1.1 About IFIP Technical Committee Eight (TC8)

IFIP TC8 is the IFIP Technical Committee dedicated to the field of Information Systems. It was established in 1966, and aims to promote and encourage the advancement of research and practice of concepts, methods, techniques, and issues related to information systems in organizations.

The declared scope of TC 8 scope is the planning, analysis, design, construction, modification, implementation, utilization, evaluation, and management of information systems that use information technology to support and coordinate organizational activities including

- effective utilization of information technologies in organizational context

- interdependencies of information technologies and organizational structure, relationships and interaction
- evaluation and management of information systems
- analysis, design, construction, modification and implementation of computer-based information systems for organizations
- management of knowledge, information, and data in organizations
- information systems applications in organizations such as transaction processing, routine data processing, decision support, office support, com -puter-integrated manufacturing, expert support, executive support, and support for strategic advantage plus the coordination and interaction of such applications
- relevant research and practice from associated fields such as computer science, operations management, economics, organization theory, cognitive science, knowledge engineering, and systems theory

1.2 About IFIP Working Group 8.2: The Interaction of Information Systems and the Organization

The International Federation for Information Processing Working Group 8.2 (WG 8.2) was established by IFIP in 1977 as a working group concerned with "the interaction of information systems and the organization." WG 8.2 conducts working conferences, publishes books through IFIP, and publishes a semi-annual newsletter (OASIS). In addition, the working group maintains a listserv, a Web site and holds business meetings.

The aims of the working group are the investigation of the relationships and interactions among four major components: information systems, information technology, organizations, and society. The focus is on the interrelationships, not on the components themselves. Its scope is defined in terms of information systems, organizations, and society as follows:

- *Information systems*: includes information processing, the design of systems, organizational implementation and the economic ramifications of information.
- *Information technology*: includes technological changes such as microcomputers, distributed processing, and new methods of communications.
- *Organizations*: includes the social group, the individual, decision making and the design of organizational structures and processes.
- *Society*: includes the economic systems, society's institutions and values of professional groups.

1.3 How to Join WG 8.2

One can become involved in the working group as a correspondent, a friend or a member. If you would like to be placed on our mailing list, just write to our secretary (preferably by e-mail) and asked to be placed on our mailing list. You will receive newsletters and conference notices. The Web page also has information relating to forthcoming events and conferences. You can also subscribe to our listserv. (Visit the WG 8.2 Web site at http://www.ifipwg82.org/.)

You can become a friend of the group by attending one of our working conferences or business meetings. Working conferences are held about every 12 to 18 months. Business meetings are typically conducted twice yearly, once in conjunction with a working conference, and once in conjunction with the International Conference on Information Systems (ICIS).

Typically, a friend who has participated in two out of three consecutive business meetings is eligible for election as a member. By this election process, the members of the group nominate new members, who must then be confirmed by TC8.

2 THE SCHOOL OF INFORMATICS, UNIVERSITY OF MANCHESTER

Formerly the School was known as the Department of Computation, at the University of Manchester Institute of Science and Technology (UMIST). This summer is marked by a major event in the world of academe in the North West of England—namely, the merger of two illustrious universities based in Manchester: the Victoria University of Manchester (VUMAN) and UMIST. The new institution will be known as the University of Manchester. Two IT-related departments are part of this new milieu: Computer Science (from VUMAN) and Computation (UMIST). The latter will be renamed the School of Informatics and will concentrate on the development of its historical strengths on the applied side of the discipline of Computing. For this reason, it will take up a position in the Humanities Faculty of the new university, alongside Business, Education, Accounting, and other cognate disciplines. Information Systems will be a powerful force within this new Faculty, bringing together a large cadre of well-known scholars within the IS discipline, many of whom have played, and continue to play, a prominent part in the work of WG8.2.

3 SALFORD CITY COUNCIL

The City of Salford is one of the various independent municipalities that make up the conurbation of Greater Manchester, lying on the north west side of the conurbation. There is a long tradition of collaboration between the City and local universities in various areas of research, especially regarding information systems and the application of IT. For nearly 10 years, there has been a particularly close relationship between the IT Services department at the City, and researchers at Salford University, Manchester Business School, and UMIST. Action research on eGovernment has been a strong feature of this collaboration, culminating in national recognition for the City as a pathfinder authority in this field and the establishment of a CRM Academy at Manchester Business School, in partnership with the School of Informatics at UMIST.

PREFACE

This volume includes the papers and panel descriptions refereed for presentation at an International Federation of Information Processing (IFIP) Working Group 8.2 conference entitled "Relevant Theory and Informed Practice: Looking Forward from a 20 Year Perspective on IS Research." The conference was held at the University of Manchester in Manchester, England, on July 15-17, 2004.

It was during the working group business meeting following the IFIP WG 8.2 working conference in 2001 on "Realigning Research and Practice in Information Systems Development" in Boise, Idaho, that a conference call was approved for a new conference dealing with the alignment of research practice and IS development. Those who proposed the conference had been involved with WG 8.2 and other TC 8 working groups over many years. The initial incentive for developing the theme of the conference dates back to 1997 at the Philadelphia conference. It was observed that while some were celebrating the end of the "methods wars" because some types of qualitative work had become acceptable for publication in mainstream journals, the work by others who were exploring questions outside the managerial, organizational, or technological mainstream, or who were employing innovative research approaches, was still being excluded from the discourse. Thus it was concluded that further attention to the question of research approaches was required.

In a real sense, however, the seeds of this conference were sown in 1984 at the first Manchester conference, when in the proceedings introduction Enid Mumford made a declaration that continues to express a concern of the working group.

> The members of the W.G. 8.2 are dedicated to stimulating and maintaining a debate on the interrelationships between information systems, organizations and society; and to influencing IFIP members, and information scientists, teachers, trade unions, and the user of information systems, to think carefully about the organizational and societal consequences of the systems they are developing and using.
>
> One of our areas of interest is research methodology and we have been looking critically at the kinds of research associated up to now with information science, and discussing the need for new approaches.
>
> Our concern that traditional research methods can not adequately investigate social needs and problems...

So, as part of the working group's living tradition, wherein about every six years it makes an assessment of IS research methods, the call for this conference went out. And how the community responded!

1 THE REVIEW PROCESS

The papers in this volume are those that survived a rigorous review and selection process. We were gratified, at times even overwhelmed, by the record number of 113 submissions we received for the conference, with papers in a number of categories: full research papers, practice-oriented papers, and panels. Three of the papers were by authors specifically invited to provide a more panoramic view of the working group's progress and of IS research methods in general. In keeping with WG 8.2 standards, all papers (including the invited papers) were subjected to a rigorous reviewing process involving a minimum of four independent readings, with some manuscripts having double that number. An associate editor (AE) assigned to each paper solicited two independent, blind reviews. The AE reports and recommendations were then considered by the four program cochairs at a meeting in Atlanta in late November 2003, with each cochair taking lead responsibility for an equal allocation of roughly 28 papers. Any borderline or contentious cases were referred to one or more fellow cochairs, and discussed by the group as a whole. The usual care was taken to avoid conflicts of interest in the assignment of associate editors and reviewers. In a few instances, papers were sent out for yet additional review consideration by the general chairs or others with the appropriate topical or methodological expertise.

The whole review process was underpinned and orchestrated using the AIS-ICIS conference software, enabling this complex process to operate smoothly despite severe time-space problems. The Web system allowed authors to submit papers electronically, associate editors and reviewers to be assigned by the cochairs, reviews and reports to be garnered and evaluated, decisions and recommendations to be made and recorded, and accept/reject letters to be dispatched. Although there were moments when we felt (with an acute sense of irony) in the midst of yet another IS failure, in the end all worked very well, and it is doubtful whether a manual system could have supported the process without considerably more blood, sweat, and tears.

We are deeply appreciative of the efforts made by the WG 8.2 community to see through the review process to successful conclusion, and the associate editors, in particular, who were the lynchpins of the whole process. Despite the exacting deadlines (little over a month was available), virtually all of the reviews and reports were received by the time of our Atlanta cochairs congress, enabling us to focus on our main tasks of making the final selection of papers and drawing up the preliminary program structure.

We were gratified that 8.2's reputation for rigorous review produced high quality submissions. Of the 113 submissions received, we accepted 33 full research papers (representing an acceptance rate of 29 percent for full papers) and 6 panel proposals. Outside of this favored selection, there were many other interesting and valuable submissions that we felt could make a very useful contribution to the conference. Rather than limit this conference to a predetermined acceptance rate, we were anxious to include these as well, in the interests of both building the community and stimulating lively debate. Accordingly, the cochairs decided at their Atlanta meeting to create a new category of position papers, referred to more colloquially as "bright ideas." Authors were invited to submit a 2000 word précis, summarizing their main points in a pithy and provocative fashion; 11 such pieces are featured in the final program.

Considering the final count for all papers, both full (33), position paper (11), and panels (6) yields a more egalitarian overall acceptance rate of 44 percent. This compares interestingly with the three previous research method conferences. According to our best records, the initial gathering from September 1-3, 1984, (called a *colloquium* because of sensitivities within IFIP itself, as we are informed by those who were present) had 18 papers and only 44 non-presenting participants. Virtually all papers submitted were presented and published, nearly a 100 percent acceptance rate. It seems the community of radicals in our then-new discipline were few in number. By 1990 and the Copenhagen conference, a total of 23 papers (including 4 invited papers and 2 panels) were presented and published. Since there were 59 papers submitted, the overall acceptance rate had fallen to 42 percent. Interestingly, the conference itself was among the most heavily attended in the working group's history. There were nearly 200 participants, attesting to the interest in the topic. The 1997 Philadelphia conference had 28 pieces, including 2 panels and 2 invited papers, from a field of roughly 60 submissions, yielding approximately a 45 percent acceptance rate.

So, for the present conference, it is with some confidence that we can claim that the quality of the reviewing process, the useful and thoughtful reviews received by authors for both accepted and rejected works, and the rigor of that process, are in keeping with the best traditions of the working group. Moreover, we are told by members of the publishing community that the WG 8.2 conferences hold a much higher standard for acceptance than is the norm for other working group conferences. We are proud to maintain that standard. And for the record, the working group itself now numbers more than 200 acknowledged members, 453 friends, plus correspondents and others who participate via our books and our newsletter, OASIS. Perhaps the cost of being a revolutionary is no longer so high!

2 THANKS TO OUR SPONSORS

The organization of such an IFIP working conference can be a daunting process. For starters, given the bylaws and procedures of the parent organization, the working group itself is not allowed to raise funds and maintain accounts, except for very small balances to fund direct operating activities such as publishing newsletters. IFIP working groups are voluntary organizations without paid staff or deep financial pockets. Accordingly, each conference is treated as a relatively independent fiscal entity wherein the conference organizers and book editors are taking on all of the responsibility to fund and manage the event. This insulates IFIP and WG 8.2 and scares the devil out of the organizers. Sponsorship in various forms is both essential and greatly appreciated. In the case of the Manchester 2004 event, our gratitude and thanks go to the following organizations for financial and material support to the conference:

- The School of Informatics, University of Manchester—£10,000 plus moral and administrative assistance.
- The IT Services, Salford City Council—£3,000 that provided the initial seed money to the conference.
- Georgia State University Department of Computer Information Systems for housing members of the conference committee and for hosting the November 2003 meeting of the conference chairs.
- The AIS for providing the use of the on-line reviewing system. Special thanks to Eph McLean and Samantha Spears at the AIS for their assistance.
- The home institutions of the conference chairs for funding the needed travel, administrative support, and indirect costs of managing the conference. Thank you the to Florida International University's Chapman Graduate School of the College of Business Administration, and The School of Informatics at the University of Manchester.

3 ACCOLADES AND ACKNOWLEDGMENTS

It is customary to thank everyone from our families and teachers to our colleagues and support staffs, parakeets, and other ways of maintaining sanity under pressure. Customary as it may be, we none-the-less truly are grateful. We heartily thank Janice DeGross for her much-needed acerbic dose of keeping us on schedule and her superb work making all the authors' various, and often times tardy, contributions into a book. Thanks also to the local organizing committee led by Peter Kawalek of the University of Manchester for aggressive negotiations and sensitive choice of venues for all of the conference events. Also, we acknowledge Kath Howell at the University of Manchester for stepping in to tidy up a host of problems and last minute crises. We doff our hats to the 8.2 officers, Julie Kendall, Michael Myers, and Nancy Russo, WG 8.2 Webmaster Kevin Crowston, and the Honorary Conference Chairs, Richard Baskerville and Frank Land, for their supportive advice on sensitive matters. Rod Padilla, technical support manager at Georgia State University, kept the reviewing site running at the most critical times and customized it for our use.

Thanks, finally, to the conference chairs of the previous research methods conferences and to the past WG 8.2 chairs. We asked each of them to provide personal remembrances of the research conferences and the history of the working group to help us in framing our own remarks and continuing narrative of the working group's activities, for we see this event as part of a living and evolving intellectual and social history. It is a history in which we are grateful to have had the chance to play a part.

We are most thankful for each other, for our ability to work together, to complement each other's strengths and tolerate one another's foibles. But finally, of course, we really are most thankful to you, the members of the IFIP Working Group 8.2 community, for bringing this all about.

Bonnie Kaplan
Duane Truex
Dave Wastell
Trevor Wood-Harper

CONFERENCE CHAIRS

General Chairs
Richard Baskerville
Georgia State University

Frank Land
London School of Economics

Program Chairs
Bonnie Kaplan
Yale University

Duane P. Truex, III
Florida International University and
Georgia State University

David Wastell
University of Manchester

A. Trevor Wood-Harper
University of Manchester and
University of South Australia

Organizing Chairs
Peter Kawalek
University of Manchester

Bob Wood
University of Manchester

ASSOCIATE EDITORS

REVIEWERS

Pekka Abrahamsson, VTT Electronics - Embedded Software, Finland
Frederic Adam, University College Cork, Ireland
Alison Adam, University of Salford, UK
Chandra Amaravadi, Western Illinois University, USA
Kim Viborg Andersen, Copenhagen Business School, Denmark
Bryant Antony, Leeds Metropolitan University, School of Management, UK
Doug Atkinson, Curtin University of Technology,
David Avison, Essec, France
Michel Avital, Case Western Reserve University, USA
Lars Baekgaard, Aalborg University, Denmark
Jørgen Bansler, Technical University of Denmark, Denmark
Stuart Barnes, Wellington University, New Zealand
Frances Bell, University of Salford, UK
Niels Bjørn-Anderson, Copenhagen Business School, Denmark
Ahmed Bounfour, University of Marne La Vallee, France
Laurence Brooks, Brunel University, UK
Christopher Bull, Manchester Metropolitan University, UK
Tom Butler, University College, Cork, Ireland
Bong-Sug Chae, Kansas State University , USA
Mike Chiasson, University of Calgary, Canada
Melissa Cole, Brunel, UK
Fred Collopy, Case Western Reserve University, USA
Stephen Corea, Warwick Business School, UK
Joe Cunningham, University College, Cork, Ireland
Wendy Currie, DISC, UK
Christopher Davis, University of South Florida St Petersburg, USA
Bill Doolin, Auckland University of Technology, New Zealand
Kristin Eschenfelder, University of Wisconsin-Madison, USA
Frances Fabian, A.B. Freeman School of Business, USA
Walter Fernandez, Australian National University, Australia
Per Flensburg, Växjö University, Sweden
Uri Gal, Case Western Reserve University, USA
Susan Gasson, Drexel University, UK
Matt Germonprez, Case Western Reserve University, USA
Ake Gronlund, Orebro University, Sweden
Arvind Gudi, Florida International University, USA
Noriko Hara, Indiana School of Library and Information Science, USA
Erling Havn, Technical University of Denmark, CTI, Denmark
Karin Hedstrom, University of Orebro, Sweden
Jukka Heikkila, University of Jyvaskyla, Finland
Ola Henfridsson, Viktoria Institute, Sweden
Helle Zinner Henriksen, Copenhagen Business School, Denmark, Denmark
David Hinds, Florida International University, USA

Lionel Honoré, Université de Nantes, France
Gordon Hunter, The University of Lethbridge, Canada
Julio Ibarra, Florida International University, USA
Pertti Järvinen, University of Tampere, Finland
Nimal Jayaratna, Curtin University, Australia
Katrin Jonsson, Umea University, Sweden
Sten Jönsson, Gothenburg School of Economics, Sweden
Michelle Kaarst-Brown, Syracuse School of Information Studies, USA
Jannis Kallinikos, London School of Economics, UK
Eija Karsten, University of Turku, Finland
Seamus Kelly, University College Dublin, Ireland
Gaye Kiely, University College, Cork, Ireland
Heinz Klein, Temple University, USA
Ralf Klischewski, University of Hamburg, Germany
Ned Kock, Texas A&M University , USA
Lynette Kvasny, Penn State University, USA
Jeannie Ledington, University of Canberra, Australia
Jonathan Liebenau, London School of Economics and Political Science, UK
Angela Lin, University of Sheffield, UK
Rikard Lindgren, Viktoria Institute, Sweden
Jan Ljungberg, University of Gothenburg, Sweden
Cheri Long, Florida International University, USA
Jessica Luo, Case Western Reserve University, USA
Magnus Mähring, Stockholm School of Economics, Sweden
Angela Mattia, Virginia Commonwealth University, USA
Donald McDermid, Edith Cowan University, Australia
Tom McMaster, University of Salford, UK
Emmanuel Monod, University de Nantes, France
Ramiro Montealegre, University of Colarado, USA
Eric Monteiro, Norwegian University of Science and Technology, Norway
Enid Mumford, University of Manchester, UK
Bjorn Munkvold, Agder University College, Norway
Lisa Murphy, University of Alabama IS Group, USA
Alistair Mutch, Nottingham Trent University, UK
Karen Neville, University College, Cork, Ireland
Petter Nielsen, University of Oslo, Norway
Sue Nielsen, Griffith University, Australia
Jacob Nørbjerg, Copenhagen Business School, Denmark
Torbjörn Nordström, Umea University, Sweden
Philip O'Reilly, University College, Cork, Ireland
Niki Panteli, University of Bath, UK
Graham Pervan, Curtin University of Technology, Australia
Athanasia Pouloudi, Athens University of Economics and Business, Greece
Philip Powell, University of Bath, UK
Marlei Pozzebon, HEC Montreal, Canada
Sandeep Purao, Pennsylvania State University, USA

Jeria Quesenberry, The Pennsylvania State University, USA
Julie Rennecker, Case Western Reserve University, USA
Helen Richardson, University of Salford, UK
Suzanne Rivard, HEC Montréal, Canada
Knut Rolland, Norwegian University of Science and Technology, Norway
Duska Rosenberg, Royal Holloway University of London, UK
Matti Rossi, Helsinki School of Economics, Finland
Bruce Rowlands, Griffith University, Australia
Sundeep Sahay, University of Oslo, Norway
David Sammon, University College, Cork, Ireland
Harry Scarborough, Warwick Business School, UK
Rens Scheepers, University of Melbourne, Australia
Ulrike Schultze, Southern Methodist University, USA
Gamila Shoib, School of Management, United Kingdom
Silvia Silas, Florida International University, USA
Rahul Singh, ISOM, USA
Mikael Söderström, Umeä University, Sweden
Carsten Sorensen, London School of Economics and Political Science, UK
Valerie Spitler, University of North Florida, USA
Jan Stage, Aalborg University, Denmark
Andrea Hoplight Tapia, Pennsylvania State University, USA
David Targett, Imperial College London, UK
Mark Thompson, Cambridge University , UK
Virpi Tuunainen, Helsinki School of Economics, Finland
Werner Ulrich, University of Fribourg, Switzerland
Betty Vandenbosch, Case Western Reserve University, USA
Richard Vidgen, University of Bath , UK
Erica Wagner, Cornell University, USA
David Wainwright, University of Northumbria, UK
Jonathan Wareham, Georgia State University, USA
Ulrika Westergren, Umeä University, Sweden
Chris Westrup, University of Manchester, UK
Mikael Wiberg, Umeä University, Sweden
Francis Wilson, University of Salford, UK
Youngjin Yoo, Case Western Reserve University, USA

1 YOUNG TURKS, OLD GUARDSMEN, AND THE CONUNDRUM OF THE BROKEN MOLD: A Progress Report on Twenty Years of Information Systems Research

Bonnie Kaplan
Yale University

Duane P. Truex III
Florida International University and Georgia State University

David Wastell
School of Informatics University of Manchester

A. Trevor Wood-Harper
School of Informatics, University of Manchester, and School of Accounting and Information Systems, University of South Australia

1 INTRODUCTION

It is now 20 years since the first Manchester conference on information systems research methodology. Since that auspicious gathering, the reputation, the reach, and the impact of the IFIP WG 8.2 scholarly community has extended into mainstream IS journals, conferences, and academic institutions world-wide. Twenty working conferences in eight nations have published almost 400 papers covering all manner of social theories and IS topics. The first gathering had bold and radical ambitions. Its provocative title, "Information Systems Research—A Doubtful Science?," challenged the prevailing orthodoxy that the research methods of "normal science" should be the only methods defining proper research in our field. The gauntlet was thrown and the research methods theme, albeit with evolving twists, was revisited again at the 1990 Copenhagen IFIP 8.2 Working Conference, and again in 1997 in Philadelphia. The Copenhagen gathering was organized around the assumption that radically different research approaches existed in the IS research community; the call was made for methodological pluralism and the debate extended around the philosophical traditions grounding IS research methods.

The Philadelphia conference organizers acknowledged the battle for recognition fought in the previous decade by IS researchers deploying nontraditional research methods, and that these approaches were finally being recognized. They argued that mature disciplines allowed both qualitative and quantitative research traditions to co-exist. In contrast to the more tentative position of the earlier conclaves, the Philadelphia conference invited authors to "assertively and unapologetically" apply and refine qualitative research approaches. Where the Copenhagen conference called for methodological pluralism, Philadelphia dealt more with the diversity in approaches within the qualitative research community. Each of the conferences took into account the role of previous research methods conferences in shaping the debate, and each was a product of the larger disciplinary discourse about research and the evolution of the discipline.

Following in this tradition, and 20 years after the first WG 8.2 research methods conference, we thought it timely to look back and take stock of the working group's impact upon the practice of information systems development (ISD) and use in organizations and other social contexts. This conference is the result. We variously wondered: How relevant has our work really been? To what extent have we made an impact on IS practice? To what degree have our theories been enhanced by drawing on practice? Has the positivist mold really been broken or is the victory pyrrhic? How has the past informed our developing and future research approaches?

We invited researchers and practitioners, both members of the now "old guard" and the new "young Turks," to continue WG 8.2's tradition of debating method, exploring the relevance of our research, and examining interrelationships between information systems, organization, and society. We solicited both empirical and theoretical papers that examined or empirically used IS research methods. The return of the conference to its geographical origin provided an auspicious opportunity both to celebrate the iconoclastic idealism of its forebears and to take stock of how the discourse on research questions, themes, and methods has evolved in the interim. We were delighted with the response to the Call for Papers, with a record number of submissions being received. As noted in the Preface, a final selection of 33 full research papers was made, together with 6 panels and a novel category of short position papers.

The remainder of this chapter provides an overview of these various submissions, beginning with the full research papers. These have been clustered into a number of themes, which are reflected in the broad structure of the conference program. The first section contains the three invited papers, which attempt in their various ways to outline a set a panoramic views of our field, challenging the community to look critically at our prescriptions, practices, and rhetoric. The further groupings of papers reflect a diverse response to the various imperatives of this agenda. Section 2 contains papers addressing foundational issues bearing on the core identity of the field itself, whereas those in the subsequent section exemplify the critical interpretive tradition that has come to flourish within the field, emblematizing (perhaps) our success in challenging the research orthodoxy of two decades ago. This is followed in section 4 by a clutch of papers on another alternative to the paradigm of normal science, namely action research, which has gained a strong following within the field. A concern with the use of social theory within IS research is the characteristic that unites the papers in section 5; these papers emphasize new developments in theory, either of an integrative nature or regarding the potential of some relatively unused, but possibly fruitful, new approach. The papers in section 6 focus on the second sense of *method* which preoccupies us within our field (i.e., the process for building systems rather than for conducting research). Methodological issues regarding information systems development, and the problems that beset IS projects in real organizational contexts, reflect the linked concerns of these papers. Following this overview of the research papers, we give a brief overview of the panels and position papers, then move on in our final remarks to the adumbration of cross-cutting themes and issues that emerge when reflecting on the conference content as a whole.

2 OVERVIEW OF THE PAPERS AND OTHER SUBMISSIONS

2.1 Panoramas

The invited papers open the book, laying out panoramic views and consequent challenges of the field. In the opening paper, Lee makes an attempt to collectively hoist us by our own petard. The expression means to be thrown up in the air by an explosive charge (the petard) placed under castle walls to gain access. The WG 8.2 community has been storming the castle gates since its earliest days, with its methods conferences at the vanguard of the campaign. Lee's paper challenges us to apply some of our own prescriptive medicine to the conduct of our research. Using three articles from the previous WG 8.2 methods conferences, Lee uses the rhetorical ploy of replacing key phrases such as "information systems development" with "IS research" or "research on the practice of research" to test the fit of our own prescriptions on our process of making research. Lee imagines a gathering of influential IS researchers to design "helpful interventions into our own research community," employing our prescriptions for others to our own process of research production and community building. The gathering would examine the process by which we, through our journal editing activities and tenure and promotion decisions, come to decide that which is enduring and of high quality. Lee challenges us to turn our lenses of analyses on our own community of practice.

In the second paper, Sawyer and Crowston survey all of the previous WG 8.2 proceedings. They identify six characteristic research themes: (1) an orientation toward social theories, (2) dominant conceptualizations of information and technology and a common level of analysis, (3) an orientation toward the use of intensive research methods, (4) use of critical and analytical perspectives in research, (5) an openness to a range of research settings, and (6) an open discourse on the study of IS in organizations and society. They argue that WG 8.2 has not been as influential as it might have been in shaping the wider disciplinary discourse, and offer ideas how it might be more influential in the future. They see two opportunities to capitalize on current strengths and past traditions: by better conceptualizing information and communication technologies (ICT) and by leading in further developing socio-technical theories of IS. They illustrate how future 8.2 scholars may resolve dualistic tendencies and understand ICTs as simultaneously social and technological.

In the third paper, Boland and Lyytinen recall how previous WG 8.2 methods conferences were "characterized by our fervent struggles to define the *correct way of doing research,*" lamenting that despite the progress made, "we now find ourselves in another quagmire, rooted in a questioning of our identity." Boland and Lyytinen rise to Lee's challenge in their analysis of the Group's identity "as found in our theories, our method and our reflexive practices," and also to the concerns of Sawyer and Crowston in suggesting the design of a better interface of the socio-technical world we as IS scholars inhabit. Boland and Lyytinen argue against the current and misguided predilection of defining the core of our field by what objects we study and how we align ourselves with "the things that *should be* part of our identity." They propose instead that we consider the process through which we construct a common identity, and that we consider researchers as *designers* of a process and of a shared identity. To illustrate the point, they apply a kind of structurational analysis reflexively to their own research.

2.2 Reflections on the IS Discipline

This section picks up themes of reflexivity, challenge, and identity. These papers are distinguished by discussion of crises and dichotomies within IS as a discipline. Ramage begins by challenging us with profound questions of IS identity and institutional acceptance in an engagingly playful way. His analysis provides, in the words of Marcon et al., "an altered frame of mind" as he argues for a model of information systems in terms of cyborgs, a metaphor for the human-technical mix of our times, and a description of situations where the social and the technical merge and blur. IS, he reminds us, is inevitably interdisciplinary, straddling the divide between social and technical perspectives. Ramage embraces the inherent nature of IS as cyborg formed by an unholy fusion of many disciplines, facing continual struggle for self-identity and legitimization: "to live as a cyborg is not to be comfortable, it is to be challenging and challenged." His discussion suggests an explanation for the continual IS crises: Cyborgs are threatening. Whether individual researchers or entire disciplines, cyborgs break societal norms. The very existence of the cyborg breaks down power achieved and maintained through categorization and dichotomization. Cyborgs are seen as double-headed monsters, to be persecuted or rendered invisible.

Rowe, Truex, and Kvasny also address the constant challenge to the credibility of IS as an academic field, calling for an end to the cognitive legitimacy crisis on political, economic, and scholarly grounds. Their concern is with establishing clear boundaries between our research concerns and those of others, so as to "mark a territory that is uniquely our own." They argue that our focus on evaluation and control are the distinctive characteristics of IS that differentiate us from other fields. They ask what we mean by an *IT-enabled solution* and even by *information system* while noting that *ontology*, too, has multiple, and therefore unclear, definitions in our field. They invoke three French sociologists (Crozier, Bourdieu, and Latour) to focus debate on the ontological grounding of our field, thus highlighting the contribution of French scholars and encouraging us to continue to explore the relevance of their work.

Ramage pushes us to reject either/or language in favor of both/and, thus avoiding entrapment in a single understanding that works only for a particular time. The same might be said for other dichotomies within IS, including rigor/relevance, qualitative/quantitative, and theory/practice. Introna and Whittaker look at these divides, raising the question of power while tracing the shift away from practice in *MIS Quarterly*. Taking a Foucauldian stance, they analyze editorial statements and other claims to tackle the important question of what the politics of truth means for research and publishing in academic journals. Their analysis traces growing tension between research and practice, or relevance and rigor, as *MIS Quarterly* pursued a policy of raising its stature as an academic journal. Methodological emphasis also shifted, with editorial calls for positivistic and theory-driven papers followed by attempts to make the journal more pluralistic. Under the current leadership, *MIS Quarterly* claims to have moved beyond "methods wars" and dismissed methodology as a problem. Nevertheless, the strong emphasis on empirical work makes it difficult for critical and speculative papers to get an audience in *MIS Quarterly*. Consequently, Introna and Whittaker argue that journal publication is more an indication of compliance with a journal's regime of truth than an indication of quality, and urge a questioning attitude toward attempts to institutionalize rankings of journals.

Jones also rises to the role of intellectual inquisitor by examining publication practice in the light of publication theory. He compares the *doxa* of a generally accepted set of required elements for good IS research against what actually is done in practice. He tests a set of papers judged to be best of breed in our field, finding that we do not practice what we preach. He posits four accepted principles: that good research should follow the scientific method, should fulfill certain criteria, should be relevant, and should employ multiple methods. His analysis of the best papers from the International Conference on Information Systems and *MIS Quarterly* either illustrates a kind of "do as we say and not as we do" hypocrisy in both the positivistic and the interpretive communities, or that something more complicated constitutes good research. This suggests that methodological checklists "may be more likely to encourage ritualistic adherence than improved [research] practice." Thus Jones invites us to conduct a kind of methodological reality check much like that envisioned by Lee.

The remaining authors in this section attempt to resolve splits within the discipline. Marcon, Chiasson and Gopal also attempt to push us toward an altered frame of mind. Like Boland and Lyytinen, they address questions of disciplinary core and identity, offering a rethinking to address the crisis of relevance we now face, thus turning a problem into an opportunity for renewal. They argue for reclaiming wider meanings of *critique* from the way the term commonly is understood in IS, either as its deployment in *critical* social theory or as in methodological *critique*. In exploring the connection between critique and crisis, they point out how critique can be a critical turning point. They advocate engendering crisis as way to move the field toward a holistic integration of research, teaching, and consulting.

Bell and Adam also are concerned with divisions, specifically that between ethics and IS. They see this as problematic both for education and practice. Ethics is taught (when it is taught) as separate from ISD. Both students and practitioners have trouble applying ethical codes or ISD methodologies to messy real-life situations. Bell and Adam find fault with rationalist rule-based decision models and the prominence of quantitative studies. Their discussion of ISD methodologies implicitly returns to IS education. Educators could use case studies that explicitly incorporate ethics into system development and rich descriptions of how ethical dilemmas actually are handled. They look forward to a body of qualitative research in this area as a useful resource for educators, transcending the limited view of ethical reasoning provided by quantitative research.

Purao and Truex also address the issue of practice and relevance, this time seen through another set of divergent paths in IS. They contrast software engineering's focus on creating information technology artifacts with research concerning organizational impact and change. In an attempt to remedy the lack of impact on the practice of research in either area, Purao and Truex combine insights from software engineering and social theories. They attempt to integrate the two streams by shifting from traditional development practices to a continuous redevelopment process. Drawing on emergent systems development, they propose a set of requirements for new representation techniques to take account of both the engineering of the IT artifact as well as the emergent nature of organizational context in which the IT artifact will be deployed. They call upon the WG 8.2 community to take up the challenge of reconciling these important and complex questions. They also sound a warning cry, that in having two research communities on parallel paths with little cross fertilization or mutual awareness, we miss the opportunity for great intellectual synergy and waste a great deal of creative energy in duplicated effort.

2.3 Critical Interpretive Studies

Papers in this section all consider different aspects of critical social theory (CST) in its relation to IS, both its use in our discipline or in more generic issues of interpretive methodology. CST is operationalized by anchoring it to various interpretive research traditions and techniques, or by positing linkages with other methods and theories.

The paper by Howcroft and Trauth considers the critical theoretic tradition in IS research by reflexively applying a critical lens to critical IS research. The authors find "little in the critical literature that differentiates critical IS research from other critical arenas." While avoiding issues of research technique, on the grounds that such questions "can detract from the more central problem of how we chose to interpret and represent social reality," they argue that a researcher should be concerned with a trio of tasks (namely, insight, critique, and transformative redefinition) when conducting critical IS research.

Greenhill's paper continues in a reflexive bent. It a self-conscious account of the process of conducting a detailed case study while employing thick description. The paper illustrates this with a description of the description and of the process. The author's goal was to contextualize the research by focusing on the method and the process of the research itself.

In a paper dealing with integrating CST and method, Cukier, Bauer, and Middleton offer an approach to operationalizing Habermas's validity claims in critical discourse analysis. They illustrate this in a technology-enabled learning case study, analyzing distortions of communication in the discourse of adoption and use of this system.

Klecuń brings a new wrinkle to the growing endeavor to integrate actor-network theory with other social theories. Like Marcon, Chiasson, and Gopal, she looks outside the traditional writing on critical social theory, the perspective of the Frankfurt School. She makes a more difficult leap toward a critical postmodernism through the late work of Foucault, hinting at how to scope a research program using the related and relative strength of each approach.

The paper by Pozzebon responds to the invitation for authors in non-hermeneutic traditions of interpretivism to suggest other criteria for research quality than those made by Klein and Myers in their influential *MIS Quarterly* article. Pozzebon enriches the discourse on the critical interpretative perspective by exploring the link between interpretation and CST. She propounds four criteria for critical interpretative research: authenticity, plausibility, criticality, and reflexivity.

Focusing on interpretive work as a class (case research in particular), Barrett and Walsham address the basic question of the nature of a research *contribution* and then proceed to illustrate how IS research employing interpretive case studies can contribute to the advance of knowledge in our field. Their approach is interesting because they examine research contributions as statements in a network of ideas. Following Latour's second rule of method—that is, to examine the transformations of statements—they situate the particular case research findings within that network. Thus a contribution can be judged by its impact on the network and upon the types of transformations it undergoes as the network develops. They study the process of network construction via an examination of how others cite a familiar and well-cited interpretive case study. Barrett and Walsham's description of their own process represents an intriguing response to Lee's opening challenge. It embodies a sort of doubly reflexive approach to research in which the goal (i.e., the contribution) of the research is examined by reflecting on how that contribution has itself been incorporated into the larger contextual network.

2.4 Action Research

Directly addressing the need for real-world relevance, action research has been postulated as the method for researchers to "rub" theory with practice. There has been an increase in its use in Europe and Australasia, but not significantly in North America and Asia. This has been due to different emphases on the nature of the main role of the researcher in the change process. One sees the academic as mainly acting as a social scientist in order to understand and explain practice. The other depicts the academic as a joint collaborator acting both as a researcher and also as a practitioner to improve practical outcomes in the organizational situation. In this section, the three papers explicitly or implicitly outline the role of the researcher, and attempt to answer some of the critiques of action research.

The first paper regards the researcher as a collaborator, with Oates reflecting on the move of social science from a *linguist turn* to an *action turn*. The author then defines a newer form of action research that places less emphasis on contribution to theoretical knowledge, and stresses participation and individual personal growth in organizations. This newer version of action research is then contrasted with more traditional forms in the literature. The paper presents a confessional account of a study and outlines five quality issues raised by this type of action research: relational praxis, reflexive-practical outcome, plurality of knowing, significant work, and new and enduring consequences or infrastructure. Because of the recent interest in action-based methods, the discussion about how to address these issues is important for the IS research community.

Germonprez and Mathiassen appreciate the researcher's role as a collaborator but want to bring more rigor associated with social science to the action-based research process. They explore the roles of conventional research methods that could contribute to the use of action research in multi-method approaches and the means by which they facilitate the creation of multi-contribution projects. The authors suggest and outline two approaches for the integration of action research. The first is a planned strategy where the main method is action research supplemented with other methods. The second is an emergent one in which more conventional methods are employed initially, and action research is then used to understand and explain the ongoing results that unfold in many projects. Finally, the authors argue that both combinations can decrease significantly the risk a researcher takes in using action research. If these approaches were adopted, action research would be more attractive for conventional IS researchers and doctoral students.

The final paper in this section, by Holwell, is based on work using Checkland's soft systems methodology for the UK National Health Service, with the academic again operating as a collaborator. This version of action research uses an intellectual framework of linked ideas through an intervention process applied to a situation. This process is a learning cycle, in which lessons can be generated about theory, method, and application. The author addresses the main criticisms leveled at action research: that it lacks both generalizability and also external validity from a single site and single focus study. The research program, described in three phases, covered a 4-year period, involved 20 organizations, and included 10 discrete, single action research interventions. Three concepts that are important in beginning to counter the main positivistic critics of action research that arise from these studies are recoverability, iteration, and themes.

2.5 Theoretical Perspectives in IS Research

The papers in this section are distinguished by a dominant concern with the use of theory in IS research, either by reviewing established practices or outlining a novel development of general significance to the field, by refining, integrating or even disinterring existing theoretical endeavor. A good starting point for synopsizing these contributions is the excellent panorama of the use of social theory in the WG's deliberations provided by Flynn and Gregory. Like Sawyer and Crowston, they look at past conference proceedings. In a remarkable effort of scholarship, all 381 papers in the 17 conferences since 1984 have been assayed. The prevalence of empirical social theory in this collective oeuvre apparently runs at 46 percent, with 175 papers manifesting a significant interest in theory either as an analytic device or the object of validation and development. There are oscillations over the period, but a trend toward greater prominence of social theory can be descried. Other highlights are the steadily increasing use of qualitative and interpretive methods, together with some fascinating research demographics, such as the proclivity of male researchers, especially in North America, to adopt the positivist paradigm.

Intriguingly, Flynn and Gregory draw up a "hit parade" of the top 10 social theories in WG 8.2 research. Actor-network theory (ANT), perhaps unsurprisingly, comes out as number 1, followed closely by structuration theory. Despite this, we are clearly a cosmopolitan community, with the top 10 accounting for less that 20 percent of the papers sampled. Interestingly, CST is not as prominent in the list as might have been expected, although Foucault comes in at number 3. The clutch of papers in this section, and in the conference more generally, is consistent with its forebears. Both ANT and structuration theory figure prominently in the section, with two of the papers attempting in different ways to conjoin the two perspectives.

Brooks and Atkinson propose a proprietary synthesis which they dub *StructurANTion*. The neologism symbolizes the complementarity seen by the authors. Whereas structuration theory provides an account of the interactive dynamics whereby social structures are held together (through the recursive enactment of socially constructed rules) and thus emphasizes stability, ANT provides a complementary narrative addressing the dialectics of socio-technical transformation in terms of the reconfiguration of networks of human and nonhuman actants. Brooks and Atkinson's framework is intended to be a practical as well as an ornamental edifice, and they illustrate its deployment in an action research project in the UK National Health Service focused on the development of patient-oriented services for cancer care.

Rose, Lindgren, and Henfridsson are also concerned with theoretical unification, again involving ANT and structuration theory. Their primary concern is fundamental for our discipline, namely whether we are to be eternally condemned as the users (and abusers?) of theory from elsewhere, rather than the builders of theory in our own right. The authors propose the idea of "adaptive theory making" as a distinctive role for applied disciplines such as IS. Our concern with practical intervention inevitably enjoins a degree of healthy pragmatism in which we draw on the most appropriate theory available from a diverse array of sources, and assemble it into a multifaceted whole. The concept of *structure* is taken as an example, and an attempt made to integrate elements from three theoretical discourses: structuration theory, ANT, and Chomsky's linguistic work on deep versus surface structure. A case study is used to illustrate the approach in action, this time in the manufacturing domain (Volvo's attempt to design a competence management system).

ANT figures for a third time in the paper by Wagner, Galliers, and Scott. Here the aim is methodological development regarding the deployment of ANT. The authors propose the use of narrative methods for conducting interviews with human actants in ANT studies. In these interviews, the researcher focuses on the meanings and explanations given by participants to the unfolding events in IT-enabled organizational change. They contend that there is a natural sympathy between ANT and the narrative approach. Again, a case study is presented for illustrative purposes. It describes the vicissitudes of an enterprise research planning implementation in a university setting, in which the use of the best practice ideal as a means of translating interests in support of the project was only partially successful, with significant concessions ultimately being forced in terms of local customs and traditional practices.

Activity theory is the subject of the paper by Korpela et al. Although on Flynn and Gregory's hit parade (at number 6), the authors argue that, since its first appearance at the Copenhagen conference in 1991, activity theory has received little attention within IS, despite being influential in cognate fields (such as computer-supported collaborative work) and having an impressive intellectual pedigree stretching back 80 years. The authors pose the somewhat rhetorical question of whether this is a "dead horse" worth flogging, or a paradigm deserving of renewed attention. They clearly believe the latter, arguing that certain features of activity theory make it particularly suited for a key role in IS research and development, such as its focus on work systems and its concern with worker emancipation. In order to promulgate its greater use, they argue that a practical development methodology is necessary, embodying activity theory concepts and principles. Such a framework is described (ActAD) and an illustrative case study set in a perinatal intensive care unit is provided.

The work of Orlikowski also features in the top 10, at number 5 in the Gregory and Flynn list of favorites. It is the subject of the next of our papers, by Davidson and Pau, who see unfulfilled potential in Orlikowski's work with Gash on technological frames of reference (TFM). Despite many citations, little real use of TFM in IS research practice has transpired. After tracing the genealogy of TFM, eight studies involving its use are analyzed, from which a number of potential refinements and possible enhancements are delineated. Some of these are substantive (e.g., developing a set of generic frame attributes, moving beyond a limited concern with frame incongruence as the source of change resistance); others are methodological (e.g., the deployment of TFM in an action research mode to enhance its relevance to practice and its fertility as a source of new IS theory).

The last of the papers in this section marks a departure by introducing theoretical ideas that are relatively new in our domain. Although social theory in general has been widely used in the 8.2 community, it is arguable that psychological theory has been neglected, reflecting our general preoccupation with social and organizational issues rather than those at the individual level. Adams and Avison draw on an eclectic mix of psychological theory that deals with cognitive blocks and biases, and how these are influenced and reinforced by the technical characteristics of development tools and methods. These are serious concerns, as the adoption of a particular methodology will inevitably have a decisive bearing on the conduct of the design process and the lineaments of the resulting IS artefact. The paper makes a useful contribution by providing a theoretical framework in which to understand these biases and the possible malignant effects they may have.

2.6 Systems Development: Methods, Politics, and Users

This section addresses the theory/practice relationship and disciplinary concerns from another standpoint, that of systems development. IS has been recognized as a discipline for more than 30 years. During that time it has rapidly evolved to reflect changes in the application of information and communication technologies to a variety of situations. Today, information systems development still remains one of the major areas for enabling IS researchers to understand how theories can be applied to complex practice. In this section, seven papers raise issues and challenge our assumptions about the systems development process itself, methods in use, political aspects, and users' participation and behavior.

Tan, Lim, Pan, and Chan analyze two in-depth case studies of a governmental institution and a commercial establishment using Montealegre's process model of capability to explore how enterprise system adoption can be strategized for the purpose of dynamic capability development. In doing this, the authors develop a process model for enterprise system adoption that captures the essence of the interdependencies between enterprise systems and dynamic capability development. They conclude that enterprise systems can be strategic partners in the capability development process in organizations.

Lings and Lundell discuss the difficulties of transferring a research method into a commercial context. They argue that method transfer is a special case of knowledge transfer. Based on four case studies of method transfer, a framework is postulated with implications for method development. Four main themes in the framework are the importance of a clear conceptual framework for a method, support for learning, usability within a defined context, and acceptability to stakeholders.

The next two papers take up the challenge of integrating critical social theory with practice. In the first, an action research project by Waring provides an opportunity to engage in a critical approach to systems analysis with the intention of exploring the politics of organizational life through the medium of integrated information systems projects in the UK National Health Service. The paper describes how the emancipatory principles of Habermas can be used to develop an innovative approach to participative process and information flow modeling. One of the conclusions is that complex social, organizational, and political issues endemic within organizations inhibit *true* discourse and therefore constitute a barrier to effective ICT introduction and the integration of information systems.

In the following paper, Tapia argues that a critical orientation is necessary to understand ICT-enabled workplace culture and employee behavior. This study is based in a dot-com organization where a group was resistant to organizational authority as the result of comparable companies beginning to close. The paper shows that ICT-based innovation may lead to increased deviant or resistant behavior in staff. Furthermore, it concludes that the social environment of the dot-com bubble has allowed several myths to propagate and affect human behavior in similar organizations.

The paper by Puri and Sahay picks up on the themes of participation and power. Puri and Sahay concentrate on how to understand the knowledge politics in using and designing a geographical information system (GIS) for land management in an Indian context. They argue that power and politics are inseparable from the systems development process and also that knowledge of the local context can complement scientific knowledge that is needed to develop and use the GIS. The link between participation and knowledge for meaningful use of GIS is crucial and communicative action can lead to better design and technology acceptance by end-users.

User participation in developing information systems also is addressed by Aanestad, Henriksen, and Pors. This has always been an important area for the 8.2 community. This paper discusses three user-led projects that utilize generic technologies. These developments were not formal processes but involved users significantly in influencing the direction and outcome of the projects. The importance of continuing redesign, tailoring, and adaptation when using these technologies is learned from these cases.

Finally in this section, Bansler and Havn give an account of improvisation in action as an attempt to make sense of information systems development in organizations. They report a longitudinal field study of the development of a Web-based groupware application in a multinational corporation. In analyzing the dynamics of this situated process, they argue that improvisation and bricolage play a vital role in the development of the project. In conclusion, they suggest that this case provides an opportunity to reconceptualize IS development.

2.7 Panels and Position Papers

The panels and position papers unsurprisingly echo many of the themes and motifs of the full research papers.

Issues regarding research method preoccupy two of the panels. The first, "Twenty Years of Applying Grounded Theory in Information Systems," discusses the promise grounded theory holds for our field and its wide use as a method for generating IS theory from qualitative data. The panel addresses a set of concerns, including its problematc relationship with positivism and the all-too-common lack of rigor in its deployment, often to endow spurious legitimacy on any form of coding or indeed informal content analysis. Action research (section 4) is the subject of the panel entitled "Building Capacity for E-Government: Contradictions and Synergies in the Dialectics of Action Research." The promise of action research as a tool for building relevant IS theory is directly addressed by this panel, with the spotlight thrown on the tensions that this creates in the research process as the imperative to solve a practical organizational problem conflicts with the requirement to deliver results of general theoretical interest to the wider research community.

Theoretical matters are addressed by two of the panels, picking up and developing some of the themes intoned in section 5. The nature of agency in socio-technical systems is tackled by the panel on "New Insights into Studying Agency and Information Technology" Is agency an exclusively human attribute, is it primarily a technological capability, or is it an inseparable property of the interaction of the two? Several of the theoretical perspectives featured in the full papers are drawn into the debate, with structuration theory, ANT and critical theory all making an appearance. The potential of activity theory as a theoretical substrate underpinning a critical approach to IS practice is addressed by the panel on "Researching and Developing Work Activities in Information Systems: Experiences and the Way Forward." This panel carries forward the general arguments for a participative, work-oriented approach to ISD laid out in the full research paper coauthored by several of the panelists (Korpela et al.). The focus of the panel is on developing methodological guidelines for implementing the proposed approach.

The two remaining panels address issues facing the conference on a more general front. The panel "Crossing Disciplinary Boundaries: Reflections on Information Systems Research in Health Care and the State of Information Systems" asks, in a wide-ranging way, to what extent the aspirations of relevant research have been realized in the domain of health care. The commonalities, and differences, in the experience of researchers in the medical informatics community and in our own field are addressed. The differential treatment of ethics in the two areas might be highlighted given the relative neglect of this topic in our field, as Bell and Adam lament (see section 2). In general, the flow of knowledge and experience could be in either direction, with the medical informatics and IS communities learning from each other's travails. The last and only invited panel, "The Great Quantitative/Qualitative Debate: The Past, Present, and Future of Positivism and Post-Positivism in Information Systems," provides a more panoramic and dialectical discussion of IS methodology. This panel addresses many of the core issues debated within the conference, which have been leitmotifs of the larger methodological discourse over the last two decades. As a microcosm of the conference itself, the panel will review *progress* in terms of two key deliberations: the opposition between quantitative and qualitative research methods and the ascendancy of post-positivistic approaches (critical and interpretive). The modernist notion of progress will itself be debated, and a plea entered for greater diversity and pluralism in our research practice, an injunction of earlier methods conferences echoed in several submissions presented here.

Turning finally to the position papers, a number of resonances are notable, with the papers falling into three distinctive clusters, mirroring the structure of the main program. Unsurprisingly, research methodology preoccupies many of these short pieces. The inherent tensions in action research are trenchantly pointed up by Breu, Hemingway, and Peppard. An illustration is given showing how the unequal exercise of power on behalf of the practitioner "side" severely compromised the rigor of the research interest in an industrially based project. The relationships between critical research and both positivism and interpretivism are addressed by Wilson and Greenhill, who add their voice to the general call for methodological pluralism. Of note in their stance is the adoption of a realist ontology to complement their overall concern with emancipation. The limits of positivism and the poverty of scientism are the subject of Jain's polemic. His philippic is based on a rejection of Cartesian dualism and an advocacy of non-dualist positions, drawing on the philosophies of Kant, Hegel, Heidegger, and Zen. Bednar's paper also addresses method, though here the primary concern is with IS development rather than research. The analysis of contextual dependencies, within a double-loop sense-making process, is proposed as a tool for addressing gender-related issues, at both the micro and macro levels, within the systems development process.

Foundational issues come to the fore in the second group of position papers. Ethical concerns resurface in the paper by Stolterman and Fors, whose main thesis challenges us to reflect critically on core issues at the heart of our discipline by asking questions regarding the purpose of IS development and research, as well as issues of methodology and ontology. They call for improving the quality of life (the idea of the "good life") as a design aim, echoing the ideological stance of extant IS methodologies such as socio-technical systems design. Foundational issues also are addressed by Stephens, who challenges the often uncritical definition of the concept of *information* adopted within the IS field and cognate domains. He considers it vital that this concept is reproblematized, and a broader view taken of information as a "complex phenomena embracing such issues as propriety, regulation, ethics, accessibility, and even aesthetics." Like Introna and Whittaker (section 2), Webb also visits the problematic nature of truth and its relationship to the academic publication process. Webb, in an interesting comparison of a conference and a journal report of the same study, finds different truth standards applied to inductive and deductive generalizations.

The final clutch of position papers addresses aspects of theory and its deployment in IS research. The methodological enhancement of ANT is the concern of Pouloudi, Gandecha, Atkinson, and Papazafeiropoulou, just as it is in the full paper by Wagner et al. (section 5). Here stakeholder analysis is advocated as a tool for identifying actants within an overall analytical approach based on ANT. The paper by Ng and Tan is also of interest from a theoretical and methodological perspective. It deploys an interesting admixture of ethnography and symbolic interactionism in an intriguing analysis of users' adherence to in-house, legacy systems in the face of an ERP implementation. The use of symbolic interactionism is of particular interest as this theoretical lens has *prima facie* much to offer to IS research, although arguably underutilized to date. Further theoretical novelty is provided by Whyte, who draws on work in the innovation studies tradition to develop a new theoretical lens for analyzing the IT artifact that draws together both a supply and a use side perspective. Whyte argues that these viewpoints traditionally have been kept separate in our field. A case study of virtual reality technology is provided to demonstrate the new lens in action. The rapid development of information technology, outstripping the pace at which we can understand its meaning and effects, is Olsson and Russo's concern. They are interested in so-called "nomadic information systems," and provide a case study of a context-aware exemplar of this relatively new technology (CABdriver) to illustrate the effectiveness of adaptive structuration theory as an analytic lens to examine its impact.

3 CONCLUDING REFLECTIONS

In putting together our final thoughts for this twentieth anniversary of the Manchester Conference, we asked the past chairs of the Working Group, and of the previous three methods conferences, to reflect on the Group's influence and significance in the various methodological turns taken by the IS community in the past 20 years or so. We are very grateful for their whole-hearted cooperation in this retrospection. Their remarks were remarkably similar. All pointed to the stimulation and enjoyment provided by participation in the Group's work, to the pleasure of finding a collection of people who took ideas seriously, self-critically challenging themselves to find new ways of thinking. This has been a hallmark of the Working Group throughout its history, turning our gatherings, according to one chair, into life-changing inspirational events. The past chairs reminisced with pride about the growth of WG 8.2 to be the largest IFIP working group. All were proud of our role in making qualitative methods a respected part of IS research and relished the coming together of varied but kindred spirits within our community as both fun and intellectually stimulating. All encouraged us to "stick at it," to keep our thinking fresh and our meetings provocative.

All agree that the 1984 Manchester conference had thrown down a gauntlet and challenged the traditionalist orthodoxy with approaches that were not common at that time in the discipline. The conference had challenged the *scientific* method in information systems research and called for greater pluralism and diversity. Its success was marked by the growing legitimacy of the linguistic and qualitative turns in IS research. As WG 8.2 grew and asserted itself, held meetings in conjunction with the International Conference on Information Systems, and nurtured researchers who began publishing in, and later joining editorial boards of, respected journals, what we originally thought of as "methods wars" were over, and the battle won. Victory was effectively proclaimed at the 1997 conference and journals such as *MIS Quarterly* claim to have left such concerns behind.

But is the war over? Can we now rest on our laurels? Have not the tyros of 1984 become the old guard of today; no longer rebels laying siege to the bastions of research conservatism, have they now become gatekeepers, sometimes seeming to police the very methods they helped establish as legitimate? Will we continue to see more flowers blooming and intellectual diversity flourishing, or are storm clouds brewing? Is history really at an end? The papers in this volume suggest not, that there is indeed much unfinished business. While not generally addressing the qualitative-quantitative debate, the present collection of papers reflect remarkably similar concerns to those exercising the 1984 gathering: a sense of crisis in IS, concern over the relevance/rigor and research/practice splits, debate over the role of theory, challenges to the very legitimacy of IS as a discipline. Can the methods wars be over while these issues persist?

With such questions in mind, and with the encouragement from the past chairs and authors in this collection, we offer further challenges to our community based on our observations of the contributions to this conference. The papers offer an exciting mix of ideas and combinations of theories, continuing the WG 8.2 traditions both of critical reflection and methodological eclecticism. We are delighted that the reviewers and associate editors kept to the Group's tradition by turning a critical eye on the papers, treating nothing and no one as a sacred cow. Everyone, the highly regarded as well as the newcomers, benefitted from their scrutiny and thoughtfulness. Not one paper made it through the review process unscathed. All were asked to revise, and many revised more than once, so papers became better and better. The reviewers proved themselves critical in the best senses of the word and in the best traditions of WG 8.2, and we, the program chairs, tried to do the same. In the interest of seeing more diversity in thought and approach, we occasionally overrode reviewers' decisions when we thought a paper had something particularly interesting to say, and decided to give the author(s) another chance to say it better. Some of those turned out to be excellent thoughtful papers, to challenge us to think, and then think some more. We also decided to try a new idea which eventually developed into a forum for the "bright ideas" or position papers. This innovation, we hope, will further stimulate radical thought and add to the spirit of debate.

We have taken to heart the past chairs' admonitions not simply to congratulate ourselves on the successes of the Group, but to continue to pioneer new ideas. In the spirit of pushing our community further forward, we note some potentially problematic tendencies in the papers in this volume. One is a continuing confusion over, and sometimes conflation of, IS development *method* and IS *research method*. These are distinctive problem domains, and clear differentiation is required. Whereas the objective of the former is to design and deploy working artifacts through the use of information systems development methods, the prerogative of the latter is the production of IS theory through appropriate research techniques. Methods for designing and building systems (e.g., structured methods, soft systems) are not the same as methods for developing social theory (e.g., the hypothetico-deductive process of normal science, grounded theory, etc.). This conference covers both dimensions of method, but in clearly demarcated categories.

We have also attuned ourselves to the general sense of defensiveness that sometimes seems to imbue our field, that we need to guard ourselves against encroachment or imperialism from other, less-diffident disciplines. The conference papers reflect two sets of possibly contradictory threads marking current discussion and controversy within IS resulting from this. Perhaps in the spirit of "attack being the best form of defense" or the theory of preemptive strikes, one thread calls for IS to be a reference discipline for other fields (i.e., to be imperialistic itself). This thread is as apparent in this volume as it perhaps has been in earlier times. The other is a persistent feeling that we must bring in theory from outside, especially from esoteric French and German social theorists and philosophers. We note a continuing dilettante tendency in the conference papers to borrow theory from Latour, Habermas, Foucault, and assorted others. We have nothing against French and German theory, but why these individuals in particular? Are there no other social theorists or philosophers worthy of attention? Is this interest promoted by the siren allure of the obscure and the exotic? Of course, there are examples outside this clique in the erudition mustered here, but perhaps not the degree of diversity suggestive of a healthy, open spirit of inquiry and scholarship. The list is more notable in that few French and Germans even participate in WG 8.2, much as we wish they would.

A number of papers propose novel combinations of these and other theories, creating "new" theory through something like hybridization. Indeed, we half-jokingly entitled a conference session "ANT-plus" at one point in our deliberations. As with any hybrids, we wait to see which will be robust examples of hybrid vigor. But hybrids are variations on their parents. We would like to see more of our own theory developed in addition to borrowing from others. Must theory come from outside to be acceptable? One way a corpus of home-grown theory might be cultivated is through grounded theory. Although we applaud using qualitative approaches for theory generation, we are concerned that grounded theory is sometimes written about as a theory, rather than as a *method* for generating for theory. Moreover, there seems to be a tendency to conflate grounded theory with qualitative research; labeling what one does as grounded theory seems to serve the purpose of legitimating it, when a less grandiose appellation (such as inductive reflection) may actually be more accurate. Similar concerns are taken up by one of the panels. We hope our observations will prevent co-optation of qualitative approaches into some increasingly innocuous, and therefore acceptable, form.

We point to a rather different concern when we examine the make-up of IFIP Working Group 8.2. Just as a plurality of methods and theories enlivens the field and leads to increased understanding, so would a good mix of people along a variety of dimensions. For example, there have been relatively few participants outside the United States, the United Kingdom and her former colonies, and northern Europe in general. Alas, the same is true for this conference. While we clearly see the contribution, for instance, that the Scandinavian social democratic tradition has made to both theory and practice in our field, would that other perspectives were brought into the fold from different political and social backgrounds removed from those with which we have become so comfortable. We would like to see further enrichment deriving from consideration of a broader range of working contexts than those of the business world, and we are gratified to see papers in this volume based on experiences in hospital, governmental, and educational settings. We hope WG 8.2 becomes an even more welcoming place for participation from those studying IT in settings outside business organizations. Similarly, although we tend to draw on ideas from a variety of other disciplines, we infrequently work with people in those disciplines or invite them to our conferences. As one past chair remarked, when we have expanded our horizons by inviting "outsiders," we changed the networks and dynamics of the Working Group. Here's to making more outsiders insiders, or at least welcomed guests.

We also note a tendency toward introspection. We applaud self-reflection. As is evident in this chapter, we are certainly guilty of it ourselves. But we also need to be wary. At what point does legitimate concern with methodology, or the state of the discipline, become an immobilizing obsession? Is debate about methodology a displacement activity that removes and detaches us from genuine but daunting engagement with the world of practice that we purport to influence? Perhaps we had best get on with our work, doing our research as best we can, *sans* such posturing over theory and method. But that, too, is part of what WG 8.2 is about. We invite all to join in the fun and fray of this anniversary celebration and, through this conference and the future work we hope it stimulates, continue the tradition of keeping our meetings and our field, fresh, vigorous, and provocative.

ABOUT THE AUTHORS

Bonnie Kaplan is on the faculty of Yale University's School of Medicine and is a faculty affiliate of the Information Society Project at the Yale Law School, and a Senior Scientist at Boston University's Medical Information Systems Unit. She previously held faculty appointments in Information Systems. She specializes in change management, benefits realization, and people's reactions to new technologies in health care. She consults for academic, governmental, private, and business organizations in health care. Her publications have appeared in such journals as *Methods of Information in Medicine, Journal of the American Medical Informatics Association, International Journal of Medical Informatics, Artificial Intelligence in Medicine; Science, Technology and Human Values*, and *MIS Quarterly*. She chairs the International Medical Informatics Association Working Group-13: Organizational and Social Issues, and chairs the Yale University Interdisciplinary Bioethics Project Research Working Group on Technology and Ethics. She is a recipient of the American Medical Informatics Association President's Award and a Fellow of the American College of Medical Informatics.

Duane Truex researches the social impacts of information systems and emergent ISD. He is an associate editor for the *Information Systems Journal*, has coedited two special issues of *The Database for Advances in Information Systems*, and is on the editorial board of the *Scandinavian Journal of Information Systems*, the *Journal of Communication, Information Technology & Work*, and the *Online Journal of International Case Analysis*. His work has been published in *Communications of the ACM, Accounting Management and Information Technologies, The Database for Advances in Information Systems, European Journal of Information Systems, le journal de la Societé d'Information et Management, Information Systems Journal, Journal of Arts Management and Law, IEEE Transactions on Engineering Management*, and 40 assorted IFIP transactions, edited books, and conference proceedings. He is a member of the Decision Sciences and Information Systems faculty in the Chapman Graduate School, College of Business, at Florida

International University, and is an associate professor on leave from the Computer Information Systems Department, Robinson College of Business, at Georgia State University.

David Wastell is Professor of the Information Society in the School of Informatics at the University of Manchester. His current interests are in business process re-engineering, electronic governance, IS research methods and the human factors design of complex systems. He has published around 100 journal articles and conference papers in Information Systems, human factors, health informatics and research methods, after an early research career in cognitive and clinical psychophysiology. He is on the editorial board of *European Journal of Information Systems* and *Information and Management*, has co-organized one previous IFIP conference (WG 8.6), and has co-authored two previous edited collections. He has considerable consultancy experience, especially in the public sector.

Trevor Wood-Harper is Professor of Information Systems at the School of Informatics at the University of Manchester and is Professor of Management Information Systems at the University of South Australia, Adelaide. Trevor has held visiting chairs at University of Oslo, Copenhagen Business School, and Georgia State University. He has coauthored or coedited 16 books and proceedings and over 200 research articles in a wide range of topics including the Multiview Methodology, information systems evolution for developing countries, electronic government, action research, ethical considerations in systems development, fundamentals of information systems and doctoral education. He has successfully supervised 25 doctoral students and acted as an external examiner for more than 75 Ph.D. theses in the UK, South Africa, Norway, Sweden, and Australia.

Part 1:

Panoramas

2 DOCTOR OF PHILOSOPHY, HEAL THYSELF

Allen S. Lee
Virginia Commonwealth University

Abstract As doctors of philosophy who are specialists in information systems, we routinely perform diagnoses of, and write prescriptions for, individuals, groups, organizations, societies, and their artifacts. The proverb "physician, heal thyself" requires that we ourselves, along with our scholarly artifacts, societies, organizations, and groups, undergo the same manner of diagnosis to which we subject others, and that we have a taste of our own medicine. This essay uses three published papers of Working Group 8.2 of the International Federation for Information Processing—from the 1984 Manchester meeting, from the 1990 Copenhagen meeting, and from the 1997 Philadelphia meeting—as a source of rich material with which to illustrate the difference in our diagnoses and prescriptions if we were to do unto ourselves what we do unto others.

1 INTRODUCTION

We are doctors of philosophy who are specialists in information systems. In this role we perform diagnoses of, and we write prescriptions for, individuals, groups, organizations, societies, and their artifacts. At the same time we need to ask ourselves: Must we practice what we preach? Must we ourselves, along with our scholarly artifacts, societies, organizations, and groups, undergo the same manner of diagnosis to which we subject others, and must we have a taste of our own medicine? The answer is that we must: We doctors of philosophy may not exempt ourselves from our own scrutiny or our own medicine—lest we violate the scientific requirement of consistency in our research and the ethical requirement of the golden rule in our conduct.

Some might argue that our research has indeed been inconsistent and our conduct hypocritical. It can appear that we doctors of philosophy readily and routinely train a critical eye on others, but not ourselves. Others might argue that a finding of inconsistency and hypocrisy is premature. There is no *a priori* reason that the optimal or only time to begin diagnosing and prescribing for ourselves is already in the past. And because the discipline of information systems is, at most, 50 years old, one can also argue that the information systems discipline is ready, only now, to turn a critical eye on itself.

Although sharing a specialty in information systems, we doctors of philosophy are a diverse lot. Mirroring the diversity among ourselves is the diversity of the research that we publish in our journals, such as *European Journal of Information Systems*, *Information Systems Research*, *Information Systems Journal*, and *MIS Quarterly*. Of the many different segments of our information systems research community, one that offers itself as especially promising material for a revelatory case study is the one that calls itself Working Group 8.2 of the International Federation for Information Processing. Given its self-conscious and reflective stance on research methods and its comfort with critical social theory, Working Group 8.2 is more likely than any other segment of the information systems research community to be able to understand the scientific and ethical necessity to heal itself and not just others. If the case cannot be made that we doctors of philosophy of WG 8.2 are ready, willing, and able to do unto ourselves what we have been doing unto others, then there would be little hope that the same case can be made for the information systems research community overall.

WG 8.2's self-conscious and reflective nature has long been evident in its existence and has manifested itself in the form of a persistent concern over research methods. Not only does the topic of research methods provide the theme for the current conference in Manchester, but it was also the theme for three of WG 8.2's past conferences—1997 in Philadelphia, 1990 in Copenhagen, and in 1984, also in Manchester. WG 8.2's regularly recurring reflection on research methods is a manifestation of its awareness of and sensitivity to the *process* of scientific research, apparently seen as distinct from, and no less important than, any *content* that the process produces.[1] In this light, if there are any doctors of philosophy in the overall information systems research community who are ready, willing, and able to do unto themselves what they do unto others, they are likely to be found among the doctors of philosophy in WG 8.2.

The following argument examines three past instances in which WG 8.2 focused on diagnosing and prescribing for others, and also what would have been different in these instances if WG 8.2 had focused, in addition, on diagnosing and prescribing for itself. This difference will serve to illustrate what we—people who are members of the 8.2 community and the information systems discipline overall—can do to practice what we preach, with the result that we can satisfy both the scientific requirement of consistency in our research and the ethical requirement of the golden rule in our conduct. The argument will begin with a fundamental point from Thomas Kuhn's history of science—that a community of scientific researchers has, and is shaped by, its own sociology, not unlike any other community that these researchers themselves would typically investigate.

---

[1]"A great deal hinges on whether science is viewed as a body of propositions or as the enterprise in which they are generated, as product or as process" (Kaplan 1964, p. 7). At the same time, no choice need be made. Science can be viewed as both process and product. Arguably, viewing it as process is more important because science as product is determined by science as process.

2 SCIENTIFIC RESEARCHERS ARE RESEARCH SUBJECTS TOO

Kuhn's *Structure of Scientific Revolutions* (1996) and his related studies are instances of research in history and sociology. In his empirical, historical investigation of natives who called themselves "scientists"—including those who called themselves physicists and biologists and whose shared cultural beliefs, rituals, politics, and superstitions are fascinating but beyond the scope of this single paper to examine—Kuhn often refers to their sociology. One example is

> Some of the principles deployed in my explanation of science are irreducibly sociological, at least at this time. In particular, confronted with the problem of theory-choice, the structure of my response runs roughly as follows: take a *group* of the ablest available people with the most appropriate motivation; train them in some science and in the specialties relevant to the choice at hand; imbue them with the value system, the ideology, current in their discipline (and to a great extent in other scientific fields as well); and finally, *let them make the choice.* If that technique does not account for scientific development as we know it, then no other will. There can be no set of rules of choice adequate to dictate desired *individual* behaviour in the concrete cases that scientists will meet in the course of their careers. Whatever scientific progress may be, we must account for it by examining the nature of the scientific group, discovering what it values, what it tolerates, and what it disdains.
>
> That position is intrinsically sociological...(Kuhn 1970, pp. 237-238).

In examining scientists in this way, Kuhn is casting them in the role of research subjects and, therefore, rendering them as objects of study.

One can argue that if physicists have a sociology, then other researchers do too. Accepting the generalization that the people who call themselves social scientists and information systems researchers are themselves research subjects similarly to how the self-proclaimed natural scientists are research subjects, we may conclude that the former are no more immune to sociological, historical, and other scientific investigation than any other natives.

In some of our past annual meetings, we members of a society which we call 8.2 have offered research methods and perspectives for how we would diagnose, and prescribe for, others—in particular, people in the midst of their organization, their organization's information technology, and the many phenomena emerging from the mutually transforming interactions between the organization and the information technology. In what follows, there are three instances of research methods or research perspectives that members of 8.2 have entertained in presentations at their earlier conferences. I have chosen the three papers so as to represent each of the earlier three meetings, as well as to include authors whose prominence, recognition, and active research programs extend to the present. Following each instance, in turn, is a scenario for how— if ready, willing, and able—we members of WG 8.2 or the information systems research community overall could likewise diagnose and prescribe for ourselves. The examples build on earlier ones (Lee 2000) and are in keeping with their spirit.

2.1 Three Knowledge Interests for ~~Information Systems Development~~ *Information Systems Research and Publishing*[2]

As early as at its 1984 meeting in Manchester, WG 8.2 recognized and embraced critical theory. Lyytinen and Klein (1985) introduced the critical theory of Jürgen Habermas to WG 8.2. The following passage from their paper well conveys its overall spirit and identifies some ramifications of critical social theory for how information systems researchers diagnose and prescribe for information systems development (pp. 225-226).[3]

> The Implications of Three Knowledge Interests for Information Systems Development
>
> Because information systems development is currently dominated by approaches based on the idea of purposive-rational action, the underlying knowledge basis of many of its methodologies is [the] technical knowledge interest. This appears to be true even of those methodologies which take a broader social perspective such as socio-technical system approaches and implementation research. Variations can only be found in the scope of inquiry, its conceptual basis and applied inquiring methods.
>
> Our understanding of the process and content of information systems development and its supporting methodologies can be improved considerably if it is recognized that it includes not only [the] technical knowledge interest, but also practical and emancipatory knowledge interests.
>
> First, restricting attention to [the] technical knowledge interest influences how problems are defined and understood. They are perceived as given and as totally independent of the investigator. Because of this narrow focus, methodologies are unable to explain how people, through social learning, create new meanings and concepts to cope with new situations.
>
> Second, a concentration on technical knowledge interests conceals the real processes of information systems development and their dependency on communicative action. In the majority of information systems design methodologies, design groups see users as "producers of information," as "primary problem solvers" and as "opponents in an implementation game." Information systems development as a process of communicative action through ordinary language is hardly known and rarely studied. In consequence, methods to assist the sharing of different opinions and problems, and the role of ordinary lan-

[2]The use of strike-outs, followed by italicized words replacing the stricken words, is intentional. As the subsequent text will make clear, I use it to indicate how I am mapping lessons about people, organizations, and information technologies from each of the three earlier 8.2 papers to the current situation of ourselves as researchers regarding our own information technology (i.e., research methods) and our own organization (i.e., our 8.2 community).

[3]The citations in the quoted materials are suppressed.

> guage in this process, have not been developed and studied. Because of this, most methodologies cannot handle the participation issue or examine it theoretically.
>
> Third, existing methodologies appeal to value-neutrality and instrumental reason. They define all information systems problems in terms of means and ends, and the most efficient way of pursuing these. This selecting implies a tyranny of means over ends. There is little consideration of values and goals, and the design process is seen as "an act of faith." There is no attempt to legitimate goals through developing a rationally grounded consensus among the stakeholders.

Lyytinen and Klein's use of critical theory to diagnose and prescribe for information systems development is also suggestive of how they and other information systems researchers can use critical theory to diagnose and prescribe for themselves and their work (i.e., information systems research). Consider

> <u>The Implications of Three Knowledge Interests for ~~Information Systems Development~~ *Information Systems Research and Publishing*</u>
>
> Because ~~information systems development~~ *information systems research and publishing* is currently dominated by approaches based on the idea of purposive-rational action, the underlying knowledge basis of many of ~~its methodologies~~ *the methods used in information systems research* is [the] technical knowledge interest. This appears to be true even of those ~~methodologies~~ *research methods* which take a broader social perspective such as *those used in* socio-technical system approaches and implementation research. Variations can only be found in the scope of inquiry, its conceptual basis and applied inquiring methods.
>
> Our understanding of the process and content of ~~information systems development~~ *information systems research and publishing* and its supporting ~~methodologies~~ *research methods* can be improved considerably if it is recognized that it includes not only [the] technical knowledge interest, but also practical and emancipatory knowledge interests.
>
> First, restricting attention to [the] technical knowledge interest influences how *research* problems are defined and understood. They are perceived as given *to the information systems researcher by what the research discipline itself (its literature, its journals, its conferences) considers to be significant research* ~~and as totally independent of the investigator~~. Because of this narrow focus, ~~methodologies~~ *research methods* are unable to explain how ~~people~~ *information systems researchers*, through social learning, create new meanings and concepts to cope with new situations.
>
> Second, a concentration on technical knowledge interests conceals the real processes of ~~information systems development~~ *information systems research and publishing* and their dependency on communicative action. In the majority of information systems ~~design methodologies~~ *research methods*, ~~design groups see users~~ *editors and reviewers see researchers* as "producers of information"

> *and* as "primary problem solvers" *while researchers see editors and reviewers* as "opponents in ~~an implementation~~ *the publication* game." Information systems ~~development~~ *research and publishing* as a process of communicative action through ordinary language is hardly known and rarely studied. In consequence, methods to assist the sharing of different opinions and problems *among information systems researchers*, and the role of ordinary language in this process, have not been developed and studied. Because of this, most ~~methodologies~~ *research methods* cannot handle the ~~participation issue~~ *research-and-publication process* or examine it theoretically.
>
> Third, existing ~~methodologies~~ *research methods* appeal to value-neutrality and instrumental reason. They define all information systems *research* problems in terms of means and ends, and the most efficient way of pursuing these. This selecting implies a tyranny of means over ends. There is little consideration of values and goals *of the information systems researchers*, and the ~~design~~ *research-and-publication* process is seen as "an act of faith." There is no attempt to legitimate *research* goals through developing a rationally grounded consensus among the stakeholders, *who include researchers, editors, reviewers, tenure-and-promotion committees, and research funding agencies*.

The above rendering of "The Implications of Three Knowledge Interests for Information Systems Development" into "The Implications of Three Knowledge Interests for ~~Information Systems Development~~ *Information Systems Research and Publishing*" is not a determinative mapping of features from one domain to another, but just one possible way of suggesting similarities between the two domains. Just as diagnosing and prescribing occur in the former, they can occur in the latter. Other possible renderings would lead to the same conclusion, which is that we, as information systems researchers, have much work remaining to be done in diagnosing ourselves as researchers, as well as prescribing what we should be doing, or doing better, in the research-and-publication process of our information-systems research community.

2.2 ~~Information Systems Use~~ *Information Systems Research and Publishing* as a Hermeneutic Process

A trait of WG 8.2 that has distinguished it within the information systems research community is its singular devotion to qualitative research, particularly interpretive research. An interpretive approach that has received much attention from 8.2 members is hermeneutics. At the 1990 meeting of 8.2 in Copenhagen, Boland (1991) presented a paper on, among other things, the hermeneutic interpretation of accounting data. Boland offered the following conclusion about information systems use as a hermeneutic process (p. 454).

> Viewing information system use as a hermeneutic process opens a new set of research concerns. From a hermeneutic view, attention would shift from how well an information system represented a situation to how well it enabled the reader to appropriate possibilities for being within the situation and themselves.

> Attention would shift from identifying the user's essential, foundational and enduring set of information needs to identifying how different information availability enabled the juxtaposition of quantity and quality, the shifting back and forth from numbers and calculation to persons and values. Attention would shift from the information systems as a device for data output to the information system as an environment for acting out interpretations —a space for actively appropriating meaning about our situations and ourselves.

It is an easy step from Boland's prescription of hermeneutics for viewing information systems *use* to the same or same-styled prescription of hermeneutics for viewing information systems *research.*

> Viewing ~~information system use~~ *information systems research and publishing* as a hermeneutic process opens a new set of research concerns. From a hermeneutic view, attention would shift from how well ~~an information system represented~~ *information systems research represents, explains, or interprets* a situation to how well it ~~enabled the reader~~ *enables information systems researchers* to appropriate possibilities for being within the ~~situation~~ *research-and-publication process* and *being* themselves. Attention would shift from identifying the ~~user's~~ *researcher's* essential, foundational and enduring set of ~~information needs~~ *needs regarding theory and data* to identifying how ~~different information~~ *different research methods* ~~enabled~~ *could enable* the juxtaposition of ~~quantity and quality~~ *positivist research and interpretive research*, the shifting back and forth from ~~numbers and calculation~~ *survey, field, experimental, or archival data and data analysis* to ~~persons~~ *the researchers themselves* and *their* values. Attention would shift from ~~the information systems as a device for data output~~ *information systems research as a process for developing theory about people and technologies that we researchers observe in organizational settings* to ~~the information system~~ *information systems research as additionally involving a parallel process in which we researchers strive to engage in research as a meaningful and rewarding activity unto itself, that is,* as an environment for acting out interpretations *of the social, political, technological, and even scholarly dimensions that both enrich and diminish us in the day-to-day work that we do as researchers and that allow us to enjoy being ourselves*—a space for actively appropriating meaning about our situations and ourselves.

Because we information systems researchers are people or research subjects no less than the people in organizations whom we observe, and because the research methods that we information systems researchers use are no less an information technology than the electronic information technologies that routinely interact with the people and organizations about whom we theorize (Lee 2003, pp. 312-314), the theories and methods that we researchers have already accumulated are also ready and available for us to use in diagnosing, and prescribing for, ourselves—including our own social, political, technological, and even scholarly activities. Furthermore, in a situation where the theories or methods of any of us researchers fail *when applied to ourselves*, the failure itself would constitute hard evidence unfavorable to the given theory or method and thereby contribute to invalidating it. In the process of further developing or otherwise improving the theory or method so that it might succeed instead of fail in this situation, we information systems researchers would be diagnosing shortcomings in our own activity of researching, as well as prescribing and taking remedies for how to overcome them.

2.3 ~~Research on Practice~~ *Research on the Practice of Research*

At the 1997 meeting of 8.2 in Philadelphia, Markus gave the opening keynote address. She rightly noted the great progress in the development of qualitative research methods and their acceptance since the time of the 1984 meeting in Manchester (Markus 1997, p. 12).

> As I look back over the proceedings of the Manchester and Copenhagen meetings and the research published in IS over the same time frame, it is clear to me that qualitative research has won at least one major championship—academic acceptance, both within the IS field and within the larger domain of academic management studies. Today, most high-status members of the IS community acknowledge that qualitative research methods occupy an important niche along with formal modeling and quantitative empirical methods (survey and experiments). Qualitative studies and methodological essays dealing with qualitative methods increasingly appear in our conferences and journals. Some research articles employing qualitative methods figure among the seminal studies read by IS doctoral students. An increasing number of IS doctoral programs teach qualitative research methods and legitimize the use of qualitative methods in dissertation research. Further, members of our field whose work is largely or exclusively qualitative in methods have been granted promotion and tenure in their respective institutions, signifying that academics from other management disciplines also accept the legitimacy of qualitative IS research. Members of our field have been appointed to the editorial boards of journals in other fields. Other signs of acceptance can be noted.

Markus did not broach the possibility in which we qualitative information systems researchers would apply our increasingly accepted qualitative research methods in studies of ourselves and our research work, but she did acknowledge another area no less deserving of further, serious study, namely, practical research (p. 18).

> A third new [arena] I think we should enter is the appreciation of practical knowledge....By practical research, I mean academic research that seeks primarily to describe, qualify or measure, evaluate or interpret practice...I am deliberately contrasting practical research with theoretical research, which seeks primarily to build or test academic theory... .Therefore, by appreciation of practical research, I mean that we as an academic field should (collectively)

> consume, reward, and contribute more heavily to a literature about what is going on in practice than we do today.

She continued (p. 22)

> If you agree that practical IS research fills important needs, you may still need to be convinced that it is something that we qualitative IS researchers should include in our portfolio of activities. I hardly need to convince you that this is something we can do: with some shift in mindset, perhaps, our methods are ideally suited to practical research. Instead, I'll try to show why we need to do it, because it requires skills we have and because it is something that no other group of professionals has the skills and incentives to do.

However, before plying our qualitative research skills in doing what Markus calls practical research, it would behoove us first to recognize that we, as researchers, are also practitioners, where we engage in the practice of research. To diagnose and prescribe for other practitioners, we would also need, eventually, to use our own methods to diagnose and prescribe for ourselves in our own roles as practitioners—i.e., practitioners of research. Ruling out the necessity of eventually diagnosing and prescribing for ourselves in our role as research practitioners would be as unscientific and unethical as ruling out this necessity in any of the other roles we play. If Markus were to agree with this conclusion, how might she suggest we proceed?

> A third new [arena] I think we should enter is the appreciation of ~~practical knowledge~~ *knowledge about the practice of research*... By ~~practical research~~ *research on the practice of research*, I mean academic research that seeks primarily to describe, qualify or measure, evaluate or interpret ~~practice~~ *the practice of research*...I am deliberately contrasting ~~practical research~~ *research on the practice of research* with ~~theoretical~~ *methodological* research, which seeks primarily to *develop research methods that research practitioners can use instrumentally to* build or test academic theory....Therefore, by appreciation of ~~practical research~~ *research on the practice of research*, I mean that we as an academic field should (collectively) consume, reward, and contribute more heavily to a literature about what is going on in ~~practice~~ *the practice of research* than we do today....
>
> If you agree that ~~practical IS research~~ *research on the practice of IS research* fills important needs, you may still need to be convinced that it is something that we qualitative IS researchers should include in our portfolio of activities. I hardly need to convince you that this is something we can do: with some shift in mindset, perhaps, our methods are ideally suited to ~~practical research~~ *research on the practice of research*. Instead, I'll try to show why we need to do it, because it requires skills we have and because it is something that no other group of professionals has the skills and incentives to do.

In the eyes of an anthropologist whose research subjects are the natives in a village who practice this or that craft, we information systems researchers would be these natives and our scientific research would be the craft that we, the natives, practice. In the same way that our information systems theories, shaped by our research methods, either bear or should be made to bear a diagnostic and prescriptive relationship to information systems practitioners and their craft, our information systems theories and methods either bear or should be made to bear a diagnostic and prescriptive relationship to ourselves and our craft of research. To proceed as the Markus-inspired text suggests would lead us, as researchers, not only to learn from undergoing the same manner of scrutiny with which we routinely diagnose IS practitioners, but also to learn from having a taste of the same medicine that our published research routinely dispenses under the heading of "ramifications for practitioners" or "recommendations for practice."

3 READY, WILLING, AND ABLE

The titles appearing in the table of contents of the anthology published for the current, 2004 meeting in Manchester would, in themselves, constitute telling signs of whether any doctors of philosophy in the information systems research community feel ready, willing, and able to heal ourselves. For example, instead of a title such as "The Critical Theory of Jürgen Habermas as a Basis for a Theory of Information Systems" (Lyytinen and Klein 1986) there would be "The Critical Theory of Jürgen Habermas as a Basis for a Theory of Information Systems Researchers"; instead of "Information Systems Use as a Hermeneutic Process" there might be "Information Systems Research and Publishing as a Hermeneutic Process"; and instead of "The Qualitative Difference in Information Systems Research and Practice" there would be, perhaps, "The Qualitative Difference in Information Systems Researchers and Practitioners." The absence of the latter sort of titles would provide a sign that we information systems researchers are not yet ready, willing, or able to diagnose and prescribe for ourselves as we have diagnosed and prescribed for others. However, if such titles were to appear, the results would be fascinating.

At the same time, it would not be appropriate to fault any current or past research article in our information systems discipline for not having pondered explicitly on our own praxis or for not having taken our own research methods in diagnosing and prescribing for ourselves. The unit of analysis or unit of action is not the individual research article. Every individual research article exists only in a larger context, and it is the context that would, or would not, make it conducive and feasible for an individual researcher to infuse the prescription, heal thyself, into the development and writing of his or her individual research article. Change must occur at institutional levels before, or concurrently with, change at the level of individuals, so as to set the stage for changes in the practice of research by individual researchers and the content of their individual research studies.

I foresee the needed change as unfolding concurrently in three processes. I am referring to the processes to which how Berger and Luckmann (1966, p. 61) refer when explaining their concept of the social construction of reality

- *Society is a human product*
- *Society is an objective reality*

- *Man is a social product*

The three concepts refer respectively to the processes that Berger and Luckmann call *externalization*, *objectivation*, and *internalization*. We members of the 8.2 research community have already been applying these phenomenological or social-constructionist concepts in our research on people, organizations, and information technologies for years. Now, in applying the same concepts to ourselves, I argue

- *Information systems research is a product of information systems researchers*
- *Information systems research itself is an objective reality*
- *An information systems researcher is a product of information systems research*

These three concepts also refer respectively to the processes of *externalization, objectivation,* and *internalization*. This conceptualization of the social context in which a research article comes into being suggests that the prescription heal thyself ought to be diffused concurrently in (1) the process of externalization, in which information systems researchers produce information systems research and which involves not only individual-level activities such as a single researcher's work in crafting a dissertation or a submission to a journal, but also social-level activities such as the enactment of preexisting shared norms and practices (the research culture) in the manuscript review process or the tenure-and-promotion process; (2) the process of objectivation, in which information systems research that has already been produced comes to take a life of its own, continues to exist even when its authors leave the research community, and, like any other objective reality, comes to be accepted as "given" by new information systems researchers who encounter it for the first time; and (3) the process of internalization, in which information systems researchers (especially those newly entering the research community) learn and appropriate the given information systems research, as well as come to have their research thinking and research behaviors shaped by it. Until and unless such interventions occur, any attempt by an individual researcher to follow the prescription heal thyself in an individual research paper would be met with disapproval or misunderstanding from the overall community of information systems researchers—whether the individual researcher's effort were to involve the crafting of a new paper or the revision of an old one. In short, it is not only the individual researcher or individual research paper, but also the research context, that requires change and intervention.

What, precisely, might some helpful interventions in the three processes be? Being but an individual researcher existing in a larger research context, I alone cannot design and carry out any such intervention. However, if there happened to be a physical, same-time, same-place gathering of information systems researchers—including those who wield influence—these individuals and I could initiate a dialogue amongst ourselves, where we would conspire to design and carry out what we believe would be helpful interventions into our own research community. To this group, I would give a few examples, for illustrative purposes, of what the interventions could be.

Regarding the process of internalization, one possible starting point would be the developmental process by which information systems doctoral students become fully fledged information systems researchers. Just as people who study to become psychotherapists often must undergo psychotherapy themselves as part of their education, doctoral students in information systems could undergo experiences as experimental subjects in positivist research and as organizational members in interpretive research where, of course, the research topic would fall in the domain of people, organizations, and information technologies. In their subsequent reading of the research write-up, the doctoral students could see and feel themselves objectified and they could assess the appropriateness or helpfulness of the practical recommendations given to them for how they may improve themselves and what they ought to do. In similar fashion, information systems researchers who are advisors and teachers of doctoral students could serve as research subjects in studies conducted by doctoral students (although, perhaps, not those from the same institutions as the advisors and teachers), where the information systems researchers could then ponder how they have been conceptualized (e.g., as effective or ineffective users of technology, or as actors who use technology to thwart the aspirations of other organizational members) as well as how appropriate and helpful they find the practical recommendations given to them.

Regarding the process of externalization, researchers could be given a particular setting in which they would feel safe to produce research that embodies specific aspects of the prescription heal thyself, such as those which I suggested above in my applications of ideas rooted in the works of Lyytinen and Klein, of Boland, and of Markus. One possibility would be for a safe context to be provided by Working Group 8.2—i.e., a future conference that would accept papers that do such things as the following: It would be fascinating to attend the presentation of papers in which we 8.2 members (1) provide thick descriptions about what we have experienced as "intellectual and social domination" (Lyytinen and Klein 1985, p. 225), whether at the hands of the larger information systems research community or at the hands of other 8.2 members, (2) describe, in our own words, our "concern to have free, open communications and the conditions that enable these to take place" (Lyytinen and Klein 1985, p. 225) both within 8.2 and in the overall information systems research community, and (3) listen to ourselves describe what emancipation would mean for us if we were to say "The purpose of such inquiries is our emancipation" (Lyytinen and Klein 1985, p. 225). It would be fascinating to attend paper presentations and panel discussions in which we members of 8.2 describe our success or failure in shifting from a focus on how well our theory represents, explains, or interprets one or another situation of technology-organization interaction to a focus on how well the research-and-publication process enables us, as researchers, to appropriate possibilities for scholarly development and fulfillment within the research-and-publication process and for being our personal and professional selves (cf. Boland 1991, p. 454). It would be fascinating, even if only in the long run, to listen to an 8.2 keynote speaker proclaim that our reflective research, in which we doctors of philosophy diagnose and prescribe for ourselves so as to heal ourselves, "has won academic acceptance, both within the information-systems research community and within the larger domain of academic management studies" (Markus 1997, p. 12). Such papers could not simply be written and submitted to any information systems conference today with the expectation of being accepted. However, a conference of WG 8.2 dedicated to the theme of "Doctor of Philosophy, Heal Thyself" would provide a safe setting welcoming of such research.

The process of objectivation would be largely the responsibility of those information systems researchers who wield influence as journal editors, conference program chairs, journal and conference reviewers, external reviewers in tenure-and-promotion cases, and highly respected members of our research community in general. It is these scholars of influence who determine the research papers and research ideas that are accepted and approved. It is also these scholars who determine the research papers to which subsequent papers must refer and cite. In this way, they are primarily responsible for selecting the research articles and books that come to endure and continue to shape future research, even when their authors cease doing research or otherwise leave the research community. One plausible setting in which these scholars may wield their influence would be a conference that they organize and whose theme would be "Doctor of Philosophy, Heal Thyself." As the organizers, program chairs, and reviewers serving this conference, they could explicitly solicit research on this theme, provide criteria by which such research submissions would be judged, and eventually encourage the further development of some of these papers for submission to mainstream information systems journals.

It is possible that we members of the information-systems research community do not yet feel ready, willing, and able to heal ourselves. However, there remains the entire future for us to prepare to feel this way. Perfection achieved today is always ideal, but progress will suffice—progress toward satisfying the scientific requirement of consistency in our research and the ethical requirement of the golden rule in our conduct.

REFERENCES

Berger, P. L., and Luckmann, T. *The Social Construction of Reality: A Treatise in the Sociology of Knowledge*, Garden City, NY: Doubleday, 1966

Boland, R. J. "Information System Use as a Hermeneutic Process," in H.-E. Nissen, H. K. Klein, and R. Hirshheim (Eds.), *Information Systems Research: Contemporary Approaches & Emergent Traditions*, Amsterdam: North-Holland, 1991, pp. 439-464.

Kaplan, A. *The Conduct of Inquiry*, San Francisco: Chandler Publishing, 1964.

Kuhn, T. S. "Reflections on My Critics," in I. Lakatos and A. Musgrave (Eds.), *Criticism and the Growth of Knowledge*, London: Cambridge University Press, 1970, pp. 231-278.

Kuhn, T. S. *The Structure of Scientific Revolutions* (3rd ed.), Chicago: University of Chicago Press, 1996.

Lee, A. S. "Building and Testing Theory on New Organizational Forms Enabled by Information Technology," in B. Sundgren, P. Mårtensson, M. Mähring, and K. Nilsson (Eds.), *Exploring Patterns in Information Management: Concepts and Perspectives for Understanding IT-Related Change* (Essays in Honor of Mats Lundeberg, Professor of Information Management), Stockholm: The Economic Research Institute, Stockholm School of Economics, 2003, pp. 305-320.

Lee, A. S. "Challenges to Qualitative Researchers in Information Systems," in E. M. Trauth (Ed.), *Qualitative Research in Information Systems: Issues and Trends*, Hershey, PA: Idea Group Publishing, 2000, pp. 240-270.

Lyytinen, K. J., and Klein, H. K. "The Critical Theory of Jürgen Habermas as a Basis for a Theory of Information Systems," in E. Mumford, R. Hirschheim, G. Fitzgerald, and A. T. Wood-Harper (Eds.), *Research Methods in Information Systems*, Amsterdam: North-Holland, 1985, pp. 219-236.

Markus, M. L. "The Qualitative Difference in Information Systems Research and Practice," in A. S. Lee, J. Liebenau, and J. I. DeGross (Eds.), *Information Systems and Qualitative Research*, London: Chapman & Hall, 1997, pp. 11-27.

ABOUT THE AUTHOR

Allen S. Lee a senior editor and former editor-in-chief of *MIS Quarterly*. He is professor of information systems and associate dean for research and graduate studies in the School of Business at Virginia Commonwealth University. Currently he is also a visiting professor at the London School of Economics and at Queen's University Belfast, and a visiting scholar at Indiana University. As a research methodologist, Allen has maintained a research stream on research approaches such as interpretivism, positivism, case studies, and action research. He served as a program chair for the 1997 meeting of IFIP WG 8.2 in Philadelphia and the general chair for the 2001 meeting of 8.2 in Boise.

3 INFORMATION SYSTEMS IN ORGANIZATIONS AND SOCIETY: Speculating on the Next 25 Years of Research

Steve Sawyer
Pennsylvania State University

Kevin Crowston
Syracuse University

Abstract The community of scholars focused on information systems in organizations and society (the IFIP 8.2 community) has grown in number, voice, and influence over the last 25 years. What will this community contribute during the next 25 years? We speculate on two possible areas: more articulate conceptualizations of information systems and more detailed socio-technical theories of their effects. For both of these possibilities, we project forward from the historical trajectory of the IFIP 8.2 community's involvement. Like all speculative scholarship, our argumentation is more about imagining possible directions than arguing the superiority of one particular view relative to all others. This considered speculation is directed at both stirring the community's collective mind and advancing the value of this community's work to interested others.

1 INTRODUCTION

The community of scholars focused on information systems in organizations and society[1] has grown in number, voice, and influence over the last 25 years. In this paper

[1]This is known here as the IFIP 8.2 community. By IFIP 8.2 we mean the International Federation on Information Processing's (IFIP) Technical Committee 8 (information systems) Working Group 8.2 (information systems in organizations and society). The community has grown to be about 300 members and friends with about 60 percent who are not in North America.

we presume to look ahead and ask: What will this community contribute during the next 25 years? Responding to this question, in this paper we propose two areas where the IFIP 8.2 community of scholars might make a particular contribution: better conceptualizations of information systems and more detailed socio-technical theories of their effects on organizations and society.

To provide context we begin with a brief introduction to the intellectual geography of the IFIP 8.2 community. In doing this we identify six contributions of the contemporary IFIP 8.2 scholarship. We then introduce and discuss two candidates for the IFIP 8.2 community's consideration as particularly promising opportunities for future research. Our two candidates (conceptualizing IS and socio-technical theorizing) hold particular promise. Clearly, however, the members of the IFIP 8.2 community will engage (and lead) in numerous other opportunities such as methodological innovations, the broadening of viable research domains for IS research, adapting and using social theories to study IS, and championing broader (non-Popperian) views of science in IS research (e.g., Dutton 1999).

2 THE INTELLECTUAL GEOGRAPHY OF THE IFIP 8.2 COMMUNITY

The IFIP 8.2 community roots can be traced to meetings held in the early 1980s by a dedicated group of scholars, which led to the milestone 1984 Working Conference held in Manchester, UK. The scholars at these early meetings, many of whom also attended the 1984 conference, came primarily from Europe and the UK, from a variety of disciplines, and shared an interest in the mutually interdependent nature of the effects and influences of information systems, organizational operations, and social life. Contemporary IFIP 8.2 efforts and its charter reflect the energy of the early 1980s to legitimize scholarly attention on the study of information systems in organizations.[2]

The specific efforts of the early members of IFIP 8.2 suggests that they had a different vision of the IS field than did the early ICIS organizers (e.g., Keen 1980). The scholars who helped to found IFIP 8.2 explicitly linked their interests to the burgeoning presence of information systems in organizational and societal contexts. They espoused through their works a willingness to use what are now known collectively in the IS community as intensive research methods.[3] The early members of the IFIP 8.2 community focused on theory-building research and did their work in a wide range of settings (homes, hospitals, schools, public sector units, and small businesses). These scholars can now be found in business schools, information science schools, and a wide range of computing- and information-oriented schools.

See http://www.ifip.org and http://www.ifipwg82.org for more information.

[2]This growing interest was also signified by the first International Conference on Information Systems (ICIS) in 1980. The ICIS and IFIP8.2 communities overlap and their growth is interconnected.

[3]Intensive research methods include field-based studies employing case study and ethnographic techniques, hermeneutic approaches, and textual analyses. See Lee (1997) and Weick (1984).

Conversely, the mainstream of the ICIS community has tended to focus on large, for-profit, organizations (e.g., the Fortune 500).[4] The ICIS community has tended to embrace experimental (or quasi-experimental) research approaches and emphasize theory-testing research. Since 1980, the ICIS community has grown to number 2,300 people, with more than half located in North America.[5]

Both the ICIS and IFIP 8.2 communities are weakly tied to information science, sociology, communications, medical informatics, and public administration disciplines—where studies of the effects of information systems and organizations are also conducted (e.g., Ellis et al. 1999). This disconnect suggests that creation and maintenance of community identity, an important part of the last 25 year's effort by both the IFIP 8.2 and ICIS communities, may have come at the expense of connecting to like-interested research communities.

3 SUMMARIZING THE IFIP 8.2 COMMUNITY'S CONTRIBUTIONS

As a starting point to understanding the distinctive characteristics of the IFIP 8.2 community's contributions, we draw on the 23 edited proceedings of the IFIP 8.2 working conferences as an archive of the community's research output. While the IFIP 8.2 community's intellective production and intellectual contributions extend far beyond these specific books, they provide a window into the community's heart. Jones (2000) and Sawyer and Chen (2002) have, for different purposes, both reviewed this collection. Beyond that, Lee and Liebenau (1997) provide some interim reflections (on method, particularly). The editorial comments in each of these edited books highlight the ongoing discourse, current issues, and place in the community's trajectories of interest.

From these analyses of the IFIP 8.2 collected volumes, we identify and discuss six characteristics: (1) an orientation towards social theories, (2) dominant conceptualizations of information and technology and a common level of analysis, (3) an orientation toward the use of intensive research methods, (4) use of critical and analytical perspectives in research, (5) an openness to a range of research settings, and (6) an open discourse on the study of IS in organizations and society.

3.1 An Orientation Toward Social Theories

By social theory, we mean here theories that take into account the relations among individuals or collections of individuals. Social theories further reflect the presence of enduring social relations, their effects, or the nature of these effects. The IFIP 8.2 literature often draws on social theories, for example, those of Latour, Giddens, Habarmas, and others (Jones 2000). Jones notes that over the last 25 years a variety social theories have been drawn into the IFIP 8.2 community and adapted for specific

---

[4]This seems to reflect that the primary academic home for many ICIS scholars is a business school.

[5]Here we use the subscription numbers to the ISWORLD listserv as a surrogate measure of this community. See http://www.isworld.org.

attention to IS (e.g., Walsham 1997). He also notes that the use of social theory is uncommon (found in less than 5 percent of papers) in the broader IS literature, which is dominated by individualistic theories of human behavior (Lamb and Kling 2003; Sawyer and Chen 2002).

3.2 Dominant Conceptualizations of Information and Technology and Common Levels of Analysis

Sawyer and Chen's (2002) analysis of the IFIP 8.2 literature indicates that this body of research has several common conceptualizations. For example, *information* is often conceptualized as embedded in a web of people and their uses of ICT. That is, information arises from discourse and cannot be removed from the people and context in which it occurs. Technology (*ICT*) is most often conceptualized as ensembles of artifacts and people and represented in functional, or use-based, analyses. This functional analysis is often done by juxtaposing findings of use with design intentions of the ICT.

The common *levels of analysis* in IFIP 8.2 research are organizational or social and there is an explicit linking between individual actions and the larger social context. In this view, *people* are social actors (Lamb and Kling 2003), having some, but not complete, individual agency and living and acting within a shifting set of social norms that, while fluid, shape behavior in both subtle and direct ways. The contextual nature of this work means that the level of analysis is explicitly tied to a larger context. As Avgerou (2002) notes, the *social and organizational contexts of ICT use* are often conceptualized as institutions. By institutions we mean here enduring social structures that can be both stable *and* changeable (e.g., Agre 1999, 2000, 2003; Scott 2002).

3.3 An Orientation Toward Intensive Methods

The IFIP 8.2 community is one of the primary forums in IS for field work, ethnography, hermeneutics, action research, critical analysis, and a range of research methods that do not rely on inferential statistics, simulations, computer modeling, etc. Exemplars and discussions of these intensive methods have been showcased in the 1985, 1991, and 1997 edited proceedings (see Jones 2000). Lee and Liebenau (1997, p. 5) further note that "all Working Conferences of IFIP 8.2 are about qualitative research."

3.4 Critical and Analytical Perspectives

The IFIP 8.2 research often takes a critical or analytic perspective, eschewing normative angles. The *critical perspective* is one that examines the roles, values, uses, and purposes of ICT without automatically and uncritically accepting the goals and beliefs of the groups that commission, design, or implement specific ICTs (e.g., Wastell 2002). An *analytical perspective* is one that leads to developing theories about ICTs in institutional and cultural contexts, or to doing empirical studies that are organized to contribute to such theorizing (e.g., Richardson 2003). In contrast, a *normative perspective* to research is one whose aim is to recommend alternatives for professionals who design, implement, use, or make policy about ICTs.

3.5 Openness to a Range of Research Settings

The IFIP 8.2 community's research reports on the roles of information systems in a wide range of settings: developing countries, education, military, small and medium enterprises, faith-based organizations, formal and informal communities, online milieus, public sector/government and larger social settings.[6] This broadened view of acceptable settings for studying the roles, functions and purposes of ICT provides verdant territory for theorizing on the institutional nature of ICTs (as we discuss in more detail in the next section).

3.6 An Open Discourse

This final characteristic reflects the IFIP 8.2 community in its openness to diverse views on, and approaches towards, IS research. Over the past 25 years, there has been explicit attention to recruiting new members via workshops and working conferences. Concurrently, there has been an explicit educational agenda to provide alternative conceptualizations, research methods, and domains to study IS. The IFIP 8.2 scholarly discourse has also focused on understanding and theorizing on IS, and not focused on falsification and hypothesis testing (which often characterizes IS research). In this acceptance of intellectual plurality and anomaly, the IFIP 8.2 community reflects more a Kuhnian,[7] rather than a Popperian, view of science.

In summary, and like others, we assert that these six characteristics of the IFIP 8.2 research, and the evolving and growing community of scholars producing the work, have influenced the larger field of IS to accept these contributions (Markus 1997). In contrast, Jones notes that very little of the larger IS community research draws on the theory base that the IFIP 8.2 community uses. Sawyer and Chen note that the IFIP 8.2 conceptualizations of technology and people represent a small percentage of the work published in premier IS research outlets. They also note that much of this work speaks more to the organizational and social context than it does about the nature and form of the ICT and IS in these contexts.

The IFIP 8.2 community's research may be influencing the larger IS discourse, but we speculate that this influence is limited (or at least should be larger and more profound than its current levels). The last 25 years of work in IFIP 8.2 has been driven in part by twin desires. One desire has been to shape the larger discourse. A second desire has been to provide a forum for those who take on critical stances, social theories and intensive approaches to IS work. The successful efforts to legitimate IFIP 8.2 as a community within IS over the past 25 years provides us the current opportunity to focus more on shaping the future discourses on IS.

With this as context, we return to our formative question: Over the next 25 years, what should the community of IFIP 8.2 scholars contribute to IS research and, more broadly, society? Underlying this question is the belief that our community needs to be

---

[6]As mentioned above, much of the IS community has generally focused on private sector corporations (exemplified by *Fortune* 500 companies).

[7]Or perhaps a Bourdiean view....

more than an intellectual home for those taking on different approaches to the conduct of IS research, examining unique and underexplored domains, and privileging social theories.[8] Communities have aspirations and these aspirations must both excite the members and extend beyond self-preservation. For the IFIP 8.2 community, it seems these aspirations take form as opportunities to continue shaping the academic discourse on, and society's understanding of, the roles and values of ICT.

4 TWO OPPORTUNITIES FOR THE IFIP 8.2 COMMUNITY

We focus in the balance of this paper on two particular opportunities. These represent the confluence of our community's intellectual trajectories, scholarly strengths, gaps in the larger discourse on information systems, and the needs of society. The first opportunity is to examine more critically and publicly the meanings and features of ICT. An important contribution of the larger IS community (and we include the IFIP 8.2 scholars as a part of this larger group) is to help science and society understand what it means to have and use ICT. Orlikowski and Iacono (2001) argue that ironically the IS field has paid little attention to conceptualizing its core artifact. Sawyer and Chen (2002) note that the IFIP 8.2 community specifically has focused more on the ways that social contexts shape the development, deployment, and use ICT than on specific meanings and features of the ICTs being studied.

The IFIP 8.2 community's research focus (and its strength) has been cross-level, context-embracing, and, while imprecise as a term, "wide-angle" scholarship on the nature and meanings of social and organizational structures in which ICT exist, and in the intertwined roles of ICT within organizations and society. Raising the level of attention on the meanings and nature of computing in these settings is the right next step because it specifically fits the IFIP 8.2 community's intellectual trajectories, the group's research strengths, fills gaps in the larger IS discourse, and addresses the needs of society.

That is, the IFIP 8.2 community's orientation to theory-building use of intensive methods, and attention to a broad discourse provides the basis for moving forward our theorizing on the nature and meaning of ICT. We speculate that the IFIP 8.2 community's theory-building orientation means that members are likely to be interested in the role of theory development around ICT and have methodological skills to pursue this effort. Further, we assert that this theory building orientation will be aided by the community's social norm of open discourse, meaning that those who take on ICT theory-building will find a venue and community to both critique and support their efforts.

The second opportunity we see for the IFIP 8.2 community is to continue to extend and better articulate our theories and concepts of the socio-technical nature of ICT.[9]

---

[8]In saying this, we make explicit the need to both continue being an institution that balances shared interests and individual autonomy. We are suggesting opportunities to develop shared interests, not prescribing particular futures.

[9]Mumford (2000) noted that the early work by scholars in IFIP 8.2 relative to socio-technical theorizing has faded from more recent discourse.

Work in this area often draw from related literature in science and technology studies (for example, Bijker 1995; Giddens 1984; Hughes 1987; Latour 1987) to explore the ways in which technologies are formed, influenced, and adapted by the social milieus in which they exist. However, Baskerville and Myers (2002) have argued that IS may now be in a position to articulate its own theories and thus become a reference discipline for others interested in the use and effects of ICT. And, we seeing glimpses of socio-technical theories explicitly developed toward ICT from friends of IFIP 8.2 (e.g., Kling and McKim 2001).

The IFIP 8.2 community's interests in social theories and its role in understanding and explaining the roles that ICT play in organizations and society suggest that there is interest in developing better socio-technical theories of ICT. Our community's theory-building orientation, expertise in intensive methods, and the norms of open discourse further suggest that our community is well-positioned to take on socio-technical theorizing. Moreover, when combined with the first opportunity (to better theorize on ICT), it may be that the two opportunities can, for some in the IFIP 8.2 community, be self-reinforcing.

4.1 Opportunity One: Better Conceptualizing ICT

The first opportunity for the IFIP 8.2 community is to focus on better developing the concepts and constructs of ICT. We focus on conceptualizing ICT because, for at least two reasons, the study of IS is also the study of ICT. First, ICT is central to essentially all contemporary IS and certainly to those that have more than a local effect. Second, much of the growth in the use and value of IS can be attributed to the increased power of the underlying ICT.

4.1.1 Conceptualizing ICT

In suggesting a focus on conceptualizing ICT, we build on Orlikowski and Iacono (2001) who identified five general approaches to conceptualizing ICT in the scholarly discourse on IS: (1) as a tool or set of features, (2) as an ensemble or a set of functions, (3) via some proxy, (4) as a computation, or (5) not at all (a nominal approach).[10] In Appendix A we briefly discuss each of the characterizations of ICT relative to what it means for conceptualizing an IS. In the rest of this section we outline current conceptualizations of ICT in the IFIP 8.2 community, explain why this opportunity is worth collective interest, and suggest a path forward.

4.1.2 Current Conceptualizations of ICT in the IFIP Community

Sawyer and Chen's (2002) analysis of the IFIP 8.2 literature reveals that 55 percent of the papers published in the IFIP 8.2 conference proceedings between 1984 and 2000

---

[10]Other approaches such as Taylor's (1982, 1986) value-adding model of ICT or Henderson and Cooprider's (1990) functional model of ICT are also viable bases for this theorizing. We focus on Orlikowski and Iacono's (2001) approach because it reflects the most recent and most broad-based effort to categorize approaches to theorizing ICT.

represent ICT as either a proxy or as presence/absence. In other words, a bit more than half of the papers have little to say about the details of the ICT.[11] Simply, the focus has not been on ICT; it has been on the context around ICT. The remaining 45 percent of the papers characterize ICT (and the larger IS) as either collections of features or as functional ensembles. We therefore focus on these two later approaches and especially on the linkage between these views as a starting point for building richer conceptualizations of ICT.

Functional conceptualizations of ICT differ from feature-based views in the level of abstraction (Henderson and Cooprider 1990; Taylor 1982, 1986). Feature-based conceptualizations detail the design intentions of an ICT and focus on the direct or intended effects of particular elements of an IS or the ICT in an IS. Functional conceptualizations focus on the ways in which ICT get used. Functional models of ICT provide a means to better assess new technologies and compare across existing technologies. By explicitly linking the ICT and people's uses, the focus of the approach is on empirical studies of use and the intended and unintended consequences of use. For example, and building on Taylor (1986), Sawyer and Tapia (2003) document five functional uses of ICT: to support personal or institutional production, to allow for control, to provide access to information, to enable coordination, or to entertain. They do not, however, link this functional conceptualization to any particular set of features, leaving open the means to translate their functional view into specific feature sets.

4.1.3 Towards Richer Conceptualizations of ICT

Given the underlying differences among feature-based and functional conceptualizations of ICT (and IS), it is often difficult to reconcile the findings from these studies. Studies that present functional depictions of computing packages do not easily map to specific features. Likewise, specific features of an IS are not easily mapped to their functional uses (e.g., Taylor 1982, 1986). Scholarship that explicitly focuses on these relationships among uses by users leads to identifying key elements of a functional view of ICT (i.e., Palen 2003). Further, studies that explore the differences in intention and use from a feature-based view of ICT (as is common in studies of human-computer interaction) are steps toward reconciliation. Whatever the path, the opportunity is to develop a means to bridge these two approaches. A trajectory of research that will help bring together the findings from these two approaches may lead to additional conceptualizations of ICT, to a means of translating or relating differing views, and to discourse on the assumptions, methods, and problems tied to each of the two current conceptualizations. A more robust conceptualization of ICT, one that links features of a particular form to their functional uses, would enable professionals in practice to make better decisions on the value and effects of new IS. Likewise, conceptualizations of ICT that allow us to combine or bridge feature and function perspectives will lead to stronger theories of use, value, and outcome.

---

[11]There was virtually no proof-of-concept work published in the IFIP 8.2 literature. This is not surprising given the community's focus on the social aspects of IS rather than building IS.

4.1.4 Why this Opportunity?

There are two reasons why the IFIP 8.2 community is particularly well positioned to engage this work. First, the adoption of a critical and analytic perspective has led scholars in this community to conceptualize technology in use, rather than taking as authoritative the intended purposes of the technology. Second, the IFIP 8.2 community's use of intensive methods provides access to the types of rich data that will be needed for the work of conceptualizing activity that theory-building around ICT will require. In doing this we anticipate that out theorizing on ICT will broaden (perhaps extending current Orlikowski and Iacono depictions) our means of representing what an IS is and means.

4.2 Opportunity Two: Developing Socio-Technical Theories of IS Effects

The second opportunity we see for attention by the IFIP 8.2 research community is to more explicitly pursue the arrangements, interactions, and elements of socio-technical theorizing relative to IS. Weick (1995) argues that theorizing is a process, while theory is an outcome of that process. He further notes that theorizing is often not visible, but should be.

4.2.1 Why this Opportunity?

Scholars generally accept that IS are socio-technical, although there seems to be little systematic attention paid to the arrangements, interactions, and elements of this socio-technical relationship (Bostrom and Heinen 1977a, 1977b; Mumford 2000). The acceptance stands unquestioned and underdeveloped, separated from the main discourses of IS and mostly outside the main discourses of socio-technical scholars. The IFIP 8.2 community, known for its open discourse, uses of social theory, and theory-building orientation, seems well-positioned to pursue more explicit theorizing on the relations among the social and the technical.

4.2.2 Towards Richer Conceptualizations of ICT Effects

There are a number of different conceptualizations of socio-technical research, many of which have been successfully applied to studies of ICT, such as actor-network theory, structuration theory, social shaping of technology, social construction of technology, and institutional approaches.[12] Socio-technical theories focus on the multiple, intricate, and evolving links among social and technical elements and helps to highlight how these elements exist in relation (not independently) of one another. As we outline below, Bijker's (1995) four principles of socio-technical change theory help illustrate the generic goals and the theoretical tensions that exist in socio-technical

---

[12]For more information on these approaches, see MacKenzie and Wacjman (1999).

research: (1) the *seamless web* principle, (2) the principle of *change and continuity*, (3) the *symmetry* principle, and (4) the principle of *action and structure*.

(1) The *seamless web* principle states that any socio-technical theory of ICT should not *a priori* privilege technological or material explanations ahead of social explanations, and vice versa.
(2) The principle of *change and continuity* argues that socio-technical theories of ICT must account both for change and for continuity, not just one or the other.
(3) The *symmetry* principle states that the successful working of a technology must be explained as a process, rather than assumed to be the outcome of superior technology.
(4) The *actor and structure* principle states that socio-technical theories of ICT should address both the actor-oriented side of social behavior, with its actor strategies and micro interactions, and the structure-oriented side of social behavior, with its larger collective and institutionalized social processes.

While Bijker's principles provide a set of ideals for socio-technical research to strive for, in practice they illustrate tensions to be managed in the research process. Specific studies of ICT and IS using the socio-technical perspective will vary depending on

- the degree to which they focus on technological processes and technological features versus social processes and social objects versus balancing both (*seamless web*)
- the degree to which they focus on processes of change versus continuity with the past and maintenance of existing processes versus balancing both (*change and continuity*)
- the depth of description and explanation in the socio-technical change process, ranging from *thick* or *rich* descriptions of an ICT to relatively *thin* associations between outcomes and various success factors (*symmetry*)
- the degree to which they focus on action-oriented aspects of social behavior, such as actor strategies and interests, versus structural aspects of social behavior such as institutional processes (*action and structure*)

A socio-technical perspective on ICT leads us to argue that ICT are simultaneously a technological and social phenomenon, and that theories of ICT should address both of these aspects. A socio-technical perspective also provides a means to organize and critique existing research. The IFIP 8.2 community has been aggressively pursuing this high-level goal (although, as we note in the previous section, with more attention to the social and relational elements of socio-technical systems).

4.2.3 Two Examples of Socio-Technical Theories of ICT

In the rest of this section, we highlight two examples which represent some of the nascent work being done in the IFIP 8.2 community to move beyond declaring that things are socio-technical. The work of Crowston (2000) to advance a theory of process and Avgerou's (2002) work where she advances a broad-scale depiction of IS as innovation processes help us illustrate socio-technical theorizing. Both should be seen as starting points.[13] Simply, these two examples are illustrative of socio-technical theorizing, not exclusive.

4.2.3.1 Crowston: Theories That Link IS, Individuals and Processes

Crowston (2000) argues that focusing on processes (conceptualized as sequences of activities) provides insight into the linkages among individual work and the larger work setting. Individuals perform the various activities that comprise the work done by the organization. A process focus is often used to analyze organizations. However, process theories can provide a means to link the activities in nonwork settings. For example, Frissen (2001) reports on the uses of cellular phones to manage the joint schedules of family members. This effort is detailed as a series of cases, each of which can be developed as a sequence of activities that demand coordination (some of which is enabled by the uses of ICT).

In process theories, uses of ICT are enacted by individuals who, through their actions, change the conduct of their work in response to the availability of these technologies. Therefore, ICT has to be conceptualized as combination of the technology and the process to which it is applied, what Orlikowski and Iacono refer to as "technology as embedded system" (2001, p. 126). It might be that ICT use makes individuals more efficient or effective at the activities they have always performed. For example, a manager using a spreadsheet to analyze a decision may be able to reach a conclusion more quickly or to consider more alternatives, thus improving the speed or quality of the process without changing the activities involved. A real estate agent might be able to search the database of house listings for each client every day instead of just occasionally. Yet, the same ICT used in the same process but in a different context (e.g., for a highly qualified and experienced agent dealing only in large mansions) might lead to fewer (if any) benefits.

In order to understand how changes in individual work might have a larger effect, Crowston argues that we must examine the ways that individual work changes relative to the process of which they are a part. To understand how individual work changes affect the process, it is necessary to examine the constraints on assembling activities that limit the possible arrangements and rearrangements of activities into processes. To identify these constraints, Crowston focuses in particular on the implications of dependencies between tasks for process assembly. For example, *producer/consumer* dependencies restrict the order in which activities can be performed, since the activity that consumes some resource (a piece of information, for example) can not be performed by the activity that produces it. Activities that are not involved in a dependency can be more easily rearranged. Beyond constraining the order of activities, dependencies often require additional activities to manage them. For example, Malone and Crowston (1994)

---

[13]There are, of course, other candidates for this discussion whose work may be known by members of IFIP 8.2 such as Ackerman's (2000) exploration of the social and technical in computer-supported cooperative work, Mansell's (2002) work on communication technologies and social change, Agre's (2003) work on the relationships among technical architectures and institutional structures, and Kling and McKim's (2000) socio-technical interaction networks.

note that the *producer/consumer* dependency not only constrains the order of the activities (a *precedence* dependency), but may also require additional activities to manage the *transfer* of the resource between or to ensure the *usability* of the resource.

Precedence requires that the producer activity be performed before the consumer activity. This dependency can be managed in one of two ways: either the person performing the first activity can notify the person performing the second that a resource is ready, or the second can monitor the performance of the first. Here ICT may have an effect by providing a mechanism for cheap monitoring.

Transfer dependencies are managed by a range of mechanisms for physically moving resources to the actors performing the consuming activities (or vice versa). For example, inventory management systems can be classified here. Using ICT may alter both the mechanisms of transfer dependencies or the actors involved in transfer. *Usability* can be managed by having the consumer specify the nature of the resources required or by having the producer create standardized resources expected by the user (among other mechanisms). The roles of ICT in usability might be to help specify or standardize resources.

More generally, process models highlight the value of ICT as a means of change because many of the coordination activities in processes are primarily information processing. These activities are particularly susceptible to ICT-related changes. For example, increased information flow between the user and producer of a resource may make it easier for the user to convey specific needs, allowing the producer to create a more precisely tailored product. A process perspective is broadly applicable. For example, a process perspective can help us to explore and understand knowledge work and work processes, information flows such as exchanges of digital goods, and transorganizational interactions.

4.2.3.2 Avgerou: ICT and Global Diversity

In her treatise on IS and global diversity, Avgerou (2002) also depicts ICT as embedded in ensembles—although these ensembles are depicted as IS of a national or multinational scale. In contrast to the deterministic nature of process theorizing, she further promotes the variegated nature of the relations among an IS and its context as leading to "non-deterministic, non-essentialist view(s) of ICT..." (p. 231). Avgerou defines IS as innovation processes that unfold over time. This process of unfolding occurs through discourses and actions by mangers, technologists, users, and others. The process is erratic in that path dependencies are unclear; the innovation effort is intimately tied to events that unfold in particular situations and that have historical roots. These processes tend to reshape the thinking about, and the uses of, the IS and to the disenfranchisement of some participants.

In this forms of socio-technical theorizing, *social contexts* form the basis in which *innovation processes* unfold. This unfolding reflects a relationship among *actors* connected in a *network* of relations that both enable and constrain action. The *temporal* nature of this process and the relations among the actor network and that process are primarily driven by the context (and not some *a priori* force) in ways that can be locally justified and realized, but may seem chaotic and unpredictable if the contexts of action are not appreciated.

This form of socio-technical theorizing demands a critical analysis of the multiple institutions that make of the context and a depiction of the temporal process of innovation noting key events (both large and small scale) that together represent the way that the IS is depicted in discourse and through action. This analysis typically crosses several levels (e.g., connecting individual, organizational, and regional contexts and behaviors) and sets the technical elements of the IS within social contexts (focusing on collective uses and less on individual user behaviors[14]). The IS innovation process is thus represented as multiple paths, with the discourses and events used to both relate these paths and to depict the seemingly erratic process of negotiating multiple interests and agency over time. The critical orientation of her work is reflected in her highlighting that IS innovation efforts and "managerialist" thinking are often tightly linked, leading to disruptive processes and events.

Avgerou's critical theory of IS as an innovation process provides a means to reexamine normative conceptualizations of IS. We speculate that this might lead to alternative conceptualizations of IS and their effects (such as what Trauth and Jessup [2000] did regarding group decision support systems). A critical theory of IS will also provide a means to explore both large-scale innovations (such as ICT infrastructure studies) and small-scale deployments (such as work group and organizational implementations). Taking a non-managerialist perspective is also likely to raise questions on the power and effect of ICT (e.g., Clement 1994)

4.2.3.3 Comparing the Approaches

Crowston and Avgerou take different approaches to theorizing on the socio-technical nature of ICT. Crowston's process model is analytic. Avgerou takes a critical approach to developing a contextual model of IS as an innovation process. However, both showcase principles of socio-technical change theory. Crowston reflects the seamless web principle by connecting the actions of individuals through process and showing how ICT can also link to process. Avgerou reflects the seamless web principle by developing the process of technical development with the pressures of the social and organizational context. Crowston's process theory approach reflects the principle of change of continuity though the mechanisms of process and coordination, while Avgerou's approach depicts the temporal nature of IT as a process of innovation. Both Crowston and Avgerou make it clear that technology is path dependent and its use is negotiated (in context or as it pertains to process decisions). Finally, both Crowston's and Avgerou's theorizing reflect the actor and structure principle. Crowston notes that processes are the larger structure that relates actors. Avgerou highlights the institutional nature of context and how these shape both individual behavior and the temporal process of IS innovation.

Crowston's process theory and Avgerou's critical innovation process theory advance socio-technical theorizing. However, what about their characterizations of ICT? Both reflect ensemble or functional conceptualizations, although the specific characteristics of this ensemble or functional view remain underdeveloped. So, these represent excellent starting points for advancing conceptualizations of ICT. We further speculate that the IS (and other related academic) community's interest in socio-

---

[14]Individual level phenomena can be central, as Avgerou (2002, Chapter 7) notes in the study of Cyprus' efforts to take on new IS in small manufacturing enterprises. The point is that individual behavior can only be understood within the contexts in which it occurs.

technical theories of ICT is tied in part to the ability of these theories to provide additional insight into the social and the technical elements. It seems that too much of the insight of current socio-technical theories is on the social; the technical elements of these theories lags.

5 SUMMARY

Predicting the future is tricky, since we've not been there yet.[15] Speculating on the future is more commonplace, and we do that here to engage the IS community, and our colleagues in the IFIP 8.2 community more specifically, to think about what we can do. In this paper we limited ourselves to speculating on two possible areas where we believe IFIP 8.2 community of scholars are well-positioned to advance our understanding of ICT in organizations and society: to advance conceptualizations of ICT and to better develop socio-technical theories of ICT. Moreover, we note that these two opportunities can be linked. For example, members of the IFIP 8.2 community might take on conceptualizations of ICT and use these as the basis for more robust socio-technical theories. Conversely, a socio-technical lens can be used to advance our conceptual understanding of ICT.

Better conceptualizations of ICT, and more powerful socio-technical theories, are likely to be needed as we engage in the debates on the design, development, implementation, policies of use, and implications of presence for a host of emerging technologies. For example, the growing interest in deploying broad-scale sensor networks, increasingly self-reliant and semiautonomous computing applications, powerful biometrics, a range of nano-technologies, and ever more pervasive and ubiquitous computing devices and computing applications are, as Winner (1986) notes, artifacts that embed politics.[16]

Theory often travels faster and further than does the empirical work on which it is often painstakingly constructed and we should strive to become theory "exporters" as Baskerville and Myers (2001) have suggested. We speculate that collective attention to both better developing (and perhaps expanding) our conceptualizations of ICT and theorizing on their socio-technical nature is the type of intellectual export that will attract scholars in related fields like information science, sociology, communications, medical informatics, and public administration units to our work. Implied in this speculation is that the effort in the next 25 years by the IFIP 8.2 community is to connect to like-interested research communities.

Like all speculative scholarship, our argumentation is more about imagining possible directions than advocating for one particular view in opposition to all others. To this point, one strength of IFIP 8.2 scholarship is its diversity. So, in suggesting these two opportunities, we also decry attempts to constrain activities to these (or any other particular) directions. The innovative nature of the IFIP 8.2 research community is central to the value of the community's work. Our suggestions should be seen as a path that some in this community should take, leaving others to take other paths, or cut new

[15]As both Neils Bohr and Yogi Berri, among many, have noted....

[16]In particular, see Winner's chapter, "Do Artifacts Have Politics?" (pp. 19-39).

paths. This considered speculation is directed at both stirring the community's collective mind and advancing the value of this community's work to interested others. In 2024, what do we want the future members of our community to say on the eve of the 40th (and 20th) anniversary of the previous Manchester conference(s)?

ACKNOWLEDGEMENTS

Comments from Lynette Kvasny, Jim Jansen, Carleen Maitland, Andrea Hoplight Tapia, Duane Truex, and three anonymous reviewers have helped us to substantially improve this paper.

REFERENCES

Ackerman, M. "The Intellectual Challenge of CSCW: The Gap Between Social Requirements and Technical Feasibility," *Human-Computer Interaction* (15:2/3), 2000, pp. 181-205.

Agre, P. "The Architecture of Identity: Embedding Privacy in Market Institutions," *Information, Communication and Society* (2:1), 1999, pp. 1-25.

Agre, P. "Infrastructure and Institutional Change in the Networked University," *Information, Communication and Society* (3:4), 2000, pp. 494-507.

Agre, P. "Peer-to-Peer and the Promise of Internet Equality," *Communications of the ACM* (45:2), 2003, pp. 39-42.

Avgerou, C. *Information Systems and Global Diversity*, Oxford: Oxford University Press, 2002.

Baskerville, R., and Myers, M. "Information Systems as a Reference Discipline," *MIS Quarterly* (25:1), 2002, pp. 1-14.

Bijker, W. *Of Bicycles, Bakelites, and Bulbs: Toward a Theory of Socio-technical Change*, Cambridge, MA: MIT Press, 1995.

Bostrom, R., and Heinen, S. "MIS Problems and Failures: A Socio-Technical Perspective Part 1: The Causes," *MIS Quarterly* (1:3), 1977a, pp. 17-32.

Bostrom, R., and Heinen, S. "MIS Problems and Failures: A Socio-Technical Perspective Part 2: The Application of Socio-Technical Theory," *MIS Quarterly* (1:4), 1977b, pp. 11-27.

Brynjolfsson, E., and Hitt, L. "Beyond the Productivity Paradox," *Communications of the ACM* (41:8), 1998, pp. 49-55.

Clement, A. "Computing at Work: Empowering Action by 'Low-Level Users'," *Communications of the ACM* (37:1), 1994, pp. 52-63.

Crowston, K. "Process as Theory in Information Systems Research," in R. Baskerville, J. Stage, and J. I. DeGross (Eds.), *Organizational and Social Perspectives on Information Technology*, Boston: Kluwer Academic Publishers, 2000, pp. 149-166.

Dutton, W. "The Web of Technology and People: Challenges for Economic and Social Research," *Prometheus* (17:1), 1999, pp. 5-20.

Ellis, D.; Allen, D.; and Wilson, T. "Information Science and Information Systems: Conjunct Subjects, Disjunct Disciplines," *Journal of the American Society of Information Science* (50:12), 1999, pp. 1095-1107.

Frissen, V. "ICTs in the Rush Hour of Life," *The Information Society* (16), 2000, pp. 65-77.

Giddens, A. *The Constitution of Society: Outline of the Theory of Structure*, Berkeley, CA: University of California Press, 1984.

Henderson, J., and Cooprider, J. "Dimensions of I/S Planning and Design Aids: A Functional Model of CASE Technology," *Information Systems Research* (1:3), 1990, pp. 227-254.

Hughes, T. *Networks of Power*, Baltimore, MD: The Johns Hopkins University Press, 1983.

Jones, M. "The Moving Finger: The Use of Social Theory in WG 8.2 Conference Papers, 1975-1999," in R. Baskerville, J. Stage, and J. I. DeGross (Eds.), *Organizational and Social*

Perspectives on Information Technology, Boston: Kluwer Academic Publishers, 2000, pp. 15-32.

Keen, P. "MIS Research: Reference Disciplines and Cumulative Traditions," in E. McLean (Ed.), *Proceedings of the First International Conference on Information Systems*, Philadelphia, PA, 1980, pp. 9-18.

Kling, R., and Scacchi, W. "The Web of Computing: Computer Technology as Social Organization," *Advances in Computers* (21), 1982, pp. 1-90.

Kling, R., and McKim, G. "Not Just a Matter of Time: Field Differences in the Shaping of Electronic Media in Supporting Scientific Communication," *Journal of the American Society for Information Science* (51:4), 2000, pp. 1306-1320.

Lamb, R., and Kling, R. "Reconceptualizing Users as Social Actors in Information Systems Research," *MIS Quarterly* (27:2), 2003, pp. 197-235.

Latour, B. *Science in Action*, Milton Keynes, UK: Open University Press, 1987.

Lee, A. "*MIS Quarterly* Special Issue—Call For Papers: Intensive Research in Information Systems: Using Qualitative, Interpretive, and Case Methods to Study Information Technology," 1997 (available online at http://www.people.vcu.edu/~aslee/misq-sp-.htm).

Lee, A., and Liebenau, J. "Information Systems and Qualitative Research," in A. Lee, J. Liebenau, J. I. DeGross (Eds.), *Information Systems and Qualitative Research*, London: Chapman & Hall, 1997, pp. 1-8.

Malone, T., and Crowston, K. "The Interdisciplinary Study of Coordination," *Computer Surveys* (26:1), 1994, pp. 87-119

Mansell, R. *Inside the Communication Revolution: Evolving Patterns of Social and Technical Interaction*, London: Oxford University Press, 2002.

Markus, M. "The Qualitative Difference in Information Systems Research and Practice," in A. S. Lee, J. Liebenau, and J. I. DeGross (Eds.), *Information Systems and Qualitative Research*, London: Chapman & Hall, 1997, pp. 11-27.

McKenzie, D., and Wajcman, J. *The Social Shaping of Technology* (2nd ed.), Philadelphia: Open University Press, 1999.

Morrison, J., and George, J. "Exploring the Software Engineering Component in MIS Research," *Communications of the ACM* (38:7), 1995, pp. 80-91.

Mumford, E. "Socio-Technical Design: An Unfulfilled Promise or a Future Opportunity," in R. Baskerville, J. Stage, and J. I. DeGross (Eds.), *Organizational and Social Perspectives on Information Technology*, Boston: Kluwer Academic Publishers, 2000, pp. 33-46.

Orlikowski, W., and Iacono, S. "Desperately Seeking the 'IT' in IT Research—A Call to Theorizing the IT Artifact," *Information Systems Research* (12:2), 2001, pp. 121-124.

Palen, L. "Beyond the Handset: Designing for Wireless Communications Usability," *Transactions on Human Computer Interaction* (9:2), 2002, pp. 125-151.

Richardson, H. "CRM in Call Centres: The Logic of Practice," in M. Korpela, R. Montealegre, and A. Poulymenakpu (Eds.), *Organizational Information Systems in the Context of Globalization*, Boston: Kluwer Academic Publishers, 2003pp. 69-84.

Sawyer, S. and Chen, T. "Conceptualizing Information Technology and Studying Information Systems: Trends and Issues," in M. Myers, E. Whitley, E. Wynn, and J. I. DeGross (Eds.), *Global and Organizational Discourse about Information Technology*, Boston: Kluwer Academic Publishers, 2002, pp. 109-131.

Sawyer, S., and Tapia, A. "The Computerization of Work: A Social Informatics Perspective," in J. George (Ed.), *Social Issues of Computing*, New York: Oxford, 2003, pp. 93-109.

Scott, W. *Institutions and Organizations*, Thousand Oaks, CA: Sage Publications, 2001.

Simon, H. "Organizations and Markets," *Journal of Economic Perspectives* (5:2), 1991, pp. 25-44.

Trauth, E., and Jessup, L. "Understanding Computer-Mediated Discussions: Positivist and Interpretive Analyses of Group Support System Use," *MIS Quarterly* (24:1), 2000, pp. 123-156.

Taylor, R. "The Value-Added Model," Chapter 4 in *The Value-Added Processes in Information Systems*, Philadelphia: Ablex Publishing Corporation, 1986, pp. 48-70.

Taylor, R. "Value-Added Processes in the Information Life Cycle," *Journal of the American Society for Information Science* (31:5), 1982, pp. 341-346.

Walsham, G. "Actor-Network Theory and IS Research: Current Status and Future Prospects," in A. S. Lee, J. Liebenau, and J. I. DeGross (Eds.), *Information Systems and Qualitative Research*, London: Chapman & Hall, 1997, pp. 466-480.

Wastell, D. "Organizational Discourse as Social Defense: Taming the Tiger of Electronic Government," in M. Myers, E. Whitley, E. Wynn, and J. I. DeGross (Eds.), *Global and Organizational Discourse About Information Technology*, Boston: Kluwer Academic Publishers, 2002, pp. 179-196.

Weick, K. "Theoretical Assumptions and Research Methodology Selection," in F. W. McFarlan (Ed.), *The Information Systems Research Challenge*, Boston: Harvard Business School Press, 1984, p. 115.

Weick, K. "What Theory is Not: Theorizing Is," *Administrative Science Quarterly* (40), 1995, pp. 385-390.

Winner, L. *The Whale and the Reactor: A Search for Limits in an Age of High Technology*, Chicago: University of Chicago Press, 1986.

ABOUT THE AUTHORS

Steve Sawyer is a founding member of the Pennsylvania State University's School of Information Sciences and Technology. Steve holds affiliate appointments in Management and Organizations; Labor Studies and Industrial Relations; and Science, Technology and Society. Steve does social and organizational informatics research with a particular focus on people working together using information and communication technologies. Steve can be reached at sawyer@ist.psu.edu.

Kevin Crowston joined the School of Information Studies at Syracuse University in 1996. Before moving to Syracuse, he was a founding member of the Collaboratory for Research on Electronic Work at the University of Michigan and of the Center for Coordination Science at MIT. Kevin has published articles and book chapters in the area of information systems and new organizational forms. His current research interests include empirical studies of coordination-intensive processes in human organizations; theoretical characterizations of coordination problems and alternative methods for managing them; and design and empirical evaluation of new kinds of computer systems to support people working together. A specific example of the final interest is the application of document genre to the World-Wide Web. He can be reached at crowston@syr.edu.

Appendix A

FIVE CONCEPTUALIZATIONS OF ICT

The *feature or tool view* is the most common (or received) view of ICT. A tool view means that ICT is characterized as a collection of features that will operate as they were designed to behave. The roles of the ICT/features are seen as primarily technical in nature and direct in their effect. These feature-based approaches to studying ICT focus on the values, effects, and impacts of particular (and identifiable) technical aspects of an ICT. The resulting IS is thus an aggregation of the various features. When ICT are depicted as collections of features, people are characterized as both individuals and as social units, and the level of analysis is typically institutional. Feature-based ICT research is most often conducted as either a theory-building or intensive research effort. The ICT being studied is depicted as a moderating or mediating factor.

Conversely, the *functional or ensemble view* of ICT is one where specific artifacts and people are interdependently connected through roles, uses of information, and actions. In the functional view there is often an explicit attention on the ways of using a particular ICT. This conceptualization of ICT leads to so-called web models of IS in which people, their roles, their uses of and needs for information, and the computing elements (hardware and software) are connected together and embedded in a larger social milieu (Kling and Scacchi, 1982). When ICT are conceptualized as functions or ensembles, people are depicted as social actors. The research is typically structured as a cross-level analysis, but at varying aggregate levels. The ICT being studied is often depicted in terms of functionality provided through use.

The *proxy* view of ICT is that some (often quantified) surrogate can capture or measure the value of ICT. For example, a researcher might use IS budgets as a proxy for the level of ICT in use. Proxy views of IS and ICT focus on making clear the ways in which the measure highlights the value of the ICT or IS. The *computational* view of ICT focuses on the computational power or abilities of an artifact. This approach highlights the construction of a computational artifact, where that artifact instantiates an idea or theory. Finally, *nominal* treatments of ICT are those where a study mentions ICT without actually including it in the study.

4 INFORMATION SYSTEMS RESEARCH AS DESIGN: Identity, Process, and Narrative

Richard J. Boland, Jr.
Kalle Lyytinen
Case Western Reserve University

Abstract

Information systems research has moved beyond the antagonistic dualisms that dominated its discourse over the last 20 years. Our community is now largely inclusive of diverse research traditions, and without a strong dogma. But it is also experiencing an identity crisis. Scholars are asking what the information systems field is or should be, and where it is going or should go. In this paper, we argue that such questions, although understandable as sense-making devices, are fundamentally misdirected because they ask about the *things* that should be part of our identity rather than the *process* through which we should construct it. As an alternative to this search for identity through the identification of things with which to align, we propose that a better way forward is to appreciate that researchers are designers. Viewing researchers as designers allows our identity to emerge from the unique and critical processes whereby we both reflect and shape the socio-technical world, as well as establish our position in an intellectual field. Viewing the researcher as designer leads to a questioning of the structurational processes in which researchers are, at the same time, both representing the socio-technical world (it is our medium) and shaping it through our knowledge generation (it is our outcome). Our ongoing choice of theories, methods, artifacts, and subjects becomes a fateful, existential choice of our identity, for which we should assume responsibility in the reflexive monitoring of our research conduct. The narratives we accept or resist in making our studies of information systems constitute our identity, as well as that of our field and of our subjects. The way forward is to take responsibility for maintaining a dynamic balance in the existential choices through which we bring ourselves as researchers, our research subjects, and the socio-technical world into being.

1 INTRODUCTION

Since the first IFIP 8.2 conference on research methods in 1984, much has changed in the field of information systems research. The 1984 conference was characterized by our fervent struggles to define the *correct way of doing research* and to raise the concerns of previously silenced voices in those debates. The conference and the resulting book (Mumford *et al.* 1985) offered a stage for alternative views on what it meant to do our research. As a result, the conference opened the gates of the information systems domain and allowed the demons and gods of continental philosophy to argue freely in its research terrain. This resulted in debates between a series of bi-polar oppositions, such as subjective versus objective, quantitative versus qualitative, positivist versus interpretivist, and critical versus noncritical modes of information systems inquiry (see, for example, Boland 1985; Jenkins 1985; Klein and Lyytinen 1985; Lyytinen and Klein 1985; Sandberg 1985). The tension of these oppositions fueled much of the research discourse in information systems for the next 10 to 15 years as reflected in the conferences that followed (Lee et al. 1997; Nissen et al. 1991).

But now, after 20 years and the turn of the millennium, these issues have become largely passé. Whereas information system scholars at the 1984 conference argued within an "either/or" dichotomy of research approaches for their favored one, we now find a more secure appreciation of a "both/and" approach that is inclusive and values the mutually informing capabilities of multiple research methods and traditions (see, for example, Baskerville and Myers 2002; Mingers 2001). This same attitude is visible in our theorizing, and we are pleased to note that the information systems field has not fallen prey to the "one best theory" mentality that can be seen in some other management disciplines (e.g., Robey 1996). Instead, the information systems field currently offers a relatively open space for multiple modes of inquiry with an almost bewildering diversity of theoretical perspectives, research topics, data collection strategies, and analysis methods.

Many of the nagging research questions that were on the intellectual agenda of our IFIP 8.2 community 20 years ago have been resolved in our theory and practice. *Apres la lutte*, we can say that we have won and that there is a sense of movement and accomplishment in the field. But at the same time, there is also a growing sense that we are floundering. Many of us ask: who are we as a discipline? What is the artifact and domain we are supposed to study? Are we doomed to be a second tier discipline, borrowing theory and methods from others? Are we weak because we do not have our own unique theories and methods? (For examples, see Baskerville and Myers 2002; Benbasat and Zmud 2003; Orlikowski and Iacono 2001; Weber 2003.) In short, although we have progressed beyond the simple dualisms that characterized our inquiries at its dawning, we now find ourselves in another quagmire, rooted in a questioning of our identity.

In this paper we analyze the question of identity as found in our theories, our methods, and our reflexive practice. From that analysis, we propose a way forward in thinking about the intellectual mission and the future role of information systems research. We propose that doing our research be viewed as a design act in which we are located at a boundary between making representations of the socio-technical world as it is, and changing it into what it might be. We are always moving between describing it and constructing it, between seeing it and shaping it, between reflecting upon it, and acting upon it and thereby facing moments of existential choice in our identity construction. Our identity is made and remade through those existential choices of what we do. It is a process that we live in language, not a thing we should be or should align with. We will first discuss the concept of identity and characterize the way we see the information systems field currently addressing the question of its identity. We will then pose an alternative approach to identity construction based on viewing the researcher as a designer. This shift in approach moves us from a search for a fixed constellation of things that define our field toward a reflexive awareness of our engagements in the research process, informed by structuration theory. The role of narrative in constructing identity is then elaborated and related to our conduct of research. Finally, our unique responsibility for existential choices in our research process is highlighted with an illustration of how structuration helps us understand a current research project of our own. Finally we formulate some implications for a way forward in constructing an identity as information system researchers.

2 INFORMATION SYSTEMS AND IDENTITY

Identity refers to a set of characteristics that makes a thing recognizable or known. The word comes from the Latin *idem,* "the same." The essential attribute of identity is consistency and persistence in character: it is that which remains the same under changing circumstances. Identity as continuity over time is also one's sense of self. Within the information systems field, our research identity refers to a wide range of phenomena that display consistency, including the uniqueness of our research objects, our personal identity as researchers, the specific identities of our theories and methods, and, finally, the identities of actors and artifacts that play roles in our theories.

If there were an enduring and firm foundation for knowing about information systems, and for defining who we are as researchers and what our research object is, the question of identity and existence could not be raised with much effect. Paradoxically, then, the success we have had in moving beyond the hope for one true form of research, which has resulted in being more open to a variety of theories and methods, has paved the way for recognizing our current problematic of identity. Most of the challenges to our identity have been addressed by calling for strong core theories in our field in order to define our "own territory" (Benbasat and Weber 1996), or for a careful characterization of "IT artifacts" (Orlikowski and Iacono 2001), or for making our work more relevant and unique to the diverse set of information systems research stakeholders (Klein and Lyytinen 2003). These are all calls for actions that are expected to strengthen our identity in the sense of creating a consistency of character in our discipline. Yet, they address the quest for identity as a search for a fixed "sameness" in relationships to things like core theories, artifacts, actors, or audiences. This is another, although higher-level example, of being trapped into bipolar oppositions in defining what is proper information systems research and what is not. What we are seeking to explore instead is how something remains the same while it is continuously changing. In other words, we seek to characterize *the recursive, dialectic process of our identity construction as it takes place within a social field as a set of knowledge production and dissemination practices* (Klein and Lyytinen 2003) *that define what information system researchers are.*

We believe that our field's fresh openness to the social nature of scientific knowledge has also given us the ability to approach the identity question with a heightened reflexivity on our own role in constructing the socio-technical world of information systems. Our actions as researchers of information systems engage us in their social construction and thereby involve us, knowingly or unknowingly, as designers of them. We argue, therefore, that we should approach the issue of identity through the eyes of designers who project their stories into the world (Suchman 2002) as narratives and at the same time build their identities.

As a theoretical sensitizing device, we will employ a structurational view of knowledge and social practice (Giddens 1984). It is important to note that we are applying structurational analysis *to* our research, not *in* it—i.e., we view researchers as engaged in the reflexive monitoring of their own conduct in light of the signification structures offered by structuration theory. Thus, our epistemological position applies structuration and the social constructionist imagination to our own engagement in the world as information systems researchers. We ask: how do we and how will we create identities for ourselves, our research subject, and the world we produce and reproduce through our situated practice? Consequently, it is not a question of what identity we should produce, but a question of how we will engage in the reflexive production of it through our action. In other words, we propose that our identity can be more productively thought of as a question of agency and its process of structuration. We believe that approaching our identity as a process of identity construction lays bare the existential choices we have to make in performing our research. It is these moments of choice that we should look to as we try to understand how to go forward in building an information systems research identity. How are we making ourselves and our world as we perform our research of it? How might it be made and how might we be made as part of it (by reflecting upon and constituting it at the same time)?

If there is to be some notion of "thingness" associated with our identity, it should be as an active agent, an engaged constructor of the socio-technical world—in other words, as a designer. We as information system researchers, perhaps more than other organizational scholars, should have an appreciation for the centrality of design in shaping the socio-technical world. We teach design and we study how others create designs but we are hard pressed to place ourselves into this active position in our field. Yet, design is an inescapable element of our being agents in the socio-technical world, especially if we think of design in the way that Herbert Simon developed it in his classic *The Sciences of the Artificial* (Simon 1996). He persistently argued that managers and organization scholars, along with other professionals, should properly be considered as designers in the engineering sense, rather than as natural scientists.

The natural scientist is concerned with the study of what is, but the engineer or the manager is concerned with creating what might be—the designed artifact. The way forward for us as information systems researchers is thus not only in perfecting our methods or theories about the world, but also in perfecting our design capabilities for and with that world. We build our identity simultaneously as researcher–scientists who reflect the socio-technical world, and as researcher–designers who exercise agency in constructing it. As information system researchers, we are like the two faced Roman god,

Janus, living in the juncture of the present and the future, between what is and what could be, combining both of these features in our agency. Our first face sees the world as it is by looking back to the past, and by refining our theories and methods of what can be known of what is and why. Our second face sees the world as it could be by looking ahead to the future and shaping it by refining our cognitive and practical capabilities of what can be created and how. We are both scientists and designers, simultaneously, and are always located at this dynamic juncture. It is through reflexivity that we engage this duality in our professional life, and it is through reflexivity that we encounter and make our existential choices.

3 INFORMATION SYSTEMS RESEARCH AS LANGUAGE GAMES THAT NARRATE THE WORLD

Viewing information systems research as a two-faced process that is always looking both at the past and toward possible futures, heightens our awareness of language and language use. Our discourse within information systems research constitutes a language game (Apel 1980; Deetz 1996; Pitkin 1972; Wittgenstein 1953) in which meaning is constructed and projected into the world as we both act in it and observe it. Within this language game we build our understanding of the socio-technical world by being scientists who reflect on its form of life, and at the same time by being designers who imagine new forms of life. Our understanding of this world is echoed in our evolving concepts, terms, vocabularies, key words, models, and so forth that are used in these linguistic moves. Hence, by participating in that language game, our understanding is intimately related to our ongoing uses of language, or discourse moves (Deetz 1996), and through speech acts that narrate that world. We explain the world to ourselves and others by punctuating the "blooming, buzzing" stream of experience into sequences of events in the form of a story.

Our use of language through speech acts and the ongoing narration of experience constitute the basic stuff of our daily academic work: they are the papers, books, software, articles, algorithms, and presentations that we compose and digest. These uses of language position us in a social field as a scholar, a consultant, a change agent, etc., and thus continually enact and constitute our identities. Our projections of understanding about the world and about our identity in that world are within this specific *form of life*. In our case, it is a form of life that involves socio-technical assemblages of computers, networks, people, applications, organizations, representations, and practices of informing (Boland and O'Leary 1991), as well as our role in both shaping and reflecting upon these socio-technical worlds.

Our uses of language and the understandings we create are not innocent. Instead, by engaging in our form of life, we not only construct our identity and position in the field, but we also trigger identity transformations in the subjects that we study. Hence, the process of identity formation is ubiquitous and inevitable. All we ask is that researchers recognize and take responsibility for critical moments of identity formation in the discourse that shapes the information systems field. We do this through the reflexive monitoring of conduct that is the central dynamic of structuration. By attending to our reflective monitoring of conduct, we shift our attention from asking what things we should be, or be aligned with, to asking how we are becoming who we are through our discourse moves.

Our language games are both the medium and outcome of our research practice, and our scholarly work is an identity constructing enterprise—a design project. By engaging in the design of ourselves and the world through our research, we build an identity of who we and our subjects are and might be. The ongoing flow of observing, interviewing, surveying, programming, and experimenting in our research work draws upon and shapes both our own identities as researchers and the identity of our research subject. Hence, one way forward in doing information systems research and constructing the identity of the field is to avoid static thinking of what our theories, methods, or objects of study should be, and to instead reflexively and critically engage in the social construction of ourselves, our field, and our object of study. Next we discuss how we can use some of Giddens' (1984) concepts of structuration theory to engage in this process.

4 INFORMATION SYSTEMS RESEARCH AS AN ONGOING PROCESS OF STRUCTURING THE INFORMATION SYSTEMS FIELD

We are not committed to structuration theory as true believers of it. Yet, we do find that Giddens' (1984) modes of structuration offer helpful ways of classifying and thinking about how information systems researchers become active designers of both selves and of a socio-technical world (see Figure 1).

Empirical research, as traditionally conceived, can be defined as the practice of observing and intervening into information systems as they are, through a theory-based inquiry that draws upon modes of signification created and made available in our information systems research practice (bold arrow 5). Constructive information systems research (Madnick 1995; March and Smith 1995; Nunamaker et al. 1991), on the other hand, can be defined as the practice of observing current practice (arrow 5), imaging and building novel technical artifacts based on information and computational theories (arrow 2) that offer new capabilities (resources) and ways of thinking about organizational computing (signification structures). The role of our research in constituting these structures which we draw upon in our observations and interventions is evident (arrow 2): we utilize and quote earlier theories and artifacts that we have learned, built, and internalized as well as the words and concepts passed on by our ancestors; we use software packages and instruments (resources) that are critical in our data collection and analysis or design practice; and we subjugate ourselves to the norms and expectations of good academic citizenship by following rules of blind refereeing, pro bono reviewing, and avoidance of plagiarism. These structures are enforced and enacted in our every day work as academic citizens as shown through the cycle produced through arrows 1 and 2.

All information systems research practice *simultaneously* constitutes a design practice to the extent it becomes an activity that creates over time new modes of signification, new modes of power (resources), and new modes of legitimation in the socio-technical world of information systems practice. This happens either *directly* by designing and transferring technology artifacts and solutions that are appropriated in practice through technology transfer, or *indirectly* by formulating new signification structures or modes of legitimation

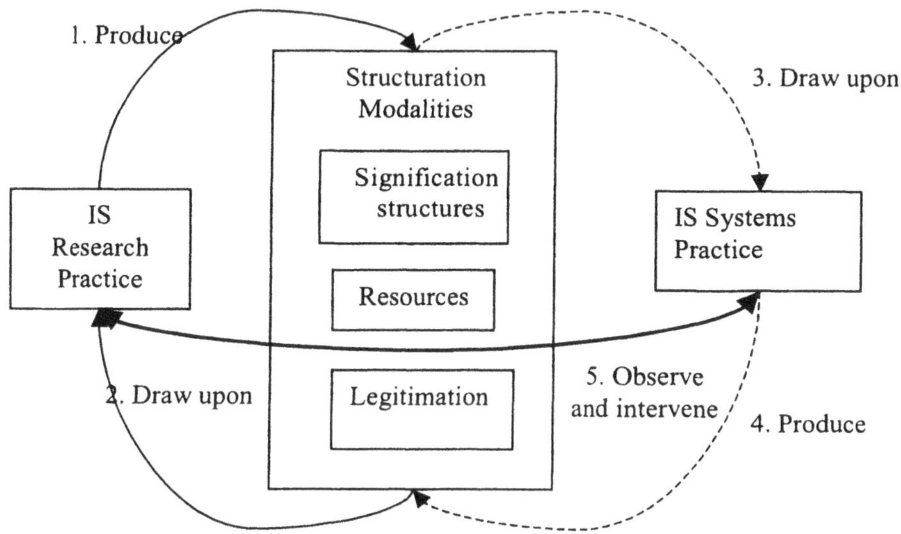

Figure 1. Structuration Processes and Information Systems Research

that are adopted in practice. This process of designing the world through continuous structuration of the social and technical field is depicted by arrows 1 and 3 in Figure 1. Stocks of signification structures, new computing and design resources, and new legitimation modes produced in information systems discourse *may* become actively drawn upon in information system practice through the processes of learning, appropriation, and adoption.[1] We emphasize the word *may* because actors in information systems practice normally draw upon a large number of different modes of signification and resources, and therefore the link between research outcomes and transformed information systems practice is not fixed or deterministic. In contrast, it may demand engagements in various types of knowledge creation and transformation networks that do the structuration work. Many of these are fragile and currently poorly understood (Klein and Lyytinen 2003).

Academic research practice is, in turn, central to establishing our role as designers of the socio-technical world in multiple ways. First, our discourses both draw upon (arrow 4) and reinforce (arrows 2 and 3) conventional sets of computational resources, but they also create powerful new immutable mobiles of resources that are appropriated and drawn upon in practice. These resources may sometimes become so powerful as to reconstitute or transform large parts of information systems practice. The invention of CASE tools, ERP systems, and collaborative tools within our discourse is a case in point. Second, academic research practice also establishes our role as designers indirectly in that we are always designing the symbols circulating in our social field (arrows

---

[1]To clarify the different ways in which modalities of structuration are drawn upon and made available, we use different types of arrows to distinguish the duality of structure in research discourse from that of informations system practice.

1 and 2). We as scholars explore existing modes of signification and transfer them to other parts of information systems practice, thereby transforming or reinforcing them. At some rare moments, if we are lucky, we may invent new modes of signification, thus giving new meanings to old practices or creating new practices that open wholly new socio-technical worlds to explore. When this happens, the terms of our research become the terms in which managers and information systems practitioners come to see and discuss their own organizational lives in the context of the socio-technical projects in which they are engaged. Similarly, assumptions of power and its possibilities for being exercised are being inhibited or encouraged by the continual design and invention of new tools and methods, which offer unprecedented capabilities to shape the behaviors of that socio-technical world. We also design indirectly by highlighting certain socio-technical organization features and hiding others—by serving some goals and needs in organizations, but overlooking others.

Finally, we cannot and do not conduct our research without exercising values in selecting, operationalizing, and giving order to particular features of a socio-technical landscape, and valuing them in certain ways as effective, appropriate, or harmful, which we then encourage others to take seriously as being real components of their world. Hence, the normative and power implications of our selection of research subjects, theories, and methods are inescapable and make us, both directly and indirectly, designers of the socio-technical world.

The way forward in our research, then, is to become more reflexive about the wake of design artifacts, signifiers, power relations, and norms that we strengthen or weaken in the process of making and reporting our research. This way forward recognizes our identity as scholars that fill the world with new meanings, offer judgments about what is good or bad, effective or not, and endow actors with new powers and capabilities. It is all of these at the same time that defines the identity of an information systems scholar as a designer. Hence, in thinking about the identity of the information systems field, it is critical to recognize the ongoing, multilayered structuration processes through which we design the very same socio-technical world that we study. This is also an identity construction process through language games that shape the forms of life we inhabit as well as of those whom we study. In other words, the thing in itself that we study is not as important as the (structurational) way we study it, because the way we study it brings both it and ourselves into being, and makes them significant.

If we try to study a thing *an sich*,[2] we are endlessly caught up in debates about what the true shape, qualities, and characteristics of the thing are, and how it is different from other things. If we accept that our research is a design project, and that the quality of our work and construction of our identity is found in that process of design in which structurational elements of power (resources), meaning (signification modes), and legitimation (norms and values) are shaped for an organizational information systems practice, then we can open a new space for dialogue in exploring our normative positions as designers—a space in which movement that appreciates both the accomplishments of the past and the creation of new possibilities is valued. Similarly, if we accept that the research space we create provides a set of structurational elements to the

[2]*An sich*—the thing-in-itself.

very subjects we study for their own practices and identity, our focus turns from studying them as things or objects to enabling their identity construction by providing them a stimulating field of values and norms, a potent vocabulary, and new resources with which to play.

5 INFORMATION SYSTEMS RESEARCH IDENTITY AS A NARRATIVE

Identity as a process, for ourselves as researchers, for technologies, and for organizations and their actors, is largely constructed through the narratives of the language games in which we participate. As noted, we establish ourselves as powerful and significant agents in the double-hermeneutics of the socio-technical world we design: we build meta-narratives (arrow 1) of the narratives that are available in the world (arrows 2 and 4), and already full of meaning. But our meta-narratives also enter that world as texts (arrow 3) and thereby affect the way in which meaning and understanding emerges in that world. We as researchers punctuate the stream of our life experience into scripts of events in which both social and technical agents of all kinds, with motivations and causal powers, play out specific strategies (Callon 1992; Latour 1993, 1996; Law 1992). The narratives we make of our experience with these heterogeneous networks of artifacts and humans makes things sensible to us and also to those we study.

Narratives have a beginning, middle, and end, which we as narrators keep in motion. A story begins when confusion arises about the world we engage and project in our designs. The story then tells about our struggle to make the world more intelligible, or more desirable. The story ends with our victory (or loss) and with new vocabulary elements that are subsequently entered into the broader discourses of the socio-technical world. In the beginning of the story, there is an order to a situation. That order is disturbed and finally restored or reconceived at the end—either claiming a success or failure (a tragedy) for the system. It is not just researchers, but also managers, clients, and customers who are constructing understandings of self and navigating organizational space by making sense of their flow of experience within similar narrative structures. Those narratives often take the form of a journey, in which they face hardships, overcome obstacles, suffer defeats, receive assistance, and seek a purpose. In our subjects' identity construction process, our research (meta)narratives become part of their organizational/technological narratives, or at least become a part of the narrative space in which they make sense of their experience. The plot and moral of our stories are not written into them, but read into them in a process which Bruner (1990) refers to as subjunctive. When we tell a story to ourselves, or as we hear a story being told, we are always also "reading between the lines" and inferring its meanings, intentions, plots, and significance for ourselves.

In our narratives we are trapped by our subjunctive power and the modes of signification that are re-created and drawn upon in these narratives. If we choose to make our research more relevant and realistic, we risk reinforcing the narratives that already structure organizational life in postmodern society, thereby subsuming ourselves within existing stories and worlds. If we chronicle a narrative that tells a different story of the future or negates the world as it is, we face the risk of not being read and of pitting ourselves against a powerful, dominant narrative, thereby marginalizing ourselves as designers. If we choose the narrative of a highly specialized and abstract theory, we risk that its appropriation is difficult and its value for providing meaning for action becomes questioned so that no subjunctive power can be applied to it. This leaves our designs lifeless and devoid of passion. Finally, if we choose to align ourselves with a particular research tradition and its ways of understanding the world, we are reinforcing its struggle to become a dominant narrative, thereby excluding other design possibilities. In following any of these paths, we are relinquishing our own identity construction to a residue of the past—as found in the sedimented layers of social practices or research traditions that we happen to be attending to at any given moment.

For instance, we may reinforce the narrative of organization as sentient being with values that shape its reasoned course of action, or we may reinforce a narrative of humans and organizations as information processing systems, or we may reinforce the narrative that technologies and tools are the central feature in a design solution, but in any case we run the risk of losing our own identity by serving the structurational narratives of another. As a researcher–designer we should ask if a particular information processing view (contingency, transaction cost, resource based, etc.) or organizational narrative (beneficent parent, rogue pirate, abusive spouse, etc.) is useful to us or our readers in how we want to engage in design. The adoption of others' narratives is both powerful and dangerous. We should realize that adopting any one of them is appropriate only if done in moderation and with a sense of irony, even though at the same time we are taking it seriously in the reflexive monitoring of our own conduct. Again, our responsibility as reflexive designers is to keep the stock of our signification structures and legitimation devices in motion—by sometimes following a strong current in one direction and allowing it to carry us along with curiosity as to where it will lead, and sometimes working against that current, to establish our own voice and an alternative branch of thought. This keeps the design scope of our research open to the world. The choice of when to allow the current narrative or our pet theory to carry us along and when to resist it and work against it is an existential, design choice and is the critical moment of how we create an identity as reflexive researchers and designers.

As Wittgenstein (1953) writes, we are always "bewitched" by the vocabulary of concepts, images, and narratives of our research that we cannot escape. The only hope we have in overcoming the illusions and blindness of our language is what Wittgenstein offers us: to understand both the limits and the traps of any language and its vocabulary. Yet, our and others' ability to make sense of a stream of experience and to act at all depends upon that same vocabulary. We cannot speak without it. We can only think within it and we can only act through it. But our vocabulary does change, and we can make changes to it through reflexivity and by being critical toward it, by acting on it, and by pushing it toward its limits (into a new structuration space). Never all at once, but piece by piece, we can and do reconstruct and deconstruct our vocabulary and thereby our identity and our world through simultaneously designing and learning about it. This forms an important critical goal (norm) for us as information systems researchers *qua* designers—a productive reshaping of our vocabulary driven by a dialogue of assimilation and accommodation, of dialectic tensions, and of structural oppositions within our forms of life, all within narratives of moving forward on a journey of self discovery in designing our socio-technical world. In short, we cannot escape our vocabulary, but we can and must resist it if are to act as designers at all. Perhaps resisting our received vocabulary is necessary to make research a fully human endeavor, which helps us keep our identities consistent yet living and continually changing. Only in this way can we overcome the traps of thingness in building up our identity where we close our language and reflect a given and fixed world and thereby become forever bewitched by just one vocabulary that we admire or endorse.

6 AN ILLUSTRATIVE CASE STUDY OF REFLEXIVE MONITORING OF IS RESEARCH AND CONSTRUCTION OF OUR IDENTITIES

By way of example and to clarify the concepts of language games, structuration, and narrative in a concrete research context we shall discuss our situation in a research project we have recently begun. Our project studies the impacts of using three-dimensional digital representation technologies in architecture and construction. These technologies are the result of significant research and design that developed computational methods for managing, rendering, and manipulating 3D representations that simulate buildings (or any other physical object). We have initially conceived the adoption and use of these technologies as a *path creation process* whereby multiple actors across different communities of practice engage in deviant and innovative work processes using such representational capabilities (qua resources) in designing and constructing a building. Our assumption is to view this as an active re-creation, destruction, and modification of legitmation, signification, and power structures through which architectural and construction agency is exercised due to the availability of new representational and computational technologies. These technologies of representation embed new information about construction objects and their features, reshaping the way in which a building is conceived, represented, and analyzed. In particular, we are interested in how Frank Gehry's architectural practice has been affected by the exploitation of such software capabilities and how it has changed the language of his architecture and the resources and legitimation structures that need to be mobilized to construct his buildings.

As part of our engagement in this project we recently received a letter from a well established architect who has worked in Gehry's office for over 10 years and who is very knowledgeable of the state of the architectural design and of the ideas that underlie the agency of being an architect. In this letter he said in part,

> What I personally find interesting about your research is that you are resurrecting the status of architecture as an instrument of knowledge. I think that's a marvelous side effect. You are reestablishing the position and weight of architecture that architecture once held in the world... (Fineout 2003).

Our research project may or may not succeed in the domain of journal publication or in achieving the noble goal of restoring architects as master builders, but while writing this essay we recognized that our project is not just an innocent intervention that follows the cycle of observation–theory building– and validation (shown in Figure 1), but, in addition, it inevitably involves a process of identity construction for both our subjects and ourselves (through circulating across all paths in Figure 1). Hence, we have become designers of the socio-technical world unknowingly by engaging in this project. This particular subject happened to recognize it immediately as he had struggled for a while with his relationship between agency and representation in architecture. Many other research subjects may not realize it as easily, including ourselves until we started to write this essay and read Fineout's letter again.

If we analyze our research project in light of Figure 1, we can observe the following. First by embarking on this project we are redefining ourselves as information system scholars by beginning to study something that is not a part of information systems research as it is currently conceived, and thus expanding its signification structures to topics and theories that previously were not part of our identity. We are not studying how to manage technology resources, or how to align strategies of architects' offices or constructors with specific new technological capabilities. We are also not interested in specific attributed causal impacts between 3D technology adoption (your favorite TAM version) and its organizational impacts (like work satisfaction, profitability, centralization, etc.). Instead our primary focus is to understand how 3D technologies as developed in one domain (aerospace industry) become serendipitously adopted within another industry (architecture and construction) and how this adoption is driven by specific dreams of what it means to be an architect and what structures such agency draws upon during this transformation process.

By studying the process of constructing a new architectural agency through new means of representation, and the new resources that come with it, we also establish a new understanding of our identity as information system researchers. We enter as a new agency into the field of construction and architectural practice by offering concepts, theories, and models which different actors in that field can use to make sense of and narrate their activity differently, and thereby learn to deconstruct and reconstruct their identity.

This is not a painless process and throughout our interventions and interviews we have learned how multiple contractors engaging in 3D-based construction projects have struggled to redefine their identity: how craftsmen have changed their skill sets and capabilities to become competent users of 3D representations, how roles and responsibilities of workgroups and trades have become renegotiated, and how we have made them narrate their experience in light of the (meta)narratives that guide our storytelling.[3] Later, we might offer new theoretical concepts, models, and language to talk about and build agency in architecture and construction. If we offer new resources that can be appropriated, and articulate new values or norms that emerge from discussions of what good construction and architecture means, we *may* become even more deeply involved in the design of the socio-technical world of 3D technologies in architecture and construction.

---

[3]These include theories of information processing, actor network theory, distributed cognition and knowledge sharing, organizational learning, and economics of knowledge and production networks.

7 IMPLICATIONS

What are the implications of this view of information systems research that emphasizes its constitutive process of design over its outcomes as our major concern? We propose that there is an implication for balance, which calls for both a heightened resistance, and for heightened responsibility. First, sensitivity to balance opens us to a deconstructive questioning of what is too well represented and what is not represented well enough in our research, theories, methods, and subjects. Our activities as designers can easily result in blind spots and aspects in the world that we do not engage and thereby silence. For instance, the way information systems research (except for certain socio-technical studies) has ignored labor is distressing, despite early attempts by Scandinavian researchers (see, for example, Ehn 1988; Sandberg 1985). We seem to act as if information and intelligence are solely the province of managers and knowledge workers. We act as if craftsmen (machinists, tool and die makers, sheet metal workers, plumbers, electricians, painters, carpenters, etc.) do not use information with intelligence and do not create knowledge in ways that are just as important, if not more important, than software designers, marketing directors, or plant managers. Labor is powerfully involved in the construction and use of the socio-technical world, and is a peculiar blind spot in our research, which we should resist.

Second, for us as researchers and designers, the question becomes: How do we structure our research space to overcome imbalances in the ways we constitute our research subjects and subsequently shape the socio-technical world? Do we seek a full, vibrant, dynamic space in which boundaries are stretched in multiple directions, favorite vocabularies are alternately strengthened and challenged, familiar situations are re-narrativized as exotic, and a sense of playfulness keeps our work ironic and surprising? We should continuously ask: What kinds of organizational and technological space are we creating for the subjects of our research and for society at large? Again, the problem is to resist the default path of searching for things in the form of theories, models, and tools, and instead seek a richer vocabulary of images, concepts, and narratives that can enable us and our research subjects to reflexively engage in design and create more desirable states of affairs.

Third, we have to be prepared for unavoidable struggle that follows from seeking the balance: within any one theory or method, we have a problem of needing to both believe and not believe in it—to both live in a tradition and to overcome it (Gadamer 1975). The seeking of balance is never simple or straightforward. Balance comes from both strengthening a position and from challenging it, from pushing work further in one area as well as from opening new horizons and proposing new questions to address. What will give balance in a situation is always dependent on an interpretive act within a tradition which comes down to what (meta)narratives we follow and believe and how we subjugate to them. Hence, we are back to our earlier assertion that we always face an existential choice in creating our research, our identity, and the identity of our research subject.

Fourth, seeing information systems research as a process of social construction and structuration may lead some to conclude that we are resigned to complete relativism and are giving up the foundations of a rational base for doing research. Aren't we also negating the value of theories and rigor in conducting research if any narrative can be deployed and mobilized in information system practice? We deny such allegations as the power of our research agency comes *exactly* from the specific sets of resources, signification structures, and legitimation modes that are drawn upon and recreated in our research practice. Moreover, these structures are always reflexively monitored (we are doing it just now!) through the signification structures and legitimation modes in which such concepts as rigor, theory-based, causality, and validity are embedded within our discourses and are mobilized to weave specific (meta) narratives. The modalities of signification that we employ continually redefine our identity and our position in the field—they are not separate from who we are as information system researchers, and should be the result of justified and authentically informed existential choices that we and our ancestors have made.

Finally, and most importantly, the challenge we pose in this essay is that *not making existential choice a part of our reflexive awareness leaves us as inauthentic, passive observers of the world.* It confines us to be researchers who pretend that there are firm foundations for a preferred theory and method to study a world given as a set of things, a preferred research object, and a preferred identity for ourselves. But, as we argue, there is no such foundation and a realization of that is just what has made us flounder over the last years as we came to value the pluralistic and relatively open space of discourse we find ourselves in today. To fix our identity into a specific set of relationships with specific things would be a step back into history. Yet, to move forward in that new space with no foundation and no boundaries is a problem that leads us to ask what is our identity, what is consistent and stable?

In this paper we suggest that our identity lies in the consistency of the fundamental processes in which we engage as designers. This demands that we reflexively embrace responsibility for the existential moments of choice which give rise to our narratives of self and other; the theories, methods, and subjects those stories embody; and our continuing process of constructing identity for self and others. These choices are the medium and outcome of our form of discourse and engagement with them is our unavoidable fate.

REFERENCES

Apel, K. *Towards a Transformation of Philosophy*, London: Routledge and Kegan Paul, 1980

Baskerville, R., and Myers, M. "Information Systems as a Reference Discipline," *MIS Quarterly* (26:1), 2002, pp. 1-14.

Benbasat, I., and Weber, R. "Research Commentary: Rethinking 'Diversity' in Information Systems Research," *Information Systems Research* (7:4), 1996, pp. 389-399

Benbasat, I., and Zmud, R. "The Identity Crisis Within the IS Discipline: Defining and Communicating the Disciplines's Core Properties, *MIS Quarterly* (27:2), 2003, pp. 183-194.

Boland, R. "Phenomenology: A Preferred Approach to Research in Information Systems," in E Mumford, R. Hirschheim, G. Fitzgerald, and A. T. Wood-Harper (Eds.), *Research Methods in Information Systems*, Amsterdam: North-Holland, 1985, pp. 193-203.

Boland, R., and O'Leary, T. "Technologies of Inscribing and Organizing: Emerging Research Agendas," *Accounting, Management and Information Technologies* (1:1), 1991, pp. 1-7.

Bruner, J. *Acts of Meaning*, Boston: Harvard University Press, 1990.

Callon, M. "The Dynamics of Techno-Economic Networks," in R. Coombs, P. Savotti, and V. Walsh (Eds.), *Technological Change and Company Strategies: Economic and Sociological Perspectives*, San Diego: Harcourt Brace Jovanovich, Publishers, 1992.

Deetz, S. "Describing Differences in Approaches to Organization Science: Rethinking Burrell and Morgan and Their Legacy," *Organization Science* (7:2), 1996, pp. 191-207

Ehn, P. *Work-Oriented Design of Computer Artifacts*, Ph.D. Dissertation, Stockholm: Arbetslivscentrum, 1988.

Fineout, M. Private communication to Richard Boland and Kalle Lyytinen, February 22, 2003.

Gadamer, H.-G. *Truth and Method*, G. Barden and J. Cumming, Trans., New York: Seabury Press, 1975.

Giddens, A. *The Constitution of Society*, Cambridge, MA: Polity Press, 1984.

Jenkins, A. M. "Research Methodologies and MIS Research," in E Mumford, R. Hirschheim, G. Fitzgerald, and A. T. Wood-Harper (Eds.), *Research Methods in Information Systems*, Amsterdam: North-Holland, 1985, pp. 25-47.

Landry, M., and Banville, C. "Disciplined Methodological Pluralism for MIS Research," *Accounting, Management and Information Technologies* (2:2), 1992, pp. 77-97.

Klein, H., and Lyytinen, K. "The Powerty of Scientism in Information Systems," in E. Mumford, R. Hirschheim, G. Fitzgerald, and A. T. Wood-Harper (Eds.), *Research Methods in Information Systems*, Amsterdam: North-Holland, 1985, pp. 131-162.

Klein, H., and Lyytinen, K. "Knowledge Creation and Transformation in Networks: The Case of Relevancy of IS Research," unpublished working paper, Weatherhead School of Management, Case Western Reserve University, 2003.

Latour, B. "Ethnography of a 'High-Tech' Case: About Aramis," in P. Lemannier (Ed.), *Technological Choices: Transformations in Material Culture Since the Neolithic*, London: Routledge and Kegan Paul, 1993.

Latour , B. *Science in Action: How to Follow Engineers and Scientists Through the Society*, Boston: Harvard University Press, 1992.

Latour, B. "Social Theory and the Study of Computerized Work Sites," in W. J. Orlikowski, G. Walsham, M. R. Jones, and J. I. DeGross (Eds.), *Information Technology and Change in Organizational Work*, London: Chapman & Hall, 1996, pp. 295-307.

Law, J. "Notes on the Theory of the Actor-Network: Ordering, Strategy, and Heterogenity," *System Practice* (5:4), 1992, pp. 379-393.

Lee, A. S., Liebenau J., and DeGross J. I. (Eds.). *Information Systems and Qualitative Research*, London: Chapman & Hall, 1997

Lyytinen, K., and Klein, H. "Critical Social Theory of Jurgen Habermas (CST) as a Basis for the Theory of Information Systems," in E. Mumford, R. Hirschheim, G. Fitzgerald, and A. T. Wood-Harper (Eds.), *Research Methods in Information Systems*, Amsterdam: North-Holland, 1985, pp. 219-236.

Madnick, S. "Information Technology: The Reinvention of the Linkage between Information Systems and Computer Science," *Decision Support Systems* (13:4), 1995, pp. 373-380

March, S., and Smith, G. "Design and Natural Science Research in Information Technology," *Decision Support Systems* (13:4), 1995, pp. 251-266

Mingers, J. "Combining IS Research Methods: Towards a Pluralist Methodology," *Information Systems Research* (2:3), September 2001, pp. 240-259

Mumford, E.; Hirschheim, R., Fitzgerald, G.; and Wood-Harper, A. T. (Eds.). *Research Methods in Information Systems*, Amsterdam: North-Holland, 1985.

Nissen, K-E.; Klein, H.; and Hirschheim, R. (Eds.). *Information Systems Research: Contemporary Approaches and Emergent Traditions*, Amsterdam: North-Holland, 1991.

Nunamaker, J.; Chen, M.; and Purdin, T. "Systems Development in Information Systems Research," *Journal of Management Information Systems* (7:3), 1991, pp. 89-106.

Orlikowski, W. J., and Iacono, C. S. "Desperately Seeking the 'IT' in IT Research—A Call to Theorizing the IT Artifact," *Information Systems Research* (12:2), June 2001, pp. 121-134.
Pitkin, H. F. *Wittgenstein and Justice,* Berkeley, CA: University of California Press, 1972.
Robey, D. "Diversity in Information Systems Research: Threat, Promise and Responsibility," *Information Systems Research* (7:4), 1996, pp. 400-408.
Sandberg, A. "Socio-Technical Design, Trade Union Strategies and Action Research," in E. Mumford, R. Hirschheim, G. Fitzgerald, and A. T. Wood-Harper (Eds.), *Research Methods in Information Systems*, Amsterdam: North-Holland, 1985, pp. 79-92.
Simon, H. *The Sciences of the Artificial* (3rd ed.), Boston: MIT Press, 1996.
Suchman, L. "Organizing Alignment: A Case of Bridge-Building," *Organization* (7:2), 2000, pp. 311-327.
Weber, R. "Editor's Comments: Still Desperately Seeking the IT Artifact," *MIS Quarterly* (27:2), pp. iii-xi.
Wittgenstein, L. *Philosophical Investigations*, Oxford: Blackwell, 1953.

ABOUT THE AUTHORS

Richard J. Boland, Jr. is a professor of Information Systems at the Weatherhead School of Management, Case Western Reserve University. Previously he was Professor of Accountancy at the University of Illinois at Urbana-Champaign. He has held a number of visiting positions, including the Eric Malmsten Professorship at the University of Gothenburg in Sweden in 1988-89, and the Arthur Andersen Distinguished Visiting Fellow at the Judge Institute of Management Studies at the University of Cambridge in 1995. His major area of research is the qualitative study of the design and use of information systems. Recent papers have concerned sense making in distributed cognition, hermeneutics applied to organizational texts, and narrative as a mode of cognition. Richard was founding editor of the research journal *Accounting, Management and Information Technologies* (now *Information and Organization*) and is coeditor of the Wiley Series in Information Systems. He serves on the editorial Board of six journals, including *Accounting, Organizations and Society*. His most recent book is *Managing as Designing* (Stanford University Press, 2004) coedited with Fred Collopy. He can be reached at Boland@Case.edu.

Kalle Lyytinen is Iris S. Wolstein Professor in the Department of Information Systems at Case Western Reserve University and an adjunct professor at the University of Jyväskylä, Finland. He serves currently on the editorial boards of several leading IS journals including *Journal of the AIS, Information Systems Research, Information & Organization, Requirements Engineering Journal,* and *Information Systems Journal*. He has published over 150 scientific articles and conference papers and edited or written eight books on topics related to system design, method engineering, implementation, software risk assessment, computer supported cooperative work, standardization, and ubiquitous computing. He is the former chairman of IFIP 8.2. and was one of the original contributors to the proceedings of the conference in 1985. His research interests include information system theories and research methods, computer-aided system design and method engineering, system failures and risk assessment, computer supported cooperative work and nomadic computing, and the innovation and diffusion of complex technologies and the role of institutions in such processes. Kalle can be reached at kjl13@cwru.edu.

Part 2:

Reflections on the IS Discipline

5 INFORMATION SYSTEMS— A CYBORG DISCIPLINE?

Magnus Ramage
Open Systems Research Group
Open University

Abstract This paper argues for a model of information systems in terms of cyborgs: a boundary-crossing mixture of the technical and the social. The argument for this model is substantiated from the personal experience of the author, presented as examples of being a cyborg researcher within a disciplinary context. Lessons for information systems are drawn.

Keywords: Cyborgs, interdisciplinarity, reflective practice, sociology of technology

1 INTRODUCTION: ON CYBORGS, CYBORGNESS AND DISCIPLINES

In this paper, I shall put a case for regarding information systems as a cyborg discipline. To do this, I shall begin in this section by discussing the concepts of cyborgs and of disciplines.

The term *cyborg* is a shortening of the phrase *cybernetic organism*, used by Clynes and Kline (1960) to refer to a combination of human and machine that would be able to function in the harsh physical environment of space travel. The science-fictional resonances of such a concept—and its parallels in the hybrid monsters of literature—made the term well-known. Cyborgs have become widely adopted in popular culture—a typical example is Arnold Schwarznegger's character in the movie *The Terminator*—to mean something that is part machine and part human.

However, the term cyborg is used in a wider sense by Haraway (1991), writing within the sociology of science with a strong feminist and postmodernist tone. Haraway's argument is that a defining property of the cyborg is that it straddles the social and technical domain: it is part human and part machine. This clearly challenges the requirement of modernist society to have everything categorized, to belong to a well-understood domain. It is thus both a metaphor for the human-technical mix of our times, and a description of situations where the social and the technical merge and blur.

The use of the word *cybernetics* in the term cyborg is interesting, given the strong links of that field with the field of systems thinking (and thus of information systems). Cybernetics as a field (Heims 1991; Wiener 1948) was explicitly concerned in its early days with the study of messages, information, and feedback within a range of domains, but especially machines and humans, and the way that knowledge about one might appropriately be applied to the other. However, it began around the time of two key events, with which it was closely linked: the birth of the digital computer and the contribution of American science and technology to the cold war. Both of these events led to cybernetics being popularly regarded in Wstern culture as being concerned with issues such as artificial intelligence, robotics, and the space race. (It is for this reason that the prefix "cyber-" has been used to denote a range of computer-related areas, such as cyberspace.)

For Haraway (1991), the concept of the cyborg is "an ironic political myth" (p. 149), it is "a condensed image of both imagination and material reality." Haraway's cyborg is both an ontological statement, about the way things *are*, that "we are all chimeras, theorized and fabricated hybrids of machine and organism." However, it is also a statement about how things *could be*, that by breaking down the boundaries between human and machine, we break down boundaries and categories in general, and allow the questioning of the hierarchies of power that depend on boundary. It is metaphorical in the sense that it uses an image of one thing to describe something else, but it has a close parallel to the thing being described; as Haraway (2000, p. 82) says in a different context, it "is not merely a metaphor that illuminates something else, but an inexhaustible source of getting at the non-literalness of the world."

Richard and Whitley (2000) have considered the concept of the cyborg as applied to IS. They equate it to the concept of the *hybrid agent* taken from the application of actor-network theory within IS; they describe the cyborg as "neither human nor machine, but [a] hybrid construct of the two that is fleeting, precarious and always mutating." However, they are skeptical about its usefulness as a term, suggesting it "has become too fashionable and politicized to be of much to the IS community at present." I hope to show in this paper ways in which the concept can be useful to the IS community.

There seems a clear parallel between the cyborg boundary-crossing and information systems, in that IS as a field of study inevitably straddles both the social and the technical domains. It is hardly a new statement to say that IS is inevitably interdisciplinary, nor is it new to look at ways to straddle the divide between social and technical perspectives. In various ways this has been a key theme in much research in IS and cognate fields (e.g., Checkland and Holwell 1998).

However, this has been primarily considered from an *internal* perspective (i.e., from within the IS community). Looked at from outside IS, what one sees is precisely the double-headed monster (cf. Law 1991) that breaks societal norms, is therefore threatening and must be persecuted.

This may sound extreme. Yet it fits with the experience of many in the IS community. This resembles the argument of Jones (1997), who talks about the concept of an academic discipline. He suggests, drawing on the work of Foucault, that the common use of the term *to discipline* in the sense of *to punish* is relevant to the way that academic disciplines police their boundaries.

For example, a department of computer science might regard the institution of an information systems program as encroaching upon its territory in an inappropriate way: not only is it covering the same intellectual ground (the study of computers) but, much worse, it is doing so in a way that understands that ground quite differently. It is thus breaking the rules of the game as they see it: "They are playing at not playing a game. If I show them I see they are, I shall break the rules and they will punish me. I must play their game, of not seeing I see the game" (Laing 1970).

Worse, for some, the whole information systems field is invisible. I have experienced this in attending a workshop on the theme of how information systems and software engineering could come together, where a senior professor of software engineering argued that the question was simply a category error, as software engineering was an academic discipline, while an information system was just a thing. Although well-respected and well-informed in his own field, he had no conception of information systems as a discipline.

Referring to information systems as a cyborg discipline carries risks as well as insights. Perhaps the most striking is the technocratic emphasis of the term *cyborg*: beyond Haraway's work, its main connotations are around machines rather than people. Given the prevalence of technological imagery and concerns within information systems, and its constant confusion with information technology, this could be problematic. In a way, though, this confusion gives strength to the cyborg concept in its ambiguity and fluidity. To live as a cyborg is not to be comfortable, it is to be challenging and challenged. As Haraway (2000, p. 129) says, the cyborg concept "does unexpected things and accounts for contradictory histories while allowing for some kind of working *in* and *of* the world."

A further danger is found in the organismic nature of the cyborg concept (cf. Morgan 1986), with its overtones of analysis leading to a single perspective, ignoring the politics of, and conflict between, multiple points of view. In fact, it is precisely the ambiguity and boundary-crossing nature of the cyborg that makes it a useful model. Undoubtedly the concept of the cyborg is that of an organism—the origins of the term imply as much—but the focus of the concept is to blur the boundaries between the cyborg's different parts. It is systemic in the sense that it has emergent properties that go beyond those of its components, and in the sense that it can only be understood through the relationships between those components (cf. Bateson 1972), not in the sense that it has a single purpose or goal. Indeed, as Letiche (1999:150) remarks, a key feature of the cyborg concept is that it embodies "*différance*—complex relationships of individual, mechanical, natural, synthetic and cultural activity that would lead to indeterminate identity and dynamic interaction." We do need to beware of reification, however: while the cyborg of science fiction may be a thing, the cyborg concept describes something fluid and changing.

What I hope to do in this paper is to illustrate the experience of IS as a cyborg discipline—to argue for its cyborgness—by describing my own encounters, as an IS academic, with boundary crossing in various academic departments.

I present my personal experiences here not because they are of interest in their own right, but as a set of typical examples which illustrate the case I am trying to make. I intend this to be within the spirit of the reflective practitioner (Schön 1983). It is also relevant from the feminist perspective that partly informed Haraway in her discussion of cyborgs, which validates personal experience as a mode of discourse.

I shall discuss these experiences in two different settings. First was my time as a doctoral student at the University of Lancaster, within their Computer Science department. Second was a period I spent at the University of Durham as a researcher on a consciously interdisciplinary project. The accounts are necessarily personal; I hope nothing in them is taken as criticism of particular individuals. Both accounts were written while I was in the situation, so the "now" in each story is some years in the past.

2 CYBORG TALE 1: AS A DOCTORAL STUDENT

My first experiences come from a piece I wrote in 1996 (but never published) that reflected on my experiences as a doctoral student. Although based in a department of computer science, I was working within the field of computer-supported cooperative work (CSCW). This field arose within computer science, separately from information systems, as an offshoot of human-computer interaction. However, there are many parallels with IS (Kuutti 1996) and the issues around its cyborg status are similar.

Shapiro (1994) lists about 15 disciplines from which CSCW has taken some input that have gone to shape its discussions. The number of these contributing disciplines makes the field immensely richer, as well as considerably more complex. It does lend the field a slightly uneasy air, however.

Could the nature of CSCW be any different? Surely not. One of its characteristics is that the very subject of its discourse is itself a cyborg: a mixture of the technical and social, a mixture of computers and people and networks and organizations. Given this cyborg nature, it would be strange if the research and practice of the discipline was not itself a mixture of the disciplines that have studied these things. Of course, there are places where the combination of people and technology is studied with purely technical interest with little concern for the effects upon people (such as in some computer science departments and IT consultancy firms); again, there are some places where the technology is ignored and only its social effects considered (such as by some sociologists), or only the effects upon the individual psyche (such as by some psycho dynamic psychologists). Such perspectives do tell us useful things: how to build better ISDN networks, what are the societal dangers of the Internet, what to celebrate and what to be wary of in electronic communication. But the perspective of CSCW is different from these: it considers instead how people work together (cooperative work) *and* how computers can change this (computer support). Combining these two aspects to more effectively design socio-technical systems has been much of the effort of CSCW.

To ask this question in a slightly different way: Who does CSCW? Is it an enterprise for members of well-defined disciplines (computer science, sociology, social psychology, management, etc.), who come together on multidisciplinary projects to study and develop new computer systems? Is it an enterprise for researchers who remain within their own traditional disciplines, while learning something of the knowledge of researchers from other disciplines, so that they covertly become interdisciplinary within themselves? Or is it an enterprise for those who are less interested in disciplinary boundaries than in relevant information, from whatever source it may come? The answer to each question is yes. All three models have been followed in various projects within CSCW. All three represent cyborg research, research that crosses boundaries of disciplines. In the first case, the research team as a whole is a cyborg. In the second, the cyborgness is somewhere between the team and the individual. In the third case, the researcher him/herself is a cyborg. People in the third category often seem to move from one discipline to another, as new opportunities arise, or they end up establishing jobs and departments that reflect their new cyborg status (such as the universities which now have CSCW departments and research centers independent from other departments, although more often they seem to exist as a kind of virtual department, where collaboration between people is more significant than structures).

However, cyborgs are not popular with society. Establishing a new department is one move sometimes undertaken to calm this insecurity, although it is a move that does little but reinforce the disciplinary walls by setting up new disciplines. In this way, for example, computer science was formed as a new discipline in the 1940s by a mixture of electronic engineers and mathematicians, but rather quickly set up its own disciplinary structures and now is as much a participant in the fractured academic culture as the older disciplines. In the context of CSCW, this doesn't matter as such (as its focus is not the breaking down of barriers for the sake of doing so) but it does remove some of the creative tension that exists between the different constituent disciplines of the field.

This last point brings us to the other reason why CSCW is a cyborg discipline: because it is useful for it to be so. If the only place where people from different disciplines, all looking at people working together via computers, could meet was at an annual conference, or the occasional project with disciplinary boundaries fully up ("you are the computer scientist on this project, you are the anthropologist and you are the organization theorist"), it would be rather dull and a lot less fruitful. One of the big risks in CSCW are the "paradigm wars" seen in various kinds of social science—groups of true believers in one way of conducting research or another, who come together not so much to engage in dialogue as to fight each other with the same old arguments. If the risks of this are so strong at the moment, imagine what it would be like if the members of those paradigm communities never spoke to each other except at conferences.

This leads me to label CSCW as a cyborg discipline: of itself, by its nature, it is a cyborg between the technical (of various kinds) and the social (of various kinds). On the one hand, this is simply its nature, a description of what it is. On the other, it is usefully so, and much productive research and practice has been conducted as a result.

But of more interest to me is the fact that I see myself (writing in 1996) as a cyborg researcher. In what way do I mean this? In the sense, as with CSCW itself, of sitting between the technical and the social. Thus my background and interests are a mixture of the technical (computer science, mathematics) and the social (psychology, sociology, management, philosophy). Likewise, my worldview sits between these, not pure computing (no C++ code was to be found in the eventual thesis, and while I find computers to be a useful tool, to me they are part of an overall process of organizational change); but also not pure social science (references to Habermas, Weber or Garfinkel were kept to a minimum, and the aim of the work is essentially pragmatic).

Why should this be so? Partly, as with my first reason to be a cyborg given above, because it is what I am. I have this mixture of disciplines within me, I don't find it possible or interesting to confine my thoughts within a single disciplinary matrix, and my thoughts by their nature move swiftly from one set of ideas to another, like a bee resting upon different flowers and (hopefully) spreading pollen from one to another.

This, it might be thought, is my problem. If I want to be a cyborg, then that's fine in my own time, but if I want to write a Ph.D. I should knuckle under the disciplinary norms of computer science and write a thesis that is pure computing (whatever that means). I have not done this, choosing instead to write in a style that is a mixture of social and technical not just because it suits my temperament, but also because I think it is good science—it is appropriate to this context.

This is principally, of course, because of the nature of CSCW that I discussed above. By its nature, CSCW is a mixture of people, computers, and organizations, a study of the facilitation of human cooperation by technology. Therefore it will be appropriate to get a handle on what is the nature of work, cooperation, and organizations and to be aware of what things affect people and how, as well as being aware of the technology and how to make it better. My task here is the evaluation of systems (some part of which are based on computers) that support cooperative work, and to evaluate these effectively requires a knowledge of the full context of the work.

It might be helpful to briefly consider what happens when one does *not* consider the organizational and human context of work when designing and evaluating computer systems to support it. A good example from my research concerned a university accounting system (Ramage 1999). One reason why this was a failure was the change in organizational culture that occurred in the finance department around the same time as the accounting system was introduced. The new culture was one strongly focused on financial targets, as favored by the government at the time, and also on a highly structured information hierarchy, where as little information flowed through the hierarchy (i.e., from the finance director to the departmental budget-holders, or vice-versa) as possible. This led to a large degree of resentment, which made budget-holders considerably more resistant to the system, but also to inflexibility in that the system (a package written for several universities) was not changed to meet the information needs of the budget-holders. The resulting problems with the system are an indicator of one reason why it would have been useful to consider properly the whole organizational system sooner.

The complementary challenge to the earlier example about computer science might also be that if I want to write a thesis on the human influences of technology, then I should be writing it in a sociology or management department, and confine myself to the norms of those cultures. After all, do such places not frequently specialize in such things? This I would similarly refute, saying that these areas alone are equally inadequate to the systemic study of technology in use. (Of course, one can also do perfectly good work on the sociological or organizational aspects of technology, and plenty has been done; my point is not to denigrate that work, but rather to say that I prefer to use a wider angle of lens.)

And so it is that I chose the difficult middle way of being a cyborg, sitting between the technical and the social in a department of computing, writing about computers but being concerned for their effects upon people and organizations, an unholy and unclean mixture—but a necessary one.

3 CYBORG TALE 2: AN INTERDISCIPLINARY RESEARCH PROJECT

The second experiences around cyborg research arise from the problems of communication occurring in an interdisciplinary research project. The project, Software as a Business Asset, ran from 1997 to 2000, with three academic staff (two software engineers and one organizational analyst), one researcher (me), and one doctoral student. The results from the project have been extensively written-up elsewhere (e.g., Bennett et al. 1999; Brooke and Ramage 2001).

The project arose from the need of two different groups of people at the University of Durham. First was a group of software engineers who had developed various methods for dealing with the maintenance of existing software, through understanding the code thoroughly, through performing mathematical transformations on it so it did the same things on different hardware, and through patching it up in various different ways. Their methods were successful, they got grants, student scholarships, and consultancies without problem, and industry used their work. However, somehow there wasn't as much effect of the work as there could be. Somehow businesses took it up and used it, but it got snarled up in politics and structure and process. Somehow they knew that they needed a concern for organizational issues. Elsewhere in the university, a lecturer was doing research and teaching MBA students about people, change and information systems. She had a view of how to help businesses make strategic decisions about their information systems. She used a method that involved looking at various possible futures for the business and thinking those over before you did anything much to the technology. She had some contact with the software engineers down the hill already—so few people at the business school were interested in computers that she needed all the company she could get. So she knew some of their problems, and they realized together that some good work could be done here.

Language was a major issue throughout the project. It seems at times that almost any word which might be used by one group of academics to mean one thing will be used by other academics to mean something completely different. This caused quite a bit of misunderstanding on a number of different occasions.

A particular feature of this was the precision with which words are used. It's not that software engineers actually use words more precisely and exactly than organizational analysts, but they often seem to think they do, and this was a continual issue of tension.

An example of a particular word which turned out to be used rather differently by the two communities was been *tool.* A tool, says the dictionary, is an implement which assists people to do their work more effectively or efficiently. Human beings are, it is often said, "tool-making animals." But what do those tools constitute? Clearly, in everyday situations, a tool is something like a hammer or a chisel. For a software engineer, however, a tool refers to a piece of software which enables them to get their work done more efficiently—for example, in analyzing the structure of a piece of program code. In organizational analysis, by contrast, a tool is more abstract and usually refers to some way of helping people to interact or think more effectively. So when the organizational analyst referred to "the organizational scenarios tool," it made perfect sense in her context that this tool was a way of structuring ideas. However, the software engineers found this such a strange thing to refer to as a tool that they kept writing little notes, in papers intended for their community, to the effect that this wasn't really the kind of tool that you might expect when you heard that word normally.

We constantly came up against the question of whether this sort of interdisciplinary work can actually take place at all, in any meaningful way. In particular, we became aware very early on that the two sides of the project were working from very different intellectual paradigms. In the terms of Burrell and Morgan (1979), software engineers work from a positivist paradigm whereas the organizational analyst works from an interpretivist paradigm. Formally speaking, these paradigms are incommensurable—that is, it is not possible to resolve the differences between them at an intellectual level. Our constant task was to try to resolve them at a practical level, which was sometimes successful and sometimes not.

The differences in paradigms became apparent at our first full project meeting, about six weeks after the project started. How should we plan the work of the project? What model of research should we use in doing this? Should we expect to build a complete picture of how to handle legacy systems and then try it out in industry, or should we aim to combine the development of our method with trials of small parts of it in industry? The project proposal, principally written by one of the software engineers, reflected the first approach, one which is common in engineering. At the meeting, however, we found ourselves moving more in the direction of an approach based on action research, the more iterative form of research.

Ironically, as the project developed, we moved back to the more engineering-based model; this is partly to do with the lack of industrial involvement in the project, but must surely also derive from the location of the bulk of the project team in a Computer Science department. It was only toward the end of the project, as we conducted the work reported in Brooke and Ramage (2001), that we began once again to take up an action research approach.

The question of paradigms also arose with respect to the relationship between the two parts of the model: organizational and technical change analysis (Bennett et al. 1999). Which part of it should be primary? The organizational change aspects occur first, but the *output* (itself somewhat of an engineering term) from them must be in a form suitable for use by the software change tool.

These were some of the tensions to be found between the two perspectives during the SABA project. Yet we did make a conscious and continual effort to work together as a single team, to do work that was not just multidisciplinary but interdisciplinary, and to try to go beyond the boundaries of our home disciplines. That is, we tried to create a cyborg enterprise together.

4 TREATING INFORMATION SYSTEMS AS A CYBORG DISCIPLINE

Straddling the disciplinary divide is not a luxury in the study of information systems, but rather a necessity. For a full understanding both social and technical perspectives are necessary, and this can be seen from either side of the divide. From the social perspective, one can see that people interact with technology, it impacts on their lives and their work, but that the detail of the technology makes considerable difference to the nature of that impact. From the technical perspective, one can see that the way in which one's carefully crafted and highly efficient technology is used depends on a whole range of factors that go beyond the value of it as a technology, and thus if one wants it to be used fully (or at all) one must be aware of those factors.

However, to portray oneself, either as an individual or as a group, as conducting information systems work, is to set oneself up as a cyborg entity, and thus due for persecution by the rest of the academic community.

How can we deal with this? There are various solutions with varying likelihood of happening.

Least likely, we can strive to have institutional acceptance of cyborgs (as individuals or disciplines) as a general category. For the reasons outlined above about the challenging nature of cyborgs, this is difficult. An example of this not happening can be seen in the troubles of IS in establishing itself as legitimate as a category within the UK's Research Assessment Exercise, where (despite considerable efforts) in 2001 it existed only as a subsection of the Library and Information Sciences category.

More productive is to put, in particular contexts, the pragmatic case of defending the value of the cyborg nature of IS. Arguments like the one at the start of this section can be made to demonstrate the usefulness of IS, and the necessity of its twin perspectives.

It might be argued that the above is just another way of discussing interdisciplinarity. While it is true that I have drawn on the interdisciplinary character of information systems above, talking of IS in terms of cyborgs adds a different character to the nature of the interdisciplinarity. Haraway (1991) argues clearly that the boundary-crossing nature of cyborgs is to be celebrated, not simply tolerated: "Cyborg imagery can suggest a way out of the maze of dualisms in which we have explained our bodies and our tools to ourselves. This is a dream not of a common language, but of a powerful infidel heteroglossia."

As with Haraway's use of the concept of the cyborg, my use of the term is metaphorical in the sense that it is an image, but I use it to cast light upon the boundary issues in IS, to raise questions about the nature of the discipline. In this sense, the question mark in the title of the paper is deliberate. The goal of looking at IS as a cyborg discipline is not to build "metrics of cyborgness" in particular papers or projects, but precisely to raise questions about the nature of the discipline and the extent to which it crosses boundaries and the implications of that boundary-crossing.

Weber (2003) asks how the IS discipline might establish an identity. I would suggest that it is in this way that looking at the cyborg concept can help. By considering the ways in which our discipline is neither precisely technical, nor social, does not derive its identity from one academic field or another but from a fusion of many—and thus in creating a new way of looking at the world that goes beyond the technical and the social. Exploring what this might mean in practice is a deeper question, but the question of the identity of information systems as a discipline is not simple.

If the concept of IS as a cyborg discipline has use, two final implications follow. First, this boundary-crossing is embedded into the nature of the discipline so firmly that it cannot be escaped—it must rather be embraced. This brings liberation from the strictures of the technical/social divide—it is to reject the language of "either/or" in favor of that of "both/and."

Second, this means that the continual struggle for self-identity, seen in IS research and scholarship over so many years, is both inevitable and a necessary part of the discipline. It is only by asking ourselves who we are that we can begin to grasp the fluid nature of what it means to be both human and machine in our perspectives, and only by continuing to ask that question that we can avoid getting trapped into a single understanding that only works for a particular time. To consider the technical and the social in one, at once, to cross the boundaries of both—that is the cyborg nature of information systems.

ACKNOWLEDGMENTS

This paper has been formulated over several years through conversations (and sometimes readings of pieces of text) with Michael Twidale, Fides Matzdorf, Keith Bennett, Carole Brooke, Karen Shipp, and Rebecca Calcraft. I am also grateful to the four anonymous reviewers for this conference, whose comments helped me to shape the paper.

REFERENCES

Bateson, G. 'The Cybernetics of 'Self': A Theory of Alcoholism," *Steps to an Ecology of Mind*, San Francisco: Chandler, 1972.

Bennett, K., Ramage, M., and Munro, M. "A Decision Model for Legacy Systems," *IEE Proceedings—Software* (146:3), 1999, pp. 153-159.

Brooke, C., and Ramage, M. "Organizational Scenarios and Legacy Systems," *International Journal of Information Management* (21:5), 2001, pp. 365-384.

Burrell, G., and Morgan, G. *Sociological Paradigms and Organizational Analysis*, London: Heinemann, 1979.

Checkland, P., and Holwell, S. *Information, Systems and Information Systems: Making Sense of the Field,* Chichester, UK: John Wiley & Sons, 1998.

Clynes, M. E., and Kline, N. S. "Cyborgs and Space," *Astronautics,* September 26-27, 1960, pp. 75-76.

Haraway, D. J. "A Cyborg Manifesto: Science, Technology, and Socialist-Feminism in the Late Twentieth Century," in *Simians, Cyborgs and Women: The Reinvention of Nature,* New York: Routledge, 1991, pp. 149-181.

Haraway, D. J. *How Like a Leaf: An Interview with Thyrza Nichols Goodeve*, New York: Routledge, 2000.

Heims, S. J. *Constructing a Social Science for Postwar America: The Cybernetics Group, 1946-1953,* Cambridge, MA: MIT Press. 1991.

Jones, M. "It All Depends What You Mean by Discipline...," in J. Mingers and F. Stowell (Eds.), *Information Systems: An Emerging Discipline?,* London: McGraw-Hill, 1997, pp. 97-112.

Kuutti, K. "Debates in IS and CSCW Research: Anticipating System Design for Post-Fordist Work," in W. J. Orlikowski, G. Walsham, M. R. Jones, and J. I. DeGross (Eds.), *Information Technology and Changes in Organizational Work*, London: Chapman & Hall, 1996, pp. 287-308.

Laing, R. D. *Knots*, London: Tavistock Publications, 1970.

Law, J. (Ed.). *A Sociology of Monsters: Essays on Power, Technology and Domination*, London: Routledge, 1991.

Letiche, H. "Donna Haraway: The Schöne, Schöne, Dona(u)," *Organization* (6:1), 1999, pp. 149-166.

Morgan, G. *Images of Organization,* Beverly Hills, CA: Sage Publications, 1986.

Ramage, M. *The Learning Way: Evaluating Co-operative Systems,* Unpublished Ph.D. Thesis, University of Lancaster, UK, 1999.

Richard, M., and Whitley, E. "Addressing the Cyborg: A Useful Concept for Information Systems Research?," Working Paper 89, Department of Information Systems, London School of Economics, 2000.

Schön, D. A. *The Reflective Practitioner: How Professionals Think in Action,* New York: Basic Books, 1983.

Shapiro, D. "The Limits of Ethnography: Combining Social Sciences for CSCW," in *Proceedings of the Conference on Computer Supported Cooperative Work* (Chapel Hill, NC, 22-26 October 1994), New York: ACM Press, 1994, pp. 417-428.

Weber, R. "Editor's Comments: Still Desperately Seeking the IT Artifact," *MIS Quarterly* (27:2), 2003, pp. iii-xi.

Wiener, N. *Cybernetics: Or Control and Communications in the Animal and the Machine*, Cambridge, MA: MIT Press, 1948.

ABOUT THE AUTHOR

Magnus Ramage works as a lecturer in information systems at the Open University, within their Systems department, a cyborg group that combines interests in information systems, sustainable development, and organizational change, united by perspectives from systems thinking. He teaches and researches on the evolution of information systems (both planned and unplanned), on systems thinking and its history, on information systems evaluation, and on issues around organizational learning and change. As indicated in the paper, he received his PhD in computer science from University of Lancaster, and has also worked at the University of Durham. Magnus can be reached at m.ramage@open.ac.uk.

6 CORES AND DEFINITIONS: Building the Cognitive Legitimacy of the Information Systems Discipline Across the Atlantic

Frantz Rowe
University of Nantes

Duane P. Truex III
Florida International University and Georgia State University

Lynette Kvasny
Pennsylvania State University

Abstract The issue of the legitimacy of Information Systems is important for researchers in this field because other disciplines have begun to lay claim to research topics often thought to belong to the domain of IS research, and the field itself is under challenge in academic intuitions around the world (Avison 2002). Benbasat and Zmud's (2003) opinion is that IS has gained socio-political legitimacy but not cognitive legitimacy in large measure because the object of study in much IS research is not clearly delineated. In part, they are defining a disciplinary boundary issue and beginning to define criteria by which our field may be distinguished from reference disciplines or other related disciplines. Therefore, to gain more cognitive legitimacy, a clearer understanding of what we mean by "an information system" and of the central issues driving its creation and use is needed if it is at the core of that which we study. This paper advances that discourse by examining the role of a handful of French scholars, many of whom are not well known out of French academic circles, but whose thoughts on the issue are useful in furthering the debate on the ontological grounding of our field.

Keywords: Cores of the discipline, ontology, information system definition, IT enabled solutions, social theories

1 INTRODUCTION

This paper begins from the position that a general recognition and broad acceptance of a set of underlying core issues differentiating our field is an important element in asserting the legitimacy of Information Systems as a discipline. Is it important to seek agreement as to the core constructs underpinning our discipline? Stated more precisely, why is it necessary to constantly seek to justify our own existence?

This is not a new question. More than a decade ago Banville and Landry (1989) questioned if the field might be *disciplined*, other scholars examined the development of distinctiveness in the field using various bibliometric analyses to see if IS as a field had made a break from our own reference disciplines (Culnan 1986; Culnan and Swanson 1986). Still other scholars considered if or how the field might fare with a diffuse pluralistic core wherein many flowers were allowed to bloom (Robey 1996). More recently, scholars have been considering the lack of theorizing about technology in IS and ICT research (Orlikowski and Iacono 2001; Sawyer and Chen 2002), while others have revisited the issue of whether we have become our own, unique, reference discipline (Alter 2003; Baskerville and Myers 2002).

Recent threads on ISWorld attest to more recent challenges being made to the continuance, structure, and integrity of IS programs worldwide. Influential scholars and journal editors suggest that there is serious disagreement about the efficacy in the field and that the field is in crisis (Karahanna et al. 2003; Hirschheim and Klein 2003; Weber 2003). *Harvard Business Review* published an article that taps into a wellspring of resentment against the enterprise IS function and created a firestorm in the popular press (Carr 2003).

So, while there may not yet be a consensus as to the nature of the problem, there seems to be continuing concern that, at the core, there may be something amorphous and ill defined about our field. Benbasat and Zmud (2003) lament the ambiguity of the core or our field and argue that the field lacks a cognitive legitimacy as a result. The general notion is that identifying a core to the discipline helps explain that which is unique about the discipline, thus differentiating it from reference fields or other functional disciplines and helping to establish legitimacy. Albert and Whetten (1985) argue that, to claim legitimacy as a separate field of endeavor, a discipline must establish (1) the central character it is studying, (2) its distinctiveness, and (3) its temporal continuity. We return to these three points in section two following. But first we address the question of why we should be concerned with establishing commonly accepted notions about the core of our discipline.

In our view there are political, economic, and scholarly reasons for addressing the issue of core notions of the field. The first, or political, argument holds that since faculty representing other disciplines sit on tenure and promotion committees and serve on curricular committees and other university-wide bodies that decide upon resource allocation, it is wise to relate to them in ways common to their own discipline. Since the sister functions—marketing, finance, economics, management, accounting, and the like—have a commonly accepted core around which their own disciplinary work occurs, it is appropriate that we do the same.

Gordon Davis has developed this idea, saying that ours is an applied field tied strongly to the organizational functions of information systems and information management, and that the vitality of our academic field is tied closely to the vitality and boundaries of the IS function (Karahanna et al. 2003). Within that function, there exist specialized skills and specialized knowledge that help differentiate the field; dealing with these core issues defines the purview of the function and hence the academic discipline. Davis then generalizes back to our sister academic functions in suggesting that there are "strongly shared activities" at the boundary of these functions. These are the activities where IS interfaces with and become key tools to another discipline such as marketing or production. At these interfaces are opportunities for shared research and for communication about core issues. But the boundaries, clearly differentiating our research concerns from other disciplines mark a territory that is uniquely our own.

The second, or economic, reason is closely associated with the political argument and is manifest in the concerns over declining enrollments in academic IS programs across the globe. Are structural changes in IS development and operations putting the field itself at risk? Will there be places for our graduates to work? This question seems to motivate students in choosing majors. We need to respect their concern and learn what the economic consequences may be for changing demand patterns for IS graduates. A clear understanding of and ability to articulate the *core* of the field may help isolate us from concerns that the field has died because programming has moved off shore, or infrastructures are embedded in enterprise-level applications, or other cyclical changes arising from social and technical change. At the core there should be more than a single technology, method, or research concern.

The third, or scholarly, rationale is more complex, and is being considered on a number of fronts. Benbasat and Zmud argue that IS has gained socio-political legitimacy but not cognitive legitimacy in large measure because the object of study in much IS research is not clearly delineated. They define a boundary condition in which they suggest a kind of delineation by proximity to actual systems artifacts. Hirschheim and Klein (2003) argue that to save the field we need to take corrective or transformative action to solve communications problems within the scholarly community, and between the field and the practitioner community with whom we run the risk of being considered irrelevant. To these authors, it becomes important to reach an agreement of the existence and organization of a common body of knowledge central to the field. Agreement with or acceptance of this common body of knowledge then becomes a criterion of membership in the field (Hirschheim and Klein 2003). For some, this is an issue of language. The late essayist Neil Postman (1988) said that knowledge of any discipline required defining, learning, and managing the language of that discipline. Steven Alter (2003), addressing the communications gaps described by Hirschheim and Klein, suggests that the field needs a better way to talk about itself and the core objects of interest. Like others, he argues "that in order for a business professional to understand an information system it is necessary to understand the work system that the information system serves." In other words, Alter is struggling with defining a means where the field might have a kind of "Sysperanto" or useful language for describing systems and other central concepts to the discipline. His notion of a useful language includes providing practical help in identifying, observing, and conceptualizing information systems and their operations and place in organizational contexts.

It should be noted that a few other scholars suggest that there is no need to define a core of the field, arguing rather that we speak via the systems we build and install

(Karahanna et al. 2003). Yet even this position entails core assumptions, even if unexpressed, about a core, and it is one in which the technical artefact is itself central.

In this paper we do not posit a single core, rather that there may be a *core set* of issues. We wish to make a contribution to that discourse itself by introducing and illustrating how French research in IS, philosophy, and sociology can advance the debate on cognitive legitimacy of IS. Specifically we address the central notion that to achieve higher cognitive legitimacy for the IS discipline, we must establish the *central character of what we are studying; its distinctiveness and possibly its temporal continuity*. This manuscript examines how French authors help consider the question of the central character of our field and then briefly discusses the themes of distinctiveness and temporal continuity, leaving a fuller treatment of those themes for a later paper.

The balance of this paper proceeds as follows: Section 2 examines three themes that warrant claims to a field's intellectual legitimacy. Section 3 addresses the question of an ontology raised inevitably by considering the nature of the IS as an object of study. The fourth section examines how five French scholars characterize or cause the IS object to be characterized. Section 5 summarizes and suggests a plan for further research.

2 THREE THEMES SUPPORTING THE FIELD'S CLAIMS TO LEGITIMACY

We now return to the question of how a field might solidify claims to its legitimacy by considering the argument by Albert and Whetten (1985) that we must first establish the central character we study, then we must establish our field's distinctiveness and finally achieve a measure of temporal continuity.

2.1 Theme 1: The Field's Central Character

One of the central claims to IS disciplinary distinctiveness is the focus upon the control and evaluation of IT in organizations, or more precisely what Benbasat and Zmud (2003) call "IT-enabled solutions." This claim is supported by a study examining 1,018 articles in major English and French publication venues from 1977-2001.[1] This study identifies six major problem areas addressed by the papers in these publication years and venues. Those are (1) *gestion stratégique* (strategic management), (2) *économie, divers* (various economic issues), (3) *conception* (design), (4) *gestion des projects* (project management), (5) *evaluation* (evaluation), and (6) *animation* (or roughly, appropriation and change management (Desq et al 2002). Their study illustrates how for both English and French authors the issue of *evaluation* is an important focus

---

[1]Those venues included from the Anglophone world two relatively older venues—*MIS Quarterly* and the proceedings of the International Conference on Information Systems (1980 onward)—and relative Francophone newcomers, two French journals, *Technologies Information et Société* (TIS) (f. 1996) and *Systémes d'Information et Management* (f. 1996), and two conferences,l'Association Information et Mangement (AIM) (f. 1997) and les journées nationals des IAE (f. 1984)

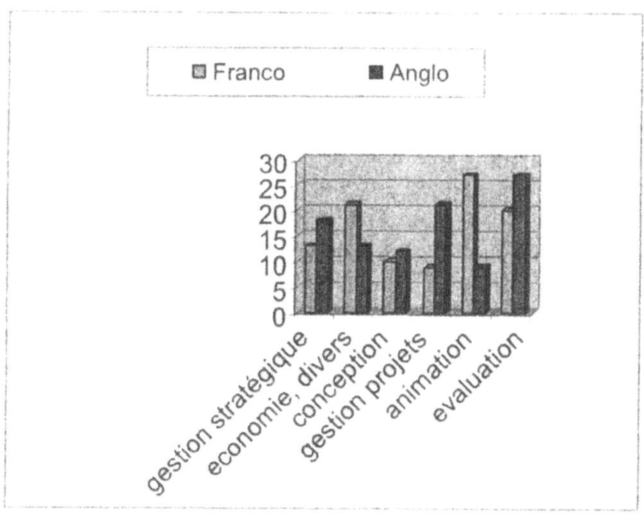

Figure 1. Problems

and that a generalized notion of *control* is the dominant theme of 25 years of research (see Figure 1). One result stands clear: the theme of evaluation of info systems represents 25 percent of the work (Shapiro 1998). The more general theme of *control* (animation, evaluation, and personnel management) dominates the field with 45 percent, far ahead of design (28 percent and strategy 23 percent.

The theme of evaluation was found to be a relatively stable construct over time even though historically the issue was analyzed at the individual level, whereas in recent years it is applied more at the organizational or interorganizational level. In recent work, the dependent variables are more likely to address potentials (e.g., competitive advantage, flexibility) than actual results.

In this manuscript, we use the concepts of control and evaluation in a slightly broader sense. We view evaluation as a process that occurs at the various stages of information systems evolution (i.e., design, use, and impacts; Soh and Markus 1995). The later concern includes organizational, managerial, and other stakeholder impacts. In the domain of IS management, Willcocks and Lester (1993) say that

> evaluation is about establishing by quantitative and/or qualitative means the worth of IS to the organization. Evaluation brings in to play notions of costs, benefits, risk and value. It also implies an organizational process by which these factors are accessed, whether formally or informally.

We like this definition because it identifies how evaluation is undertaken in any organizational setting as a matter of course in routine interfacing with an information system. The evaluation process is done in both formal and informal modes by managers as well as by other organizational stakeholders. So, given the embedded nature of IS artifacts in human organizational settings, the process of evaluation requires both managerial and technological evaluation tools and methods.

IS design methods also incorporate procedures for managerial control and technical evaluation, taking into account goals, anticipating some of the impacts, and finally dealing with the specification methods for information needs (Purao et al. 2002). In the realm of system use, control and evaluation occur for two main reasons: to anticipate or learn exactly how many people will really use the system, and to understand how (or if) they will appropriate it. The situated use of the IS often implies some measure of transformation from use anticipated by the system's designer and builder's to that employed at the user level. Thus one other meaning of evaluation is concerned with diffusion and infusion issues (Saga and Zmud 1994). This lag between conception and the adoption in organizational settings means that it is only *ex post* to system design, development, and deployment, and typically much later after the appropriation in the organization that we can assess the net economic benefits of the system. In fact it has been suggested that it is only after competitors have made (or not made) similar investments that we can assess the net economic benefits of the system. The context of evaluation is not only organizational, but includes the competitive structure of the industry in which the firm is located (Soh and Markus 1995).

2.2 Theme 2: Distinctiveness of the Field

Following from the claim above, that evaluation and control are issues at the core of IS research, one of the key claims to the distinctiveness of our discipline lies in the way we assess IT-based systems. In this regard, IS research is a big tent allowing many approaches to this process.

Nevertheless our discipline is distinct in the way it helps develop evaluation methods at each stage of evolution of the IT artefact (proposal, development, implementation, post-implementation, routine operations (Willcocks and Lester 1993), and at the same time takes into account the role, importance, and interaction of social actors, the structures of organizations, strategies, and tasks among a host of other issues (Benbasat and Zmud 2003; Marciniak and Rowe 1997).

2.3 Theme 3: The Field's Temporal Continuity

Temporal continuity deals with the relative maturity and consistency of views in a field. It is more difficult to argue our field's legitimacy on this dimension because of the relative youth of IS and because the field is very dependent on dynamic technological innovations (Reix and Rowe 2002). Previous attempts to delineate the evolving independence of our field have focused on citation analyses of IS papers, examining the degree to which they cite authors and sources in reference disciplines versus citing other IS works (Culnan 1986; Culnan and Swanson 1986). More recently, we have seen claims that the growing continuity of distinctively IS research provide evidence of a temporal continuity to the field (Baskerville and Myers 2002). The notion of process or organizational control has been a part of this literature almost from the beginning (cf. the "Minnesota experiments") and remains key in the literature to this day. As discussed above, the study of 25 years of IS research suggests a reasonably constant set of issue. This provides evidence of the temporal continuity needed in legitimizing a field.

3 THE DEFINITION OF AN INFORMATION SYSTEM AS AN OBJECT OF STUDY: THE QUESTIONS OF AN ONTOLOGY

Following from these observations, it is our contention that, in order to achieve higher cognitive legitimacy, our own discipline must clearly articulate the core concepts it is using. This is not a trivial exercise because it is both raises ontological questions and presents the political challenges of any definitional issue. Postman (1988) reasons that the knowledge of any discipline is defined by knowledge of the language of the discipline. This explains why definitional battles are prominent in many disciplines. Terminological development itself is a convoluted process. In our own field, this point is well illustrated by Robert Gray's (2003) historical review of the distinction between the commonly used and accepted terms *data* and *information*. Thus if we claim that we are principally dealing with issues of control in the development of IT-based solutions, we are left with the question of what do we mean by IT-enabled solutions. What do we mean by and define as the central object of our interest and study in the field? What is an IT-enabled solution? What is an information system? Once we get a clear picture of these questions, then we can better tackle larger questions, such as how do we conceptualize the IT artifact (Orlikowski and Iacono 2001; Sawyer and Chen 2002)?

This paper takes aim at this issue from the contributions made to this discourse by French IS researchers and sociologists who have influenced other IS researchers in and out of France. Defining an information system is an ontological issue. The discourse on ontologies is long and rich within the literature of the IFIP WG 8.2 community. However, because the term is now being used in very different ways within our own field, as for example between the so-called design community, the requirements engineering community, and the IS community, it is necessary to briefly explain our use of the term as well as where within the ontology discourse we place ourselves.

Ontology is the part of philosophy aimed at studying being as being. It can be centered on the fact of being (existentialism) or on the nature of being (essentialism). The relationship between essence and existence is the fundamental problem of any ontology. In IS, as in other disciplines, we can consider two different problems.

1. The building of an ontology—a problem that has been taken very seriously by many researchers in data base management and in IS development (Hirschheim et al. 1995). The challenge is to reduce the elements of an objective reality to a limited number of notions, as general as possible (see, for instance, the two-layered ontology of Parsons and Wand 2000) and to describe the structure of the universe from these notions and their relationships.
2. The ontology can be defined as the "exploitation of being of being structures and we would rather define metaphysics as the questioning of the existence of being" (Sartre 1943, pp. 358-359). Sarte also stated that, "In an ontology things take ontic attributes. But conscience can overcome the fact of being to the sense of being" (p. 30).

Therefore, we could say that what IS scholars consider is the study of the information system in its intimate and deep nature, as opposed to its appearance or its attributes. Using Sartre's own words, we are dealing with the ontological or metaphysical problem as opposed to the ontic problem of an information system (the latter being considered as the ontological problem in software engineering).

This paper does not allow a full treatment of the ontological discourse nor does it allow for the identification of the full population of productive French IS academics nor of French sociologists and philosophers who have influenced other (curiously often non-French) IS scholars. So in the balance of this paper, we will present a small, illustrative sample of French IS scholars and point to but three French sociologists (Crozier, Bourdieu, and Latour, the last two being well known to our community). As a set, we use these scholars to illustrate the question of how one may view the core of the IS research ontological debate from a French perspective.

4 FRENCH SCHOLARS ON THE INFORMATION SYSTEM DEFINITION

We now turn our attention to specific French IS scholars, some of whom are, regrettably, only occasionally available in English translation. Within this initial community we can find five conceptions of an information system arising from different positions on the ontological spectrum. This contradicts the idea that there is a single French ontological view of the IS core in contrast with say an American or German ontological view of an IS core. Those views expressed in the French IS literature are

- derived from a rationalist and software engineering viewpoint; the nature of an information system is a formal code and an artifact of a different nature than the socio-technical system it controls (social or natural)
- derived from general system theory; the nature of an information system is not different from that of the socio-technical system in which it is embedded
- derived from a pluralist view the nature of an information system is human and social
- derived from a critical sociologist view where structures are reinforced by culture and language and constrain more (Bourdieu) or less (Crozier) actions
- derived from a historical view of science and technology (Latour)

4.1 Peaucelle and Rolland: Information Systems as a Codified Language—A Rationalist Software Engineering View

The prevailing conceptualization comes from computer science and the software engineering literature. One of its leading figures, former IFIP 8.1 chair, Rolland wrote (1986)

> An information system is an artefact, an artificial object, grafted on a natural object which can be an organization, an industrial process, an embarked command. It is designed for memorizing a set of images of the real object at different times in its life; these images must be accessed by partners of the organization for decision purposes in the best conditions.

For Peaucelle (1981), translator of Davis and Olson's (1985) book, the notion of an information system is implicitly restricted to formal systems, those dealing with data according to specified rules (p. 8). He views an information system as a formal code and the outcome of an intersubjective process, one which has a unique and collective fixed meaning that is justified by organizational routines.

> An information system is a communication language of the organization, build [sic] consciously by its members to represent, in a reliable and objective manner, rapidly and economically, some aspects of its activity, past or future. Sentences and words of this language are data and their signification comes from the rules of their making by men or machines. The representation mechanisms, special to this type of language, prove efficient in the routinization of organizations acts (pp. 24-25).

Peaucelle and Rolland share the view that an information system is an artefact (Weber 1987), in the first sense recalled by Alter (2003). Consequently, their nature is different from that of organizations which can be considered either as socio-technical systems with an important informal component (Peaucelle 1981), or as natural objects (Rolland 1986).

By their insistence on a possible objective coding and fixed meaning, information systems appears as the tool of rationalization of the organization. It aims at designating transactions and formal processes. It naturalizes the organization and derives from a realist ontology.

4.2 LeMoigne and Melèse: The Appropriation of General Systems Theory and its Evolution Toward Constructivism

LeMoigne is well known in France as one of the men who has adapted Von Bertralanffy's general systems theory (LeMoigne 1977) while also offering a strong criticism of the analytical and Cartesian method for designing information systems (LeMoigne 1996). He is also one of the first authors in France to write a book on information systems and not on data processing or "informatique de gestion," in which he concluded

> the information system of an organization can and must only be in charge of one essential function, that of memorizing—collecting, storing and retrieving—all information generated by its transactions with its environment and, incidentally, some others that members of the organization wish to share for some time (LeMoigne 1973, p. 195).

However, he has not explicitly proposed a formal definition of an information system. In one of his most well known contributions on this topic (LeMoigne 1986) first presented at the IFIP Conference in Toronto in 1977 with Maurice Landry, we can only find it in a figure (Landry and LeMoigne 1977). There they convey the idea that an organizational information system is the memorization system of the organization's information. His definition also constitutes both an intermediary and a regulation mechanism between the system of operations and the control system of the former cybernetics representation of the functioning of an organization. As such, we did not contemplate a revolution in terms of IS thinking that might call for abandoning Davis's (1974) definition of an information system, a definition that LeMoigne explicitly associates with a cybernetic view of the organization. Nevertheless, LeMoigne's view is that the goal of an information system cannot be the control of a rational norm (Dehaene 1992). More interestingly, LeMoigne criticizes the MIS paradigm for not doing justice to the generation and memorization of information processes. LeMoigne especially insists on these two problems: "The new paradigm of information systems must take into account the capacity of the organization to represent itself, its behaviors and its transformations and not anymore its capacity to control them" (1986, p. 27).

In fact, an organization is organizing itself by its information processes, reciprocally in an auto-referential process (Wilensky 1983). Information systems inform the organization which forms (makes) them. The way information systems are constructed is, therefore, generated by the activities of the organization and by the data modeling, which are particular to each organization. Secondly, this representation must be appropriated and co-memorized by several actors (LeMoigne and Pascot 1979).

LeMoigne and his team were very influential in the French IS world both with academics and practitioners when they developed the MERISE design method. For his team, information systems modeling enriches the modeled reality, not only because it is incorporated in it and as such simplifies it, but because even if it were not incorporated in it, it potentially contains more diverse realities than what has been modeled (Tabourier 1986, p. 32). In view of his team, LeMoigne's golden triangle of intellectual debt is attributed to Morin, Piaget, and Simon (GRASCE 1999, p. 5). From an epistemological viewpoint, he openly declares himself to be a social constructivist. For him IS and many other kinds of engineering activity are social constructions and cannot be studied with positivist lenses.

LeMoigne (1996) refers to a remarkable theoretical essay by Jacques Mélèse, another advocate of the systems approach, where Mélèse (1979, p. 36) defines an information system as "the set of processes which, by exchange of significations, allow, globally and locally in the firm, consistency, equilibration, backup and innovation." Mélèse insists that relevant information is not only codified and formal, but is often informal and qualitative, and that each organizational unit has to represent itself and its environment and allow other units to do so (i.e., to set up an informational situation in a participative way).

As Mélèse added, "perceived information should be most interpretable and locally usable" (p. 37). For that matter, formal communication of codified information between units and levels should be limited, because of noise and distortion. Conversely, in each unit, informational learning and sense making should be developed. He finally added, "Make of all organizational level, places of information association and sense emerging" (p. 37). Therefore, we consider Melèse as an interpretivist, in the sense close to Karl Weick's characterization of the objects of research, but also a social constructivist (LeMoigne 1996, p. 131) who was strongly influenced by Beer, Morin, Piaget, and Bateson.

What is important and common to both Melèse and LeMoigne is that to a certain extent they explicitly refuse to tackle the ontology of an information system even though we find the definition, cited above, in Melèse's work. The important point here is that, for them, an information system is a socio-technical system and one isomorphic to that of the organization. It is not just a management information system, but a socially constructed organizational information system (LeMoigne and Van Gigch 1990).

4.3 Reix and Rowe: A Pluralist View—The Nature of an Information System Is Human and Social

In their introduction to IS research, after a review of the history of the IS discipline and prominent definitions, Reix and Rowe (2002) offered the following definition: "A set of social actors who memorize and transform representations, via information technologies and operating modes " (p. 7). They founded their definition on previous work in IS as well as on Bourdieu and Crozier and argued that this definition is also the result of an eidetic reduction (Husserl 1950) and takes into account

- The fundamental human and social nature of any information system, by putting upfront the social actor. An information system is not just an abstract objective representation or the fixed outcome of intersubjectivity, but it always remains subject to interpretation, social games, and conflicts (Bourdieu 1980; Crozier and Friedberg 1977). This view contrasts with that of Lamb and Kling (2003). While both views agree upon the central role of the social actor, Lamb and Kling see the information system as a product and external representation of man's activity.
- The fundamental functions of memorizing (LeMoigne 1973) and transforming representations (LeMoigne 1990).
- The possibility of working with or without information technology and with or without some modus operandi (Bourdieu 1980).

The first point also appears as the logical outcome of the idea of information being constructed as meaningful for a user and needing interpretation (Gray 2003). If we talk about information system, it does not mean they exist out there, as external things; we are talking about an intellectual construct, which from a phenomenological viewpoint demands a human and social interpretation at some point. Moreover, this definition, with the second additional point, opens the possibility of considering activities such as cognitive processes, informal talks and the use of tacit knowledge, for instance in communities of practice, as phenomena linked to the concept (Michaux and Rowe 2004).

The fact of considering the concept as encompassing informal phenomena departs from the rationalist and empiricist position developed in software engineering. However, from an epistemological viewpoint, Monod (2002, p. 44) considered this definition as reflecting a pluralist view. For him, in addition to its insistence on representation processes tending toward interpretivism, this definition also stresses social games tending toward the sociology of conflicts, while the technology supporting information processes and its operating modes tends toward a positivist view. The authors fully agree with Kallinikos (2002, p. 289) that "the formation of the premises governing the human-technology interaction must be analyzed with reference to the constitutive properties of technology and the distinctive forms by which various technologies emerge as standing possibilities of one type or another," as it has been shown on empirical works (Rowe and Struck 1999).

4.4 Bourdieu's Critical Realism and Influence on IS Scholars

We see at least four reasons for using Bourdieu in Information Systems. First, as a sociologist of practice, he clearly distinguishes between *opus operatum* and *modus operandi*, between prescription and activity. Second, he expressly fought the language philosophers (Ricoeur, Austin, and Searle) for language is rarely performative in itself; it is the social status of the speaker which gives legitimacy and meaning to language propositions. Third, Bourdieu (1980) sharply criticizes the reductionism of most quantitative surveys. In his own work, he adopts and advocates for mixed methods to investigate social phenomenon. Finally, and above all, Bourdieu helps us think about control, power, and domination, which are at the heart of the first core of the IS discipline. In doing so, we can better theorize the practices of social actors.

For Bourdieu, societal structures are socially defined and maintained. They have great persistence and are very difficult to change. As such, they have enormous influence over human behavior. One objective of Bourdieu's theoretical framework is to uncover the buried organizational structures and mechanisms that are used to ensure the reproduction of social order. His framework helps us understand how changes arising from information technology may actually reinforce existing power structures and help perpetuate the social order. For Bourdieu, change (including technological change) is a self-regenerative mechanism required for the maintenance of stratified organizational hierarchies. So where, one the one hand, static structures can be figured out and conquered over time, on the other hand, changing structures keep actors off balance, and thus lead them to apply familiar strategies in unfamiliar contexts, reinforcing old structures, behaviors, rules, and order. It is this reuse of learned dispositions (habitus) in new settings that make existing class positions self-sustaining.

Referring to Bourdieu, IS researchers examine the nature of those structures and the impact they have on the introduction and use of IS artifacts (here symbolic meaning) in societal settings. On the one hand, Shultze and Boland (2000) use Bourdieu to help understand the roles of information gatekeepers. Schultze (2001a, 2001b) uses Van Maanen's (1988) notion of confessional tales to frame the narratives in her ethnographic fieldwork in a way especially attuned to Bourdieu's call for reflexivity in intensive research. On the other hand, Kvasny's research program examining the digital divide in African American communities uses Bourdieu's concepts of capital, habitus, and field as theories for understanding the IS practices of individuals, groups, and organizations. In particular, she is interested in how IT reproduces social inequality (Kvasny 2002a, 2002b; Kvasny and Keil 2003; Kvasny and Truex 2000, 2001).

To these authors, one advantage Bourdieu provides over that of other European critical social theorists such as Giddens and Habermas or over postmodern theorists like Derrida is that Bourdieu's own empirical research offers some guidance as to how to go about using his theoretical framework. Whereas Bourdieu uses empirical work to develop theory, thus making his theory more convincing and easier to apply, other social theorists have little to say about empirical research and methodology. Thus this French variation on critical social theory combines an interest in the practical concerns of examining social order with the more cerebral act of theorizing about that order.

In another attempt to add further empirical work inspired by Bourdieu's CST in the domain of information technologies, Helen Richardson (2003) studied customer relationship management technology utilization. Richardson, in examining social relations around CRM system use, discovered the application of symbolic violence as a mode of domination and illustrated how the relationships between agencies and structures (social and technology enforced and supported) manifest and reinforce themselves in the logic of practice.

While Bourdieu has been used on both sides of the Atlantic by IS scholars, it turns out that references to two other prominent French thinkers differ in that respect. Crozier has been essentially used by IS scholars in France, while Latour was mostly used outside of France.

4.5 Crozier: The Uncertainty Zone Enlarged by Social Actors

After showing that, even in bureaucracies, actors circumvent the rules and find some degree of leeway (Crozier 1963), Crozier and Friedberg (1977) tended to generalize this conduct and develop its theorizing: in order to avoid domination, actors tend to increase their "uncertainty zone" (i.e., their power to act as they want) which implies that their conduct is not totally constrained. Therefore, "power resides in the degree of leeway [freedom; marge de liberté] that each partner has in a power relationship" (Crozier and Friedberg 1977, p. 60). The circular character of this definition is noted by Caillé ("De finalité par nature, il devient moyen de fins indéterminées," 1981). The first stake is power as the possibility of chosen action in a collective setting. But it is surprising that Caillé posits that a legitimate goal of power becomes a means servicing indeterminate ends. For in the same article, Caillé criticizes the more substantive sociology of interest of Bourdieu. Finally, it is clear that with Bourdieu, the social actor is more structurally constrained than with Crozier. Many in the French IS literature have cited Crozier, probably because his theory provides greater openness to human agency than do the poststructuralist theories of Bourdieu, Foucault, and Giddens.

If we do the genealogy of Crozier's use in IS, its paradoxically by Peaucelle that we should begin. In fact, after a doctorate in computer science, Peaucelle spent three years with Crozier's research center. Ballé and Peaucelle (1973) then published a book, *The Power of Data Processing*. In his preface, Crozier (1973) writes

> As with any important techno-organization innovation, computerization does not raise the problem of functioning or implementing some model, but that of change, from a socio-technical system to another one whose characteristics cannot be predicted precisely, and which in any case cannot be considered as and fully achieved, but as a stage in an evolution....The situation is generally the reverse of what is generally thought. At the beginning there are all the more many problems as participation of employees is actively searched (pp. 11-14).

Crozier also strongly influenced Pavé who showed the utopia—due to the hyper-functionalism of computer scientists—of firms made of transparent human relationships. He also advocated that there could not be any correspondence between computers and organizational structures because of the attitude of employees maintaining their uncertainty zone. Then Morley (1993) used these works, with the exception of Pavé, to show that the effect of participation on the complexity of projects depends on the choice of the users, the attribution of roles, and the selection of phases during which they participate. More recently Besson and Rowe (2001) sketch the role of leeway margins in the dynamics of enterprise systems projects. Many in the French IS literature have cited Crozier, probably because he opens a window for thinking action other than with the sociologists of domination (Bourdieu, Foucault, and Giddens).

4.6 Latour and Agency: A Constructivist Ontology

The ideas of the French sociologist and scholar of science and technology, Bruno Latour, have captured the interest of a generation of IS scholars (Baskerville and Myers 2002; Bijker 1994; Bijker and Law 1994; Walsham 1995, 1997), an assertion readily confirmed by glancing through the published proceedings of the last eight IFIP WG 8.2 transactions (Jones 2000). Much of this work utilizes Latour's actor network theory to help understand the interaction of both human actors and nonhuman actants. In studying these relationships, researchers hope to see how this interaction creates and shapes the social contexts in which the IT object exists. It is a form of evaluation of the IS artefact and its social setting. Thus one could argue that the ontology behind the relatively equal agency of any action in a network is relativistic and very socially constructivist. The network is a linguistic and relationally constructed object always in the process of being constructed. However, from an ontological point of view, this is not the end of the story with Latour, for he does not accept the dichotomous view of the reality that has inspired the previous discussion.

An underlying and critical concept in Latour's thinking, and one that set his work apart, is that he rejects the separation of the natural world from the social world, or what he terms the "modern constitution" (Latour 1993). This constitution has as its first guarantors the idea that "Nature is transcendent but mobilizable (immanent), and that Society is immanent but it infinitely surpasses us (transcendent)" (ibid, p. 141). Rather, Latour holds for a "nonmodern constitution" in which he sees the "nonseparability of the common production of societies and natures" (ibid, p. 141). It is an ontological blended middle ground.

5 CONCLUSION: THE CONTRIBUTION OF FRENCH SCHOLARS TO THE QUESTION OF THE CORE OF THE IS DISCIPLINE

If we examine the French scholars who have had some influence on IFIP 8.2 since the first Manchester conference in 1984, the scholars discussed in sections 4.1 through 4.3 have not been cited often. On the other hand, work inspired by Bourdieu has been present in the past three conferences. Latour's ideas have been more widely used and cited. In fact Latour was an invited plenary speaker, as well as a published author at the 1995 Cambridge WG 8.2 conference (Latour 1996).

For the moment, we suggest that one contribution of this paper is to raise these questions while highlighting the contribution of French scholars to the debate regarding one of the central objects of study in our field.

We see three contributions of French scholars to information systems and the central issue of control.

1. They help us see how central the concepts of control and evaluation are in our field.
2. To underline the major difference between a software engineering ontological view (realist) and that of a management and social scientist ontological view (be it constructivist, critical realist, pluralist, or that of Latour). By definition, the latter assesses the relevance of an information system, as an artefact, with respect to its human and organizational context and not just with respect its capabilities and specifications for some tasks (i.e., from a logical viewpoint).
3. They give some theoretical and methodological advice as to the study of the exercise of power and control through the contribution of Bourdieu, Crozier, and Latour.

In continued research we will explore the nature and influence that French thinkers and researchers have contributed both directly and indirectly to this debate.

REFERENCES

Albert, S., and Whetten, D. A. "Organizational Identity," *Research in Organizational Behavior* (7), 1985, pp. 263-295.

Alter, S. "A General, Yet Useful theory of Information Systems," *Communications of the AIS* (1:13), 2003.

Avison, D. "The UK Information Systems Perspective: A Personal View," *Systeme d'Infomation et Management* (3:7), 2002, pp. 49-54.

Ballé, C., and Peaucelle, J. L. *The Power of Data Processing*, Paris: Editions d'organisation, 1973.

Banville, C., and Landry, M. "Can the Field of MIS Be Disciplined?," *Communications of the ACM* (32:1) 1989, pp. 48-60.

Baskerville, R. L., and Myers, M. D. "Information Systems as a Reference Discipline," *MIS Quarterly* (26:1), 2002, pp. 1-14.

Benbasat, I., and Zmud, R. "The Identity Crisis within the IS Discipline: Defining and Communicating the Discipline's Core Properties," *MIS Quarterly* (27:2), 2003, pp. 183, 193.

Besson, P., and Rowe F. "ERP Project Dynamics and Enacted Dialogue: Perceived Understanding, Perceived Leeway, and the Nature of Task-Related Conflicts," *DataBase* (32:4), 2001, pp. 47-66.

Bijker, W. E. "The Social Construction of Fluorescent Lighting, or How an Aircraft was Invented in its Diffusion Stage," in W. E. Bijker and J. Law (Eds.), *Shaping Technology/Building Society: Studies in Sociotechnical* Cambridge MA: The MIT Press, 1994, pp. 75-102.

Bijker, W. E., and Law, J. (Eds.). *Shaping Technology / Building Society: Studies in Sociotechnical Change*, Cambridge MA: The MIT Press, 1994.

Bourdieu, M. *Le sens pratique*, Paris: Editions de Minuit, 1980.

Caillé, A. "La sociologie de l'intérêt est-elle intéressante?" *Sociologie du travail* (23:3), 1981, pp. 257-274.

Carr, N. "IT Doesn't Matter," *Harvard Business Review*, May 2003, pp. 1-10.

Crozier, M. "Le phénmomèn bureaucratique," Paris: Seuil, 1963.

Crozier, M. "Préface," in C. Ballé and J. L. Peaucelle, *Le pouvoir informatique dans l'entreprise*, Paris: Les Editions d'Organisation, 1973, pp. 11-19.

Crozier, M., and Friedberg, E. *Actors and System: The Politics of Collective Action*, Boston: Ginn and Co., 1977.

Culnan, M. J. "The Intellectual Development of Management Information Systems, 1972-1982: A Co-citation Analysis," *Management Science* (32:2), 1986, pp. 156-172.

Culnan, M. J., and Swanson, E. B. "Research in Management Information Systems, 1980-1984: Points of Work and Reference," *MIS Quarterly* (10:3), 1986, pp. 288-303.

Davis, G. B. *Management Information Systems: Conceptual Foundations, Structure, and Development*, New York: McGraw-Hill Book Company, 1974.

Davis, G. B., and Olson, M. H. *Management Information Systems: Conceptual Foundations, Structure, and Develompent* (2nd ed.), New York: McGraw-Hill Book Company, 1985.

Dehaene, P. "Organization, Project and Strategy as Symbols," in *Proceedings of the Third Conference Economique et Intelligence Artificielle (CECOIA III-CEMIT)*, Tokyo, JASMIN, Volume I, 1992, pp. 243-247.

Desq, S.; Fallery, B.; Rodhain, F.; and Reix R. "25 ans de recherches en systèmes d'informations ," *Systèmes d'Information et Management* (7:3), 2002, pp. 5-33.

GRASCE. *Entre systémique et complexité chemin faisant: mélanges en l'honneur du Professeur Jean-Louis LeMoigne*, Paris: Presses Universitaires de France, 1999.

Gray, R. "A Brief Historical Review of the Development of the Distinction Between Data and Information in the Information Systems Literature," in J. Ross and D. Galletta (Eds.), *Proceedings of the 9th Americas Conference on Information Systems*, Tampa, FL, 2003, pp. 2843-2849.

Hirschheim, R., and Klein, H. K. "Crisis in the IS Field? A Critical Reflection on the State of the Discipline," *Journal of the Association for Information Systems* (4:5), 2003, pp. 237-293.

Hirschheim, R.; Klein, H. K.; and Lyytinen, K. *Information Systems Development and Data Modeling: Conceptual and Philosophical Foundations*, Cambridge, England: Cambridge University Press, 1995.

Husserl, E. *Idées directrices pour une phénoménologie* (3rd ed.), Paris: Gallimard, 1950 (1st ed., 1913).

Jones, M. R. "The Moving Finger: The Use of Social Theory in WG8.2 Conference Papers, 1975-1999," in Baskerville, J. Stage, and J. I. DeGross (Eds.), *Organizational and Social Perspectives on Information Technology*, Boston: Kluwer Academic Publishers, 2000, pp. 15-31.

Kallinikos, J. "Reopening the Black Box of Technology: Artefacts and Human Agency," in L. Applegate, R. Galliers, and J. I. DeGross (Eds.), *Proceedings of the 23rd International Conference on Information Systems*, Barcelona, Spain, 2002, pp. 287-294.

Karahanna, E.; Davis, G. B.; Mukhopadhyay, T.; Watson, R., and Weber, R. "Embarking on Information Systems' Voyage to Self-Discovery: Identifying the Core of the Discipline," Panel Presentation at the 24th International Conference on Information Systems, Seattle, WA, 2003.

Kvasny, L. "A Conceptual Framework For Examining Digital Inequality," in R. Ramsower and J. Windsor (Eds.), *Proceedings of the 8th Americas Conference on Information Systems*, Dallas, TX, August 2002a, pp. 1798-1805.

Kvasny, L. *Problematizing the Digital Divide: Cultural and Social Reproduction in a Community Technology Initiative*, Unpublished Ph.D. Thesis, Department of Computer Information Systems, Robinson College of Business, Georgia State University, 2002b.

Kvasny, L., and Keil, M. "The Challenges of Redressing the Digital Divide: A Tale of Two Cities," in L. Applegate, R. Galliers, and J. I. DeGross (Eds.), *Proceedings of the 23rd International Conference on Information Systems*, Barcelona, December 2003, pp. 817-828.

Kvasny, L., and Truex, D. "Defining Away the Digital Divide: A Content Analysis of Institutional Influences on Popular Representations of Technology," in B. Fitzgerald, N. Russo, and J. I. DeGross (eds.), *Realigning Research and Practice in Information Systems Development: The Social and Organizational Perspective*, Boston: Kluwer Academic Publishers, 2001, pp. 399-414.

Kvasny, L., and Truex, D. "Information Technology and the Cultural Reproduction of Social Order: A Research Program," In R. Baskerville, J. Stage, and J. I. DeGross (Eds.), *Organizational and Social Perspectives on Information Technology*, Boston: Kluwer Academic Publishers, 2000, pp. 277-294.

Lamb, R., and Kling, R. "Reconceptualizing Users as Social Actors in Information Systems Research," *MIS Quarterly* (27:2), 2003, pp. 197-235.

Landry, M., and LeMoigne J. L. "Towards a Theory of Organizational Systems: A General System Perspective," in B. Gilchrist, *Proceedings of the IFIP Congress '77*, Amsterdam: Elsevier-Science, 1977, pp. 801-805.

Latour, B. *We Never Have Been Modern (Nous n'avons jamain été modernes)*, C. Porter (trans), Hemel Hempstead, England: Harvester Wheatsheaf, 1993.

Latour, B. "Social Theory and the Study of Computerized Work Sites," in W. J. Orlikowski, G. Walsham, M. R. Jones, and J. I. DeGross (Eds.), *Information Technology and Changes in Organizational Work*, London: Chapman & Hall, 1966, pp. 295-307.

LeMoigne, J. L. "La conception des systèmes d'information organisationnels: de l'ingénierie informatique à l'ingéniérie systémique," in J. A. Bartoli and J. L. LeMoigne (Eds.), *Organisation intelligente et systéme d'information stratégique*, Paris: Economica, 1996, pp. 25-52.

LeMoigne, J. L. *La Modelisation des sytèmes complexes*, Paris: Dunod, 1990.

LeMoigne, J. L. *La theorie du systeme general, theorie de la modelisation*, Paris: Presses Universitaires de France, 1977.

LeMoigne, J. L. *Les systèmes d'information dans les organisations*, Paris: Presses Universitaires de France, 1973.

LeMoigne, J. L. Vers un système d'information organisationnel? Revue Française de Gestion, Novembre-Décembre, 1986, pp. 20-31.

LeMoigne, J. L., and Pascot, D. (Eds.). *Les processus collectifs de mémorisation*, Aix-en-Provence: Librairie de l'Université d'Aix-en-rovence, 1979.

LeMoigne, J. L., and Van Gigch J. P. "The Design of an Organization Information System: Intelligent Artifact for Complex Organizations," *Information and Management* (19), 1990, pp. 325-331.

Marciniak, R., and Rowe F. *Systèmes d'information et dynamique des organisations*, Paris: Economica, 1997.

Mélèse, J. *Approche systémique des organisations*, Paris: Editions d'organisation, 1979.

Michaux, V., and Rowe, F., Complémentarité entre système d'information informatisé, communautés de pratiques et vigilance dans la haute fiabilité: le cas d'une compagnie d'assistance, *Systèmes d'Information et Management* (9:1), 2004.

Monod, E. "Epistémologie de la recherche en systèmes d'information," in F. Rowe (Ed.), *Faire de la recherche en systèmes d'information*, Paris: Vuibert, 2002, pp. 21-56.

Morley, C. "Information Systems Development Methods and User Participation: A Contingency Approach," in D. Avison, J. E. Kendall, and J. I. DeGross (Eds.), *Human, Organizational, and Social Dimensions of Information Systems Development*, Amsterdam: North Holland, 1993, pp. 127-142.

Orlikowski, W. J., and Iacono, C. S. "Research Commentary: Desperately Seeking the 'IT' in IT—A Call to Theorizing the IT Artifact," *Information Systems Research* (1:2), 2001, pp. 121-134.

Parsons, J., and Wand, Y. "Emancipating Instances from the Tyranny of Classes in Information Modeling," *ACM Transactions on Database Systems* (25:2), 2000, pp. 228-268.

Peaucelle J.-L. *Les systèmes d'information – la représentation*, Paris: Presses Universitaires de France, 1981.

Postman, N. *Conscientious Objections: Stirring up Trouble about Language, Technology and Education*, New York: Vintage Books, 1988.

Purao, S.; Rossi, M.; and Bush, A. "Towards an Understanding of the Use of Problem and Design Spaces During Object-Oriented System Development," *Information and Organization* (12:4), 2002.

Reix, R., and Rowe, F. "La recherche en systèmes d'information: de l'histoire au concept," in F. Rowe (Ed.), *Faire de la recherche en systèmes d'information*, Paris: Vuibert, 2002, pp. 1-21.

Richardson, H. "CRM in Call Centres: The Logic of Practice," in M. Korpela, R. Montealegre, and A. Poulymenakou (Eds.), *Organizational Information Systems in the Context of Globalization*, Boston: Kluwer Academic Publishers, 2003, pp. 68-83.

Robey, D. "Diversity in Information Systems Research: Threat, Promise, and Responsibility," *Information Systems Research* (7:4), 1996, pp. 400-408.

Rolland, C. "Introduction à la conception des systèmes d'information et panorama des méthodes disponibles," *Revue Génie logiciel* (4), June, 1986, pp. 7-62.

Rowe, F., and Struck, D. "Cultural Values, Media Richness and Telecommunication Use in an Organization," *Accounting, Management and Information Technologies* (9:3), 1999, pp. 161-92.

Saga, V., and Zmudm R. "The Nature and Determinants of Information Technology Acceptance, Routinization and Infusion," in L. Levine (Ed.), *Diffusion, Transfer and Implementation of Information Technology*, Amsterdam: North-Holland, 1994, pp. 67-86.

Sartre J.-P. *L'être et le néant: essai d'ontologie phénoménologique*, Paris: Gallimard, 1943.

Sawyer, S., and Chen, T. "Conceptualizing Information Technology in the Study of Information Systems: Trends and Issues," in E. Wynn, E. R. Whitley, M. Myers, and J. I. DeGross (Eds.), *Global and Organizational Discourse About Information Technology*, Boston: Kluwer Academic Publishers, 2002, pp. 109-131.

Schultze, U. "A Confessional Account of an Ethnography about Knowledge Work," *MIS Quarterly* (23:1), 2001a, pp. 1-39.

Schultze, U. "Reflexive Ethnography in Information Systems Research," in E. Trauth (Ed.), *Qualitative Research in IS: Issues and Trends*, Hershey, PA: Idea Group, 2001b, pp. 78-103.

Schultze, U., and Boland Jr., R. J. "Knowledge Management Technology and the Reproduction of Knowledge Wwork Practices," *Journal of Strategic Information Systems* (9), 2000, pp. 193-212.

Shapiro, S. "Places and Spaces: The Historical Interaction of Technology, Home, and Privacy," *The Information Society* (14), 1998, pp. 275-284.

Soh, C., and Markus L. "How IT Creates Business Value: A Process Theory Synthesis," in J. I. DeGross, G. Ariav, C. Beath, R. Hoyer, and C. Kemerer (Eds.), *Proceedings of the Sixteenth International Conference on Information Systems*, Amsterdam: Amsterdam, The Netherlands, 1985, pp. 29-41.

Tabourier Y. *De l'autre côté de MERISE*, Paris: Editions d'Organisation, 1986.

Van Maanen, J. *Tales from the Field: On Writing Ethnography*, Chicago: University of Chicago Press, 1988.

Walsham, G. "Actor Network Theory and IS Research," in A. S. Lee, J. Liebenau, and J. I. DeGross (Eds.), *Information Systems and Qualitative*, London: Chapman & Hall, 1997, pp. 446-480.

Walsham, G. "The Emergence of Interpretivism in IS Research," *Information Systems Research* (6:4), 1995, pp. 376-394.

Weber, R. "Editors Comments: Theoretically Speaking," *MIS Quarterly* (27:3), 2003, pp. iii-xiii.

Weber, R. "Towards a Theory of Artifacts: A Paradigmatic Base for Information Systems Research," *Journal of Information Systems*, Spring 1987, pp. 3-19.

Willcocks, L., and Lester, S. "How Do Organizations Evaluate and Control Information Systems Investments? Recent UK Survey Evidence," in D. Avison, J. E. Kendall, and J. I. DeGross (Eds.), *Human, Organizational, and Social Dimensions of Information Systems Development*, Amsterdam: North-Holland, 1993, pp. 15-40.

Wilensky, R. *Planning and Understanding: A Computational Approach to Human Reasoning*, Reading, MA: Addison Wesley, 1983.

ABOUT THE AUTHORS

Frantz Rowe is a professor in Information Systems at the University of Nantes. He has an MA in Economics and an ME in Civil Engineering from the University of Lyon, an MS from the University of California, Berkeley, and a Ph.D. from the University of Paris in Information Systems. He cofounded the French Association in Information Systems in 1991 and was appointed Editor in Chief of *Systemes d'Information et Management* in 1996. With the support of the National Foundation for Management (FNEGE in French), he edited a book, *Doing Research in IS*, in 2002. His research interests include the transformations of work and organizations linked to DSS, new media and in particular to the integration of enterprise systems. He can be reached at frantz.rowe@sc-eco.univ-nantes.fr.

Duane Truex researches the social impacts of information systems and emergent ISD. He is an associate editor for the *Information Systems Journal*, has coedited two special issues of *The Database for Advances in Information Systems*, and is on the editorial board of the *Scandinavian Journal of Information Systems*, the *Journal of Communication, Information Technology & Work*, and the *Online Journal of International Case Analysis*. His work has been published in *Communications of the ACM, Accounting Management and Information Technologies, The Database for Advances in Information Systems, European Journal of Information Systems, le journal de la Societé d'Information et Management, Information Systems Journal, Journal of Arts Management and Law, IEEE Transactions on Engineering Management*, and 40 assorted IFIP transactions, edited books, and conference proceedings. He is a member of the Decision Sciences and Information Systems faculty in the Chapman Graduate School, College of Business, at Florida International University, and is an associate professor on leave from the Computer Information Systems Department, Robinson College of Business, at Georgia State University. He can be reached at duane.truex@fiu.edu

Lynette Kvasny is an assistant professor of Information Sciences and Technology at the Pennsylvania State University. She received her Ph.D. from the Department of Computer Information Systems at Georgia State University, Robinson College of Business in Atlanta. Her research focuses on the appropriation of information and communication technology by historically underserved groups and institutions. Her research has appeared in *The Data Base for Advances in Information Systems*, the proceedings of IFIP Working Group 8.2 conferences, and the the proceedings of the International Conference on Information Systems. Her research is funded by a Faculty Early Career Development Grant from the National Science Foundation. Kvasny can be reached at lkvasny@ist.psu.edu.

7 TRUTH, JOURNALS, AND POLITICS: The Case of the *MIS Quarterly*

Lucas Introna
Lancaster University

Louise Whittaker
University of the Witwatersrand

Abstract In this paper, we want to demonstrate the way in which regimes of truth at the *MIS Quarterly* (MISQ) have made it possible for certain types of research to be published there, and others not. The importance of this claim lies in the fact that publication in MISQ is often seen as an indication of status. Furthermore, publication in MISQ also plays an important role in decisions about tenure and promotion. However, the aim of the paper is not to rid MISQ of regimes of truth—this is not possible. The paper will argue, with Foucault, that all institutions always already have their politics of truth. The production of truth is always intimately tied to relations of power which itself depends upon truth for its sustenance. The aim of the paper is to show this intimate connection between truth and power. In particular, in the case of MISQ, we want to question the often-implied legitimacy and status that the MISQ has over and against other high quality journals in the field. Foucault argues that power is most effective when it hides itself. This paper is an attempt to make its face more public and open to scrutiny.

1 INTRODUCTION

There is no doubt that academic journals play an important role in shaping a discipline. Equally, it is generally known that many decisions about the academic careers of university faculty are intimately tied to publication and participation in the so-called leading journals of the field. It would, therefore, be reasonable to argue that these journals should be free from politics; that these journals ought to allow for a level playing field in which networks of power and influence are limited, if not illuminated. Yet this does not seem to be the case as the recent widely published row in economics illustrates (Jacobsen 2001). There have been some, like Habermas (1979, 1984, 1987), who have argued that it is only in an apolitical space of rational debate that a discipline can flourish. They argue that all participants must be equal in making and defending truth claims. Although we may want this to be true—and we are not sure we do—we do not believe this will ever be the case. We want to argue, using Foucault (1977), that truth is always a very mundane human affair. As such, human institutions are directly implicated in its production and circulation. With truth we mean statements or claims about the state of affairs *that we want others to accept as true*. We, like Foucault, are not here interested in absolute truths that are supposed to exist, or not. Every individual may believe what he or she wants—this of no concern to us in this context. However, when someone makes truth claims—about the validity of certain types of research over and above other types, for example—with the expectation that others or we should accept them as such, then truth becomes directly implicated as a human affair. As such, our concern with academic journals as spaces where truth claims are proffered is not directly a concern with the validity (or not) of a particular ontology and epistemology. Rather, our concern is with the way in which any particular ontology and epistemology becomes constituted as more legitimate than others equally valid—especially when it has a material bearing on issues of tenure and promotion. It is important to say from the outset that the claim that a particular regime of truth is operating is not necessarily a claim that there exists an intentional strategy or some sort of a *conspiracy* or plot to this effect.

We will structure the discussion as follows: first, we want to explain the intimate connection between the production of truth and the mechanisms of power that sustain this production using the Foucauldian notion of regimes of truth; second, we want to outline the way in which regimes of truth functioned and function in the *MIS Quarterly* (MISQ); and finally, we want to discuss some of the implications of our analysis.

2 KNOWLEDGE AND POWER: THE CONSTITUTION OF REGIMES OF TRUTH

In opposition to the modern view that knowledge gets produced in a zone where power is suspended, Foucault argues that any attempt to separate power and knowledge is futile since the production of knowledge is political all the way down. To separate knowledge and power would be to claim that we could separate statements of fact from the values and mechanisms that constitute them as such. Latour (1987, 1993) has convincingly argued that what we find *in practice* is that "facts do not speak for themselves." Facts are produced as facts because we value them as such. It is institutional mechanisms and practices that give facts a voice in the first instance. They become constituted as facts through processes, procedures, and discursive practices that produce them and are likewise produced by them. For example, in the modern scientific regime of truth we value scientific method and, therefore, we judge its products to be facts. We do not value intuition and, therefore, we judge it products to be speculation. One could say that facts are merely legitimized value choices accorded that status through the prevailing institutional mechanisms. Thus, for every *recognized* fact (or set of facts) one could always, in principle, find the institutional mechanism that accords it that status and which is itself dependent on that status. For example, the valuing of profit becomes sedimented as facts in the income statement and balance sheet of the company, which are themselves necessary to sustain that value. Through the rituals of accounting practices and stock exchanges, the income statement and balance sheet become constituted as truth at the expense of equally valid alternative values, such as environmental or employee concerns. Nietzsche (1967) writes in *Will to Power*: "But what is truth? Perhaps a kind of belief that has become a condition for life" (p. 248)—or as Foucault later articulates it more clearly, truth simply means beliefs that have become a condition for that institution to sustain itself as that which it believes it is.

Every attempt to secure knowledge in a zone outside of power will itself become a resource for power. For we should, as Foucault argues, "admit rather...that there is no power relation without the correlative constitution of a field of knowledge, nor any knowledge that does not presuppose and constitute at the same time power relations" (1977, p. 486). The linking of power and knowledge through discourse gives rise to what he calls *regimes of truth*. A regime of truth is the institutional infrastructure for the production and circulation of *truth claims*. Truth is produced in and through institutionalized discursive practices—in our discussion below, for example, we will look at editorial statements and other claims made in the MISQ as such discourse. Discourse here is understood as a particular way of talking, of making statements or claims, about the state of affairs. Truth, as pointed out above, is understood as claims about the world that are proffered as valid claims within a particular regime of truth—claims those that make them *expect us to accept* as valid. Foucault (1977) argued that each institution or society has its "regime of truth, its 'general politics' of truth." A particular regime of truth is constituted through a set of mechanisms and discursive practices which legitimizes claims and is itself dependent on the legitimacy of those claims. In *Power/Knowledge* Foucault (1977, p.131) mentions the following mechanisms and practices that constitute a particular regime of truth:

- *The types of discourse that it accepts and makes function as true*. Clearly, not all discourse in institutions functions as truth claims. For example when we present this paper at an academic conference we are proffering truth claims in a way we are not when chatting to a colleague in the corridor about the paper.
- *The mechanisms and instances that enable one to distinguish true and false statements*. Once we occasion certain discursive moments as instances of truth making, we must set in place the mechanisms for identifying true statements from false statements. How will these be verified?
- *The means by which each is sanctioned*. Truth is not produced if it is not sanctioned. Presenting the paper to our peers at an academic conference is one of the practices to sanction the truth claims we are proffering.
- *The techniques and procedures accorded value in the acquisition of truth*. In distinguishing true from false statements, and in sanctioning these, careful attention is given to the techniques and procedures used in the construction of these truth claims.
- *The states of those who are charged with saying what counts as true*. In institutions, the charge of acknowledging truth claims is carefully distributed and controlled. The auditor is charged to make claims about the financial status of the

organization in a way that the marketing manager is not. However, to make these statements the auditor must be at least a certified accountant. Likewise, we normally require professors as examiners of Ph.D. dissertations.

In regimes of truth "[truth] is linked in a circular relation with systems of power which it induces and which extend it" (Foucault 1977, p. 133) Power, through micro-practices and mechanisms of meaning, membership, and discipline (Clegg 1989), structures and restructures discourse, a way of talking about the world, in a discontinuous and diffused manner. Owing to the non-egalitarian and diffused nature of the relationship, such discourse gives rise to a particular regime of truth. The regime of truth in turn produces discursive resources in support of the very power relations that constitutes it. Power and truth are co-constitutive.

If claims of truth are always already within an existing sets of power relations, as Foucault claims, then we can only exchange one regime of truth for another. In knowledge production, which is a fundamentally social enterprise, we can not escape power (as proposed by Habermas [1984, 1987] in his ideal speech situation). Every ideal speech situation will always already assume a regime of truth for its very existence. Of course, we could say that in some cases power relations can become systemically asymmetrical as in a dictatorial state or monopolistic market, although Foucault will deny this. He would argue that even the king is only king because of a whole network of alliances that must constantly be serviced and secured. Even the big positive research programs in the natural sciences—that supposedly produce objective, value free truth—must make deals and produce the appropriate truth to secure funding, publications, and so forth.

Let us consider the notion of a regime of truth more closely by developing contrasting examples of the regimes of truth in different institutions. In Table 1, we present a comparison of the *publication of a paper in a peer-reviewed journal* (in scientific institutions) with *the publishing of the annual report of a public company* (in the capitalist enterprise), with *the delivery of a sermon in a church service* (in the Christian church). The table aims to show that each of these institutions have an identifiable set of institutional mechanisms and practices to produce what will be considered truth in that particular institution. Although these mechanisms and practices differ widely from institution to institution, they serve essentially the same purpose: namely, to ensure the production of truth in a manner that would sustain the institution as that which it claims to be.

In the table, we can observe the interplay between power and knowledge (in and through truth claims). Through a set of mechanisms, techniques, and sanctions, the truth is produced and confirmed as such. The mechanisms and practices are constituted through relations of power in such a manner that the truth produced would maintain and sustain these very relations of power. It is this relation between power and truth that stabilizes the institution (one can think of Kuhn's [1970] studies here; however, from this perspective we should say that his paradigm shifts are more shifts in power than shifts in epistemology/ontology). Any regime of truth, irrespective of its power relations, is always under threat. As Kuhn indicated, such shifts in power might be evolutionary or revolutionary, but regimes of truth are never fixed. Nevertheless, they do not necessary represent a conspiracy or a plot of some sort, as mentioned above. Their

Table 1. Regimes of Truth in Different Institutions

Regime of Truth	***Institutions of Science***	***Capitalist Enterprise***	***Christian Church***
Types of discourse which it accepts and makes function as true	*Publishing a paper in a peer reviewed journal,* Defending a Ph.D.; presenting a conference paper, etc.	*Publishing the annual company report,* AGM, annual employment review, etc.	*Delivering the sermon*, administering the sacraments, counseling a member of the church, etc.
Mechanisms and instances for distinguishing true and false statements	Scientific argument and proof, (dis)agreements in viva, using canonical texts/authority, etc.	Review by the auditors, economic argument (efficiency, profitability), appealing to canonical texts (Porter, etc.) or consultants, etc.	Review of sermon by the church elders, use of canonical text for authority, appealing to a higher church authority (for example, the bishop), etc.
The **means by which each is sanctioned**	Review by peers, publication in journal, citation in subsequent papers, citation, and journal indices.	Report presented to the board of directors, delivered at the AGM as the official financial position of the company, reaction of stock exchange, etc.	Sermon delivered as part of liturgy, starts with (or follows) the reading from bible, sermon starts or ends with "so says the Lord."
Techniques and procedures accorded value in the acquisition of truth	Scientific method/research method	General accepted accounting practices (GAAP), audit process, strategic planning, etc.	Biblical exegesis, interpretations of church edicts, etc.
The **states of those who are charged** with saying what counts as true	Reviewers must be recognized experts in their field, editors must be seen to be objective, acting on behalf of academic community.	Auditor must be a chartered auditor, managing director acts *ex officio* on behalf of the shareholders, etc.	Must be a licensed minister or religion, and an appointed leader in a congregation

origins and sustenance are often due to contingent events that are seized upon as resources for the play of power. The "logic is clear, the aims decipherable, and yet it is often the case that no one is there to have invented them" (Foucault 1977, p. 486). The local tactics may link together and combine into overall strategies that create the illusion of grand design but are in fact outcomes of very local contingent actions.

Every claim to truth whatsoever always implies a regime of truth for its force or validity. Some regimes of truth may be subtle and mobile and others more explicit and fixed. For example, in a family, husband and wife have the ability to make truth claims in a way that children cannot. This is because of the relations of power constituted by their access to resources of power, such as the sanction by society of their role as parents and guardians. Yet, this regime of truth is continually open for dispute and maneuvering in a way that a court of law is not. Thus, no claim to truth can be made outside of a regime of truth—not even in the intimacy of a family. The production of truth is never outside of power.

It is also the case, however, that in some instances, there are *blocks* in which power and truth, as claimed through resources of communication and objective capacities, constitute *regulated and concerted systems*, or disciplines, in an enlarged sense of the word. In this regard, Foucault (1994) refers to monastic, penitential, medical, and technical disciplines as examples. Certainly the general understanding of information systems as an academic discipline would fall within this definition. For even given its, in some quarters notorious, lack of consensus as to what rightfully constitutes the discipline, there are nonetheless powerful regulatory processes that ensure aptitudes and particular types of behavior, across and within institutions in which the discipline is taught. Indeed, it has been a specific objective of academics within the discipline, since the mid-1960s at least, to "be a profession" (Dickson 1982, p. v), as much as medicine or engineering are professions.

Within a discipline, the specificities of power relations are made clearer, perhaps, than in less regulated regimes of truth. And yet they cannot by any means be said to be fixed or unidirectional, since it is easy to see that "all forms of dependence offer some resources whereby those who are subordinate can influence the activities of the superiors" (Giddens 1984, p.16). Foucault, in fact goes further than this: "Power is exercised only over free subjects, and only insofar as they are 'free.'" Power relations exist where individuals have a field of possibilities, however sparse these may be, and even though the power relations are inscribed in the more or less permanent structures underpinning that field. In any relationship of power, "a whole field of responses, reactions, results and possible inventions may open up" (Foucault 1994, p. 340). This implies, then, that intertwined as they are with power, the mechanisms used to produce truth in any discipline are inherently contingent. Through intentional and unintentional moves, these regimes of truth are continually shifting, opening spaces for certain types of research to become legitimate and others not.

Within information systems as a discipline, the *MIS Quarterly*, as a leading journal, is one of those mechanisms used to validate both the truth claims made by the discipline, and the discipline itself. Further, in so far as the MISQ is a regulated and concerted "conduct of conduct" or leader of behavior in the production of truth, it can itself be said to be a discipline, in Foucault's enlarged sense of the word. In the sections that follow, we want to analyze some of the mechanisms that constitute the regimes of truth of the MISQ, with a view to showing, first, that these do in fact operate and, second, just how contingent they are. Our analysis is by no means comprehensive or complete. It merely suggests some outlines of such an analysis.

3 REGIMES OF TRUTH AT THE MISQ

The MISQ was established in 1977 with Gary W. Dickson as Editor-in-Chief. Dickson's tenure lasted six years. Thereafter, editors were appointed for three-year terms (see Appendix A for the list of past and present editors). For the analysis, we focus on the period between 1981 and 2003 since we have access to journals only from 1981. Dickson's "Apology of a Retiring Founding Editor (6:4)[1], does, however, give us some insight into the workings of the journal in the first five years.

Obviously, there are many elements that constitute the regime of truth at the MISQ, or any academic journal for that matter. We as academics are quite familiar with these. They are the editors (past and present), associate editors, the mission or scope of the journal, editorial statements, the review process, keyword classification systems, and so forth. In the regime of truth no one person pulls all the strings. One may be tempted to think the editors do have many of the strings in their hands. Nevertheless, as Foucault argued, the king is only king as long as he can maintain the alliances that legitimize him as king and are themselves dependent on the legitimacy of the king as king. Editors have to continually defend their own legitimacy in the face of previous editors, associate editors, authors, and readers—to name but a few. Thus, to really get a picture of the regimes of truth at MISQ, we would need to trace a multiplicity of relations, statements, codes, processes, and mechanisms operating there. When we do this, we will see that these are multiple, mundane, and contingent. Nevertheless, they do weave together a fairly coherent network of resources for the execution of power that allows certain papers to be judged as true, valid, and legitimate and others not.

It is beyond the scope of this paper to do such a detailed analysis. We are further conscious that, even were it not, "the analysis of power relations...cannot be reduced to the study of a series of institutions" (Foucault 1994, p. 345). Power relations are rooted in the whole network of the discipline. What Foucault does suggest is that a "certain number of points be established." We will attempt to trace some of the power relations constituting this regime of truth by addressing some of these points in the MISQ. In particular we will focus on the types of objectives pursued, in the form of the scope of the journal; and the instrumental modes of action, or rules, in the form of the definition of legitimate research method, as these have shifted over the past 20-odd years.

3.1 The Shifting Scope of the MISQ: From Managerial Issues and Practice to Technology and Organizational Research Problems

When the MISQ was established in 1977, its scope was described as "Our major goals are to be *managerially oriented* and to offer something of benefit to the practitioner. At the same time, we intend to provide a vehicle for *researchers working in the information systems field to communicate with each other* and with practitioners." This objective is not surprising as the MISQ was at the time a cosponsored project of The Society for Management Information Systems—later to become the Society for

---

[1]We will only refer to the volume and issue number (volume:issue) in referencing the editorial comments of the editors. The editorial comments are usually only three to five pages so the text being referenced should be easy to locate.

Information Management—and the University of Minnesota's Management Information Systems Research Center (MISRC). In fact, as Dickson noted upon his retirement, "we wanted a journal in which MIS academics could publish while our primary source of funding was a society of information systems practitioners" (6:4). This dual focus of managerial orientation as well as research forum arose from the task of specifying a product that could satisfy both groups. Initially, this dual focus was maintained by establishing a two-section journal, and by appointing both practitioners and academics as consulting and associate editors. As it evolved, it became articulated as the tension between rigor and relevance.

William King, on his appointment, suggested that his objective was to improve on the "quality and credibility of the journal" (7:1), with the expressed intention of improving its scientific credibility and its use in tenure and promotion decisions (7:2). At the same time, he was anxious to "recognize the importance of *both* research and applications." However, it is certainly the case that King succeeded in raising the profile of a struggling journal,[2] not least by nurturing credibility for MISQ as an *academic* journal, personally lobbying universities to recognize it in promotion criteria (9:4), and getting it listed on databases like ISI (8:4). He also dropped the practitioner-only consulting board for "turnaround-time on review" reasons (9:1), and the number of academics on the associate editors board increased. The journal became steadily more academic in its focus, in that the in-coming third editor, Warren McFarlan, suggested that "sound research-based articles" were now at the "very core of the journal's editorial philosophy." (This despite the fact that neither the editorial policy board, nor the official editorial policy per se, had changed at all over this time.) And while McFarlan seems to have confined himself in editorial statements to reflecting on practically based tools, approaches, and insights, it was in his tenure that Izak Benbasat was appointed (without comment) as Senior Associate Editor of Theory and Research (11:1). Under Benbasat's "primary editorial responsibility for all papers classified as Theory and Research," quantitative, positive research started to dominate the journal.[3] The split in the editorial board between Research and Theory and Application was dropped in the final year of McFarlan's editorship and a single all-academic board established. Nonetheless, much as Dickson had done, McFarlan closed his editorship with a comment on the tension between theory and practice, and a plea to practitioners and academics to "establish a dialogue" (12:1).

James Emery, too, highlighted the need to "provide a forum for materials addressing the information systems field in both theory and practice...[hopefully] thereby, to unite the efforts of those teaching and doing research in this area with those applying information systems to organizational problems" (13:3). This, however, was to be done with an emphasis of adding to the body of knowledge, as the screening of application

---

[2]A "hand-to-mouth" editorial existence forced the late production of the first two issues of volume 7 (8:4).

[3]A detailed classification of all papers published in the period under discussion is beyond the scope of this paper, for reasons of length, but a review of all empirical research papers published in this period shows them to be positivist, and overwhelming quantitative.

papers became more rigorous through the application of the criteria of generalizability, delimitations of scope, and reference to the existing literature (13:4). Where application papers were published that did not "satisfy the standard criteria for research"—and these were now presumed to be understood—their inclusion was explicitly justified (14:4). Unsurprisingly, having become a recognized academic journal, relevance to the practitioner community now emerged as specifically problematic. Emery closed his tenure by suggesting that "practitioners must also bear part of the burden of translating theory into practice" (14:4).

In the tenure of Blake Ives, this tension became even more apparent. He argued,

> The business school is a professional school with obvious linkages to the business professions. Drawing a parallel to the business school, a colleague recently asked what we would think of medical schools whose research did not address the treatment of disease? Too often business school research addresses problems of little relevance or, equally damaging, fails to be tested in the world of practice. (16:1)

The managerial concern of MISQ was also highlighted by his incoming Senior Editor of Theory and Research, Gerardine DeSanctis: "One of the most distinguishing characteristics of the *MIS Quarterly* as compared to other journals devoted to information systems is its emphasis on management." Nevertheless she explicitly broadened the definition of management to "include management of public and private organizations, government and labor organizations, and professional and social societies." She equally made it clear that "[a] managerial emphasis does not mean that every paper will have direct implications for practicing management, nor does it mean that every paper in its entirety will appeal to all readers...we do not apologize for publishing papers that fail to reap so-called 'real world relevance.' The *potential* for relevance is what matters" (17:1, emphasis added).

In 1995, when Bob Zmud took over as Senior Editor, the MISQ was unbundled as an automatic subscription benefit with the Society for Information Management (SIM) membership. The institutional arrangement that had demanded both Theory and Application in the first instance was dissolved, and a major practitioner audience was lost. With this change in mind, Zmud suggested that a move away from practitioners concerns had now become possible although he did not support such a move: "this changed relationship also raises the possibility that the *Quarterly*, given a potential of fewer practitioner readers, could redirect its direction toward the academic community and away from the practitioner community. I wish to state as strongly as possible that this is not my intention"(19:1). This stated commitment became less apparent as he sought to improve the theoretical foundations of the papers in MISQ. As for the scope of papers appropriate for the journal, some minor but important shifts are also noticeable. He saw the MISQ editorial objective to be "the development and communication of knowledge concerning both the management of information technology and the use of information technology for managerial and organizational purposes" (19:2). He clarified "the operative terms...are: management, information, and information technology....Notice that who exactly is involved in the act of managing is left open." This emphasis on *managing* rather than *managers* is important as it opened up the horizon for research that is less managerialist. Toward the end of his first year as Senior Editor, Zmud made a strong appeal for theory as the foundation of all MISQ papers: "Adequate theory development is a fundamental requirement for all manuscripts submitted to the *Quarterly*....All manuscripts submitted to the *Quarterly*—this includes both 'Theory and Research' and 'Applications' articles—must have a theory section" (19:3). He also indicated what he did *not* want to see in MISQ: "pragmatic descriptions of information systems applications, methodologies, or practices; formal descriptions of information systems applications, methodologies, or practices; replications of prior studies; and criticisms of prior studies." These topics, especially the first two, were very much "the stuff" for MISQ in the 1970s and 1980s. As the theory focus of MISQ grew, Zmud introduced an interesting new distinction between *weak relevance* and *strong relevance*. He explained: "An article can be said to demonstrate weak relevance when the research questions being examined *touch on* organizational phenomena that are clearly of interest to practice, and I expect all articles submitted to *MIS Quarterly* to minimally demonstrate weak relevance" (20:3, emphasis added). He further clarified that strong relevance is accomplished when "(1) using the concerns of practice (rather than scientific literature) as the primary motivation behind the research effort being described in an article, and (2) clearly discussing the meaningfulness of the article's scientific contributions to the executive audience" (20:3). In his last year as Editor-in-Chief, Zmud made an rather unusual appeal for pure theory papers: "I wish to encourage the *MIS Quarterly* readership to consider submitting manuscripts whose primary contribution lies with the theory being developed and articulated" (22:2). Zmud's commitment to move MISQ toward an academically and intellectually respectable position was most clearly demonstrated when he suggested in his final Editor's Comments that the Research/Application divisions, long at the heart of MISQ's dual focus, be abandoned. As Zmud left office, MISQ had become a journal with a strong emphasis of theory in which most papers had only weak relevance. This is a very significant distance away from the happy marriage of its early years.

In his inaugural editorial comments, Allen Lee confirmed this direction: "I am more than happy to accept the responsibility...to secure *MIS Quarterly*'s place among the *best research journals in the academic field of MIS*." (23:1, emphasis added). Maybe it is a Freudian slip but there is no mention of the other half of the relationship here. He also announced the consolidation of the Theory and Research department and the Application department as suggested by Zmud. The new department was to be called Research and publish "Articles that are 'pure theory,' articles that empirically test or illustrate theory and articles that apply existing theory." In his second editorial comments (23:2), Lee argued for a further broadening of the editorial scope to include social and political dimensions: "authors can make their papers more likely to be compelling and significant if they additionally attend to the *social and political dimensions* of their craft," and not only on the "traditional 'objective' (cognitive and intellectual) dimensions of research" (23:2). The push toward theory-driven work that is academically and intellectually rigorous and legitimate at the expense of that which practitioners may want and find relevant surfaced again toward the end of Lee's second year. In his editorial comments, the actual commitment of MISQ was made abundantly clear:

> Must all instances of information systems research [that which MISQ publishes] be directly consumable by and have immediate relevance to

> managers, executives, consultants, and other practitioners? Must all information systems research be targeted at practitioners? Absolutely not!" (24:3)

He explained—quite correctly in our view—that the relations of power in academic institutions that govern decisions about tenure and promotions (its regime of truth) are not commensurable with those expected in business practice. In our view, he identified the most important problem that plagued MISQ from the start: The regimes of truth of academia are not commensurable with the regimes of truth of business practice. He therefore proposed that a new and different journal be established for the now-alienated and illegitimate half brother—the *MISQ Executive*: "This line of reasoning has motivated me to work with Jeanne Ross and Michael Vitale in pursuing the possibility of starting a new *MIS Quarterly* publication (other MISQ publications are *MISQ Discovery* and *MISQ Review*) whose mission would be the publication of *research that would appeal immediately to managers, executives, consultants, and other practitioners*. The organizational change would involve the *instituting of new editorial and reviewing norms, emphasizing practicality and relevance*.... Plans are still at a preliminary stage for the new publication, tentatively named *MISQ Executive*" (24:3, emphasis added). In fact, as it was launched in 2001, *MISQ Executive* seemed to look very much like the original MISQ: "targeting two audiences: information systems practitioners and information systems academics" (26:2). With this move, MISQ finally shed its dual focus to become an exclusively academic journal.[4] This was not seen as a problem, as there will "always be a place for the theory-oriented, basic research of the kind that now predominates in *MIS Quarterly*" (24:3).

Taking up his tenure, the current editor, Ron Weber, similarly commented that, "Clearly we need to publish the best research undertaken within the information systems discipline," in defense of the MISQ's position "among the very top journals" in all three of the fields of information systems, computer science and management (26:1). He further extended the research scope of the journal to include the managerial and organizational implications of "the full gamut of topics that command the attention of *researchers* in the information systems field" (26:1, emphasis added). A significant institutional change occurred under the editorship of Weber, with the establishment of an alliance between the Association for Information Systems (of which Weber is a past President) and the MISQ. This alliance was driven by economic imperatives. Once again, the costs of maintaining a journal had become prohibitive for the independent journal, such that MISQ faced an uncertain future. In this case, however, the alliance created a tension not between theory and practice, but rather between two journals that "seek to publish high-quality research papers," namely MISQ and the *Journal of the AIS* (26:3). The effect that this may or may not have on the scope of each journal will be interesting to observe. Weber has, further to this development, editorialized on the need for IS researchers to identify and research the "deep, substantive problems" in the field (27:1). He argues that MISQ has prioritized rigor *over* relevance, but now relevance has

---

[4]A small, but telling, detail in the academic evolution of the journal is the switch from only the month and year of publication appearing in the footnote of each page, to a volume, issue/month and year footnote, more typical of purely academic journals.

taken on a different meaning. Weber is apparently precisely not concerned with relevance to practitioners (as the abolition of executive overviews in the same issue indicates), but with "the most important problems," those that have longevity. In this editorial, Weber appears to address the nature of information systems as a discipline, MISQ being the leading research outlet in that discipline, while acknowledging in this, and his following editorial (27:2) that this has been an ongoing debate since at least the mid-1980s. He further determines the purpose of the MISQ as being to defend, or perhaps establish, depending on one's position, the core of our discipline, and not to "stretch the boundaries of our discipline to an extreme" (27:2). Thus the current editor has quite clearly articulated MISQ's role as defender and protector of the *research* scope of IS as a discipline.

In this narrative, we have traced the shifting editorial scope of the MISQ. We aimed to demonstrate that this shift happened through many individual statements in the editorial comments, and perhaps also gradual changes in the structure and constitution of the editorial boards, as well as the MISQ's institutional arrangements. We would claim that this shift was not a plot or a strategy in the minds of the editors as such. Nevertheless, papers that were very legitimate bearers of truth in the 1970s and 1980s would not get a second look by the editors today. Likewise, most executives reading MISQ today would probably now agree with the quote from Blake Ives in (16:1) in his editorial comments of 1992: "They say nothing in these articles and they say it in a pretentious way." In the following section, we will consider how research methodology as the legitimate techniques and procedures accorded value in the acquisition of truth also shifted.

3.2 The Battle of Method at MISQ: From Positivism to Pluralism (of Some Sort)

In the first 10 years of the MISQ, method was not seen as an issue—mostly because rigor in research was not the only criteria for the acceptance of papers. Many papers that were seen as highly relevant were published irrespective of their theory and their research methods. We can thus see that the points of power relations are not only contingent, but also intertwined, as the *scope* issue directly affected the *method* issue. It was furthermore evident to most in the field that the positive methods of the natural sciences were the way to do scientific research. Thus, the issue of method, where it arose at all, was self-evident and quickly subsided again. It was with appointment of Benbasat as the first Senior Associate Editor of Theory and Research in 1987, that method became articulated as an issue, as the first papers actually reflecting on positivist research methods were published (Benbasat, Goldstein, and Mead 1987; Culnan 1987; Culnan and Swanson 1986). As noted above, the editor at that time, McFarlan, did not discuss questions of method at all. It was first in the tenure of James Emery that the positive methodological criteria for a paper in the Theory and Research category were clarified:

> It should be based on a set of well-defined hypotheses, unbiased and reproducible procedures for collecting evidence that supports or refutes the hypotheses, and sound analytical procedures for drawing appropriate

> conclusions from the evidence. The research often involves the collection of considerable quantitative data through such means as laboratory experiments or survey instruments. The data are then subjected to statistical analysis to draw the appropriate inferences from the research (13:3).

Emery did, however, hint that there might be an alternative: "High quality research need not be limited to work generating large quantities of data that can be statistically validated, however: a well constructed case study can also meet the tests of rigorous research." The emphasis on positivism is not surprising as it is clearly evident in the work of Benbasat and Ginsberg as Senior Associate Editors of Theory and Research. In the tenure of Blake Ives, DeSanctis—also as Senior Associate Editor of Theory and Research— made a plea for strong theoretical and empirical work that would provide "fresh theoretical ideas about causal relationships in order to facilitate understanding and prediction of events in the world" (DeSanctis, 17:1). This is in *opposition* to the dominance of managerial frameworks: "Frameworks are useful systems for identifying and organizing variables for scientists to study; but they do not lend good insight into cause/effect relationships, nor do they articulate the properties and behaviors underlying phenomena" (DeSanctis, 17:1). She also started to open the door for other research approaches: "On the empirical side, we welcome research based on positivist, interpretive, or integrated approaches. Traditionally, *MIS Quarterly* has emphasized positivist research methods. Though we remain strong in our commitment to hypotheses testing and quantitative data analysis, we would like to stress our interest in research that applies interpretive techniques, such as case studies, textual analysis, ethnography, and participant/ observation" (DeSanctis, 17:1). In spite of this declaration of openness and a restatement of this commitment of "openness with regard to the research methods" by Zmud (19:3), they seemed not to receive other approaches as submissions. Zmud explained: "We truly believe the *Quarterly* is not biased against manuscripts that adopt other than positivistic perspectives and methods. If it seems that positivistic articles tend to dominate the *Quarterly*, a sound explanation does exist—the majority of submitted manuscripts adopt such a perspective. We strongly encourage a variety of research approaches" (19:3). These are very interesting statements. They show us that the editors do indeed not have all the strings of the regime of truth in their hands. They can only work with the resources available. If they do not get non-positivist submissions, they cannot enact a shift in the regime of truth, even if they want to.

In an attempt to establish a new, more pluralistic, regime of truth it was decided to create a special issue on "Intensive Research in Information Systems: Using Qualitative, Interpretive, and Case Methods to Study Information Technology"—under the editorship of Lynn Markus and Allen Lee. The first articles of this special issue appeared as volume 23, issue one, the first issue of Lee's tenure. Other attempts were also made to increase the legitimacy of non-positivistic research such as the publication of Michael Myers' site on "Qualitative Research in Information Systems" in *MISQ Discovery* (21:2). One can only speculate about why it required such explicit effort for the editors to change the regime of truth with regard to method. We would venture to say that we must not forget that the MISQ itself operated in a regime of truth of management schools (where tenure and promotion decisions were made). Maybe this regime of truth still valued positivist research and, therefore, the authors continued to produce positivist research in spite of the attempts of the editors to preach pluralism. The regime of truth of methodological pluralism is sealed in the editorial statements of Allen Lee at the start of his second year. He concluded, after reviewing some examples of published work in MISQ, that the "*MIS Quarterly* welcomes: the research would (1) be positivist, quantitative, and mathematical, (2) involve a rationalistic/economic decision-making framework, and (3) conduct hypothesis testing," also that the MISQ welcomes "(1) qualitative, interpretive, and case research, (2) a historical framework, and (3) theory building" (24:1). More pluralistic than this one cannot get—besides allowing speculative philosophical essays (like this one).

Recently there has been continued evidence of such pluralism—there were, in our estimation, three interpretive papers *about* research in 2001 and 2002 (Baskerville and Myers 2002; Orlikowski and Barley 2001; Schultze and Leidner 2002)—while the special issue articles edited by Bob Zmud on "Redefining the Organizational Roles of Information Technology in the Information Age" appearing in September 2002 and June 2003 contained a number of papers that address more interpretive issues. Interestingly, Weber has explicitly dismissed methodology as a concern, because he is concerned very much with the problem of the problem (that is, scope). He suggests that "we need to change some deeply ingrained mindsets,' specifically, the way in which "our identity as researchers is a function of the extent to which we can flex our muscles using a particular research approach or method." He identifies this mindset as "a vestige of the old 'methodology wars' that occurred long ago in our discipline," and concludes "it is time to forget and move on" (27:1). Clearly the methodology wars have, in the view of the editor, been won, or least there is some kind of truce declared.

We hope to have shown here how the regime of truth with regard to method has change over the tenure of the last four editors of MISQ. It ought to be clear that work, legitimate early on in the 1970s and 1980s both with respect to scope and method, will no longer be legitimate, or not. Nevertheless, there is still a strong emphasis on empirical work. Thus, critical and speculative work will still find it hard to get an audience in MISQ. This analysis of the regimes of truth of the MISQ is still very superficial. We have not analyzed the editorial board, the relationships between the editorial board members, the way in which Senior Editors are chosen, the actual reviewing practices, and so forth. All of these may, to a lesser or greater degree, have shifted the regime of truth of the MISQ in a particular direction rather than another. We want to conclude the paper by looking at some of the implications of the analysis.

4 TOWARD A BETTER POLITICS OF KNOWLEDGE

We first want to comment on a philosophical issue then say more about academic publishing. If we admit that truth is not some essential ontological claim—as the moderns assumed—but rather a claim for legitimacy of epistemological categories, does this mean we are now adrift in a sea of relativism in which anything goes? We would claim not. For as much as we want our truth claims to be recognized as such, we will have to appeal to the regime of truth that operates. We can claim all sorts of things about MISQ, but in as much as we want to publish our work there, we will have to address ourselves to the regime of truth that operates there, likewise for all other journals and discourses of science. Every form of relativism (or fundamentalism) will eventually be mediated by the regimes of truth that operate in the institutions where it seeks legitimacy. We do not have guarantees nor an ultimate foundation or meta-narrative that we can appeal to for the legitimacy of our truth claims *outside of power*. Nevertheless, this does not mean truth is an arbitrary process. Likewise, from an ethical point of view, every person will have to mediate and reconsider their values subject to those regimes of truth where they seek legitimacy. Thus, contrary to Feyerabend (1993), *anything* can simply not go—absolute relativism, like fundamentalism, is simply not a political option in the production of truth. Bad research will be rejected wherever it is presented. The issue is rather that good and valid research (depending on your criteria) may find a home in some journals and not in others. If all journals were equally relevant in promotion and tenure decisions, then this might have been the end of it—unfortunately it is not. We now want to return to the issue of academic publication. What does the politics of truth mean for research and publishing in academic journals?

It is important to realize—and tell our students—that academic publication is first and foremost a political rather than a pure epistemological issue. Obviously epistemological considerations are also important, but they can never be seen separately from the regime of truth within which the ultimate truth claims will be made. We claim that all research must be *reasonable* (as opposed to *rational*). And by this we mean it should be congruent with the regime of truth in which it locates itself. If you do positive research and want to publish in a positivist journal, you must adhere to that regime of truth. Likewise, the interpretivist must be congruent with that regime of truth. Ironically, even the post-modernist must appeal to a regime of truth for legitimacy. This implies that the researcher must make a significant effort to know the regime of truth to which she will eventually appeal. This is not easy. Foucault has taught us that power is only effective if it hides itself. Thus, the power relations that make up the regime of truth will be reluctant to spell out in detail what constituted legitimate claims as that would limit their room to manoeuver. It is, therefore, often necessary not so much to know the scope and aims of the journal as to know the particular dispositions of those who make the important decisions. Furthermore, the journals also seek legitimacy within the academic community. They themselves will, therefore, have to address themselves to *that* regime of truth. To the extent that the modern positivist regime of truth is still dominant, they will have to be seen to be scientific—at some level of analysis. Knowledge production is a complex set of power relations within and between regimes of truth that the experienced researcher must learn to negotiate. Publication in academic journals is an experience in disciplinary power, in Foucault's terminology. As we all know, we must be seen to address the reviewer's comments, even when we fundamentally disagree with them. Thus, a publication in MISQ, or any other journal, is not necessarily so much an indication of quality as it is an indication of compliance with the regime of truth that operates there. Indeed we often bemoan the quality of the work published in MISQ. Yet, the editors cannot fail to agree to publish work that obviously conforms to the disciplinary processes that make up the MISQ—scope, method, reviewing, etc.

What we have said about MISQ is equally true for other journals such as *Information Systems Research*, *Information and Organization*, *Journal of Management Information Systems*, and so forth. We would argue that it would be very difficult if not impossible to publish a purely positivist paper in *Information and Organization*—even if the editor might wish to. We need to admit that since knowledge production—in the form of publications in this case—is a fundamentally social process, questions of epistemology cannot be separated from questions of politics. This is not bad—indeed it is the very condition of its ongoing possibility. However, it is dangerous, as Foucault has argued. It is our duty to continually disclose this politics of truth and to subject it to ongoing scrutiny. It is when power becomes hidden *and* systematically asymmetrical that it becomes bad. It is our belief that the systemic dominance of the MISQ in university rankings is bad for the field—our evident institutional silence on this politics of truth is even worse. It is our moral duty to continually challenge any attempt to institutionalize rankings of journals that tend to favor certain regimes of truth and not others—especially as this may have material consequences for those equally legitimate researchers on the outside.

REFERENCES

Baskerville, R., and Myers, M. "Information Systems as Reference Discipline," *MIS Quarterly* (25:1), 2002, pp. 1-14.

Benbasat, I., Goldstein, D. K., and Mead, M. "The Case Research Strategy in Studies of Information Systems," *MIS Quarterly* (11:3), September 1987, pp. 369-386.

Clegg, S. R. *Frameworks of Power*, London: Sage Publications Ltd., 1989.

Culnan, M. J. "Mapping the Intellectual Structure of MIS, 1980-1985: A Co-Citation Analysis," *MIS Quarterly* (11:3), 1987, pp. 341-349.

Culnan, M. J., and Swanson, E. B. "Research in Management Information Systems, 1980-1984: Points of Reference," *MIS Quarterly* (10:3), 1986, pp. 289-302.

Dickson, G. "The Apology of a Retiring Founding Editor," *MIS Quarterly* (6:4), 1982, pp. v-viii.

Feyerabend, P. *Against Method* (3rd ed.), London: Verso, 1993.

Foucault, M. "Truth and Power," in *Power/Knowledge: Selected Interviews and Other Writings 1972-1977*, C. Gordon (Ed.), New York: Pantheon Books, 1977.

Foucault, M. "The Subject and Power," in *Power: Essential Works of Foucault 1954-1984, Volume Three*, James D. Faubion (Ed.), London: Penguin, 1994, pp. 325-348.

Giddens, A. *The Constitution of Society: Outline of the Theory of Structuration*, Oxford: Blackwell, 1984.

Habermas, J. *Communication and the Evolution of Society*, London: Heinemann Press, 1979.

Habermas, J. *The Theory of Communicative Action, Volume 1*, London: Heinemann Education, 1984.

Habermas, J. *The Theory of Communicative Action, Volume 2*, Cambridge: Polity, 1987.

Jacobsen, K. "Unreal Man," *The Guardian*, April 3, 2001, p. 12.

Kuhn, T. S. *The Structure of Scientific Revolutions* (2nd ed.), Chicago: University of Chicago Press, 1970.

Latour, B. *Science in Action: How to Follow Scientists and Engineers Through Society*, Cambridge, MA: Harvard University Press, 1987.

Latour, B. *We Have Never Been Modern*, New York: Harvester, 1993.

Nietzsche, F. W. *The Will to Power*, translated by R. Hollingdale and Walter Kaufmann, London: Random House, 1967.

Orlikowski, W. J., and Barley, S. R. "Technology and Institutions: What Can Research on Information Technology and Research on Organizations Learn from Each Other," *MIS Quarterly* (25:2), 2001, pp. 145-165.

Schultze, U., and Leidner, D. "Studying Knowledge Management in Information Systems Research: Discourses and Theoretical Assumptions," *MIS Quarterly* (26:3), 2002, pp. 213-242.

ABOUT THE AUTHORS

Lucas Introna is a reader in Organisation, Technology, and Ethics in the Centre for the Study of Technology and Organisation at Lancaster University. His research interest is the social dimensions of information technology and its consequences for society. In particular he is concerned with the way information technology transforms and mediates social interaction with specific reference to the moral dimension. He was associate editor of *Information Technology & People* (1996-2000) and is coeditor of *Ethics and Information Technology*. He is a founding member of the International Society for Ethics and Information Technology (INSEIT) and an active member of IFIP WG 8.2, The Society for Philosophy in Contemporary World (SPCW), and a number of other academic and professional societies. His most recent work includes a book, *Management, Information and Power* published by Macmillan, and various academic papers in leading journals and conference proceedings on a variety of topics such as theories of information, privacy, surveillance, information technology and post-modern ethics, autopoiesis and social systems, and virtual organisations. He holds degrees in Management, Information Systems and Philosophy. Lucas can be reached at l.introna@lancaster.ac.uk

Louise Whittaker is a senior lecturer in Information Management at the Graduate School of Business Administration, University of the Witwatersrand in Johannesburg. Her research interests include the organizational effects of information technology, the evaluation of information systems, and information systems and organizational strategy. She is particularly interested in a phenomenological approach to understanding these issues. She completed her B.Com. and M.Com. degrees at the University of Witwatersrand and her Ph.D. at the University of Pretoria. Louise is a member of IFIP WG 8.2. Louise can be reached at whittaker.l@wbs.wits.ac.za

Appendix A

MISQ Editors from 1977 to 2004

Year	Vol.	Editor-in-Chief	Senior Editors
1977- 1982	1-6	Gary W. Dickson	
1973-1985	7-9	William R. King	
1986	10	F. Warren McFarlan	
1987-1988	11-12	F. Warren McFarlan	Izak Benbasat
1989	13	James C. Emery	Izak Benbasat
1990-1991	14-15	James C. Emery	Michael Ginsberg
1992	16	Blake Ives	Michael Ginsberg
1993-1994	17-18	Blake Ives	Gerardine DeSanctis
1995	19	Robert Zmud	Izak Benbasat, Gerardine DeSanctis, Allen Lee, Blake Ives
1996	20	Robert Zmud	Lynda Applegate, Izak Benbasat, Sirkka Jarvenpaa, Allen Lee
1997	21	Robert Zmud	Lynda Applegate, Sirkka Jarvenpaa, Allen Lee, Blake Ives, Kalle Lyytinen, Ron Weber
1998	22	Robert Zmud	Lynda Applegate, Blake Ives, Sirkka Jarvenpaa, Kalle Lyytinen, Rick Watson, Ron Weber
1999	23	Allen Lee	Lynda Applegate, Cynthia Beath, Sirkka Jarvenpaa, Kalle Lyytinen, Rick Watson, Daniel Roby, Ron Weber, Robert Zmud
2000	24	Allen Lee	Cynthia Beath, Michael D. Myers, Daniel Roby, Rick Watson, V. Sambamurthy, Jane Webster, Kwok-Kee Wei, Ilze Zigurs, Robert Zmud
2001	25	Allen Lee	Cynthia Beath, Michael D. Myers, Daniel Roby, Rick Watson, V. Sambamurthy, Jane Webster, Kwok-Kee Wei, Ilze Zigurs, Robert Zmud
2002	26	Ron Weber	Ritu Agarwal, Allen Lee, Michael Myers, V Sambamurthy, Peter Todd, Jane Webster, Kwok-Kee Wei, Ilze Zigurs
2003	27	Ron Weber	Ritu Agarwal, Deborah Compeau, Allen Lee, Michael Myers, Rajiv Sabherwal, V Sambamurthy, Carol Saunders, V. Storey, Peter Todd, Jane Webster, Kwok-Kee Wei
2004	28	Ron Weber	Ritu Agarwal, Deborah Compeau, Allen Lee, Lars Mathiassen, Michael Myers, Rajiv Sabherwal, Carol Saunders, Veda Storey, Peter Todd, Bernard Tan, Dov Te'eni

8 DEBATABLE ADVICE AND INCONSISTENT EVIDENCE: Methodology in Information Systems Research

Matthew R. Jones
University of Cambridge

Abstract The range of legitimate methods in IS research has expanded considerably over the past 20 years, a process to which IFIP Working Group 8.2 is seen to have made an important contribution. This has probably made it even harder, however, for IS researchers to know what constitutes good methodological practice. This paper addresses this issue from two angles: first through a critical analysis of claims made in the IS literature regarding the characteristics of good research; and second through an examination of the use of methodology, as reported in a number of IS research papers. The characteristics of good research considered are that it should follow the scientific method; that it should fulfil certain criteria; that it should be relevant; and that it should employ multiple methods Each of these is shown to have limitations. With respect to methodology in practice, the analysis indicates a remarkable lack of consistency in the reporting of IS research. The implications of these findings are discussed.

1 INTRODUCTION

One of the particular contributions of IFIP Working Group 8.2, as the call for papers for this conference describes, is seen to have been in enlarging the range of research methodologies considered legitimate in the Information Systems (IS) research field. Although itself not perhaps the key legitimating institution, Working Group 8.2 has provided a forum for discussion of, and reflection upon, the methods appropriate to IS research, and is considered as being in the vanguard of the adoption of new, especially qualitative and interpretative, methods, as it has been in the use of social theory (Jones 2000b).

While increasing the number of recognized methodologies and theories in the field may have enriched the way in which IS are studied, it does not mean that better insight has necessarily been gained on IS phenomena, as this may be considered to depend, at least in part, on the way in which methodologies are used and related to the theories employed. Indeed, the proliferation of methodologies and theories may actually make it harder to judge whether research has been conducted well, as greater diversity and specialization within the field increasingly limits any particular individual's knowledge, let alone experience, of more than a small proportion of available methods and theories, leaving them unable to comment effectively on other approaches. Unless, however, it is decided that this means that there is no basis for saying whether a particular piece of research has been done well or not, some concept of good methodological practice would still seem necessary.

Although, as has been noted, the appropriate use of both methodology and theory are important to good research, it would seem feasible to give proper consideration to only one of them in a paper of this sort. The current focus, therefore, reflecting that of the 1984 Manchester conference, will be on methodology, that is, what constitutes good practice with respect to the use of methodology and how can it be identified?

In this paper, this question will be approached from two angles, firstly in terms of espoused theory and secondly in terms of theory in practice. Initially, therefore, various statements from the IS literature, both direct and indirect, about the characteristics of good methodological practice will be identified, discussed, and their assumptions critically explored. Then the question will be examined empirically, considering how methodology is actually used in a number of papers that may be seen to be examples of good practice in IS research. Finally the results of this analysis and its implications for IS research practice will be discussed.

2 CHARACTERISTICS OF GOOD RESEARCH METHODOLOGY: IN PRINCIPLE

An examination of the IS literature suggests the existence of at least four different, but sometimes connected, views on the characteristics of good research from a methodological perspective. These are that good research should follow the scientific method; that good research should fulfil certain criteria; that good research should be relevant; and that good research should employ multiple methods. The first of these is generally considered to be the mainstream view in IS research, as the continuing dominance of positivist papers in the literature may be seen to demonstrate (Nandhakumar and Jones 1997; Orlikowski and Baroudi 1991; Vessey et al. 2002). It is, therefore, to this that attention is initially directed.

2.1 Good Research Follows the Scientific Method

This view is promoted in many standard works on research methodology in the social sciences as well as business and management (the likely reference point for IS researchers in the absence of specific IS research textbooks). Yin (1993, p. xvi), for example, argues that "case studies that follow the procedures from 'normal' science are likely to be of higher quality than case studies that do not," while Cooper and Schindler (1998, p. 15) write that "good [business] research follows the standards of the scientific method," and King, Keohane, and Verba (1994, p. 7) argue that good research, is by definition, scientific, following the same logic of inference whether "quantitative or qualitative in style." This is endorsed in the IS literature by writers such as Emery (1989, p. xi) who argued in an *MIS Quarterly* editorial that IS research papers should meet "the strict criteria for rigorous scientific research," Remenyi and Williams (1995, p. 191), who state that "to [undertake research] satisfactorily...the researcher should comply with the 'scientific method,'" and Lee (1989, p. 33), who sought to demonstrate that IS case studies can satisfy "the standards of the natural science model of scientific research."

Notwithstanding the substantial body of studies in the sociology of science that show that research practice in the natural sciences may bear little relation to such standards (e.g., Collins 1985; Latour and Woolgar 1979; Pickering 1992), advocates of "the scientific method" often seek to present characteristics that *scientific* management or IS research should display. Sekaran (1992, p. 10), for example lists these as "purposiveness, rigor, testability, replicability, precision and confidence, objectivity, generalizability and parsimony." More concretely, Emery (1989, p. xi) argued that scientific IS research

> should satisfy the traditional criteria for high quality scholarly research. It should be based on a set of well-defined hypotheses, unbiased and reproducible procedures for collecting evidence that supports or refutes the hypotheses, and sound analytical procedures for drawing appropriate conclusions from the evidence. This research often involves the collection of considerable quantities of quantitative data through such means as laboratory experiments or survey instruments [and] the data are then subjected to statistical analysis to draw the appropriate inferences from the research.

Good research, it would seem, is necessarily positivist in epistemology and may be identified by evidence of its adherence to such precepts of the scientific method.

2.2 Good Research Fulfils Certain Criteria

While apparently widely accepted in the IS research field, the syllogism that good research follows the scientific method, the scientific method is positivist, therefore, good research is positivist has also been the focus of considerable criticism, not least at conferences of IFIP Working Group 8.2. Indeed, as Fitzgerald et al. (1985, p. 3) note, the first Manchester conference was originally entitled "Information Systems Research—A Doubtful Science?" and specifically sought to call into question the idea that the "the scientific research methodology is the only relevant methodology for information systems research."

Evidence of the success of this challenge may be found in an *MIS Quarterly* editorial (DeSanctis 1993) less than five years after Emery's pronouncement. This suggested that the association of good research with positivism was not a required one, and that interpretive studies could also be of high quality (i.e., publishable in a strong scholarly journal devoted to good science). This breaking of the alleged positivist monopoly on good IS research, however, did not necessarily lead to a complete abandonment of the notion that good research fulfils certain criteria (and may be judged by whether it demonstrates this in its reports). Rather it lead, in some circles at least, to calls for the development of alternative "criteria for judging qualitative, case and interpretive research in IS" (Lee et al. 1995, p. 367), the most widely recognized response to which may be the "principles for conducting and evaluating interpretive field studies in information systems" of Klein and Myers (1999). Thus while they are careful to emphasize that their principles are "not like bureaucratic rules of conduct," nor should their use be considered mandatory, Klein and Myers' claim that they offer an alternative to "inappropriate (positivist) criteria" (p. 81) may be seen to suggest that they fulfil a comparable function to these criteria, albeit from a markedly different philosophical perspective.

This idea, that research reports exhibiting certain characteristics are necessarily indicative of superior research practice than reports lacking them, however, may be seen as relying on a number of assumptions: first, that consensus can be reached on appropriate criteria, if not for all research, then at least for particular types; secondly, that evidence of having met these criteria is sufficient to establish the merits of a particular piece of research; and finally, that the presence of claims concerning these criteria accurately describes research practice. The first of these assumptions is clearly not restricted to claims based on the content of research reports. The latter two, however, may be seen to be vulnerable to the criticism that they treat research reports as adequate descriptions of research practice, rather than as accounts constructed for particular purposes (such as getting published).

This situation may also be considered in terms of Giddens' (1976) concept of the double hermeneutic of social research, whereby researchers' findings become part of actors' understandings of their settings. Thus there is the possibility that the criteria may come to shape research practice, either explicitly as researchers adapt their reports to ensure they demonstrate the required characteristics, or implicitly, as researchers unreflexively adopt the criteria as measures of quality.

Of course, this may be precisely the objective of those promulgating such criteria, but it is critically dependent on the second of the assumptions above, that demonstrating conformance to the criteria is enough, in itself, to establish the quality of a piece of research. As quality management research (e.g., Harari 1997) discusses, however, there is a risk that fulfilment of criteria can become the objective in itself, without reference to the wider aims of the research and that researchers simply use the criteria as a template for reporting their work without this significantly influencing their research practice.

2.3 Good Research Is Relevant

Similar concerns about technique being seen as more important than substance may also be seen to lie behind the relevance vs. rigor debates in the IS literature (Benbasat and Zmud 1999; Davenport and Markus 1999; Keen 1990; Turner et al. 1990; Zmud 1996), where critics have claimed that the emphasis on methodological rigor in IS research has been at the expense of addressing relevant problems and engaging with practitioner audiences. Although directed at IS research as a whole, the main target for complaints has been what is seen to be an excessive attention to methodological refinement, especially in positivist studies of the sort recommended by Emery (1989), that may be exacerbated by academic career incentives (Applegate and King 1999). Lee (1999) also argues that positivist IS research, in seeking to emulate the natural sciences, has a particular tendency to be driven by theory rather than practice.

More typically, however, *relevance* itself would not appear to have any specific methodological implications; rather it is simply a matter of whether the research "demonstrate[s] a meaningfulness regarding its application to the significant problems being faced by today's organizations and their members" (Zmud 1996, p. xxvii) in terms of both the topics and the audiences it seeks to address. Relevance and rigor, it is thus generally claimed, need not be incompatible and good research can, and should, strive for both (Keen 1990; Zmud 1996). From a methodological perspective, therefore, relevance focuses on the inputs and outputs of IS research, i.e., whether the research topics are meaningful to practitioners and the results are presented in an accessible style, rather than the process of research itself (except to the extent that this may detract from meeting these objectives). It is, therefore, unable to serve as a source of guidance on good methodological practice.

2.4 Good Research Is Multi-Methodological

The development of alternative criteria for judging non-positivist IS research may be seen as one particular response to the calling into question of the scientific research methodology as the only appropriate research methodology for IS research. More generally, as the organizers of the first Manchester Conference proposed, "this might be thought to argue strongly for an acceptance of a pluralism of methods in this area of research" (Fitzgerald et al. 1985, p. 4). Similar arguments have also been advanced by Allen and Ellis (1997), Landry and Banville (1992), Mingers (2001), and Smithson (1990).

Pluralism in this context is often seen to refer to the use of multiple methods in one piece of research. Fitzgerald et al., for example, talk of "the possibility that the combination of two or more research approaches might lead to progress" (p. 5) (albeit in the context of a discussion of the institutional barriers to pluralism). Thus the limitations of the positivist, scientific method, it is suggested, may be overcome by supplementing it with other methods. This theme has been taken up in a number of subsequent studies (Gable 1994; Kaplan and Duchon 1988; Lee 1991; Markus 1994; Mingers 2001; Sawyer 2000, 2001; Trauth and Jessup 2000) that have sought to demonstrate the feasibility of such multi-method research and to propose it in some cases, normatively, as a model for IS research practice.

In considering the merits of these claims it would seem necessary first to establish what is meant by multi-method research, as this will affect what is feasible (and hence whether it is sensible to advocate it as something to which IS researchers should aspire). The key issue here would seem to be whether the *methods* it is proposed to combine refers to the underpinning ontological and epistemological beliefs, sometimes referred to as *paradigms* (Burrell and Morgan 1979), or simply relates to the mechanics of the research process, such as the gathering of quantitative or qualitative data, or the use of surveys or case studies, which, in themselves, carry no necessary philosophical assumptions. Of course, in practice, these two aspects of research methods are often linked. Thus positivist researchers tend to favor experimental studies and surveys, involving large numbers of subjects, whereas interpretivists typically favor interviews and observations of smaller numbers of sites. As Jones (2000a) argues, however, many such associations are a matter of convention, rather than being necessary, i.e., both positivist and interpretivist researchers can, and do, use qualitative and quantitative data, case studies, and surveys, if in quite different ways.

Accepting this distinction between the philosophy and practice of methods, therefore, there would seem no particular issues with the feasibility of multi-method research combining qualitative and quantitative data gathering. Research combining, say, positivism and interpetivism within the same study, however, would seem to require the simultaneous maintenance of contradictory beliefs. Hence Parker and McHugh (1991) argue that for an individual researcher to pursue such a multi-paradigm combination of methods would involve either that core beliefs can be changed as an act of will, or that the researcher is capable of *authentically* feigning alternative beliefs, or that they have multiple personalities which dominate at different times. It would also seem unclear that such differences can be adequately reconciled within a research team to permit integration (rather than just accommodation) of different perspectives. Thus, as Jones (2000a) has shown, most, if not all, IS studies that claim to demonstrate such integration do not, in practice, involve multiple paradigms, or retain unresolved philosophical differences. The feasibility of such multi-paradigm research would, therefore, seem unproven.

Despite, as Mingers (2003) acknowledges, the paucity of examples of such research, however, it might still be the case that good research should be multi-method. Mingers (2001, p. 243), for example advocates multi-method research on two grounds: firstly, that all research settings are so complex and multidimensional that they would "benefit from a range of methods" (although it is not specifically indicated whether this need involve different research paradigms or be undertaken by the same researcher); and secondly, because research is typically a process that involves different phases, that "pose different tasks and problems for the researcher" for which different methods may be more useful. Mingers also reports further advantages cited by Tashakkori and Teddlie (1998) that strong pluralism enables triangulation of results, encourages creativity through the discovery of paradoxical findings, and enables the scope of studies to be widened.

Although these arguments may be attractive in suggesting the potential to achieve a richer understanding of IS phenomena, they need to be weighed against some practical and philosophical objections to multi-method research to decide whether it is something that should be considered a mark of good IS research practice. Mingers (2001) identifies some of these practical concerns in terms of the funding and assessment of research, the training of researchers, and local research cultures. It might also be questioned whether it is the best deployment of resources for individual researchers to be expected to learn and apply multiple methods.

Philosophically, the argument that different methods are needed to illuminate different aspects of reality rests on the assumption that there is a single reality, independent of the observer, or the observation process, that is being studied. As Seale (1999) discusses in the context of methodological triangulation, this view would be challenged by idealist and constructivist researchers and relies on the inductivist fallacy that valid conclusions can be reached from specific instances. While multi-method research may have some potential advantages, therefore, it is not without its drawbacks and may also not offer the improved insights that its proponents claim.

The reasons for this may perhaps be better understood in terms of the types of pluralism identified by Watson (1990). Thus the claims of Mingers (2001), that different approaches provide only partial access to a complex reality and that a mixture of methods is therefore necessary and ultimately reconcilable, represent only one of a number of possible pluralist positions, each of which may be seen to have rather different methodological implications.

Pluralism of hypotheses, for example, is perhaps closest to the position of IS advocates of multi-method research in suggesting that there is a single reality, but that different opinions are possible about it. Such opinions, it is suggested, may be incompatible, but incompatibility will disappear as truth is discovered. This may be seen to suggest a contingent approach to methodology choice (Mingers [2001] describes this as *complementarism*), adopting the principle of "horses for courses," with a number of valid methods, the choice of which will depend on the particular research question being addressed. This assumes, however, that research questions have an intrinsic character for which certain methods are best suited and that all researchers are pursuing a common understanding.

Archic pluralism, on the other hand, does not assume a single reality, but argues that it is constituted by the inquirer, so that each person has their own reality. Individual perspectives, however, are seen to reflect essential possibilities of reason, so mutual intelligibility and dialogue are possible, even if, as with translation between languages, all concepts and nuances cannot be conveyed. Such a dialogical model, moreover, does not require that there is agreement; rather, debate between methods can stimulate independent development. Good research would therefore be seen as striving for improvement on its own terms, but simultaneously seeking informed engagement with other approaches.

Perspectival pluralism is perhaps the furthest from the assumptions of multi-method advocates in suggesting that individuals do not experience the same world and thus that each of us has our own reality. Different individuals' views are therefore incommensurable and the only possibility for research is to follow Feyerabend's (1975) maxim of "anything goes" and "let many flowers bloom," as Fitzgerald et al. put it, with no common basis for defining good research.

It does not follow, therefore, that an acceptance of methodological pluralism in a research field necessarily means that good research should be multi-method. This is not to argue that IS researchers should not undertake multi-method studies or that one particular form of pluralism is correct, but simply to point out that pluralism *per se* does not have specific methodological implications and that the methodological quality of IS research papers may not perhaps be best assessed on the basis of whether they employ multiple methods or not.

3 CHARACTERISTICS OF GOOD RESEARCH METHODOLOGY: IN PRACTICE

It appears that the IS literature provides several different views on what good research methodology should involve (and how it may be identified), each of which have their limitations. In the absence of universally accepted or, arguably, even plausibly effective guidance on the methodology of IS research, it would seem interesting to see which, if any, of these precepts are actually adhered to by IS researchers and, if they are not complied with, what characteristics of good research methodological practice may be identified from IS research studies.

3.1 Research Methodology

A grounded analysis (Strauss and Corbin 1990) was undertaken of a number of papers seen to represent good IS research practice, namely "best papers" from *MIS Quarterly* and from the International Conference on Information Systems, as identified on the AIS Website. To enlarge the sample, empirical papers from the 1990, 1995, and 2000 IFIP WG8.2 conferences that included an explicit discussion of methodology were also analyzed, yielding a total of 32 papers. These are listed in Appendix A.

All of the papers were examined and coded, focusing on the methodologies employed. This technique uses a form of content analysis where the data are read and categorized into concepts that are suggested by the data rather than imposed from outside (Agar 1980). This is known as open coding (Strauss and Corbin 1990), and it relies on an analytic technique for identifying possible categories and their properties and dimensions. Once all of the data were examined, the concepts were organized by recurring theme. These themes became prime candidates for a set of stable and common categories, which linked a number of associated concepts. This is known as axial coding (Strauss and Corbin 1990), and it relies on a synthetic technique of making connections between subcategories to construct a more comprehensive scheme. The data were then reexamined and recoded using this proposed scheme, the goal being to determine the set of categories and concepts that covered as much of the data as possible. This iterative examination yielded a set of broad categories and associated concepts that described the characteristics of the research methods employed (cf. Orlikowski 1993). The categories are listed in Table 1.

For categories 5 to 17, no entry was made where no data were given in the paper. For example, if a paper did not mention the research approach, then the entry was left blank, even if it appeared that the paper was positivist or interpretivist. For these categories, any qualifiers offered by way of explanation for how these issues were dealt with in practice were also noted. For example, a response rate to a survey might be qualified by explaining that this was "generally accepted as sufficient for path analysis" (Nelson and Cooprider 1993).

3.2 Results

Perhaps the most striking feature of the analysis was the almost complete absence of consistency in the way papers report their research methods. This extended to the name of the research methods section, or even whether such a section was included in the paper. Nine of the 32 papers had no research methods section, although five of these were theoretical/methodological papers. Three of the four economics papers included no separate discussion of methodology. Of the papers with a section discussing research

Table 1. Characteristics of Research Methodology Discussion in IS Research Papers

Category	Category name	Defintion
1	Author	Name of authors (for reference)
2	Year	Year of publication
3	Venue	*MIS Quarterly*, ICIS,or IFIP WG8.2
4	Version	Extended abstract, conference paper or journal version (where published)
5	Section heading	What name was given to the section, if any, discussing research methods
6	Proportion of methods discussion	The length of the methods discussion as a percentage of the total length of the paper
7	Research approach	Positivist, interpretive, critical, or other
8	Level of analysis	Economy, institution, organization, group, or individual
9	Sample	The number and type of sites or instances studied
10	Respondents	The types of individuals from whom data was sought
11	Response rate	The proportion (or number) of sites or individuals approached who supplied data used in the study
12	Measurement	The types of data gathered
13	Constructs	The categories of responses
14	Pilot	Whether a pilot study was conducted and what role it served in the study
15	Reliability	Measure of reliability used
16	Data analysis	Method of data analysis employed
17	Other claims	

methods, the titles (and the frequency with which they were used) included "Research Method(s)" (5); "Methodology" (3); "Research Approach" (2); "Research Design" (2); "Research Methodology" (2); and "Research Methodology and Data Collection" (1). Other titles of sections in which research methods were discussed included "Constructing the Methodology," "The Learning Audit Methodology," and "The Study."

The proportion of the paper devoted to methodology varied between 0 percent and 38 percent (for a paper reporting the methodology itself) with 8 of the 23 papers that included a research methods section devoting less than 5 percent of the paper to their discussion and the average being 11 percent. Some of the longer research methods sections included large tables of variables and lists of equations, but a few provided quite detailed accounts of, and justifications for, the research process adopted.

The research approach, where specified, was variously described in terms of the reference discipline (e.g., organization studies), the form of data collection (e.g., case study, field experiment, quasi-experiment), the philosophical stance (e.g., interpretive and constructivist), and with reference to particular research traditions (e.g., ethnomethodological and grounded theory).

The level of analysis was specifically identified in only four papers, although for many this was evident from the description of research conducted. More information was usually given on the sample, although the numbers of units of analysis and the duration of contact with them (where appropriate) was not always specified. For some of the more quantitative papers, the numbers of units was very high, e.g., 16,5875 prices (Wood and Kauffman 2000), 370 firms (Hitt and Brynjolfsson 1996). The qualitative and interpretive papers, on the other hand, typically studied one or two cases. Only a few papers commented on the sample selection, e.g., convenience (Nelson and Cooprider 1993), random (Guimaraes 1995), theoretical sampling (Orlikowski 1993).

About half the studies reported the number of organizations or individuals approached or interviewed (where appropriate), but in a few cases just their organizational roles, e.g., senior HR manager or developer, or their credentials, e.g., knowledgeable individuals, were given. The numbers of interviewees varied between 4 and 159. Only the four studies using surveys reported response rates, while one study reported only the number of valid responses.

Most papers provided information on the methods used to gather data, but only one study reported using predefined measures. Rather, most simply described the type of interviewing or observation and the types of secondary documentation studied. Three papers reported the use of predefined constructs from previous studies, two more described their constructs as field-driven, and four referred to specific theoretical approaches as informing the analysis. Only three of the papers reported pilot studies and the same number gave reliability measures (two Cronbach Alpha's and one Cohen's Kappa).

Five of the research methods sections included discussion of data analysis. The detail with which this was described however, varied from simply qualitative to substantial descriptions of coding procedures. Other claims made in the description of research methodology included reference to the number of pages of data collected and the size of the research team.

Most studies described their research methods without qualifiers. Where qualifications were offered, this was typically by reference to accepted practice in reference disciplines or research of the particular type, often substantiated with appropriate references. For example, Wilson and Howcroft (2000) justified their use of a case study with the argument that it is "particularly relevant given the size and diversity of the organization...[and also] recommended where there is a desire to gain insight into emerging topics, but there is no need to control behavioral events or variables" backed by references to Benbasat et al. (1987) and Yin (1993). A number of such references were cited in several papers. Other claims were more vague. For example Nelson and Cooprider (1993) explain that the key informant method is "frequently adopted" and that "there was no reason to suspect systematic bias." Some authors also offered explanations for limitations of their methodology or sought to counter possible alternative interpretations of findings. Toraskar (1990), for example, explained that it had been intended to collect live data of participants using the evaluation method, but this had proved infeasible so it was necessary to resort to an anecdotal approach, while Guimaraes (1995), commented that "while no relevant changes extraneous to the BPR project were apparent, this possible threat to the results validity cannot be completely discarded."

Given the nature of the data collected, and the small sample size, it was not possible to undertake a statistical comparison of the papers from the different venues. Apart from the fact that the IFIP WG8.2 papers had already been selected for their discussion of research methods, it was not evident that they were significantly different from the best papers in terms of their reporting of methodology (if not in terms of their methods and theoretical interests; Sawyer 2002).

4 DISCUSSION

What would seem clear from this analysis is that there is considerable variation in how research methods are reported in IS papers, or indeed whether they are reported at all in some cases. As the variety of terms used to describe the sections of papers discussing research methods and their very different contents also indicates, there would not seem to be a common understanding even of what is meant by the term *research methodology*. While some authors (e.g., Mingers 2001) have attempted to identify some of the different interpretations of the term, and others (e.g., Remenyi et al. 1998) have offered their own definitions, what would seem clear is that it is difficult to envisage the achievement of consensus on the characteristics of good research methodology without agreement on even to what the term applies.

From an examination of the best papers, it would also seem that few, if any, of them could be considered to demonstrate their adherence to the scientific method. Of course this could be seen as indicating that even the *best* IS research papers are not of high quality (or that the process for selecting best papers or conference submissions is faulty). While there might be some who would argue that this is a valid conclusion, it would seem more reasonable to suggest either that this indicates the inappropriateness of demonstrating compliance with the scientific method as the sole criterion for assessing IS research practice, or that good research cannot be evaluated solely in terms of whether it reports certain things in its publications. This second conclusion would also seem to apply equally to other criteria/principle-based approaches to evaluating research. Indeed as Klein and Myers (1999) themselves demonstrate, a number of well-reputed IS papers do not fulfil their principles.

The results of the analysis would seem to confirm Mingers' (2003) findings regarding the frequency of use of multi-method research. Only two papers reported the use of qualitative and quantitative data and none were evidently multi-paradigm. In one sense this could be seen as illustrating the cultural and institutional barriers to multi-method research, but it could equally be taken to show that advocating that all IS research should be multi-method research, even as an aspiration (Mingers 2001), may be unrealistic.

Looking at the findings more critically, however, what would seem surprising is how little information on research methods is actually given in many of the papers. This is not just that some economists don't seem to consider that methodology is something that needs to be discussed in their papers, or that other papers devote almost all their attention to reporting findings rather than how they were obtained, but that even when papers do report on their research methods (or approaches, designs, or whatever), significant aspects of the research process, which might have bearing on the findings, are not mentioned at all.

This is not to propose that there should be a checklist of items that every paper should report, since one of the arguments of this paper is that such a checklist is unlikely to be universally accepted, and, if promulgated with sufficient authority, may be more likely to encourage ritualistic adherence than improved practice. At the same time, however, there are papers in the research methods literature that discuss what a researcher might be expected to tell their readers about the mechanics of the research process (as opposed to the more philosophical principles of Klein and Myers). For example, Taylor and Bogdan (1998) suggest that qualitative researchers should report their methodology, the time and length of the study, the nature and number of settings and informants, the research design, the researchers own frame of mind, the researcher's relationship with the informants, and the checks made on the data. While such recommendations are likely only to apply to particular types of research and to be guidance rather than prescriptions, they may nevertheless serve as stimuli for reflection in the preparation of descriptions of research methods.

Whether or not IS researchers reflect on such guidance and seek to respond to it in their papers, though, it would be wrong to consider what is reported in research papers as bearing a one-to-one relationship with research practice. Thus even the most detailed, reflexive account of research practice (e.g., Schultze 2000), while it may tell us a lot more than a four line comment that a particular method was used (with appropriate references), is still able to address only a tiny part of the research experience. Our evaluation of the method in both cases involves trust that the author conducted the research appropriately. No amount of appendices or instruments or datasets available from the author can overcome this.

If what a paper tells us is insufficient to judge definitively the quality of the research methods, but it is also accepted that the papers analyzed do meet some suitable quality standard in terms of their methodology, then what does this suggest about good research methods practice? Three general features of the papers studied may be identified. First, it would seem from the variability of presentation and the limited information supplied on research methods that this is not considered a critical element of good research, or perhaps that deficiencies in the conduct or reporting of research can be balanced by strengths elsewhere in the paper (especially given constraints on the length of papers in these venues). Second, it would appear that the greater length devoted to discussion of methods in some papers reflects a greater attention paid to the resources expended in carrying out the research, for example, citing the large numbers of interviews undertaken or the elaborate procedures adopted to validate and analyze findings, and extended discussion of the precedents for the approach used. That is, longer discussions of research methods tend to include more evidence and more elaborate arguments to support their claims. Third, to the extent that conventions have been adopted in the reporting of IS research methods, these would seem to relate more to the language and style of the reporting than to the content. Thus there would appear to be certain keywords and references that recur in research methods descriptions, although it is not possible to judge whether this constitutes evidence of the existence of some limited templates of the sort it has been suggested might be accentuated by attempts to define criteria for research reports.

These findings may be seen to suggest that in research methods, as in other aspects of the research process, the measure of good practice may be its outputs. That is, good research methods produce convincing accounts, however cursory their description, and longer discussions are about making claims more credible. Thus, as Geertz (1988, p. 4) writes with respect to anthropologists, the ability to

> get us to take what they say seriously has less to do with either a factual look or an air of conceptual elegance than it has with their capacity to convince us that what they say is a result of their having actually penetrated (or, if you prefer, been penetrated by) another form of life, of having, one way or another, truly "been there."

The researcher's task is thus

> to demonstrate, or more exactly to demonstrate again, in different times and with different means, that accounts of how others live that are presented neither as tales about things that did not actually happen, nor as reports of measurable phenomena produced by calculable forces, can carry conviction (Geertz 1988, p. 141).

Similar arguments are made by Van Maanen (1989) in drawing attention to the importance of writing in organization studies, and by Golden-Biddle and Locke (1997) in their discussion of researchers' stories.

What the findings also highlight, in the apparent absence of accepted standards of reporting of research methods, are some of the means that are employed by IS researchers in seeking to establish the methodological credibility of their work. One important method would seem to be references to certain texts. Jones (2000b), for example, showed that Burrell and Morgan (1979) were cited in 24 papers, Glaser and Strauss (1968) in 16 papers, and Yin (1993) in 27 papers at IFIP Working Group 8.2 conferences between 1979 and 1999 (this may be compared to 34 papers citing Giddens, the most frequently referenced social theorist, and 15 papers citing Foucault). In actor network terms, as Latour (1987) describes, this may be interpreted as strengthening the network through drawing in more actors and strategically orienting them to support the author's case. No matter, for example, that Benbasat et al. (1987) and Yin (1993) advocate the scientific method, they can be used in an interpretative paper to justify the use of case studies in exploratory research. Similar intertextual reinforcement would appear to be at work in positivists' reuse of others' instruments and constructs. Perhaps the greatest show of strength, though, as some of the economists demonstrate, is to not discuss research methods at all. By black-boxing the method, its authority is removed from debate.

Interestingly, numbers, such as Cronbach's Alpha scores or response rates, that are generally seen as highly authoritative, appear, in these papers, to be insufficiently persuasive on their own, requiring reinforcement with claims of their acceptability. Numbers appear to carry greater weight, however, when used to denote Herculean efforts in data collection or analysis. Qualifications also appear to play a significant role in establishing credibility by seeking to preemptively disarm critics. This would seem to take two forms: either appeals to some general authority, such as something being frequently accepted, or more specific acknowledgment of potential weaknesses, evoking the critics' clemency and sympathy.

While a more detailed analysis of these rhetorical strategies would no doubt be fruitful, for the present purposes they would seem sufficient to suggest some of the possible elements of apparently convincing IS research methods accounts. This is not to propose that these could, or should, be translated into guidance/recommendations/requirements for good research methods reporting. This is not just because the primary concern of this paper is with research methods practice rather than its output, but because, as has been emphasized, the outcomes of such guidance may even be counter-productive. The point to be made, rather, is that, from a methodological perspective, the IS research literature appears neither to provide effective guidance for research practice, nor evidence of standards of good methodological practice.

5 CONCLUSIONS

An examination of the guidance offered in the IS literature on the qualities of good research methodology suggests that none of the available recommendations, if not prescriptions, have sufficient acceptance to form a satisfactory basis for evaluating the merits of research papers. Were they to somehow acquire such authority, moreover, they would risk ossifying research methods reporting according to particular templates, without necessarily improving research practice.

At the same time, however, an analysis of how IS research methods are reported in the literature in practice reveals almost no consistency, even at the basic level of terminology. Although this analysis did identify some features of these discussions, these appeared to have more to do with constructing convincing accounts than with demonstrating sound method. This need not be considered a cynical position, however. As Geertz (1988, p. 145) argues,

> all this is not to say that descriptions of how things look to ones subjects, efforts to get texts exact and translations veridical...and rigorous examination of one's assumptions are not supremely worth doing for anyone who aspires to [report on research]. It is to say that doing these things does not relieve the burden of authorship, it deepens it.

The choice, therefore, is not between slavish following of boilerplates or complete fiction. Paying careful attention to how we describe our research, in the light of our acknowledged philosophical position, and awareness of the artifice involved in this process are both necessary.

REFERENCES

Agar, M. H. *The Professional Stranger: An Informal Introduction to Ethnography*, New York: Academic Press, 1980.

Allen, D., and Ellis, D. "Beyond Paradigm Closure in Information Systems Research: Theoretical Possibilities for Pluralism," in R. Galliers, C. Murphy, H. Hansen, R. O'Callaghan, S. Carlsson, and C. Loebbecke (Eds.), *Proceedings of the 5th European Conference on Information Systems*, Cork, Ireland: Cork Publishing Company, 1997, pp. 760-776.

Aanestad, M., and Hanseth, O. "Implementing Open Network Technologies in Complex Work Practices: A Case from Telemedicine," in R. Baskerville, J. Stage, and J. I. DeGross (Eds.),

Organizational and Social Perspectives on Information Technology, Boston: Kluwer Academic Publishers, 2000.

Applegate, L. M., and King, J. L. "Rigor vs. Relevance: Careers on the Line," *MIS Quarterly* (23:1), 1999, pp. 17-18

Benbasat, I.; Goldstein, D. K.; and Mead, M. "The Case Study Research Strategy in Studies of Information Systems," *MIS Quarterly* (11:4), 1987, pp. 369-386.

Benbasat, I., and Zmud, R. W. "Empirical Research in Information Systems: The Practice of Relevance," *MIS Quarterly* (23:1), 1999, pp. 3-16.

Burrell, G., and Morgan, G. *Sociological Paradigms and Organizational Analysis*, Portsmouth, NJ: Heinemann, 1979.

Broadbent, M.; Weill, P.; O'Brien, T.; and Neo, B. S. "Firm Context and Patterns of IT Infrastructure Capability," in J. I. DeGross, S. Jarvenpaa, and A. Srinivasan (Eds.), *Proceedings of the 17th International Conference on Information Systems*, Cleveland, OH, 1996, pp. 174-194.

Calloway, L. J., and Ariav, G. "Developing and Using a Qualitative Methodology to Study Relationships Among Designers and Tools," in H-E. Nissen, H. K. Klein, and R. Hirschheim (Eds.), *Information Systems Research: Contemporary Approaches and Emergent Traditions*, Amsterdam: Elsevier, 1990.

Ciborra, C.; Patriotta, G.; and Erlicher, L. "Disassembling Frames on the Assembly Line: The Theory and Practice of the New Division of Learning in Advanced Manufacturing," in W. J. Orlikowski, G. Walsham, M. R. Jones, and J. I. DeGross (Eds.), *Information Technology and Changes in Organizational Work*, London: Chapman & Hall, 1996.

Collins, H. *Changing Order: Replication and Induction in Scientific Practice*, London: Sage Publications, 1985.

Cooper, D. R., and Schindler, P. S. *Business Research Methods*, London: McGraw-Hill, 1998.

Davenport, T. H., and Markus, M. L. "Rigor vs. Relevance Revisited: Response to Benbasat and Zmud," *MIS Quarterly* (23:1), 1999, pp. 19-23,

Davies, L. J. "Researching the Organizational Culture Contexts of Information Systems Strategy: A Case Study of the British Army," in H-E. Nissen, H. K. Klein, and R. Hirschheim (Eds.), *Information Systems Research: Contemporary Approaches and Emergent Traditions*, Amsterdam: Elsevier, 1990.

DeSanctis, G. "Theory and Research: Goals, Priorities and Approaches," *MIS Quarterly* (17:1), 1993, pp. vi-viii.

Elam, J.; Walz; D. R.; Curtis, B.; and Krasner, H. "Measuring Group Processes in Software Design Teams," in H-E. Nissen, H. K. Klein, and R. Hirschheim (Eds.), *Information Systems Research: Contemporary Approaches and Emergent Traditions*, Amsterdam: Elsevier, 1990.

Emery, J. C. "Editor's Comments," *MIS Quarterly* (13:3), 1989, pp. xi-xii

Feyerabend, P. *Against Method*, London: Verso, 1975.

Fitzgerald, G.; Hirschheim, R.; Mumford, E.; and Wood-Harper, A. T. "Information Systems Research Methodology: An Introduction to the Debate," in E. Mumford, R. Hirschheim, G. Fitzgerald, and A. T. Wood-Harper (Eds.), *Research Methods in Information Systems*, Amsterdam: North-Holland, 1985, pp. 3-9.

Gable, G. G. "Integrating Case Study and Survey Research Methods: An Example in Information Systems," *European Journal of Information Systems* (3:2), 1994, pp. 112-126.

Gallivan, M. J. "Contradictions Among Stakeholder Assessments of a Radical Change Initiative: A Cognitive Frames Analysis," in W. J. Orlikowski, G. Walsham, M. R. Jones, and J. I. DeGross (Eds.), *Information Technology and Changes in Organizational Work*, London: Chapman & Hall, 1996.

Gasson, S., and Holland, N. "The Nature and Processes of IT-Related Change," in W. J. Orlikowski, G. Walsham, M. R. Jones, and J. I. DeGross (Eds.), *Information Technology and Changes in Organizational Work*, London: Chapman & Hall, 1996.

Geertz, C. *Works and Lives: The Anthropologist as Author*, Cambridge: Polity Press, 1988.

Giddens, A. *New Rules of Sociological Method*, London: Hutchinson, 1976.

Giddens, A. *New Rules of Sociological Method* (2nd ed.), Cambridge: Polity Press, 1993.

Glaser, B., and Strauss, A. *The Discovery of Grounded Theory*, London: Weidenfeld and Nicholson, 1968.

Golden-Biddle, K., and Locke, K. D. *Composing Qualitative Research*, London: Sage Publications, 1997.

Grunden, K. "MOA-S: A Scenario Model for Integrating Work Organization Aspects into the Design Process of CSCW Systems," in R. Baskerville, J. Stage, and J. I. DeGross (Eds.), *Organizational and Social Perspectives on Information Technology*, Boston: Kluwer Academic Publishers, 2000.

Guimaraes, T. "Assessing Employee Turnover Intentions Before/After BPR," in W. J. Orlikowski, G. Walsham, M. R. Jones, and J. I. DeGross (Eds.), *Information Technology and Changes in Organizational Work*, London: Chapman & Hall, 1996.

Hamilton, D., and Atchison, M. "The COMIS Plan: IT-Mediated Business Reengineering in Telecom Australia During the 1960s," in W. J. Orlikowski, G. Walsham, M. R. Jones, and J. I. DeGross (Eds.), *Information Technology and Changes in Organizational Work*, London: Chapman & Hall, 1996.

Harari, O. "Ten Reasons Why TQM Doesn't Work," *Management Review* (86:2), 1997, pp. 38-44.

Hess, C. M., and Kemerer, C. F. "Computerized Loan Origination Systems: An Industry Case Study of the Electronic Markets Hypothesis," *MIS Quarterly* (18:3), September 1994, pp. 251-275.

Hitt, L. M., and Brynjolfsson, E. "Productivity, Business Profitability, and Consumer Surplus: Three Different Measures of Information Technology Value," *MIS Quarterly* (20:2), 1996, pp. 121-142.

Hitt, L., and Brynjolfsson, E. "The Three Faces of IT Value: Theory and Evidence," in J. I. DeGross, S. L. Huff, and M. C. Munro (Eds.), *Proceedings of the 15th International Conference on Information Systems*, Vancouver, British Columbia, Canada, 1994, pp. 263-277.

Jasperson, J.; Carte, T. A.; Saunders, C. S.; Butler, B. S.; Croes, H. J. P.; and Zheng, W. "Review: Power and Information Technology Research: A Metatriangulation Review," *MIS Quarterly* (26:4), 2002, pp. 397-459.

Jones, M. R. "Mission Impossible? Pluralism and Multiparadigm IS Research," *Information Systems Review* (1:1), 2000a, pp. 217-232.

Jones, M. R. "The Moving Finger: The Use of Social Theory in WG8.2 Conference Papers, 1975-1999," in R. Baskerville, J. Stage, and J. I. DeGross (Eds), *Organizational and Social Perspectives on Information Technology*, Boston: Kluwer Academic Publishers, 2000b, pp. 15-31.

Kaplan, B., and Duchon, D. "Combining Qualitative and Quantitative Methods in Information Systems Research: A Case Study," *MIS Quarterly* (12:4), 1988, pp. 571-586.

Kauffman, R. J., and Wood, C. A. "Follow the Leader? Strategic Pricing of E-Commerce," in W. J. Orlikowski, S. Ang, P. Weill, H. C. Krcmar, and J. I. DeGross (Eds.), *Proceedings of the 21st International Conference on Information Systems*, Brisbane, 2000, pp. 145-151.

Keen, P. G. W. "Relevance and Rigor in Information Systems Research: Improving Quality, Confidence, Citation and Impact," in H-E. Nissen, H. K. Klein, and R. Hirschheim (Eds.), *Information Systems Research: Contemporary Approaches and Emergent Traditions*, Amsterdam: Elsevier, 1990, pp. 27-49.

King, G.; Keohane, R. O.; and Verba, S. *Designing Social Inquiry*, Princeton: Princeton University Press, 1994.

Klein, H. K., and Myers, M. D. "A Set of Principles for Conducting and Evaluating Interpretive Field Studies in Information Systems," *MIS Quarterly* (23:1), 1999, pp. 67-94.

Kumar, K.; van Dissel, H. G.; and Bielli, P. "The Merchant of Prato—Revisited: Toward a Third Rationality of Information Systems," *MIS Quarterly* (22:2), 1998, pp. 199-226.

Landry, M., and Banville, C. "A Disciplined Methodological Pluralism for MIS Research," *Accounting, Management and Information Technology* (2:2), 1992, pp. 77-97.

Latour, B. *Science in Action*, Milton Keynes, UK: Open University Press, 1987.

Latour, B., and Woolgar, S. *Laboratory Life: The Social Construction of Scientific Facts*, London: Sage Publications, 1979.

Lee, A. S. "Electronic Mail as a Medium for Rich Communication: An Empirical Investigation Using Hermeneutic Interpretation," in J. I. DeGross, R. P. Bostrom, and D. Robey (Eds.), *Proceedings of the 14th International Conference on Information Systems*, Tampa, FL, 1993, pp. 13-21.

Lee, A. S. "Integrating Positivist and Interpretive Approaches to Organizational Research," *Organization Science* (2:4), 1991, pp. 342-365.

Lee, A. S. "Rigor and Relevance in IS Research: Beyond the Approach of Positivism Alone," *MIS Quarterly* (23:1), 1999, pp. 29-34.

Lee, A. S. "A Scientific Methodology for MIS Case Studies," *MIS Quarterly* (13:1), 1989, pp. 32-51.

Lee, A. S.; Baskerville, R. L.; Liebenau, J; and Myers, M. D. "Judging Qualitative Research in Information Systems: Criteria for Accepting and Rejecting Manuscripts," in J. I. DeGross, G. Ariav, C. Beath, R. Hoyer, and C. Kemerer, *Proceedings of the 16th International Conference on Information Systems*, Amsterdam, 1995, p. 367.

Lee, S.; Goldstein, D. K.; and Guinan, P. J. "Informant Bias in Information Systems Design Team Research," in H-E. Nissen, H. K. Klein, and R. Hirschheim (Eds.), *Information Systems Research: Contemporary Approaches and Emergent Traditions*, Amsterdam: Elsevier, 1990.

Leidner, D. E., and Jarvenpaa, S. L. "The Use of Information Technology to Enhance Management School Education: A Theoretical View," *MIS Quarterly* (19:3), 1995, pp. 265-291.

Majchrzak, A.; Rice, R. E.; Malhotra, A.; King, N.; and Ba, S. "Technology Adaptation: The Case of a Computer-Supported Inter-Organizational Virtual Team," *MIS Quarterly* (24:4), 2000, pp. 569-600.

Mark, G. "Some Challenges Facing Virtually Collocated Teams," in R. Baskerville, J. Stage, and J. I. DeGross (Eds.), *Organizational and Social Perspectives on Information Technology*, Boston: Kluwer Academic Publishers, 2000.

Markus, M. L. "Electronic Mail as the Medium of Managerial Choice," *Organization Science* (5:4), 1994, pp. 502-527.

Mingers, J. "Combining IS Research Methods: Towards a Pluralist Methodology," *Information Systems Research* (12:3), 2001, pp. 240-259.

Mingers, J. "The Paucity of Multimethod Research: A Review of the Information Systems Literature," *Information Systems Journal* (13), 2003, pp. 233-249.

Mukhopadhyay, T.; Kekre, S.; and Kalathur, S. "Business Value of Information Technology: A Study of Electronic Data Interchange," *MIS Quarterly* (19:2), 1995, pp. 137-156.

Nandhakumar, J., and Jones, M. R. "Designing in the Dark: The Changing User-Developer Relationship in Information Systems Development," in K. Kumar and J. I. DeGross (Eds.), *Proceedings of the 18th International Conference on Information Systems*, Atlanta, GA, 1997, pp. 75-87.

Nandhakumar, J., and Jones, M. R. "Too Close for Comfort? Distance and Engagement in Interpretive Information Systems Research," *Information Systems Journal* (7), 1997, pp. 109-131.

Nelson, Kay M. and Cooprider, Jay G. "The Relationship of Software System Flexibility to Software System and Team Performance," in J. I. DeGross, R. P. Bostrom, and D. Robey (Eds.), *Proceedings of the 14th International Conference on Information Systems*, Tampa, FL, 1993, pp. 23-32.

Ngwenyama, O. K., and Lee, A. S. "Communication Richness in Electronic Mail: Critical Social Theory and the Contextuality of Meaning," *MIS Quarterly* (21:2), 1997, pp. 145-167.

Orlikowski, W. "CASE Tools as Organizational Change: Investigating Incremental and Radical Changes in Systems Development," *MIS Quarterly* (17:3), 1993, pp. 309-340.

Orlikowski, W. J., and Baroudi, J. J. "Studying Information Technology in Organizations: Research Approaches and Assumptions," *Information Systems Research* (2:1), 1991, pp. 1-28.

Parker, M., and McHugh, G. "Five Texts in Search of an Author: A Response to John Hassard's 'Multiple Paradigms and Organizational Analysis,'" *Organization Studies* (12:3), 1991, pp. (3), 451-456.

Pickering, A. *Science as Practice and Culture*, London: University of Chicago Press, 1992.

Remenyi, D., and Williams, B. "Some Aspects of Methodology for Research in Information Systems," *Journal of Information Technology* (10), 1995, pp. 191-201.

Remenyi, D.; Williams, B.; Money, A.; and Swartz, E. *Doing Research in Business and Management*, London: Sage Publications, 1998.

Robey, D. "Theories that Explain Contradiction: Accounting for Contradictory Organizational Consequences of Information Technology," in J. I. DeGross, G. Ariav, C. Beath, R. Hoyer, and C. Kemerer, *Proceedings of the 16th International Conference on Information Systems*, Amsterdam, 1995, pp. 55-63.

Sawyer, S. "Analysis by Long Walk: Some Approaches to the Synthesis of Multiple Sources of Evidence," in E. Trauth (Ed.), *Qualitative Research in IS: Issues and Trends*, Hershey, PA: Idea Group Publishing, 2001.

Sawyer, S. "Conceptualizing Information Technology in the Study of Information Systems: Trends and Issues," in E. Wynn, E. A. Whitley, M. Myers, and J. I. DeGross (Eds.), *Global and Organizational Discourse About Information Technology*, Boston: Kluwer Academic Publishers, 2002, pp. 109-132.

Sawyer, S. "Studying Organizational Computing Infrastructures: Multi-Method Approaches," in R. Baskerville, J. Stage, and J. I. DeGross (Eds.), *Organizational and Social Perspectives on Information Technology*, Boston: Kluwer Academic Publishers, 2000, pp. 213-222.

Schultze, U. "A Confessional Account of an Ethnography About Knowledge Work," *MIS Quarterly* (24:1), 2000, pp. 3-41.

Seale, C. *The Quality of Qualitative Research*, London: Sage Publications, 1999.

Sekaran, U. *Research Methods for Business*, New York: Wiley, 1992.

Smithson, S. "Combining Different Approaches," in H-E. Nissen, H. K. Klein, and R. Hirschheim (Eds.), *Information Systems Research: Contemporary Approaches and Emergent Traditions*, Amsterdam: Elsevier, 1990, pp. 365-369.

Strauss, A., and Corbin, J. *Basics of Qualitative Research*, London: Sage Publications, 1990.

Subramani, M., and Walden, E. "The Dot Com Effect: The Impact of E-Commerce Announcements on the Market Value of Firms," in P. De and J. I. DeGross (Eds.), *Proceedings of the 20th International Conference on Information Systems*, Charlotte, NC, 1999, pp. 193-207.

Tashakkori, A., and Teddlie, C. *Mixed Methodology: Combining Qualitative and Quantitative Approaches*, London: Sage Publications, 1998.

Taylor, S., and Bogdan, R. *Introduction to Qualitative Research Methods*, Chichester, UK: Wiley, 1998.

Te'eni, D. "Review: A Cognitive-Affective Model of Organizational Communication for Designing IT," *MIS Quarterly* (25:2), 2001, pp. 251-212.

Toraskar, K. "How Managerial Users Evaluate Their Decision Support: A Grounded Theory Approach," in H-E. Nissen, H. K. Klein, and R. Hirschheim (Eds.), *Information Systems Research: Contemporary Approaches and Emergent Traditions*, Amsterdam: Elsevier, 1990.

Trauth, E. M., and Jessup, L. M. "Understanding Computer-Mediated Discussions: Positivist and Interpretive Analysis of Group Support System Use," *MIS Quarterly* (24:1), 2000, pp. 43-79.

Turner, J. A.; Bikson, T. K.; Lyytinen, K.; Mathiassen, L.; and Orlikowski, W. "Relevance Versus Rigor in Information Systems Research: An Issue of Quality," in H-E. Nissen, H. K. Klein, and R. Hirschheim (Eds.), *Information Systems Research: Contemporary Approaches and Emergent Traditions*, Amsterdam: Elsevier, 1990, pp. 715-745.

Van Maanen, J. "Some Notes on the Importance of Writing in Organization Studies," in J. I. Cash and P. R. Lawrence (Eds.), *The Information Systems Research Challenge: Volume 1*, Boston: Harvard Business School Press, 1989, pp. pp 27-33.

Vessey, I.; Ramesh, V.; and Glass, R. L. Research in Information Systems: An Empirical Study ofDiversity in the Discipline and its Journals," *Journal of Management Information Systems* (19:2), 2002, pp. 129-173.

Vidgen, R., and McMaster, T. "Black Boxes, Non-Human Stakeholders and the Translation of IT through Mediation," in W. J. Orlikowski, G. Walsham, M. R. Jones, and J. I. DeGross (Eds.), *Information Technology and Changes in Organizational Work*, London: Chapman & Hall, 1996.

Watson, W. "Types of Pluralism," *The Monist* (73:3), 1990, pp. 350-365.

Wilson, M., and Howcroft, D. "The Role of Gender in User Resistance and Information Systems Failure," in R. Baskerville, J. Stage, and J. I. DeGross (Eds.), *Organizational and Social Perspectives on Information Technology*, Boston: Kluwer Academic Publishers, 2000.

Yap, A. Y., and Bjørn-Andersen, N. "Energizing the Nexus of Corporate Knowledge: A Portal Toward the Virtual Organization," in R. Hirschheim, M. Newman, and J. I. DeGross (Eds.), *Proceedings of the 19th International Conference on Information Systems*, Helsinki, Finland, 1998, pp. 273-286.

Yin, R. K. *Applications of Case Study Research*, London: Sage Publications, 1993.

Zmud, R. W. "Editor's Comments," *MIS Quarterly* (23:3), 1996, pp. xxxvii-xxxix.

ABOUT THE AUTHOR

Matthew R. Jones is a University Lecturer in Information Management in the Judge Institute of Management and Department of Engineering at the University of Cambridge. His research interests are concerned with the relationship between information systems and social and organizational change, and theoretical and methodological issues in information systems research. He has published widely in these areas. He can be reached at mrj10@cam.ac.uk.

Appendix A

IFIP Best Paper Recipients

2000 Kauffman, Robert. J., and Wood, Charles A. "Follow the Leader? Strategic Pricing of E-Commerce" (extended abstract only)

1999 Subramani, Mani, and Walden, Eric. "The Dot Com Effect: The Impact of E-Commerce Announcements on the Market Value of Firms"

1998 Yap, Alexander Y., and Bjørn-Andersen, Niels. "Energizing the Nexus of Corporate Knowledge: A Portal Towards the Virtual Organization"

1997 Nandhakumar, Joe, and Jones, Matthew. "Designing in the Dark: The Changing User-Developer Relationship in Information Systems Development"

1996 Broadbent, Marianne, Weil, Peter, O'Brien, Tim, and Neo, Boon Siong. "Firm Context and Patterns of IT Infrastructure Capability"

1995 Robey, Dan. "Theories that Explain Contradiction: Accounting for Contradictory Organizational Consequences of Information Technology" (published in *Information Systems Research* (10:2), 1999, pp. 167-185)

1994 Hitt, Lorin, and Brynjolfsson, Eric "The Three Faces of IT Value: Theory and Evidence" (published in *MIS Quarterly* (20:2), 1996, pp. 121-142)

1993 Lee, Allen, "Electronic Mail as a Medium for Rich Communication: An Empirical Investigation Using Hermeneutic Interpretation" (published in *MIS Quarterly* (18:2), 1994, pp. 143-157)

1993 Nelson, Kay M. and Cooprider, Jay G. "The Relationship of Software System Flexibility to Software System and Team Performance" (published as "The Contribution of Shared Knowledge to I/S Group Performance" in *MIS Quarterly* (20:4), 1996, pp. 409-432)

MISQ Paper of the Year Recipients

2002 Jasperson, 'Jon (Sean); Carte, Traci A.; Saunders, Carol S.; Butler, Brian S.; Croes, Henry J. P.; and Zheng, Weijun. "Review: Power and Information Technology Research: A Metatriangulation Review," *MIS Quarterly* (26:4), 2002, pp. 397-459.

2001 Te'eni, Dov. "Review: A Cognitive-Affective Model of Organizational Communication for Designing IT," *MIS Quarterly* (25:2), 2001, pp. 251-212.

2000 Majchrzak, Ann; Rice, Ronald E.; Malhotra, Arvind; King, Nelson; and Ba, Sulin. "Technology Adaptation: The Case of a Computer-Supported Inter-Organizational Virtual Team," *MIS Quarterly* (24:4), 2000, pp. 569-600.

1999 Klein, Heinz K., and Myers, Michael D. "A Set of Principles for Conducting and Evaluating Interpretive Field Studies in Information Systems," *MIS Quarterly* (23:1), 1999, pp. 67-94.

1998 Kumar, Kuldeep; van Dissel, Han G.; and Bielli, Paola. "The Merchant of Prato—Revisited: Toward a Third Rationality of Information Systems," *MIS Quarterly* (22:2), 1998, pp. 199-226.

1997 Ngwenyama, Ojelanki K., and Lee, Allen S. "Communication Richness in Electronic Mail: Critical Social Theory and the Contextuality of Meaning," *MIS Quarterly* (21:2), 1997, pp. 145-167.

1996 Hitt, Lorin M., and Brynjolfsson, Eric. "Productivity, Business Profitability, and Consumer Surplus: Three Different Measures of Information Technology Value," *MIS Quarterly* (20:2), 1996, pp. 121-142.

1995 Mukhopadhyay, Tridas; Kekre, Sunder; and Kalathur, Suresh. "Business Value of Information Technology: A Study of Electronic Data Interchange," *MIS Quarterly* (19:2), 1995, pp. 137-156.

Leidner, Dorothy E., and Jarvenpaa, Sirkka L. "The Use of Information Technology to Enhance Management School Education: A Theoretical View," *MIS Quarterly* (19:3), 1995, pp. 265-291.

1994 Hess, Christopher M., and Kemerer, Chris F. "Computerized Loan Origination Systems: An Industry Case Study of the Electronic Markets Hypothesis," *MIS Quarterly* (18:3), September 1994, pp. 251-275.

1993 Orlikowski, Wanda. "Case Tools as Organizational Change: Investigating Incremental and Radical Changes in Systems Development," *MIS Quarterly* (17:3), September 1993, pp. 309-340.

Empirical Papers at WG8.2 Conferences 1990, 1995, and 2000

1990 Calloway, Linda Jo, and Ariav, Gad. "Developing and Using a Qualitative Methodology to Study Relationships Among Designers and Tools," in H-E. Nissen, H. K. Klein, and R. Hirschheim (Eds.), *Information Systems Research: Contemporary Approaches and Emergent Traditions*, Amsterdam: Elsevier, 1990.

Davies, Lynda J. "Researching the Organizational Culture Contexts of Information Systems Strategy: A Case Study of the British Army," in H-E. Nissen, H. K. Klein, and R. Hirschheim (Eds.), *Information Systems Research: Contemporary Approaches and Emergent Traditions*, Amsterdam: Elsevier, 1990.

Elam, Joyce; Walz; Diane R.; Curtis, Bill; and Krasner, Herb. "Measuring Group Processes in Software Design Teams," in H-E. Nissen, H. K. Klein, and R. Hirschheim (Eds.), *Information Systems Research: Contemporary Approaches and Emergent Traditions*, Amsterdam: Elsevier, 1990.

Lee, Soonchul; Goldstein, David K.; and Guinan, Patricia J. "Informant Bias in Information Systems Design Team Research," in H-E. Nissen, H. K. Klein, and R. Hirschheim (Eds.), *Information Systems Research: Contemporary Approaches and Emergent Traditions*, Amsterdam: Elsevier, 1990.

Toraskar, Kranti. "How Managerial Users Evaluate Their Decision Support: A Grounded Theory Approach," in H-E. Nissen, H. K. Klein, and R. Hirschheim (Eds.), *Information Systems Research: Contemporary Approaches and Emergent Traditions*, Amsterdam: Elsevier, 1990.

1995 Ciborra, Claudio; Patriotta, Gerardo; and Erlicher, Luisella. "Disassembling Frames on the Assembly Line: The Theory and Practice of the New Division of Learning in Advanced Manufacturing," in W. J. Orlikowski, G. Walsham, M. R. Jones, and J. I. DeGross (Eds.), *Information Technology and Changes in Organizational Work*, London: Chapman & Hall, 1996.

Hamilton, Doug, and Atchison, Martin. The COMIS Plan: IT-Mediated Business Reengineering in Telecom Australia During the 1960s," in W. J. Orlikowski, G. Walsham, M. R. Jones, and J. I. DeGross (Eds.), *Information Technology and Changes in Organizational Work*, London: Chapman & Hall, 1996.

Gallivan, Michael J. "Contradictions Among Stakeholder Assessments of a Radical Change Initiative: A Cognitive Frames Analysis," in W. J. Orlikowski, G. Walsham, M. R. Jones, and J. I. DeGross (Eds.), *Information Technology and Changes in Organizational Work*, London: Chapman & Hall, 1996.

Gasson, Susan, and Holland, Niki. "The Nature and Processes of IT-Related Change," in W. J. Orlikowski, G. Walsham, M. R. Jones, and J. I. DeGross (Eds.), *Information Technology and Changes in Organizational Work*, London: Chapman & Hall, 1996.

Guimaraes, Tor. "Assessing Employee Turnover Intentions Before/After BPR," in W. J. Orlikowski, G. Walsham, M. R. Jones, and J. I. DeGross (Eds.), *Information Technology and Changes in Organizational Work*, London: Chapman & Hall, 1996.

Vidgen, Richard, and McMaster, Tom. "Black Boxes, Non-Human Stakeholders and the Translation of IT through Mediation," in W. J. Orlikowski, G. Walsham, M. R. Jones, and J. I. DeGross (Eds.), *Information Technology and Changes in Organizational Work*, London: Chapman & Hall, 1996.

2000 Aanestad, Margunn, and Hanseth, Ole. "Implementing Open Network Technologies in Complex Work Practices: A Case from Telemedicine," in R. Baskerville, J. Stage, and J. I. DeGross (Eds.), *Organizational and Social Perspectives on Information Technology*, Boston: Kluwer Academic Publishers, 2000.

Grunden, Kerstin. "MOA-S: A Scenario Model for Integrating Work Organization Aspects into the Design Process of CSCW Systems," in R. Baskerville, J. Stage, and J. I. DeGross (Eds.), *Organizational and Social Perspectives on Information Technology*, Boston: Kluwer Academic Publishers, 2000.

Mark, Gloria. "Some Challenges Facing Virtually Collocated Teams," in R. Baskerville, J. Stage, and J. I. DeGross (Eds.), *Organizational and Social Perspectives on Information Technology*, Boston: Kluwer Academic Publishers, 2000.

Wilson, Melanie, and Howcroft, Debra. "The Role of Gender in User Resistance and Information Systems Failure," in R. Baskerville, J. Stage, and J. I. DeGross (Eds.), *Organizational and Social Perspectives on Information Technology*, Boston: Kluwer Academic Publishers, 2000.

9 THE CRISIS OF RELEVANCE AND THE RELEVANCE OF CRISIS: Renegotiating Critique in Information Systems Scholarship

Teresa Marcon
University of Western Ontario

Mike Chiasson
University of Calgary

Abhijit Gopal
University of Western Ontario

Abstract Information systems as a discipline has recently been under pressure to justify its existence as a core subject within the management curriculum. There has also been recent pressure about the relevance of the IS research agenda. These are pressures felt at the more general level of business education as well, and calls have been made for business scholars to take a more holistic approach to scholarship as well as to make more explicit links to the practice of business. We take the position in this paper that the pressures can be addressed in one way by renegotiating the notion of scholarly critique. Specifically, we re-connect the idea of critique to that of crisis and attempt to show how crisis has the potential to reengage the IS scholar with praxis and help bring the often disparate projects of research, teaching, and consulting into an integrated scholarly enterprise.

Keywords: Critique, crisis, praxis, relevance

1 INTRODUCTION

The information systems discipline appears poised at a critical juncture. Given recent market trends and a back-to-basics mentality, most IS scholars are familiar with the skeptical attitude in business schools, among both colleagues and students, toward the discipline and its relevance to business education (Avison 2003). Coupled with recent angst within the field about the questionable relevance of IS research,[1] we might venture to say that the discipline faces a crisis of sorts. The referent world which defines us, the world of business practice, appears not to feel a reciprocal attachment. It certainly consumes in copious quantities the kinds of services we offer—education, research, consulting—but often does not obtain these from us. All in all, the situation appears to merit more than passing attention from the IS community at this time.

Why does the skepticism arise and why is it restricted to the IS discipline? Richard Mowday's (1997) presidential address to the 1996 Academy of Management meeting suggests that the problem applies to business scholarship in general. According to the Faculty Leadership Task Force of the AACSB (cited in Mowday 1997), business faculty lack real world experience, are slow to adopt new technologies, and are resistant to change. The result is a perception that business academics are increasingly irrelevant to the business world.

The practice of business is clearly the *raison d'etre* of both business and IS scholarship. As Keen (1991) points out, IS research "is intended to influence action in some domain." Indeed, following Mowday, we might say that business (and IS) scholarship is practically oriented in all its guises, including teaching, research, and consulting. The unique confluence of IS with rapid changes in technology and business practices clearly makes a lack of practical engagement untenable. But how do we address this issue?

We believe that a familiar device, made unfamiliar, has considerable potential for us in this regard: scholarly critique. Critique has developed increasingly specialized connotations in our field and a recovery of its wider meanings might help us reengage with the practical world. Moreover, we believe that critique, conceived in this light, allows us to address in a holistic way our entire scholarly enterprise across the increasingly separated domains of research, teaching, and consulting. What we suggest in the following pages is an altered frame of mind, a rethinking, that might help us address the crisis of relevance we now face.

2 THE MEANING OF CRITIQUE IN IS

Our interest lies in examining the ways in which critique is presently understood in IS and the ways in which other possible understandings have been obscured. We attempt to disturb the obviousness of current conceptualizations of critique to encourage a rethinking that expands the range of possibilities for scholarly critique and to bolster the call for a widespread engagement with praxis in IS scholarship. We describe two principal ways of understanding critique that currently dominate the IS field: critical social theory (CST) and methodological critique. We explore these approaches to critique not to challenge their legitimacy or value but to uncover alternative ways of understanding and engaging in critique.

---

[1]We are not revisiting the rigor-relevance debate. We do not see the two terms here as being in opposition. We are pointing only to the clear importance that the idea of relevance to practice has within our field.

Critique in IS is most often associated with the theoretical approaches of the Frankfurt School of Critical Theory and particularly the work of Jürgen Habermas (Brooke 2002). This tradition bears the distinctive mark of the Enlightenment, privileging the rise of reason over metaphysics and embodying the modernist belief that reason will light the way to a more just society (Collins 1994). It yields a critique of human practices formulated on the basis of an ideological position that reflects the Enlightenment values of rationality and equality and the goal of promoting social change through emancipation. The goal of emancipation, in particular, is central to current conceptions of CST in IS (Boudreau 1997; Howcroft and Truex 2001). Indeed, the CST movement, which relies heavily on Habermasian perspectives, has largely become synonymous with the critical in IS.

A second way in which the term *critical* is commonly understood is as methodological critique. Critique in this sense encompasses the scholarly practice of systematically and objectively evaluating existing research to build upon the work of previous researchers. It does so by identifying opportunities for expansion or refinement at the theoretical or methodological level. The critical review of existing literature sets the stage for further exploration and provides the motivation for new research. Additionally, through discussion of potential problems and limitations, scholars are encouraged to cast a critical eye on their own efforts, reflecting on methodological limitations and exploring alternative explanations. Commonly regarded as good scholarship, such critical reflections and the practice of writing critical reviews of the literature are skills that are considered central to IS scholarship (e.g., Webster and Watson 2002).

While methodological critique retains an important role in scholarly practice,[2] members of the IS community have recently attempted to expand the common understanding of critique beyond its traditional associations. In the title of the leading article of the June 2002 issue of the *Journal of Information Technology* on critical IS research, Carole Brooke poses the question: "What does it mean to be critical in IS research?" Brooke argues that, until recently, our interest in critique has been limited to Habermas' theory of communicative action. She warns us against becoming "locked into a Habermasian discourse" and proposes that "IS research must continue to push beyond this thinking in order to enrich our work" (p. 49). Brooke argues for a Foucauldian perspective as one possible avenue to critique. Similarly, Doolin and Lowe (2002) conclude that "the definition of 'critical' used thus far in IS research is too limiting" (p. 69). They make a case for recasting this notion in broader terms as an act of revelation: "to reveal is to critique" (p. 74). They point to the critical potential of actor-network theory (e.g., Latour 1999) which, in exposing the contingency of the world, reveals "how things could have been otherwise" (p. 75). These efforts broaden the accepted view of critique by encompassing post-structuralist theories that eschew an explicitly emancipative aim or the articulation of a particular ideological position as an alternative to the current order.

---

[2]This is not to say that methodological critique could not be questioned as leading, for example, to rather formulaic applications of critique. However, a critique of methodological critique is beyond the scope of this paper.

Expanding the critical playing field along such lines might be seen as a radical departure from the explicitly transformational intent of CST (Boudreau 1997; Howcroft and Truex 2001). Yet, this is not a new idea. Indeed, a decade before Brooke, Lyytinen (1992) proposed a need to supplement a Habermasian approach in IS with the work of other critical theorists (p. 176) such as Foucault and Giddens. Within the neighboring field of organization studies there is also evidence of a movement toward a more encompassing view of critical research. Kincheloe and McLaren (2000) and Alvesson and Deetz (1996), for example, have sketched the continuity of thought and purpose, and also the differences, between Frankfurt School theories and critical discourses centered on feminist, post-structuralist, and postmodernist positions. This broadening of the concept of the critical pursues similarities and complementarities among a range of approaches from neo-Marxism to deconstruction.

Efforts to push beyond current thinking and to cast critique in a different light within IS and related fields are certainly valuable in our view. But has this questioning gone far enough? And have these voices been heard? Despite the emergence and reemergence of calls for expanding the theoretical bases for critical research and despite the strength of the theoretical apparatuses that have been brought to bear, the idea of the critical in IS has remained in large measure tied to the central tenets of the Frankfurt School (e.g., Orlikowski and Baroudi 1991). Perhaps because of the central position it has historically occupied in the IS field and particularly in the CST scholarly community, critical theory of the Frankfurt School variety has become the standard against which other approaches are measured and deemed as either *critical* or *not critical.* Efforts to expand the range of theoretical approaches suitable for critical IS research often measure themselves against this standard and demonstrate congruent, if not identical, aims. For example, Brooke explores how certain post-structuralist approaches fit into a broadly emancipative project in her argument for the relevance of a Foucauldian approach to the project of critique.

While such considerations are important and even necessary, tethering the idea of critique to particular theoretical positions, even if to an expanding list, may stand in the way of a deeper appreciation of the idea of the critical. Recent arguments in favor of post-structuralist approaches to critique run the risk of producing yet another static division that might well lull us into some unexamined conclusions. Is it the case that only some theoretical choices afford the opportunity for critical engagement (e.g., Walsham 1993)? Are critical theory, post-structuralist, or postmodernist methodologies critical, whereas interpretivism and functionalism are not? We believe that such efforts at anchoring the idea of critique in selected methodologies and delimiting what *counts as* critique tend to obscure possibilities and limit further engagement with critique as a broader orientation to IS scholarship.

Whether a theoretical approach or a particular piece of research is critical is a question that merits consideration in its own right. Marcon and Gopal (2003), for example, have argued for the critical potential of ethnomethodology (Garfinkel 1967) based on how this approach "seeks to reveal the way in which *taken for granted* social practices maintain the appearance of things" (Rawls 2002, p. 54). Despite this critical potential, ethnomethodology is generally regarded as falling under the umbrella of interpretivism (Burrell and Morgan 1979). While the classification of ethnomethodology within the interpretative tradition is open to challenge (e.g., Lemert 1979), the case of ethnomethodology illustrates the danger in attempting to demarcate the territory of the critical by relying on broad classifications of research traditions such as critical theory[3] or poststructuralism. While such labels may be useful at times, we believe it is important to retain an awareness of the diversity of positions they encompass and of the differences and similarities between research approaches that are subsumed or lost in such broad divisions. Limiting the possibility of engaging in critique based on particular theoretical approaches or prescribed definitions of what counts as critique (e.g., Howcroft and Truex 2001) turns our attention away from a broader conceptualization of critique as a common orientation toward IS scholarship.

Similarly, when critique is restricted to the identification of theoretical lacunae and opportunities for methodological improvement in existing research, as is perhaps more common in functionalist work, we risk disengaging from the idea of the critical as a common praxical project concerned with human activities and with knowledge of the practical world. To consider critique merely as an aspect of sound scholarship that allows us to build cumulative traditions (Keen 1980) focuses our gaze inward, and overlooks the potential for turning our critical gaze outward toward the human practices around IS that find expression in organizations, and our own crucially important reflexive engagement with this world.

Rather than adding to the list of theories that might be considered critical or limiting the critical to methodological reflection, our own project follows a different path. We attempt to expand the concept of critique by exploring its broader meanings. We seek to rediscover broader understandings of critique that have informed critical scholarly work across research, teaching, and consulting. Such understandings allow for an ongoing and widespread engagement with the everyday world of human practice, raising the potential to meaningfully inform the practices of the people for whom IS research is produced and those of our own scholarly community. In particular, we explore the connection between critique and *crisis* and attempt to show how the production of crises through critique may serve to guide our scholarly efforts toward a common, reflexive, and intellectually fruitful engagement with the world of practice.

3 CRITIQUING THE MEANING OF CRITIQUE

In our discussion, we have explored the manner in which the term *critical* is understood in IS, privileging its denotative meanings within the field. Although these conceptions of the critical may have the appearance of the obvious or natural, denotation is only one way in which language functions to give shape to interpretation. As Barthes (1974) notes,

> denotation is not the first meaning, but pretends to be so; under this illusion, it is ultimately no more than the last of the connotations (the one which seems both to establish and close the reading), the superior myth by which the text pretends to return to the nature of language, to language as nature (p. 9).

---

[3]We could argue that the name critical theory itself is a black hole—it sucks all critical light into it through its label.

What appear as the obvious meanings of critique in IS reflect the specialized ways in which the term has come to be understood in the field. Such meanings can be overlaid with a web of connotations, those relations and connections within and between texts that form "nebulae of signifieds" (p. 8), allowing for a spreading out and a broadening of interpretation of what it means to engage in critique.

To expand the idea of the critical, we begin with notions of critique in academic circles and in popular culture. Consider, for example, the association of the practice of critique with the literary critic, the social critic, or the film critic. In *Keywords: A Vocabulary of Culture and Society*, Williams (1985) notes that criticism is related to the Greek word for "judge" and thus has carried the primary meaning of "passing judgment." In particular areas of art and literature, Williams argues that the term has relegated the critic to the role of an expert with the ability or *taste* to differentiate good from bad. This restricts critique to the realm of opinion or acculturated taste, associated with necessarily individual values and ethics. In considering the usage of the term critical across disciplines and over time, Williams also notes that

> judgment depended, of course, on the social confidence of a class and later a profession. The confidence was variously specified, originally as learning or scholarship, later as cultivation and taste, later still as sensibility….At various stages, forms of this confidence have broken down, and especially in [the 20th century], attempts have been made to replace it by objective...methodologies, providing another kind of basis for judgment. What has not been questioned is the assumption of "authoritative judgment."

This latter movement provides a link to the practice of methodological critique in the social sciences and points to the implicit authority granted to those who strive for an objective scholarship in a society that has in large part placed its faith in science and relegated critique to local and subjective opinion.[4]

Dictionary definitions of the term critical encompass a plurality of popular and specialized meanings. As Barthes suggests, these definitions reflect both denotative and connotative aspects. *The American Heritage Dictionary of the English Language* (2000), for example, defines the term critical as

1. Inclined to judge severely and find fault.
2. Characterized by careful, exact evaluation and judgment: a critical reading.
3. Of, relating to, or characteristic of critics or criticism: critical acclaim; a critical analysis of Melville's writings.
4. Forming or having the nature of a turning point; crucial or decisive: a critical point in the campaign.
5. a. Of or relating to a medical crisis: an illness at the critical stage.

---

[4]This perceived difference between ideologically based and methodological (or objective) critique has been challenged in epistemological reflections which collapse the differences between fact and value (e.g., Bohman 1993).

   b. Being or relating to a grave physical condition especially of a patient.
6. Indispensable; essential: a critical element of the plan; a second income that is critical to the family's well-being.
7. Being in or verging on a state of crisis or emergency: a critical shortage of food.
8. Fraught with danger or risk; perilous.

The emphasis on critique as judgment, and particularly negative judgment, is evident in the first of these definitions which relates criticism to finding fault, reflecting the "oppositional" approach sometimes associated with the project of social critique (e.g., Grey and Mitev 1995, cited in Burrell 2001, p. 14). The negative connotations of the term critique also become evident when we consider antonyms in popular usage which oppose critique to verbs like encourage, flatter, and praise (*Merriam-Webster OnLine* 2003). Yet, another set of antonyms opposes the critical to that which is "cursory, shallow, superficial" (*Merriam-Webster OnLine* 2003), shifting the focus to the potential for depth and discernment that is contained in critical reflection.

Other ways of understanding the term critical point to the most salient, decisive, or urgent moments in particular events, evoking images of conflict, harm and impending danger (definitions 4 to 8 above). A recurrent idea in definitions of the term critical is the turning point, a moment when someone's fate hangs in the balance and outcomes possibly involving life and death are decided (definitions 4 and 5 in particular). Arriving at a turning point suggests a progression toward a moment of danger, a point where resources are low and options few as in the case of a state of emergency (definition 7) or where necessity rather than choice sets the terms for action (definition 6). Such meanings of the word critical highlight a different set of connotations which unite the idea of critique with a companion notion: crisis.

While crisis is inevitably only one of the many possible connotations of the term critical, we single out this aspect to explore its potential as a device to rethink critique in IS. Critique and crisis stood in a proximal relation to each other in ancient Greece when the term crisis encompassed both terms, meaning "discrimination and dispute, but also decision, in the sense of final judgment or appraisal" (Koselleck 1988, p. 103). The word crisis derives from the Greek word *krinein*—to separate, decide and judge (*The Houghton Mifflin Canadian Dictionary* 1982).[5] Over time, critique and crisis did separate, acquiring distinct meanings as usage evolved and was transformed at the intersections between fields. Crisis assumed specialized meanings, for example in medicine, where it came to denote the turning point in a serious illness. Criticism, a practice originally associated with revelation in religious texts, was transported into

---

[5]In evoking an etymological thread, it is not our intent to return to the root of the word crisis in order to hone in on the essential or original meaning of critique. Theorists from Saussure (Culler 1976) to Derrida (1972) and Foucault (1970) have taught us that the meaning of words is neither stable nor definite, but rather embedded in systems of opposition and relations that are subject to constant shifts, translations, and reinterpretations over time. Rather, our purpose is to *reunite* critique and crisis as a way of thinking about the practice of critique and bring a unity of purpose to all aspects of scholarly practice.

other disciplines and enrolled in support of reason in the early 18th century (for an excellent history, see Koselleck 1988). Current scholarship across the social sciences tends to distinguish critique from crisis. Whereas crisis, the "economic crisis" or the "crisis of capitalism" (O'Connor 1987), for example, has acquired a largely autonomous or even deterministic status that suggests the inevitability of natural disasters (e.g., a hurricane), critique has been largely relegated to the realm of the subjective, ethical, or moral, as in the practice of ideological critique, or to methodological critique as objective judgment. Resonances with such understandings of the critical can be found in the broader societal realm where critique is often associated with opinion (Williams 1985) and sometimes also technical skill, as in critical thinking (Thayer-Bacon 1998).

The connection between critique and crisis implicit in current usage surfaces in more explicit form in the writings of Continental philosophers. Continental philosophy is a rich and complex tradition that spans several centuries and distinct strands of thought that cannot comfortably be collapsed into a unified position (Critchley 2001). As an undercurrent to this tradition, we can conceive of Continental philosophy as a style of thinking that, in its deep concern with praxis, relates critique to the present "through the consciousness of crisis" (Kompridis 2000, p. 40). The theme of crisis assumes various forms in the work of Continental philosophers from German idealism to the present: for example, in Marx's crisis of the capitalist state, Husserl's crisis of the European sciences, Heidegger's forgetfulness of Being, and Foucault's and Derrida's crises of the human sciences. Within this seemly disparate assortment of theoretical perspectives, from historical materialism to phenomenology and deconstruction, the production of crisis through critique marks the path to the touchstone of Continental philosophy: praxis (Critchley 2001).[6]

In praxis lies the promise of an IS scholarship that addresses the needs and interests of the academic community and of those who attempt to understand and use information systems and information technologies in action. The connection between critique and crisis which we have attempted to make evident in exploring the history and connotations of the term critical may serve as a device to help us rethink critique in the context of IS scholarship and to renew our connection with praxis. Following the tradition in Continental philosophy that set the philosopher to the task of "promot[ing] a reflective awareness of the present as being in crisis" (Critchley 2001, p. 73), we suggest that, through critique, the work of IS researchers might aim to engender crisis in its various audiences (including itself), a sense of "instability or even discomfort that is a distinctive feature of genuine intellectual undertakings" (Kingwell 2002, p. 7). In the following section we explore the relevance of crisis to the IS academic enterprise and the unity of purpose that it may engender across the academic practices of research, education, and consulting.

---

[6]The term *praxis* has been associated with a variety of meanings from the time of the Greeks to the present (for a brief history, see, Bottomore 1991). We use the term in its broadest sense of signaling a (not indifferent) concern with the realm of ongoing practice and the human activities in which ordinary people engage.

4 THE RELEVANCE OF CRISIS

What is the relevance of engendering crisis through critique? In considering this question, we must account for the role of and the benefits to both IS researchers (the authors) and their audiences (the readers).[7]

In focusing on the production of crisis, academics seek to foster an *awakening* among those for whom they write: managers, employees, students, and members of their scholarly community. When such attempts succeed in producing crisis in the audience, they bring into play the possibility of genuine engagement and the opportunity for a fruitful exchange. Awakening goes beyond attempts to present new perspectives that "stimulate critical thinking" or "restructure the mental models managers [and students] apply in their practice" (Benbasat and Zmud 1999, p. 5), although such aims are part of the process. Much has also been said about the need to communicate research findings through appropriate publication channels and teaching materials that are accessible and of interest to practitioner audiences (e.g., Benbasat and Zmud 1999; Lyytinen 1999). Although such suggestions are certainly valuable and even necessary, engendering crisis shifts the focus from research and teaching that can be communicated and implemented to research and teaching that are compelling because of their ability to disturb the obviousness of *what everybody knows* (Garfinkel 1967). Crises create a sense of immediacy, urgency, and even peril that admits the possibility of insight, of an "open[ing] to light" (Heidegger 1993).

The possibility of insight applies also to IS academics. As we engage our audiences through the crises we attempt to produce, we need to remain open to the possibility of being informed by them. Often, a combination of dialogues across research, teaching, and consulting allows us to identify enduring issues. The compelling call of crisis fosters our own engagement with the problems we choose to tackle and the audiences we address.

Beyond the possibility of a genuine dialogue, the production of crisis may serve as a device to orient the efforts of IS academics in their research, teaching, and consulting toward areas of practice of significant interest to practitioners, students, and fellow academics, being at the center of their concern, anxiety, and discomfort and thus perhaps most in need of and in readiness for critical attention. Considering the ability to engender crises in their audiences through writing, teaching, and consulting may guide IS researchers toward the "deep, substantive, prototypical problems" (Weber 2003, p. iv) faced by academics, managers, or users of information systems and provide a process to tackle what Weber in his recent *MIS Quarterly* editorial statement has labeled a "dark art" (p. vi): the problem of choosing research problems.

---

[7]As Michel de Certeau (1984) has aptly noted, the reader is not a passive recipient. The distinction between author and reader could easily be dissolved, given the multiple connections between researchers, the people they study, teach or interact with in their consulting role. Consider, for example, the extent to which subjects in a research setting inform and shape the writings of academics. We merely draw a distinction between authors and readers as an analytical device to assist us in exploring the various facets of the practice of producing crises through critique.

Perhaps most importantly, when critique engenders crisis, the awareness of crisis compels IS scholars to reflect on the directions and outcomes of their intervention in the world. When uncoupled from the notion of crisis, critique runs the risk of being unreflexive at the level of engagement with the wider world in which it is embedded. Walzer (2002) has argued that "criticism is most properly the work of 'insiders,' men and women mindful of and committed to the society whose policies or practices they call into question—who *care about* what happens to it" (p. xi). The "caring critic" (p. xii) is someone who assumes the stance of the independent observer (to the extent to which this is ever possible) while avoiding indifference (Lynch 1997) and maintaining a genuine and sustained interest in the successes and failures of the community s/he studies and seeks to inform.

A critical scholarly practice that seeks to engender crisis does not come to an end with the publication of the research report or the delivery of a lesson in the classroom. Rather, when critique engenders crisis and leads to practical outcomes, it calls for an evaluation of its effects (Davenport and Markus 1999) and also of the researcher's and instructor's practices. As members of the IS academic community, we can learn from our successes and also our failures. If, through the production of crises, we succeed in our efforts to alter practice and foster new understanding and practices through teaching, we must attend not only to the desired effects but also to the unintended consequences (Giddens 1984) of our interventions. Similarly, failure to engender a crisis should lead us to question our assumptions, our choice of problems, and the manner in which we have framed and communicated these: "critique entails the mutual transformation of both subject and object—entails changing oneself as well as the world. By engaging in critique we are engaging in self-critique" (Kompridis 2000 p. 43).

Seeking to engender crisis is relevant also to critical dialogue within the IS research community. Indeed, the information systems literature contains several examples of critical commentary that have left the IS community feeling discomfort and doubt, if not yet in crisis. Among such works are classic pieces from the early years of IS scholarship, for example, Ackoff's (1967) critical reflections on the manner in which management "misinformation" systems are put to use in the world of practice and Churchman and Schainblatt's (1964) detailed exploration of the unexamined assumptions that have (mis)guided attempted collaboration between MIS researchers and practitioners. More recently, Orlikowski and Iacono (2000), in reminding the IS community that the IT artifact—purportedly at the core of the discipline—is rarely engaged by scholars, have indeed succeeded in engendering a crisis in the field, as suggested by the inclusion of a discussion panel on theorizing the IT artifact in the 2002 International Conference on Information Systems (Boland 2002).[8]

We might also suggest that the practice of engendering crisis through critique in empirical research has in some measure coexisted, although in relative obscurity, with

---

[8]Contrast this form of critical engagement with a body of research with the practice of the critical review of literature, which has as its principal target the movement toward a cumulative tradition. The former is a form of critical reflection that is not uncommon in the information systems literature and yet rarely labeled as critical, another indication of the restricted manner in which critique is understood in the IS community.

dominant conceptions of critique in IS. For example, the theme of crisis is salient in Montealegre and Keil's (2000) in-depth case study of the de-escalation of a highly visible and ambitious IT project at the Denver International Airport. The crisis playing out in the lifeworld provides Montealegre and Keil with a compelling setting with substantial economic and political significance and an altogether good story, full of twists and turns and practical lessons. Yet, a second crisis is also interwoven in their paper which frames de-escalation in terms of a human "commitment to a failing course of action" (p. 418). IT failure unfolds in the context of a crisis in the life of the manager who is unwillingly caught up in a series of decisions and external contingencies moving toward an unfavorable outcome. Although, in our view, Montealegre and Keil do not make extensive use of crisis but limit their concluding remarks to promoting a form of reflexive monitoring among managers to avoid a sunk-cost syndrome, their work stands as an example of a critique of human practices that contains within it an inherent pathos and dramatic tension. This lived crisis calls both authors and readers to a sympathetic engagement with the participants in the situation.

Across the multiple sites of contact between academics and their audiences, the production of crises has the potential to open up debate, discussion, and the exchange of ideas. Most importantly, we see crisis as a device that, through its compelling call to reflexivity, may promote an academic practice that maintains a continuous engagement with the world in which it takes place. Crisis fosters a reflexive scholarship that considers its own achievements, not only according to the methodological and publication standards of the academy (e.g., Applegate and King 1999), but also in light of its accountability to the participants in the research setting, the consulting relationship, or the classroom and in terms of the practical outcomes it engenders or fails to achieve.

It is, of course, entirely possible to set out to engender crises without an enduring commitment to a continuing engagement with praxis. Yet, a genuinely reflexive and caring critique based on the production of crises would seem to ask more of us. A reflexive and caring critical scholarship demands a commitment to work toward understanding and the articulation of alternatives or solutions (although perhaps temporary, partial, local, and imperfect ones). This commitment cannot be comfortably circumscribed, including particular activities such as research, while excluding others, such as teaching. A commitment to praxis through a reflexive, critical orientation demands that we attend to all of our activities with a unity of purpose and engagement that derives its energy from a genuine concern. Beyond mere words, integration across all of our activities keeps our gaze focused on the world of practice and is a manifestation of a genuine and enduring commitment to making a difference.

5 CONCLUSIONS

There are two senses of crisis in the title of our paper. In the first sense, the crisis of relevance, we encounter crisis as an unintended consequence: the appearance of a crisis in the IS community that it had not set out to produce—even if its appearance has everything to do with how the community has conducted itself in the past. To help address that crisis, we have turned our attention in this paper to crisis itself—the second sense in our title, the relevance of crisis—as an intentional means of informing and influencing praxis. We (re)unite crisis with the established scholarly practice of critique to invigorate that practice and to recover its potential as a means to provoke change. The premise we adopt in juxtaposing these two senses of crisis is that while the first sense is the one that dominates the use of the word, the recovery of crisis (in the second sense) as a companion notion to critique might arm our scholarly endeavors with the potential for real—and ongoing—change.

We also see crisis as the means to evaluate critique, to assess its praxical influence. Seen in this way, it should be evident that for an account to "be critical" it need not adhere to particular epistemological, ontological, or axiological tenets as much as it needs to be intentional (in Husserl's sense), directed to some definable end. The relevance sought by the IS community flows from the world of practice and we offer critique/crisis as a device that establishes an umbilical connection to that world, a means to evaluate whether our critical commentaries on practice achieve their intended consequences while allowing us to reflexively monitor and reorient the critique that falls short of its mark in a dialogical (and dialectical) relationship with that world.

Here we must confront what would amount to a most embarrassing critique of our own position, if it were to hold true: that what we have posited is nothing more than common sense, that scholarly IS critique is already result-oriented and that the means to achieve results through critique are amply clear; in other words, that our position contributes little if anything. We offer three reasons, based on what we note above, why this argument might not hold. First, as the notion of crisis within our field is decidedly one-sided, connoting the unintended consequence, the idea of intentional crisis emanating from critique is rarely, if ever, articulated. One might argue that it does not need articulation, that crisis is always already implicit in critique, that critique is by definition provocative. While we can certainly agree that critique is geared toward disturbing the status quo and inciting change, there is little evidence within the IS community that such disturbances within the field of practice are monitored, evaluated, and aligned; our aim has been to visit this very issue and propose a tangible means—crisis—to make the implicit *explicit* as a means to reflexively monitor the intended consequences of critique.

Second, we have tried to show the relevance of critique—in tandem with the outcome orientation of crisis—to scholarship in a broad sense rather than restricted to research alone; that is to say, as a coordinated means to address research, teaching, and consulting activities by imbuing all of them with a common result orientation. The crisis of relevance in our field is not confined to research alone and any meaningful attempt to confront the crisis will need an initiative appropriately coordinated among these three activities.

Third, and perhaps most controversially, we have tried to show how critique coupled with crisis can take critical work beyond the confines of CST in the IS field. This is not to say that CST is less than suited to carrying the banner of critique or even that critique from this direction has failed to achieve its purpose. Indeed, critical theorists might well be in the best position to incorporate a crisis-orientation in their repertoire, given their inherently reflexive stance and their attention to history. What we are advocating here is that the larger IS field consider critique—the very device that the critical theorists realize is invaluable—and its companion notion, crisis, for the value they offer within the firmament of business academia.

We see the above three factors both as pointers to what we hope to have accomplished in this paper and as our means of distinguishing ourselves from an orientation based exclusively on CST. By drawing on the idea of crisis, we have tried to bring back into play within the critical project an orientation that has long informed the Continental philosophical tradition in general (Critchley, 2002). This orientation is, in fact, evident in other fields—the work of Shiva (1993) and Bourdieu (1990) provide ready examples of scholars who have used a variety of points of engagement to sustain a critical dialogue with their focal worlds. These examples point to the possibilities in our own field to forge a more enduring and even symbiotic connection to the world of praxis on which we thrive.

Where and how, then, can crises be fruitfully engendered? We offer, below, some broad suggestions that we believe might start us in the direction of bringing critique/crisis into play in our academic practice.

- Taking a holistic view of our scholarship rather than viewing research, teaching, and consulting as compartmentalized activities that do not inform each other. This allows us to seek a synthesis that begins with a rich and fertile ground of exposure to the real life concerns of practitioners and students.
- Learning from our failures. We might argue that our inability to engender a crisis is almost as instructive as the successful production of a crisis. Failure to touch a nerve alerts us to a lack of connection, a point of disjuncture which might require reframing or reconsideration of our choice of problem.
- Deliberately using crisis as a rhetorical device, a means of drawing the reader into a dialogue in which a genuine two-way exchange of ideas might take place and new insights arise. This is applicable to the classroom, to the conversation with a client, and to written discourse.
- Committing to a long term interest in the organizations we study and the problems we try to solve. Maintaining an ongoing engagement after the initial research project is over, for example, can alert us to unintended consequences which may evolve over time.
- Engaging in critical dialogue in our journals though a crisis engendering critique that calls us collectively to reflect on our assumptions and practices. Such reflection could be facilitated by journal policies that invite responses and debate after a crisis engendering article is published.

We do not intend these suggestions as prescriptions but rather as ideas that might help spark consideration of the ways in which different contexts of praxis might be approached when we adopt a result-oriented critical perspective, as well as the ways in which we might meaningfully conduct the conversations that follow the crises we are able to engender. To illustrate the kind of crisis and its follow-up to which we refer, we return to the work of Orlikowski and Iacono (2000), who were able to create crisis within the IS community by pointing out how the IT artifact had become obscured in research in the field. They made members of the community look up and pay attention to the profound contradiction brewing in their midst: the drift away from what they had set out to study in the first place. They then participated in conversations (at ICIS 2002) with community members from diverse subfields to consider ways in which academic practice could be changed. We will let their effort stand as a good if (inevitably) imprecise example of the critique/crisis nexus that we have sought to articulate.

REFERENCES

Ackoff, R. L. "Management Misinformation Systems," *Management Science* (14:4), December 1967, pp. 147-155.

Alvesson, M., and Deetz, S. "Critical Theory and Postmodernism Approaches to Organizational Studies," in S. R. Clegg, C. Hardy, and W. R. Nord (Eds.), *Handbook of Organization Studies,* Thousand Oaks, CA: Sage Publications, 1996, pp. 191-217.

American Heritage Dictionary of the English Language (4th Ed.). Houghton Mifflin Company, 2000 (accessed online at http://www.dictionary.com, January 10, 2004).

Applegate, L. M., and King, J. L. "Rigor and Relevance: Careers on the Line," *MIS Quarterly* (23:1), March 1999, pp. 17-18.

Avison, D. "Information Systems in the MBA Curriculum: An International Perspective," *Communications of the Association for Information Systems* (11), 2003, pp. 117-127.

Barthes, R. *S/Z*, R. Miller (trans.), New York: Hill and Wang, 1974.

Benbasat, I., and Zmud, R. W. "Empirical Research in Information Systems: The Practice of Relevance," *MIS Quarterly* (23:1), 1999, pp. 3-16.

Bohman, J. *New Philosophy of Social Science: Problems of Indeterminancy*, Cambridge, MA: The MIT Press, 1993.

Boland Jr., R. J. "Taking the IS Artifact Seriously in IS Research: Theory Development From Multiple Perspectives," in L. Applegate, R. Galliers, and J. I. DeGross (Eds.), *Proceedings of the 23rd International Conference on Information Systems*, Barcelona, Spain, 2002, pp. 910-911.

Bottomore, T. (Ed.). *A Dictionary of Marxist Thought.* Cambridge, MA: Basic Blackwell Inc., 1991.

Boudreau, M.-C. "Report on the Discussion Panel on Assessing Critical Social Theory research in Information Systems," 1997 (available online at http://www.people.vcu.edu/~aslee/Philadelphia-CST.htm, accessed January 15, 2004).

Bourdieu, P. "In Other Words. Essays Towards a Reflexive Sociology," Cambridge, England: Polity Press, 1990, pp. 123-139.

Brooke, C. "What Does it Mean to be 'Critical' in IS Research?," *Journal of Information Technology* (17), 2002, pp. 49-57.

Burrell, G. "Ephemera: Critical Dialogue on Organization," *Ephemera: Critical Dialogue on Organization* (1:1), 2001, pp. 11-29.

Burrell, G., and Morgan, G. *Sociological Paradigms and Organizational Analysis*, London: Heinemann, 1979.

Churchman, C. W., and Schainblatt, A. H. "The Researcher and the Manager: A Dialectic of Implementation," *Management Science* (11:4), February 1964, pp. B-69-B-87.

Collins, R. Four Sociological Traditions, New York: Oxford University Press, New York, 1994.

Critchley, S. *Continental Philosophy: A Very Short Introduction,* Oxford, England: Oxford University Press, 2001.

Culler, J. D. "Saussure's Theory of Language," in *Saussure*, Hassocks, England: Harvester Press, Hassocks, 1976, pp. 18-52.

Davenport, T. H., and Markus, L. M. "Rigor vs. Relevance Revisited: Response to Benbasat and Zmud," *MIS Quarterly* (23:1), March 1999, pp. 19-23.

de Certeau, M. *The Practice of Everyday Life*, Berkeley, CA: University of California Press, 1984.

Deetz, S., and Kersten, A. "Critical Models of Interpretive Research," in L. L. Putnam and M. E. Pacanowsky (Eds.), *Communication and Organizations: An Interpretive Approach*, Beverly Hills, CA: Sage Publications, 1983, pp. 147-171.

Derrida, J. *Positions*, Chicago: University of Chicago Press, 1972.

Doolin, B., and Lowe, A. "To Reveal is to Critique: Actor-Network Theory and Critical Information Systems Research," *Journal of Information Technology* (17), 2002, pp. 69-78.

Foucault, M. *The Order of Things: An Archeology of the Human Sciences*, London: Rutledge Classics, 1970.

Garfinkel, H. *Studies in Ethnomethodology*, Englewood Cliffs, NJ: Prentice-Hall, Inc., 1967.

Giddens, A. *The Constitution of Society*, Berkeley, CA: University of California Press, 1984.

Grey, C., and Mitev, N. "Management Education: A Polemic," *Management Education* (26:1), 1995, pp. 73-90.

Heidegger, M. "The Question Concerning Technology," in D. F. Krell (Ed.), *Martin Heidegger: Basic Writings*, New York: HarperCollins Publishers, Inc., 1993.

The Houghton Mifflin Canadian Dictionary of the English Language Houghton Mifflin Canada, Markham, Ont., 1982.

Howcroft, D., and Truex, D. "Editorial Statement," *The DATA BASE for Advances in Information Systems* (32:4), 2001, pp. 14-18.

Keen, P. G. W. "MIS Research: Reference Disciplines and a Cumulative Tradition," in E. R. McLean (Ed.), *Proceedings of the First International Conference on Information Systems*, 1980, pp. 9-18.

Keen, P. G. W. "Relevance and Rigor in Information Systems Research: Improving Quality, Confidence, Cohesion and Impact," in H-E. Nissen, H. K. Klein, and R. Hirschheim (Eds.), *Information Systems Research: Contemporary Approaches and Emergent Traditions*, Amsterdam: North-Holland, 1991, pp. 27-49.

Kincheloe, J. L., and McLaren, P. L. "Rethinking Critical Theory and Qualitative Research," in N. K. Denzin and Y. S. Lincoln (Eds.), *Handbook of Qualitative Research*, Thousand Oaks, CA: Sage Publications, 2000, pp. 138-157.

Kingwell, M. *Practical Judgments. Essays in Culture, Politics, and Interpretation*, Toronto: University of Toronto Press, 2002.

Kompridis, N. "Reorienting Critique. From Ironist Theory to Transformative Practice," *Philosophy & Social Criticism* (26:4), 2000, pp. 23-47.

Koselleck, R. *Critique and Crisis: Enlightenment and the Pathogenesis of Modern Society*, New York: Berg, 1988.

Latour, B. *Science in Action*, Cambridge, MA: Harvard University Press, 1999.

Lemert, C. C. "De-Centered Analysis," *Theory and Society* (7:3), May 1979, pp. 289-306.

Lynch, M. "Silence in Context: Ethnomethodology and Social Theory," *Human Studies* (22), 1997, pp. 211-233.

Lyytinen, K. "Empirical Research in Information Systems: On the Relevance of Practice in Thinking of IS Research," *MIS Quarterly* (23:1), March 1999, pp. 25-28.

Lyytinen, K. "Information Systems and Critical Theory," in M. Alvesson and H. Willmott (Eds.), *Critical Management Studies*, London: Sage Publications, 1992, pp. 159-180.

Marcon, T., and Gopal, A. "Irony, Critique and Ethnomethodology: Fundamental Tensions," Paper Presented at the Third Critical Management Studies Conference, Lancaster, UK, 2003.

Merriam-Webster Online Dictionary and Thesaurus, 2003 (available online at http://www.m-w.com; accessed October 1, 2003).

Montealegre, R., and Keil, M. "De-Escalating Information Technology Projects: Lessons from the Denver International Airport," *MIS Quarterly* (24:3), September 2000, pp. 417-447.

Mowday, R. T. "Reaffirming Our Scholarly Values," *Academy of Management Review* (22:2), 1997, pp. 335-345.

O'Connor, J. *The Meaning of Crisis. A Theoretical Introduction*, Oxford, England: Basil Blackwell Inc., 1987.

Orlikowski, W. J., and Baroudi, J. J. "Studying Information Technology in Organizations: Research Approaches and Assumptions," *Information Systems Research* (2:1), 1991, pp. 1-28.

Orlikowski, W. J., and Iacono, C. S. "Research Commentary: Desperately Seeking the 'IT' in IT Research—A Call to Theorizing the IT Artifact," *Information Systems Research* (12:2), 2000, pp. 121-134.

Rawls, A. W. "Editor's Introduction," in H. Garfinkel, *Ethnomethodology's Program. Working out Durkheim's Aphorism*, Lantham, MA: Rowman & Littlefield Publishers Inc., 2002, pp. 1-64.

Shiva, V. *Monocultures of the Mind*, Penang, Malaysia: Third World Network, 1993.

Thayer-Bacon, B. "Transforming and Redescribing Critical Thinking: Constructive Thinking," *Studies in Philosophy and Education* (17), 1998, pp. 123-148.

Walsham, G. *Interpreting Information Systems in Organizations*, Chichester, England: John Wiley & Sons, 1993.

Walzer, M. *The Company of Critics*, New York: Basic Books, 2002.

Weber, R. "Editor's Comments: The Problem of the Problem," *MIS Quarterly* (27:1), 2003, pp. iii-ix.

Webster, J., and Watson, R. T. "Analyzing the Past to Prepare for the Future: Writing a Literature Review," *MIS Quarterly* (26:2), 2002, pp. xiii-xxiii.

Williams, R. *Keywords: A Vocabulary of Culture and Society*, New York: Oxford University Press, 1985.

ABOUT THE AUTHORS

Teresa Marcon is a doctoral student at the Richard Ivey School of Business, University of Western Ontario, Canada. Her interests fall at the intersection of contemporary theory and information technologies in society. Her current work attempts to explore this nexus in conceptual and empirical directions. She can be reached at tmarcon@ivey.uwo.ca.

Mike Chiasson received his Ph.D. in information systems and his postdoctoral in health promotion from the University of British Columbia, Canada. His research is increasingly focused on agency-structure issues in the development and implementation of information systems, in both traditional and nontraditional (e.g., entrepreneurial, professional, and legal) contexts. His specific IT areas include Web-based systems for e-commerce and knowledge management systems. He is currently investigating the effect of emerging IS on entrepreneurial organizations, criminal and civil cases involving e-commerce and software development, and knowledge management systems in law firms. Mike can be reached at chiasson@ucalgary.ca.

Abhijit Gopal received his Ph.D. from the University of Georgia. He is an associate professor at the Richard Ivey School of Business, University of Western Ontario, Canada. His interests lie in the interstitial and peripheral locations of information systems use with particular emphasis on the manner in which information technologies shape and are shaped by political and social conditions and considerations. He has published both in leading journals in the field and in marginal ones. Abhijit can be reaced at agopal@ivey.uwo.ca.

10 WHATEVER HAPPENED TO INFORMATION SYSTEMS ETHICS? Caught between the Devil and the Deep Blue Sea

Frances Bell
Alison Adam
Information Systems Institute
University of Salford

Abstract This paper explores the development of information systems and computer ethics along separate trajectories over the 20 years since the first Manchester Conference, and ponders how things might have been and could be different. Along each trajectory, the challenge of aligning theory and practice has stimulated much research. We evaluate some of this research with respect to this alignment, discuss ethical theories and behavior, and explore the role of education in the development of practitioners who can and do behave ethically. We recommend the inclusion of the ethics of care, and more research into the teaching and learning of ethics as part of the personal journey of students, teachers, and practitioners.

Keywords: Computer ethics, information systems, information systems development methodologies, education

1 INTRODUCTION

Many interesting observations can be made regarding the development of the Information Systems discipline since the Manchester IFIP 8.2 conference in 1984 (Mumford et al. 1985). One of the most important of these relates to the way in which IS, as a discipline, has developed a largely separate trajectory from that of information or computer ethics. Things might have been different. With the increased interest in participative approaches, and the recasting of IS development as a thoroughly social and cultural enterprise from the 1980s onward, we might have expected an information systems ethics to have developed as part of the parent discipline. However, this does not appear to have happened, prompting our question of what has happened to information systems ethics. Instead there appears to have been a split between the development of IS as a discipline and information or computer ethics. Henceforth we shall refer to the discipline of IS as DIS, and computer ethics (or information ethics) as CE.

As it is overly ambitious to take on the whole scope of the DIS and CE disciplines in one research paper, we are concentrating on some of the ramifications of this split in terms of the implications for IS education and practice. While recognizing that there are many information systems management ethical issue (e.g., relating to data protection, privacy, etc.), we center our argument, in the main, on information systems development. Although the maturation of DIS and CE as separate disciplines has allowed each of them considerable space to develop theoretically and empirically, we do not believe that it is ultimately a good thing for them to be so separate in scope, and we conclude with some suggestions as to how the interests of the IS community and those affected by IS might be better served by a closer intellectual relationship between the two disciplines.

Information systems is often characterized as a young discipline concerned with the reflexivity between theory and practice. The scope of IS has greatly increased over the last 20 years as information technology has pervaded first the workplace and, lately, the home and social life in general, at least in Western societies, with the growing convergence of information and communications technologies. CE is an equally young discipline addressing similar issues concerning the social and ethical contexts of information systems and of information and communications technologies (ICT). However, given the relative fluidity of definitions of the respective disciplines, and given the intersection of interest, it is surprising that the two disciplines have grown up quite so separately. As Walsham (1996) notes, papers in IS journals often mention ethical issues yet they rarely focus on such topics in terms of explicit ethical concepts and systems of ethics, nor do they tend to cite CE research explicitly, although there have been some attempts to integrate ethical reasoning into systems methodologies which will be discussed below. However, there are a number of mainly quantitative studies of IS professionals' attitudes to ethical issues that have been reported in the IS and business studies literature (e.g., Khazanchi 1995; Kreie and Cronan 2000). Walsham's criticism of the lack of ethical development in IS does not appear to extend to this work, as such. Nevertheless, as discussed below, statistical surveys of ethical beliefs tend to deflect interest away from ethical theory and theorizing, and focus unethical behavior into ethical decision making (Adam 2001b).

Revisiting the place of ethics in the IS research agenda in this millennium, noted earlier by Walsham, we find it instructive to review briefly how far publication on ethics has permeated the leading IS journals, taking *Information Systems Research* and *Communications of the ACM* as our sample. We note that in the research commentaries (where previous research in an area is reviewed and future areas laid out) published by *Information Systems Research*, a leading IS journal, none published since January 2000 mention ethics (Alavi and Leidner 2001; Ba et al. 2001; Basu and Kumar 2002; Lyytinen and Yoo 2002; March et al. 2000; Orlikowski and Iacono 2001; Sambamurthy and Zmud 2000; Straub and Watson 2001; Wand and Weber 2002), although ethics was mentioned in an earlier commentary (Benbasat and Weber 1996). Specific journals, such as *Ethics in IT* and *Information, Technology and People*, explore the ethics of IS research and practice but they are likely to be addressing those with an existing interest in ethics. This leads us to question the place of ethics in the IS research agenda.

Communications of the ACM is a leading publication read and written by practitioners and academics, and it has shown sustained concern with the ethics of practice. The topics covered in articles on ethics, published in the last 25 years, include ethical decision-making, codes of ethics, ethics education, and issues raised by specific problem domains or technologies. An early example was the self-assessment procedure published in 1982 (Weiss 1982), and revised in 1990 (Weiss et al. 1990), an educational article encouraging the reader to explore ethical scenarios with reference to the ACM Code of Conduct. The 1982 article was the only one specifically related to ethics published in *CACM* in the 1980s, but as the decade progressed, the subject of ethics began to crop up in editorial material and letters and, since the publication of the second self-assessment procedure in 1990, there has been a steady stream of articles relating to ethics. Almost all of these articles make some reference to ethical theories and offer guidance to the reader on applying ethical theory generally, or in specific situations, such as given scenarios or in the context of particular technologies. Several articles cover specific codes of conduct and practice (Anderson et al. 1992; Anderson et al. 1993; Farber 1989; Gotterbarn et al. 1997; Gotterbarn et al. 1999), while others refer to them in offering guidance. Some articles offer explicit methodological support on ethical practice, usually decision-making (Collins et al. 1994; Huff and Martin 1995; Mason 1995; Wood-Harper et al. 1996), while others situate their discussion of ethics in a given area of practice or in the use of a specific technology, often the subject of other articles in the same issue (Berdichevsky and Neunschwander 1999; Bowen 2000; Johnson 1997; Sipior and Ward 1995; Wagner 1993). Survey-based research into ethics has been presented to try to understand how individuals make ethical decisions (Kreie and Cronan 1998, 2000; Loch and Conger 1996; Moores and Dhillon 2000; Pearson et al. 1997), and one article deals with ethical issues in Internet research (Duncan 1996). While we may regret that all of the research is survey-based, and question some of the more formulaic guidance, we recognize *CACM*'s commitment to dealing with ethics in the context of practice, education, and research. An interesting editorial piece publicizes action taken against an author who plagiarized in an article published in an ACM publication (Denning 1995). Although it is difficult to generalize from one publication, we tentatively hypothesize that *CACM*'s engagement with a practitioner audience, in contrast to other more academically focused IS journals, may partially account for its apparently greater interest in ethical issues.

Similarly CE research rarely references mainstream IS research. Some CE writing is philosophical and abstract (e.g., Floridi 1999). Yet there are many practical case studies and good philosophical approaches to relevant writing (e.g., Tavani, 2004). However, quantitative studies, so popular in the North American management literature, are rare in the CE literature; philosophical analysis and case studies are much more common. This implies a certain amount of incommensurability in the ethics research paradigms of DIS and CE that may go part of the way toward explaining the split.

A number of implications spring from this separation. As well as different paradigms for research on information systems ethics and the tendency to focus on ethical decision making, a significant implication is the difficulty of integrating ethical practice into IS development. This is manifest initially in terms of IS education and later in relation to the development, and use, of IS in the workplace. Both DIS and CE struggle, on their separate trajectories, to align theory and practice. The IS community's response to the challenge of integrating social aspects into IS development is evident in much of the IS development methodology research over the last 30 years. In the remainder of the paper, we consider the *status quo*, in terms of what practitioners do and what students learn, and continue by reflecting on the implications of characterizing moral behavior as decision making. We consider some of the problems of the separation of CE and DIS within the context of the rise of CE as an identifiable, separate area of research and the significance of professional codes of conduct in delineating the boundaries of an emerging computing-IT-IS profession. We continue by discussing three areas: first, codes of ethics as teachable constructs; second, the related perils of focusing on rules, decisions, and goal-centered activity as a reflection of moral life, a view of ethics that is encouraged by the prominence of quantitative studies of IS professionals' views on ethics; third, efforts to integrate ethics more formally into systems development methodologies.

2 WHAT PRACTITIONERS DO AND WHAT STUDENTS LEARN

As well as incorporating ethics in methodologies and professional codes, we can include ethics in the curriculum of putative IS developers (e.g., on undergraduate courses) with the hope of encouraging practitioners to become moral agents. However, it is not clear how ethical development can be taught effectively as part of the IS curriculum unless a more adequate means of integrating ethics into mainstream IS is developed. Currently ethics is often taught as a separate subject within the IS curriculum. While this may be better than not teaching ethics at all, it suggests to novice IS students and professionals that ethics and practice are separate and may encourage them to compartmentalize elements of their education. Our experience in teaching a core undergraduate module in ethics to a range of IT, IS, and computer science students suggests that this may be happening. Although most students enjoy this module, there is always a small group of students, admittedly usually drawn from the computer science and software engineering end of the spectrum, who struggle to see the significance of the material. Their concerns seem to run deeper than questions of teaching quality, relating, rather, to what they, and we, expect to see as a suitable topic in the IS curriculum. Despite the efforts of professional bodies to emphasize the importance of ethics in the curriculum, it often looks like something of an optional extra, a theory that is irrelevant to practice. We argue that, unless we find ways of integrating ethics into core elements of the IS curriculum, it will always seem marginal and not something that graduates would expect to apply in their professional practice. In addition, were we able to achieve a better integration into the curriculum, it would undoubtedly affect what is taught as ethics. The emphasis might move away from hacker ethics and Internet pornography, admittedly important topics, to information systems management and development, for example, to ethical applications of systems development methodologies in the workplace, a more mainstream concern.

This signals one of the most problematic aspects of the separation of ethics and IS education and practice that manifests itself in the way that students may avoid asking difficult questions about applying ethics to practice. This is inevitable; they cannot frame such questions if they have no obvious means of connecting ethics and practice. There is already evidence that students or new graduates find it difficult to integrate ethical awareness into the workplace. Alarmingly, a recent study at Nottingham Trent University found that final-year business students who had attended work placements were less likely to show ethical awareness than students in their second year (*Times Higher Educational Supplement*, December 6, 2002, p. 8). Although this might suggest that earlier ethics training is needed, we argue that it is just as likely that novices in the workplace have little idea of how to integrate ethical practice into their work even if they have been taught ethics. This problem is unlikely to be stemmed by earlier training, unless it explicitly addresses practical methods of application and integration.

3 MORAL BEHAVIOR AS DECISION MAKING OR WHAT?

What is it that we do when we act morally as IS professionals? We think it is doubtful that we apply rules in a conscious way in acting morally. The concept of a rule implies that there is a potentially correct answer, a decision or set of decisions that can be chosen. Therefore, there are two further issues to consider: first, the construction of techniques that can be used to choose rules to decide between ethical alternatives, and second, the related characterization of ethical behavior in IS in terms of making a decision based on a rule or rules. On the first point, an emphasis on rule following and concomitant decision making as encapsulating moral behavior in relation to IS can lead researchers into developing difficult and potentially convoluted ethical methods that are difficult for teachers to teach, and for students to apply.

Possibly partially in response to such albeit somewhat intangible difficulties, some authors have interpreted the ambiguities involved in terms of a need to resolve conflict between different courses of action. For instance, Mason describes such conflict resolution as supersession, where the moral agent selects the ethical principle or principles that is the most compelling in a particular case. Supersession requires an individual to make decisions even when using ethical principles agreed by a collectivity, such as a professional body. Depending on the source for the judgement of which principle is the most compelling, this process involves selecting the higher order ethical principle, and the ability to defend the reasoning by which the superseding principle is chosen (Mason et al. 1995).

There are a number of problems with such a process. It describes an ideal situation that is very unlikely to exist in real life, whereas a decision may have to be made swiftly and against a messy backcloth of conflicting parameters that do not readily map onto sets of ethical principles amenable to priority rating. If we accept, as has been argued elsewhere (Adam 2001a), that moral reasoning in CE tends to follow traditional utilitarian and deontological moral reasoning, then criticisms of traditional ethical theories may be relevant to CE. Critics (Adam 2001a; Tong 1993) have commented on the way that traditional systems of ethics, particularly those based on Kantian theories, presuppose a rational, individual moral agent, who can select among a set of abstract principles of justice that are available *a priori* those principles which should apply in a given situation. The individual moral agent is the free man of traditional liberal theory, making decisions on justice and rights unfettered, in his decision making, by the nexus of societal relations. However a number of writers on ethics (Robinson 1999; Tong 1993) have, more recently, emphasized power relations in relation to ethics. One might not be able to choose freely among competing alternatives, but may be constrained heavily in one's choice depending on one's position in the hierarchy.

A related issue is the emphasis placed on the role of the professional at the expense of other workers, such as information owners, users, or other stakeholders (Orlikowski and Baroudi 1989). This is not just a limitation in the scope of professional codes but is more generally part of the disguised power relations embedded in traditional liberal ethics. There is a particular tension here regarding the traditional ethics that act as a cornerstone for computer and professional ethics, as its focus on the individual moral agent is based on the assumption that all have an equal chance to speak. Given the movement toward user participation in IS, the focus of CE could be at odds with user-centered participative approaches. This is especially important in relation to the newer critical movement in IS (Hirschheim and Klein 1994) where, following Habermas, the will towards emancipation emphasizes the concept of the ideal speech situation. Yet within the critical IS field, we have barely begun to construct the critical IS ethics that could make use of these ideas (Adam and Bell 2003).

An additional difficulty lies in the tacit and persistent acceptance, embedded in approaches such as supersession, echoing scientific management, of management activity as rational pursuit of a goal through decision making. Such a view has been extraordinarily tenacious, stretching from Taylorism, through Simon's (1976) later work on the scientific management of human problem solving, even into views of intelligence as rational problem solving encapsulated in artificial intelligence and knowledge management. Later approaches toward the characterization of decision making in scientific management include the theory of reasoned action and, more broadly, rational choice theory (Archer and Tritter 2000). These theories impose a mathematical model on the business of making decisions, often with weightings. Only with the advent of anthropological and interpretive approaches to recording management and workplace behavior has the reliance on rationalistic, goal-seeking decision making diminished. Nevertheless such approaches have been extraordinarily tenacious especially in normative areas such as methodologies for system design and for ethical analysis.

Taking such criticisms on board for CE leads us to doubt the practical use of ethical methodologies such as supersession and reliance on the power of the code of ethics and rationalist goal-centered approaches in ethical decision making (Mason et al. 1995). Furthermore, this leads us to question the teaching of ethical methods in CE education especially those built round the objective of designing methods to enable following professional codes and applying sanctions if codes are not followed. Apart from the whole question of the ineffectiveness of professional codes in an industry that is largely unregulated, this view places far too much reliance on formal processes of rational decision making, ignoring the question of how far moral activity is directed into activities other than decisions. On the other hand, unless we are to admit some sort of methodology such as that put forth by Mason et al., we may have little by way of practical suggestions as to how to teach ethical principles, other than abstract approaches that struggle to connect case studies to ethical theory.

4 CODES OF CONDUCT AND PRACTICE

The rise of CE as a discipline involves the intersection of several vectors. It can be seen as part of the professionalization strategy of the emerging computing-information technology profession (Adam 2001a) where the notion of a *social contract* is important. As citizens, we are all bound by a social contract to act in particular ways toward each other and with regard to the instruments and institutions of the state. The social contract takes on additional burdens beyond those expected of the individual member of society when applied to professions. A profession has particular duties to its users and a wider public, not only to do them no harm, but also, more positively, to act in their interests according to the dictates of the profession. The computing industry, in subscribing to codes of ethics, attempts to enter into a similar social contract. Yet it can be argued that the computing profession hardly matches any of the traditional indicators of professional status (e.g., standard education, professional autonomy, regulatory bodies).

Computing codes of conduct and practice, as an explicit representation of the social contract, have two goals: to capture the essence of the profession's commitments and responsibilities as a basis for ethical decision-making and to convince the public that the profession is capable of self-regulation (Walsham 1996). This can be characterized as a deal between a profession and society: accountability of the profession and its members in return for the trust, confidence, and respect of the public (and the accompanying increased social and economic rewards) (Mason et al. 1995). Early versions of the codes of bodies such as the ACM tended to be regulatory, but in the 1990s these codes become more normative in nature. Gotterbarn (1997) sees normative codes as reflecting some sort of consensus of traditions and a growing sense of maturity in a profession.

Although in this discussion we recognize the role of the ethical code in formalizing the social contract between a profession and the public, we note that much of that social contract is tacit, not written down and not strictly enforceable. This is especially the case in the computing-IT-IS profession where most practitioners practice perfectly well without reference to professional membership and one need not be licensed in order to practice. This reinforces Gotterbarn's point. It is not so much that the codes of ethics strictly lay down rules for the profession to follow, rather that they reflect the maturity of the social contract into which the profession has developed. Codes of ethics are generally seen to provide useful sets of principles and duties but several IS researchers have clearly alluded to the difficulties that practitioners may have difficulty in applying them (Anderson et al. 1993; Mason et al. 1995; Walsham 1996).

Teaching codes of conduct, including an awareness of how and why these have changed through the years, is therefore important. Although codes are by no means the only topics taught on CE courses, they do form a convenient peg on which to hang the topic of professionalism. Yet too sharp a focus on ethical codes may prove problematic. Rules cannot stand alone as simple prescriptions or proscriptions for action; in professions, they are bedded into practice by a variety of means including education. We have the legal profession to help us apply legal rules and mathematics teachers to help us apply mathematical rules, at least until we become expert in the application of mathematical principles. Therefore, we should not be surprised if ethical rules are difficult to apply, and we should expect that ethics be taught by teachers who can understand ethical theories and apply them to the information systems context.

This serves to emphasize that although codes can help rule out unacceptable decisions, they are not prescriptions for action. This is partly because, in real-life contexts, different principles and duties may be in conflict, but also because, especially in more recent form, ethical codes tend to display the "open texturedness" we expect from a good rule, where all the states to which the rule applies are not written down in advance (they cannot be). Rather the rule is subject to interpretation in each new case, in the same way that legal rules are constantly reinterpreted in new legal cases, thus building up the body of case law. In summary, the point we make here is that teaching codes of conduct, especially if they are taught as part of a separate professional studies or ethics module, tends to move the focus of information systems ethics teaching and practice away from real life practical action toward more abstract rules.

5 TEACHING AND LEARNING COMPUTER ETHICS

Although professional bodies attach importance to the inclusion of ethics on approved course syllabi, there is a lack of consensus about the effectiveness of ethics education in improving information systems practice, and this is an issue that runs through business and management literature more generally. Indeed the considerations raised in the previous section reinforce doubts about CE education, especially in terms of an over-reliance on the idealized decision- making processes that traditional systems of ethics offer. Wright (1995) claims that education is the best means of developing good ethical behavior in the modern business environment. However, a recent statistical survey found that in terms of the ethical values examined in this survey, there were no significant differences between business students who had taken an ethics course and those who had not (Peppas and Diskin 2001). This tends to reinforce the findings of the Nottingham Trent University study cited earlier. Additionally, these findings are borne out, at least to some extent, by the studies referenced in Wright's literature survey on learning ethical behavior and judgement where results were mixed, to say the least. This is a disturbing finding, suggesting that our attempts to incorporate ethics into the curriculum may be to little avail. However, Peppas and Diskin (2001) suggest that further research is needed into how ethics can be learned, suggesting that case studies could help simulate the experience of exposure to business circumstances and may, therefore, be more effective than teaching abstract principles, avoiding the difficulties that students experience in trying to apply systems of ethics (Johnson 1994).

The difficulty of turning abstract ethical principles into teachable moral procedures is, perhaps, to be expected for any discipline, not just for IS and computing. We contend that the teaching of ethics demands teachers capable of post-formal reasoning, who can facilitate the development of such reasoning by their students, and who appreciate

> that the intersections of expert knowledge, imagination, and ethical decision are governed by a postformal stage of reasoning, that is, a way of thinking allowing for multiple and contradictory views of truth, for bridging across belief systems, and for bringing to the foreground subjective and self-referential thought (Lee 1993).

Post-formal thinkers do not rely solely on propositional knowledge but can also use self-referential and subjective knowledge in their consideration of wider issues, guided by the compassion, responsiveness, and responsibility that are the hallmarks of an ethics of care. They can connect hypothetical situations to their own experiences, and still consider the other. Achieving this is a tall order.

Our understanding of ethical decision-making is constrained by the style of empirical study that preponderates in the research literature on business and IS ethics (Adam 2001b). In many studies of ethical decision making, the main research tool is the questionnaire followed by statistical analysis. This is a standard quantitative research approach that predominates in North American management research. It also reinforces an approach to business and management, echoed in IS and computing, that focuses on decision making as the primary thing that managers do. However, for ethical analysis, the effect is to take a "snapshot" of the ethical event, to focus on actual decisions that respondents would take, or at least the decisions they say they would make under the circumstances outlined in the questionnaires (Kreie and Cronan 1998). Apart from our concerns about the value of such an exclusive concentration, this also raises the age-old problem that we do not know how to correlate what people say with what they do. More importantly, this style of research has three important consequences. First, it assumes that there is a "right" answer that is clear from the brief description of the case. Second, it forces considerations of moral behavior into the end process of an ethical decision, de-emphasizing or ignoring the complexities of the process and context within which the decision was made. Third, it ignores the way that much, if not the majority, of moral behavior is not concerned with making decisions. Making good decisions may not be all there is to being "good." Hence the effectiveness of ethics education is as much about providing the opportunity to reflect on accountable, ethical practice as about learning codes or theories and emphasizing decisions based on such theories. Post-formal thinkers will be equipped to grapple with the age- old ethical dilemmas and their current manifestations in the context of globalization, people working, learning and socializing via the Internet, as well as face-to-face—namely, power, trust, identity, and many other issues. In the next section, we examine the possible role of methodology in learning the practice of an important activity in IS, systems development.

6 METHODOLOGIES AS NORMATIVE ETHICAL DEVICES

In the education and practice of systems analysts, information systems development methodologies (ISDMs) are typically used as normative devices to encourage "good practice" (Klein and Hirschheim 2001). Where ISDMs operate as static rule-systems that require developers to operate in a standardized manner, they offer poor support for learning compared with second-order learning processes that provide a framework within which individual learning can take place (Floyd 1987). The combination of the structure of a documented process with reflection and tutor feedback can make a methodology a good framework for students learning systems development, even if they do not use the methodology once they graduate. However the concept of *good* in an IS development methodology rarely maps on to what is understood by good in a moral sense. We regard this separation as problematic. It is as if the goodness of ISDMs is to be understood in functionalist terms rather than moral terms and further underlines both the apparent separation of ethics from other parts of life and the lesser status of ethics within disciplines. Witness the way that business ethics exists as a separate and somewhat lesser status discipline from management, a situation that parallels the separation and status of CE and DIS.

Research into the use of methodologies by practitioners indicates that in many cases they are not used (Chatzoglou 1997); that where they are used, they are adapted to the exigencies of the problem and development situations (Fitzgerald 1997); and that in some cases the goal of using them may be displaced to legitimize the development process, what Wastell calls a social defense (Fitzgerald 1996; Wastell 1996). Therefore, we can see that practitioner respond to methodologies by a combination of ignoring them and using them in their own way, or for their own purposes—in short, they have difficulties in aligning theory and practice. Commercial methodologies are often adopted by virtue of a decision to purchase a computer-aided software engineering (CASE) tool. Many of these are focused on a bounded, often technical, rationality, with recognition of the need to align with business needs and be usable by end-users, but with little or no recognition of wider ethical issues raised by consideration of stakeholders beyond rational views of client, developer, and user, for example, rational unified process (Kruchten 2000). As we explore later, even methodologies that take account of social aspects do not explicitly include tools and techniques to support ethical analysis.

Hirschheim and Klein (1995) define an information systems development methodology as "an organized collection of concepts, methods, beliefs, values and normative principles supported by material resources." By including "beliefs, values and normative principles" in their definition, Hirschheim and Klein suggest that the adoption of a particular information system development methodology (ISDM) may have an effect on the analyst's treatment (or not) of ethical issues. Methods based on a technical rationality (e.g., SSADM— Goodland and Slater 1995), pay minimal attention to ethical issues (Rogerson et al. 2000; Walsham 1993). Paradigmatic analysis has revealed the extent to which different methodologies facilitate the consideration of ethical and social issues (e.g., ETHICS—Mumford 1996; soft systems methodology—Checkland and Scholes 1990; Multiview—Wood-Harper et al. 1985; the collective resource approach—Ehn and Kyng 1987; and critical action research— Hirschheim and Klein 1995; Jonsson 1991; Walsham 1993). Even those methodologies that encourage the analyst to raise ethical issues offer limited support for the resolution of these issues. In his exploration of the support offered by SSM to the analyst as moral agent, Walsham (1993) provisionally concludes that the degree of support offered by SSM depends on the analyst's own actions and the particular adaptation of SSM adopted by the analyst.

In the case of information systems development, as opposed to information systems management or strategy, there is a small body of literature on the role of ethics. An important strand of this work looks at the philosophies that underpin various systems development methodologies, and claims that they favor various value orientations, apparent as design ideals (Hirschheim and Klein 1989; Iivari et al. 1998; Klein and Hirschheim 2001). The speech act-based approach developed separately in North America and Scandinavia, the latter stream being strongly influenced by the critical social theory of Habermas. While Iivari et al. (1998) identify a means-end orientation in the research based on this approach, with the IS designer adopting an emancipatory role, they point out that the emphasis on the intersubjective use of rational communication can be used to increase organizational effectiveness. Soft systems methodology, with its use of *Weltanschauungen*, does offer the opportunity to consider alternative viewpoints (including ethical ones), but its ethical approach depends on how it used, and specifically how "accommodation" is achieved between these viewpoints in plans for action (Checkland and Scholes 1990; Iivari et al. 1998; Walsham 1996). It seems, therefore, that methodology is no guarantor of ethics; the change agent(s) and the problem situation also affect the process of making ethical decisions.

Let us examine two analyses of how specific methodologies could be extended to improve their support for ethical analysis. In the first example, Wood-Harper et al. (1996) take the view that there is a dominant ethical belief that can help to predict and understand group behavior in a given situation, while also considering the various, possibly conflicting, stakeholder ethical views. The analyst must choose a methodology, and decide how to analyze and resolve conflicting ethical viewpoints. In order to do this, the analyst should understand and be able to apply ethical theory. They offer a five step ethical analysis approach, which they claim might be integrated into any systems development methodology, and they then retrospectively map that approach on to soft systems methodology as it was actually applied in their case study.

In the second example, Rogerson et al. (2000) attempt to map the Australian Computer Society's Code of Ethics (ACSCE) on to structured systems analysis design method on the basis that such a mapping on to a technically oriented systems method is a good test of the possibilities for enrichment of methodologies with ethical analysis. Their initial approach covers a mapping of ACSCE on to SSADM modules, and an example of how ACS articles might be used to derive (ethical) product criteria for SSADM products.

In neither of these examples has an ethical methodology been identified then explored in practice, but the mapping between a practice case study and a methodology, in the light of ethical theories, may be a useful activity in ethics education. Systems development methodologies have been claimed to offer a useful learning framework for novice system developers, making explicit activities and decisions that more experienced developers may treat as a matter of course. Case studies that use the ethical SSADM product criteria or follow the approach suggested by Wood-Harper et al. or give rich descriptions of how ethical dilemmas were handled (as messy processes rather than correct decisions) will be a useful resource for educators. Using such case studies in ethics education has the attraction of offering a safe "sand pit" where novices can experiment with ethical thinking, digging and building without doing any damage. However, we need case studies that capture the process and complexities of dealing with ethical issues in a range of contexts for IS, systems development, systems integration and implementation, management, and use of information systems, in the home, the workplace, and other social settings.

7 CONCLUSION

In this paper we have raised a number of issues relating to the connection, or lack of connection, between DIS and CE. These disciplines have shared concerns: they seek to align social and technical concerns, and theory and practice; such goals are inherently difficult; they seek to influence practice through education; and research methods are moving from quantitative to qualitative. We note the relatively separate trajectories of the two disciplines; this is especially notable when we consider the relative lack of interest in ethical issues displayed by flagship IS journals. This is also hard to understand when DIS and CE address similar subject matter and when other disciplines, such as science and technology studies, have taken a turn toward the ethical.

Our primary concern is with the integration of ethics into IS education, an endeavor that remains problematic for reasons relating to the curriculum in relation to the dominant theories espoused by teachers, and to the inherent difficulty of learning to reason ethically in practice. The split between CE and DIS has exacerbated these problems. Furthermore, the emphasis on moral behavior as decision making, which is displayed more generally in business ethics, but is also found in CE, leads us to develop ever more convoluted decision-making techniques that are difficult to apply and emphasize the individual, Kantian, rational-moral agent in favor of considering a network of moral relationships. This appears to be the traditional Tayloristic view of meaningfully activity as rational pursuit of a goal.

We recommend that, as is usually the case, where ethics is taught as a separate subject, the curriculum be changed to integrate ethics with practice, for example, in systems analysis or in reflection on information systems management and development. This would also emphasize a more relational approach to ethics with the inclusion of an ethics of care that encourages the connection of individual experience to consideration of that of others, within the context of self-referential and subjective thought. We recognize how difficult it will be to achieve the development of such ethical thinking in undergraduates (and in us, the teachers), and look forward to a body of qualitative research in this area, that can transcend the limited view of ethical reasoning as judged by decisions that follow from a rational process, provided by the body of quantitative research to date.

The very few attempts to integrate ethical analysis into systems development methodologies appear to be at an early stage of development and have not really spawned a research tradition. Given the emphasis on emancipation and communicative rationality within critical IS, we argue that further empirical work in this field would provide more appropriate case studies of integration of ethics into IS, particularly where these can be integrated into political and social contexts as suggested by Benhabib (1992). We leave you to surmise which is the devil, and which is the deep blue sea.

REFERENCES

Adam, A. "Computer Ethics in a Different Voice," *Information and Organization* (11:4), 2001a, pp. 235-261.

Adam, A. "Gender and Computer Ethics," in R. Spinello and H. Tavani (Eds.), *Readings in Cyberethics*, Sudbury, MA: Jones and Bartlett, 2001b, pp. 63-76.

Adam, A., and Bell, F. "Critical Information Systems Ethics," in C. H. C. Gilson, I. Grugulis, and H. Willmott (Eds.), *Proceedings of the CMS3—Third International Critical Management Studies Conference*, Lancaster University Management School, Bailrigg, Lancaster, UK, 2003 (available online at http://www.mngt.waikato.ac.nz/research/ejrot/cmsconference/2003/proceedings/exploringthemeaning/Adam.pdf).

Alavi, M., and Leidner, D. E. "Research Commentary: Technology-Mediated Learning—A Call for Greater Depth and Breadth of Research," *Information Systems Research* (12:1), 2001, pp. 1-10.

Anderson, R. E.; Engel, G.; Gotterbarn, D.; Hertlein, G. C.; Hoffman, A.; Jawer, B.; Johnson, D. G.; Lidtke, D. K.; Little, J. C.; Martin, D.; Parker, D. B.; Perrolle, J. A.; and Rosenberg, R. S. "ACM Code of Ethics and Professional Conduct," *Communications of the ACM* (35:5), 1992, pp. 94-99.

Anderson, R. E.; Johnson, D. G.; Gotterbarn, D.; and Perrolle, J. "Using the New ACM Code of Ethics in Decision-Making," *Communications of the ACM* (36:2), 1993, pp. 98-107.

Archer, M., and Tritter, J. *Rational Choice Theory*, London: Routledge, 2000.

Ba, S. L.; Stallaert, J.; and Whinston, A. B. "Research Commentary: Introducing a Third Dimension in Information Systems Ddesign the Case for Incentive Alignment," *Information Systems Research* (12:3), 2001, pp. 225-239.

Basu, A., and Kumar, A. "Research Commentary: Workflow Management Issues in E-Business," *Information Systems Research* (13:1), 2002, pp. 1-14.

Benbasat, I., and Weber, R. "Research Commentary: Rethinking 'Diversity' in Information Systems Research," *Information Systems Research* (7:4), 1996, pp. 389-399.

Benhabib, S. *Situating the Self, Gender, Community and Postmodern Contemporary Ethics*, Cambridge, MA: Polity, 1992.

Berdichevsky, D., and Neunschwander, E. "Toward an Ethics of Persuasive Technology: Ask Yourself Whether Your Technology Persuades Users to Do Something You Wouldn't Want to Be Persuaded to Do Yourself," *Communications of the ACM* (42:5), 1999, Pp. 51-58.

Bowen, J. "The Ethics of Safety-Critical Systems," *Communications of the ACM* (43:4), 2000, pp. 91-97.

Chatzoglou, P. D. "Use of Methodologies: An Empirical Analysis of Their Impact on the Economics of the Development Process," *European Journal of Information Systems* (6:4), 1997, pp. 256-270.

Checkland, P., and Scholes, J. *Soft Systems Methodology in Action*, Chichester, England: Wiley, 1990.

Collins, W. R.; Miller, K. W.; Spielman, B. J.; and Wherry, P. "How Good Is Good Enough?," *Communications of the ACM* (37:1), 1994, pp. 81-91.

Denning, P. J. "Plagiarism in the Web," *Communications of the ACM* (38:12), 1995, pp. 29-29.

Duncan, G. T. "Is My Research Ethical?," *Communications of the ACM* (39:12), 1996, pp. 67-68.

Ehn, P., and Kyng, M. "The Collective Resource Approach to Systems Design," in G. Bjerknes, P. Ehn, and M. Kyng (Eds.), *Computers and Democracy: A Scandinavian Challenge*, Aldershot, England: Avebury, 1987, pp. 17-57.

Farber, D. J. "NSF Poses Code of Networking Ethics," *Communications of the ACM* (32:6), 1989, pp. 688-688.

Fitzgerald, B. "Formalized Systems Development Methodologies: A Critical Perspective," *The Information Systems Journal* (6:1), 1996, pp. 3-23.

Fitzgerald, B. "The Use of Systems Development Methodologies in Practice: A Field Study," *Information Systems Journal* (7:3), 1997, pp. 201-212.

Floridi, L. *Philosophy and Computing: An Introduction*, London: Routledge, 1999.

Floyd, C. "Outline of a Paradigm Change in Software Engineering," in G. Bjerknes, P. Ehn, and M. Kyng (Eds.), *Computers and Democracy: A Scandinavian Challenge*, Aldershot, England: Avebury 1987, pp. 191-210.

Goodland, M., and Slater, C. *SSADM Version 4: A Practical Approach*, New York: McGraw Hill, 1995.

Gotterbarn, D. "Software Engineering: A New Professionalism," in C. Myers, T. Hall, and D. Pitt (Ed.), *The Responsible Software Engineer: Selected Readings in IT Professionalism*, London: Springer-Verlag, 1997, pp. 21-31.

Gotterbarn, D.; Miller, K.; and Rogerson, S. "Software Engineering Code of Ethics," *Communications of the ACM* (40:11), 1997, pp. 110-116.

Gotterbarn, D.; Miller, K.; and Rogerson, S. "Software Engineering Code of Ethics is Approved," *Communications of the ACM* (42:10), 1999, pp. 102-107.

Hirschheim, R., and Klein, H. K. "Four Paradigms of Information Systems Development," *Communications of the ACM* (32:10), 1989, pp. 1199-1216.

Hirschheim, R., and Klein, H. K. *Information System Development and Data Modeling. Conceptual & Philosophical Foundations*, New York: Cambridge University Press, 1995.

Hirschheim, R., and Klein, H. K. "Realizing Emancipatory Principles in Information-Systems Development: The Case For Ethics," *MIS Quarterly* (18:1), 1994, pp. 83-109.

Huff, C., and Martin, C. D. "Computing Consequences: A Framework for Teaching Ethical Computing," *Communications of the ACM* (38:12), 1995, pp. 75-84.

Iivari, J.; Hirschheim, R.; and Klein, H. K. "A Paradigmatic Analysis Contrasting Information Systems Development Approaches and Methodologies," *Information Systems Research* (9:2), 1998, pp. 164-193.

Johnson, D. *Computer Ethics*, Englewood Cliffs, NJ: Prentice-Hall, 1994.

Johnson, D. G. "Ethics Online," *Communications of the ACM* (40:1), 1997, pp. 60-65.

Jonsson, S. "Action Research," in H.-E. Nissen, H. K. Klein, and R. Hirschheim (Eds.), *Information Systems Research: Contemporary Approaches and Emergent Traditions*, Amsterdam: North-Holland, 1991, pp. 371-396.

Khazanchi, D. "Unethical Behavior in Information Systems: The Gender Factor," *Journal of Business Ethics* (14), 1995, pp. 741-749.

Klein, H. K., and Hirschheim, R. "Choosing between Competing Design Ideals in Information Systems Development," *Information Systems Frontiers* (3:1), 2001, pp. 75-90.

Kreie, J., and Cronan, T. "How Men and Women View Ethics," *Communications of the ACM* (41:9), 1998, pp. 70-76.

Kreie, J., and Cronan, P. "Making Ethical Decisions," *Communications of the ACM* (43:12), 2000, pp. 66-71.

Kruchten, P. *The Rational Unified Process: An Introduction*, Reading, MA: Addison-Wesley, 2000.

Lee, D. M. "The Place of Wisdom in Teaching," *Learning and Individual Differences* (5:4), 1993, pp. 301-317.

Loch, K. D., and Conger, S. "Evaluating Ethical Decision Making and Computer Use," *Communications of the ACM* (39:7), 1996, pp. 74-83.

Lyytinen, K., and Yoo, Y. "Research Commentary: The Next Wave of Nomadic Computing," *Information Systems Research* (13:4), 2002, pp. 377-388.

March, S.; Hevner, A.; and Ram, S. "Research Commentary: An Agenda for Information Technology Research in Heterogeneous and Distributed Environments," *Information Systems Research* (11:4), 2000, pp. 327-341.

Mason, R. O. "Applying Ethics to Information Technology Issues," *Communications of the ACM* (38:12), 1995, pp. 55-57.

Mason, R. O., Mason, F. M., and Culnan, M. J. *Ethics of Information Management*, Thousand Oaks, CA: Sage Publications, 1995.

Moores, T., and Dhillon, G. "Software Piracy: A View from Hong Kong," *Communications of the ACM* (43:12), 2000, pp. 88-93.

Mumford, E. *Systems Design: Ethical Tools for Ethical Change*, London: Macmillan, 1996.

Mumford, E., Hirschheim, R., Fitzgerald, G., and Wood-Harper, A. T. *Research Methods in Information Systems*, Amsterdam: North-Holland, 1985.

Orlikowski, W. J., and Baroudi, J. J. "The Information Systems Profession: Myth or Reality," *Office, Technology and People* (4), 1989, pp. 13-30.

Orlikowski, W. J., and Iacono, C. S. "Research Commentary: Desperately Seeking the 'IT' in IT Research—A Call to Theorizing the IT Artifact," *Information Systems Research* (12:2), 2001, pp. 121-134.

Pearson, J. M.; Crosby, L.; and Shim, J. P. "Measuring the Importance of Ethical Behavior Criteria," *Communications of the ACM* (40:9), 1997, pp. 94-100.

Peppas, S. C., and Diskin, B. A. "College Courses in Ethics: Do They Really Make a Difference?," *International Journal of Education Management* (15:7), 2001, pp. 347-353.

Robinson, F. *Globalizing Care: Ethics, Feminist Theory, and International Relations*, Boulder, CO: Westview Press, 1999.

Rogerson, S.; Weckert, J.; and Simpson, C. "An Ethical Review of Information Systems Development," *Information Technology & People* (13:2), 2000, pp. 121-136.

Sambamurthy, V., and Zmud, R. W. "Research Commentary: The Organizing Logic for an Enterprise's IT Activities in the Digital Era—A Prognosis of Practice and a Call for Research," *Information Systems Research* (11:2), 2000, pp. 105-114.

Simon, H. A. *Administrative Behavior*, New York: The Free Press, 1976.

Sipior, J. C., and Ward, B. T. "The Ethical and Legal Quandary of Email Privacy," *Communications of the ACM* (38:12), 1995, pp. 48-54.

Straub, D. W., and Watson, R. T. "Research Commentary: Transformational Issues in Researching IS and Net-Enabled Organizations," *Information Systems Research* (12:4), 2001, pp. 337-345.

Tavani, H. *Ethics and Technology: Ethical Issues in Information and Communication Technology*, New York: John Wiley & Sons, 2004.

Tong, R. *Feminine and Feminist Ethics*, Belmont, CA: Wadsworth, 1993.

Wagner, I. "A Web of Fuzzy Problems: Confronting the Ethical Issues," *Communications of the ACM* (36:6), 1993, pp. 94-101.

Walsham, G. "Ethical Issues in Information Systems Development: The Analyst as Moral Agent," in D. Avison, J. E. Kendall, and J. I. DeGross (Eds.), *Human, Organizational, and Social Dimensions of Information Systems Development*, Amsterdam: North-Holland, 1993, pp. 281-294.

Walsham, G. "Ethical Theory, Codes of Ethics and IS Practice," *Information Systems Journal* (6:1), 1996, pp. 69-81.

Wand, Y., and Weber, R. "Research Commentary: Information Systems and Conceptual Modeling—A Research Agenda," *Information Systems Research* (13:4), 2002, pp. 363-376.

Wastell, D. "The Fetish of Technique: Methodology as a Social Defense," *Information Systems Journal* (6), 1996, pp. 25-40.

Weiss, E. A. "A Self-Assessment Procedure Dealing with Ethics in Computing," *Communications of the ACM* (25:3), 1982, pp. 181-195.

Weiss, E. A.; Parker, D. B.; Swope, S.; and Baker, B. N. "Self-Assessment Procedure 22: The Ethics of Computing," *Communications of the ACM* (33:11), 1990, pp. 110-132.

Wood-Harper, A. T.; Avison, D. E.; and Antill, L. *Information Systems Definition: The Multiview Approach*, Oxford: Blackwell Scientific, 1985.

Wood-Harper, A. T.; Corder, S.; Wood, J. R. G.; and Watson, H. "How We Profess: The Ethical Systems Analyst," *Communications of the ACM* (39:3), 1996, pp. 69.

Wright, M. "Can Moral Judgement and Ethical Behavior Be Learned? A Review of the Literature," *Management Decision* (33:10), 1995, pp. 17-28.

ABOUT THE AUTHORS

Frances Bell's research interests include virtual organizing and the use of Information and Communications Technologies (ICT) in Education. She has worked on a Fifth Framework project to support a Virtual Enterprise in Rural Tourism, and is currently working on a Minerva project, Collaboration Across Borders, that is building staff and student networks for international collaboration in education. She has published in journals and in conferences such as UKAIS, ALT-C, and EUNIS. Frances can be reached at F.Bell@salford.ac.uk.

Alison Adam is Professor of Information Systems and Head of School at the Information Systems Insitute, University of Salford. Her research interests include gender and IS, critical IS and computer ethics. She is involved in two European Social Fund projects, based at the ISI in Salford, researching women in the IT industry (WINIT and WINWIT). She is currently completing a book on gender and computer ethics. She can be reached on a.adam@salford.ac.uk

11 SUPPORTING ENGINEERING OF INFORMATION SYSTEMS IN EMERGENT ORGANIZATIONS

Sandeep Purao
The Pennsylvania State University

Duane P. Truex III
Florida International University and Georgia State University

Abstract Research related to information systems development has roughly followed two diverse paths. The first, pursued by the software engineering community, is aimed at creating techniques for the efficient engineering of IT artifacts. The second, pursued by communities such as IFIP 8.2, attempts to understand and anticipate the impact of IT on organizations or upon one another by incorporating social science theories. The two views are in conflict because the former treats information systems merely as reflections of requirements, whereas the latter views them as agents of change. As a result, the two streams have suffered from increasing emphasis on minutiae and are, at worst, in danger of losing their relevance. Recent research in emergent systems development and developers' engagement in problem and design spaces suggest a possible approach to integrating the two streams. In particular, we argue that novel R-forms (representation techniques) can proactively facilitate the engineering of information systems in emergent organizations. Using insights from research in both streams, we develop a set of requirements that can guide the development of new R-forms that may take into account both the engineering of the IT artifact as well as the emergent nature of organizational context in which the IT artifact will be deployed.

Keywords: Emergent systems design, deferred systems design, social theories, information systems development (ISD), representational forms

1 INTRODUCTION

Much current research in software engineering focuses on models and methods for creating IT artifacts that are intended to function in organizational contexts. An implicit assumption in this research stream is the availability of stable or slowly evolving requirements (Robinson et al. 2003) and the need to reflect these accurately in the object system being constructed (Welke 1980). Improved methods and models following this assumption cannot sufficiently address demands posed by increasingly turbulent business environments, which require not only changes to the IT artifacts but also business practices. Such traditional ideals of information systems development are being questioned by researchers in the IFIP WG 8.2 community (e.g., emergent systems development—Bello et al. 2002; Truex, Baskerville, and Klein 1999; amethodical systems development—Truex, Baskerville, and Travis 2000; deferred systems development—Patel 1999; and improvisation and bricolage—Bansler and Havn 2002). The IFIP Working Group 8.2 focus may be described as making sense of the complex interrelationship between the IT artifact and the organizational context in which it is embedded. Much of this work has been conducted as *post hoc* studies of impacts following the implementation of information systems. Resulting insights from this community, while often telling and profound, have not been translated into proactive, actionable techniques for building more effective information systems. Thus, the two related research communities, each ostensibly having the mission to help design and introduce effective information systems into organizational settings, have missed important opportunities to interact and work on this shared goal. Without greater awareness and integration of the work in the two domains, there is an increasing risk that each will continue its emphasis on minutia accompanied by decreasing relevance of the research outcomes. This paper builds its argument in the context of repeated calls for integrating insights from these two research streams (see, for example, Probert and Rogers 1999).

A specific objective of the paper is to investigate a possible path to realizing this integration in the form of representation techniques that can assist in codesigning information systems and emergent organizational forms. The remainder of the paper is organized in four sections. Section 2 traces research in the two streams identified above with a view to highlighting the need for integration. In section 3, we envision how modeling techniques may be used to facilitate this integrated perspective (i.e., for codesigning information systems and emergent organizations following developer behaviors observed in practice). Section 4 develops desired requirements for such representational techniques. We conclude in section 5.

2. PRIOR RESEARCH

2.1 A Software Engineering Perspective

The software engineering perspective views information systems as technical artifacts that need to be developed, built, and deployed. It, therefore, requires a focus on techniques and models that facilitate design and production of software artifacts with a limited consideration of the environments in which the artifact will be deployed. Pressman and Associates (2003) define software engineering as a discipline that encompasses the process associated with software development, the methods used to analyze, design, and test computer software, the management techniques associated with the control and monitoring of software projects, and the tools used to support process, methods, and techniques. Like many other engineering disciplines, the environment (e.g., an organization) is seen as the source of requirements, often simplified to organizational processes, behaviors, or roles (ignoring important issues such as power, conflict, or structure). The goal of software engineering is, thus, converting these requirements to a format appropriate for realization in the software artifact in a manner that leads to few bugs, on time delivery, and greater maintainability. More enlightened approaches from the requirements engineering research community (see, for example, Nuseibeh 2001; Nuseibeh and Easterbrook 2000) have argued for a closer connection between real-world *goals* and *software specifications* favoring a multidisciplinary, human-centered process for requirements engineering (Nuseibeh and Easterbrook 2000, p. 38). Such linkage has also been explored for the purpose of understanding nonfunctional requirements (Mylopoulos et al. 1992) and for expressing the mappings between different viewpoints during the specification and capture of requirements (Nuseibeh et al. 1994).

Methods and modeling techniques in software engineering, however, continue to assume relative stability and consensus in an organization. The techniques often simplify, generalize, and abstract the organizational domain to develop archetypes, which are then translated into functional descriptions (Gane and Sarson 1979) or use cases (OMG 2003). These plan-driven methodologies (Boehm 2002) are often not appropriate for turbulent environments, where business practices or technology is undergoing rapid change. Levine et al. (2002), for example, suggest that requirements for many systems are vague and likely to change. Truex, Baskerville and Klein (1999) suggest that developing a set of specifications that are clear, consistent, and complete before starting the design may be a fools errand. As a response, the definition of requirements specifications has been expanded to include evolution over time and across software families (Zave 1997). However, the linkage between the organizational subsystem and the information system is still tenuous, largely restricted to treating the former as the source of requirements.

One large-scale effort addressing the formalization of concepts for organizational information systems has been the FRISCO project (1997, 2001), representing the culmination of a decade-long multinational research effort. The FRISCO report develops an elaborate system of definitions to map underlying conceptions of things in a domain to representations of those concepts (FRISCO 1997, p. 34). The superstructure of definitions in FRISCO has been critiqued by Hesse and Verrijn-Stuart (2000) for reasons such as over-formalization, circular definitions, and ambiguity. Similarly, Stamper (2000), one of the signatories to the FRISCO report, has argued forcefully for the combination of signs and norms as the basis for a better conceptualization of organizational information systems in his critique of the FRISCO project and in reports on the MEASUR project (1994). More recently, the notion of complete requirements specifications drawn from largely static organizational perspectives is being challenged by lightweight or agile methodologies (Cockburn 2001), extreme programming (Beck 2000), SCRUM (Schwaber and Beedle 2002), and Crystal (Cockburn 2001). Another technique proposed to manage software evolution in changing environments is traceability (i.e., linking requirements to the artifacts being created; Ramesh and Jarke 2001). However, even these approaches underplay the organizational context, providing no explicit mechanisms to take into account concerns such as conflict or structural changes in the organization that may be brought about by the introduction of the information system.

As this overview suggests much work in software engineering is motivated by the idea that changes introduced late in the software development cycle are costly and should be avoided as much as possible. Information systems, thus, are seen as reflections of requirements instead of as agents of change for the intended environment. As reflections of complete requirements, the systems are expected to be relatively stable, resembling the computational view of the IT artifact described by Orlikowski and Iacono (2001).

2.2 A Social Science Perspective

A social science perspective, on the other hand, considers information systems as agents of change and not mere technical artifacts. Practitioners sharing this worldview consider the insertion of information systems in organizations as opportunities to study organizational consequences of the introduction of the artifact—following a nominal view of the IT artifact (Orlikowski and Iacono 2001). A number of these researchers using structuration theory also investigate the dualistic interplay of the organization and technology (DeSanctis and Poole 1994; Jones 1999; Lyytinen and Ngwenyama 1992; Orlikowski 1992; Orlikowski and Robey 1991; Truex, Baskerville, and Travis 2000; Walsham 2002; Walsham and Han 1991). Others using critical social theoretic approaches examine the roles of power differential and how IT emancipates or imprisons social actors. Still others, guided by actor network approaches examine the development and interactions of human and nonhuman networks. The thrust of this perspective is, therefore, on understanding and explaining the underlying logic of social organizations when they come into contact with information systems.

Jones (2000), based on a content analysis on the IFIP 8.2 working group proceedings, identifies predominant social science theories as Giddens' (1979, 1984, 1987) structuration theory, Habermas' (1984) theory of communicative action, and the actor-network theory (Latour 1987, 1993, 1996, 1998)—which attests to the complex relationship between the organizational context and the IT artifact. Further bolstered in terms of their appropriateness by Holmström and Truex (2001), these represent appropriate theoretical lenses to study organizational changes. Giddens' structuration theory, for example, is considered well-suited for exploring the ephemeral boundaries between IT and organization, characterizing IT as both enabler and constrainer of social action (DeSanctis and Poole 1994; Orlikowski 1992, 1996, 2000; Orlikowski and Robey 1991). Other social theories address related questions, but from different points of view. Critical social theory as applied to IS research addresses issues of conflict and of power differentials in ISD and IS use (Apel 1980; Bourdieu 1991; Habermas 1981, 1984). Actor-network theory addresses the complex relationships between human and non-human actants in rich social and technological and social domains (Callon 1986, 1991; Latour 1987, 1993, 1996, 1998). The challenge for these research streams is to go beyond the idea of reactively studying information systems as change agents to proactively improving specific ways of engineering systems that can contribute to the desired changes in the environment.

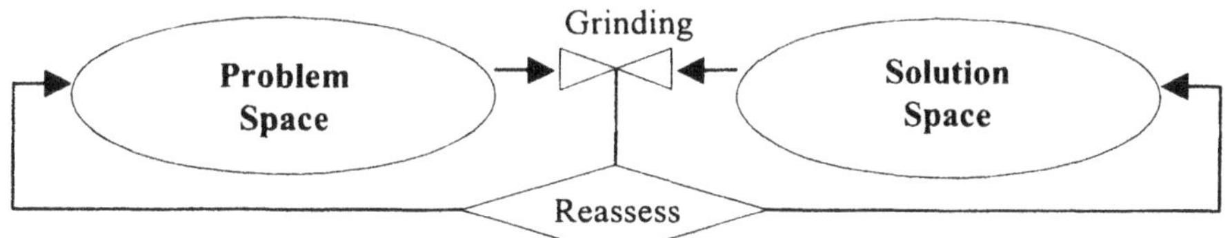

Figure 1. Heterogeneous Engineering (Adapted from Bergman et al. 2001)

2.3 Reconciling the Perspectives

Both the software engineering and organizational IS communities have been slow at recognizing that the unidimensional views embraced within their own research stream may be inadequate to deal with the difficult problems associated with developing and deploying complex information systems that must function within emergent organizations. Even recent articles defining the boundaries of IS research and of the conditions that define IS research often narrowly define the field, leaving aspects of design and human research out of the description (Benbasat and Zmud 2003; Weber 2003a, 2003b). A few recent writings, however, attest to the increasing recognition of the limits of these unidimensional approaches. Bergman et al. (2001), for example, suggest the term *heterogeneous engineering*, that is, requirements analysis as an iterative process moving between an existing solution space to problem space, then back again to the future solution space, a process they metaphorically call *grinding* (see Figure 1).

They recognize at a macro scale the interplay between the information system and the organization captured at the micro level (Guindon 1990; Mathiassen and Munk-Madsen 2000; Purao, Rossi, and Bush 2002) with the ideas of *problem* and *design* spaces. We note that this view presents an epistemological challenge in that it separates the social context (the organization) and the technical component (the IT artifact). An alternative is to see the two as parts of a whole, wherein the IT artifact becomes a part of the organizational fabric and where organizational assumptions are woven into the IT artifact. Viewing the two objects separately suggests a dualism versus seeing the duality of the organization and the IT artifact. A possible approach to this quandary is seen in the line of work loosely called emergent systems development or deferred systems development that is challenging traditional notions of systems development (Bansler et al. 2000; Bansler and Havn 2002; Bello et al. 2002; Patel 1999; Truex, Baskerville, and Klein 1999; Truex, Baskerville, and Travis 2000). This work suggests a fundamental shift from traditional development practices to a continuous redevelopment process (Truex, Baskerville, and Klein 1999).

The need to reconcile the two perspectives is also seen in the work of Liu (2000), and Liu et al. (2002a), which suggests use of semiotics for information systems engineering. Liu et al. (2002b) also argue for the need for codesigning information systems and organizational processes, presenting the hypothesis that "an organic integration of IT into both processes will allow both systems to evolve naturally" (p. 254), emphasizing that the codesigning of business and IT systems is an important research issue. De Moor (2002) suggests a requirements elicitation process that may allow surfacing of concerns and constraints to inform such codesign. Eatock et al. (2002) demonstrate the possibility of using simulation to study the problem, whereas Beeson et al. (2002) describe a case study focused on identifying and modeling links between information systems and business. These efforts are also indicative of the early research efforts in this emerging research area.

As a group, these authors further underscore the need to reconcile two research streams. Although the notions of "IT artifact as a reflection of requirements" and "IT artifact as change agent" may, prima facie, seem incommensurate, the above writings point to several possibilities for reconciling the two perspectives. Specifically, the line of work on deferred system development challenges the software engineering perspective away from its focus on the notion of reflecting requirements while recognizing the emergent nature of organizations. A specific approach to addressing this is to support the engineering of software in emergent organizations in the form of representations that may encompass both perspectives.

3. SUPPORTING THE ENGINEERING OF INFORMATION SYSTEMS IN EMERGENT ORGANIZATIONS

The engineering of software for emergent organizations presents several interlinked challenges. On one hand, the very idea of engineering an IT artifact requires us to confer upon it an ontological status that is separate from the organization in which it will be deployed. On the other hand, considering the organization as emergent requires that we consider the mutually interdependent nature of the IT artifact and the organization, removing the distinct ontological status accorded to it for the purpose of engineering. This mirrors the epistemological challenge described earlier, that is, *engineering in emergent organizations* requires us to simultaneously separate the social context (the organization) and the technical component (the IT artifact), as well as consider it as a duality, where the IT artifact is a part of the organizational fabric with organizational assumptions woven into the IT artifact.

Thus, we struggle with how we can facilitate such an integrated, yet discrete, perspective on IT artifacts and organizations. We believe that a prerequisite to reconciling these viewpoints is to consider each—the IT artifact as well as the organization—as malleable and being in a constant state of flux. The process of emergence (Truex, Baskerville, and Klein 1999), then, must be seen as one that is aimed at codesigning both the information systems and the organization. We note that codesigning refers to a process, not a specific outcome, that is, the process of codesigning may result in a series of temporary equilibria, which hold in balance the IT artifact and the organization. Facilitating a process that allows such codesigning also requires that we take into account developer behaviors observed in practice (Purao, Rossi, and Bush 2002), which point to cycles between the problem and design space (i.e., the organization and the IT artifact). The cycles are punctuated by apparent resting points or emergent regularities that allow the organization to function normally, that is, at times of relative regularity when the distance between the design and problem spaces may be comfortably slight.

3.1 The Role of Representations

As in other professional disciplines, both ISD research communities embrace the need to create representations of extant and changed *object* systems. Unlike architecture or engineering, however, representations play an even more vital role for information systems development for two reasons. The artifacts developed by information systems developers are virtual as opposed to real, making representations extremely important as a mode of communication. And the environments where these artifacts are intended to be deployed are themselves malleable, making the representations of interaction between the two even more important.

A modeling technique, in essence, externalizes representations that cannot be "held in the head" by a developer, allowing the developer to extend the short-term memory to notations that spill over into the physical world. The world being modeled, in all but the most trivial examples, is simply too complex; often it is too complex to be modeled by a single person. This requires then the sharing and codevelopment of the representation, requiring close interaction and communication between those creating the representation. The representations must, therefore, support effective communication between different participants in the development process. A representational technique can be used, for example, to envision a system, and its behavior, before specifying the system (for example, scenario-based design; Carroll 1995). Representation techniques are also important because of the significant role they play in problem-solving and development as aids to cognition (Bødker 1998; Suchman 1990). Better representation has been shown to directly contribute to better solutions by bringing to light solution approaches heretofore not evident (Larkin and Simon 1987). Finally, representation techniques are also important because they suggest to the developer what is most important and relevant, and what may be ignored. The absence of a technique for codesigning the organization and the artifact, therefore, implicitly suggests that organizational context is not important. Clearly this is an untenable position.

3.2 Representations to Support Engineering of Software

A large number of representation techniques have been proposed for the engineering of software. A representative example is found in the unified modeling language (UML), which suggests modeling the structural, behavioral, dynamic, and functional properties of IT artifacts. A survey by Wieringa (1998) classifies and discusses these techniques in detail. Several other techniques have been proposed over the years including the so-called traditional techniques such as the data flow diagram (Gane and Sarson 1979), entity-relationship modeling (Chen 1976) and numerous extensions thereof, as well as newer modeling techniques for object-oriented modeling such as UML and OML (Opdahl et al. 2000). Reviews and classifications of these techniques appear in Rossi and Brinkkemper (1996) and Siau et al. (1997). In addition, considerable work has been carried out in understanding appropriateness of modeling notations (Kim et al. 2000), modeling the modeling notations (Rossi et al. 1992), and evaluation of modeling notations (Wand and Weber 1993).

In their research manifesto for conceptual modeling Wand and Weber (2002) suggest four levels of focus for research activities. These include grammar, methods, scripts, and context. They indicate that while work has been carried out in the first two levels, much less work has been carried out for the last two. Work in the first stream has dealt with issues such as faithful representation, ontological expressiveness, and overlaps. Work in the second stream has dealt with issues of appropriate application of the grammar. Work in the third and fourth streams has dealt with the idea of using the grammars to generate scripts, and the use of conceptual modeling techniques in different contexts, different individuals, and their empirical investigation. The agenda presented by Wand and Weber (2002) has, at the core, the goal of accurate reflection of the universe of discourse (see section 2.1). They indicate, for example, their key motivation as "how can we model the world to better facilitate...information systems?' (Wand and Weber 2002, p. 363). The arguments we have presented so far require us to go beyond this perspective, treating the organization itself as emergent.

3.3 Representations to Describe Emergent Organizations

Few representation techniques have been proposed for modeling organizations in the context of designing information systems. Where such techniques have been proposed, they have been infused with metaphors borrowed from software engineering, imposed on the organizational practices and processes. For example, as early as 1979, data-flow-diagrams were proposed to model current business functions (Gane and Sarson 1979). Techniques such as BIAIT and PSL/PSA attempted to match the complex relationships between data, process, and owners of both in organizations. More recently, the IDEF suite of process maps (Mayer et al. 1995) has been proposed. These techniques have not directly addressed purely organizational concepts, instead focusing on only those aspects that will eventually be represented in an information system.

In contrast to these software engineering techniques are graphical approaches like rich pictures as devised by Checkland (1998, 1999) and adopted in Multiview by Avison and Wood-Harper (1990), and others in which cartoon-like drawings provide high-level representations identifying stakeholders, their concerns, and some of the structure underlying the work context. This technique remains informal, providing no direct mapping from the rich picture to other representation forms and models of data, work-flow, human interaction, or organizational process. Returning to textual representations, Boland (1985, 1991) and others have recognized the importance of the concept *metaphor* when deployed as a grammatical construct used in describing organizational context and ambiguity.

However, bridging from rich organizational contexts to the information system being developed has proved an elusive goal for several reasons. Organizations themselves are ephemeral and emergent. Processes and the interaction of human and machine actors are never really fixed. They are, in short, prototypical dynamic organisms, devilishly difficult to model and replicate. Scenarios are limited because each provides a representation of an organizational moment frozen in time. This technique offers no direct way to stitch the scenarios together to represent a continuously changing organizational tapestry. While scenarios offer the advantage of "ready to go" models, like all contingency approaches, anticipating when to deploy one scenario versus another and knowing in what particular order remains an inexact science.

3.4 Representations that Support Developer Engagement with Both Domains

The usefulness of a representation technique is moderated by how well it supports behaviors that developers may utilize in their development processes. A few prescriptive descriptions (e.g., Kruchten 1998) have attempted to specify correct modes of using representation techniques. However, a more practice-oriented view is necessary to fully understand how representation techniques may help developer behaviors. Further, it is necessary to understand how developers engage with both the envisioned artifact as well as the environment where the artifact is intended to be deployed even when there are no formal representation techniques at hand. There is some evidence that a systems analyst's mental models, experience, and training help address this challenge (Fitzgerald 1997; Lee and Truex 2000) but this dependence on non-systematic factors does not lead to the kind of replicability sought by creators of formal methods. Having representations to help in the process is therefore thought to advance the discipline in a more orderly fashion.

A recent study by Purao, Rossi, and Bush (2002) identifies specific behaviors that developers use for engaging with problem and design spaces, which loosely mirror the ideas of the organization, and the IT artifact. Their study reveals that developers do, indeed, treat both spaces as malleable and engage with each either in turn or simultaneously, and represent elements of each that affect their decisions in the other. They found evidence in developer behaviors that may be interpreted as indicating strong support for the duality between the information system and the domain of influence, which they term the problem space and the design space. Micro-cycles between the two spaces appear as back-and-forth engagements in the two spaces, which facilitate simulation and expansion of the design space. Problem space behaviors represent reinterpreting the problem space based on decisions in the design space. These three directions—engineering of software, describing the emergent organization, and accommodating developer behaviors—suggest the need for novel representational forms (or R-forms; Welke 1980) that can support engineering the emergence of both, the IT artifact as well as the organization in which it will be deployed.

4. ENGINEERING THE EMERGENCE: AN R-FORMS MANIFESTO

We argue that new R-forms (modeling and representation techniques; Welke 1980) are necessary to support the codesigning of information systems and organization (i.e., to support the three directions described above). Clearly, such a technique must not sacrifice the first direction, engineering of software, which has been traditionally the primary consideration for representational forms. Rather, it should strive to incorporate constructs or means to support the second dimension, portraying organizational emergence. Finally, it must facilitate and support the third direction, known developer behaviors such as cycles of engagement with the two spaces. Desirable attributes of such R-forms are difficult to identify because they require operationalizing micro-level behaviors and contextually mandated adjustments of the type outlined above. Such operationalizing will, of necessity, involve some over-determination reflecting the researchers' biases. Therefore, instead of presenting prescriptions that will dictate specific structures or procedures that may shape such R-forms, we outline a set of requirements that could guide research interested in the discovery and assessment of such R-forms. Our recommendations are guided by the following key shortcomings of existing R-forms.

- An implicit, exclusive focus on facilitating the engineering of the IT artifact
- An implicit mode of representing frozen moments in time, favoring a snapshot view instead of an evolving perspective
- A codification of universal descriptions of work practices and processes instead of recognizing the particulars and context-laden specializations
- A fixation on issues such as *correctness* and *completeness* favoring the assumption of a closed set of the functionality of the IT artifact
- An implicit operationalization of the duality between the organization and the IT artifact instead of recognizing the dualism that the two represent
- Treating only the IT artifact as malleable during the development process and the organization as rigid

It is necessary to examine whether we have a paradox in the making specifically with our call for R-forms to address the third shortcoming indicated above. For example, it has been argued (Truex, Baskerville, and Klein 1999; Truex, Baskerville, and Travis 2000) that standard, universal, one-size-fits-all solutions are not appropriate for organizational life, which is complex and can often present a series of unique challenges. Current prescriptions for R-forms, on the other hand, require classification, generalization, and specification of constructs that may be embedded in the IT artifact. Is it possible, then, to reconcile the concept of uniqueness and organizational emergence against that of classification, embedded in the current R-forms, which allow IT artifacts to be engineered? A fully developed and cogent answer will, in fact, take the shape of a novel R-form. One possible approach to reconciling the two is suggested by Parsons and Wand (2000) in the context of database design, where they separate the specification of the schema layer and the instance layer, allowing varying mapping between the two. Work related to schema evolution in object-oriented databases (Franconi et al. 2000) also suggests that it is, in fact, possible to accord instances a first-class citizenship. These efforts suggest that it is possible to construct IT artifacts that respect the particularities of organizational life.

In particular, realizing such an R-form will require that we take account of a number of organizational concepts familiar to the IFIP WG 8.2 community. Those concepts include (1) the relative persistence of certain social structures when reinforced by technologies supporting dominant values and power relationships, (2) the interplay and relative agency of both human and organizational actors, (3) how systems may be used to consolidate power and control or to emancipate, and (4) the process of negotiation enabling constant changes in organizational process and meanings. Doubtless there are other such notions requiring attention. IS researchers have identified theories and have advanced the discourse on these topics. We have found ways to examine and describe these concepts in the interplay of ISD and IS use in organizational life, but not to model them effectively or parsimoniously. It is difficult to capture these concepts because they represent ephemeral and dynamic social processes. Even so, they should be represented in our descriptions of systems.

4.1 R-Forms: A Set of Requirements

Based on the discussion above, we outline a set of requirements for R-forms. Where possible, we indicate ongoing work that may be leveraged to help creation of such R-forms.

- **Separating universals and particulars**. The R-form should allow modeling of categories and classes as well as instances that may not fit into the categories. A similar case is made by Parsons and Wand (2000) in their suggestions for a new mode of database design that separates schemas and instances.
- **Portraying multiple perspectives.** The R-form should allow multiple perspectives to deal with the underlying complexity of the phenomenon— the IT artifact as well as the organizational environment in which it will function. A similar use of multiple perspectives is seen in the now *de facto* standard for modeling object-oriented software, UML (OMG 2003).
- **Capturing organizational characteristics.** The R-form should allow capturing organizational actors, relationships among them, and fuzzy notions such as power. Techniques analogous to these are found in the portrayal of roles in workflow diagrams (van der Aalst et al. 2003), and those for illustrating webs of power and influence and interaction (Kling 1987; Kling and Scacchi 1982). Early work related to rich pictures represents another direction that can be useful here.
- **Balancing formal and informal descriptions.** The R-form should not only possess sufficient expressive power to capture, informally, the organizational characteristics, but should also contain sufficient formalisms to capture the details needed for engineering the software solutions. The balance between formal and informal descriptions is also necessary to ensure ease of use for the developers.
- **Bridging the duality between the IT artifact and the organization.** The R-form should allow mechanisms to ensure that elements from the IT artifact can be mapped against those in the organization and vice versa to ensure internal consistency. A similar requirement for consistency is enforced by UML modeling conventions (OMG 2003). Such conceptual bridging should ensure that both, the IT artifact itself, as well as the organization, is treated as malleable, allowing codesign of both. One technique addressing this goal is suggested in recent work by Purao, Truex, and Cao (2003).
- **Allowing design to continue beyond deployment.** The R-form should recognize that emergence must continue past deployment of the information system, and that the process of designing should continue to encompass what is traditionally referred to as the maintenance stage. A possible approach to achieving this suggested by Chua et al. (2003).
- **Supporting progress without commitment.** The R-form should allow capturing incomplete decisions allowing progress without commitment. A similar idea is put

forward in the context of use cases (see survey by Hurlbut 2003), which allow specifications to evolve without requiring commitment.

- **Managing complexity without defining it away.** It should allow means to simultaneously represent the whole of complex settings while allowing users to zero in on elements requiring immediate focus. This is analogous to the decomposition allowed in traditional data-flow diagraming. On the one hand, the analyst should be able to represent the whole of the system, perhaps in a rich picture form and follow with a logical mapping through successive layers of detail to the desired level and objects—either technical or organizational—of interest.

Interesting conceptual advances have been made in a number of fields that may lead to possible solutions that meet the above requirements. We have already pointed out several advances in the discussion above. A complete explication of these is beyond the purview of this paper, and continues to be part of our ongoing research agenda. A few promising directions we are considering include the ideas of patterns (Fowler 1997), metaphors (Lakoff and Johnson 2003; Madsen 1994), and scenarios (Carroll 1995) and in the work of those exploring deferred systems development approaches.

5. CONCLUDING REMARKS

Combining insights from software engineering and social theories, we have argued that new R-forms are necessary to support the codesign and coevolution of problem and design spaces. In doing so, we are presenting a counterargument to that presented by Wand and Weber (2002). Their agenda focuses solely on the concern of accurate reflection of requirements, which they indicate as "how can we model the world better to facilitate the development of information systems?" The agenda we have argued for in this paper suggests expanding the set of concerns to address the question, how can we model the envisioned IT artifact as well as the context in which it will be deployed better to support the codesign and coevolution of both, the IT artifact as well as the organizational context?

We have positioned our arguments in a manner that does not detract from the notion of engineering IT artifacts, which we believe continues to be an important dimension of the systems development process. This dimension, however, does not take account of the symbiotic relationship between the IT artifact and the organizational context—a relationship that is better described as emergence. In our arguments, we have attempted to integrate these orthogonal dimensions to argue for novel R-forms that may better support engineering of IT artifacts in emergent organizations.

As of this writing there are no fora dedicated to the goal of reconciling these important and complex questions. Thus we turn to the IFIP WG 8.2 community, which in it earliest days chose to cast the gauntlet to the establishment with a call for the IS research community and for the establishment within that community to consider more expansive and less familiar ways by which one could consider and research IS phenomenon and open minds as to the kinds of ontological and epistemological questions that would be relevant to such studies. The IFIP WG 8.2 community has demonstrated its ability to embrace new and challenging approaches to IS research. We hope they are willing and able to take the challenge we have outlined and offer both advice as to how to advance the discourse and a forum in which it may be developed.

REFERENCES

Apel, K.-O. *Towards A Transformation of Philosophy*, Translated by G. Adey and D. Frisby, Edited by John O'Neill, London: Routledge & Kegan Paul, 1980.

Avison, D. E., and Wood-Harper, A. T. *MULTIVIEW: An Exploration in Information Systems Development*, Oxford: Blackwell Scientific Publications, 1990.

Bansler, J. P.; Damsgaard, J.; Scheepers, R.; Havn, E.; and Thommesen, J. "Corporate Intranet Implementation: Managing Emergent Technologies and Organizational Practices," *Journal of the Association of Information Systems* (1), December 2000.

Bansler, J. P., and Havn, E. "Improvisation and Bricolage in Information Systems Development: A Field Study," Working Paper 11-11-02, 10, Technical University of Denmark, Copenhagen, 2002.

Beck, K. *eXtreme Programming Explained*, Reading, MA: Addison Wesley, 2000.

Beeson, I.; Green, S.; Sa, J.; and Sully, A. "Linking Business Processes and Information Systems Provision in a Dynamic Environment," *Information Systems Frontiers* (4:3), 2002, pp. 317–329.

Bello, M.; Sorrentino, M.; and Virili, F. "Web Services and Emergent Organizations: Opportunities and Challenges for IS Development," in S. Wrycza (Ed.), *Proceedings of the 10th European Conference on Information Systems*, Gdansk, Poland, 2002, pp. 439-449.

Benbasat, I., and Zmud, R. W. "The Identity Crisis within the IS Discipline: Defining and Communicating the Discipline's Core Properties," *MIS Quarterly* (27:2), June 2003, pp. 183-194.

Bergman, M.; King, J.; and Lyytinen, K. "Large Scale Requirements Analysis as Heterogeneous Engineering," in C. Floyd and R. Klischewski (Eds.), *Social Thinking— Software Practice*, Cambridge, MA: MIT Press, 2001.

Bødker, S. "Understanding Representation in Design," *Human-Computer Interaction* (13:2), 1998, pp. 107-125.

Boehm, B. "Agile and Plan-Driven Methodologies: Oil and Water," Presentation at Agile Universe 2002, August 5, 2002 (available online at http://www.agilealliance.com/articles/articles/agileAndPlanDrivenMethods.pdf; accessed December 30, 2003).

Boland, Jr., R. J. "Phenomenology: A Preferred Approach to Research on Information Systems," in E. Mumford, R. Hirschheim, G. Fitzgerald, and A. T. Wood-Harper (Eds.), *Research Systems in Information Systems*, Amsterdam: North-Holland, 1985, pp. 193-201.

Boland, R. "Information Systems Use as a Hermeneutic Process," in H-E., Nissen, H. K. Klein, and R. Hirschheim (Eds.), *Information Systems Research: Contemporary Approaches & Emergent Traditions*, Amsterdam: North-Holland, 1991, pp. 439-458.

Bourdieu, P. *Language and Symbolic Power*, Cambridge, MA: Harvard University Press, 1991.

Callon, M. "Some Elements of a Sociology of Translation: Domestication of the Scallops and the Fishermen of St Brieuc Bay," in J. Law (Ed.), *Power Action and Belief: A New Sociology of Knowledge*, London: Routledge, 1986, 196-233.

Callon, M. "Techno-Economic Networks and Irreversibility," in J. Law (Ed.), *A Sociology of Monsters: Essays on Power, Technology and Domination*, London: Routledge, 1991, 132-161.

Carroll, J. M. *Scenario Based Design*, New York: Wiley & Sons, 1995.

Checkland, P. *Systems Thinking, Systems Practice*, New York: John Wiley & Sons, Ltd., 1999.

Checkland, P., and Holwell, S. *Information, Systems, and Information Systems: Making Sense of the Field*, Chichester, England: John Wiley & Sons, 1998.

Chen, P. P. S. "The Entity-Relationship Model: Towards a Unified View of Data," *ACM Transactions on Database Systems* (1:1), 1976, pp. 9-36.

Chua, C. E. H.; Purao, S.; and Storey, V. "An Approach for Developing Maintainable Software," in J. Ross and D. Galletta (Eds.), *Proceedings of the 9th Americas Conference on Information Systems*, Tampa Florida, 2003, pp. 1911-1921.

Cockburn, A. "Agile Software Development Ecosystems," in A. Cockburn and J. Highsmith (Eds.), *The Agile Software Development Series*, Reading, MA: Addison Wesley Longman, 2001.

De Moor, A. "Language/Action Meets Organizational Semiotics: Situating Conversations with Norms," *Information Systems Frontiers* (4:3), 2002, pp. 257–272.

DeSanctis, G., and Poole, M. S. "Capturing the Complexity in Advanced Technology Use: Adaptive Structuration Theory," *Organization Science* (5:2), 1994, pp. 121-147.

Eatock, J.; Paul, R. J.; and and Serrano, A. "Developing a Theory to Explain the Insights Gained Concerning Information Systems and Business Process Behavior: The ASSESS-IT Project," *Information Systems Frontiers* (4:3), 2002, pp. 303–316.

Fitzgerald, B. "The Use of Systems Development Methodologies in Practice: A Field Study," *Information Systems Journal* (7), 1997, pp. 201-212.

Fowler, M. *Analysis Patterns: Reusable Object Models*, Reading, MA: Addison Wesley, 1997.

Franconi, E.; Grandi, F.; and Mandreoli, F. "A Semantic Approach for Schema Evolution and Versioning in Object-Oriented Databases," *Computational Logic 2000*, London, July 24-28, 2000.

FRISCO. "A Framework of Information Systems Concepts: The FRISCO Report," 1997 (available online at http://www.wi.leidenuniv.nl/~verrynst/fri-full-7.pdf).

FRISCO "A Framework of Information System Concepts: The Revised Frisco Report," Draft January 2001 (available online at http://www.wi.leidenuniv.nl/~verrynst/Draft-Jan-01.zip).

Gane, E. C., and Sarson, T. *Structured Systems Analysis: Tools and Techniques*, Englewood Cliffs, NJ: Prentice-Hall, 1979.

Giddens, A. *A Contemporary Critique of Historical Materialism*, Volume 2, Berkeley, CA: University of California Press, 1987.

Giddens, A. *Central Problems in Social Theory: Action, Structure and Contradiction in Social Analysis*, Berkeley, CA: University of California Press, 1979.

Giddens, A. *The Constitution of Society: Outline of the Theory of Structuration*, Cambridge, MA: Polity Press, 1984.

Guindon, R. "Designing the Design Process: Exploiting Opportunistic Thoughts," *Human-Computer Interaction* (5:2-3), 1990, pp. 305-344.

Habermas, J. "Reason and the Rationalization of Society," Chapter 3 in *The Theory of Communicative Action*, Boston: Beacon Press, 1981, pp. 273-333.

Habermas, J. *The Theory of Communicative Action*, Volume 1, Boston: Beacon Press, 1984.

Hesse, W.; and Verrijn-Stuart, A. A. "Towards a Theory of Information Systems: The FRISCO Approach," in *Proceedings of the 10th European-Japanese Conference on Information Modeling and Knowledge Bases*, Saariselkä, Finland, 2000.

Holmström, J., and Truex, D. "What Does it Mean to be an Informed IS Researcher? Some Criteria for the Selection and Use of Social Theories in IS Research," in S. Bjørnestad, R. E. Moe, A. I. Mørch, and A. L. Opdahl (Eds.), *Proceedings of the Information Systems Research in Scandinavia 2001 Conference*, Volume 2, Bergen, Norway, August 11-14, 2001, pp. 313-326.

Hurlbut, R. *A. Survey of Approaches For Describing and Formalizing Use Cases*, Wheaton, IL: Expertech, Ltd., 2003 (available online at http://www.iit.edu/~rhurlbut/xpt-tr-97-03.html; accessed August 20, 2003).

Jones, M. "The Moving Finger: The Use of Social Theory in WG 8.2 Conference Papers, 1975-1999," in R. Baskerville, J. Stage, and J. I. DeGross (Eds.), *Organizational and Social*

Perspectives on Information Technology, Boston: Kluwer Academic Publishers, 2000, pp. 15-31.

Jones, M. "Structuration Theory," in W. L. Currie and R. D. Galliers (Eds.), *Re-Thinking Management Information Systems*, Oxford: Oxford University Press, 1999, pp. 103-135.

Kim, J.; Hahn, J.; and Hahn, H. "How Do We Understand a System with (So) Many Diagrams? Cognitive Integration Processes in Diagrammatic Reasoning," *Information Systems Research* (11:3), 11:3), September 2000, pp. 284-303.

Kling, R. "Computerization as an Ongoing Social and Political Process," in G. Bjerknes, P. Ehn, and M. Kyng (Eds.), *Computers and Democracy: A Scandinavian Challenge*, Aldershot, England: Avebury, 1987, pp. 117-136.

Kling, R., and Scacchi, W. "The Social Web of Computing: Computer Technology as Social Organization," *Advances in Computers* (21), 1982, pp. 2-90.

Kruchten, P. *The Rational Unified Process: An Introduction*, Reading, MA: Addison-Wesley, 1998.

Lakoff, G., and Johnson, M. *Metaphors We Live By*, Chicago: University of Chicago Press, 2003.

Larkin, J., and Simon, H. "Why a Diagram Is (Sometimes) Worth Ten Thousand Words," *Cognitive Science* (11), 1987, pp. 65-99

Latour, B. "From the World of Science to that of Research?," *Science*, April 1998, pp. 14-19.

Latour, B. *Science in Action*, Cambridge, MA: Harvard University Press, 1987.

Latour, B. "Social Theory and the Study of Computer Work Sites," in W. J. Orlikowski, G. Walsham, M. R. Jones, and J. I. DeGross (Eds.), *Information Technology and Changes in Organizational Work*, London: Chapman & Hall, 1996, pp. 295-307.

Latour, B. *We Never Have Been Modern (Nous n'avons jamain été modernes)*, Translated by Catherine Porter, Hemel Hempstead, England: Harvester Wheatsheaf, 1993.

Lee, J., and Truex, D. "Exploring the Impact of Training in ISD Methods on the Cognitive Structure of Novice Information Systems Developers," *Information Systems Journal* (10:4), October 2000, pp. 347-368.

Levine, L.; Baskerville, R.; Loveland Link, J. L.; Pries-Heje, J.; Ramesh, B.; and Slaughter, S. "Discovery Colloquium: Quality Software Development @ Internet Speed," SEI Technical Report CMU/SEI 2002-TR-020, ESC-TR-2002-020, Software Engineering Institute, Carnegie-Mellon University, 2002.

Liu, K. *Semiotics in Information Systems Engineering*, Cambridge, England: Cambridge University Press, 2000.

Liu, K.; Clarke, R. J.; Andersen, P. B.; and Stamper, R. K. *Organizational Semiotics: Evolving a Science of Information Systems*, Boston: Kluwer Academic Publishers, 2002a.

Liu, K.; Sun, L.; and Bennett, K. "Co-design of Business and IT Systems—Introduction," *Information Systems Frontiers* (4:3), 2002b, pp. 251-256.

Lyytinen, K. J., and Ngwenyama, O. K. "What Does Computer Support for Cooperative Work Mean? A Structurational Analysis of Computer Supported Cooperative Work," *Accounting, Management and Information Technology* (2:1), 1992, pp. 19-37.

Madsen, K. H. "A Guide to Metaphorical Design," *Communications of the ACM* (37:12), December 1994, pp. 57-62.

Mathiassen, L., and Munk-Madsen, A. *Object Oriented Analysis & Design*, Aalborg, Denmark: Marko Publishers, 2000.

Mayer, R. J.; Benjamin, P. C.; Caraway, B. E.; and Painter, M. K. "A Framework and a Suite of Methods for Business Process Reengineering," in V. Grover and W. J. Kettinger (Eds.), *Business Process Change: Reengineering Concepts, Methods, and Technologies*, Harrisburg, PA: Idea Group Publishing, 1995, pp. 245-290.

Mylopoulos, J.; Chung, L.; and Nixon, B. "Representing and Using Non-Functional Requirements: A Process-Oriented Approach," *IEEE Transactions on Software Engineering* (18:6), June 1992, pp. 483-497.

Nuseibeh, B. "Weaving Together Requirements and Architectures," *IEEE Computer* (34:3), 2001, pp. 115-117.

Nuseibeh, B., and Easterbrook, S. "Requirements Engineering: A Roadmap," *Software Engineering* (20:10), 2000, pp. 760-773.

Nuseibeh, B.; Kramer, J.; and Finkelstein, A. C. W. "A Framework for Expressing the Relationships between Multiple Views in Requirements Specification," *IEEE Transactions on Software Engineering* (20:10), 1994, pp. 760-773.

OMG. "Unified Modeling Language, Standard Specification Version 2.0," Object Management Group, 2003 (available online at http://www.omg.org/uml, accessed November 30, 2003).

Opdahl, A. L; Henderson-Sellers, B.; and Barbier, F. "An Ontological Evaluation of the OML Metamodel," in E. D. Falkenberg, K. Lyytinen, and A. A. Verrijn-Stuart (Eds.), *Information Systems Concepts: An Integrated Discipline Emerging*, Boston: Kluwer Academic Publishers, 2000, pp. 217-232.

Orlikowski, W. J. "The Duality of Technology: Rethinking the Concept of Technology in Organizations," *Organization Science* (3:3), 1992, pp. 398-429.

Orlikowski, W. J. "Improvising Organizational Transformation Over Time: A Situated Change Perspective," *Information Systems Research* (7:1), 1996, pp. 63-92.

Orlikowski, J. "Using Technology and Constituting Structures: A Practice Lens for Studying Technology in Organizations," *Organization Science* (11:4), July-August, 2000, pp. 404-428.

Orlikowski, W. J., and Iacono, C. S. "Research Commentary: Desperately Seeking the 'IT' in IT Research—A Call to Theorizing the IT Artifact," *Information Systems Research* (12:2), 2001, pp. 121-134.

Orlikowski, W. J., and Robey, D. "Information Technology and the Structuring of Organizations," *Information Systems Research* (2:2), June 1991, pp. 143-169.

Parsons, J., and Wand, Y. "Emancipating Instances from the Tyranny of Classes in Information Modeling," *ACM Transactions on Database Systems* (25:2), 2000, pp. 228-268.

Patel, N. V. "Developing Tailorable Information Systems through Deferred System's Design," in W. D. Haseman and D. L. Nazareth (Eds.), *Proceedings of the 5th Americas Conference on Information Systems*, Milwaukee, WI, 1999, pp. 4-6.

Pressman, R. S., and Associates. "Glossary of Software Engineering Terms," 2003 (available online at http://www.rspa.com/spi/glossary.html; accessed September 30, 2003).

Probert, S. K., and Rogers, A. "Understanding Hard and Soft IS Development Methods: Paradigmatic Rigidities or Different Ends of a Spectrum?," in W. D. Haseman and D. L. Nazareth (Eds.), *Proceedings of the 5th Americas Conference on Information Systems*, Milwaukee, WI, 1999, pp. 650-652.

Purao, S.; Rossi, M.; and Bush, A. "Towards an Understanding of the Use of Problem and Design Spaces During Object-Oriented System Development," *Information and Organization* (12:4), 2002, pp. 249-281.

Purao, S.; Truex, D.; and Cao, L. "Now the Twain Shall Meet: Combining Social Sciences and Software Engineering to Support Development of Emergent Systems," in G. Ross and D. Galletta (Eds.), *Proceedings of the 9th Americas Conference on Information Systems*, Tampa, FL, 2003, pp. 2738-2744.

Ramesh, B., and. Jarke, M. "Towards Reference Models for Requirements Traceability," *IEEE Transactions on Software Engineering* (27:1), 2001, pp. 58-93.

Robinson, W. N.; Pawlowski, S. D.; and Volkov, V. "Requirements Interaction Management," *ACM Computing Surveys* (35:2), 2003, pp. 132-190.

Rossi, M., and Brinkkemper, S. "Complexity Metrics for Systems Development Methods and Techniques," *Information Systems Journal* (21:2), 1996, pp. 209-227.

Rossi, M.; Gustafsson, M.; Smolander, K.; Johansson, L-A.; and Lyytinen, K. "Metamodeling Editor as a Front End Tool for a CASE Shell," in P. Loucopoulos (Ed.), *Advanced Systems Engineering, CAiSE'92*, Manchester, UK, May 12-15, 1992, pp. 546-567.

Schwaber, K., and Beedle, M. *Agile Software Development with Scrum*, Upper Saddle River, NJ: Prentice Hall, 2002.

Siau, K.; Wand, Y.; and Benbasat, I. "The Relative Importance of Structural Constraints and Surface Semantics in Information Modeling," *Information Systems Journal* (22:2/3), 1997, pp. 155-170.

Stamper, R. K. "Information Systems as a Social Science: An Alternative to the FRISCO Formalism," in E. D. Falkenberg, K. Lyytinen, and A. A. Verrijn-Stuart (Eds.), *Information Systems Concepts: An Integrated Discipline Emerging*, Boston: Kluwer Academic Publishers, 2000, pp. 1-51.

Stamper, R. "Social Norms in Requirements Analysis—An Outline of MEASUR," in M. Jirotka and J. A. Goguen (Eds.), *Requirements Engineering: Technical and Social Aspects*, New York: Academic Press, 1994, pp. 107-139.

Suchman, L. A. "What is Human-Machine Interaction," in S. P. Robertson, W. Zachary, and J. B. Black (Eds.), *Cognition, Computing, and Cooperation*, Norwood, NJ: Ablex Publishing Corporation, 1990, pp. 25-55.

Truex, D. P.; Baskerville, R.; and Klein, H. K. "Growing Systems in an Emergent Organization," *Communications of the ACM* (42:8), August 1999, pp. 117-123.

Truex, D. P.; Baskerville, R.; and Travis, J. "Amethodical Systems Development: The Deferred Meaning of Systems Development Methods," *Accounting Management and Information Technologies* (10), 2000, pp. 53-79.

Van der Aalst, W. M. P.; ter Hofstede, A. H. M.; and Weske, M. "Business Process Management: A Survey," in W. M. P. van der Aalst, A.Hm. M. ter Hofstede, and M. Weske (Eds.), *Proceedings of the International Conference on Business Process Management*, 2003, pp. 1-12.

Walsham, G. "Cross-Cultural Software Production and Use: A Structural Analysis," *MIS Quarterly* (26:4), 2002, pp. 359-379.

Walsham, G., and Han, C. K. "Structuration Theory and Information Systems Research," *Journal of Applied Systems Analysis* (17), 1991, pp. 77-85.

Wand, Y., and Weber, R. "On the Ontological Expressiveness of Information Systems Analysis and Design Grammars," *Journal of Information Systems* (3:4), 1993, pp. 217-237.

Wand, Y., and Weber, R. "Research Commentary: Information Systems and Conceptual Modeling—A Research Agenda," *Information Systems Research* (13:4), 2002, pp. 363-376.

Weber, R. "Editor's Comments: Still Desperately Seeking the IT Artifact," *MIS Quarterly* (27:2), June 2003a, pp. iii-xi.

Weber, R. "Editor's Comments: Theoretically Speaking," *MIS Quarterly* (27:3), September 2003b, pp. iii-xiii.

Welke, R. J. "A Context Approach to Information Systems Development Synthesis," ISRAM Working Paper W-8007-1.0, McMaster University, Hamilton, Ontario, Canada, 1980.

Wieringa, R. "A Survey of Structured and Object-Oriented Software Specification Methods and Techniques," *ACM Computing Surveys* (30:4), 1998, pp. 459-527.

Zave, P. "Classification of Research Efforts in Requirements Engineering," *ACM Computing Surveys* (29:4), 1997, pp. 315-321.

ABOUT THE AUTHORS

Sandeep Purao received his Ph.D. in management information systems from the University of Wisconsin-Milwaukee in May 1995. His research focuses on various aspects of information system design and development in organizational settings. His current projects include reuse-based design of information systems, evaluation of information systems, co-design of information systems and organizations, and measurement of information systems. His work has been published in *Information Systems Research*, *ACM Computing Surveys*, *Communications of the ACM*, *IEEE Transactions on Systems, Man and Cybernetics*, *Information & Organizations*, *Decision Support Systems*, and other venues.

Duane Truex researches the social impacts of information systems and emergent ISD. He is an associate editor for the *Information Systems Journal*, has coedited two special issues of *The Database for Advances in Information Systems*, and is on the editorial board of the *Scandinavian Journal of Information Systems*, the *Journal of Communication, Information Technology & Work*, and the *Online Journal of International Case Analysis*. His work has been published in *Communications of the ACM, Accounting Management and Information Technologies, The Database for Advances in Information Systems, European Journal of Information Systems, le journal de la Societé d'Information et Management, Information Systems Journal, Journal of Arts Management and Law, IEEE Transactions on Engineering Management*, and 40 assorted IFIP transactions, edited books, and conference proceedings. He is a member of the Decision Sciences and Information Systems faculty in the Chapman Graduate School, College of Business, at Florida International University, and is an associate professor on leave from the Computer Information Systems Department, Robinson College of Business, at Georgia State University.

Part 3:

Critical Interpretive Studies

12 THE CHOICE OF CRITICAL INFORMATION SYSTEMS RESEARCH

Debra Howcroft
Manchester University

Eileen M. Trauth
Pennsylvania State University

Abstract Accompanying the development and diffusion of information technologies throughout organizations and society is the research challenge to examine the relationship between information systems and the organizations and societies within which they are embedded. As the field of information systems matures, it is fitting that consideration be given to the ways in which such an examination is carried out. Thus, there is a research need not only to examine and assess the ways in which information systems are used and affect people; there is also a research need to examine and assess the research approaches that are used to carry out these assessments. This paper examines the enactment of the critical tradition in IS research and the possibilities for new insights that can arise from shifting the lens from positivist or interpretive to critical. This consideration leads to a discussion of issues that arise from the choice of critical IS research, followed by some recommendations for addressing these issues.

Keywords: Critical theory, critical management studies, critical information systems, gender, information systems development, research methods, information systems research

1 INTRODUCTION

Accompanying the development and diffusion of information technologies throughout organizations and society is the research challenge to examine the relationship between information systems and the organizations and societies within which they are embedded. Consideration should also be given to the ways in which such an examination is carried out. Thus, there are two research needs: (1) to examine and assess the ways in which information systems are used and affect people; and (2) to examine and assess the *research approaches* that are used to carry out these assessments. This recognition is reflected in the number of research papers, books, and conferences devoted to the topic of research approaches in IS. This paper contributes to this literature by considering the contribution and effect of employing a particular research tradition—the critical tradition—in the conduct of information systems research.

The paper is structured in the following way. Following a review of the critical tradition in social science research and its enactment in IS research to date, we consider the new insights that can arise from shifting the lens from the positivist or the interpretive to the critical. This consideration leads to a discussion of issues that arise from the choice of critical IS research and some recommendations for addressing them.

2 WHAT IS A CRITICAL EPISTEMOLOGY?

In the social sciences the term *critical* is used describe a range of related approaches, including critical theory (Horkheimer 1976), critical operational research (Mingers 1992), critical accounting (critical perspectives on accounting), critical ethnography (Forester 1992),and critical management studies (Alvesson and Willmott 1996). Each of these is subject to its own disciplinary connotations (Mingers 2000). However, a commonality across all of them is their dependence upon the critical theory of the Frankfurt School (Hammersley 1995).

Yet, despite some areas of commonality, there are some fairly distinct styles in the way research is performed (geographically, institutionally, and disciplinarily), resulting in a diversity of intellectual activity, some of which is indeed oppositional (e.g., realism vs. relativism, class politics vs. gender politics). Hence, there exists a broad range of epistemological/ontological positions, which fall under the critical umbrella, and which draw upon a variety of social theories and social thinkers. These include, for example, the Frankfurt School of critical theory (Horkheimer 1976), actor-network theory (Latour 1991), Marxism (Marx 1974), feminist theory (Wajcman 1991), and the work of Bourdieu (1990), Dooyeweerd (1973), Foucault (1979), and Heidegger (1953).

In contrast to the diversity within the social sciences, information systems research has been dominated by the Frankfurt School generally (Brooke 2002a), and more particularly, the work of Habermas (Doolin and Lowe 2002), with a core of authors committed to this area (Cecez-Kecmanovic 2001; Cecez-Kecmanovic et al. 1999; Hirschheim and Klein 1994; Klein and Hirschheim 1993; Lyytinen 1992; Lyytinen and Hirschheim 1988, 1989; Lyytinen and Klein 1985; Ngwenyama 1991; Ngwenyama and Lee 1997). Some have argued that the relative dominance of the Habermasian approach is unnecessarily limiting (Doolin and Lowe 2002) and have called for enrolling other critical social theorists whose work could be of relevance to IS (Brooke 2002b).

In order to outline a critical epistemology, we will draw upon five key themes emanating from the critical management studies (CMS) literature and resonating with IS research. The first theme, emancipation (Alvesson and Willmott 1992), is fundamental in a range of intellectual traditions be it Habermasian, feminist, or Marxist research. A thread running through all of these perspectives is a commitment to freeing individuals from power relations around which social and organizational life are woven (Fournier and Grey 2000). The objective is to focus on "the oppositions, conflicts and contradictions in contemporary society, and seeks to be emancipatory; that is, it should help to eliminate the causes of alienation and domination" (Myers and Avison 2002, p. 7).

The second theme, critique of tradition, seeks to disrupt rather than reproduce the status quo. Whereas traditional accounts seek to justify organizational and technological imperatives as natural and/or unavoidable, CMS challenges rather than confirms that which is established and encourages dissent rather than accepting surface consensus. This critique of tradition (Mingers 2000) endeavors to upset existing patterns of power and authority. Critical research questions and deconstructs the taken-for-granted assumptions inherent in the status quo, and interprets organizational activity (including information systems) by recourse to a wider social, political, historical, economic, and ideological context (Doolin 1998).

The third theme, non-performative intent (Fournier and Grey 2000), concerns the rejection of the provision of tools to support and assist managerial efficiency through reengineering minimum inputs for maximum outputs. This stands in contrast to non-CMS research, which aims to develop knowledge that contributes to the production of maximum output for minimum input (means-ends calculation). Similar claims are made on behalf of technology in general and information systems in particular, which are seen as augmenting the power of managerial decision-making.

The fourth theme, critique of technological determinism, challenges the discourse surrounding socio-economic change—be it post-industrial society, information society, or globalization (Avgerou 2002)—which assumes that technological development is autonomous and that societal development is determined by the technology (Bijker 1995). It disrupts the inner logic of technology as a given, something that is assumed to provide an effective and reliable vehicle for social and organizational change (Williams and Edge 1996). Critique of the technological determinist tradition highlights both its explanatory inadequacy and its ideological function of furthering the vested interests in technical change (Russell and Williams 2002).

The final theme, reflexivity, highlights a methodological distinction between critical management research and other management research. Whereas mainstream management studies are positivistic, critical management research engages in a critique of objectivity (Mingers 2000) which questions the validity of objective, value-free knowledge and information that is available, noting how this is often shaped by structures of power and interests. Like interpretive research, critical research engages in philosophical and methodological reflexivity (Fournier and Grey 2000).

3 EMPLOYING A CRITICAL APPROACH IN IS RESEARCH

It can be argued that the social nature of activities associated with the development, implementation, and use of IS and the management of people who carry out these activities leads naturally to considerations of social and political power. This consideration of power, in turn, encourages critical analysis (e.g., Beath and Orlikowski 1994; Franz and Robey 1984; Markus 1983). Yet even though there is a body of IS research concerned with issues of power and politics, it has not necessarily been identified as critical IS. Nevertheless, there has been a considerable shift in the research landscape since the publication of the seminal paper by Orlikowski and Baroudi (1991) which noted the dearth of critical IS research. While the end of the 20th century witnessed the gradual emergence of a critical stream in IS research, albeit generally in the Habermasian tradition, the last few years has seen the effort to focus more explicit attention on critical perspectives. This is evidenced in an increasing number of publications, conference streams, special issues, and academic electronic networks concerned with discussing critical IS. Continuing in that vein, the discussion below considers alternative insights into the assessment of information systems uses and impacts that can arise from critical IS research. Just as Trauth and Jessup (2000) showed that different results emerge from an assessment of IT when the lens shifts from positivist to interpretive, we show how further insights can result from shifting the lens to critical.

3.1 Critical Examination of Information Systems Development

The area of information systems development (ISD) is vast and so this section will briefly highlight some of the differences in the various epistemological perspectives. The positivist epistemology characterizes systems development as one of rationality whereby actions are justified on rational grounds and the appropriate organizational rituals are adhered to (Boland and Pondy 1983). Such a perspective adopts a unitary model of organizations and is based on the assumption that information systems are designed to contribute to specific ends, ends that can be articulated, are shared, and are objective. Once built and installed, the system, itself an "icon of rationality" (Franz and Robey 1984), will improve the efficiency or effectiveness of decision-making processes, thus supporting managerial practices. IS developers are seen as systems experts (Hirschheim and Klein 1989), rational thinkers whose profession is based on their ability to solve abstract, complex problems, with computers being programmed to solve their problems. The tools and methodologies associated with this process also possess an aura of rationality, often based on mathematical and logical processing techniques as opposed to reliance on human intuition, judgement, and politics. The construction of the information system represents the mapping of organizational reality onto a more concrete machine-oriented level, and this process of translation is seen as enabling organizational reengineering or transformation. When performing these translations, developers are expected to follow structured techniques, which, in a Tayloristic fashion, facilitate the division of labor, provide an audit trail, and produce a so-called maintainable system. The resultant information system is thus seen as the "embodiment of rationality" (Newman 1989).

Emerging as a reaction to the shortcomings of the positivist paradigm and its ability to come to grips with the complexity of systems development and the organizations within which they are embedded, an interpretivist analysis of ISD has enjoyed increasing popularity. Here, the focus is on the interpretations of actors and how these actors socially construct organizations and information systems. The process of systems development involves sense-making of the social interaction among various actors, resulting in consensus and a set of agreed objectives. With this perspective, we witness a shift in focus from technical development to interpretation of social action. The role of the systems developer is that of facilitator (Hirschheim and Klein 1989), working alongside users in order to assist them in articulating their preferred views. Hence, this stream of research is often associated with the emergence of a number of development approaches that are concerned with the socio-technical fit (ETHICs— Mumford 1983) or understanding problem construction from multiple perspectives (e.g., soft systems methodology—Checkland 1981). Such approaches favor strong user participation, allowing a system to emerge based upon what the social actors construct as being a good system, although problems with these approaches have been widely reported (Franz and Robey 1984; Howcroft and Wilson 2003; Newman 1989). Interpretivist research has helped advance IS research and offers an alternative to the positivist orthodoxy, but, like its positivist counterparts, interpretivism is also reliant upon a "regulation theory of society" (Burrell and Morgan 1979), which lacks analysis of the relations of power and control that regulate and legitimize socially constructed meanings. In this respect, interpretivism often neglects to analyze many aspects of context, such as the dominance of one social interest or single economic imperative, that shape organizational form and processes.

In contrast, critical research on ISD is more reflective in that it considers organizations and information systems in their wider social context, attending to issues such as power, domination, conflict, and contradiction. Actors within ISD are affected both by wider macro-socio-economic forces that constrain their opportunities and scope, and by factors that shape their everyday interactions within their local organizations. In this respect, ISD takes place within a complex network of structures, with the developer often placed in an oppositional role faced with the opposing interests of both management and end-users. Arguably, the Scandinavian tradition has helped to lay the foundations for a more critically informed approach to IS development and use, enabling other researchers to tread along a similar path. Scandinavian research projects operate within a different paradigm from the contemporary North American MIS tradition (Iivari and Lyytinen 1998) in that they subscribe to the notion of increasing workplace democracy, the intention being that all employees should have influence over their work situation and participate in decision-making forums (Bjerknes and Bratteteig 1995). In the late 1960s, a number of experimental projects consisting of trade union and employer organizations focused on the goals of job satisfaction and higher productivity (Ehn 1988). However, many of these projects faced severe difficulties during systems implementation as the differing interests of management and employees were evidently irreconcilable. Indeed, Scandinavian and British participatory projects were criticized at the time by Marxist theorists in Scandinavia for promoting fundamentally capitalist values; that is, increasing productivity and curtailing worker resistance (Asaro 2000, p. 268). Over time, the notion of joint decision making and worker influence has virtually disappeared (Kyng 1998) and it appears that the socio-technical approach "was a product of a particular socio-political regime" (Avgerou 2002, p. 55) that was popular during times of labor shortages when efforts were made to retain employees. Nevertheless, the Scandinavian tradition, particularly with regard to its attitude toward participatory practices, has had considerable influence in the IS literature and has contributed to the legitimization of ISD as a political project.

Critical information systems research overcomes the limitations of positivist research, with its rejection of the unitary model of organizations and the technologically determinist view of information technology. It also overcomes some of the weaknesses of the interpretivist approach by calling upon greater contextual awareness that may help explain why certain interpretations (rather than others) dominate and are seen to represent organizational reality. As mentioned, much of the critical information systems research that is concerned with ISD draws upon the Habermasian perspective, striving for emancipation through free and undistorted communication, yet how the emancipatory ideal may be applied in the context of ISD is "very much in the making with no examples or strict methodologies available" (Alvesson and Willmott 1996). For this research area to advance, practical examples would certainly add credibility.

3.2 Critical Examination of Gender and Information Technology

While the previous discussion considered the application of critical IS research to information system development and use, this section considers the application of critical IS research to the users of information systems and technology. Specifically, it considers the use of critical IS research to examine power relations and under represented voices in the context of gender and IT use. This example was chosen because the issues involved in this topic are consistent with the themes represented in critical research that were discussed in section 2.

When the positivist epistemology is applied to the topic of gender and IT, the objective is typically to discover *whether* and *where* there are gender differences—in women's vs. men's use (adoption, acceptance, etc.) of IT (e.g., Gefen and Straub 1997) or in women's participation rate in the IS profession (e.g., Carayon et al. 2003; Truman and Baroudi 1994). Further, the theory underlying positivist gender research is often essentialism. That is, observed gender differences are understood to arise from the dichotomizing of male/ female roles that, in turn, are assumed to derive from bio-psychological differences (Wajcman 1991). The motivation for conducting this research is generally to advance managerial objectives. For example, it might be to consider gender as a factor of production in better harnessing diversity in pursuit of effectiveness and productivity (e.g., Gallivan 2003; Igbaria and Baroudi 1995; Igbaria and Chidambaram 1997; Venkatesh and Morris 2000). Problems of inequalities are seen in terms of wasted resources, and increased equality is promoted in the hope of optimizing efficiency.

The main criticisms of this approach to studying the topic of gender and IT relate to the scope and the particular point of view that becomes privileged in the research. Positivist investigations of gender and IT remain on the surface of observable and documentable differences. In so doing, they are vulnerable to charges of superficial and unproblematic treatment of the topic. Further, by offering only managerialist perspectives, positivist gender and IT research privileges one perspective over others. Hence, the gendered aspects of IT use are not considered from the perspective of those experiencing it.

These limitations are addressed in interpretive studies of gender and IT use. Research that employs the interpretive lens to examine this topic focuses on developing a better understanding of *how* these gender differences in IT use have come about. The objective is to add context to the observations about the relationship between gender and IT. Thus, this research is more likely to invoke theories of social construction or individual differences (Trauth 2002) in developing theoretical explanations for gender differences. The motivation is to better understand the social influences underlying inequality (i.e., observable differences) between the genders. While some of this research privileges managerialist goals, not all of it does. Some interpretive gender and IT research desires to advance our understanding of the relationship between gender and IT from the point of view of women IT users. Thus, an interpretive examination of gender and the IS profession might explore the influence of national culture on the social construction of gender identity as it relates to the IT workforce (Trauth 1995; Trauth et. al 2003). However, a criticism of the interpretive approach is that the focus is on *understanding* the societal influences, not *questioning* them. It is directed at coping with the dynamics of inequality, not questioning the legitimacy of underlying social influences or changing them.

In contrast, the objective of critical research about gender and IT is to advocate a position as to *why* gender inequality exists. In doing so, it challenges the status quo and the dominant discourses about gender and IT. Critical social theory, postmodernism, and feminist theory (Adam 2002; Adam and Richardson 2001), for example, are used to inform the search for the underlying causes of gender inequality. Thus, this approach to gender and IT research moves away from traditional themes found in positivist and interpretive gender research such as profitability, efficiency, effectiveness, and gender identity. Instead, it focuses on concepts such as power, control, resistance, and inequality. The motivation for this research is more activist than positivist or interpretive research: to challenge power relations that reproduce inequality. Thus, a critical perspective on gender and IT might concentrate on the gendered nature of the workplace and technological skills (e.g., Adam and Richardson 2001; Wilson 2002). However, a criticism that could be made of critical gender and IT research is that the research itself is subject to power relations and dominant discourses.

The use of critical IS research to study this topic addresses the limitations of both positivist and interpretive research on gender and IT. In doing so, alternative explanations and theories become available. Critical information systems overcomes positivist limitations by offering alternatives to a managerialist perspective and managerialist theories. It goes beyond interpretive understanding of societal influences to explore power relations, marginality, and dominant discourses in the organizational and societal context. While critical information systems research regarding gender and IT opens up new avenues for exploring the topic of gender and IT, it also brings with it issues and challenges that could be limiting factors. These issues are addressed in the next section.

4 IMPLICATIONS FOR IS RESEARCH AND PRACTICE

Consistent with a critical stance on IS research, we now apply a critical lens to critical IS research. In doing so, we raise issues regarding the use of critical research to examine the relationship between information systems and the organizations or societies within which they are embedded. We then offer some challenges and recommendations to the IS community regarding the conduct of critical information systems research.

4.1 Issues with the Choice of Critical Research

We begin with a consideration of issues that arise from the choice of a critical approach to the study of IS topics. Using a framework that was developed to assess the choice of qualitative research (Trauth 2001), we consider theoretical, methodological, and political issues associated with the choice of critical IS research.

One theoretical issue is the need for greater theoretical diversity. As noted earlier in this paper, the thrust of critical IS research to date has been in the Habermasian tradition. As Adam (2002) pointed out in her argument for exploring emancipation through feminist theory, broadening the theoretical base of critical IS research will yield both epistemological and practical benefits. Another theoretical issue is that in order to do justice to a critical approach, IS research requires levels of analysis that go beyond the organizational level. In order to better understand the interplay of the social and the technical, we need to expand our level of understanding to include other elements such as macro-economic, societal, cultural, and industry-specific factors.

A range of methodological issues emanate from the decision to conduct critical research. One is whether there are any methodological principles that are specific to this kind of research or whether one can simply use any methodology suitable for the study of IS in general. This issue is discussed in section 4.2 below. A second issue, with notable exceptions such as Klecun-Dabrowska (2002), Kvasny (2002) and Richardson (2003), is the dearth of empirical studies. However, when empirical critical IS research *is* conducted, other issues arise. One set of issues relates to the research subject and data collection. It can be very problematic to elicit the kinds of data that further a critical investigation of the topic. For example, in the researcher's effort to elicit comments from the participant about emancipation, she or he might be encouraging the participant to verbalize comments about personally painful experiences or organizationally inappropriate behavior that the participant might prefer not to confront. But if the researcher is successful in bringing these experiences into the participant's consciousness, another issue arises: coping with the aftermath of this consciousness raising. At the end of the interview session, the researcher walks away but the subject may undergo a personal transformation because of the interview experience. In its mildest form, the participants may have a new awareness about issues in their personal or organizational lives. In some circumstances, however, the transformation the subjects experience might be psychologically or emotionally traumatic. The challenge for the researcher has both methodological and ethical dimensions.

Another set of methodological issues relates to data analysis. If interpretive research must overcome the charge of bias, the challenge for critical research is even greater. The reduction of bias is given much attention by positivists, since it is assumed that such a reduction will move one closer to the truth. Another perspective is that all research is inherently biased and that the only difference is that the critical researcher is more likely to acknowledge (or even defend) their explicit bias. Given the dominant discourse, how does one cope with the charge of bias when the research epistemology consciously adopts a particular stance (i.e., in favor of emancipation and in critique of tradition and technological determinism) in interpreting the data?

Political issues are deeply embedded in critical research. Since one of the objectives is to critique the status quo, the vested interests of the dominant discourse can be threatened. For example, critical IS research that adopts perspectives other than managerial and is interested in understanding why some problem or situation occurs from alternative points of view can pose a serious challenge to journals and academic departments whose publishing norms and reward systems are limited to a managerialist (i.e., a management information system) perspective on IS problems. A second political issue is that the search for relevant social theories may result in ones that are new to the IS community, the journals, and the reviewers. Further, they may even contradict current theories-in-use.

4.2 Challenges and Recommendations for the Conduct of Critical IS Research

When considering the conduct of critical IS research studies, one might assume that qualitative techniques would seem the most appropriate choice. Drawing a battle line between questions of technique can be misleading since it detracts from the more central problem of how we choose to interpret and represent social reality. Of greater concern is the issue of being reflective about the social and organizational underpinnings of one's own recommendations and practices. As noted elsewhere,

> Method is thus not primarily a matter of "data management" or the mechanics and logistics of data production/processing, but is a reflexive activity where empirical material calls for careful interpretation—a process in which the theoretical, political and ethical issues are central (Alvesson and Skoldberg 2000, p. 5).

The central premise of critical research is the balance between being informed by critical theoretical ideas and a political agenda, and maintaining an empirical sensitivity toward and interest in the discovery of repression. Having too much of a theoretical focus can lead to intellectual elitism and insensitivity to the richness of empirical settings and the lived experience of people at "the front line." Having too much empirical work can lead to a narrow focus which ignores the macro, economic, and social context, and becomes engrossed in surface phenomena, thus losing the critical edge. In an attempt to show how we can strive to achieve this balance, we draw on the work of Alvesson and Deetz (2000) in offering some guidelines on the conduct of critical research. Three overlapping tasks that the critical researcher may wish to consider are insight, critique, and transformative redefinition.

4.2.1 Insight

Insight refers to understanding how various forms of knowledge, objects, and events are formed and sustained, highlighting hidden or less obvious aspects of social reality.

The task here is to investigate at a local level and then relate these empirical themes to wider economic, social, historical, and political forces. Insight is about the process of producing meaning from the data, and understanding the conditions (e.g., socio-economic context) which frame how we make sense of the data. This serves as an important reminder of the value of engaging with organizational practices and people within them as our subject of interest. Indeed, one of the greatest criticisms leveled at critical research is the lack of extended empirical studies (Alvesson and Deetz 2000; Alvesson and Willmott 1996; Boudreau 1997).

Critical theory and postmodernism's strong critique of empiricism does not mean that reflective empirical work is not a worthwhile activity. To ground theories of bureaucracy, capitalism, managerialism, and technological determinism in organizational contexts and the lives of organizational members can only aid our understanding of these issues. At the same time, we should be attendant to some of the difficulties of empirical work. For critical studies, the metaphor of researcher as *mirror* is replaced by *lens* noting the role of the researcher as positioned and active (Alvesson and Deetz 2000). An example of this can be found in the work of Richardson (2003), who used the work of Bourdieu (1990) in her analysis of call center workers in the UK to highlight significant contradictions between system objectives and outcomes in practice. The paper throws the spotlight on the individuals working at the call center front line and tells stories often left untold in studies of IS adoption and use.

For critical research, the focus is on the ability to challenge guiding assumptions, values, social practices, and routines of an observation. Rather than seeing the world from the native's point of view by viewing events, actions, and values from the perspective of the people being studied, critical researchers aim to balance their interest in the level of meaning, with an awareness of less explicit ideological and structural forces (see, for example, Lukes 1974). This is in contrast to what Bhaskar (1979) has described as the "linguistic fallacy," the claim adopted by many interpretivists that subjects, concepts, meanings, and accounts of their actions cannot be criticized. This fallacy is based on the assumption that reality is expressed in the language of social actors. The experiences of the subjects of study are neither the primary nor the only focus of interest. The lens shifts from exclusive focus on individuals, situations, and meaning to the systems of relations, which make such meanings possible. This is not to suggest that experiences should be ignored; rather, they must be balanced with attendance to issues of an ideological nature that may frame the experiences and prescribe meaning. This enables researchers to deal with the conditions which give rise to the meaning and interpretations of social actors, an element that is often absent in much interpretivist research (Fay 1975). Ultimately, insight can lead to the production of competing discourses and even counter-discourses.

4.2.2 Critique

Critique challenges many of the assumptions, beliefs, ideologies, and discourses that permeate IS phenomena. Political, economic, and social forces are inscribed in organizational arrangements (Alvesson and Willmott 1996) and technological artefacts (Akrich 1992; Winner 1985). To assume that this is natural or represents the one best way is insufficient, since these phenomena arise as a result of ideological and historical forces that privilege certain groups. Addressing this privileging of certain discourses and constructions is, therefore, a crucial aspect of critical research. Whereas insight focuses on local specifics, with critique the lens shifts to general characterizations relating to wider social concerns and often the larger global community. Critique explicitly relates to power constraints, repression, ideology, social asymmetries, and technological determinism that give priority to certain ways of viewing the world. If we were to study IS phenomena at the micro, meso, and macro levels (Drummond 1996), it may provide us with a deeper understanding of information systems. Doing so would entail examining the process of systems development and use, and their relationship to organizational context. These, in turn, reflect and reproduce the major social inequalities in society (Knights and Murray 1997). At the macro level, understanding such issues as managerial and labor processes and IT markets could provide further understanding of information systems at the organizational level. An example of such work includes research by Klecun-Dabrowska (2002), who applied a critical perspective to telehealth in the UK. Klecun-Dabrowska examined the historical conditions and structures that shape telehealth, including the organization of health care in Britain, the health policy process, strategies for the employment of ICTs, and visions of an information society.

4.2.3 Transformative Redefinition

A third task refers to the development of critical, relevant knowledge and practical understandings that facilitate change. In this respect, the third element is the most difficult. While it is important that critical information systems researchers aim to disrupt the dominant ethos in mainstream literature, there are problems with building a research stream that is based only on critique and negation, but does not put anything forward. Research that is perceived as anti-technology and anti-management will have limited appeal and adds little toward the goal of social change. Even though critical researchers seek to avoid telling people what to do, transformative redefinition aims to suggest an alternative and radically different view of the world, which emphasizes change but in a more positive way. This is difficult to achieve as efforts to progress things forward (such as the attempt by Deetz [2003] to overcome the problems of work-place communication in the hope of improving collaborative decision-making) run the risk of co-optation. The challenge of smoothing over irreconcilable differences could result in exemplification of what we initially set out to criticize. Co-optation is best avoided by careful consideration of *who* one chooses to be involved with within the organization, rather than merely espousing utopian ideals.

In terms of transformative redefinition, perhaps here we can draw on some of the discussions that have taken place within CMS concerning the different perspectives on our role as educators (see Grey and Mitev 1995). This breaks down essentially into two opposing positions. On the one hand, there is the militant position, which is committed to the victims of corporate power and leverages our role as academics in the form of "hell raising and muckraking" (Adler 2002). This position is one of complete disengagement with management practice, where the task is not to reform management but to undermine it and expose it through critique. The other position is that of humanist, which is based on a respectful engagement with students, based on the premise that managers are people too. It aims to develop, promote, and transform more humane forms of management by engaging in a dialogue with management practitioners in the hope that they could become potential allies in the future (e.g., Alvesson and Willmott 1996). As IS researchers, many of us are engaged in the teaching of students and thus have the potential for considerable influence. Given that our voices have an element of legitimacy with students, by encouraging and enabling them to question many of the prevailing assumptions within our field (such as an acceptance of organizational hierarchy, managerialism, globalization, the primacy of markets, and technological determinism), a more mature understanding may arise that emerges as a refreshing antithesis to the orthodoxy.

This issue clearly links in with the notion in IS research of research for whom that asks the question who is our target audience? This is a highly contentious and lengthy debate[1] centered on the issue of the relevance of our research. When IS research exposes inappropriate and unjust work and power relations that practitioners are unwilling to accept, this does not imply that such research is not relevant (Lee 1999). The problem of relevance (or engagement) is that the debate has been conducted in a one-sided way and is centered on the assumption that we should engage with (IT) practitioner managers and the interests of business, rather than concern ourselves with the managed—the people lower down the organizational hierarchy who are on the receiving end of managerial practice. If we are to be critical in terms of local action and practice, then consideration should be given to those with whom we choose to engage. As pointed out elsewhere (Truex 2001), less powerful groups such as trade unions or community groups should also have access to the skills and insights that we have to offer. As researchers, we can choose whether we want to give advice to management or to people who are on the receiving end of management decisions. Relevance is not only about whether or not we engage in a dialogue with management, but also about whether or not we choose to broaden our focus of study to include a much more inclusive organizational constituency. Perhaps we need to consider the possibility of telling different stories.

As critical researchers, our role in transformative definition can also extend beyond the local environment to broader issues relating to the publication of our work. Critically oriented publications in both academically oriented and practitioner-oriented outlets can reach a broad audience while simultaneously challenging the status quo and critiquing many taken-for-granted assumptions regarding technology and organizations. Publications of this nature can generate additional insights into recurring problems such as information systems failures, the conflictual nature of user-developer relations, or the productivity paradox. While publications of this nature cannot solve such problems, the insights that arise from this can be enlightening and potentiality transformative. They can at least contribute to a greater understanding of the nature of organizational relations and avoid the scapegoating of people and technologies that fail to live up to expectations

5 CONCLUSION

The purpose of this paper is to address the themes of both this conference and the 1984 IFIP WG 8.2 conference to "think carefully about the organizational and societal consequences of the systems they are developing and using" (see the Call for Papers for

[1]See Lee (1999) for an overview.

this conference). This has been accomplished by considering the *research approaches* by which we examine and assess the ways in which information systems are used by and affect people. We employ critical IS research methods to carry out this assessment. In the course of examining research approaches, we also address another theme of this conference: concern with the ability of traditional research methods to adequately address these organizational and societal dimensions. In 1984, traditional would have been understood to be quantitative, positivist studies; in 2004 the definition of traditional can, arguably, be broadened to include qualitative, interpretive studies. In our consideration of critical IS research we revisit this theme of methodological limitations by illustrating the benefits of broadening the definition even further to include critical research as one of the mainstream IS research methods.

This paper also contributes to the broader IS research community in two ways. First, it can serve as a useful resource for people trying to understand critical IS. Second, it contributes to the development of a distinct, critical IS research literature. There is little in the general critical literature that differentiates critical IS research from other critical arenas; we are provided with insights about emancipation, for example, but with no mention of technological determinism. In terms of social theory, technology is an off-stage phenomenon, that has not been seriously integrated (Sorensen 2002) and the field of sociology has never steadfastly nurtured an interest in technology (Button 1992). Given the nature of the IS field and our object of study, we need suitable ways of conceptualizing how we integrate the material (technology) into the analysis of human societies. Critical IS can provide a way of helping with this conceptualization.

REFERENCES

Adam, A., and Richardson, H. "Feminist Philosophy and Information Systems," *Information System Frontiers* (3:2), 2001, pp. 143-154.

Adler, P. "Critical in the Name of Whom and What?," *Organization* (9:3), 2002, pp. 387-395.

Akrich, M. "The De-scription of Technical Objects," in W. E. Bijker and J. Law (Eds.), *Shaping Technology/Building Society: Studies in Sociotechnical Change*, Cambridge, MA: MIT Press, 1992.

Alvesson, M., and Deetz, S. *Doing Critical Management Research,* London: Sage Publications, 2000.

Alvesson, M., and Skoldberg, K. *Reflexive Methodology*, London: Sage Publications, 2000.

Alvesson, M., and Willmott, H. "On the Idea of Emancipation in Management and Organization Studies," *Academy of Management Review* (17:3), 1992, pp. 432-464.

Alvesson, M., and Willmott, H. *Making Sense of Management: A Critical Introduction*, London: Sage Publications, 1996.

Avgerou, C. *Information Systems and Global Diversity*, Oxford: Oxford University Press, 2002.

Asaro, P. "Transforming Society by Transforming Technology: The Science and Politics of Participatory Design," *Accounting, Management and Information* (10), 2000, pp. 257-290.

Beath, C. M., and Orlikowski, W. J. "The Contradictory Structure of Systems Development Methodologies: Deconstructing the IS-User Relationship in Information Engineering," *Information Systems Research* (5:4), 1994, pp. 350-77.

Bhaskar, R. *The Possibility of Naturalism: A Philosophical Critique of the Contemporary Human Sciences*, Brighton, UK: Harvester, 1979.

Bijker, W. E. "Sociohistorical Technology Studies," in T. Pinch (Ed.), *Handbook of Science and Technology Studies*, Thousand Oaks, CA: Sage Publications, 1995, pp. 229-256.

Bjerknes, G., and Bratteteig, T. "User Participation and Democracy: A Discussion of Scandinavian Research on System Development," *Scandinavian Journal of Information Systems* (7:1), 1995, pp. 73-98.

Boland, R. J., and Pondy, L. R. "Accounting in Organizations: A Union of Natural and Rational Perspectives," *Accounting, Organizations and Society* (8), 1983, pp. 223-234.

Bourdieu, P. *The Logic of Practice*, Cambridge, MA: Polity Press, 1990.

Brooke, C. "Critical Perspectives on Information Systems: An Impression of the Landscape," *Journal of Information Technology* (17), 2002a, pp. 271-283.

Brooke, C. "Editorial: Critical Research in Information Systems," *Journal of Information Technology* (17), 2002b, pp. 45-47.

Burrell, G., and Morgan, G. *Sociological Paradigms and Organizational Analysis*, London: Heinemann, 1979.

Button, G. "The Curious Case of the Vanishing Technology," in G. Button (Ed.), *Technology in Working Order: Studies of Work, Interaction and Technology*, London: Routledge, 1992.

Carayon, P.; Hoonakker, P.; Marchland, S.; and Schwarz., J. "Job Characteristics and Quality of Working Life in the IT Workforce: The Role of Gender," in E. Trauth (Ed.), *Proceedings of the 2003 SIGMIS Conference on Freedom in Philadelphia: Levering Differences and Diversity in IT*, New York: ACM Press, 2003, pp. 58-63.

Cecez-Kecmanovic, D. "Doing Critical IS Research: The Question of Methodology," in E. M. Trauth (Ed.), *Qualitative Research in IS: Issues and Trends*, Hershey, PA: Idea Group Publishing, 2001, pp. 141-163.

Cecez-Kecmanovic, D.; Moodie, D.; Busuttil, A.; and Plesman, F. "Organizational Change Mediated by E-Mail and Intranet: An Ethnographic Study," *Information Technology and People* (12:1), 1999, pp. 9-26.

Checkland, P. B. *Systems Thinking, Systems Practice*, Chichester, England: John Wiley & Sons, 1981.

Deetz, S. "Developing Critical Communication Theories for Collaboration: Using Workplace Participation as an Example," in *Proceedings of Critical Management Studies 3*, Lancaster, UK, 2003 (available online at http://www.mngt.waikato.ac.nz/research/ejrot/cmsconference/2003/abstracts/abstracts_communication.asp; accessed December 2003).

Doolin, B. "Information Technology as Disciplinary Technology: Being Critical in Interpretive Research on Information Systems," *Journal of Information Technology* (13), 1998, pp. 301-311.

Doolin, B., and Lowe, A. "To Reveal Is to Critique: Actor-Network Theory and Critical Information Systems Research," *Journal of Information Technology* (17), 2002, pp. 69-78.

Dooyeweerd, H. "Introduction," *Philosophia Reformata* (38), 1973, pp. 5-16.

Drummond, H. *Escalation in Decision-Making: The Tragedy of Taurus*, Oxford: Oxford University Press, 1996.

Ehn, P. *Work-Oriented Design of Computer Artefacts*, Stockholm: Arbeslivscentrum, 1988.

Fay, B. *Social Theory and Political Practice*, London: Allen & Unwin, 1975.

Forester, J. "Critical Ethnography: On Fieldwork in a Habermasian Way," in M. Alvesson and H. Willmott (Eds.), *Critical Management Studies*, London: Sage Publications, 1992.

Foucault, M. *Discipline and Punish: The Birth of the Prison*, Harmondsworth, England: Penguin, 1979.

Fournier, V., and Grey, C. "At the Critical Moment: Conditions and Prospects for Critical Management Studies," *Human Relations* (53:1), 2000, pp. 7-32.

Franz, C. R., and Robey, D. "An Investigation of User-Led Systems Design: Rational and Political Perspectives," *Communications of the ACM* (27:12), 1984, pp. 1202-1209.

Gallivan, M. "Examining Gender Differences in IT Professionals' Perceptions of Job Stress in Response to Technological Change," in E. Trauth (Ed.), *Proceedings of the ACM SIGMIS CPR Conference*, New York: ACM Press, 2003, pp. 10-23.

Gefen, D., and Straub, D. W. "Gender Differences in the Perception and Use of E-Mail: An Extension to the Technology Acceptance Model," *MIS Quarterly* (21:4), 1997, pp. 389-400.

Grey, C., and Mitev, N. "Management Education: A Polemic," *Management Learning* (26:1), 1995, pp. 73-90.

Hammersley, M. *The Politics of Social Research*, London: Sage Publications, 1995.

Heidegger, M. *Being and Time*, New York: State University of New York Press, 1953.

Hirschheim, R., and Klein, H. K. "Four Paradigms of Information Systems Development," *Communications of the ACM* (32:10), 1989, pp. 1199-1216.

Hirschheim, R., and Klein, H. K. "Realizing Emancipatory Principles in Information Systems Development: The Case for ETHICS," *MIS Quarterly* (18:1), 1994, pp. 83-109.

Horkheimer, M. "Traditional and Critical Theory (1937)," in P. Connerton (Ed.), *Critical Sociology*. Harmondsworth, England: Penguin, 1976.

Howcroft, D., and Wilson, M. "Paradoxes of Participatory Design: The End-User Perspective," *Information & Organization* (13:1), 2003, pp. 1-24.

Igbaria, M., and Baroudi, J. J. "The Impact of Job Performance Evaluations on Career Development Prospects: An Examination of Gender Differences in the IS Workplace," *MIS Quarterly* (19:1), 1995, pp. 107-123.

Igbaria, M., and Chidambaram, M. "The Impact of Gender on Career Success of Information Systems Professionals," *Information Technology & People* (10:1), 1997, pp. 63-86.

Iivari, J., and Lyytinen, K. "Research on Information Systems Development in Scandinavia: Unity in Plurality," *Scandinavian Journal of Information Systems* (10:1/2), 1998, pp. 135-185.

Klecun-Dabrowska, E. *Telehealth and Information Society: A Critical Study of Emerging Concepts in Policy and Practice*, Unpublished Ph.D. Thesis, Department of Information Systems, London School of Economics and Political Science, 2002.

Klein, H. K., and Hirschheim, R. "The Application of Neohumanist Principles in Information Systems Development," in D. Avison, J. E. Kendall, and J. I. DeGross (Eds.), *Human, Organizational and Social Dimensions of Information Systems Development*, Amsterdam: North Holland, 1993, pp. 263-279.

Knights, D., and Murray, F. "Markets, Managers and Messages: Managing Information Systems in Financial Services," in B. Bloomfield, R. Coombs, D. Knights, and D. Littler (Eds.), *Information Technology and Organizations: Strategies, Networks and Integration*, Oxford: Oxford University Press, 1997, pp. 36-56.

Kvasny, L. *Problematizing the Digital Divide: Cultural and Social Reproduction in a Community Technology Initiative*, Unpublished Ph.D. Thesis, Department of Computer Systems, Robinson College of Business, Georgia State University, 2002.

Kyng, M. "Users and Computers: A Contextual Approach to Design of Computer Artifacts," *Scandinavian Journal of Information Systems* (10:1/2), 1998, pp. 7-44.

Latour, B. "Technology Is Society Made Durable," in J. Law (Ed.), *A Sociology of Monsters: Essays on Power, Technology and Domination*, London: Routledge, 1991, pp. 103-131.

Lee, A. S. "Rigor and Relevance in MIS Research: Beyond the Approach of Positivism Alone," *MIS Quarterly* (23:1), 1999, pp. 29-34.

Lukes, S. *Power: A Radical View*, London: MacMillan, 1974.

Lyytinen, K. "Information Systems and Critical Theory," in M. Alvesson and H. Willmott (Eds.), *Critical Management Studies*, London: Sage Publications, 1992.

Lyytinen, K., and Hirschheim, R. "Information Systems and Emancipation: Promise or Threat?," in H. Klein and K. Kumar (Eds.), *Systems Development for Human Progress*, Amsterdam: North Holland, 1989, pp. 115-139.

Lyytinen, K., and Hirschheim, R. "Information Systems as Rational Discourse: An Application of Habermas's Theory of Communicative Action," *Scandinavian Journal of Information Systems* (4:1/2), 1988, pp. 19-30.

Lyytinen, K., and Klein, H. K. "The Critical Theory of Jurgen Harbermas as a Basis for a Theory of Information Systems," in E. Mumford, R. Hirschheim, G. Fitzgerald, and A. T. Wood-Harper (Eds.), *Research Methods in Information Systems*, Amsterdam: North Holland, 1985, pp. 219-236.

Markus, M. L. "Power, Politics, and MIS Implementation," *Communications of the ACM* (26:6), 1983, pp. 430-444.

Marx, K. *Capital*, London: Penguin, 1974.

Mingers, J. "Technical, Practical and Critical OR: Past, Present and Future?" in M. Alvesson and H. Willmott (Eds.), *Critical Management Studies*, London: Sage Publications, 1992, pp. 90-112.

Mingers, J. "What Is it to Be Critical? Teaching a Critical Approach to Management Undergraduates," *Management Learning* (31:2), 2000, pp. 219-237.

Mumford, E. *Designing Human Systems: The ETHICS Method*," Manchester Business School, Manchester, England, 1983.

Myers, M. D., and Avison, D. E. *Qualitative Research in Information Systems: A Reader*, London: Sage Publications, 2002.

Newman, M. "Some Fallacies in Information Systems Development," *International Journal of Information Management* (9), 1989, pp. 127-143.

Ngwenyama, O. K. "The Critical Social Theory Approach to Information Systems: Problems and Challenges," in H-E. Nissen, H. K. Klein, and R. Hirschheim (Eds.), *Information Systems Research: Contemporary Approaches and Emergent Traditions*, Amsterdam: North-Holland, 1991, pp. 267-280.

Ngwenyama, O. K., and Lee, A. S. "Communication Richness in Electronic Mail: Critical Social Theory and the Contextuality of Meaning," *MIS Quarterly* (21:2), 1997, pp. 145-167.

Orlikowski, W. J., and Baroudi, J. J. "Studying IT in Organizations: Research Approaches and Assumptions," *Information Systems Research* (2:1), 1991, pp. 1-28.

Richardson, H. "CRM in Call Centres: The Logic of Practice," in M. Korpela, R. Montealegre, and A. Poulymenakou (Ed.), *Organizational Information Systems in the Context of Globalization*, Boston: Kluwer Academic Publishers, 2003, pp. 69-84.

Russell, S., and Williams, R. "Social Shaping of Technology: Frameworks, Findings and Implications for Policy with Glossary of Social Shaping Concepts," in R. Williams (Ed.), *Shaping Technology, Guiding Policy: Concepts, Spaces and Tools*, Cheltenham, UK: Edward Elgar, 2002, pp. 37-132.

Sorensen, K. "Social Shaping on the Move? On the Policy Relevance of the Social Shaping of Technology Perspective," in R. Williams (Ed.), *Shaping Technology, Guiding Policy: Concepts, Spaces and Tools*, Cheltenham, UK: Edward Elgar, 2002, pp. 19-35.

Trauth, E. "Odd Girl Out: An Individual Differences Perspective on Women in the IT Profession," *Information Technology & People* (15:2), 2002, pp. 98-118.

Trauth, E. (Ed.). *Qualitative Research in IS: Issues and Trends*, Hershey, PA: Idea Group Publishing, 2001.

Trauth, E. "Women in Ireland's Information Industry: Voices from Inside," *Eire-Ireland* (30:3), 1995, pp. 133-150.

Trauth, E., and Jessup, L. "Understanding Computer-Mediated Discussions: Positivist and Interpretive Analyses of Group Support System Use," *MIS Quarterly* (24:1), 2000, pp. 43-79.

Trauth, E., Nielsen, S.; and von Hellens, L. "Explaining the IT Gender Gap: Australian Stories for the New Millennium," *Journal of Research and Practice in Information Technology* (35:1), 2003, pp. 7-20.

Truex, D. P. "Three Issues Concerning Relevance in IS Research: Epistemology, Audience and Method, *Communication of the AIS*, 2001.

Truman, G. E., and Baroudi, J. J. "Gender Differences in the Information Systems Managerial Ranks: An Assessment of Potential Discriminatory Practices," *MIS Quarterly* (18:2), 1994, pp. 129-141.

Venkatesh, V., and Morris, M. "Why Don't Men Ever Stop to Ask for Directions? Gender, Social Influence, and Their Role in Technology Acceptance and Usage Behavior," *MIS Quarterly* (24:1), 2000, pp. 115-139.

Wajcman, J. *Feminism Confronts Technology*, Cambridge, MA: Polity Press, 1991.

Williams, R., and Edge, D. "The Social Shaping of Technology," *Research Policy* (2), 1996, pp. 865-899.

Wilson, M. "Making Nursing Visible? Gender, Technology and the Care Plan as Script," *Information Technology & People* (15:2), 2002, pp. 139-158.

Winner, L. "Do Artifacts have Politics?," in D. MacKenzie and J. Wajcman (Eds.), *The Social Shaping of Technology*, Buckingham, England: Open University Press, 1985, pp. 26-38.

ABOUT THE AUTHORS

Debra Howcroft is a senior lecturer in information systems at Manchester School of Accounting and Finance, the University of Manchester. Her research interests focus upon the social and organizational aspects of information systems. For more information, visit Debra's website at http://man.ac.uk/accounting/Staff/Academic/D_Howcroft.htm. She can be reached at debra.a.howcroft@man.ac.uk.

Eileen M. Trauth is a professor of Information Sciences and Technology and Director of the Center for the Information Society at The Pennsylvania State University. She received her Ph.D. in information science from the University of Pittsburgh. Her current research interest is the influence of socio-cultural factors and public policy on the development of information economies. Eileen has conducted extensive research into Ireland's information economy as a Fulbright Scholar, and was the recipient of an Irish Science Foundation Walton Visitor Award in 2003. Eileen is also the recipient of an NSF IT Workforce grant to investigate the under-representation of women in the American IT labor force. She is the author of *The Culture of an Information Economy: Influences and Impacts in the Republic of Ireland* (Kluwer, 2000), co-author of *Information Literacy: An Introduction to Information Systems* (Macmillan, 1991), editor of *Qualitative Research in IS: Issues and Trends* (Idea Group Publishing, 2001) and coeditor of *Seeking Success in E-Business: A Multidisciplinary Approach* (Kluwer, 2003). Eileen has conducted cross-cultural research in Europe and Australia and has published articles on socio-cultural and organizational influences on IT, qualitative research methods, global informatics, information policy, information management, and the development of IS professionals. She serves on the editorial boards of several international journals.

13 THE RESEARCH APPROACH AND METHODOLOGY USED IN AN INTERPRETIVE STUDY OF A WEB INFORMATION SYSTEM: Contextualizing Practice

Anita Greenhill
Manchester School of Management

Abstract This paper contextualizes the research approach followed in an interpretive Information Systems study that explored the development of a Web information system. The paper provides details of the methodological approach followed in the project, the assumptions used, and the research techniques utilized when conducting a specific interpretive study of IS development. The aim of this paper is to provide a detailed presentation of the range of methods chosen to detail the rigor maintained in the research process associated with interpretive IS research. This paper illustrates how methods that are described as visual methodological techniques can be applied to an Information Systems study. With the presentation of this detail, it is hoped other IS researchers can reflect on their own practices when conducting interpretive research projects in comparison with those chosen for this project. The paper bridges the theoretical background of existing literature to illustrate how the exploration of method and research practice contributes to the examination of a chosen research topic.

Keywords: Method, interpretive study, space, visual techniques

1 INTRODUCTION

This paper extends the debate regarding appropriate methodological frameworks for the philosophy of science. Instead of weighing the suitability of individual methods, it details one study and its methodological practice. The study was qualitative, interpretive, critical, and asked, how do developers spatially perceive their Web information system? This focus also casts the project as an ontological exploration of the topic. The methodology pursued by the study makes a contribution to theory testing within the field of Information Systems but did not aspire to become a methodology for researching the spatial aspects of Information Systems. This paper presents the ontological and epistemological basis for the study. It reveals the assumptions that were embedded in the study and how these, in turn, informed methodological decisions regarding the project. This conceptualizes the use of specific methods by providing examples of data collected during original research. This paper contributes to the field of Information Systems research by illustrating the way interpretive methods enable detailed insight of Information Systems development.

The focus of the original study was on how information systems are spatially perceived by Web information system developers. The study focused on a team of Web publishers in a large government department in Australia who maintained an electronic publishing system. The team was developing a new Web information system for their publishing requirements. The research project provided insight into the internal operations of a large government department and its capacity to deal with the spatial forms that new technologies enable. The findings included an in-depth presentation of Web usage characteristics of the department. The primary informants of the study were described collectively as the **Web development team**. The research site was their workspace. The benefit of the original research lies with its detailed discussion of Web information systems development. The original study also included analysis of data gathered over a six-month period from in-depth observation and interviews that were conducted with team members regarding the new Web information system. The research offers descriptions of the socio-cultural and power relations in the organization that were reflected in the observed spatial configurations. The relationships conducted in the physical and virtual spaces of the organization being studied were described. None of these findings are presented in this paper. What is focused upon here is the relationship between the research project and the methods used to examine the topic of research. The paper specifically justifies the choice of the method finally utilized within the interpretive study.

2 THE CASE METHOD AS A RESEARCH STRATEGY

As a research strategy, the case method is used in situations where the purpose of research and its objectives are to find answers questions of *how* or *what* and to focus upon understanding the dynamics of an individual setting (Klein and Myers 1999; Yin 1989). The research incorporated descriptions and analysis from six months of in-depth observation and interviews concerning the organization's new Web information system. The research also described the power relations that were reflected by the spatial configuration of this unit through physical and virtual space allocations. The case study, as the name suggests, utilized a theory to explore a single case. In this way, the research detailed stories of what happened over a period of time between the people observed by the researcher. The study was empirical; however, it did not attempt to draw any positivist conclusions with the attendant requirements for identifying cause and effect in order to draw definitive findings. Interpretive research provides a mechanism to explore what those who were observed *perceive* to be an information system. The case study helps to understand if different socio-cultural experiences influence these perceptions. The research did not attempt to suggest that the way the research was interpreted is the only possible reading of this organization's new Web information system. The research undertook a thorough process of data gathering and, through a particular research lens, of detailing detail a particular situation.

It is important to stress that the examination of the information system was spatially oriented with an emphasis upon spatial configurations and not on other perspectives or phenomena. Emphasis was on locating the spatial dimensions to information sharing and the study sought to interpret the differences and similarities in the use of space.

3 CONTEXTUALIZING THE STUDY BY DEFINING INFORMATION SYSTEMS

Information Systems, as a discipline, is cross disciplinary. A broad research foundation provides the IS researcher with the best and worst aspects of its reference disciplines. As a result, there is no methodological uniformity within the discipline and intense debate over the methods employed to conduct information systems research continues (Benbasat et al. 1987; Checkland and Howell 1998; Kaplan 2001) and extends to the legitimacy of the discipline.

The range of perspectives that *are* IS necessitate the researcher to repeatedly and regularly contextualize their research by providing a definition of an information system within their work. For example, the study discussed in this paper used a combination of two definitions of IS to situate the research to the field as a whole. The first came from Keen (1980, p. 713), who states that information systems research should include the "study of effective delivery and usage of IS in organizations (and society)." The second definition used came from Galliers and Land (1987, p. 901), who identify information systems as being associated with the relations between "the organization and the people they serve." The context of this research project was clearly an IS study, as it considered the organization-wide issues of information system (a Web information system) and following Galliers and Land's extended definition, was also an exploration of the relationships held between the people and the system being studied.

The empirical aspects of the study examined space in the context of a Web information system.

> [Space] is not simply inherited from nature, or passed on by the dead hand of the past, or autonomously determined by "laws" of spatial geometry as per conventional location theory. Space is produced and reproduced through human intentions, even if unanticipated consequences also develop, and even as space constrains and influences those producing it (Lefebvre in Molotch 1993, p. 887).

There is a growing number of studies concerning information systems and the World Wide Web (Eschenfelder and Sawyer 2001), the Internet (Bradley and Briggs 1995), and virtual reality (Robey et al. 2003; Truex and Baskerville 1998). The emphasis in the research considered here, however, differs from the assumptions of these previous studies. This research considered the spatial and socio-cultural work practices of individuals involved in the development of an information system. The work extends current spatially focused research in information systems by examining the usefulness of space for exploring new technologies, including the World Wide Web for information systems development.

4 CONTEXTUALIZING THE RESEARCH

The initial assumption of the research concentrated on the creation and utilization of space by the organization in the context of an information system. To this end the study considered whether a spatial consideration of information systems development was useful in its development. To achieve this, a detailed examination of organizational interactions was conducted to understand how information was developed by individuals for publication on the Web; how individuals differed in their socio-cultural position, technological context, and physical placement within the organization. The study also explored the similarities and differences between the social processes involved in the construction of space by asking how groups of people construct a shared understanding of a space. The research considered the combination of spatial, socio-cultural, and work practices of individuals involved in the development of an information system as these all contributed to its shared understanding.

This research aimed to produce an understanding of the context of information systems development (Walsham 1993, p. 58). The context for this research was an exploration of the social construction of space as it impacts on the development of a Web Information System. This research adapted Lefebrve's (1991) theory regarding the social production of space to explore whether spatial constructions are important to the development of workplace-based information systems. A spatial approach does not place primary emphasis upon the human aspects of the information system. What is privileged instead is the human construction of space. This emphasis is derived from the foundational assumption of the social construction of reality; we know the world only as we perceive it (Berger and Luckmann 1966). Social constructionism argues that our perceptions are based on learned interpretations of the world around us and that these learned interpretations are social. Berger and Luckmann also argue that we learn when we are engaged in social interaction. The main vehicles for conveying social meaning are symbols, cultural myths, the structure of our institutions and our rules for congruent action (pp. 55-56). These vehicles of meaning together construct our worldview, our sense of ourselves, our identity and purpose, as well as our ideologies. Our selves, our societies, and our institutions are consequently in a continual state of change through these interactions.

5 CONSTRUCTIVIST RESEARCH

This research project is located within the constructivist tradition. Constructivist ontologies aim to not reduce an observed research situation to generalizations but attempt to unfold context-specific patterns and tendencies (Karsten 2000). Constructivist research is always "irretrievably hermeneutic" (Giddens 1984). The subjects of the study, including the researcher, are influenced by, and influence, the research.

Interpretations cannot be value-free but are subjective, multiple, and complex. The concern for bias in interpretive research is significantly different from that of scientific and realist approaches. In the social sciences, the subjective experience and presence of social agency including the researcher's own are focal points for the understanding of social phenomena (Boland 1991; Burrell and Morgan 1979; Giddens 1984; Hirschheim et al. 1995; Lee 1994; Trauth and Jessup 2000).

Interpretive epistemologies were utilized in the research (Klein and Myers 1996) to examine a Web information system (Van de Ven and Poole 1990; Walsham 1993). The research focused upon the observable cultural and historical aspects of the site with emphasis given to the spatial context of the developers and users of the information system. Central to this research was the identification of the organization's physical and virtual spatial environment. Therefore, it was necessary to draw heavily upon those methodologies associated with the visual domain of enquiry. Emmison and Smith (2000) describe visual enquiry as "the study of visibility, mutual interaction and semiotics as they relate to objects, buildings and people as well as to the study of images." They continue, "visual enquiry is no longer just the study of the image, but rather the study of the seen and observable." The application of visual methodological techniques to the study is expanded further in this paper.

6 CONTEXTUALIZING THE RESEARCH PROCESS

As this was a qualitative and critical study, particular assumptions were ever-present. These assumptions concerned the way the investigators understood the nature of reality, knowledge, human nature, and methodology, and the way that they interpret the research site (Pihlanto 1994). The assumptions in this research were founded around a subjectivity in which truth only exists through interlocking contextual understandings, and where reality is a subjective state of sensibility. These assumptions are based on the observation "that people create and associate their own subjective and intersubjective meanings as they interact with the world around them" (Orlikowski and Baroudi 1991, p. 24, citing Chua 1986). This research was an exploration of how truths can be found in social life which is itself socially constructed. If this assumption is pursued, then the social construction of these truths can inform both the participant and the researcher of their effect and presence. To clarify the research approach utilized in the original study, Orlikowski and Baroudi's framework has been adapted to illustrate the conduct of the research process (Table 1).

The research processes and their associated assumptions outlined in Table 1 informed and impacted upon the exploration of the research site. For example, contextual and ideographic methods were applied to enable the researcher to observe social practice to understand the Web development team's dynamic and conflictual development process. By applying a social construction of meaning with the data collected, a number of targeted questions could be asked of the development team. It was intended that the research would reveal how this team spatially perceived their Web Information System. This would then inform the Web development team and their employer of how an informed outsider saw their organizational processes. This was appealing to the department as there was an economic agenda regarding the funding of the Internet development.

Table 1. Description of the Properties and Patterns of the Study

Beliefs About	Explanation	This Research
Physical and social reality		
Ontology	Whether social and physical worlds are objective and exist independently of humans, or subjective and exist only through human action	Subjectivist, constructionist at least when regarding social reality
Human Rationality	The internationality ascribed to human action	Agency as ability to do otherwise
Social Relations	Whether social relations are intrinsically stable and orderly, or essentially dynamic and conflictual	Dynamic and often conflictual
Knowledge		
Epistemology	Criteria for constructing and evaluating knowledge	Interpretive, intersubjective, Hermeneutic
Methodology	Which research methods are appropriate for generating valid evidence	Contextualist, ideographic methods
Relationship between theory and practice	The purpose of knowledge in practice	To inform and to enlighten

7 METHODOLOGICAL APPROACHES

A critical approach was applied to each theme in the project including space, organizations and information systems, the developers, information systems and work spaces, the Web, and virtual reality. The study challenged the limited understandings of the spatial qualities of information systems and the limited regard of space by developers and users in relation to information systems development. Orlikowski and Baroudi (1991, p. 24, citing Chua 1986) explain that

> Critical studies aim to critique the *status quo*, through the exposure of what are believed to be deep-seated, structural contradictions within social systems, and thereby to transform these alienating and restrictive social conditions.

Space is a socially constructed phenomenon. If the socio-cultural aspects of real-life space are ignored, then virtual space and all spaces of occupation become problematic. This was an issue for this research as it examined the spaces associated with a Web information systems and whether physical restrictions are experienced in virtual reality based systems as they are in real-life systems. As a result, an aspect of the study was to examine if the development and usage of a Web information system would also be experienced in the same way.

Specific methodological choices had to be made in the study for their capacity to meet the demands of the theory. Theory, in this way, determined and directed method. Lefebvre (1991) emphasizes that space should be examined using the broad dichotomy of social and abstract space. Social space is, therefore, further categorized as *spatial practice*, *representations of space*, and *representational space*. The researcher can include observation of textures, shapes, groupings of objects, creative expressions like drawings and even memories (Bachelard 1969, p. 6; Game 1995, p. 201). Consequently, emphasis is placed upon the spatial dimensions of information sharing, interplay, and grouped according to the overarching concepts of social space. Data was gathered relating to each individual category of social space and then analyzed separately. To capture spatial data it was necessary for methodological choices to be broad enough to encapsulate this triplicate analysis. Each data collection configuration was identified and discussed in relation to the appropriate theoretical lens in the analysis phase of the research project.

8 CHOOSING THICK DESCRIPTION AS A METHODOLOGICAL APPROACH TO DATA COLLECTION

The most significant methodological practice followed throughout the study was thick description. Thick description is underutilized in contrast to ethnography within qualitative approaches to information systems exploration (Myers 1999; Trauth 2000; Trauth and Jessup 2000). The significance of ethnography, for the field of Information Systems and for qualitative analysis generally, is in its ability to present an intensive study of a particular social phenomenon (Trauth 2000; Trauth and Jessup 2000). However, the research conducted for this study was not a ethnography or an interpretive in-depth case study, as it is described by Walsham (1993). Instead it drew upon the research foundations articulated in anthropology by utilizing an ethnographic approach to data collection in a manner that Geertz (1973) describes as "thick description." Schultze (2001, p. 90) states that "the ethnographic method requires the researcher to closely observe, record, and engage in the daily life of people in the field and write about it in descriptive detail." In this respect, both ethnography and thick description utilize a similar impetus. A broad range of data-gathering techniques are drawn upon as well as the intuitive and interpretive skills of the researcher in order to conduct research. The significant difference between ethnography and thick description is the duration of time spent at the research site (Klein and Myers 1999). Ethnography calls for extensive interaction over time to fully immerse the researcher in the culture in order to reduce the bias of observations and to give a less tainted account of the subjects of observation (Atkinson 1990; Daniel and Peck 1996, p. 21; Hughes et al. 1999). Thick description also aims at rich and full descriptions of a particular research site that is not linked to a time frame. Instead it has the objective of providing a thorough, careful, and accurate exploration of the lived experiences of the research subjects as they have been observed by the researcher (Atkinson 1990, p. 8; Geertz 1996). For example, data gathered regarding the informant's daily activities and utilizing a thick description focus appeared as such:

> Crystal walks over to Cassandra's cubicle. Cassandra's cubicle backs onto Crystal's but as an indication of her higher organizational rank Cassandra occupies a cubicle to herself. Cassandra's cubicle is her own personal space, rectangular in shape and fitted out with the departmental standard grey and beige furnishings. Crystal enters Cassandra's cubicle and asks if they can talk about the day's activities. Crystal in her debriefing keeps Cassandra informed on what she and Penny have discussed and agreed to do today. Crystal tells Cassandra that Penny will contribute to the resolution of the file size issue and the construction of the office map.

The depth and detail of the observations recorded by this method offers an account that maintains an unbiased interpretation of the events observed solely through the researcher's own expertise and sensitivity to the research subject. To achieve this objectivity, the study presented a narrative of the spatial interactions of the Web development team. A chronology provided a thick and detailed description of their daily activities and experiences. The narrative did not disentangle time or space; instead it provided an immediate description of the worker's daily life. To convey the depth and complexity of ethnographic studies, the description was offered in the narrative form truest to the *ethical imperative* of the writer and the multitude of representations the subject reveals in a research context (Barthes 1983). In this way, the researcher's narrative of interactive events at the research site provides additional detail to the observations.

9 REFLECTING ON CHOICE OF QUALITATIVE METHOD

Trauth's discussion of the influences on the choice of qualitative methods in information system research is utilized here. Trauth (2001, pp. 4-11) poses five influences regarding the choice of qualitative analysis. The five influences are the research problem, the researcher's theoretical lens, the degree of uncertainty surrounding the phenomenon, the researcher's skills, and contemporary academic politics.

9.1 Research Influence 1: The Research Problem

The initial assumption for the research concentrated on the creation and utilization of space by the organization within the context of an information system. To this end the study considered whether *a spatial consideration of information systems development was useful during the development of that information system*. To achieve this, a detailed examination of organizational interactions was conducted to understand how information was developed by individuals for publication on the Web; how individuals differed in their socio-cultural position, technological context, and physical placement within the organization. The *objective of the research was to understand the significance of space within information systems development.* The study extended current spatially focused research in Information Systems by examining *the usefulness of space for exploring new technologies*. Significantly, for this analysis, a clear picture

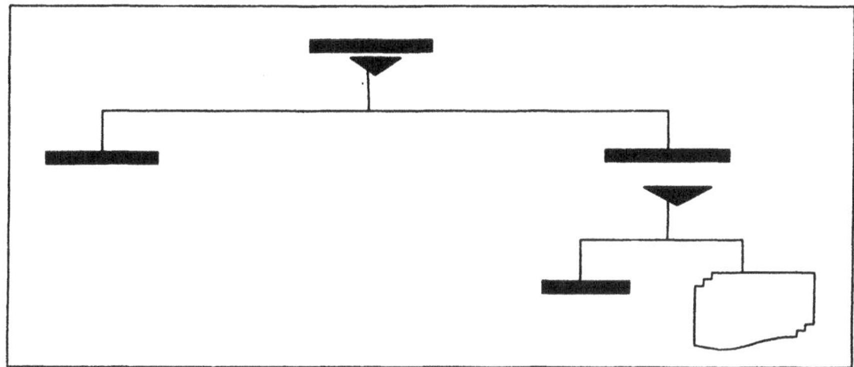

Figure 1. Diagram Taken from System Specification Document

of the Web development team's understanding of the system was provided by their technical documentation. Figure 1 came from an illustration included in the functional specifications documents that had been drawn up by an outside contractor and presented to the Web development team. This particular example provides a detailed pictorial representation of the Web information system subsystem that was to be developed to store and maintain information about news, events and other system generated events.

The drawing was used by the designers to illustrate to the team where they understood specific information would be positioned within the new Web site. This diagram is also a site map. It uses pages and lines to indicate general connectivity between the pages of the Web site. It symbolizes the connectivity of buttons and hypertext links. This is represented as the small arrows in the picture and indicates links between related pages. Each page has a series of links. Spatial references were collected and observed during all aspects of the research but it was only with the analysis and collation of this data that the significance of spatiality to the development process was revealed.

9.2 Research Influence 2: Researcher's Theoretical Lens

Lefebrve's primary argument is that space and the social construction of space are not always easily observed. However, he claims that they are important considerations for research. An appreciation of social space can, therefore, only be achieved through the exploration of the subjective position (Lefebvre 1991, p. 6). The thick description approach to ethnography was adopted because it is through analysis of the data gathered about the languages, images, institutional practices, and other social practices discerned in the primary research that questions about space and the development of a Web information system could be explored. Another example drawn from the research is the interview transcript and complimentary discussion offered by the researcher:

> I think of it in a similar way as there being a pool of documents. I think of it more hierarchical in a way. You have a large pool of documents all sitting in

> a big pond at the bottom, sort of at a level. You have the home page right at the top, and then you have these intermediate pages in between, which are like menu pages that help to sort and give access to the big pool at the bottom....they're dynamically done, so I might choose to create a intermediary menu page that's on the theme...and then by tagging all the documents [the pages] on the bottom, they automatically appear on the menu....It's never completely hierarchical but that's a general way to get a grip on the information.

This example shows the usefulness of thick description and non-participant observation as an appropriate way to explore the research site's socio-cultural, technical, and historical makeup.

To adequately address the central research problem regarding space in information systems development it was also necessary for a critical ethnographic approach (Bentley et al. 1992, p. 123). Lefebrve states that space is an integral component within the circuits of capitalism, and therefore contributes to the perpetuation of workers' inequality. It was necessary, in the framework of this claim, to explore the existing power structures of the research site. Immersive techniques allow critical explorations to uncover the forms of these interactions (Schultze 2001, p. 80). The immersed researcher can provide the depth of analysis necessary to explore social phenomena such as space. The researcher, at appropriate periods of observation, sketched the spatial surroundings of the working environment. One example is presented in Figure 2.

The use of an ethnographic technique was chosen because this method was most suited to revealing "the story behind the space" of a Web information system from the developer's perspective. This method enabled the researcher to observe the daily events of a group of people working on developing a Web information system. Research activities included observing the people and the events and included mundane elements of a working team's daily lives. For example, detailed observation from the data collection existed in this form:

> I look at my watch. It is 9:00 a.m. The office was quickly filling with people. They all began observing their morning work preparation rituals: starting up their computers, planning their day's activities, greeting those around them, collecting coffees, sharing tales and salutations, "Good morning Mike, how was your weekend?" Nigel spotted me, and after extending the usual welcoming remarks, introduced me to a number of the people in the Web development team. Nigel was the acting project manager. He is a thin, wiry man whose age and length of experience is very hard to ascertain. His indefiniteness is heightened by the impact of his very white bleached hair with dark roots, his weathered complexion, and his crisp mannerisms.

This method of data collection provides a different understanding of the working environment than those that can be revealed by a questionnaire or other methods.

The ability to experience the team's daily life in its totality was restricted by the organization and the research focus itself. It was necessary to consider the speed with which this form of technology was evolving and being embraced by the organization. As a consequence, an elongated observation period covering a number of years risked

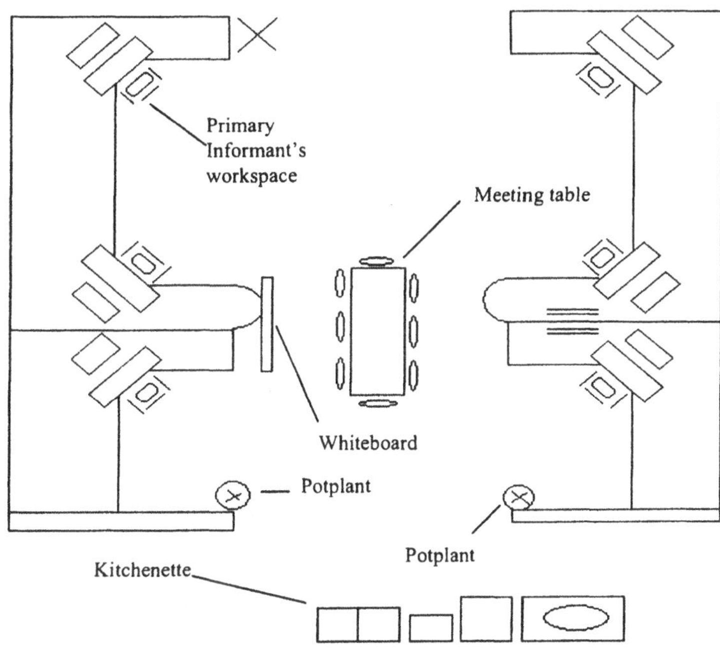

Figure 2. Diagram of the Primary Informant's Work Area

encountering new impetus and changes in direction brought with the introduction of still newer Web technologies. Utilizing a more time consuming approach would result in a lack of accounting for the immediacy of the development. The project itself was already behind schedule and the excessive demands on the host organization meant that a prolonged period of in-depth observation of a specific team member could not be sustained for more than one week. Although the organization supported the research it could not sustain a drawn-out research protocol.

9.3 Research Influence 3: Degree of Uncertainty

There was a high degree of uncertainty surrounding the phenomena under study. An example of this uncertainty can be discerned from the data that the interviews and informants themselves provided.

> The information often just comes in as, say, a raw manuscript, if you're backing a print with it. I mean, it's just a raw manuscript with the author editing it. Initially when we started off over 3 years ago now, we thought we could train the staff out there how to create HTML pages. It is high technology and they're not all DAP people, are not capable of doing the work required [to

> put it online]. You know, to take it to print, you get the raw manuscript and take it through the publishing process to look at what comes out the other end.

The exploration of the social construction of space and the perceptions of the developers and users of the Web information system represents the intangible dimensions of an information system. Other studies into space and information systems utilize geographic constructs or artifactual assessments of the information system environment. The exploration of the socio-cultural impact brought by the people engaging with the Web information system has a high degree of uncertainty associated with it. There was no existing documentation or description about any of the people studied or, in a general sense, how the developers of a Web information system spatially perceive their system. The study utilized visual data collection in the form of photography to help clarify the spatial dimensions of organizational practices.

The approach to data collection required an open mind and acceptance of the fact that the informants would reveal their own particular socio-cultural perspectives and biases. In this way, the *what* of the theory drove the *how* associated with the method of this research project (Järvinen 1999; Trauth 2001). To minimize the degree of uncertainty of the phenomenon being studied, appropriate methods had to be utilized to explore *how* the developers and users spatially perceived a Web information system. This was done by drawing on the researcher's own interpretation of the research site and by employing a critical eye to delve beyond the immediately obvious interactions. This approach contrasts with a quantitative exploration of a Web information system's spatiality. A quantitative method would need to establish the reality of such a statement. To do this, the researcher might measure the volume of traffic on the Web sites or measure the distance between individuals working on the Web information system and then assess the extent that this is a significant space to traverse. However, the focus of this research project was to explore the sociality of an information system through an ethnographic approach aiming to detail spatial perceptions, and their existence. Using a variety of qualitative techniques as an interpretive foundation the finalized study was able to reveal that the Web development team did hold specific perceptions about the Web pages contained on their Web site. These perceptions predominantly existed as an awareness of the page as a bounded box and these visualizations were revealed through descriptive passages relating to the Web information system itself. The information contained on the Web site was also described as contained and bounded. For example, Crystal used metaphor to explain how she mentally visualizes the site.

> There's usually some sort of categorizing and clustering of things, there's usually some sort of overarching framework that you can use that means it can be contained with one page. In terms of the Internet and what's happening with the Internet, this is a really old sort of interface.

> I think of it in a similar way as there being a pool of documents. I think of it more hierarchical in a way. You have a large pool of documents all sitting in a big pond at the bottom, sort of at a level. You have the home page right at the top, and then you have these intermediate pages in between, which are like menu pages that help to sort and give access to the big pool at the bottom....

> they're dynamically done, so I might choose to create a intermediary menu page that's on the theme...and then by tagging all the documents [the pages] on the bottom, they automatically appear on the menu... .It's never completely hierarchical but that's a general way to get a grip on the information.

This example provides some insight into the visual image that Crystal had about the Web and the Web information system itself. Chorafas and Steinmann (1995, pp. 206-7) see visualization as a means to discovering users' understanding of technology. In this way, drawing upon ethnographic methods allowed the researcher to go beyond understanding the technology to explore the how element of spatial perception.

9.4 Research Influence 4: Researcher's Skill

The foundations for the research topic and interest in the area of virtual space grew out of earlier research in the field of anthropology and sociology. An extensive history associated with the use qualitative methods led to the decision to use an array of qualitative methods that was not restricted to a narrow range of method. Having studied many methods, with significant practical experience of qualitative studies, and experience as a market researcher, I had practiced a variety of methodological techniques over the previous eight years. The methods I had used ranged from the development and usage of nationwide survey instruments, to psychological and behavioral market research practices, to undercover research evaluation of retail business practices, as well as many one-to-one in-depth interviewing and group interviewing sessions. At no time was the prospect of using interpretive methods to explore the notion of perceived space intimidating. I also had an established interest in computing, multimedia and World Wide Web related technologies.

9.5 Research Influence 5: Contemporary Academic Politics

I have a Bachelor of Arts with Honors and this provides me with a strong theoretical and analytical understanding relating to the social practices of computer usage. The academic political climate in Australia in the late 1990s also played a significant role in my choice to explore other research projects that examine the social impact of networking. The degradation of the Humanities and Arts Faculties in terms of guaranteed employment, funding, and job certainty for new academics (Keen 2001) seemed to lead naturally to the more secure field of information systems, where jobs were relatively abundant with a degree of job security. From this stable academic environment, a research focus regarding computers and society focusing on Web Information Systems was concretized.

In the context of the wider academic political environment there is also acknowledgment by other disciplines that a need exists for more critical research regarding Information Systems and embracing Web technology by the mainstream. The intended audience for this research was fellow researchers and academics situated within the Information Systems field, although it was, and is, hoped that others may also show interest in this cross-disciplinary field.

10 DATA COLLECTION AND ANALYSIS IN AN INTERPRETIVE STUDY

Data collection in thick description often incorporates a number of techniques to capture the breadth of activity being observed. The data collection for this research project used a single case study with a variety of collection techniques over two specific periods. To ensure that a chain of evidence could be adhered to, the research conduct was divided into phases.

Exploratory research was conducted in the first phase of research. This was undertaken with a series of semi-structured interviews with managerial level staff within the department. Three employees who worked directly under these managers were also interviewed to provide alternative sources and stories to those presented by the managers. This was done to include the many levels of the workplace hierarchy. All of the informants interviewed in phase I were identified as people responsible for the maintenance and development of the organization's Web sites. The questions put to them encompassed a range of issues including questions about personal interactions, work practices, and specific spatial understanding. The initial interviewing provided clues as to where further detail could be built upon to explore the core research question associated with Information Systems development. More visually oriented techniques such as the field diary, hand-drawn sketches, and digital photographs were expanded upon in later research phases. Table 3 provides details of all of the types of data or information sources that were gathered in the context of this particular study. It shows the variety of types of information that can be collected from a research site.

Significantly, in all research studies data analysis is undertaken to examine, categorize, and assess what has been revealed by a particular set of data. This research project has, in a manner similar to that of Heaton (1998), utilized interpretive methods to gather data through nonparticipant observation, a research diary, interviews, and documentation. The choice of these methods was influenced by the desire to explore the socio-cultural understandings of the informants and how they reflect their own understandings in the design of an information system. The research diary, taped debriefings, sketches, and other narrative descriptions and observations were kept as integral elements of thick description of the research site. The structure of the field diary is shown in Table 4.

The field diary was a significant tool in this study as it enabled the researcher to clearly and consistently observe and document the interactions that were occurring at the time of the study. *Ad hoc* data gathering could not achieve the detail that a consistent field diary can achieve. It is also important to note, as van Maanen (1988, p. xi) has said, that discussions relating to method and particularly ethnographic methods should always

> explicitly consider (1) the assumed relationship between culture and behavior (the observed); (2) the experience of the fieldworker (the observer); (3) the representational style selected to join the observed (the tale); and (4) the role of the reader engaged in the active reconstruction of the tale (the audience).

Table 2. Phases of the Research Conduct

Phase	Description	Timetable
I	Formation of the theoretical framework.	May 1998
II	Cooperation with DAP begins. First exploratory interviews begin. Interviews conducted with Alison, Kent, Brock, Diane, George, and Nigel. Written material collection begins. Material includes booklets and pamphlets released to general public, internal documents including corporate structure, aims and objects, internal policies corporate image, documents on technical training programs, print-outs from intranet internal announcements, general site layout of intranet, Internet information printouts. Analysis of potential leads for second phase study occurs.	October 1999
III	Collection of second round interviews begins. Interviews with the manager of the Web development team. Portia (Nigel interviewed in phase I).	December 1999
IV	Collection of interviews continues. Participant observation (one week period) shadowing Crystal. More documentation collected including meeting agendas, newspaper articles from the site, systems development documents, organizational charts and maps, documents on proposed amendments to the systems, copies of e-mails, change request forms, systems development time lines and costs, information flow diagrams both formal and informal, progress reports, more Web pages relating to site development and construction, functional specification documents of new system, secondary research site background documents from the Web. Further interviews with Web development team including Roger, Penny, and Crystal. Interviews conducted with two Ping employees: Manny (the lead technical developer) and Cain (back end or systems developer).	April 2000
V	Final data collection ready for analysis. Pattern matching. Coding and exploration of transcripts. Coding and exploration of field notes and diary. Analysis of written documentation.	July 2000
VI	Confirmation of interpretation and analysis with research site. E-mail interactions.	September 2000
VII	Research results ready for reporting.	May 2001

Table 3. Information Sources Used in the Study

Information Source	Form of Materials	Period Collected
Formal documentation	Printed material ranging from memoranda to minutes of meetings to formal reports and Web site pages.	Phase II , Phase IV
Archival and historical records	Organizational charts, financial and personal records, maps, graphs and service statistics.	Phase I, Phase II, Phase IV
Informal documentation	Hand-drawn material ranging from sketches on white boards to scribbles and notes to other employees or self.	Phase II, Phase IV
Interviews	Open-ended and semi-structured interviews coupled with probing questions.	Phase II, Phase IV
Direct or participant observation	Absorbing and noting details, actions, and subtleties of the field environment.	Phase IV
Physical artifacts	Office and room construction, chairs tables, office cubicles, etc.	Phase II, Phase III, Phase IV

Table 4. Layout of Field Diary: Tuesday, April 4

Time	Action	Interpretation
9:20 a.m.	Spoke for a while.	
9:31 a.m.	Turned computer on.	Takes a while to up load. Mixture of messages. Their own e-mails and chat group.
9:42 a.m.	E-mail problem	Problem is with someone she has trained—Cascade I style sheet (enabled).
9:43 a.m.	Telephones woman to check out problem.	Sarah is not quite sure why, then gets an idea. Will relink pages. Cascade needs linking. File size is too long—strip down file size (html). She leaves it to test or keep playing with it at a later date. It works; problem is solved. She gives more advice to change the different directories.

Consistency in the data gathering techniques followed by the researcher regardless of their qualitative or quantitative foundations should adhere to these considerations. In this particular study, these sentiments were adopted and choosing visually, spatially oriented methods of data collection were selected to reinforce the spatial dimensions of the theoretical focus of the study.

11 CONCLUSION

This paper has contextualized a recent research project by outlining its method and process. The definitions given by Keen (1980) and Galliers and Land (1987) are used to contextualize the foundational assumptions that define an information system. A combination of these two definitions was used to include and situate human relationships within the information system studied. The methodological approach, assumptions used, and the research techniques are presented to reveal the way in which a qualitative, interpretive, and critical research project has been conducted. Considerable explanation has been provided about the epistemological, ontological, and theoretical positions pursued in this study. This situates the research in contrast to other research of similar topics. This paper clarifies how and why theoretical choices are made in research. Theory often informs the methodological practices the research must use to explore an interpretive study at a chosen research site. This paper provides an overview of one form of qualitative research in information systems. It argues that the primary research question is the hook on which all research approaches and methodological decisions rest. Because the research was ontological and theory testing was emphasized, ethnographic analysis in the form of thick description has been presented as the most appropriate method with which to conduct this research. A variety of data gathering techniques including visual techniques and the necessity of a field diary for consistency are discussed to illustrate the maintenance of rigor in such studies.

REFERENCES

Atkinson, P. *The Ethnographic Imagination: Textual Constructions of Reality*, London: Routledge, 1990.

Bachelard, G. *The Poetics of Space*, Boston: Beacon Press., 1969.

Barthers, R. "Barthes: Selected Writings," in S. Sontag (Ed.), Douglas, Isle of Man: Fontana Press, 1983.

Benbasat, I.; Goldstein, D. ; and Mead, M. "The Case Research Strategy in Studies of Information Systems," *MIS Quarterly* (11:3), September 1987, pp. 369-387.

Bentley, R. H.; Randall, J.; Rodden, D.; Sawyer, T.; Shapiro, D.; and Sommerville, I. "Ethnographically-Informed Systems Design for Air Traffic Control," in J. Tuner and R. Kraut (Eds.), *Proceedings of the Computer Supported Cooperative Work Conference*, Lancaster, UK, 1992.

Berger, P. L., and Luckmann, T. *The Social Construction of Reality: A Treatise in the Sociology of Knowledge*, New York: Anchor Books, 1966.

Boland Jr., R. "Information System Use as a Hermeneutic Process," in H-R. Nissen, H. Klein, and R. Hirschheim (Eds.), *Information Systems Research: Contemporary Approaches and Emergent Traditions*, Amsterdam: North-Holland, 1991, pp. 439-458.

Bradley, M. P., and Briggs J. S. "An Internet Information System for GP's," *Information Research: An Electronic Journal*, 1995 (available online at http:// InformationR.net/ir/3-3/paper41.html).

Burrell, G., and Morgan G. *Sociological Paradigms and Organizational Analysis*, Aldershot, England: Gower Publishing Company Ltd., 1979.

Checkland, P., and Howell, S. *Information, Systems and Information Systems: Making Sense of the Field*, Chichester, UK: Wiley, 1998.

Chorafas, D. N., and Steinmann, H. *Virtual Reality: Practical Applications in Business and Industry,* Englewood Cliffs,NJ: Prentice-Hall, 1995.

Chua, W. "Radical Developments in Accounting Thought," *Accounting Review* (61), 1986, pp. 583-598.

Daniel, E. V., and Peck J. M. "Culture/Contexture: An Introduction," in E. V. Daniel and J. M. Peck (Eds.), *Culture/Contexture: Explorations in Anthropology and Literary Studies*, Berkeley, CA: University of California Press, 1996.

Emmison, M., and Smith, P. *Researching the Visual*, London: Sage Publications, 2000.

Eschenfelder, K., and Sawyer, S. "Web Information Systems Management: Proactive or Reactive Emergence," in N. L. Russo, B. Fitzgerald, and J. I. DeGross (Eds.), *Realigning Research and Practice in Information Systems Development: The Social and Organizational Perspective*, Boston: Kluwer Academic Publishers, 2001, pp. 163-181.

Galliers, R., and Land, F. F. "Choosing Appropriate Information Research Methodologies," *Communications of the ACM* (30:11), 1987, pp. 900-902.

Game, A. "Time, Space, Memory, with Reference to Bacheland" in M. Featherstone, S. Lash, and R. Robertson (Eds.), *Global Modernities: From Modernism to Hypermodernism and Beyond*, London: Sage Publications, 1995, pp. 192-208.

Giddens, A. *The Constitution of Society*, Cambridge, MA: Polity Press, 1984.

Geertz, C. *Thick Description: Toward an Interpretive Theory of Culture*, New York: Basic Books Inc., 1973.

Geertz, C. "The World in a Text: How to Read Tristes Tropiques," in E. V. Daniel and J. M. Peck (Eds.), *Culture Contexture: Explorations in Anthropology and Literary Studies*, Berkeley, CA: University of California Press, 1996.

Heaton, L. "Talking Heads vs. Virtual Workspaces: A Comparison of Design Across Culture," *Journal of Information Technology* (13:4), 1998, pp. 259-272.

Hirschheim, R.; Klein, H. ; and Lyytinen, K. *Information Systems Development and Data Modeling: Conceptual and Philosophical Foundations*, Cambridge, UK: Cambridge University Press, 1995.

Hughes, J.; O'Brien, J.; Randall, D.; Rouncefield, M.; and Tolmie, P. "Virtual Organizations and the Customer: How 'Virtual Organizations' Deal with 'Real' Customers," in *Proceedings of the 4th United Kingdom Association of Information Systems (UKAIS) Conference*, University of York, 1999.

Järvinen, P. *On Research Methods,* Tampere, Finland: Tampereen Yliopistopaino Oy, Jevenes-Print, 1999.

Kaplan, B. "Evaluating Informatics Applications—Some Alternative Approaches: Theory, Social Interactionism, and Call for Methodological Pluralism," *International Journal of Medical Informatics* (64), 2001, pp. 39-56.

Karsten, H. "Constructing Interdependencies with Collaborative Information Technology," in R. Baskerville, J. Stage, and J. I. DeGross (eds.), *Organizational and Social Perspectives on Information Technology,* Boston: Kluwer Academic Publishers, 2000, pp. 429-450.

Keen, P. G. W. "MIS Research: Reference Disciplines and Cumulative Tradition," in E. R. McLean (Ed.), *Proceedings of the First International Conference on Information Systems*, Philadelphia, PA, 1980, pp. 9-18.

Keen, S. "Economics: From Emperor to Vassal?," *Australian Universities Review* (44), 2001, pp. 1-2, 15-17.

Klein, H. K., and Myers, M. D. "A Set of Principles for Conducting and Evaluating Interpretive Field Studies in Information Systems," *MIS Quarterly* (23:1), 1999, p. 67-93.

Klein, H., and Myers, M. "The Quality of Interpretive Research in Information Systems," Working Paper Series, Binghamton University, School of Management, Binghamton, New York, 1996.

Lee, A. S. "The Hermeneutic Circle as a Source of Emergent Richness in the Managerial Use of Electronic Mail," in J. I. DeGross, S. L. Huff, and M. C. Munro (Eds.), *Proceedings of the 15th International Conference on Information Systems*, Vancouver, Canada, 1994, pp. 129-140.

Lefebvre, H. *The Production of Space*, Oxford: Blackwell, 1991.

Molotch, H. "The Space of Lefebrve." *Theory and Society* (22), 1993, p. 887-895.

Myers, M. E. "Qualitative Research in Information Systems," *IS World*, 1999, (available online at http://www.qual.auckland.ac.nz/).

Orlikowski, W. J., and Baroudi J. J. "Studying Information Technology in Organizations: Research Approaches and Assumptions," *Information Systems Research* (2:1), 1991, pp. 1-28.

Pihlanto, P. "The Action-Oriented Approach and Case Study Method in Management Studies," *Scandinavian Journal of Management* (10:4), 1994, pp. 369-382.

Robey, D.; Schwaig, K. S.; and Jin, L. "Intertwining Material and Virtual Work," *Information and Organization* (13), 2003, pp. 111-129.

Schultze, U. "Reflexive Ethnography in Information Systems Research," in E. M. Trauth (Ed.) *Qualitative Research in IS: Issues and Trends*, Hershey, PA: Idea Group Publishing, 2001.

Trauth, E. M. *The Culture of an Information Economy*, Dordrecht: Kluwer Academic Publishers, 2000, pp. 78-103.

Trauth, E. M. *Qualitative Research in IS: Issues and Trends*, Hershey, PA: Idea Group Publishing, 2001.

Trauth, E. M., and Jessup, L. "Understanding Computer Mediated Discussions: Positivist and Interpretive Analyses of Group Support System Use," *MIS Quarterly* (24:1), 2000, pp. 43-79.

Truex, D., and Baskerville, R "Deep Structure or Emergence Theory: Contrasting Theoretical Foundations for Information Systems Development," *Information Systems Journal* (8:3), 1998, pp. 99-118.

Van de Ven, A. H., and Poole, M. S. "Methods for Studying Innovation Development in the Minnesota Innovation Research Program," *Organization Science* (1:3), 1990, pp. 13-335.

van Maanen, J. *Tales of the Field*, Chicago: University of Chicago Press, 1988.

Walsham, G. *Interpreting Information Systems in Organizations*, Chichester, UK: Wiley, 1993.

Yin, R. K *Case Study Research: Design and Methods*, Beverley Hills, CA: Sage Publications, 1989.

ABOUT THE AUTHOR

Anita Greenhill continues to carry out interpretive studies exploring the notions of space and virtuality. These research projects are cross disciplinary and bridge the fields of Information Systems, Education, Cultural Studies, Sociology, and Anthropology. Anita can be reached at A.Greenhill@umist.ac.uk.

14 APPLYING HABERMAS' VALIDITY CLAIMS AS A STANDARD FOR CRITICAL DISCOURSE ANALYSIS

Wendy Cukier
Ryerson University

Robert Bauer
Johannes Keppler University

Catherine Middleton
Ryerson University

Abstract It has been proposed that the theory and practice of information systems development could benefit from a more explicit consideration of concepts of rationality. Habermas' communicative rationality has been proposed as an approach to improve the conditions for rational discourse in systems development, thereby improving outcomes (Klein and Hirschheim 1991), and applied at the project level (Ulrich 2001) and to specific episodes of managerial communications (Ngwenyama and Lee 1997). At the same time, it is understood that societal discourses and ideologies shape the external environments of organizational decision making. A variety of approaches has been proposed to analyze these discourses including qualitative techniques for *reading* or *interpreting* texts, artifacts, and social practices (Philips and Hardy 2002). This paper examines the way in which Habermasian validity claims can provide an explicit and *ethical* standard for critical discourse analysis in order to reveal the distortions that shape the institutional environments of technology decision making. It offers an approach to operationalizing Habermas' validity claims for an analysis of media texts related to a case study involving learning technology.

Keywords: Discourse analysis, Habermas, institutional theory; ideal speech, learning technology

1 INTRODUCTION

This paper examines media discourses on a technology enabled learning project in an effort to explore ways in which these discourses may shape and reflect technology planning and decision making. Using texts that discuss the "Acadia Advantage" case (a Canadian university's program to bring notebook computing into the classroom), the paper demonstrates how Habermas' validity claims can be used as an analytical framework to guide discourse analysis. The paper contributes to the discussion of technology planning and systems development and to the techniques of discourse analysis in two ways.

(1) Discourses are important to the study of organizations and information systems because they shape the organizational and institutional environments that provide a context for planning and decision making. However, in spite of the long tradition of technology criticism that focuses attention on the construction of reality, there has been limited attention in the information systems literature to the linkages between societal level and organizational discourses.

(2) Despite the growing popularity of discourse analysis among information systems researchers (e.g., Wynn et al. 2003), there has been limited reflection to date on the methodology of discourse analysis. This paper proposes a novel approach to textual analysis grounded in Habermasian discourse ethics. It proposes that the Habermasian notion of the "ideal speech act" can serve as a standard for assessing the rationality of discourse and that Habermas' validity claims can be operationalized for textual analysis that is both rigorous and theoretically sound.

2 DISCOURSE ANALYSIS AND INFORMATION SYSTEMS

Discourse is one of the principal ways in which reality is socially constructed and has long been the subject of scholars in sociology, psychology and cultural studies (see, for example, Fairclough 1995; Fiske 1982; Hansen et al. 1998; Hirsch 1986; Inglis 1990; Jensen and Jankowski 1991; van Dijk 1991; Wodak 1989). More recently, attention has focused on exploring the role of discourses and their relevance to organizational studies and management (e.g., Grant et al. 2001, Hardy 2001, Kets de Vries and Miller 1987).

There have also been analyses of information technology discourses at the organizational level (Wynn et al. 2003). For example, Robey and Markus (1984) suggested that elements of the systems design process can be interpreted as rituals that enable actors to appear overtly rational while negotiating to achieve private interests. Orlikowski and Yates (1994) examined genre repertoires in organizational communications, and Päivärinta (2001) applies the genre concept to critical information systems development. Murray (1991) examined discourses of power among IS specialists. Bloomfield and Vurdubakis (1994) examined discourse reflected in information technology consultancy reports. Others (e.g., Boland 1985, 1991; Boland and Day 1989; Butler 1998l; Gopal and Prasad 2000; Myers 1995) have applied hermeneutic analysis to aspects of discourse analysis.

Societal discourses are part of the institutional environment of organizations, and media is an important part of societal discourse. Studies in the social sciences examine the role of media in shaping perceptions of social reality (Gerbner 1977; Lazarsfeld and Merton 1948; Lippmann 1992; McLuhan and Fiore 1968), as well as the structural forces shaping discourse (Chomsky 1989; Foucault 1980). The organizational relevance of the broader societal discourse as a part of the organization's institutional environment has, with varying degree of explicitness, been acknowledged in the organization theory literature (e.g., Czarniawska-Joerges and Joerges 1988; DiMaggio and Powell 1984; Meyer and Rowan 1977; Meyer and Scott 1983). Alvarez (1996) notes that the popularization of knowledge has to do not only with its intellectual merits but also with the political, social, and ideological position and disposition. In general, the links between societal discourses (macro level) and organizational discourses have received limited attention.

The role of discourse in management has been examined by Clegg and Palmer (1996). Abrahamson (1996, 2001) and others (e.g., Furusten 1999) examined the role of management fads and fashions. Philips and Hardy (1997) examined refugee systems, and Hardy, Lawrence and Philips (1998) examined employment services. Calás and Smircich (1991) deconstructed leadership discourse and Bowring (2000) deconstructed institutional theory. Townley (1993) used Foucauldian discourse analysis to explore human resources management. There has been some examination of the discursive practices by some professions (Clegg and Palmer 1996; Meyer and Scott 1992).

There is a well-established tradition in the social sciences focusing on exposing broad societal discourses that surround technology. For example, Ellul (1977) and Winner (1986) examine aspects of the technological imperative that has enveloped society and suppressed technology criticism. Ellul argued that "the human being who uses technology today is by that very fact the human being who serves it" (p. 325). Nardi and O'Day (1999) have examined "the rhetoric of inevitability," a language that represents technological change as unstoppable and unavoidable. Postman (1992), Rose (2003), Stoll (1995), and many others examine the ways in which technology discourses shape perceptions and behavior.

However, there has been relatively little discussion of the role of societal discourses in the information technology literature. Given the importance of ideologies and societal discourses in shaping the institutional environments of organizations, societal discourses surrounding information technology would seem to warrant further exploration (Cukier and Bauer 2002; Cukier et al. 2003). By analyzing the broader societal discourse on information technology, using a framework that is directly linked to rationality in organizational decision making, we make a contribution to narrowing this gap. The paper also provides an explicit discussion of the methodology used to explore the rationality of societal discourses on information technology.

3 PARADIGMS AND METHODS IN DISCOURSE ANALYSIS

There have been attempts to categorize approaches to discourse analysis using the concept of paradigms (e.g., Heracleous and Barrett 2001; Philips and Hardy 2002).

Such typologies distinguish, among other things, *interpretative* approaches from *critical* discourse analysis. Interpretative approaches, for the most part, are grounded in the notion that reality is socially constructed, and draw on a wide range of philosophical and linguistic theories of language and approaches to reading texts (which can include the written word or social practices). A number of scholars have applied these approaches to "reading" organizational texts (Boland 1991; Lee 1994), to analyzing metaphor or genre (Orlikowski and Yates 1994), or to deconstruction (Calás and Smircich 1991). These are modes of address that imply specific social uses of communication in relation to particular political and cultural practices (Jensen 1991). According to Burrell and Morgan's (1979) framework for paradigm analysis, the interpretative paradigm is subjective and focused on examining the status quo rather than effecting change. Consequently, it is difficult to make a link between interpretative approaches and improving practice.

In contrast, critical discourse analysis is focused on exposing the deep structures that underlie discourse, particularly power, and is grounded in normative or ethical standards. Its roots are in the Frankfurt school of neo-Marxism and it has been adapted in some forms of radical feminist analysis. Its explicit objectives are to effect radical change. Fondas (1997), for example, undertook a feminist analysis of management writings. Krefting (2001) analyzed the portrayal of women executives in mass media. Generally, the approach to discourse associated with Marxist scholars from the Frankfurt school is *critique*, the reading of texts, artifacts, or social practices to reveal underlying ideology (Hardt 1992). Foucauldian analysis of social and cultural practices (e.g., Foucault 1980) is difficult to categorize. It is inherently political and focused on exposing power relationships. However, the epistemology underlying Foucault's approach leaves little room for normative or ethical analysis, a criticism leveled at him by Habermas. The principal difficulty in linking current approaches to critical discourse analysis to practice is that they rest on large-scale structural change.

Burrell and Morgan's radical humanist paradigm has tended to be overlooked by typologists of discourse analysis or lumped in with other critical approaches. We argue, however, that its is significantly different and offers a valuable perspective which can be used to improve practice, for while radical humanism focuses, like radical structuralism, on effecting change, its means are different. The change envisaged can occur at the individual level with enlightenment and emancipation as the path. Although Habermas (1984) sprang from the Frankfurt school, his emphasis on the emancipatory power of reason distinguishes him from neo-Marxists. Habermasian discourse ethics offer a strong and unique conceptual framework for understanding communications distortions and for improving practice (Forester 1983). The challenge is finding ways to operationalize the principles into tools for discourse analysis. It is this challenge that is addressed in this paper.

4 DISCOURSE, DECISION MAKING, AND RATIONALITY

Most approaches to management and planning rest on notions of *rationality*. Often these are grounded in Weberian notions of rationality which prioritize the notion of efficiency and economic behavior. Rationality is an implicit goal of most information systems development efforts. Klein and Hirschheim (1991) systematically discuss the types of rationality that underlie different approaches to systems analysis and design. They discuss the formal and substantive form of rationality proposed by Weber and its relationship to different system development methodologies.

They maintain, however, that this is only one form of rationality. Communicative rationality is another notion of system rationality, which is focused more on the development of mutual understanding and consensus in the context of the ideal speech situation. This ideal speech situation

> is a hypothetical situation which is characterized by a) an open agenda and free access in which all claims and counter claims can be freely examined, b) no asymmetries of knowledge and power...c) a social atmosphere which encourages everyone to express their feelings, to question and examine those feelings....The opposite of rational communication is distorted communication. (Klein and Hirschheim 1991, p. 167)

The principles of ideal speech are embodied in information systems development methodologies which treat systems analysis and design as a communication and learning process. In order to achieve emancipatory rationality it is necessary to diagnose distorting tendencies in communication. There is a significant body of IS research that draws upon Habermas' work as it relates to information systems development. As noted by Päivärinta (2001), contributors in this area include Klein, Hirschheim. and Lyytinen (e.g., Klein and Hirschheim 1993; Lyytinen and Klein 1985), as well as Ngwenyama (1991) and Ulrich (2001). In addition, Ngwenyama and Lee (1997) undertook an intensive investigation of an episode in the managerial use of e-mail in a company by applying Habermasian validity claims to e-mail messages, and Truex and Klein (1991) outlined an interpretation of information systems as formalized language games based on Habermas.

While the application of the standard of communicative rationality has been discussed in the context of information systems development, it has not been applied to the broader context of organizational decision making about information technologies or to the societal discourses that shape the institutional environment in which decision making takes place. It is understood that discourse on technology mirrors power relations and structures ideology, the "shared, relatively coherent interrelated set of emotionally charged beliefs and norms," which in turn shape the way technology is understood and enacted in organizations (Feldman and March 1981). "Although planners may believe they are acting rationally in adopting new technologies, their decisions actually reflect a pervasive mystique that what can be developed, must be developed" (Attewell and Rule 1984).

Examining societal and organizational discourses provides a means of assessing communicative rationality. As the case study data presented below show, when the ideal speech situation is not realized, communication distortions exist. Forester (1989), writing in the urban planning literature, maintains that revealing such distortions can improve the rationality of communications and, in particular, planning practice. Exposing *ideologies* to the standards of *rational* discourse may provide a means of reducing their influence on decision making. Habermas maintains that *reason* may be applied to *undistort* communications and improve the human condition. By providing an explicit and *ethical* standard for assessing the validity of communications, Habermas offers a strong and unique conceptual framework that can be applied not only to analyze the distortions in discourse which reflect the dominant ideology and power structures but also to undistort communications, thereby improving practice. While the principles of Habermasian *discourse ethics* have been invoked in many different discussions of policy development and management, there are limited examples of ways in which these principles can be operationalized and very few examples of their application to the analysis of societal discourse.

5 RESEARCH QUESTIONS AND METHODOLOGY

Recognizing the importance of discourse as a means of understanding the decision-making context and environment of the Acadia Advantage project, the research questions posed in this study were as follows: What is the nature of the discourse on Acadia University's technology-enabled learning Acadia Advantage project? Are communication distortions evident in this discourse? What can be learned from the Acadia Advantage case study about the development and adoption of technology-enabled learning projects?

In order to investigate these questions, a methodology was devised using Habermas' ideal speech situation and validity claims as the framework to assess the discourse. Specifically, each validity claim was applied as an analytical lens, through which to analyze the texts describing the Acadia Advantage program. The discourses on the Acadia Advantage program considered in this paper are drawn from a total of 57 sources, representing popular, academic, and practical part discourses.

5.1 Validity Claims and the Ideal Speech Situation

Following Forester (1983), we adopt Habermas' theory of communicative action as a foundation for our analysis. Forester suggests that this theory allows for (1) empirical analysis of communicative interaction and structural settings; (2) interpretive analysis of meaning; and (3) normative analysis of systemic distortion and violation of the free discourse of humans implicit in the most ordinary communications (p. 236).

Habermas maintains that with the dissolution of a theologically based form of substantive ethics, a new form of secular, procedural morality emerges "based on moral agreement that expresses in rational form what was always intended in the symbolism of the holy" (Habermas cited in Cannon 2001, p. 101). The basis of this morality is communicative ethics. "Social integration no longer takes place directly via institutionalized values but by way of inter-subjective recognition of validity claims raised in speech acts" (Habermas cited in Cannon 2001, p. 101). Discourse ethics asserts that morality is based on a pattern inherent in the mutual understanding of a language. In this way Habermas avoids the threat of relativism by invoking a standard for communication which is universal and unconditional (Cannon 2001) yet at the same time dynamic and grounded in the social world (Duquenoy et al. 1998). A universally valid ideal speech situation may be used to assess the legitimacy of normative claims; and this in turn forms the basis for his notion of communicative or discursive ethics.

Habermas sets out four tests, or validity claims, that must exist in order for the ideal speech situation to be realized. An example of the application of the validity claims to the speech act is an examination of possible responses to a simple request. If a professor asks a student, "Would you please bring me a glass of water?," the request could be rejected on the basis of assumptions about truth, clarity, sincerity, and legitimacy[1] implicit in that speech act (Habermas 1979, 1984). Assessing the truth assumption would involve consideration of the objective facts in the speech act. If there were no water available in the building, then the truth claim would not be achieved. If the request were unclear (for example, there was no shared system of meaning between the two participants as the professor made the request in English to a student whose English was poor), then clarity was not achieved. Sincerity is assessed by considering the congruence of the expressed meaning and the speaker's agenda. For instance, was the request a genuine request for a glass of water, or was it an opportunity for the professor to demonstrate authority over the class by demanding an obedient response? Legitimacy in this case would apply to the appropriateness of the implied relationship among the parties to the speech act, e.g., is it a student's duty to serve a professor in the requested manner?

According to Habermas, rational action is the result of communicative action when actors do not violate any of the validity claims in their speech acts. This ideal speech situation results in undistorted communication and builds comprehension, trust, knowledge, and consent. In contrast, distorted communication results in misrepresentation, confusion, false assurances, and illegitimacy.

5.2 Operationalizing Habermas' Validity Claims to Assess Communicative Rationality

Habermasian communicative rationality is a useful standard for the analysis of discourse because it enables us to apply normative standards to expose the distortions in discursive practices and so improve practice. Habermas provides a way of understanding the effects of the discourses in which we participate. But Habermas does not propose a methodology for discourse analysis. Difficulties in linking discourse ethics theories with practical concerns have been explored (see, for example, Blaug 1999; Cannon 2001). Nevertheless, Habermas' validity claims do have strong appeal as a conceptual tool for empirical research. Their theoretical foundation is strong, and the validity claims are accessible, as truthfulness, clarity, sincerity, and legitimacy are easily understood.

---

[1]Note that terminology varies somewhat when describing Habermas' validity claims. In *Communication and the Evolution of Society*, the four validity claims are translated as truth, rightness, truthfulness, and comprehensibility (Habermas 1979), a translation used by Ulrich (2001) in his discursive approach to information systems development. Forester (1989) uses *comprehensibility* instead of clarity, but otherwise follows the terminology used in this paper.

Table 1. Communicative Distortions (From Forester 1989, p. 150)

Practical Level	Comprehen-sibility	Sincerity	Legitimacy	Truth
Face-to-face	Ambiguity, confusion, lack of sense	Deceit, insincerity	Meaning taken out of context	Misrepresentation
Organization	Use of jargon to exclude public	Rhetorical reassurances, false expression of concern, hiding motives	Unresponsive-ness, assertion of rationaliza-tions, domi-nance by professionals	Information with-held, responsibility obscured, need misrepresented
Political Economic Structure	Mystification, complexity	Manipulation of the public good	Lack of accountability, legitimization through empty rhetoric rather than by active participation	Policy possibilities obscured, withheld or misrepresented, ideological claims such as "public ownership is always inefficient"

In order to operationalize Habermas' validity claims, a combination of textual analysis techniques may be employed (Cukier and Bauer 2002; Cukier et al. 2003). While Forester (1989) does suggest ways in which communicative distortions may occur in face-to-face, organizational, and political/economic structures and ways of correcting communicative distortions, his work remains at the conceptual level and is not applied specifically to discourse (see Table 1).

To assess Habermas' validity claims through an examination of texts, a series of questions was developed to facilitate identification of truth, sincerity, clarity, and legitimacy claims in these texts. These questions formed the basis of a coding scheme used to identify the elements of ideal speech present in the discourse. Careful readings of the texts produced data sets coded by validity claim, facilitating content analysis that provided concise measurement of the speech acts constituting each claim. Subsequent analysis allowed for consideration of communication distortions by examining the instances where the ideal speech situation was not realized. Our focus was on understanding the range of texts within the discourse to understand overall patterns of communicative rationality, rather than on detailed analysis of specific passages.

Identification of *truth* claims within the discourse was guided by a search for objective facts. Michalos' (1986) tests for logic were helpful in developing specific questions, including what are the basic arguments? Are the issues and options clearly defined? What evidence has been provided to support these arguments? Has the relevant information been communicated without distortion or omission? Are there ideological claims which are unexamined?

Table 2. Summary of Validity Claims and Corresponding Discourse Dimensions

Validity Claim	**Result**	**Distortion**	**Speech Dimensions**
The content of the presuppositions of what is said be factual or true.	Truth	Misrepresentation	Argumentation and evidence
The speaker is honest (or sincere) in what she says.	Sincerity	False Assurance	Metaphors and connotative words
What is said is linguistically intelligible and comprehensible.	Clarity	Confusion	Rhetoric and semantic rules
What the speaker says (and hence does) is right or appropriate in the light of existing norms or values.	Legitimacy	Illegitimacy	Use of experts

Sincerity claims are identified through the use of rhetorical devices. Examining the choice of metaphors, adjectives, and connotative vocabulary used in the texts may reveal nuances not apparent on cursory reading. "Stylistic choices also have clear social and ideological implications, because they often signal opinions of the reporter about news actors and news events as well as properties of the social and communicative situation" which are not directly expressed (van Dijk 1991, p. 116). Coding for sincerity claims sought to identify instances in which metaphors or language usage could influence interpretation or understanding of the Acadia Advantage project and technology, paying particular attention to instances where metaphors or language might promote or suppress understanding or create false assurances.

Clarity was assessed in the usage of jargon, unfamiliar terminology, or incomprehensible language. Clarity is achieved when these obfuscations are absent.

In the area of *legitimacy*, coding focused on identifying texts that indicated participation in the discourse. To whom was legitimacy accorded in the texts? Who was considered an expert, and on what basis? What was assumed or implied in the discourse? How were decisions legitimized? Once coding was complete in this section, it was then possible to consider questions of absence, including which groups and viewpoints were marginalized or excluded from the discourse. What was missing or suppressed?

Table 2 summarizes the dimensions of discourse used to assess the validity claims. One of the distinct advantages of this approach is that the speech dimensions can be analyzed using both qualitative and quantitative textual analysis techniques. In combination, these add to the power of this approach to discourse analysis. Different approaches to reading text are not mutually exclusive and applying multiple perspectives to text may address the limitations of individual techniques in isolation. Kracauer (in Larsen 1991) insisted that a reading of a text necessarily involves an act of interpretation which, like other readings, is based on specific assumptions to be made explicit in the reading. "Critical discourse analysis is a particular epistemological orientation to discourse and tends to be associated with a qualitative 'reading' or artifacts" (Fairclough 1995). Critical theorists have tended to reject quantitative strategies for determining the content or meaning of media messages (Kracauer) given the importance of considering both the manifest and latent meanings. However, even Kracauer granted that quantitative studies might serve as a supplement to qualitative analysis. The sheer volume of mass media texts poses problems in terms of heterogeneity as well as quantity. We find other examples of mixing methods. Herman and Chomsky (1988), for example, employ a wide range of techniques including content analysis to demonstrate ways in which the mass media is used to manufacture consent. A major criticism often leveled at discourse analysis is that it is selective or lacks rigor (Philips and Hardy 2002). Combining qualitative and quantitative approaches is one way of responding to these criticisms.

Once the discourse has been coded, questions guiding analysis include what distortions or misrepresentations have occurred? Why have they occurred? What might undistorted communication look like? These questions are addressed below.

6 METHODS IN ACTION: THE ACADIA ADVANTAGE CASE STUDY

The texts analyzed here refer to the Acadia Advantage project. In 1996, the President of Acadia University (a 3,500 student liberal arts university in Wolfville, Nova Scotia, Canada) announced a plan to wire classrooms, promoting the development of Web-based curricula, and requiring students to lease portable computers. The Acadia Advantage became the centerpiece of the university's marketing and has been showcased by IBM in its ThinkPad University marketing efforts. It is regarded as a model for notebook or ThinkPad computing at many other post secondary institutions and has received considerable media coverage.

For text selection, we used two multidisciplinary full-text databases (EBSCO, including the Academic Search Full Text Elite segment, and ProQuest, including ABI/Inform Global, Applied Science and Technology Plus, Periodicals Abstracts II, and ProQuest Telecommunications) as well as several subject-specific electronic resources (e.g., Gale Directory of Databases, Faerber 1999). The search statement was identical in all databases: (Acadia University) AND (Technology OR Advantage). The time period searched was from 1993 to June 1998. The search, international in scope, produced a master list of 72 articles in different periodicals. Articles that did not actually relate to the Acadia Advantage technology-enabled learning project were excluded, resulting in 57 relevant articles that were used in this analysis (see Appendix A for a list of the sources).

6.1 Truth: What Are the Facts? Are the Arguments Supported with Evidence?

The analysis reveals certain patterns in argumentation (see Table 3). Many of the statements regarding the program are essentially descriptive—all students received an IBM laptop computer and software (63 mentions) which they leased for $1,200 per year for a total of $5,055 in tuition (23). The campus was wired to provide access to the Internet everywhere (18) and courses were redesigned to integrate technology (14).

Claimed positive effects for participating students included improved quality of learning, access to more information from the Internet, improved technology skills, and improved equity and access. Positive effects were also claimed for the university (e.g., industry partnerships) and for industry (e.g., increased educational markets). Negative effects claimed for students included tuition, which was the highest in the country and might be a barrier to access. Other negative effects claimed were the additional investment in infrastructure by the university and the potential negative effects for society such as privatization of education. The negative effects for faculty included the increase in workload without compensation. The principal claims are shown in Table 3.

To assess the truthfulness of the discourse, we will consider some of the claims, the evidence used to support them, and the argumentation (inductive and deductive reasoning). For instance, while proponents of the Acadia Advantage insist that it is primarily an academic initiative aimed at improving the quality of learning, little evidence is provided to support the claims. A number (8) of articles report that the program has resulted in improved student performance, some with dramatic headlines: "Acadia's Wired Students Soar to the Top of the Class" (*Halifax Daily News* 1997) or, "Technology is Getting Good Grades with Faculty and Students at Acadia University in Nova Scotia" (Pearsall 1998). Only two articles mention significant caveats in the study comparing wired and non-wired students, including the absence of controls for class size and consideration of whether or not the students were specializing in the discipline. Class sizes (96 versus 20) and teaching methods were very different (Pearsall 1998, p. 11). Even more important, the 96 non-Acadia Advantage students in the Physics course under study were mostly non-physics majors while 80 percent of the much smaller class of 20 Acadia Advantage students were Physics majors (McLaughlin 1997).

The claim that improved learning is a consequence of the use of laptop computing is frequently inferred. Many articles repeat the claim that the Acadia Advantage increased interactivity in the classroom (4), produced more learner- centered instruction (8), and promoted more practical and studio-oriented work (10). "The lecture hall is passive learning, it's boring.... We need students who are actively involved. They should be discussing, analyzing, problem solving" (Noakes 1996, p. 14). The suggestion here is that laptops alone created increased interactivity. While laptops were certainly a catalyst, increased interactivity appears to have been furthered by smaller classes coupled with extensive (and time consuming) faculty efforts to redesign courses. Acadia's technology initiative may have provided an opportunity for restructuring which produced benefits, but it does not follow that investing in laptop computers was the principal factor producing these benefits. Similarly, benefits like reduced note taking (5) and increased access to information (22) could be achieved by other means, although there is no evidence that these identified benefits are in themselves valuable.

Watters et al. (1998) claim that the Acadia Advantage program is "One example of the new class of learning support tools needed to take advantage of the reality of student centered mobile computing." However, their article offers a mere description of activities with no analysis of objectives or outcomes and is circular, arguing that the tools are needed to take advantage of the technology rather than considering the variety of means which might be used to achieve the ends of enhanced learning.

Table 3. Principal Claims Regarding Cause (AA Program) and Effects (Number of Text Segments)

Descriptive Statements	
All students receive an IBM laptop computer and software	63
Students must pay $1,200 per year to lease a computer ($5,055 tuition including computer)	23
Campus wired to provide access to Internet anywhere	18
Courses redesigned to integrate technology	14
Claimed Positive Effects for Students	
Improved quality of learning	
— More practical and studio work	10
— Improved performance of Acadia Advantage students versus traditional	8
— Learner centered—changed relationship between students and teachers	8
— More interactive learning, lectures decreased	4
— Student enthusiasm	4
— PowerPoint presentations	1
More information from the Internet	
— Increased access to information sources	22
— More communication, collaboration among students	5
Technology skills	
— Students gain experience with computers and Internet	7
Equity/Access	
— Students have equal access to computing resources	3
— Students receive identical software and hardware which is regularly upgraded and supported	2
— Technology is part of tuition therefore tax deductible	1
Claimed Positive Effects for University	
Partnerships with IBM, Microsoft, MT&T, 3Com	4
University can offload costs of upgrading technology to the students	2
Corporate donations to capital campaign	2
Leading edge	2
Contributions a marketing effort	2
Claimed Positive Effects for Industry	
Grow education market	2
Claimed Negative Effects for Students	
Tuition is the highest in the country	13
High tuition may be a barrier to accessing education	5
Only students leasing IBM computers have access to Acadia Advantage courses	2
Off campus students pay additional $25 per month for Internet access	2
Students pay $4,800 over 4 years and must return computer	2

Claimed Negative Effects for University	
Labor unrest	21
Wiring campus required additional investment by university of $16 to 20 million	6
Additional investment in "sandbox" of $300,000	1
Claimed Negative Effects for Society	
Privatization, corporatization of education	3
Funds diverted from education to IT companies	1
Claimed Negative Effects for Faculty	
Large increase in course development time borne by faculty	7
Increased communication with students adds to workload	1
Questions regarding effects and their causes	
Less costly alternatives are available to provide access to computers and Internet information sources	3
Technology no substitute for critical thinking	3
More information is not necessarily better	2
Improved performance as a result of course redesign, small class, and differences in students (e.g., majors) not technology	2

It is also claimed that the program increases access to technology for students (3). "Parents often end up buying their children computers for university, but not every student gets one. But at Acadia, the playing field is level. Students have equal access to learning" (Murphy 1998, p. 34). "For the first time in the history of Acadia," Bruce Cohoon, Director of Public Affairs at Acadia University, stated, "every person involved in the program has equal access to information and technology" (Domet 1997, p. 1). Invoking a level playing field is a powerful appeal to metaphor, as discussed below, but it is hardly an accurate description of the Acadia program. As a few articles (5) note, by requiring students to pay an additional $1,200 per year, Acadia may have created a barrier to accessible education and technology.

In summary, when Habermas' first standard, that of truth, is applied, we find unsupported claims and faulty logic, such as the inductive fallacies of faulty analogy and "false cause" (Michalos 1986). In many cases, the specific causal link between the claimed benefits for students and the Acadia Advantage program (i.e., the mandatory leasing of laptop computers) is unclear. The fact that the results attributed to Acadia Advantage might have been achieved through other means, such as significant investment in curriculum redesign, smaller teacher/student ratios, and access to other kinds of computing and Internet resources, is seldom discussed. Some of the benefits claimed are more clearly associated with access to computers and the Internet than others. However, the ThinkPad solution selected by Acadia may not have been the most affordable or equitable means of providing access to computers and the Internet. The total costs (including the investment in infrastructure and labor required to redesign the courses) or alternative approaches are almost never considered. Only three articles raise questions about the argumentation of the proponents and, in particular, question the link between claimed causes and effects (Table 3). Most of the articles simply replicate them.

6.2 Clarity: Is What Is Said Intelligible and Comprehensible?

There are a number of ways in which clarity may be undermined and confusion may be created, intentionally or unintentionally. The fallacy of jargon is committed when instead of being given accurate and comprehensible descriptions of a product, we are given technical terms that make the claim seem stronger, more important, or valuable (Michalos 1986). Several of the articles use technology terms that may obscure more than they clarify. In a number of the Acadia Advantage texts there is a detailed description of the leading edge technology components with no explanation as to why they are useful or important.

While, naturally, the assumptions of the reader's level of technical expertise may vary with the publication, there is virtually no explanation of either the meaning or importance of the technologies in any of the articles examined. Technical terms pervade the discourse, but are seldom explained, as this example shows.

> A campus wide Asynchronous Transfer Mode (ATM) switching solution enabled students and faculty to gather information electronically and to better communicate. Now in its second year, the program is being called a great success. About 4,500 "drops" have been installed all over campus, everywhere from classroom seats to libraries, cafeterias and students lounges. Drops are places where students can connect their laptops to the network via Ethernet locally and the ATM backbone. By the end of the four-year roll-out period, 8,000 of these connection points will be installed in and around campus. (Joy 1998, p. 18).

In addition, texts frequently (115 mentions) apply a wide range of adjectives with positive associations—new (11), innovative (11), wired (8), award-winning (6), hi-tech (6), ambitious (2), pioneering (4), exciting (5), etc.—which tend to reinforce a positive view of the program, and imply that new and innovative are desirable attributes. Although several articles do suggest there are problems associated with the program, adjectives that invoke skepticism— such as controversial (1), expensive (5), or scary (3)—are seldom used. Table 4 provides a summary of adjectives found in the discourse.

It may be argued the preponderance of technical jargon is not only a barrier to understanding but part of a pervasive technological mystique. This mystique can serve to elevate those who are part of the technological priesthood while excluding those who do not understand from meaningful participation in the discourse. This leaves them with essentially two alternatives: completely rejecting the propositions or accepting them uncritically, which also has implications for legitimacy, discussed below.

6.3 Sincerity: Is What Is Said What Is Meant? What Is Implied/Invoked?

The sincerity of a speech act may be assessed in several ways. Essentially, sincerity requires congruity between what is said and what is meant or the intention underlying and the intention expressed in the speech act. There is no hidden agenda in a sincere speech act.

Table 4. Connotative Language—Number of References

Term	#
Subjects	
Acadia Advantage	83
Program	82
Project	24
Initiative	6
Innovation	4
Investment	4
Revolution	3
Virtual Classroom	3
Experience, Vision, Evolution	2
Cyber-campus, Huge Attraction, Undertaking, Great Success, Opportunity, Cyber-push, Paradigm Shift, Campaign, Excellent Plan, Leader, Comprehensive Deal, Cyber-club, Hype, Best Thing	1
Adjectives—Positive	
Innovative	11
New	11
Wired	8
Hi-tech	6
First	6
Award Winning	6
Pioneering	4
Exciting	5
Electronic, Vaunted, Cyber	3
Lauded, Laptop, Significant, Sophisticated, Forefront, Future, Ambitious, Ground Breaking, Academically Driven, Unique	2
Revolutionary, Radical, Great, Effective, Latest, Large, Leading Edge, Dynamic, Ambitious, Superb, Online, Top Ranked, Technological, Highly touted, Coveted, Computer-oriented, Affordable, Up to date, Information Age, Most, Novel, Advanced, Selective, Special, Top of the Class, Recently Approved, Major, Modern, Heralded	1
Adjectives—Negative/Ambiguous	
Expensive	5
Scary	3
Mandatory	2
Compulsory, Pricey, Controversial, Multimillion dollar	1

Regardless of the denotative content of the articles, certain perspectives are implied or reinforced through connotative language, imagery, and metaphor. Specific nouns used for the project are instructive. All of the articles make repeated references to the use of laptop computers at Acadia as the Acadia Advantage (83), thereby reinforcing the benefits of the program. They often refer to it as a program (82) or a project (24) but also use terms that evoke a positive response—initiative (6), innovation (4), and investment (4).

The dominant metaphor the articles reinforce is the revolution, as in "the wired revolution" or the "techno-revolution." The Acadia Advantage is new, innovative, exciting, and pioneering, while the critics are defending old or traditional approaches. Critics of the technology are cast as "fighting a rearguard action." Many faculty members and administrators see computers as a threat to centuries-old traditions of pedagogy. "There is great irony to this," adds Tapscott, "It's not technology that's the threat. It's the status quo, if the universities don't reinvent themselves, they will be replaced" (Bergman 1998, p. 66).

Sometimes, a simple choice of words has a profound impact on the interpretation of a speech act. For example, several of the texts say or imply that Acadia *gave* computers to students without mentioning the cost. For example, "The President has been credited with putting Acadia at the technological forefront of academia with the Acadia Advantage program, in which every student will *have* a laptop computer" (McLaughlin 1998, p. 4). Or, "All first year students *are being equipped* with sophisticated software and Internet access. By 2000 all the full time undergraduate students will *have* a computer" (Sommers 1997, p. C5).

In addition, the connotative power of language can be used to reinforce the positions of certain actors and to marginalize others. It seems that language is often used to undermine the credibility of the critics of the program, particularly in the discussions of a labor dispute. "Many professors *bristled* at being told that they must employ new technology in their classrooms. 'The administration wants to require every faculty member to use the Acadia Advantage,' *complained* the faculty association president" (Bergman 1998, emphasis added). In contrast the President is cast as a *strong* and *principled leader*, an *innovator*, a *visionary*.

Proponents of the Acadia Advantage invoke metaphors and images with powerful associations regardless of their appropriateness. For example, the claim that the Acadia Advantage creates a level playing field appeals to powerful emotions and values such as improving accessibility and equity when it may be argued that the program does precisely the opposite.

One may also question the sincerity of a speech act based on evidence of unstated motives or hidden agendas. Consider, for example, the efforts to "spin" the costs of the program. One article notes that when the author suggested that the Acadia Advantage is actually a very effective way for the university to offload systems costs to students, Bruce Cohoon (Director of Public Affairs, and author of the level playing field metaphor) responded, "Let's not call it an extra expense....Tuition at Acadia is $5055 a year period. It just happens that tuition includes a $3700 laptop computer and $16 million worth of infrastructure behind it" (Pearsall 1998:12).

6.4 Legitimacy: What Is Privileged? What Is Missing?

The analysis of legitimacy centers around what was missing in the discourse. Texts concerning the Acadia Advantage provide generous information about its potential positive effects while slighting information about total costs, disadvantages, and the basis of opposition to the program. The analysis of the number of statements related to the benefits of the program (regardless of their truthfulness discussed above) compared to the statements related to the disadvantages, or the costs of the program reveals this imbalance (see Table 3). Only two articles had more statements related to the costs and disadvantages than benefits; one focused on the accessibility issue and another on the faculty association's objections. Some articles, including the academic ones, had no discussion of costs or disadvantages whatever. In most, the only cost mentioned was the $1,200 extra tuition students must pay.

Only one article mentions a $300,000 sandbox program, or the faculty time needed to develop the courses. In most cases we are told that these expenditures are good investments and paying dividends. We are also told that folding the costs of computers into tuition makes it tax deductible. In the majority of articles, benefit statements outnumbered cost statements by more than 3 to 1.

At Acadia, students who did not acquire their ThinkPad computer through the IBM leasing program were denied access to Acadia Advantage course material (Tausz 1996), a point mentioned in only two articles. Only three of the articles explain that there are other ways students could be given access to the benefits of computing and the Internet, such as allowing access to other brands of computers, other terminals, labs, used computers, or wireless technology.

There is evidence that such selective silence also relates to groups affected by the program. A total of 16 articles were specifically focused on the labor dispute and leadership review at Acadia and mentioned the Acadia Advantage. Generally, the faculty association president was quoted in this context. While the faculty repeatedly stated support for the program in principle, their concerns regarding the amount of time required to redevelop courses (estimated at 200 hours per course) are mentioned in only seven. Other concerns related to accessibility are found in only five articles. Bargaining issues, e.g., parity in wages, which were 16 percent below other universities, are mentioned only in two local stories. Not only are these perspectives underrepresented but they are marginalized through the subtle use of connotation (see above) which implies that an "award winning technological initiative" is being used as a bargaining chip, or attacked out of "blind resistance to change" rather than principles.

> In any labor conflict the union will use every lever they have. It may be disappointing, certainly in an academic environment, but is not surprising.... Really it's about change and moving the yardsticks and there are those who are resistant to that. (Lewington 1998, p. A6)

Several articles discuss the corporate partnerships but only a few texts examine what corporations actually stand to gain. One calls it an IBM deal (Tausz 1996). Another tells us

> IBM Canada Ltd., a corporate partner in the Acadia Advantage program, is promoting the institute as the place in Canada where expertise in the application of information technology to the curriculum resides....They are well ahead of other post-secondary institutions in Canada and they're willing to share what they've learned. We direct both potential and existing clients to Acadia so they can see what their peers are doing. (Sommers 1997, p. C5)

Acadia may or may not be at the forefront of effective technology-enabled learning, but it is an important IBM customer. Not only was the value of the Acadia Advantage program anticipated to be $4 million per year by 2000, the potential marketing implications were likely even more significant. Acadia policy guarantees IBM a monopoly because only students to whom Acadia has issued a ThinkPad will be enrolled in the notebook courses. Maritime Telephone and Telegraph has exclusive rights to communications and 3Com built the network. One article notes that IBM and MT&T made $4 million in in-kind donations to the University's capital campaign. Only two of the articles link the program to privatization.

Another measure of legitimacy is provided through consideration of who is included or excluded when authorities are cited. While there are legitimate reasons why some perspectives on the project may be more valid than others, it appears that certain perspectives are privileged and others marginalized. For example, almost half of the articles citing experts quote administrators. In all, 29 articles contain 62 statements by administrators regarding the program. In comparison, only 13 articles cite faculty members, a total of 31 times. Interestingly enough, more than one third of the faculty statements (12) come from a single instructor. Students, who are supposed to be the principal beneficiaries of the program, are cited in 8 articles a total of 13 times. In addition, several articles include comments from other experts (5) and vendors (3).

7 DISCUSSION

To summarize, our findings suggest that there are communication distortions in the published texts on technology in education at Acadia University. The information provided in the published texts is highly selective and at times misleading. The emphasis is on benefits rather than costs; in particular, the cost of the IT infrastructure and course redesign appear downplayed. In addition, alternatives to the program are widely ignored. In cases where there are apparent improvements in learning and teaching, there is little effort to establish causes. For example, benefits are attributed to the introduction of the technology even when reduced class size and intensive instructional redesign seem to have played a significant role. This effectively silences discussion of any way of improving pedagogical quality other than through investment in information technology. Similarly there is no acknowledgment that alternative ways of providing computing technology and Internet access exist other than by leasing IBM ThinkPads. The metaphors and images reinforce notions of the technological imperative and of progress as a value in itself. The dominant voices are those which support the program; critics are marginalized both subtly and overtly. While one might expect distortions in popular media and in trade publications funded by the private sector, we would expect more rigor and balance in the two peer-reviewed academic publications included in this study.

Table 5. Authorities Cited—Number of Articles and Statements

Authority	Articles citing	Total citations
Administrators (total)	**29**	**62**
President	8	15
Vice President	1	3
Dean of Arts	2	3
Coordinator of Acadia Advantage Program	7	20
Director of Development	2	3
Director of Computing	2	5
Director of Institution on Technology and Teaching	2	4
Director of Media Relations	5	9
Vendors (total) Microsoft, 3M, and IBM	**3**	**3**
Other experts (total) Neil Postman, Linda Harasim, Don Tapscott, David Johnston, and Tony Bates	**4**	**7**
Faculty (total)	**13**	**31**
Union Presidents (2)	6	12
Other Faculty (4 different)	7	19 (12 of them one faculty member)
Students (total)	**8**	**13**
Student Union Presidents (2)	3	6
Other Students (4)	5	7

Instead, they merely reinforce our initial comments regarding the need for a critical perspective. The relative power of the various actors engaged in this discourse, what is at stake, and the role of the suppliers of technology in this $3 trillion market cannot be ignored.

While examining reproduction of the discourse is beyond the scope of this paper, we see evidence of the rhetoric of the Acadia Advantage echoed in other forms. For example, when Canada's Sheridan College introduced its mandatory laptop program, it too talked about the Acadia Advantage and how the technology would produce a level playing field. Government documents, the popular press, and even academic journals reinforce the argumentation about advantages (regardless of the absence of evidence), the metaphors, and privilege certain voices. Vendor-sponsored conferences, publications, and Websites use these as part of their marketing strategy. The reproduction of the discourse is mirrored extensively (Cukier and Bauer 2002).

8 LIMITATIONS OF THE STUDY

This study makes contributions to the discussion of information technology and to the techniques of discourse analysis but is not without limitations. First, the scope of the analysis is very small, with its focus on a particular learning technology project over a limited time period. As such, a broader study would be needed to understand the extent to which the patterns identified are reflected in the broader societal discourse. Second, while we maintain that societal discourses as manifest in media provide part of the environment of organizational decisions about technology, the link is based on institutional theory rather than empirical study. The extent to which the media discourses regarding the Acadia Advantage shape and reflect the organizational discourse needed further investigation. In addition, as it is a retrospective analysis, it has little value for improving information technology decision making at Acadia; rather, it provides insights which might be relevant in other contexts. Finally, while the effort to operationalize Habermasian validity claims to textual analysis is defensible, the mapping of the validity claims to specific dimensions of textual analysis is problematic. For example, while metaphors are used as an indicator of sincerity, that is, the link between what is said and what is meant—we cannot impute motives. That is, one cannot suggest that the use of metaphor is a deliberate effort to deceive as they are pervasive and all but invisible.

9 CONCLUSION

Other scholars have insisted on the importance of applying standards of communicative rationality to information systems development at the organizational level. We have suggested that the same standard needs also to be applied to the broader societal discourses which form the context of organizational decision making. Institutional theory tells us that organizational behavior is shaped by the institutional environment. However, as Abell (1991) argues, a recognition of institutional isomorphism need not lead to a complete rejection of rational choice theory. We agree with Forester that Habermasian communicative rationality is an appropriate standard to apply in an effort to expose the communications distortions that suppress common sense. We suggest that further work examining the relationship between the levels of societal and organizational discourses would be a fruitful area of investigation.

The conceptual framework we have developed, building on Forester's interpretation of Habermas' communicative ethics and, specifically, Habermas' validity claims can provide a standard that can be applied systematically to the analysis of discourse in order to identify communication distortions. While not perfect, we suggest ways in which the standards of truth, clarity, sincerity, and legitimacy can be applied to texts using a combination of quantitative techniques (e.g., content analysis) as well as qualitative approaches. This approach may help respond to some of the criticisms leveled at critical discourse analysis techniques.

The example of the Acadia Advantage case was provided to offer an example of how this approach could be operationalized in order to reveal distortions in the discourse regarding learning technology.

REFERENCES

Abell, P. *Rational Choice Theory*, Brookfield, VT: Edward Elgar, 1991.

Abrahamson, E. "Management Fashions, Academic Fashions and Enduring Truths," *The Academy of Management Review* (14:27), 1996, pp. 616-619.

Abrahamson, E. "Words into Numbers," *Academy of Management Conference Report*, Washington, DC, 2001.

Alvarez, J. L. "The International Popularization of Entrepreneurial Ideas," in S. R. Clegg and G. Palmer (Eds.), *The Politics of Management Knowledge*, London: Sage Publications, 1996, pp. 80-98..

Attewell, P., and Rule, J. "Computing and Organizations: What We Know and What We Don't Know," *Communications of the ACM* (27:2), 1984, pp. 1184-1191.

Beedham, C. "Language, Indoctrination and Nuclear Arms," University of East Anglia Papers in Linguistics, 1983, pp. 15-31.

Bergman, B. "Wired Revolution," *Maclean's* (119:3), 1998, p. 66.

Blaug, R. *Democracy Real and Ideal: Discourse Ethics and Radical Politics*, New York: State University of New York Press, 1999.

Bloomfield, B., and Vurdubakis, T. "Re-presenting Technology: IT Consultancy Reports as Textual Reality Constructions," *Sociology* (28:2), 1994, pp. 455-477.

Boland, R. J. "Information System Use as a Hermeneutic Process," in H-E. Nissen, H. K. Klein, and R. A. Hirschheim (Eds.), *Information Systems Research: Contemporary Approaches and Emergent Traditions*, Amsterdam: North Holland, 1991, pp. 439-458.

Boland, R. J. "Phenomenology: A Preferred Approach to Research in Information Systems," in E. Mumford, R. A. Hirschheim, G. Fitzgerald, and A. T. Wood-Harper (Eds.), *Research Methods in Information Systems*, Amsterdam: North Holland, 1985, pp. 193-201.

Boland, R. J., and Day, W. F. "The Experience of System Design: A Hermeneutic of Organizational Action," *Scandinavian Journal of Management* (5:2), 1989, pp. 87-104.

Bowring, M. A. "De/Construction Theory: A Look at the Institutional Theory that Positivism Built," *Journal of Management Inquiry* (9:3), September 2000, pp. 135-137.

Burrell, G., and Morgan, G. *Sociological Paradigms and Organizational Analysis: Elements of the Sociology of Corporate Life*, London: Heinemann, 1979.

Butler, T. "Towards a Hermeneutic Method for Interpretive Research in Information Systems," *Journal of Information Technology* (13:4), 1998, pp. 285-300.

Calás, M., and Smircich, L. "Voicing Seduction to Silence Leadership," *Organization Studies* (12:4), 1991, pp. 567-601.

Cannon, B. *Rethinking the Normative Content of Critical Theory: Marx, Habermas and Beyond*, New York: Palgrave, 2001.

Chomsky, N. *Necessary Illusions: Thought Control in Democratic Societies*, Concord, ON: House of Anansi Press, 1989.

Clegg, S. R., and Palmer, G. (Eds.). *The Politics of Management Knowledge*, London: Sage Publications, 1996.

Cukier, W., and Bauer, R. "The (Re)Production of Technology Discourse: Technology Enabled Learning in Canada," paper presented at the European Group on Organization Studies (EGOS) Conference, Barcelona, Spain, 2002.

Cukier, W., Middleton, C. A., and Bauer, R. "A Critical Analysis of Technology Enabled Learning Discourse in Canada and Its Implications for Policy and Organizational Decision Making," in E. H. Wynn, E. A. Whitley, M. D. Myers and J. I. DeGross (Eds.), *Global and Organizational Discourse About Information Technology*, Boston: Kluwer Academic Publishers, 2003, pp. 197-221.

Czarniawska-Joerges, B., and Joerges, B. "how to Control Things with Words: Organizational Talk and Control," *Management Communications Quarterly* (2:2), 1988, pp. 170-193.

DiMaggio, P., and Powell, W. "The Iron Cage Revisited: Institutional Isomorphism and Collective Rationality in Organizational Fields," *American Sociological Review* (48), 1984, pp. 147-160.

Domet, S. "Turn On, Plug In, Hook Up: At Acadia University, the Computer Revolution Is Changing the Way Students Learn and How Professors Teach," *Quill and Quire* (63:11), 1997, pp. 1, 17.

Duquenoy, P., Thimbleby, H., and Torrance, S. "Towards a Synthesis of Discourse Ethics and Internet Regulation," School of Computing Science, Middlesex University, London, 1998 (available online at http://www.cs.mdx.ac.uk/staffpages/penny/Habermas.htm).

Ellul, J. "The Technological System," trans. Joachim Neugroschel, New York: Continuum, 1977.

Faerber, M. *Gale Directory of Databases,* Foster City, CA: Gale Group Publishers, 1999.

Fairclough, N. *Critical Discourse Analysis: The Critical Study of Language*, New York: Longman Group, 1995.

Feldman, M. S., and March, J. G. "Information in Organizations as Signal and Symbol," *Administrative Science Quarterly* (26:2), 1981, pp. 171-186.

Fiske, J. *An Introduction to Communication Studies*, London: Methuen Press, 1982.

Fondas, N. "Feminization Unveiled: Management Qualities in Contemporary Writings," *Academy of Management Review* (22:1), 1997, pp. 257-282.

Forester, J. "Critical Theory and Organizational Analysis," in G. Morgan (Ed.), *Beyond Method* Beverley Hills, CA: Sage Publications, 1983, pp. 234-346.

Forester, J. *Planning in the Face of Power*," Berkeley, CA: University of California Press, 1989.

Foucault, M. *Two Lectures: Power/Knowledge*, New York: Random House, 1980.

Furusten, S. *Popular Management Books: How They Are Made and What They Mean for Organizations*, London: Routledge, 1999.

Gerbner, G. "Comparative Cultural Indicatorsm" in G. Gerbner (Ed.), *Mass Media Policies in Changing Cultures,* New York: Wiley Publishing, 1977, pp. 199-205.

Gopal, A., and Prasad, P. "Understanding GDSS in Symbolic Context: Shifting the Focus from Technology to Interaction," *MIS Quarterly* (24:3), 2000, pp. 509-546.

Grant, D., Keenoy, T., and Oswick, C. "Organizational Discourse: Key Contributions and Challenges," *International Studies of Management and Organization* (31:3), 2001, pp. 5-25.

Habermas, J. *Communication and the Evolution of Society*, trans. T. McCarthy, Boston: Beacon Press, 1979.

Habermas, J. *The Theory of Communicative Action,* Boston: Beacon Press, 1984.

Halifax Daily News. "Acadia's Wired Students Soar to Top of Class, Study Shows, February 1, 1997, p. 4.

Hansen, A., Cottle, S., Negrine, R., and Newbold, C. (Eds.). *Mass Communications Research Methods*, New York: New York University Press, 1998.

Hardt, H. *Critical Communication Studies. Communication, History and Theory in America*, New York: Routledge, 1992.

Hardy, C. "Researching Organizational Discourse," *International Studies of Management and Organization* (31:3), 2001, pp. 25-48.

Hardy, C. Lawrence, T., and Phillips, N. "Talking Action: Conversations, Narrative, and Action in Interorganizational Collaboration," in D. Grant, T. Keenoy, and C. Oswick (Eds.), *Discourse and Organization*, London: Sage Publications, 1998, pp. 65-83.

Heracleous, L., and Barrett, M. "Organizational Change as Discourse: Communicative Actions and Deep Structures in the Context of Information Technology Implementation," *Academy of Management Journal* (44), 2001, pp. 755-778.

Herman, E. S., and Chomsky, N. *Manufacturing Consent: The Political Economy of the Mass Media*, New York: Pantheon, 1988.

Hirsch, P. M. "From Ambushes to Golden Parachutes: Corporate Takeovers and an Instance of Cultural Framing and Institutional Integration," *American Journal of Sociology* (91), 1986, pp. 800-837.

Inglis, F. *Media Theory*, Oxford: Blackwell, 1990.

Jensen, K. B. "Humanist Scholarship as Qualitative Science: Contributions to Mass Communication Research," in K. B. Jensen and N. Jankowski (Eds.), *A Handbook of Qualitative Methodologies for Mass Communication Research*, London: Routledge, 1991, pp. 17-43.

Jensen, K. B., and Jankowski, N. (Eds.). *A Handbook of Qualitative Methodologies for Mass Communication Research*, London: Routledge, 1991.

Joy, S. "Mobile Users Plug into University Network: School Plans to Have Program Extended to All 4,000 Students by 2000," *Technology in Government* (56), 1998, p. 18.

Kets de Vries, M. F. R., and Miller, D. "Interpreting Organizational Texts," *Journal of Management Studies* (24:3), 1987, pp. 233-247.

Klein, H. K., and Hirschheim, R. "The Application of Neohumanist Principles in Information Systems Development," in D. Avison, T. E. Kendall and J. I. DeGross (Eds.), *Human, Organizational and Social Dimensions of Information Systems Development*, Amsterdam: Elsevier. 1993, pp. 264-280.

Klein, H. K., and Hirschheim, R. "Rationality Concepts in Information System Development Methodologies," *Accounting, Management and Information Technologies* (1:2), 1991, pp. 157-187.

Krefting, L. A. "Re-Presenting Women Executives: Valorization and Devalorization in US Business Press," paper presented at the European Group on Organizational Studies (EGOS) Conference, Lyon, France, 2001.

Larsen, P. "Textual Analysis of Fictional Media Content," in K. B. Jensen and N. Jankowski (Eds.), *A Handbook of Qualitative Methodologies for Mass Communication Research*, London: Routledge, 1991, pp. 121-148.

Lazarsfeld, P. F., and Merton, R. K. "Mass Communications, Popular Taste, and Organized Social Action," reprinted in W. Schramm (Ed.), *Mass Communications*. Urbana, IL: University of Illinois Press, 1948, pp. 492-512.

Lee, A. S. "Electronic Mail as a Medium for Rich Communication: An Empirical Investigation Using Hermeneutic Interpretation," *MIS Quarterly* (18:2), 1994, pp. 143-157.

Lewington, J. "Acadia President's Style Assailed," *Globe and Mail*, November 18, 1998, p. A6.

Lippmann, W. *Public Opinion*, New York: MacMillan, 1992.

Lyytinen, K. J., and Klein, H. K. "The Critical Theory of Jurgen Habermas as a Basis for a Theory of Information Systems," in E. Mumford, R. Hirschheim, G. Fitzgerald, and A. T. Wood-Harper (Eds.), *Research Methods in Information Systems*, Amsterdam: North Holland, 1985, pp. 219-237.

McLaughlin, P. "The Learning Curve: St. F.X. Goes High-Tech with New $8m Network," *The Halifax Daily News*, January 19, 1998, p. 6.

McLaughlin, P. "Techno Tools: Do Students Really Need to Be Wired or Are Computers Just a Fancy Distraction from the Basics," *Halifax Daily News*, February 16, 1997, p. 4.

McLuhan, M., and Fiore, Q. *War and Peace in the Global Village*, New York: Bantam, 1968.

Meyer, J., and Rowan, B. "Institutionalized Organizations: Formal Sstructure as Myth and Ceremony, *American Journal of Sociology* (83), 1977, pp. 157-179.

Meyer, J., and Scott, R. (Eds.). *Organizational Environments: Ritual and Rationality*, Newbury Park, CA: Sage Publications, 1983.

Meyer, J., and Scott, R. (Eds.). *Organizational Environments: Ritual and Rationality* (rev. ed.). Newbury Park, CA: Sage Publications, 1992.

Michalos, A. C. *Improving Your Reasoning*, London: Prentice Hall, 1986.

Murphy, M. "Living the Acadia Experiment," *Globe and Mail*, June 22, 1998, p. C38.

Murray, F. "Technical Rationality and the IS Specialist: Power, Discourse and Identity," *Critical Perspectives on Accounting* (2), 1991, pp. 59-81.
Myers, M. D. "Dialectical Hermeneutics: A Theoretical Framework for the Implementation of Information Systems," *Information Systems Journal* (5:1), 1995, pp. 51-70.
Nardi, B. A. and O'Day, V. L. *Information Ecologies: Using Technology with Heart*, Cambridge, MA: MIT Press, 1999.
Ngwenyama, O. K. "The Critical Social Theory Approach to Information Systems: Problems and Challenges," in H-E. Nissen, H. K. Klein and R. H. Hirschheim (Eds.), *Information Systems Research: Contemporary Approaches and Emergent Traditions*, Amsterdam: North Holland, 1991, pp. 267-280.
Ngwenyama, O., and Lee, A. "Communication Richness in Electronic Mail: Critical Theory and the Contextuality of Meaning," *MIS Quarterly* (21:2), 1997, pp. 145-167.
Noakes, S. "Virtual Classroom Opens its Doors," *Financial Post*, April 24, 1996, p. 14.
NODE. "LTReport: The Mobile Campus Matures," The Node Learning Technologies Network, 2000 (available online at http://node.on.ca/ltreport/mobile).
Orlikowski, W., and Yates, J. "Genre Repertoire: The Structuring of Communicative Practices in Organizations," *Administrative Science Quarterly* (39:4), 1994, pp. 541-574.
Päivärinta, T. "The Concept of Genre Within the Critical Approach to Information Systems Development," *Information and Organization* (11), 2001, pp. 207-234.
Pearsall, K. "Higher Learning via Wired Classes: Technology is Getting Good Grades with Faculty and Students at Acadia University," *Computing Canada* (24:10), 1998, pp. 11-12.
Phillips, N., and Hardy, C. *Discourse Analysis: Investigating Processes of Social Construction*, Thousand Oaks, CA: Sage Publications, 2002.
Phillips, N., and Hardy, C. "Managing Multiple Identities: Discourse, Legitimacy and Resources in the UK Refugee System," *Organization* (4), 1997, pp. 159-186.
Postman, N. *Technology: The Surrender of Culture to Technology*, New York: Vintage, 1992.
Robey, D., and Markus, M. L. "Rituals in Information System Design," *MIS Quarterly* (8:1), 1984, pp. 5-15.
Rose, E. *User Error: Resisting Computer Culture*, Toronto: Between the Lines, 2003.
Sheridan College. " Mobile Briefing," unpublished, May 2000.
Sommers, J. A. "Acadia Students, Professors Meet at the Sandbox," *Globe and Mail*, August 26, 1997, p. C5.
Stoll, C. *Silicon Snake Oil*, New York: Anchor Books, 1995.
Tausz, A. "How Technology Firms Court the Key Education Markets," *Globe and Mail*, October 15, 1996, p. C6.
Townley, B. "Foucault, Power/Knowledge, and its Relevance for Human Resource Management," *Academy of Management Review* (18:3), 1993, pp. 518-545.
Truex, D. P., and Klein, H. K. "A Rejection of Structure as a Basis for Information Systems Development," in R. K. Stampler, P. Kerola, R. Lee, and K. Lyytinen (Eds.), *Collaborative Work, Social Communications and Information Systems*, Amsterdam: North Holland, 1991, pp. 213-235.
Ulrich, W. "A Philosophical Staircase for Information Systems Definition, Design, and Development: A Discursive Approach to Reflective Practice in ISD (Part 1)," *Journal of Information Technology Theory and Application* (3:3), 2001, pp. 55-84.
Van Dijk, T. A. "The Interdisciplinary Study of News as Discourse," in K. B. Jensen and N. Jankowski (Eds.), *A Handbook of Qualitative Methodologies for Mass Communication Research*, London: Routledge, 1991, pp. 104-120.
Watters, C., Conley, M., and Alexander, C. "The Digital Agora: Using Technology for Learning in the Social Sciences," *Communications of the ACM* (41:1), 1998, pp. 50-57.
Weick, K. E. *Sense Making in Organizations*, Thousand Oaks, CA: Sage Publications, 1995.
Winner, L. *The Whale and the Reactor*, Chicago: University of Chicago Press, 1986.

Wodak, R. (Ed.). *Language, Power and Ideology: Studies in Political Discourse*, Philadelphia: John Benjamins Publishing, 1989.
Wynn, E. H., Whitley, E. A., Myers, M. D., and DeGross, J. I. *Global and Organizational Discourse About Information Technology*, Boston: Kluwer Academic Publishers, 2003.

ABOUT THE AUTHORS

Wendy Cukier is a professor of Information Technology Management at Ryerson University in Toronto, a professor in the joint graduate program in Communication and Culture (Ryerson and York), and the Associate Dean (Academic) of the Faculty of Business. She holds M.A. and MBA degrees from the University of Toronto and a Ph.D. from York University. Her research focuses on emerging technology trends, technology discourse and gender. She is the recipient of many awards including two honorary doctorates and the Governor General's Meritorious Service Cross, one of Canada's highest civilian honours, for her advocacy work. Wendy can be reached by e-mail at wcukier@ryerson.ca.

Robert Bauer is an associate professor of Organizational Design and Behavior at Johannes Kepler University Linz, Austria. His research aims at a better understanding of different ways of knowing (including but not limited to explicit formal and every day language statements) and exploring their consequences for organizational design and behavior. He studies aspects of identity and difference on the individual, organizational and inter-organizational level as well as with respect to the philosophy of organization science. He is also a registered psychotherapist and has worked extensively as an executive coach and trainer. Robert can be reached at robert.bauer@jku.at.

Catherine Middleton is an assistant professor of Information Technology Management at Ryerson University in Toronto and also a professor in the joint graduate program in Communication and Culture (Ryerson and York). She holds a Ph.D. from York University and an MBA from Bond University. Her current research interests include consumer usage of broadband technologies, policy related to the development of broadband networks, and consumer adoption of mobile technologies. Catherine can be reached by e-mail at cmiddlet@ryerson.ca.

Appendix A

TEXTS INCLUDED IN THE ANALYSIS

Popular Discourse	
The Halifax Daily News	25
The Globe and Mail	6
Macleans	5
The Toronto Star	2
Canadian Business	1
Edmonton Journal	1
Times Colonist	1
Briarpatch	1
Financial Post Daily	1
Academic Discourse	
Communications of the ACM	2
Practical Discourse	
Systems:	
Computing Canada	2
Computer-dealer News	2
Canadian Telecom	1
Technology in Government	1
Network World Canada	1
Teaching/Administration:	
University Affairs	1
Campus Canada	1
Other:	
Quill and Quire	1
Peace Research	1
Parks and Recreation Can.	1
Total	**57**

15 CONDUCTING CRITICAL RESEARCH IN INFORMATION SYSTEMS: Can Actor-Network Theory Help?

Ela Klecuń
London School of Economics and Political Science

Abstract This paper considers the proposition that actor-network theory (ANT) might be adopted within a broader critical paradigm to conduct empirical studies. The paper outlines the main tenets of the two theories, with the critical perspective primarily represented by Foucault. The aim is not to provide an extensive discussion of critical theory and ANT but to focus on their approach to the nature, scope, and level of empirical studies, particularly in their treatment of micro/macro analysis. The paper concludes that the differences are less significant than it may appear at first and that some of ANT's ideas are close to Foucault's position. However, ANT focuses on actors and their actions as they are performed in a particular time and place and does not appear to be concerned to what extent they may be historically conditioned. Thus, ANT on its own, in the view of the author, might not offer sufficient explanations as to why the actors under study take particular actions and why some actors are excluded or marginalized from the innovation process, e.g. from the development and implementation of an IS. For these reasons this paper suggests a critical research agenda enriched by ANT insights.

Keywords: Critical theory, actor-network theory, information systems, Foucault

1 INTRODUCTION

The Information Systems field is influenced by a number of disciplines, including computer science, management and organizational studies, social science, and philosophy. Early on, many IS researchers primarily concentrated on technology, seeing technological development as following a predefined trajectory and technology as driving organizational change (Orlikowski 1992) and the majority of (published) IS research reflected a positivistic orientation (Orlikowski and Baroudi 1991). The IS discipline has significantly moved from this position, with interpretative research becoming more and more popular (Avgerou 2000; Walsham 1995). This broadening of IS research has been long on the agenda of IFIP 8.2, and is illustrated by varied contributions drawing on diverse social theories presented at the conferences since 1984.

More recently, an approach (or rather approaches) based on critical theory have been adopted by a growing (although still somewhat limited) number of researchers (Doolin and Lowe 2002; Hirschheim and Klein 1989, 1994; Jonsson 1991; Lyytinen 1992; Lyytinen and Klein 1985; Myers and Young 1997; Ngwenyama 1991; Saravanamuthu and Wood-Harper 2001; Wilson 1997).

Lately, some researchers from the IS and management fields called for broadening of the definition of critical research and for more empirically oriented studies (Alvesson and Deetz 2000; Brooke 2002a, 2002b; Doolin and Lowe 2002). Broadening the definition of critical research means including theorists who do not necessarily follow the Frankfurt School tradition. For example, the genealogical writings of Foucault, post-structuralist deconstruction associated with Derrida, and postmodernist work by Foucault and Lyotard (Kincheloe and McLaren 1994). To this list Brooke (2002a) also adds critical system thinking, critical realism, and critical postmodernism.

Yet, these calls for pluralism raise issues of paradigm incommensurability and diluting of emancipatory principles. While I agree that calling everything *critical* is unhelpful and indeed will dilute emancipatory principles, I do believe that critical theory should open itself up to new theoretical perspectives. After all, one of the fundamental tenets of critical theory is its need for self-reflexivity and self-critique. Thus, when considering what it means to be critical in IS research, I agree with Brooke (2002b) that it is the emancipatory interest rather than the detailed following of any one particular theorist that is important, providing that the underlying theoretical values and assumptions are explicated and, I would add, providing they are not incommensurable.

Nevertheless, adopting a critical approach poses two problems for IS researchers, namely the need to give attention to information and communication technology (ICT) and the difficulty of conducting empirical research. Critical theory is not specifically concerned with ICT and many, even contemporary works, remain vague on this subject. Critical theory is almost silent with respect to the use of techniques of investigation and does not prescribe how research should be conducted (Morrow and Brown 1994). In order to bridge the gap between philosophical foundations and empirical research, increasingly IS researchers are reaching out to other theories for additional insights, and actor-network theory (ANT) is becoming particularly popular (Doolin 1998; Doolin and Lowe 2002; Whitley 1999a).

This paper evaluates ANT as a good candidate to guide empirical studies within a broader critical paradigm, particularly as represented by Foucault. In doing so, the paper hopes to open a dialogue between often-separate communities of theorists and researchers representing these approaches. The paper's aim is not to provide an extensive discussion of critical theory and ANT but to focus on their approaches to empirical studies and the treatment of micro and macro levels of analysis.

The paper starts by outlining the main concepts of critical theory, focusing on works by Foucault. Then, it considers the proposition that ANT could be adopted for conducting critical empirical research. In doing so, the paper outlines the main arguments about similarities and differences of the two approaches, focusing on their treatment of the scope of empirical research and the level of analysis. The concluding section summarizes the main points of the discussion, again posing and addressing the question about the applicability of ANT to critical theory-led studies.

2 THEORETICAL PERSPECTIVES: CRITICAL THEORY AND ANT

2.1 Aspects of Critical Theory

Critical theory is not a unified theory but rather a set of loosely linked principles. It has many flavors and proponents. The central idea in critical theory is that all social phenomena are historically created and conditioned (Horkheimer 1972b; Horkheimer and Adorno 1972). Critical theory has evolved over time, often in response to historical conditions of the time. The early proponents of critical theory from the Frankfurt School sought to develop a historically grounded social theory that could help to transform the world (Held 1980). They produced grand narratives of emancipation, maintaining that social conditions, often constraining emancipation and limiting peoples' potential, are created and recreated by man. They claimed that the consciousness of man is dominated by the ideological superstructures and these may result in *alienation*. Their aim was to expose and undermine the *status quo*.

Among second generation theorists, Habermas is perhaps best known, especially in the IS community. Although, Habermas' roots are in the Frankfurt School, his writings have developed many concepts further and taken them in different directions. Habermas seeks to incorporate central notions of the hermeneutic tradition within the bounds of critical philosophy and his works are often seen as an extension of hermeneutics (Jerald 1990). His main concerns are the problems regarding the nature of communication and self-consciousness and their role in the causation of social action (Habermas 1972, 1979). In terms of macro and micro division, Habermas can be seen as leaning toward a macroscopic view of social order but he is also concerned with many micro-sociological concepts (e.g., speech performances) (Knorr-Cetina 1981).

Casting our net further away from the Frankfurt School, we could also consider theorists from the French school of thought (for example, Foucault, Derrida, and Lyotard) as well as Giddens as working in a critical tradition. This suggestion is a result of our wider, more inclusive definition of a critical tradition that focuses on emancipatory interest. However, can Foucault's work be described as having such an interest? We can argue that by providing a critique of the *status quo* Foucault undermines the inevitability of the present situation and opens up avenues for change. His genealogical method aims to expose mechanisms of power and to emphasize the contingency and fragility of the circumstances that have shaped present practices. Thus, such an approach has an emancipatory intent. Foucault clearly stated his aim during a television program titled "Human Nature: Justice versus Power" discussed in Rabinow (1984, p. 6).

> [The] real political task in a society such as ours is to criticize the working of institutions which appear to be both neutral and independent; to criticize them in such a manner that the political violence which has always exercised itself obscurely through them will be unmasked, so that one can fight them.

Increasingly, Foucault's works (particularly his later works) are considered as belonging to the critical tradition (Brooke 2002a; Kincheloe and McLaren 1994; McGrath 2003a), and this is a position taken in this paper.

This paper is interested in Foucault's writings for a number of reasons. First, although Foucault's works are detailed, historically situated accounts of different phenomena that do not aspire to produce grand theories of society, their relevance is enduring (refer, for example, Foucault 1979). Indeed, it is through such a micro approach that we can learn about the macro order (Knorr-Cetina 1981).

Second, Foucault focuses on understanding knowledge and power, two concepts I consider as key to the study of information systems. The basic claim of Foucault's genealogy of knowledge is that political power and scientific knowledge are not external to one another, and thus "regimes of truth" (truth claims in science) have political character (are characterized by power/ knowledge relations present in the scientific discourse of a particular discipline). Foucault (1979) argues that power is not necessarily bad or top-down but is relational, that it is exercised through a net-like organizations. He breaks away from the Enlightenment belief in reason, absolute truth, and totality. To some extent, so do the other critical theorists (Horkheimer and Adorno 1972) but Foucault also eschews grand narratives and emphasizes the importance of the local relations and actions. He rejects an idea of absolute or transcendental truth "outside of history" as well as of any conception of *objective* or *necessary* interests that would ground either knowledge, morality, or politics (Olssen 1996).

I believe that it is ontologically and epistemologically feasible to consider complementing a Foucauldian approach with ANT's concepts (discussed in section 3). Indeed, the influence of Foucault on ANT is acknowledged by Law (1986), who states that the concepts of translation in ANT "owe more than a little to the writing of Foucault."

What is central to this paper is Foucault's discussion of modes of analysis. He proposes a genealogical approach that aims to account for the constitution of knowledge, discourses, and domains of objects, without appeal to transcendental subjects (Foucault 1980). In another work, he elaborates the nature of criticism.

> Criticism is no longer going to be practiced in the search for formal structures with universal value, but rather as an historical investigation into the events that have led us to constitute ourselves and to recognize ourselves as subjects of what we are doing, thinking, saying. In this sense the criticism is not transcendental and its goal is not of making a metaphysics possible: it is genealogical in its design and archaeological in its method (Foucault 1984, pp. 45-46).

While Foucault alerts us to the diversity of forms and locations of power, he comes under criticism because of his lack of treatment (or inadequate treatment) of structures, the state, the school, the bureaucracy, and their relevance to the notion of power (Olssen

1996). However, as McGrath (2003b) argues, this neglect is not an inherent feature of his perspective but rather it is due to narrow interpretation of his work that does not take into account his concern with bio-power, as a normalizing and regulatory force working on whole populations. Thus, it could be argued that Foucault's theoretical conceptualizations (spanning both ontology and epistemology) might guide research (including IS research) that aspires to focus on the local, contingent, and negotiated nature of technological innovations without neglecting a broader political and economic context.

Nevertheless, Foucault (and for that matter Habermas) does not explicitly consider ICT in his work. This poses great difficulty for the IS researcher looking for a firm theoretical and methodological ground on which to place an enquiry. This is not to say that critical theory has nothing to say about the technology in general. For example, the relationship between technology and society is discussed by Winner (1977), who sees technology as augmenting, as having accumulative impacts that make some choices increasingly difficult. Ellul (1964) presents a more pessimistic, or fatalistic, account of technology (understood in a broad sense that includes technological artefacts and techniques), warning us that a process of technological change is self-generating, self-determining, and inevitable. On the other hand, Feenberg (1991), and before him Marcuse (1970), consider a technology not as autonomous but as an instrument of social control placed in the hands of the vested interests that control society. In Feenberg's words, "technology is not a destiny but a scene of struggles. It is a social battlefield, or perhaps a better metaphor would be a *parliament of things* on which civilization's alternatives are debated and decided" (1991, p. 14). Feenberg (1988) believes that we can construct a solution that avoids the excess of both utopian technophobia and uncritical acceptance of given technology as fate. He articulates this as the need to contextualize technologies taking into account more and more of the essential features of the object, bringing together its many dimensions.

Yet, critical theory comes under scrutiny because of the (perceived) dichotomy between theory and practice (Asaro 1999). I would argue that this dichotomy is not embedded in critical theory; on the contrary, critical theory sees action as an extension of theory. Thus, it is important that critical research continues to move beyond purely theoretical endeavors. Alvesson and Willmott (1992) and Alvesson and Dietz (2000) similarly argue that in the field of management it is essential to close the gap between theory and practice.

2.2 Critical IS Research

In 1992, Lyytinen appraised the state of critical IS research, pointing out its limitations, and outlining its general requirements.

> In order to move from fragmentary critical IS research to systemic "praxis"-oriented research, future studies should change their goals and research content. The inquiry needs to shift from critique into more concrete and problem-focused studies of the implications of Critical Theory for IS....In this research model, critical inquiry is concerned with the improvement of the human condition through IS, criticism of alienated and distorted practices,

> development of alternative IS forms and organizations, and with finding and enclaving an arena for emancipatory IS activity (Lyytinen 1992, pp. 171-172).

The call for systemic praxis-oriented research that is concrete and problem-focused has been taken up by a (small, so far) number of IS researchers (Cecez-Kecmanovic et al. 2002; Howcroft and Wilson 1999; McAulay et al. 2002; Myers and Young 1997; Oliver and Romm 2002; Waring 1999).

The paucity of critical theory-driven research in IS, in my opinion, is not only due to the neglect of ICT in critical theorists' writings, but also because critical theory is a meta-theory, built from a collection of many writings on different subjects. It does not prescribe specific methods for empirical research or offer detailed guidelines to follow. This poses difficulty for the researcher in terms of how to conduct the research and how to judge its quality. These problems have been tackled by a number of contemporary authors (Alvesson and Deetz 2000; Alvesson and Skoldberg 2000; Kincheloe and McLaren 1994; Morrow and Brown 1994). For example, Guba and Lincoln (1994) provide the following criteria for judging the goodness or quality of inquiry: first, its historical situatedness (i.e., that it takes account of the social, political, cultural, economic, ethnic, and gender antecedents of the studied situation); second, the extent to which the inquiry acts to erode ignorance and misapprehension; third, if and to what extent it spurs people to action, that is, to the transformation of the existing structure. These criteria are hard to satisfy and moreover they are very subjective.

Within the IS field a concern has been voiced that researchers, in their drive to account for organizational, social, and political factors, have neglected the technology itself, treating it as a black box, for example, not considering different parts of an IS, like software, standards, procedures, rules, and work activities (Monteiro 2000; Orlikowski and Iacono 2001). Similarly, Williams and Edge (1996) argue that a useful theory of the relationship between technology and society needs to address more directly the characteristics of the material world.

How might the IS researcher conduct a study that is all of the following: informed by a social theory, in our case critical theory, and taking into consideration different macro trends but also focused on the local; eschewing grand narratives and *a priori* classifications but also rejecting relativism; sensitive to social, political, and economic factors but also not forgetting or black boxing the technology itself?

Considering the dilemma facing empirical researchers, Feenberg (1999, p. 13) suggests

> we can fruitfully combine modernity theory and technology studies in an empirically informed, critical approach to important social problems. The triviality that threatens a strictly descriptive, empirical approach to such humanly significant technical phenomena as experimentation on human subjects, nuclear power, or the development of the automobile, can be avoided without falling into the opposite error of *a priori* theorizing. The alternative—global condemnation, narrow empiricism—is not exhaustive. There are ways of recovering some of the normative richness of the critique of modernity within a more concrete sociological framework that does allow entry to a few facts.

Fully agreeing with this statement, this paper suggests that critical theory can help to preserve the normative richness of the critique of modernity while actor-network theory offers a more concrete sociological framework. Indeed, such an approach has already been proposed by a number of IS researchers. For example, Whitley (1999a) suggests that the principles of critical theory can be applied to the study of IS by drawing on the work of Bruno Latour (one of the main theorists of ANT). Mitev (2003) advocates using social constructivism and ANT as the basis for a critical analysis, Monteiro (2000) suggests that ANT lends itself to empirically underpinned studies. Doolin and Lowe (2002) argue that ANT, as its stands, is well suited to conducting empirical IS research within a broader critical research project. Specifically, they claim that "actor-network theory offers a particularly effective 'alternative reading' of social interactions within organizations through its emphasis on empirical enquiry and its lack of constraining structure and ontology" (p. 72). Before the suitability of ANT for critical empirical research can be considered, its main concepts need to be outlined.

2.3 An Introduction to Actor-Network Theory

ANT rejects any *a priori* distinction between technology and society, proposing that both should be studied in the same way, through the same (or interchangeable) language and metaphors. In ANT terms, innovations are developed and adopted (or not) through the building of networks of alliances between human and nonhuman actors (Monteiro 2000). The sociology of translation, an important concept in ANT, expresses how innovations are translated or constructed and transposed from one state to another, and how different actors may be co-opted (enrolled) to support a particular innovation or project within a heterogeneous network of actors (Callon 1986). For a network to succeed or be sustained, such a transformation must become stable (stronger), even irreversible, but this stability is often difficult to sustain.

ANT might be conceptualized as both a theory and a methodology. However, as Latour tells us, ANT does not tell us *positive* things about the world (i.e., how the social world is) but rather suggests *how* to study things (Latour forthcoming). What does ANT say about conducting an empirical study? Its advice is simple: follow the actors. Only by following actors, their actions, and interactions can we say something about the situation. This means that no *a priori* expectations or theoretical concepts should guide, or rather limit, our research. For example, we should not make *a priori* distinction "between the size of actors, between the real and the unreal, between what is necessary and what contingent, between the technical and the social" (Callon and Latour 1981, pp. 291-292). According to Latour (forthcoming), "actors themselves make everything, including their own frames, their own theories, their own contexts, their own metaphysics, even their own ontologies." Thus, in commenting on how to write research findings, he tells us not to impose our own frameworks or concepts but just describe. It is a thick description that provides insights into the situation.

Typically, ANT informed research would start by identifying key actors, interests, and scenarios, and then trace them over time (Monteiro and Sahay 2000), but it is difficult for the researcher to know when to stop (i.e., where to draw the boundary of the study). Ultimately, there is no prescription on how to do it and such a process is somewhat arbitrary.

Is there anything more to the methodology of ANT? No...and yes. No, because there are no frameworks or steps to follow. Yes, because ANT provides us with certain concepts that we may find helpful while conducting research. For example, ANT elaborates on the processes of construction and deconstruction of actor networks in terms of four moments of translation (or phases): *problematization, interessement, enrolment*, and *mobilization* (Callon 1986). It also, through examples, describes how interests and expectations may be *inscribed* in actors (e.g., in an information system).

ANT is gaining popularity in IS research, partly because, through its theoretical concepts and language, it supports the researcher in being more specific about technology (e.g., through inscriptions). ANT offers a way for describing how technical design and systems are interwoven with organizational issues (Monteiro and Hanseth 1996). Furthermore, as ICT becomes more ubiquitous and embedded in our everyday activities, "we need new methodological and theoretical devices to enable us to think about hybrids of people and information technology" (Walsham 1997). ANT seems to offer just this.

However, ANT has been accused of being apolitical, lacking an evaluative stance that might help people judge the possibilities presented by the technology and consider the consequences, and not taking into account that there may be dynamics evident in technological change beyond those revealed by studying the immediate needs, interests, problems, and solutions of specific groups (Walsham 1997; Winner 1993). This is significant if we are concerned with emancipatory principles and suggest ANT as a way of operationalizing critical theory. ANT in particular, because of its symmetrical treatment of humans and nonhumans and the definition of action as distributed through different entities (human and nonhuman) in socio-technical ensembles, is open to a charge of diluting intentionality, responsibility, and accountability (Latour 1999; Stalder 2000). Furthermore, although ANT rejects *a priori* assignment of motives to actors, it can be argued that implicitly it presumes that actors are rational and goal-oriented. There is a danger that the strategic and rational aspects of ANT become over-emphasized in IS studies (Monteiro and Sahay 2000). In order to combat perceived shortcomings of ANT, in particular its neglect of influences of structures on micro-events and processes, a number of writers proposed an approach that combines ANT with ideas from other social theories (Avgerou 2002; Walsham 1997; Whitley 1999b).

Thus we see that calls to combine the critical tradition with ANT come from two quarters: those who see ANT as the overriding theory but benefitting from adopting additional insights and concepts (e.g., regarding power) and those who search for other theories, including ANT, to assist with empirical critical studies in IS.

3 CRITICAL THEORY AND ACTOR-NETWORK THEORY: CAN THEY COMPLEMENT EACH OTHER?

The previous section briefly introduced critical theory and ANT and their application in the IS field, noting that arguments have been voiced for basing IS research on both approaches (combined in some way). This section considers whether ANT can complement critical theory and enrich empirical studies in the IS field. In doing so, this section revisits some of the ontological and epistemological assumptions behind the two theories and considers to what extent, if any, they are compatible. Specifically, this section considers their position regarding macro and micro levels of analysis.

Burrell and Morgan (1985) consider critical theory as having much in common with the interpretative paradigm. They categorize critical theory as belonging to a radical humanism paradigm, placed between the subjective and the sociology of radical change axis in their framework. But increasingly, critical approaches incorporate different ontological and epistemological assumptions and exhibit methodological pluralism (Brooke 2002a). It is not the aim of this paper to discuss the pros and cons of methodological pluralism (Mingers 2001) or the relative merits of different perspectives within the critical tradition; that has been done by others (Brooke 2002a; Fitzgerald and Howcroft 1998). Here, I aim to highlight the diversity of traditions within the broadly understood critical perspective. Such diversity at least allows us to consider the possibility of incorporating (redefined?) ANT within a critical perspective.

ANT is often seen as belonging to the social constructivism tradition, and is thus used as a lens in interpretative IS research (Cordella and Shaikh 2003). This would imply that it has constructivist (relativist) ontology and interpretative epistemology, not unlike critical theory as espoused by the Frankfurt School. However, as mentioned earlier, critical theory is not a uniform theory, and Foucault's later works are seen by some as adopting a more realist position (i.e., being based on objectivist ontology) (Olssen 1996). This would suggest that Foucault's works and ANT are based on different ontological and epistemological assumptions and, if one subscribes to incommensurability of paradigms, as being incompatible. However, Latour rejects the assumption that ANT is based on constructivist ontology. Cordella and Shaikh (2003) argue that ANT has its own ontology, that it sees reality as emerging from interactions between different actors. Such an understanding of ANT would bring it closer to Foucault's critical research. Furthermore, such clear paradigm distinctions are seen as unhelpful by some (Allen and Ellis 1997; Alvesson and Deetz 2000; Fitzgerald and Howcroft 1998). Thus, I would argue, critical theory and ANT are not incommensurable and can potentially complement each other in IS research.

In conducting an empirical study, a critical researcher is faced with an important choice regarding the unit of analysis in terms of the application of emancipatory principle. Should the focus be on emancipation within organizational boundaries or, alternatively, consider how critically informed research could influence wider society? The choice of either of these models has implications for the contributions expected, and certainly the second type of research seems more difficult. Although it is customary in IS research to focus on one unit of analysis, consideration of multiple units of analysis is important for capturing different perspectives. As Markus and Robey (1988 p. 596) note, "By consciously mixing levels of analysis, researchers can explore the dynamic interplay among individuals, technology, and larger structures." From critical theory perspective Horkheimer (1972a, p. 249), commenting on critical theory, proclaims that "The theory is concerned with society as a whole." Morrow and Brown (1994) argue that a good critical research even if focusing on a particular level of enquiry should remain aware of other levels and their influence on the research area. This means that when conducting an analysis of, for example, an implementation of IS in one organization, when analyzing local dominant discourses and situated action, we need to take into consideration wider structures that may be constraining (and enabling) such actions, and influencing their consequences (i.e., class, gender, ethnicity, and so on). Indeed, many of the strands of critical theory (e.g., the Frankfurt School and feminism) consider as *a priori* certain superstructures and their influence on local circumstances. Although this is not true of Foucault's writings, it does not mean that he neglects macro structures. He is, however, against the search for an organizing principle and an explanation of all phenomena in relation to a single center (Olssen 1996). He sees history as subject to discontinuities, segmentations, and different types of relations, and maintains that there are no simple material categories (e.g., class) that explain everything. Such a perspective lends itself to empirical (but reflective) studies of local phenomena.

ANT offers a somewhat different view on the micro-macro relationship. It promotes a uniform framework regardless of the unit of analysis and it refuses to distinguish *a priori* between small and big networks (Callon and Latour 1981; Monteiro 2000). Furthermore, it refuses to assign *a priori* attribution of social interests. Latour's position that the analysis of networks suffices and that the introduction of macro-social terms would obscure the activities of the underlying actors may appear contrary to the critical theory position, particularly as espoused by the Frankfurt School. ANT suggests that it is the actors that establish (interpret and make real) the macro trends (e.g., capitalism) and thus we should not create an artificial distinction between the local and the global but that we can extend a study to ever-wider actor-networks.

This approach is criticized for its flat ontology (Monteiro 2000), and Feenberg (1999) argues that Latour reduces the terms *society* and *nature* to local actions. Furthermore, starting with actors involved with an innovation, and not referring to the traditional categories of social theory, such as class, culture, and the state, means we might not be aware of or even concerned about actors (or potential actors) who are missing. We may not be able to explain why they might have been excluded or marginalized (Feenberg 1999; Williams and Edge 1996). As Feenberg (1999) argues, "this means that stronger parties establish the definition of basic terms, culture, nature and society, and there is no appeal to a priori essence" and thus ANT contains an implicit bias toward victors (Radder 1996).

So far, this section outlines a number of important points regarding ANT's position. The reminder of the section reconsiders these one by one and examines to what extent ANT's position is (or could be) compatible with critical theory proponents. First, I stated that ANT refuses to assign *a priori* motives to actors (e.g., that managers are interested in IS only in terms of their potential for efficiency gains). However, this does not mean that ANT disregards motives or assumptions but these should be uncovered in empirical studies. For example, it doesn't ignore capitalism; it asks where the effects of capitalism can be seen. Thus this point should not be of great contention between critical theory and ANT. Indeed, Foucault does not adopt *a priori* a political or moral agenda that represents the interests of particular groups. It may be argued, however, that having in mind some *a priori* assumptions or rather suspicions (e.g., that women might be marginalized) facilitates focus on potential conflicts and mistrust of the taken-for-granted facts. Yet, such suspicions should be taken as given (as transcendental truths) and a good empirical study should stay open minded with regard to its findings.

The second point, that no *a priori* differences in macro or micro networks exist, and that researching networks is sufficient, may be understood as proposing that macro-structures do not determine micro-events, and that social processes exhibit chains of intended and unintended outcomes (Mitev 2003). Mitev suggests that, "While 'conditions of possibility' [as Knights and Murray (1997) describe external forces] frame organizational behavior, actors construct at the local level the 'external forces' that they respond to." Or in the words of Knights and Murray, quoted in Mitev (2003, p. 35), "a technological opportunity or constraint exists only in so much as people believe it exists."

This is not an interpretation I choose to favor. I agree with Kallinikos (2002) and Avgerou (2002) that the process of construction of technology is not purely local, but is a result of many processes and past experiences, either individual or group, immediate or secondary, that influence local decisions. Using critical theory language, technology is historically situated. But what is ANT's stance on this? My interpretation (based on different relevant readings, in particular Latour [1999]) is that ANT, through empirical studies, links macro and micro. Such linking takes place not just in the construction of the macro by actants (i.e., the micro elements) but rather in studying the relationships between different actants. The process of identifying the local in the global, and vice versa and thus spanning local and global (without making *a priori* distinctions) involves unpacking seemingly macro-elements down to their empirical constituents and black boxing or collapsing an entire (or a part of an) actor-network into a single actant (Monteiro 2000). Following these steps would lead an IS researcher to encompass "micromacro" perspectives in a study. Micromacro is written here as a one word to emphasize that such an approach rejects the micro-macro dichotomy; it is, however, noted that in practice boundaries of research have to be drawn and some black boxes left unopened. Such an approach would allow us to consider how wider (macro) trends (e.g., anti-trust laws, national security threats, etc.) influence local developments.

Understood in this way, ANT's position brings it closer to the approach advocated by Foucault. It differs, however, in that it is focused on actants and their actions in a particular place and time, eschewing historically based analysis. Thus, I am still left with a charge that following actors does not lend itself to explanations of why some actors take certain actions, or why they do not appear at all or are marginalized. Thus, I would suggest that we do need to bring with us an awareness of class, race, or gender, and how they are historically constituted. However, any *a priori* assumptions should be open to empirical analysis. Thus I advocate a research agenda based on a critical (Foucauldian) approach with empirical studies enriched by insights from ANT.

4 CONCLUDING REMARKS

This paper has indicated some problems facing the critical researcher in IS arising from the fragmented nature of critical theory, its cursory treatment of ICT, and the lack of guidelines on how to conduct critical research. Having in mind these problems, the paper then considered if ANT could assist the IS researcher in conducting critically inspired studies.

The paper suggests that by theorizing the nature of actor-networks and elaborating the theory of translations ANT develops concepts and vocabulary that may assist the IS researcher in investigating, for example, how networks are formed and actors are enrolled during early stages of decision-making activities regarding a new information system, and during the system implementation and subsequent use (or non-use). ANT also offers a fresh view on human and technology collectives and indicates how we may treat them. ANT is thus not only a theory but also a research methodology and has the potential to address some of the problems facing the IS researcher who seeks to draw from critical theory.

There are, however, obstacles to combining both theoretical perspectives in research. This paper briefly considered their potential ontological and epistemological differences, concentrating on their treatment of micro and macro levels of analysis. As shown, the differences, although at first appearing to be significant, might not be so. I argue ANT is close to Foucault's perspective in terms of its ontology and epistemology (e.g., both focus on local, situated actions and a rejection of grand narratives).

Yet, I still agree with ANT's critics who argue that a research methodology based on following the actors and a rejection of *a priori* constructs may lead to results that privilege the views of certain actors and do not consider those who do not immediately appear. As suggested here, Foucault's rejection of grand narratives does not necessarily mean a total rejection of *a priori* constructs (e.g., the role of the state), but instead it leads us to question such constructs through conducting detailed (we might say geological) studies. Thus, an empirical program for critically led research would aim to question the status quo through detailed studies of how things come to be, for which following the actors would be an excellent start, and *also* be sensitive to *a priori* constructs, such as gender, class, or state. Such constructs should not be taken for granted or seen as natural and enduring but act as sensitizers. Empirical study should concentrate on local, situated actions and relationships *and* see them as historically situated.

The approach proposed here does not claim to combine into a seamless whole critical theory and ANT; rather, it illustrates how ANT might be used as a way of operationalizing critical theory in practice. Thus, the paper illustrates how ANT might help critical research. Conversely we could consider how critical theory concepts may inform ANT research. The paper hopes to open a dialogue between the proponents of these two approaches.

ACKNOWLEDGMENTS

I would like to thank Tony Cornford, Edgar Whitley, and the reviewers for their very useful comments.

REFERENCES

Allen, D., and Ellis, D. "Beyond Paradigm Closure in Information Systems Research: Theoretical Possibilities for Pluralism," in R. D. Galliers, S. Carlsson, C. Loebbecke, C. Murphy, H. R. Hansen, and R. O'Callaghan (Eds.), *Proceedings of the 5th European Conference on Information Systems*, Cork, Ireland: Cork Publishing Ltd., 1997, pp. 737-759.

Alvesson, M., and Deetz, S. *Doing Critical Management Research*, London: Sage Publications, 2000.

Alvesson, M., and Skoldberg, K. *Reflexive Methodology: New Vistas for Qualitative Research*, London: Sage Publications, 2000.

Alvesson, M., and Willmott, H. (Eds..) *Critical Management Studies*, London: Sage Publications, 1992.

Asaro, P. M. "Transforming Society by Transforming Technology: The Science and Politics in Participatory Design," paper presented at the Critical Management Studies Conference, Manchester, 1999 (available online at http://www.mngt.waikato.ac.nz/ejrot/cmsconference/documents/Information%20Tech/ Pd_cms.pdf).

Avgerou, C. "Information Systems: What Sort of Science Is it?," *Omega* (28), 2000, pp. 567-579.

Avgerou,, C. *Information Systems and Organizational Diversity: The Articulation of Local and Global Rationalities*, Oxford: Oxford University Press, 2002.

Brooke, C. "Critical Perspectives on Information Systems: An Impression of the Research Landscape," *Journal of Information Technology* (17:4), 2002a, pp. 271-283.

Brooke, C. "What Does it Mean to Be 'Critical' in IS Research," *Journal of Information Technology* (17), 2002b, pp. 49-57.

Burrell, G., and Morgan, G. *Sociological Paradigms and Organizational Analysis: Elements of the Sociology of Corporate Life*, Aldershot, England: Gower, 1985.

Callon, M. "Some Elements of a Sociology of Translation: Domestication of the Scallops and the Fishermen of St Brieuc Bay," in J. Law (Ed.), *Power, Action and Belief*, London: Routledge & Kegan Paul, 1986, pp. 196-233.

Callon, M., and Latour, B. "Unscrewing the Big Leviathan," in K. D. Knorr-Cetina and A. V. Cicourel (Eds.), *Advances in Social Theory and Methodology: Towards an Integration of Micro- and Macro-Sociologies*, Boston: Routledge & Kegan Paul, 1981, pp. 277-303.

Cecez-Kecmanovic, D.; Janson, M.; and Brown, A. "The Rationality Framework for a Critical Study of Information Systems," *Journal of Information Technology* (17:4), 2002, pp. 215-227.

Cordella, A., and Shaikh, M. "Actor Network Theory and After: What's New for IS Research?," *Proceedings of the 11th European Conference on Information Systems*, Naples, Italy, 2003.

Doolin, B. "Information Technology as Disciplinary Technology: Being Critical Interpretive Research on Information Systems," *Journal of Information Technology* (13), 1998, pp. 301-311.

Doolin, B., and Lowe, A. "To Reveal Is to Critique: Actor-Network Theory and Performativity in Critical Information Systems Research," *Journal of Information Technology* (17:2), 2002, pp. 69-78.

Ellul, J. *The Technological Society*, New York: Vintage Books, 1964.

Feenberg, A. "The Bias of Technology," in R. Pippin, A. Feenberg, and C. B. Webel (Eds.), *Marcuse: Critical Theory and the Promise of Utopia*, London: MacMillan Education, 1988, pp. 225-256.

Feenberg, A. *Critical Theory of Technology*, New York: Oxford University Press, 1991.

Feenberg, A. "Modernity Theory and Technology Studies: Reflections on Bridging the Gap," paper presented at the Conference on Technology and Modernity, University of Twente, 1999 (available online at http://www-rohan.sdsu.edu/faculty/feenberg/twente.html).

Fitzgerald, B., and Howcroft, D. "Towards Dissolution of the IS Research Debate: From Polarization to Polarity," *Journal of Information Technology* (13:4), 1998, pp. 313-326.

Foucault, M. *Discipline and Punish: The Birth of the Prison*, Harmondsworth, England: Penguin, 1979.

Foucault, M. *Power/Knowledge: Selected Interviews and Other Writings 1972-77*, Brighton, England: Harvest Press, 1980.

Foucault, M. "What Is Enlightenment?" in P. Rabinow (Ed.), *The Foucault Reader*, London: Penguin Books, 1984, pp. 32-50.

Guba, E. G., and Lincoln, Y. S. "Competing Paradigms in Qualitative Research," in N. K. Denzin and Y. S. Lincoln (Eds.), *Handbook of Qualitative Research,* London: Sage Publications, 1994, pp. 105-117.

Habermas, J. *Communication and the Evolution of Society*, Boston: Beacon Press, 1979.

Habermas, J. *Knowledge and Human Interests*, Boston: Beacon Press, 1972.

Held, D. *Introduction to Critical Theory: Horkheimer to Habermas*, Oxford: Hutchinson & Co, 1980.

Hirschheim, R., and Klein, H. K. "Four Paradigms of Information Systems Development," *Communications of the ACM* (32:10), 1989, pp. 1199-1216.

Hirschheim, R., and Klein, H. K. "Realizing Emancipatory Principles in Information Systems Development: The Case for ETHICS," *MIS Quarterly* (18:1), March 1994, pp. 83-109.

Horkheimer, M. "Postscript," in *Critical Theory: Selected Essays of Max Horkheimer*, New York: Herder and Herder, 1972a, pp. 244-252.

Horkheimer, M. "Traditional and Critical Theory," in *Critical Theory: Selected Essays of Max Horkheimer*, New York: Herder and Herder, 1972b, pp. 188-243.

Horkheimer, M., and Adorno, T. W. *Dialectic of Enlightenment*, New York: Herder and Herder, 1972 (originally published in 1944).

Howcroft, D., and Wilson, M. "Paradoxes of Participatory Design: The End-User Perspective," paper presented at the Critical Management Studies Conference, Manchester, 1999 (available online at http://www.mngt.waikato.ac.nz/ejrot/cmsconference/documents/Information%20Tech/Howcroft.pdf).

Jerald, W. *The Hermeneutics of life History: Personal Achievement and History in Gadamer, Habermas, and Erikson*, Evanston, IL: Northwestern University Press, 1990.

Jonsson, S. "Action Research," in H.-E. Nissen, H. K. Klein, and R. Hirschheim (Eds.), *Information Systems Research: Contemporary Approaches and Emergent Traditions,* Amsterdam: North-Holland, 1991, pp. 371-396.

Kallinikos, J. "Reopening the Black Box of Technology: Artifacts and Human Agency," in L. Applegate, R. Galliers, and J. I. DeGross (Eds.), *Proceedings of the 23rd International Conference on Information Systems*, Barcelona, 2002, pp. 287-294.

Kincheloe, J. L., and McLaren, P. L. "Rethinking Critical Theory and Qualitative Research," in N. K. Denzin and Y. S Lincoln (Eds.), *Handbook of Qualitative Research,* London: Sage Publications, 1994, pp. 138-157.

Knights, D., and Murray, F. "Markets, Managers and Messages: Managing Information Systems in Financial Services," in B. P. Bloomfield, R. Coombs, D. Knights, and D. Littler (Eds.), *Information Technology and Organizations: Strategies, Networks and Integration,* Oxford: Oxford University Press, 1997, pp. 36-56.

Knorr-Cetina, K. D. "Introduction: The Micro-Sociological Challenge of Macro-Sociology: Towards a Reconstruction of Social Theory and Methodology," in K. D. Knorr-Cetina and A. V. Cicourel (Eds.), *Advances in Social Theory and Methodology: Toward an Integration of Micro- and Macro-Sociologies,* Boston: Routledge & Kegan Paul, 1981, pp. 1-47.

Latour, B. *Pandora's Hope: Essays on the Reality of Science Studies*, Cambridge, MA: Harvard University Press, 1999.

Latour, B. "On Using ANT for Studying Information Systems: A (Somewhat) Socratic Dialogue," in C. Averou and C. Ciborra (Eds.), *Social Study of ICT,* Oxford: Oxford University Press, Oxford, Forthcoming.

Law, J. "Editor's Introduction: Power/Knowledge and the Dissolution of the Sociology of Knowledge," in J. Law (Ed.), *Power, Action and Belief: A New Sociology of Knowledge,* London: Routledge and Kegan Paul, 1986.

Lyytinen, K. "Information Systems and Critical Theory," in M. Alvesson and H. Willmott (Eds.), *Critical Management Studies,* London: Sage Publications, 1992, pp. 159-180.

Lyytinen, K. J., and Klein, H. K. "The Critical Theory of Jurgen Habermas as basis for a Theory of Information Systems," in E. Mumford, R. Hirschheim, G. Fitzgerald, and A. T. Wood-Haper (Eds.), *Research Methods in Information Systems,* Amsterdam: North Holland, Amsterdam, 1985, pp. 219-236.

Marcuse, H. *One-Dimensional Man*, London: Sphere Books Ltd., 1970.

Markus, M. L., and Robey, D. "Information Technology and Organizational Change: Causal Structure in Theory and Research," *Management Science* (34:5), 1988, pp. 583-598.

McAulay, L.; Doherty, N.; and Keval, N. "The Stakeholder Dimension in Information Systems Evaluation," *Journal of Information Technology* (17:4), 2002, pp. 241-255.

McGrath, K. "ICTs Supporting Targetmania: How the UK Health Sector is Trying to Modernise," in M. Korpela, R. Montealegre, and A. Poulymenakou (Eds.), *Organizational Information Systems in the Context of Globalization*, Boston: Kluwer Academic Publishers, 2003a, pp. 19-34.

McGrath, K. *Organizational Culture and Information Systems Implementation: A Critical Perspective*, Unpublished Ph.D. Dissertation, Department of Information Systems, The London School of Economics, London, 2003b.

Mingers, J. "Combining IS Research Methods: Towards a Pluralist Methodology," *Information Systems Research* (12:3), 2001, pp. 240-259.

Mitev, N. "Constructivist and Critical Approaches to an IS Failure Case Study: Symmetry, Translation and Power," Working Paper Series, Information Systems Department, London School of Economics, 2003.

Monteiro, E. "Actor-Network Theory and Information Infrastructure," in C. Ciborra (Ed.), *Control to Drift,* New York: Oxford University Press, 2000, pp. 71-83.

Monteiro, E., and Hanseth, O. "Social Shaping of Information Infrastructure: On Being Specific About the Technology," in W. J. Orlikowski, G. Walsham, M. R. Jones, and J. I. DeGross (Eds.), *Information Technology and Changes in Organizational Work*, London: Chapman & Hall, 1996, pp. 325-343.

Monteiro, E., and Sahay, S. "On the Life-Blood of Actants," paper presented at the Latour's Seminar, Tromsø, Norway, 2000 (available onine at http://www.idi.ntnu.no/~ericm/life.blood.htm).

Morrow, R. D., and Brown, D. D. *Critical Theory and Methodology*, London: Sage Publications, 1994.

Myers, M. D., and Young, L. W. "Hidden Agendas, Power and Managerial Assumptions in Information Systems Development: An Ethnographic Study," *Information Technology and People* (10:3) 1997, pp. 224-240.

Ngwenyama, O. K. "The Critical Social Theory Approach to Information Systems: Problems and Challenges," in H.-E. Nissen, H. K. Klein, and R. Hirschheim (Eds.), *Information Systems Research: Contemporary Approaches and Emergent Traditions,* Amsterdam: North-Holland, 1991, pp. 267-280.

Oliver, D., and Romm, C. "Justifying Enterprise Resource Planning Adoption," *Journal of Information Technology* (17:4), 2002, pp. 199-213.

Olssen, M. "Michel Foucault's Historical Materialism," in M. Peters, W. Hope, J. Marshall, and S. Webster (Eds.), *Critical Theory, Poststructuralism and the Social Context,* Palmerston North, New Zealand: Dunmore Press, 1996, pp. 82-105.

Orlikowski, W. J. "The Duality of Technology: Rethinking the Concept of Technology in Organizations," *Organization Science* (3:3), 1992, pp. 398-427.

Orlikowski, W. J., and Baroudi, J. J. "Studying Information Technology in Organizations: Research Approaches and Assumptions," *Information Systems Research* (2:1), 1991, pp. 1-28.

Orlikowski, W. J., and Iacono, C. S. "Desperately Seeking the 'IT' in IT Research—A Call to Theorizing the IT Artifact," *Information Systems Research* (12:2), 2001, pp. 121-134.

Rabinow, P. "Introduction," in P. Rabinow (Ed.), *The Foucault Reader*, London: Penguin Books, 1984, pp. 3-29.
Radder, H. *In and About the World: Philosophical Studies of Science and Technology*, Albany, NY: State University of New York Press, 1996.
Saravanamuthu, K., and Wood-Harper, A. T. "Developing Emancipatory Information Systems," paper presented at (Re-)Defining Critical Research in Information Systems: An International Workshop, The University of Salford, 2001, pp. 91-109.
Stalder, F. "Beyond Constructivism: Towards a Realist Realism. A Review of Bruno Latour's Pandora's Hope," *The Information Society* (16), 2000, pp. 245-247.
Walsham, G. "The Emergence of Interpretivism in IS Research," *Information Systems Research* (6), 1995, pp. 376-394.
Walsham, G. "Actor-Network Theory and IS Research: Current Status and Future Prospects," in A. S. Lee, J. Liebenau, and J. I. DeGross (Eds.), *Information Systems and Qualitative Research*, London: Chapman & Hall, 1997, pp. 466-480.
Waring, T. S. "The Challenge of Emancipation in Information Systems Implementation: A Case Study in an NHS Trust Hospital," paper presented at the Critical Management Studies Conference, Manchester, 1999 (available online at http://www.mngt.waikato.ac.nz/ejrot/cmsconference/documents/Information%20Knowledge/Waring_Manchester.pdf).
Whitley, E. A. "Habermas and the Non-Humans: Towards a Critical Theory for the New Collective," paper presented at the Critical Management Studies Conference, Manchester, 1999a (available online at http://www.mngt.waikato.ac.nz/ejrot/cmsconference/documents/Information%20Tech/Habermas%20and%20the%20non-humans.pdf).
Whitley, E. A. "Understanding Participation in Entrepreneurial Organizations: Some Hermeneutic Readings," *Journal of Information Technology* (14:2), 1999b, pp. 193-202.
Williams, R., and Edge, D. "The Social Shaping of Technology," *Research Policy* (25), 1996, pp. 865-899.
Wilson, F. A. "The Truth Is Out There: The Search for Emancipatory Principles in Information Systems Design," *Information Technology and People* (10:3), 1997, pp. 187-204.
Winner, L. *Autonomous Technology: Technics-Out-of-Control as a Theme in Political Thought*, Cambridge, MA: MIT Press, 1977.
Winner, L. "Upon Opening the Black Box of Technology and Finding it Empty: Social Constructivism and the Philosophy of Technology," *Science, Technology and Human Values* (18:3), 1993, pp. 362-378.

ABOUT THE AUTHOR

Ela Klecuń is a lecturer in information systems at the London School of Economics and Political Science (LSE). She holds a Ph.D. in information systems from the LSE. Her research interests include health information systems, evaluation of information systems, and the application of critical theory and actor-network theory in the field of information systems. Ela can be reached by e-mail at e.klecun@lse.ac.uk or through her home page at http://is.lse.ac.uk/staff/klecun.

16 CONDUCTING AND EVALUATING CRITICAL INTERPRETIVE RESEARCH: Examining Criteria as a Key Component in Building a Research Tradition

Marlei Pozzebon
HEC Montreal

Abstract The collection, analysis, and interpretation of empirical materials are always conducted within some broader understanding of what constitutes legitimate inquiry and valid knowledge. In the Information Systems field, there are well-known and widely accepted methodological principles consistent with the conventions of positivism. However, the same is not yet true of interpretive research. The emergence of interpretivism in IS research was advocated by Walsham (1995) and corroborated by a series of special issues in outstanding IS journals. An example of the effort to advance the legitimacy of studies grounded in an interpretive position is the set of principles suggested by Klein and Myers (1999), which applies mostly to hermeneutics. However, because not all interpretive studies are built on a hermeneutical philosophical base, they recommended that other researchers, representing other forms of interpretivism, suggest additional principles. This paper follows in this vein, advocating the timely emergence of a critical interpretive perspective in IS research and pressing the argument that an extended version of Golden-Biddle and Locke's (1993) criteria is not only appropriate but comprehensive as initial guidelines for conducting and evaluating critical interpretive research.

Keywords: Critical interpretive research, research criteria, intensive research, qualitative research

1 INTRODUCTION

The motivation, or perhaps I should say the need, for writing this work, grew out of the moment in July 2002 when my thesis proposal defense ended. After presenting to my committee a nonorthodox perspective regarding the prevailing view in North

American universities (I hold a critical interpretive perspective which combines structuration theory and critical discourse analysis), I found myself with the obligation to better justify the validity of my work when finally defending the thesis. As a result, during the months following the thesis proposal defense, I started to compile interpretive and critical literature, looking for criteria for judging the quality of this type of research. The purpose of this text is to participate in the dialogue about our 20-year perspective on Information Systems research by presenting the provisional ideas I have developed during this time and discussing how to evaluate research carried out from a perspective that I believe is still emergent: critical interpretive.

There are several reasons why the IS field would benefit from an updated review and discussion of the existing criteria for evaluating qualitative research. The use of qualitative methods in IS research is growing rapidly. "As the focus of IS research shifts from technological to managerial and organizational issues, qualitative research methods become increasingly useful," Michael Myers (1997, p. 241) argued when announcing the creation of a special section within *MISQ Discovery's* World Wide Web archive to support qualitative research. Such increased interest in qualitative research methods is triggering the need for discussions on the criteria for evaluating qualitative research, qualitative not being unambiguously understood since qualitative does not necessarily mean intensive, or interpretive. Behind the term qualitative, a variety of philosophical assumptions and research methods coexist.

Despite the variety of approaches, most of the existing guidelines regarding the evaluation of IS qualitative research up to the 1990s are inspired by underlying philosophical assumptions espoused by a positivistic view (Lee 1989; Yin 1994). Markus and Lee (1999) focus our attention on the danger, still present, of judging interpretive research using positivist criteria, and vice versa. Recent initiatives have emerged suggesting a set of principles for the conduct and evaluation of qualitative research from an *interpretive standpoint* (Klein and Myers 1999; Schultze 2000). I could not find explicit guidelines for evaluating IS *critical research*. This was corroborated by Klein's assertion that "if one asks which methods can be taught to aspiring critical researchers, one draws almost a blank card. There appears to be no research methods literature on critical research" (1999, p. 21).

In 2004, the WG 8.2 community is celebrating 20 years of efforts toward making an impact on IS research using tools and methods that go beyond those cultivated by mainstream IS research. In keeping with my research into critical and interpretive views, I decided to conduct a review and compile a set of principles for IS researchers that, in addition to taking an interpretive view, seek to develop a critical appreciation of the way in which information technology is involved with organizational activity (Doolin 1998). Briefly, this paper has two goals: first, to reiterate the value to social investigation of a critical interpretive perspective in which social phenomena involving IS or IT are included; second, to review and extend Golden-Biddle and Locke's (1993) criteria, presenting the results as a step forward in drawing up a set of principles for guiding and evaluating critical interpretive research. By criteria, I do not mean a set of fixed standards. Any notion of criteria should be seen as enabling conditions that should only be applied contextually. The terrain upon which judgments are made is continually shifting, and should be characterized by openness, rather than stability and closure (Garrat and Hodkinson 1998).

2 WHY CRITICAL INTERPRETIVE?

My point of departure was the ISWorld.Net special section, "Qualitative Research in Information Systems," edited by Michael Myers, which aims to provide qualitative researchers in IS with useful information on the conduct, evaluation, and publication of qualitative research. This site, and the collection of references it offers, is of great value to researchers seeking to follow interpretive and/or critical work, helping them to legitimate their choices in the eyes of the mainstream IS community. Myers starts by recalling that, just as different people have different beliefs and values, there are different ways of understanding what research is. All research is based on some underlying assumptions about what constitutes *valid* research and which research methods are appropriate (Myers 1997). These beliefs and values in research have been called *paradigms of inquiry* (Denzin and Lincoln 1994), *theoretical traditions* (Patton 1990) or, simply, *orientations* (Tesch 1990). For instance, in the IS field, research has been classified according to three well-known orientations: positivist, interpretive, and critical (Orlikowski and Baroudi 1991). Although the paradigm debate is starting to provoke a sense of fatigue in many, or simply is "not a very interesting way of thinking about research program differences" (Deetz 1996, p. 194),[1] classifications according to distinct philosophical assumptions remain useful in helping researchers *position themselves clearly* and *argue for the value of their work*. Different designations have emerged, such as post-positivism and post-modernism, showing that the struggle among research groups for identity protection and legitimacy has changed its labels, but not its nature.

Recent theoretical discussions within the IS field have reinforced the benefit of *combining different perspectives*, especially the interpretive and the critical. For instance, Klein (1999) put forward the "full development of all the potential relationships between interpretivism and critical theory as one of the most fruitful avenues for future research" (p. 22). Similarly, Doolin (1998) points toward a critical interpretive perspective, arguing that interpretive researchers need to consciously adopt a critical and reflective stance in relation to the role that IT plays in maintaining social orders and social relations in organizations. Walsham (1993) advanced a similar position in his leading book about interpretivism in IS research. The research he describes has elements of both the interpretive and critical traditions and, thus, does not fit neatly into either of these categories. Indeed, he argues that constitutive process theories, such as those he espouses, are "an attempt to dissolve the boundaries between such traditions, in emphasizing not only the importance of subjective meaning for the individual actor, but also the social structures which condition and enable such meanings and are constituted by them" (p. 246).

Viewed separately, interpretivism and critical theory are far from being homogenous schools of thought. Klein and Myers (2001) recognize at least two different lines of philosophical thinking underlying the interpretive stream of thinking, drawing our attention to the fact that, even within interpretivism, not all studies should be evaluated according to the same criteria. Regarding critical theory, its foundations are often

---

[1]Deetz refers to the subjective and objective debate.

associated with two distinct schools of critical theory: the Frankfurt School of Horkheimer, Adorno, Marcuse and Fromm; and the contemporary critical theory of Habermas. Although these two approaches differ, the differences are seen as subtle (Steffy and Grimes 1986).

Even though calling for a union of critical research and interpretivism, Klein is "very skeptical if current attempts to integrate the two are founded on a clear understanding of their intrinsic connections" (p. 22). He argues that critical theory is much more theory-oriented than interpretivism, and that critical theory carries a strong legacy of Habermas' critical social theory. Yet he acknowledges a theoretical link between critical and interpretive research throughout hermeneutics: critical research emphasizes communicative orientation, which implies interest in human understanding, which, in turn, implies hermeneutics, which is the heart of interpretivism.

Klein's assertions are not incontestable, especially his claim that without an explicit reconstruction of the conceptual foundation, the union of interpretivism and critical research is merely "a matter of convenience, if not desperation" (p. 22). For instance, Doolin argues that to adopt a critical view does not necessarily mean to rely deeply on the critical theory of Habermas and of the Frankfurt School. Being critical may simply imply probing taken-for-granted assumptions inherent in the status quo by being critically reflective, while utilizing whatever theoretical framework is chosen. In combining structuration theory with critical discourse analysis in my own research, I learned from one of the leading figures in critical discourse analysis, Fairclough, that the term critical theory can be used in a "generic sense for any theory concerned with critique of ideology and the effects of domination, and not specifically for the critical theory of the Frankfurt School" (1995, p. 20). We can use the term *critical* without linking it to Habermas or the Frankfurt School.

I believe that to be critically interpretive does not require proper theoretical justifications because both approaches might just be seen as *intrinsically related.* Interpretive or constructivist approaches aim to produce fine-grained explorations of the way in which a particular social reality has been constructed. Critical approaches aim to focus more explicitly on the dynamics of power, knowledge, and ideology that surround social practices. Far from being incompatible, the boundary between interpretive and critical can be seen as a matter of degree: *many constructivist studies are sensitive to power, while critical studies include a concern for the processes of social construction that underlie the phenomena of interest* (Phillips and Hardy 2002). I conclude that IS research may be interpretive and critical without any inherent inconsistency. A number of IS researchers would suggest that it is often hard to avoid being critical when conducting interpretive research (Walsham 1993). Being critically interpretive about IT means that, in addition to understanding the context and process of IS from different interpretations arising from social interactions, researchers will avoid unreflective accounts by connecting these interpretations to broader considerations of social power and control (Doolin 1998).

The connection between interpretation and critical interpretation is nicely illustrated by Alvesson and Skoldberg's (2000) understanding of the different levels of reflection during empirical work (see Table 1). Empirical research starts from the data-constructing level (first level), where researchers make observations, talk to people and create their own pictures of the empirical phenomena. Preliminary interpretations are

Table 1. Linking Interpretation and Critical Interpretation

Aspect/Level	Focus
1. Interaction with empirical material	1. Accounts in interviews, observations of situations and other empirical materials
2. Interpretation	2. Underlying meanings
3. Critical interpretation	3. Ideology, power, social reproduction
4. Reflection on text production and language use	4. Own text, claims to authority, selectivity of the voices represented in the text

developed, the degree of which is often relatively low or somewhat unclear to the researchers themselves. This material is then subjected to further interpretation of a more systematic kind (second level), guided by ideas that can be related to theoretical frameworks or to other frames of reference. Ideally, researchers would allow the empirical material to inspire, develop, and reshape theoretical ideas. In fact, it is often the case that theoretical views allow the consideration of different meanings in empirical material. "The researcher's repertoire of interpretations limits the possibilities of making certain interpretations" (Alvesson and Skoldberg 2000, p. 250). The interpretation level that follows the interaction with empirical material is a step toward critical interpretation. Critical thinking (third level) and reflexivity (fourth level) stem from interpretive reflection.

3 DO RESEARCH METHODS IN INTERPRETIVE AND CRITICAL RESEARCH DIFFER?

Just as there are various philosophical perspectives that can inform qualitative research, so too are there various qualitative research methods. As a matter of fact, each research method represents a strategy of inquiry that moves from underlying philosophical assumptions to research design and empirical material interaction (Myers 1997). Viewed broadly, method is a mode and a framework for engaging with empirical material; method connects theoretical frameworks with the production and productive use of empirical material; method is a reflexive activity where theoretical, political, and ethical issues are central (Alvesson and Deetz 2000). Of course the choice of research methods influences the way in which the researcher collects data. Different research methods imply different skills, assumptions, and research practices. The problem related to the choice of a research method is not so much one that takes into account how many methods we employ or if those are of a quantitative or qualitative nature, but rather one that concerns the attempt to *achieve coherence over the whole process* (Schultze 2000).

Given their concern with understanding actors' meanings, **interpretive** researchers have often preferred *meaning-oriented methods,* which differ from positivist researchers' preference for *measurement-oriented methods*. In particular, from an interpretive perspective, data collection and representation have been accomplished through interviewing (Spradley 1979), ethnography (Van Maanen 1988), participant observation (Myers 1999), and case study (Walsham 1993). Walsham (1993) puts forward a view that the most appropriate method for conducting IS empirical research in the interpretive tradition is the *in-depth case study.*

Regarding **critical** research, the methodological debate is quite unclear. Myers (1999) nominates *research action* as one of critical researchers' preferred methodological approaches. Klein (1999) not only argues that there appears to be no research methods literature on critical research, but also that this lack of a recognized stock of critical methods provides the primary motivation for critical researchers to borrow interpretive approaches to data collection. Critical researchers often borrow methods like field research, historical analysis, and textual analysis from interpretive research, but utilize them in a context where theoretical ideas are used to encourage political action (Gephart 1999). The distinctions between critical research and interpretivism most clearly are not methodological in nature—both look for *meaning-oriented methods*—but are related to the recurrent commitment, or lack thereof, to critique of ideology, domination, and status quo.

In my experience conducting doctoral research from a critical interpretive perspective (Pozzebon 2003), I found in critical discourse analysis (CDA) a powerful *methodology* and *perspective* for studying social phenomena that involves ways of thinking about discourse (conceptual elements) and ways of treating discourse as data (methodological elements) which is quite distinct from most qualitative approaches (Hardy 2001; Wood and Kroger 2000). CDA has a long history in sociolinguistics (Titscher et al. 2000), is beginning to attract interest in organization studies (Grant et al. 2001), and can be seen as emergent in the IS field as well (Alvarez 2001, 2002; Heracleous and Barret 2001). CDA proved to be an example of a *compromise between my interpretive and critical claims*. On one hand, CDA reflects the constructivist epistemology underlying my research project. In order to explore the discursive production of aspects of social reality, discourse analysis is fundamentally interpretive (Phillips and Hardy 2002). On the other hand, because its techniques uncover multiple meanings and representations, and highlight multiple voices and perspectives, CDA becomes very helpful in connecting the discourses of different actors to broader considerations of their social context.

4 CRITERIA FOR CRITICAL INTERPRETIVE IS RESEARCH

The historic use of positivistic criteria for evaluating qualitative research reflects the dominance of quantitative research logic in certain social science disciplines. By the 1980s, Lincoln and Guba (1985) proposed four criteria that can be thought of as the development of slightly modified positivist criteria,[2] more aligned with the worldview of qualitative research. It is noteworthy that several interpretivists argue that such post-positivist criteria are essentially neo- positivist in nature, a sort of "realism reclothed" (Garratt and Hodkinson 1998).

---

[2]The four criteria—credibility, transferability, dependability, and confirmability—can be seen as equivalent to internal validity, external validity, reliability, and objectivity. Respecting these four criteria would guarantee the trustworthiness of findings from studies using qualitative methods.

In the IS field, the publication of Klein and Myers' paper can be seen as a response to the call to "discuss explicitly the criteria for judging qualitative, case and interpretive research in information systems" (Klein and Myers 1999, p. 68). They propose a set of principles primarily derived from anthropology, phenomenology, and hermeneutics, acknowledging that other forms of interpretivism also exist. The authors discuss the suitability of such a set of principles, arguing that some authors may feel that, in proposing them "for conducting and evaluating interpretive studies, we are going too far because we are violating the emergent nature of interpretive research, while others may think just the opposite" (p. 68). Their concluding guess is that it is better to have some principles than to have none at all. Complementarily, Garrat and Hodkinson (1998) claim that, although no prespecified criteria can ensure universally valid judgments about any type of research, writing about the ways in which our research can be judged helps "refine and develop our thinking about what doing and judging research entails" (p. 535). In addition, any notion of criteria should be applied contextually and placed continually at risk!

The principles Klein and Myers set forth are not to be mechanistically applied but are open to lively debate about interpretive research standards. Several IS researchers have relied on some of Klein and Myers' principles to validate their qualitative research. Davidson (2002), Gallivan (2001), Hanseth et al. (2001), Henfridson and Holmstrom (2002), and Trauth and Jessup (2000) are some examples. Because their set of principles applies mostly to hermeneutics and not all interpretive studies follow a *hermeneutical* philosophical base, Klein and Myers recommend that other IS authors, representing other forms of interpretivism, suggest additional principles. For instance, Gopal and Prasad (2000) propose a set of criteria particularly adapted for evaluating *symbolic interactionist* work, arguing that this research differs from other social constructionist genres, notably hermeneutics and ethnography. A number of IS interpretive researchers—like Davidson (2002), Schultze (2000), Trauth and Jessup (2000), and Walsham and Sahay (1999)—have used Golden-Biddle and Locke's three criteria for *ethnographic writing* (which does not exclude hermeneutics), as the basis for evaluating their research.

Convincing has been presented as paramount for interpretive researchers relying on ethnography. Van Maanen (1979) outlines the rhetorical effort characterizing the communication between researchers and their audience: "in large measure, our task is rhetorical, for we attempt to convince others that we've discovered something of note, made unusual sense of something, or in weak form, simply described something accurately" (p. 540). Similarly, Silverman (1997) asks, "have the researchers demonstrated successfully why we should believe them?" (p.25). In this vein, Golden-Biddle and Locke consider writing research texts to be about *convincing and persuading audiences* and about *building authorial authority*. Trying to answer such a central question from qualitative researchers—"How does ethnographic work convince?"—and positioning the *convincingness* of ethnographic texts as central, they propose three evaluation criteria: *authenticity, plausibility,* and *criticality*. Table 2 shows these three criteria in the first row. The second row presents two additional criteria, recently proposed by Schultze (2000), for evaluating research that, in addition to relying on ethnography, provides a confessional, self-reflexive, and self-revealing account of the researcher's experience.

Table 2. Interpretive Criteria for Evaluating Ethnography and Reflexive Research

Interpretive criteria (for ethnography) (Golden-Biddle and Locke 1993)	• **Authenticity:** Was the researcher there? • **Plausibility:** Does the history make sense? • **Criticality:** Does the text activate readers to re-examine assumptions that underlie their work?	
Interpretive criteria (for confessional research) (Schultze 2000)	• **Self-revealing writing:** Does the text reveal personal details about the ethnographer? • **Interlacing actual and confessional content:** Is autobiographical material interlaced with actual ethnographic material?	**Reflexivity:** Does the author reveal his/her personal role and his/her selection of the voices/actors represented in the text? (Alvesson and Skoldberg 2000)

Table 3. Assembling Criteria for Critical Interpretive Research

Criteria	**Aspects of Interpretation (Based on Alvesson and Skoldberg 2000)**
Authenticity	Interaction with empirical material
Plausibility	Sound interpretation
Criticality	Critical interpretation
Reflexivity	Reflection on text production and language use

Both Golden-Biddle and Locke's and Schultze's criteria are based on *ethnography*. Walsham has put forward *in-depth case study* as the methodological vehicle *par excellence* to carry out IS interpretive research, as he deems it appropriate "for the view of the nature of knowledge embedded in a broadly interpretive philosophy, which emphasizes the need of detailed understanding of human meanings in context" (1993, p. 247). Walsham also argued that the approach to field research for the case studies largely derives from the ethnographic research tradition, which leads me to conclude that we can differentiate in-depth case study and ethnography as a *matter of degree*. This opens the possibility of adopting Golden-Biddle and Locke's criteria for evaluating in-depth case study and other forms of intensive research. Given the lack of other studies (especially IS studies) suggesting criteria for evaluating interpretive research that it is not necessarily of hermeneutic orientation, I propose a new version of Golden-Biddle and Locke's criteria, reviewed and extended, as the basis for evaluating the quality of intensive IS research studies, especially critical interpretive research (see Table 3).

According to Golden-Biddle and Locke, the first two criteria, *authenticity* and *plausibility*, are seen as essential. The addition of *criticality* characterizes the work of a researcher who, in addition to being interpretive, is also critical (I posit criticality also as *essential*, not an optional criterion as proposed by Golden-Biddle and Locke). I also propose *reflexivity* (this time optional) as an important aspect in intensive research.

Reflexivity was inspired by Schultze's confessional research and Alvesson and Skoldberg's (2000) critical view, but also characterizes several variants of post-structural and post-modern studies.

4.1 Expressing Authenticity

Authenticity means being genuine to the field experience as a result of "being there" (Golden-Biddle and Locke 1993). Meeting this criterion assures that the researcher was there, and was genuine to the experience in writing up the account.[3] This is a moment to discuss more purposively the difference that exists between an *ethnographic work* and an *in-depth case study*. As Myers (1999) recognizes, one of the distinguishing features of ethnographic research is participant observation: "The researcher needs to be there and live in the organization for a reasonable length of time" (p. 12). In turn, researchers doing case studies strongly rely on in-depth interviews and analysis of archival documents. On-site observation, participant or not, may or may not occur, and when it does occur, its intensity often varies from low to medium, but rarely is very high (otherwise we would be inclined to talk about ethnography and not about in-depth case study). As a result, although researchers conducting in-depth case studies were there to some degree, and might even have gained a certain familiarity with the setting, the being there is not the same as the immersion that characterizes a traditional ethnographic study. Many facts the researcher will report were not directly observed but gathered during interviews and conversations with the social group under study. The closeness to the actions and events of interpretive studies is likely to be higher when the researcher works as a participant observer or action researcher, and lower when the researcher works as an outside observer and interviewer. Researchers will thus report evidence based on *their interpretations of other participants' interpretations of the phenomenon investigated* (Walsham 1995).[4]

Aware of the differences in degree of being there, I have reflected on the appropriateness of **authenticity** as a criterion for evaluating in-depth case studies as it seems, indeed, more appropriate for ethnographic work. However, in the absence of another term, I propose to provisionally retain **authenticity** for evaluating in-depth case studies, but with *nuances that respect the nature of this kind of interaction with the field* (see Table 4). For instance, instead of proving that we were there, we must prove that we had enough interaction with participants and enough access to archival documents to compensate for the lack of direct immersion during the development of the phenomena under investigation. Consequently, when researchers doing case studies

---

[3]Schultze offers an interesting comparison between Golden-Biddle and Locke's authenticity and well-known positivist criteria (reliability and validity), reminding me of the risk of constantly recreating a sort of "realism reclothed" instead of reaffirming research values of a quite different ontology (nominalist).

[4]Nandhakumar and Jones (1997) offer a provocative discussion of how a researcher's ability to obtain an understanding of actors' interpretations may be limited in a number of ways. They also put forward alternatives of how the limitations of reporting interpretations of interpretations can be overcome.

Table 4. Ways and Examples of Expressing Authenticity

Criteria	Ways to ...	Examples from IS Literature
(1) Has the author been there (in the field) *or had enough interactions with participants to compensate for the lack of direct immersion?*	(1a) Particularizing everyday life from researchers' direct immersion or *from the interaction with participants and archival documents* (Golden-Biddle and Locke 1993); demonstrating familiarity with the vernacular of the field, describing what members think about their lives in the field, etc. (Schultze 2000).	Walsham and Sahay (1999, pp. 59-60) and Schultze (2000, pp. 59-60) provide rich descriptions, with many quotes, of their presence in everyday life.
	(1b) Delineating the relationship in the field (Golden-Biddle and Locke 1993); describing how close the researchers were, whom they talked to and observed (Schultze 2000).	In addition to describing the length of their stay and the context of their fieldwork, Walsham and Sahay (1999) add further material on their role and attitudes (p. 60).
(2) Has the author been genuine to the field experience?	(2a) Depicting the disciplined pursuit and analysis of data (Golden-Biddle and Locke 1993); presenting raw data such as fieldnotes, documents, and transcribed interviews, conducting *post hoc* respondent validation (Schultze 2000).	Trauth and Jessup (2000) share the process of developing their interpretations openly with readers, rather than simply presenting it to them as a finished product (p. 69).

particularize everyday life, they are trying to provide sufficient detail, not from deep immersion in the field, but from their interaction with actors deeply immersed in the field. They are telling the reader that they are not reporting facts but their interpretations of other people's interpretations. Table 4 summarizes ways of expressing authenticity in intensive research (ethnography and/or in-depth field study) and offers examples of how IS researchers have dealt with authenticity.

4.2 Constructing Plausibility

Plausibility is defined as the ability of the text to connect to the reader's worldview (Walsham and Sahay 1999) and it addresses the rhetorical strategies used to compose a text that positions the work as relevant to the concerns of the intended audience (Schultze 2000). Whereas authenticity is concerned with the conduct of field work, plausibility addresses the write-up phase (Schultze 2000). In order to establish plausibility, researchers should be concerned with two interconnected components. First, they need to make sense, which means to deal with common concerns, establishing connections to the personal and disciplinary backgrounds and experiences of their readers. For instance, the researcher will structure the text in a way that is consistent with the academic article genre, i.e., with specified headings and the use of citations (Schultze 2000). Second, they need to offer a distinctive research contribution to a disciplinary area (Golden-Biddle and Locke 1993). A plausible intensive study will identify gaps in the literature or delineate a novel theoretical perspective to justify the research and differentiate its contribution (Schultze 2000).

This latter aspect—convincing that there is a contribution to the field—is one of the most important aspects to be considered. The value of any empirical research depends on the extent to which the author tells us something new and relevant. However, from a critical perspective, we would ask, *new and relevant for whom?* What is new for one person might not be new for another. More polemically, what is relevant strongly depends on everyone's assumptions, purposes, and expectations (Benbasat and Zmud 1999; Lyytinen 1999). Myers (1999) reminds us that it is essential for researchers to convince the reviewers and editors who serve on the editorial boards of our journals that their research contribution is new and relevant.

Plausibility also recalls the dilemma of generalization. According to Klein (1999), the ultimate goal of IS research is to produce some form of knowledge that has relevance *outside the context of the original research setting*. When the researcher assumes a positivist stance, the status of such knowledge is likely to be law-like generation. In assuming an interpretive stance, the researcher appears more conservative and talks about tendencies (Walsham 1995). The validity of drawing inferences from one or more individual cases depends not on the representativeness of such cases in a statistical sense, but on the plausibility and cogency of the logical reasoning used in describing results from the case, and in formulating inferences and conclusions from those results (Walsham and Waema 1994). Table 5 summarizes ways of constructing plausibility and offers examples of how IS researchers have constructed it.

4.3 Raising Criticality

Criticality refers to the ability of the text to entice readers to reconsider taken-for-granted ideas and beliefs (Golden-Biddle and Locke 1993). It entails the ability to propose an understanding of ourselves and others in a new and better way, including novel ways of thinking. Criticality can be achieved by challenging readers to pause and think about a specific situation, by provoking them to answer questions, and by guiding readers through novel ways of thinking (Schultze 2000). Although criticality was proposed by Golden-Biddle and Locke as a somehow optional criterion, I propose it as *essential* to critical interpretive research.

"Good research, from a critical perspective, is one that enables a qualitatively new understanding of relevant fragments of social reality, furnishing new alternatives to social action" (Alvesson and Skoldberg 2000). Critical interpretive studies should necessarily activate such a criterion in order to be able to outline and question prevailing views, to contradict conventional wisdom and multiple viewpoints, which are often in conflict. More attention should be paid not only to multiple narratives that give voice to and allow the construction of multiple worlds, but also to the role of the researcher, of his or her understanding, insights, experiences, and interpretations (link with reflexivity). Multiple narratives will not give us any single representation but they may offer us more interesting ways to think about the organization (Garcia and Quek 1997). Table 6 summarizes ways of raising criticality and offers examples of how IS researchers have been triggering it.

Table 5. Ways and Examples of Constructing Plausibility

Criteria	Ways to ...	Examples from IS Literature
(3) Does the history make sense to me?	(3a) Adhering to academic article genres, using conventional sections like method, results, discussion, and references.	Walsham and Sahay (1999) and Schultze (2000) organize their papers with sections like introduction, research methods, setting description, results, discussion, and conclusion.
	(3b) Drafting the reader (Golden-Biddle and Locke 1993); using *we* to include the authors and the reader (Walsham and Sahay 1999).	Walsham and Sahay (1999) use the *we* in several situations.
	(3c) Legitimating the atypical (Golden-Biddle and Locke 1993); showing the scope of the application of the findings (Walsham and Sahay 1999); aligning the findings with common, everyday experiences (Schultze 2000).	Walsham and Sahay (1999) show that their ideas of actor-network theory could be applied to other technologies (not only GIS) and other contexts (not only their Indian case) (p. 61).
	(3d) Justifying contestable assertions (Walsham and Sahay 1999).	Walsham and Sahay (1999) describe a rich picture and add quotes from participants in order to support contestable assertions (p. 61).
(4) Does it offer something distinctive?	(4a) Differentiating findings—a singular contribution (Golden-Biddle and Locke 1993); showing missing areas in the past (Walsham and Sahay 1999), providing the development of a novel theoretical approach (Schultze 2000).	Schultze (2000) highlights shortcomings in previous literature and her contributions with respect to substantive insights (page 33).
	(4b) Building dramatic anticipation (Golden-Biddle and Locke 1993); creating expectation.	Walsham and Sahay (1999) add a "little spice to their writing," as described in page 61.

Table 6. Ways and Examples of Raising Criticality

Criteria	Ways to ...	Examples from IS Literature
(5) Does the text motivate the readers to re-examine assumptions underlying their own work?	(5a) Carving out room to reflect (Golden-Biddle and Locke 1993); including spots in the text where the reader stops and reflect about a specific situation (Schultze 2000; Walsham and Sahay 1999).	Walsham and Sahay (1999) and Schultze (2000) do not use explicit stop signs, but both provide implicit illustration of this strategy.
	(5b) Stimulating the recognition and examination of differences (Golden-Biddle and Locke 1993); actively provoking the reader to answer questions (Walsham and Sahay 1999).	Walsham and Sahay (1999) invite readers to critically examine their own views and approaches (p. 62).
	(5c) Imagining new possibilities (Golden-Biddle and Locke 1993); using metaphors, stimulating criticality in the reader (Walsham and Sahay 1999).	Schultze (2000) challenges readers to answer questions about their own assumptions, subjectivity and objectivity (p. 33).

4.4 Experimenting with Reflexivity

Reflexivity implies reflection on text production and language use and reveals a kind of awareness of the ambiguity of language (Alvesson and Skoldberg 2000). Recalling Table 1, the level of interpretation (interpretation, critical interpretation, reflexive interpretation) each work of empirical research achieves depends, essentially, on each researcher's assumptions and purposes. As outlined by Hardy et al. (2001), work on reflexivity is well developed in areas like sociology of science but has attracted less attention in organization and management theory. I think the same could be said about IS research. Excepting the recent work of Schultze, far less attention exists in our field. In their book dedicated to reflexive methodology, Alvesson and Skoldberg stress that much good qualitative research is unreflexive, often paying much more attention to tasks such as gathering and analyzing data than to different elements of reflexivity, both during the process of research and in the final textual product. Reflexivity was defined by Clegg and Hardy as "ways of seeing which act back on and reflect existing ways of seeing" (1996, p. 4). Reflexive research often includes researchers in the subject matter they are trying to understand. Hardy et al. (2001) complement this notion: "we cannot confine our attention to the relationship between researchers and the research subject, but must also examine the relationship between researchers and the research network of which they are part" (p. 533).

Table 7. Ways and Examples of Experimenting with Reflexivity

Criteria	Ways to ...	Examples from IS literature
(6) Does the author reveal his/her personal role and personal biases and assumptions?	(6a) Self-revealing writing (Golden-Biddle and Locke 1993); describing researcher's personal role (Alvesson And Skoldberg 2000); using personal pronouns, revealing personal details about the researcher (Schultze 2000).	Schultze (2000) uses "I" in abundance in the descriptions of her own informed practices as well as in the excerpts from the field notes. She also presents herself, giving information about age, gender, race, etc.
	(6b) Interlacing actual and confessional content (Golden-Biddle and Locke 1993; Schultze 2000).	Schultze (2000) avoids overemphasis on self-reflexive and autobiographical material by describing participants' practices after describing her own practices (p. 34).
	(6c) Qualifying personal biases (Golden-Biddle and Locke 1993); describing researcher's selection of the voices/actors represented in the text (Alvesson and Skoldberg 2000); disclosing details like mistakes made (Schultze 2000).	Schultze (2000) provides examples of mistakes she made with respect to contaminating the data (p. 34).

Schultze defined the reflexive dimension of her work with two elements: self-revealing writing and the interlacing of actual ethnographic material and confessional content. A self-revealing text demands a personalized author, the use of personal pronouns to consistently highlight the point of view being represented, and the construction of the researcher as a reasonable yet fallible individual with whom the audience can identify. Regarding the second feature, confessional writing interlaces the actual ethnographic content with the confessional material, meaning that any statement about the foreign culture is also a statement about the ethnographer' and the reader's culture.

To Holland (1999), reflexivity involves reflecting on the way research is carried out and understanding how the process of doing research shapes its outcomes. This calls into discussion the responsibility of researchers to declare their biases. Hardy et al. point out that, from an interpretive standpoint, this does not mean to remove such biases, but to render them visible though personal disclosure, so that readers can take them into account. In other words, any research is seen as one representation among many possible representations, and researchers present their representations for interpretation by the reader. Table 7 summarizes ways of experimenting with reflexivity and offers examples of how IS researchers have been undertaking it.

5 CONCLUDING REMARKS

By choosing emergent ways of making sense of IS phenomena, we assume some important risks and cope with many difficulties. Critical and interpretive studies are increasing in number and are starting to be regularly published at IS conferences, in journals, and in books, but they do not yet have the wide acceptance that positivist studies enjoy. As Walsham outlines, any theoretical choice is always "a way of seeing and a way of not-seeing" (1993, p. 6). The same applies to methodological choices: each one is a way of interacting with empirical material that is guided by the researcher's background, bias, and world-view, with pros and cons. As researchers, we will always deal with some degree of uncertainty about our choices and interpretations, which are not created, shared, or applied in a social vacuum, but are involved in communication, interpersonal relations, identity construction, and convincing others (and ourselves) that our propositions are sound (Alvesson and Skoldberg 2000).

In this paper, I put forward the argument that critical interpretive research is an emerging and valuable perspective on IS research. How can we, if we so choose, determine which set of criteria to adopt in conducting and evaluating critical interpretive work? The fact that critical interpretive research is essentially constructivist and emergent does not mean that looking for evaluative criteria and judging the quality are not appropriate. Writing about ways to develop and judge any type of intensive research helps refine and develop our thinking about what conducting and evaluating intensive research leads to and also serves as a device for sharing ideas with others about these things. Most important, to discuss a set of criteria for conducting and evaluating intensive research represents a key component in building a research tradition of which we are a part. "Established approaches to doing and judging research are our collective prejudices, neither to be slavishly accepted not willfully rejected, but which should be placed continuously at risk" (Garratt and Hodkinson 1998, p. 535).

In this paper, I have also tried to trace a picture of different sets of criteria that have emerged regarding the wide *interpretive* perspective. What complicates this exercise is that it is difficult to treat equally categories like *hermeneutical* philosophical base (Klein and Myers 1999), *symbolic interactionist* work (Gopal and Prasad 2000), *ethnographic* writing (Golden-Biddle and Locke 1995) or *confessional* research (Schultze 2000). For instance, while hermeneutic can be defined as a broad theoretical tradition, ethnography can be defined as a research strategy. The two are far from being mutually exclusive. Future research can refine the above debate by clarifying distinctions regarding criteria vis-à-vis ontological or epistemological assumptions, i.e., broad paradigms, perspectives, or traditions (e.g., interpretivism); criteria vis-à-vis theoretical traditions (e.g., hermeneutic or symbolic interactionism); and criteria vis-à-vis research strategies (e.g., case study or ethnography).[5]

Analyzing the nature of criteria in qualitative research, Garratt and Hodkinson (1998) develop a provocative argument: "criteria can only be located in the interaction between research findings and the critical reader of those findings" (p. 515). They assume that most writing about the ways in which research should be judged is concerned almost exclusively with the ways in which *research was done*, and does not take into account the ways in which the *standpoint of the reader will influence their judgment* of that research. As a result, all criteria for judging research quality contain within them a defining view of what research is, and any attempt to preselect the criteria against which a piece of research is to be judged is also "predetermining what the nature

---

[5] I thank Charo Rodriguez for her insightful comments on this topic.

of that piece of research should be" (p. 525). The authors are not saying that qualitative judgments in research cannot be made but, rather, insisting that the idea of deliberately choosing any list of universal criteria in advance of reading a research report is antithetical to the process of understanding the experience. All these thoughts about research, quality of research, criteria for research, and building a research tradition, albeit partly subjective, are drawn from the evolving wisdom within the research network of which we are a part.

ACKNOWLEDGEMENTS

The author would like to thank Charo Rodriguez, Alain Pinsonneault, and two anonymous reviewers for their helpful comments in the first drafts of this paper, and HEC Montreal for its research support.

REFERENCES

Alvarez, R. "Confessions of an Information Worker: A Critical Analysis of Information Requirements Discourse," *Information and Organization* (12), 2002, pp. 85-107.

Alvarez, R. "It Was a Great System. Face-Work and the Discursive Construction of Technology During Information Systems Development," *Information Technology & People* (14:4), 2001, pp. 385-405.

Alvesson, M., and Deetz, S. *Doing Critical Management Research*, London: Sage Publications, 2000.

Alvesson, M., and Skoldberg, K. *Reflexive Methodology: New Vistas for Qualitative Research*, London: Sage Publications, 2000.

Benbasat, I., and Zmud, R. W. "Empirical Research in Information Systems: The Practice of Relevance," *MIS Quarterly* (23:1), 1999, pp. 3-16.

Clegg, S., and Hardy, C. "Some Dare Call Into Power," in S. Clegg, C. Hardy, and W. Nord (Eds.), *Handbook of Organization Studies,* London: Sage Publications, 1996.

Davidson, E. J. "Technology Frames and Framing: A Socio-Cognitive Investigation of Requirements Determination," *MIS Quarterly* (26:4), 2002, pp. 329-358.

Denzin, N. K., and Lincoln, Y. S. *Handbook of Qualitative Research*, Newbury Park, CA: Sage Publications, 1994.

Deetz, S. "Describing Differences in Approaches to Organization Science: Rethinking Burrell and Morgan and Their Legacy," *Organization Science* (7:2), 1996, pp. 191-207.

Doolin, B. "Information Technology as Disciplinary Technology: Being Critical in Interpretive Research on Information Systems," *Journal of Information Technology* (13), 1998, pp. 301-311.

Fairclough, N. *Critical Discourse Analysis—The Critical Study of Language*, London: Longman, 1995.

Gallivan, M. J. "Organizational Adoption and Assimilation of Complex Technological Innovations: Development and Application of a New Framework," *The Data Base for Advances in Information Systems* (32:3), 2001, pp. 51-84.

Garcia, L., and Quek, F. "Qualitative Research in Information Systems: Time to Be Subjective?" in A. S. Lee, J. Liebenau, and J. I. DeGross (Eds.), *Information Systems and Qualitative Research*, London: Chapman & Hall, 1997, pp. 444-466.

Garratt, D., and Hodkinson, P. "Can There Be Criteria for Selecting Research Criteria? A Hermeneutical Analysis of an Inescapable Dilemma," *Qualitative Inquiry* (4:4), 1998, pp. 515-539.

Gephart, R. "Paradigms and Research Methods," *Research Methods Forum* (4), 1999, pp. 1-11.

Golden-Biddle, K., and Locke, K. "Appealing Work: An Investigation of How Ethnographic Texts Convince," *Organization Science* (4), 1993, pp. 595-616.

Gopal, A., and Prasad, P. "Understanding GDSS in Symbolic Context: Shifting the Focus from Technology to Interaction," *MIS Quarterly* (24:3), 2000, pp. 509-546.

Grant, D., Keenoy, T., and Oswick, C. "Organizational Discourse—Key Contributions and Challenges," *International Studies of Management and Organization* (31:3), 2001, pp. 5-24.

Hanseth, O., Ciborra, C. U., and Braa, K. "The Control Devolution: ERP and the Side Effects of Globalization," *The Data Base for Advances in Information Systems* (32:4), 2001, pp. 34-46.

Hardy. C. "Researching Organizational Discourse," *International Studies in Management and Organization* (31:3), 2001, pp. 25-47.

Hardy, C., Phillips, N., and Clegg, S. "Reflexivity in Organization and Management Theory: A Study of the Production of the Research 'Subject'," *Human Relations* (54:5), 2001, pp. 531-560.

Henfridson, O., and Holmstrom, H. "Developing E-Commerce in Internetworked Organizations: A Case of Customer Involvement Throughout the Computer Gaming Value Chain," *The Data Base for Advances in Information Systems* (33:4), 2002, pp. 38-50.

Heracleous, L., and Barrett, M. "Organizational Change as Discourse: Communicative Actions and Deep Structures in the Context of Information Technology Implementation," *Academy of Management Journal* (44:4), 2001, pp. 755-778.

Holland, R. "Reflexivity," *Human Relations* (52:4), 1999, pp. 463-485.

Klein, H. K. "Knowledge and Research in IS Research: from Beginnings to the Future," in O. Ngwenyama, L. Introna, M. D. Myers, and J. I. DeGross (Eds.), *New Information Technologies in Organizational Processes: Field Studies and Theoretical Reflections on the Future of Work*, Boston: Kluwer Academic Publishers, 1999, pp. 13-25.

Klein, H. K., and Myers, M. D. "A Classification Scheme for Interpretive Research in Information Systems," in E. M. Trauth (Ed.), *Qualitative Research in IS: Issues and Trends* Hershey, PA: Idea Group Publishing, 2001, pp. 218-239.

Klein, H. K., and Myers, M. D. "A Set of Principles for Conducting and Evaluating Interpretive Field Studies in Information Systems," *MIS Quarterly* (23:1), 1999, pp. 67-93.

Lee, A. "A Scientific Methodology for MIS Case Studies," *MIS Quarterly*, 1989, pp. 33-50.

Lincoln, Y. S., and Guba, E. G. *Naturalistic Inquiry*, New York: Sage Publications, 1985.

Lyytinen, K. "Empirical Research in Information Systems: On the Relevance of Practice in Thinking," *MIS Quarterly* (23:1), 1999, pp. 25-27.

Markus, M. L., and Lee, A. S. "Special Issue on Intensive Research in Information Systems: Using Qualitative, Interpretive, and Case Methods to Study Information Technology," *MIS Quarterly* (23:1), 1999, pp. 35-38.

Myers, M. D. "Investigating Information Systems with Ethnographic Research," *Communications of AIS* (2:23), 1999, pp. 2-19.

Myers, M. D. "Qualitative Research in Information Systems," *MIS Quarterly* (21:2), 1997, pp. 241-242.

Nandhakumar, J., and Jones, M. "Too Close for Comfort? Distance and Engagement in Interpretive Information Systems Research," *Information Systems Journal*, 7, 1997, pp. 109-31.

Orlikowski, W. J., and Baroudi, J. J. "Studying Information Technology in Organizations: Research Approaches and Assumptions," *Information Systems Research* (2:1), 1991, pp. 1-28.

Patton, M. Q. *Qualitative Evaluation and Research Methods*, Newbury Park, CA: Sage Publications, 1990.

Phillips, N., and Hardy, C. *Discourse Analysis—Investigating Processes of Social Construction*, London: Sage Publications, 2002.

Pozzebon, M. *The Implementation of Configurable Technologies: Negotiations between Global Principles and Local Contexts*, Unpublished Ph.D. Dissertation, McGill University, 2003.

Schultze, U. "A Confessional Account of an Ethnography about Knowledge Work," *MIS Quarterly* (24:1), 2000, pp. 3-41.

Silverman, D. "The Logic of Qualitative Research," in J. Hassard, and M. Parker (Eds.), *Context and Method in Qualitative Research*, London: Sage Publications, 1997.

Spradley, J. P. *The Ethnographic Interview*, New York: Holt Reinhart & Winston, 1979.

Steffy, B. D., and Grimes, A. J. "A Critical Theory of Organization," *Academy of Management Review* (11:2), 1986, pp. 322-336.

Tesch, R. *Qualitative Research: Analysis, Types and Software Tools*, London: Falmer, 1990.

Titscher, S., Meyer, M., Wodak, R., and Vetter, E. *Methods of Text and Discourse Analysis*, London: Sage Publications, 2000.

Trauth, E. M., and Jessup, L. M. "Understanding Computer-mediated Discussions: Positivist and Interpretive Analyses of Group Support System Use," *MIS Quarterly* (24:1), 2000, pp. 43-79.

Van Maanen, J. "The Fact of Fiction in Organizational Ethnography," *Administrative Science Quarterly* (24), 1979, pp. 539-550.

Van Maanen, J. *Tales of the Field*, Chicago: University of Chicago Press, 1988.

Walsham, G. *Interpreting Information Systems in Organizations*, Chichester, England: John Wiley and Sons, 1993.

Walsham, G. "Interpretive Case Studies in IS Research: Nature and Method," *European Journal of Information Systems* (4), 1995, pp. 74-81.

Walsham, G., and Sahay, S. "GIS for District-level Administration in India: Problems and Opportunities," *MIS Quarterly* (23:1), 1999, pp. 39-65.

Walsham, G., and Waema, T. "Information Systems Strategy and Implementation: A Case Study of a Building Society," *ACM Transactions on Information Systems* (12:2), 1994, pp. 150-173.

Wood, L. A., and Kroger, R. O. *Doing Discourse Analysis—Methods for Studying Action in Talk and Text*, London: Sage Publications, 2000.

Yin, R. K. *Case Study Research, Design and Methods*, Newbury Park, CA: Sage Publications, 1994.

ABOUT THE AUTHOR

Marlei Pozzebon is an assistant professor at HEC Montreal. She received her Ph.D. from McGill University in 2003. Her research interests are the political and socio-cultural aspects of information technology implementation, the use of structuration theory and critical discourse analysis in the information systems field, business intelligence and social responsibility, and the role of information technology in developing countries. Prior to joining HEC, Marlei worked at three Brazilian universities and she also held positions such as consultant and systems analyst for at least 13 years. She has published papers in *Journal of Management Studies* and presented papers at the Academy of Management Conference, the International Conference of Information Systems, and the European Group on Organization Studies, among others. Marlei Pozzebon can be reached at Marlei.pozzebon@hec.ca.

17 MAKING CONTRIBUTIONS FROM INTERPRETIVE CASE STUDIES: Examining Processes of Construction and Use

Michael Barrett
Geoff Walsham
Judge Institute of Management
University of Cambridge

Abstract In this paper, we examine how contributions are established in interpretive case studies. By focusing on the introductory sections of articles, previous research in the organizational literature has recently shed light on how theorists construct opportunities for making contributions. Our theoretical approach is broader in scope, examining both the construction and use of contributions in all sections of an article. We use this approach to explore how a well-ited IS interpretive case study makes contributions. With respect to constructing contributions, our findings confirm previous strategies suggested by earlier research, and uncover the use of plural strategies. Our analysis of the use of contributions is more unexpected. For example, key theoretical and conceptual contributions that were central to the case study were rarely drawn on and incorporated by later texts. Rather, multiple audiences took up the softer, and simpler, conceptual contributions. Furthermore, our findings categorize a number of different types of referencing used by later texts in incorporating and using contributions, namely mistaken referencing, ambiguous referencing, incorporating into a new term, and related work referencing. We conclude by providing some suggestions as to how IS researchers can make better contributions from interpretive case studies

Keywords: Contributions, construction use, interpretive case studies, research methods, evaluation

1 INTRODUCTION

1.1 What Is a Contribution?

Whether one is a doctoral student preparing his or her thesis or an experienced researcher responding to reviewers' comments, the question "what is a contribution?" continues to be a challenge in written work. A key debate in the IS literature has been on the relevance of our research (Benbasat and Zmud 1999; Davenport and Markus 1999). An equally important dimension of knowledge construction in establishing a contribution has been evident in recent IS research. For example, March and Smith (1995) provided an evaluative framework for both design and natural science research in IT research. More recently, Webster and Watson (2002) provide guidance on constructing and articulating a review paper's contribution, while Klein and Myers (1999) offer a set of principles for conducting and evaluating IS interpretive field studies. This latter focus on the quality of contributions in interpretive work (Klein and Myers 1999) is perhaps not unexpected given the emergence of interpretive research (Walsham 1995a) in IS and its significant development over the last decade. In this paper, we seek to build on the current literature by carrying out needed empirical work to understand both how interpretive case studies develop contributions in their texts (cf. Locke and Golden-Biddle 1997) and to examine the subsequent use of these contributions in later texts.

Research in the organizational literature on the importance of contribution in scientific work has tended to focus on novelty or uniqueness as a key aspect of what constitutes contribution (Astley 1985; Locke and Golden-Biddle 1997). Reviewers often ask what's new or innovative about a work (Weick 1995; Whetten 1989) and scholars are urged to make unique contributions to their discipline (Mone and McKinley 1993). Furthermore, there is growing evidence of the link between novelty and uniqueness, and publication in journals (Beyer et al. 1995).

Locke and Golden-Biddle's (1997) pathbreaking work shifts this focus on the importance of contribution to examining what contribution means in practice. Their approach views contribution as the socially constructed nature of scientific knowledge (Knorr-Cetina 1981; Latour 1987), which has two key premises: (1) knowledge cannot be known separately from the "knower" and (2) knowledge is a meaning-making activity "enacted" in particular communities (Orlikowski and Baroudi 1991). This perspective suggests that knowledge is not an objective entity independent of the knower in a world composed of facts. As elaborated in the next section, their empirical work is valuable in highlighting two key processes upon which texts draw in constructing opportunities for contribution. However, as they themselves note, their grounded theory study on the *how* of contribution is somewhat limited in examining contribution only in the introduction sections of articles.

We argue that research on interpretive case studies also needs to complement this focus on the *how* of the contribution with the *what* or content of contributions. Walsham (1995b) suggests four key generalizations that case study researchers may critically develop as key findings. While he does not explicitly link these generalizations to contributions, he is clearly concerned with contributing or how "IS case study researchers reflect on the basis, conduct, and reporting of their work" (p. 80). We draw on and integrate the how and what of contributions in developing our theoretical approach later in the article.

1.2 When Is a Contribution?

Our perspective on contribution also distinguishes the *when* of a contribution. In examining this, Locke and Golden-Biddle (1997) focus again solely on the construction processes: "an idea becomes a contribution, then, when it is constructed as important by the members of a scholarly community, relative to the accepted knowledge constituted by the field's written work." They do, however, suggest future research that recognizes the importance of the subsequent use of contributions. After a text is published, it travels to a variety of audiences and is used in a variety of ways (Winsor 1993). How does a text travel, relative to its construction of contribution?

Latour's (1987) work on the rhetorical construction of knowledge claims in scientific articles in the literature is instructive in this regard. As discussed later, not only does Latour offer additional positioning strategies to those of Locke and Golden-Biddle in the construction of knowledge claims, but also exhorts us to examine the later use of claims. He sums this up in his second rule of method which "asks us not to look for the intrinsic qualities of any given statement but to look instead for all the transformations it undergoes later in other hands" (p. 59). In our theoretical developments, we draw on Latour's insights concerning the use of claims to further develop the when of contribution as well as the how of contribution (through his rhetoric on positioning strategies). In so doing, we are conscious that he is primarily concerned with scientific facts and we therefore translate his work for our purposes of examining interpretive IS case studies, although we would suggest their broader applicability to interpretive case study research in general.

In the next section, we develop our theoretical approach which integrates the how, what, and when of contribution by distinguishing the construction of contributions and the use of contributions in the literature. We then discuss the research methodology followed by a case analysis of a single well-cited interpretive case study in the literature. The final section of the paper concludes by discussing and synthesizing key findings from the analysis.

2 CONSTRUCTING AND USING CONTRIBUTIONS: A THEORETICAL APPROACH

2.1 Constructing Contributions in the Literature

Table 1 summarizes the different elements of our theoretical approach concerned with how authors construct contribution. We specify four broad strategic concepts and their associated tactical approaches. The first two concepts, structuring intertextual coherence and problematizing the context for contribution, draw from Locke and Golden-Biddle's (1997) findings from their empirical grounded theory investigations of 82 articles in *Administrative Science Quarterly* and *Academy of Management Journal*. Our third concept, positioning as translating interests, draws from Latour's (1987) ideas on the rhetorical construction of the scientific article. He suggests that scientists move from weaker to stronger rhetoric in the construction of knowledge claims as contributions in the literature. As discussed earlier, we are conscious that there are key differences between interpretive case studies and the scientific article upon which he focuses. We carefully select specific concepts and translate them in ways that might be more appropriate to interpretive case studies. In this regard, we draw on Latour's use of *soft facts*[1] and the process of *hardening of facts*, though we do not believe that his use of hard facts is appropriate for the concerns of this paper. To complement the above focus on process, the how of the contribution, implicit in the above concepts, our fourth concept develops the what or content of the contribution as four generalizations in interpretive case study research (Walsham 1995b). We now discuss each of these concepts in turn.

2.1.1 Structuring Intertextual Coherence

Locke and Golden-Biddle (1997) identified two processes, structuring intertextual coherence and problematizing context, which were in tension with each other. The first process, constructing intertextual coherence, refers to the need for texts to establish contribution by re-presenting and organizing "existing knowledge so as to configure a context for contribution that reflects the consen sus of previous work" (p. 1029).[2] At the same time, however, and this is where the tension develops, authors must be involved in a second process of problematizing the situation. In other words, they must subvert or problematize "the very literatures that provide locations and raison d'etre for the present efforts" (p. 1029).

Locke and Golden-Biddle's (1997) analysis suggested three intertextual coherences[3]: synthesized coherence, progressive coherence, or non-coherence. According to Locke and Golden-Biddle, manuscripts display synthesized coherence when they "cite and draw connections between works and investigative streams not typically cited together to suggest the existence of the undeveloped research areas" (p. 103). The implication in such texts displaying **synthesized coherence** is that researchers working in different areas are not aware that their work points to common ideas. In contrast, **progressive coherence** indicates "networks of researchers linked by shared theoretical perspectives and methods working on research programs that have advanced over time" (p. 1035), highlighting cumulative knowledge growth and construction of consensus amongst researchers. Finally, in **non-coherence** intertextual fields, they found "referenced works that are presented as belonging to a common research program but as linked by disagreement" (p. 1038).

---

[1]Latour does not define soft facts per se, but implies that they are commonly accepted statements. Unlike hard facts, these statements are not supported by statistically significant scientific evidence (cf. p. 206).

[2]Latour refers to this as the context of citation.

[3]Locke and Golden-Biddle also identify textual acts and associated rhetorical practices with each type of coherence, but this is outside the scope of this article.

Table 1. Key Concepts for Constructing Contributions

Strategic Concept	Tactical Approach
Structuring Intertextual Coherence (Locke and Golden-Biddle 1997)	• Synthesized coherence • Progressive coherence • Non-coherence
Problematizing Context for Contribution (Locke and Golden-Biddle 1997)	• Incompleteness • Inadequacy • Incommensurability
Positioning as Translating Interests (Latour 1987)	• Framing for particular audiences • Staging to highlight what audiences should find interesting to discuss, and admitting what they may find disputable • Captation or subtle control of objector's moves with due consideration of allowed margin of negotiation of soft facts • Stacking or the extension of evidence to inductively support theories
Qualitative Generalizations as Content of Contributions (Walsham 1995b)	• Concept development • Theory generation • Specific implications • Rich insights

2.1.2 Problematizing Context for Contribution

Three ways of problematizing an intertextual field were identified by Locke and Golden-Biddle: **incompleteness, inadequacy**, and **incommensurability**. By problematizing a literature as **incomplete**, the text claims that the existing literature is not finished and that the present article will further develop or specify it. A literature is problematized as **inadequate** when the text claims that the existing literature does not sufficiently incorporate different perspectives (relevant and important) and views to better understand the phenomena under investigation. Finally, in problematizing a text as **incommensurate**, an article goes further to suggest that not only does the existing literature overlook different and relevant perspectives but that the claims being made are inaccurate.

2.1.3 Positioning as Translating Interests

Our third concept, positioning as translating interests, draws on Latour's (1987) work, which examines the construction of scientific facts in the literature. His broad theory, sometimes referred to as the *sociology of translation* focuses on the process by which a knowledge claim becomes a black box (that is to say accepted as unproblematic and uncontested) or rejected. Translation consists of the ongoing process by which claims are progressively transformed as proponents seek to enrol other actors, who may accept, reject, ignore, or modify the claim depending on their own interests. In our paper, we suggest that the contribution constructed by an author is a claim to knowledge whose fate is always in later authors' hands. However, this does not mean that authors do not have some control in positioning their texts and protecting their stated contributions. In interpretive case studies, contributions as knowledge claims are always soft facts as distinct from hard facts, which may have more relevance in scientific articles. We draw on two ideas from Latour's work, namely, the hardening of (soft) facts and captation.

Latour discusses **captation** as a tactical approach in positioning strategies. He uses it to mediate the fact-builder's paradox: "how to leave someone completely free and have them at the same time completely obedient...to lay out the text so that wherever the reader is there is only one way to go" (p. 57). However, this paradox is still appropriate when translated for hardening the (soft) fact process. Latour argues that "the simplest way to spread a claim is to leave a margin of negotiation to each of the actors to transform it as he or she sees fit and to adapt it to local circumstances" (p. 208) So, a captation approach should recognize and provide a margin of negotiation in allowing later authors to translate their contribution, while simultaneously recognizing the need to retain subtle control. Control is necessary to ensure that your contribution is not lost, and still recognizable despite modification through translation by later authors. Latour goes on to note that the softer the facts the longer the networks, and the greater the potential variability in translation.

Latour suggests other positioning strategies including framing, staging, and stacking. In **framing** the contribution, the authors should carefully consider the audience: "In order to defend itself, the text has to explain how and by whom it should be read....If one wishes to increase the number of readers again, one has to decrease the intensity of the controversy, and reduce the resources" (p. 52). Furthermore, Latour suggests that, in aiming at a specific readership, authors should not only consider the kind of words used but also anticipate readers' objections in advance. Typical **staging** strategies authors should adopt are to highlight what should be discussed, what is really interesting, and that which is admittedly disputable, the latter strategy being "like taking out an insurance policy against the unexpected transformation of facts into artefacts" (p. 55). Finally, **stacking** strategies inductively draw on and extend evidence in supporting theories. Researchers seek to use their findings to move from a rather specific instance to suggest applicability in a more general field. Latour gives the example of a biologist who looks at small slices of flesh which are first "three hamster kidneys," then extended to be "hamster kidneys," then "rodent kidneys," and finally "mammal countercurrent structure in the kidney." In so doing, the biologist seeks to "prove as much as he can with as little as he can, considering the circumstances" (p. 51).

2.1.4 Qualitative Generalizations

Our fourth strategic concept is concerned with the construction of qualitative generalizations from interpretive research, which we suggest are the *content* of contributions. These generalizations are critical for researchers to develop and are often constructed toward the latter half of articles, in contrast to the introduction sections, which were the focus of Locke and Golden-Biddle's analysis. Four types of generalizations have been identified by Walsham (1995b): **development of concepts, generation of theory**, drawing of **specific implications** in particular domains of action, and contributions of **rich insights** which are broad insights not easily categorized as any of the other three types.

2.2 Use of Contributions in the Literature

As we have highlighted in the previous section, Latour emphasizes that knowledge claims or contributions are always in the hands of later authors. We drew further on his ideas in developing our theoretical approach on the use of contributions in the literature (see Table 2). We develop three elements to evaluate use of contribution along with their associated assessment criteria. The first evaluative element examines the "take up" of the **content of the contributions used by later audiences**. Relevant assessment criteria to consider are the nature and type of the contribution cited, and how it was related (or not) to one of the four qualitative generalizations. Another useful criterion is to examine any unusual cases of use by later texts, and to try to interpret reasons as to why specific contributions were adopted.

Our second evaluative element is the **incorporation and use of contribution by later audiences**. Latour argues that if (contributions as) claims to knowledge are to be effective, they must be incorporated into later statements by members of a research community. A statement (contribution) needs the next generation of papers "much like genes that cannot survive if they do not manage to pass themselves on to later bodies" (p. 38). Furthermore, Latour suggests that one needs to examine how others use and insert premises or arguments (rhetoric) from one paper into sentences of their own paper. Depending on how this modality (as he refers to it) is inserted into other sentences by others in developing their own rhetoric in papers, a given sentence becomes more of a fact or more of an artefact. Modalities, positive or negative, can further the claim positively by reinforcing the contribution or alternatively contest the claim (contribution) with negative implications for the continued life of the contribution in subsequent generations of papers. We suggest then that appropriate assessment criteria are to examine how the contributions have been incorporated and subsequently used to further other authors' claims.

A related criterion is to examine the implications of later referencing for the fate of the contribution. Latour notes a few of the many possibilities: references may be misquoted or wrong; many may have little or no bearing whatsoever on the claim and be just for display. Other, perfunctory, citations might be present but only because they are always present in the author's articles, whatever his or her other claim, to mark affiliation to a group of researchers, and finally qualified or modalized claims (as Latour calls them), where papers may be referenced which (knowingly or not) explicitly say the contrary of the author's thesis.

Our final evaluative element concerns the **success of positioning strategies**. Quite simply, by evaluating how an article is referred to by later texts, the success of the framing, staging, captation, and stacking strategies can be assessed. The assessment criteria, therefore ,examine the range of audiences that authors were successful in framing, the types of staging strategies that were successful, and the extent to which captation and stacking strategies were successful.

Table 2. Use of Contributions in the Literature

Evaluative Element	Assessment Criteria
Content of contribution(s) used by later audiences	What was the nature and type of contribution taken up and cited? How can we account for unusual cases and the take up of specific contributions?
Incorporation and use of contribution by later audiences	How have the contributions been incorporated and used (translated) to further later writer's claims? What are the implications of later referencing for the fate of the contribution?
Success of positioning strategies	What were the range of audiences that authors were successful in framing? What types of staging strategies were successful? To what extent were captation and stacking strategies successful?

3 RESEARCH METHODOLOGY

To examine the construction and use of contribution in interpretive case study research, we had to make some difficult decisions. The first decision concerned the scope of the study, and a single interpretive case study approach was chosen. We made this decision primarily because our research design required us to carry out an in-depth study of both construction and use of contributions. Our analysis, therefore, required us not only to carefully examine all sections of the chosen interpretive case study but also to include examination of a substantial body of literature that subsequently referenced this study. The second decision concerned sample selection. Our research through the use of citation indexing software suggested that either *MIS Quarterly* or *Information Systems Research* would be suitable choices based on their number of citations. We examined a wide range of articles which used an interpretive IS case study methodology and eventually selected one that was 5 to 7 years old and was very well cited. The article selected was "Steps Toward an Ecology of Infrastructure: Design and Access for Large Information Spaces" (1996) by Susan Leigh Star and Karen Ruhleder (hereafter referred to as S&R) in *Information Systems Research*. The underlying assumptions of S&R's research are consistent with those suggested by Orlikowski and Baroudi (1991) concerning the philosophical premises of an interpretive study, having a focus on understanding phenomena through the meanings that people assign to them. Furthermore, their use of theory (in their case, Bateson's theory) is as a sensitizing device to make sense of the world rather than the application of a positivist theory-testing approach

(Klein and Myers 1999). As of July 2003, when our research was conducted, there were 39 citations of this paper in the literature. Of these citations, we were able to access 35.

Both authors contributed significantly to the data analysis, particularly with respect to the initial development and subsequent testing of the evaluation scheme, which required each to code the S&R paper and a number of other articles citing this paper. Specifically, we started our data analysis by separately coding all sections of S&R using the iteratively developed theoretical framework (Table 1) for constructing contributions in the literature. We then discussed the similarities and differences in our analysis, and where we agreed we merged our findings. The data analysis for coding the use of contributions in the literature was much trickier but followed a similar process. Following open coding of a significant sample of the articles that referred to S&R, each of the authors separately coded a subset of the articles using the evaluation and assessment criteria (Table 2). We then came together to discuss similarities and differences, and to agree on how to integrate different perspectives both in our findings and in improving our data analysis approach. The first author subsequently coded the rest of the 35 articles based on the agreed scheme and concluded the analysis.

Before discussing our findings in the next section, we present the abstract from S&R as a high level summary to aid our readers in gaining an initial broad understanding of the article.

Steps Toward an Ecology of Infrastructure:
Design and Access for Large Information Spaces

S. Star and K. Ruhleder

Abstract

We analyze a large-scale custom software effort, the Worm Community System (WCS), a collaborative system designed for a geographically dispersed community of geneticists. There were complex challenges in creating this infrastructural tool, ranging from simple lack of resources to complex organizational and intellectual communication failures and tradeoffs. Despite high user satisfaction with the system and interface, and extensive user needs assessment, feedback, and analysis, many users experienced difficulties in signing on and use. The study was conducted during a time of unprecedented growth in the Internet and its utilities (1991-1994), and many respondents turned to the World Wide Web for their information exchange. Using Bateson's model of levels of learning, we analyze the levels of infrastructural complexity involved in system access and designer-user communication. We analyze the connection between system development aimed at supporting specific forms of collaborative knowledge work, local organizational transformation, and large-scale infrastructural change.

(Infrastructure; Collaboratory; Organizational Computing; Participatory Design; Ethnography; Internet; Scientific Computing)

4 THE CONSTRUCTION AND USE OF CONTRIBUTIONS IN STAR AND RUHLEDER

Our analysis in sections 4.1 and 4.2 draws on our developed theoretical approach as summarized in Tables 1 and 2 respectively. We bold the key concepts or points in our discussion below so as to aid the reader in appreciating the connections between our discussion and the earlier theoretical approach.

4.1 Constructing Contributions in Star and Ruhleder

4.1.1 Structuring Intertextual Coherence

Contrary to a tacit implication in Locke and Golden-Biddle (1997), S&R's article did not fall neatly into a single tactical approach or type of coherence. Nonetheless, the coherence categories were useful in identifying plural types of coherence used as tactical approaches in structuring intertextual coherence.

At the very beginning of the article, we identified patterns of **synthesized coherence**. In section 1, S&R develop the intertextual field as being concerned with the paradoxical relationship between technology and organizational transformation. They highlight structuration theory (Giddens 1984) as an approach in analyzing this paradox but then cite connections with "web of usability and action" approaches (after Engestrom [1990] and Kling and Scacchi [1982]) to suggest the existence of undeveloped research areas. Later on, in sections 4 and 5, S&R also adopt synthesized coherence in drawing on the above approaches; in particular, they use Bateson's ecology levels and double bind concepts to examine the logical paradox between technology and organizational transformation. Furthermore, they suggest that the added dimensions of infrastructure developed by drawing on this approach "deepens our understanding of the dual and paradoxical nature of the technology."

With respect to S&R's treatment of the dimensions of infrastructure, this is harder to categorize. We would argue that S&R use a **progressive coherence** approach as they bring together the work of a range of other scholars (Bowker 1994; Jewett and Kling 1991) linked by a shared perspective on infrastructure as a relational property and who seek to build cumulative knowledge (Locke and Golden-Biddle 1997). However, it could be argued that synthesized coherence was the tactical approach used. After all, in developing infrastructure as a "fundamentally relational concept...in relation to organized practices" (p. 113), they provide an alternative approach to the common metaphors of infrastructure as a substrate or a "thing with pre-given attributes frozen in time" (p. 112). However, as discussed below, we view this more as a problematization strategy since the article does not present an intertextual field of literature per se when making this particular point.

4.1.2 Problematizing Context for Contribution

Intriguingly, the analysis of problematizing strategies was even more nuanced and pluralistic. The categories from Locke and Golden-Biddle (1997) once again held up well but, perhaps surprisingly, we identified all three forms of problematizing in this one paper. First, as highlighted at the end of the last subsection, their direct challenge of common metaphors of infrastructure as being "neither useful nor accurate in understanding the relationship between work/practice and technology" (p. 112) represents an **incommensurable problematization**. Second, the **incompleteness problematizing** approach is evident toward the end of section 1. The authors conclude this section by emphasizing the need for more studies since they are not only important but it is "increasingly clear to us that this development effort is taking place at a moment of rare, widespread infrastructural change" (p. 114). The implication is that their study attempts to fill (or complete) this need. Finally, an **inadequacy problematization** is evident in section 3. In discussing CSCW typologies, the authors argue that existing approaches are inadequate "in analyzing the issues associated with implementation or integration... [nor] the relational aspects of computing infrastructure and work" (p. 116).

4.1.3 Positioning as Translating Interests

As structured in Table 1, our analysis identified a range of positioning strategies used by S&R in constructing opportunities for contribution across all sections of their paper. For example, their **framing** strategy addressed a wide range of audiences. In section 4, they highlight the CSCW community, system development community, widely dispersed scientists, and communication researchers. In section 7, the authors signal the importance of social, information, and computer science communities in exploring the ecology of infrastructure.

A very good example of **framing and staging** strategies used together in the paper are illustrated in section 3 of the paper. S&R start by framing the collaboratory infrastructure and the questions they are reporting on as being important and relevant to the CSCW community. Their subsequent staging strategies suggest an urgent need for interest and discussion as to the importance of the key arguments reflected in the paper concerning infrastructure and its relationship to work and communication: "Those working in the emergent field of CSCW, of which the collaboratory is a subset, have struggled to understand how infrastructural properties affect work, communication, and decision making" (p. 116).

Appreciating **captation**, or subtle control balanced with a margin of negotiation, was more challenging to ascertain within discrete parts of the paper. Pulling together various strands across the paper, we believe that captation was evident in the way the paper attempts to shift the study of information infrastructure to a study of contexts: "Information infrastructure is not a substrate which carries information on it, or in it, in a kind of mind-body dichotomy. The discontinuities are not between system and person, or technology and organization, but rather between contexts" (p. 118). In shifting the study of information infrastructure to a study of contexts, it opens the door, or provides a wide margin of negotiation in applying the concepts of information infrastructure to a number of other topics; so long (and this is where subtle control may be retained) as this is done using S&R's theory concerning contextual levels.

Finally, S&R employ **stacking** techniques to prove the value of applying Bateson's typology in examining the paradox of infrastructure. They start by examining the contexts of two generic processes common to any infrastructure, *signing on* and *hooking up*. They then extend this evidence to support their generated theory by discussing the "implications of this typology for other forms of system development" (p. 118). This stacking is further extended to the "broader implications for understanding the impact of new computer-based media and their integration into established communities." The last few lines of their abstract reproduced above also suggests such a strategy.

4.1.4 Qualitative Generalizations as Content of Contributions

As developed in Table 1, specific contributions may be manifested as one of four types of qualitative generalizations: concept development, theory generation, specific implications, and rich insights. We discuss examples of each of these in turn. First, a number of concepts are identified in S&R. A key **concept** is the "when is an infrastructure," developed as a relational concept and made explicit through defined dimensions. Another example, developed from applying Bateson's double bind to infrastructural issues is the infrastructual transcontextual syndrome (ITS) concept. Second, **theory generation** was distinguished as the novel application of Bateson's levels of communication and related first, second, and third order issues to information infrastructure. Third, the paper also highlights a number of **specific implications** in addressing first, second, and third order issues. Finally, we identified **rich insights** on the relationships between information infrastructure technology, organizations, and communities.

4.2 Use of Contribution in Star and Ruhleder

4.2.1 Content of Contribution(s) Used by Later Audiences

Our primary focus here was to examine the **nature and type of contribution cited and used** by later texts. We did not try to make the contribution fit into one of the four possible qualitative generalizations but noted if it did.

The citation and use of contributions was highly diverse and somewhat surprising. In particular, the use of Bateson's model including the concept of double binds to examine the ecology of infrastructure, although central to S&R and its analysis, was rarely discussed or incorporated by later audiences. Only 4 of the 35 articles analysed made any reference to the theory or its related key concepts such as double binds (e.g., Bishop et.al. 1999; Robey and Boudreau 1999) or orders of use issues (e.g., Bishop et al. 2000). Of these, only one article (McCarthy 2000) highlights and incorporates Bateson's theory significantly in their work.

A popular type of contribution used (nine articles) was the concept of the "when of infrastructure," along with one (e.g., use of transparency by Whyte and Macintosh [2001]) of more of its eight dimensions (e.g., Borgman 2002; Jacob 2001) associated with this relational concept. Once again, this finding is interesting and somewhat unexpected. The infrastructure concept, developed early on in S&R, is important for justifying the contextual approach undergirding Bateson's theory. However, it is Bateson's concepts that are central to the analysis and discussion sections of S&R. It was, therefore, a little surprising that the use of the earlier infrastructure concept was significantly greater than the paper's key theory and concepts.

Unlike the above contributions, which fit the concept development qualitative generalization, the largest contribution category did not neatly fit any of the four generalizations. We refer to this category as the *related work* category, which accounted for just over a third of the articles surveyed. There were two different groups. First, those articles that referenced S&R because they served as a good analogy, or early pre-Internet example, of related large-scale infrastructure. In these articles, specific contributions (whether theory, concepts, rich insights, or specific implications) are rarely ever discussed. One exception (DeSanctis and Monge 1999) recognizes the rich insights S&R offer on the coupling between technology, organization structure, and communication patterns.

A second grouping of articles highlights the general proposition or broad essence of the article, without explicit reference to any of the four specific contribution types. This general proposition refers to infrastructure as a socio-technical construct, a relational concept, and recognizes the importance of its context of use. For example, Bishop et al. (2000), while not directly referencing any specific type of contribution, cite S&R's study as being useful in examining the "context and culture of community computing in design and implementation of community services." Articles in this group recognize and adopt a "broad web of usability and action" in understanding the paradox of infrastructure. Furthermore, in two cases, the articles, in addition to citing a general proposition, also cite a specific concept, such as infrastructure as a socio-technical construct occurs when tension between local and global is resolved (Kling 2000).

The last category and smaller grouping of articles referenced specific implications or substantive concepts discussed by S&R. For example, Min and Galle (1999), in examining e-purchasing, reference S&R's substantive (third order issue) concept on network externalities and electronic participation which they discuss in relation to the Worm Community system.

4.2.2 Incorporation and Use of Contribution(s) by Later Audiences

In this subsection, we go beyond our previous examination of the nature and type of contributions. We draw on the assessment criteria in Table 2 to evaluate how **contributions are referred to and incorporated** by later audiences. We now discuss each of the categories which characterized our findings.

4.2.2.1 Mistaken Referencing: Incorporating and Using Claims Not Discussed in the Article

At one level, these articles can be construed to be mistaken in referencing concepts that are not discussed in the paper. For example, Ciborra and Andreu (2001) reference S&R in connection with the concept of boundary objects. However, S&R do not develop this concept in their article, although the concept is a well-known one developed by Star in some of her other writings. Nonetheless, in their article, Ciborra and Andreu clearly reflect the broad essence or general propositions of S&R concerning Bateson's theoretical perspective (involving three levels/orders of use) in examining the paradox of infrastructure. Their specific case examines the use of infrastructure by a community involving contexts/levels and what they term *interorganizational learning ladders*. While this paper's specific reference to boundary objects may do little to strengthen the specific contributions made in S&R's paper, the paper as a whole implicitly reinforces S&R's general propositions.

4.2.2.2 Ambiguous Referencing: Mixed Use of Negative and Positive Modalities

An interesting finding was the use of ambiguous referencing. For example, Duncker (2000) references S&R in their use of the image of infrastructure as "sinking into the background." S&R use this metaphor differently to problematize more conventional characterizations of infrastructure. Specifically, they suggest that this conceptualization "is neither useful nor accurate in understanding the relationship between work/practice and technology" (p. 112). Furthermore, this subsidiary argument in S&R as referenced by Duncker could be wrongly interpreted by later generations of texts to be S&R's central message. Therefore, this incorporation potentially invokes a negative modality to the claim made by S&R as it (likely inadvertently) supports the very opposite of what their claim is trying to say, and draws attention away from the paper's more sophisticated and central claim on the ecology of infrastructure.[4] In another twist of irony, Duncker goes on to expound in her notes at the end of the paper the transparency of infrastructure, which emerges when "the tension between local meaning and global standards is resolved" (p. 197). This description denotes a very good, accurate, and positively modalised claim to the S&R original, although its chances of being reproduced in subsequent generations of papers may be similar to that of a recessive gene given its discreet location in the paper.

Another example of ambiguous referencing mentioned briefly in an earlier section concerns the use by Min and Galle in examining e-purchasing. Using the concept of network externality and electronic participation, they develop hypotheses on the number of users participating in a purchasing community. However, S&R contrast network externality (as referring to the participation in a community) and critical mass (that focuses on the number of users). Min and Galle seem to conflate these closely related concepts in developing their hypothesis. Such ambiguous referencing could result in future translations by later audiences, which go against the intended use by S&R. In Latour's language, future translations involving negative modalities of S&R's original contribution may dilute or even serve to negate the contribution over time.

4.2.2.3 Incorporating Statements into a New Term (Neologism)

A number of authors incorporate S&R's socio-technical approach into the development of a new concept or term, sometimes referred to as a neologism. For example, using Latour's terminology, Kling (2003) modalizes the claim positively to further his socio-technical integration network (STIN) model. Similarly, Kling (2000) draws on the claim as a socio-technical approach to ICT infrastructures in developing more broadly the field of social informatics in understanding IT and social change.

---

[4]In the scientific literature, Winsor (1993) identifies a similar characterization and irony with what evolutionary biologists do with adaptionist claims in their later use of a well-cited article.

4.2.2.4 Related Work Referencing

The WCS large-scale infrastructure collaboratory, the central empirical focus of S&R, is often cited as a good analogy for scientific research communities (Bieber et al. 2002). This and other papers (e.g., Bishop et al. 2000) use S&R to legitimize their own research into digital libraries as an information infrastructure. Similarly, other papers cite the WCS as analogy for other large scale scientific infrastructures, such as BioDiversity infrastructures (Bowker 2000), while others (e.g., Baker et al. 2000) explicitly identify the WCS as a pioneering pre-Internet system, which has led to many other examples of networked information systems. Internet research (e.g., Valentine and Holloway 2002) draws on the use of the relational concept of infrastructure, recognizing the Internet as an information infrastructure, which emerges for people in practice, connected to activities and structures. Another type of related work referencing was evident in papers that examined the connections between classification and information infrastructure (Jacob 2001; Jacob and Shaw 1998; Palmer and Malone 2001).

4.2.3 Success of Positioning Strategies

This final subsection evaluates the **success of positioning strategies** by S&R in later texts using the assessment criteria developed in Table 2. First, we examine the range of audiences to whom S&R were successful in using framing strategies. We then assess the types of staging strategies that seem to be successful as well as to what extent captation and stacking have been successful.

Our analysis of the 35 articles citing S&R shows that a broad set of audiences across different academic communities have cited their work and incorporated their contributions. The main audiences include researchers from information science, supply chain/e-markets, information systems, communication researchers, systems development, social informatics, and medical informatics. The majority of researchers citing S&R were researchers in information science and social informatics, many of whom were conducting research on digital libraries. Apart from the CSCW group, which does not seem to have referenced the work to date, S&R's framing strategies seem to have been largely successful in gaining support from a broad set of researchers in translating their key contributions.

In addition to successfully framing their work to multiple audiences, S&R's staging of the importance of their pioneering pre-Internet work on large-scale infrastructure successfully led to significant related-work referencing. Researchers, in particular information scientists, viewed WCS as good examples/analogies that legitimated their work on digital libraries, the Internet, and other scientific large-scale infrastructures. A particularly successful positioning strategy was the incommensurable problematization of traditional representations of infrastructure and the development of a relational concept of infrastructure with its dimensions. A number of later texts cited this perspective or conceptualisation. In addition, the wide scope of dimensions (eight) of infrastructure as opposed to a more basic definition (as highlighted by Jacob 2001) provides more of what is interesting with specific dimensions being cited and translated by different authors. In line with Latour's prediction, infrastructure conceptualised as a softer fact allows longer networks and more variability in translation by later authors.

The concept of captation is admittedly quite difficult to evaluate, although there are signs of some success by S&R in such strategies. For example, a number of researchers from the Information Science community have subsequently examined classification as infrastructure. This demonstrates the high margin of negotiation in translating soft facts associated with infrastructure to classification while retaining some control through the use of Bateson's theory as evidenced by it being examined as a study of contexts. For instance, the analysis by Bishop et al. (2000) of classification views it as infrastructure but examines it in a contextual study using three levels as in S&R.

However, S&R's stacking strategy to show the use of Bateson's typology beyond generic and widespread processes of signing on and hooking up in infrastructure to other systems development approaches has to date not been very successful in terms of take-up by others.

5 DISCUSSION AND CONCLUSIONS

Our study examined processes of construction and use of contributions in a well-cited interpretive case study to extend our understanding as to how such texts may better develop contributions. Our theoretical approach draws on and extends previous work on how texts construct contribution by integrating the how, what, and when of contribution. We further develop the how of contribution (Locke and Golden-Biddle 1997) by appropriating (translating) Latour's (1987) positioning strategies, which include tactics authors may use in crafting their texts. We also suggest that in constructing contribution, it is important to examine later sections of a paper beyond the introduction, and to consider how authors develop the what or content of contributions as qualitative generalizations (Walsham 1995b). Finally, drawing on Latour, we develop the when of contribution by highlighting the importance of use, and recognizing contribution as a knowledge claim whose fate depends on later texts.

In our paper, we have tailored key concepts proposed by Latour and by Locke and Golden-Biddle in elaborating a theoretical framework applicable for examining interpretive case studies. However, as hinted earlier, we recognize that elements of the framework may have wider applicability beyond interpretive studies to scientific articles (cf. Latour 1987), and further research could consider how such elements might be used in conjunction with earlier frameworks (e.g., March and Smith 1995) to examine processes of construction and use in other research approaches. We also recognize the study's current limitations concerning generalizability as our unique methodology examined a single case. However, we believe that the deeper structure of construction and use of contributions in the paper can be key learning for other researchers seeking to develop their own paper's contributions. Despite these limitations, our analysis produced a number of interesting and unexpected findings which we believe are a good first step in furthering the literature on contribution in IS interpretive case studies, as well as interpretive case study research more generally. With respect to constructing opportunities for contribution, our case confirmed the usefulness of Locke and Golden-Biddle's specific types of coherence strategies. However, we go further to suggest that texts may be successful in deploying plural strategies in constructing intertextual coherence. Our analysis also shows that texts may use all three identified problematizations across different parts of the article, an intriguing finding that Locke and Golden-Biddle's limited analysis of the introduction sections of articles could not identify.

In further examining the construction of contributions by S&R, we drew on and translated positioning strategies from Latour's work on scientific facts to be relevant for soft facts in interpretive case studies. We also examined the content of the contributions in the S&R study using the four qualitative generalizations provided by Walsham (1995b). While there were fewer surprises with these construction processes, in that good examples were found in each category, it did suggest the value of these extensions of previous work being incorporated in our integrated theoretical approach.

Some key findings from our analysis of the use of contributions by later texts were quite unexpected. There was little use by later texts of central theories and concepts, which from the title and content of S&R's paper were clearly meant to be key contributions.[5] Interestingly, there may be echoes here of the rigor/relevance debate (Benbasat and Zmud 1999). In S&R, a significant level of rigor is displayed in using Bateson's more complex theoretical apparatus drawn from the communications and psychiatry community. However, the relevance as displayed by researchers in their subsequent take up seems to be lower. In contrast, the far simpler and well-crafted concept of the "when of infrastructure," clearly meant by S&R to merely set the stage for the use of Bateson's theory later on in the paper, is surprisingly used significantly more by later authors.

A large number of citations simply recognized the study as a good example or early analogy of large-scale infrastructure and confirmed the successful framing and staging identified in S&R. Exceptions were the deafening silence of feedback from the CSCW group (admittedly hard to identify) despite the most obvious framing and staging to this group. In addition, other groups less obviously targeted cited the work such as supply chain and purchasing, and medical informatics. This observation highlights that there are never any guarantees as to whether framing and staging strategies will be successful, and it is uncontrollable as to whom your contribution is ultimately taken up by and translated. What was also interesting was the frequency of use of the general propositions of the paper's socio-technical approach to infrastructure, which were drawn on more than specific qualitative generalizations. Furthermore, the stacking strategies used to prove the use of theory beyond common generic processes to systems development seemed to be largely unsuccessful to date.

Our findings categorize a number of different types of referencing used by later texts in incorporating and using contributions, namely mistaken referencing, ambiguous referencing, incorporating into a new term, and related- work referencing. Of particular note, ambiguous referencing took place when audiences translated concepts which involved incommensurate problematizations but with closely related meanings (e.g., critical mass and network externality).

---

[5]There are some parallels here to the findings by Mingers and Taylor (1992), which demonstrated that soft systems methodology (SSM) in practice had been used for a wide variety of tasks. Some elements of SSM were nearly always used while others were used less often. Finally, other techniques were often combined with SSM or modifications made to SSM in ways that extended well beyond the originators.

In conclusion, we make some suggestions as to how one can construct better contributions in interpretive case studies. First, and this is a macro-level theme of the paper, be sensitive to the how, what, and when of a contribution. For example, consider carefully the use that others might make of your contributions. Second, think through your communities/target audiences and appropriate framing/staging strategies to signal to them how different contributions may be best translated by them. Third, think about your captation strategies—how to allow a high margin of negotiation by multiple audiences in translating a wide range of soft facts while attempting to retain some subtle control of the main essence of the contributions. Fourth, in constructing opportunities for contribution, do not necessarily limit yourself to a single coherence or problematization strategy. A plural strategy may very well increase the effectiveness of your contributions. Fifth, appreciate the use of incommensurate problematizations that can act as a double-edged sword. These problematizations can be a powerful staging strategy in raising significant interest and use of a contribution. However, clarity in defining concepts is critical, especially when there is significant relatedness or similarity between concepts used in the problematization. Otherwise, as we saw with the infrastructure concept or with network externality, there is the risk of possible confusion or poor translation of the contribution in future generations of texts. Sixth, try to communicate clearly the general propositions of your research in addition to developing key qualitative generalizations as our analysis suggests that this may be equally if not more important.

It can be argued that a number of the above suggestions are tacitly used by experienced researchers such as S&R who clearly know how to construct a contribution as evidenced by the wide citation of their article. Our paper has attempted to unearth some of the deep structures (Heracleous and Barrett 2001) of these writers and their intended strategies. However, all strategies, even of the most experienced writers, are emergent and authors cannot fully control what later users do with their projected and anticipated contributions, including this paper.

REFERENCES

Astley, W. G. "Administrative Science as Socially Constructed Truth," *Administrative Science Quarterly* (30), 1985, pp. 497-513.

Baker, K. S.; Benson, B. J.; Henshaw, D. L.; Blodgett, D.; Porter, J. H.; and Stafford, S. G. "Evolution of a Multisite Newtork Information System: The LTER Information Management Paradigm," *Bioscience* (50:11), November 2000, pp. 963-978.

Bateson, G. *Steps to an Ecology of Mind*, New York: Ballantine Books, 1978.

Benbasat, I., and Zmud, B. "Empirical Research in Information Systems: The Practice of Relevance," *MIS Quarterly* (23:1), March 1999, pp. 3-16.

Beyer, J. M.; Chanove, R. G.; and Fox, W. B. "The Review Process and the Fates of Manuscripts Submitted to AMJ," *Academy of Management Journal* (38), 1995, pp. 1219-1260.

Bieber, M.; Engelbart, D.; Furuta, R.; Hiltz, S. R.; Noll, J.; Preece, J.; Stohr, E. A.; Turoff, M.; and Van de Walle, B. "Toward Virtual Community Knowledge Evolution," *Journal of Management Information Systems* (18:4), 2002, pp. 11-35.

Bishop, A. P.; Neumann, L. J.; Star, S. L.; Merkel, C.; Ignacio, E.; and Sandusky, R. J. "Digital Libraries: Situating Use in Changing Information Infrastructure," *Journal of The American Society for Information Science* (51:4), 2000, pp. 394-413.

Bishop, A. P.; Tidline, T. J.; Shoemaker, S.; and Salela, P. "Public Libraries and Networked Information Services in Low-Income Communities," *Library & Information Science Research* (21:3), 1999, pp. 361-390.

Borgman, C. L. "Challenges in Building Digital Libraries for the 21st Century," in E-P. Lim, S. Foo, and C. Khoo (Eds.), *Digital Libraries: People, Knowledge & Technology: Proceedings of the 5th International Conference on Asian Digital Libraries* (ICADL 2002), Heidelberg: Springer-Verlag, 2002, pp. 1-13

Bowker, G. "Biodiversity Datadiversity," *Social Studies of Science* (30:5), October 2000, pp. 643-683.

Bowker, G. "Information Mythology and Infrastructure," in L. Bud-Frierman (Ed.), *Information Acumen: The Understanding and Use of Knowledge in Modern Business*, London: Routledge, 1994, pp. 231-247.

Ciborra, C. U., and Andreu, R. "Sharing Knowledge Across Boundaries," *Journal of Information Technology* (16), 2001, pp. 73-81.

Davenport, T., and Markus, L. " Rigor vs. Relevance Revisited: Response to Benbasat and Zmud," *MIS Quarterly* (23:1), March 1999, pp. 19-23.

DeSanctis, G., and Monge, P. "Introduction to the Special Issue: Communication Processes for Virtual Organizations," *Organization Science* (10:6), November-December 1999, pp. 693-703.

Duncker, E. "How LINCs Were Made: Alignment and Exclusion in American Medical Informatics," *The Information Society* (16), 2000, pp. 187-199.

Engestrom, Y. "When is a Tool? Multiple Meanings of Artifacts in Human Activity," in *Learning, Working and Imagining*, Helsinki: Orienta-Konsultit Oy, 1990.

Giddens, A. *The Constitution of Society*, Cambridge: Polity Press, 1984.

Heracleous, L., and Barrett, M. "Organizational Change as Discourse: Communicative Actions and Deep Structures in the Context of IT Implementation," *Academy of Management Journal*, (44:4), 2001, pp. 755-778.

Jacob, E. K. "The Everyday World of Work: Two Approaches to the Investigation of Classification in Context," *Journal of Documentation* (57:1), January 2001, pp. 76-99.

Jacob, E. K., and Shaw, D. "Sociocognitive Perpsectives on Representation," *Annual Review of Information Science and Technology* (33), 1998, pp. 131-185.

Jewett, T., and Kling, R. "The Dynamics of Computerization in a Social Science Research Team: A Case Study of Infrastructure, Strategies, and Skills," *Social Science Computer Review* (9), 1991, pp. 246-275.

Klein, H. K., and Myers, M. D. "A Set of Principles for Conducting and Evaluating Interpretive Field Studies in Information Systems," *MIS Quarterly* (23:1), 1999, pp. 67-94.

Kling, R. "A Bit More to It: Scholarly Communication Forums as Socio-Technical Interaction Networks," *Journal of the American Society for Information Science and Technology* (54:1), 2003, pp. 47-67.

Kling, R. "Learning About Information Technologies and Social Change: The Contribution of Social Informatics," *The Information Society* (16), 2000, pp. 217-232.

Kling, R., and Scacchi, W. "The Web of Computing: Computing Technology as Social Organization," *Advances in Computers* (21), 1982, pp. 3-78.

Knorr-Cetina, K. *The Manufacture of Knowledge: An Essay on the Constructivist and Contextual Nature of Science*, New York: Pergamon Press, 1981.

Latour, B. *Science in Action*, Cambridge, MA: Harvard University Press, 1987.

Locke, K., and Golden-Biddle, K. "Constructing Opportunities for Contribution: Structuring Intertextual Coherence and 'Problematizing' in Organizational Studies," *Academy of Management Journal* (40:5), 1997, pp. 1023-1062.

March, S. T., and Smith, G. F. " Design and Natural Science Research on Information Technology," *Decision Support Systems* (15), 1995, pp. 251-266.

McCarthy, J. "The Paradox of Understanding Work for Design," *International Journal Human-Computer Studies* (53), 2000, pp. 197-219.

Min, H., and Galle, W. P. "Electronic Commerce Usage in Business-to-Business Purchasing," *International Journal of Operations & Production Management* (19:9), 1999, pp. 909-921.

Mingers, J., and Taylor, S. "The Use of Soft Systems Methodology in Practice," *Journal of Operational Research Society* (43:4), 1992, pp. 321-332.

Mone, M. A. , and McKinley, W. "The Uniqueness Value and its Consequences for Organization Studies," *Journal of Management Inquiry* (2), 1993, pp. 284-296.

Orlikowski, W. J., and Baroudi, J. J. "Studying Information Technology in Organizations: Research Approaches and Assumptions," *Information Systems Research* (2:1), 1991, pp. 1-28.

Palmer, C. L., and Malone, C. K. "Elaborate Isolation: Metastructures of Knowledge About Women," *The Information Society* (17), 2001, pp. 179-194.

Robey, D., and Boudreau, M. "Accounting for the Contradictory Organizational Consequences of Information Technology: Theoretical Directions and Methodological Implications," *Information Systems Research* (10:2), 1999, pp. 167-185.

Star, S. L., and Ruhleder, K. "Steps Toward an Ecology of Infrastructure: Design and Access for Large Information Spaces," *Information Systems Research* (7:1), 1996, pp. 111-134.

Valentine, G., and Holloway, S. L. "Cyberkids? Exploring Children's Identities and Social Networks in On-Line and Off-Line Worlds," *Annals of the Association of American Geographers* (92:2), 2002, pp. 302-319.

Walsham, G. "The Emergence of Interpretivism in IS Research," *Information Systems Research* (6:4), 1995(a), pp. 376-394.

Walsham, G. "Interpretive Case Studies in IS Research: Nature and Method," *European Journal of Information Systems* (4), 1995(b), pp. 74-81.

Webster, J., and Watson, R. T. "Analyzing the Past to Prepare for the Future: Writing a Literature Review, *MIS Quarterly* (26:2), June 2002, pp. xiii-xxiii

Weick, K. E. "Editing Innovation into Administrative Science Quarterly," in L .L. Cummings and P. J. Frost (Eds.), *Publishing in the Organizational Sciences*, Thousand Oaks, CA: Sage Publications, 1995, pp. 284-296.

Whetten, D. A. " What Constitutes a Theoretical Contribution?" *Academy of Management Review* (14), 1989, pp. 490-495.

Whyte, A., and Macintosh, A. "Transparency and Teledemocracy: Issues from an 'E-consultation'," *Journal of Information Science* (27:4), 2001, pp. 187-198.

Winsor, D. A. "Constructing Scientific Knowledge in Gould and Lewontin's 'The Spandrels of San Marco'," in J. Selzer (Ed.), *Understanding Scientific Prose*, Madison: University of Wisconsin Press, 1993, pp. 127-143.

ABOUT THE AUTHORS

Michael Barrett is a senior lecturer in Information Systems and E-Business at the Judge Institute of Management, University of Cambridge, where he also earned his Ph.D. His current research is principally concerned with processes of innovation, change, and use of information and communication technologies in the context of globalization, and the consequences for the nature of work. His work has been published in a wide range of management journals including *Information Systems Research*, *Academy of Management Journal*, and *European Journal of Information Systems*.

Geoff Walsham is a professor of Management Studies at the Judge Institute of Management, Cambridge University, UK. His teaching and research is centered on the social and management aspects of the design and use of information and communication technologies, in the context of both industrialized and developing countries. His publications include *Interpreting Information Systems in Organizations* (Wiley 1993), and *Making a World of Difference: IT in a Global Context* (Wiley, 2001). He can be reached at g.walsham@jims.cam.ac.uk.

Part 4:

Action Research

18 ACTION RESEARCH: Time to Take a Turn?

Briony J. Oates
University of Teesside

Abstract Following the linguistic turn of social sciences in the 20th century, some researchers are now taking a turn to action. They use action research but give it a broader meaning than that currently understood by many researchers in IS. This paper discusses the newer meaning of action research and indicates how it contrasts with some uses of action research reported in the IS literature. Five quality issues for the new action research are discussed: relational praxis, reflexive-practical outcome, plurality of knowing, significant work, and new and enduring consequences or infrastructure. The paper then gives a reflexive account of an IS research study that attempted to address these five issues. Finally, the paper discusses some of the broader implications for IS research of a turn to action.

Keywords: Action research, participation, metaphors, information systems development

1 INTRODUCTION

The 1984 IFIP WG 8.2 conference has been described as the occasion when some "'Young Turks'...set out to break the mold of orthodox IS research methodology" (IFIP WG 8.2 Call for Participation for the current conference). This was not just a move to promote qualitative data in IS research, but also a manifestation in IS of a phenomenon which took place last century across the social sciences, namely a change of paradigm in what has been called the linguistic turn. This turn brought the realization that there is a difference between the world itself and our interpreted experience of the world. It looked at the previously little-explored role of language in our construction of our world. In what has since become known as *interpretivism*, it was recognized that whatever reality is, it can only be accessed through social constructions such as language and shared meanings and understanding. For example, the scientific method based on positivism is itself a social construction. Interpretive studies, therefore, examine people in their social settings and try to understand phenomena through the meanings that people assign to them. The aim is "an organized discovery of how human agents make sense of their perceived worlds, and how those perceptions change over time and differ from one person or group to another" (Checkland and Holwell 1998, p. 22).

Since the 1984 IFIP 8.2 conference, interpretivism has won increasing acceptance in IS. In their 1991 survey, Orlikowski and Baroudi (1991 p. 5) categorized just 3 percent of the IS research papers they studied as interpretive. However, in 1995, Walsham found evidence of increasing use and acceptance of interpretivism in IS. Avison (1997) suggested that, in Europe at least, the emphasis is increasingly on interpretive methods, and the figure for positivist research in IS would be less than 50 percent by the turn of the century. Mingers' (2003) recent survey of journals found only 18 percent of papers were interpretive. However, this percentage would have been higher if Mingers had included action research in the interpretive paradigm, as previous surveys had, rather than in what he calls an "intervention oriented" paradigm, and his survey was restricted to only six IS journals. Even if much IS research, particularly in the U.S., continues to be based on positivism, interpretivism has become the chosen paradigm for a significant number of IS researchers.

Meanwhile, however, some social sciences researchers have taken another turn—the turn to action (Reason and Bradbury 2001b). This accepts the idea of knowledge as a social construction, but then "asks us to consider how we can act in intelligent and informed ways in a socially constructed world" (Reason and Bradbury 2001b, p. 2). Its methodology is based on action research.

Action research has also received attention in IS. It was discussed at the 1984 conference in the context of the development of the Multiview methodology (Wood-Harper 1985) and had already been used by Checkland (1981) in his development of soft systems methodology (SSM) and Mumford (1978, 1983) in her development of ETHICS. It has been used particularly, although not exclusively, for IS development methodologies (for example, Baskerville and Stage 1996; Baskerville and Wood-Harper 1996; Vidgen 2002). Other examples of its use include Scandinavian research aimed at empowering trade unions and users (Bjerknes et al. 1987), and the work of Mathiassen and his colleagues at Aalborg in Denmark (Mathiassen 1998). Comprehensive reviews of the use of action research in IS can be found in Baskerville and Wood-Harper (1998) and Lau (1997). Recent or planned special issues on action research in *Information Technology and People* (2001) and *MIS Quarterly* (forthcoming) are evidence of continuing, and perhaps increasing, interest in action research.

However, for social scientists who have taken the turn to action, action research has a broader meaning than that currently understood by many researchers in IS, in terms of scope, conceptual underpinnings, and forms of practice. This paper discusses how those who have taken the turn to action perceive action research—called here the new action research. It is structured around the five quality issues defined by Bradbury and Reason (2001), who currently offer the most detailed exposition of the emerging new action research (see also Reason and Bradbury 2001a, 2001b). This paper indicates how the new action research contrasts with some uses of action research reported in the IS literature. It suggests how IS action research could be enriched by the turn to action, and how such a turn would lead to a changed understanding of the nature of knowledge and a questioning of what is worthwhile research. The paper then illustrates this discussion via a reflexive account of an IS research study that attempted to follow the new action research principles. Finally, the paper discusses some of the broader implications for IS researchers of a turn to action.

2 THE NEW ACTION RESEARCH

The new action research is defined as

> a participatory, democratic process concerned with developing practical knowing in the pursuit of worthwhile human purposes, grounded in a participatory worldview which we believe is emerging at this historical moment. It seeks to bring together action and reflection, theory and practice, in participation with others, in the pursuit of practical solutions to issues of pressing concern to people, and more generally the flourishing of individual persons and their communities (Reason and Bradbury 2001b, p. 1).

We can compare this with Rapoport's definition of action research, which is often cited in the IS literature.

> Action research aims to contribute both to the practical concerns of people in an immediate problematic situation and to the goals of social science by joint collaboration within a mutually acceptable ethical framework (Rapoport 1970, p. 499).

It can also be compared with Kock's definition.

> A general term to refer to research methodologies and projects where the researcher(s) tries to directly improve the participating organization(s) and, at the same time, to generate scientific knowledge (Kock 1997).

The new definition places less emphasis than the other two on contribution to scientific knowledge and greater emphasis on worthwhile purposes, participation, and individual human (rather than organizational) flourishing. These aspects are all discussed in this section.

Varieties of the new action research include action science (Torbert 1991), participatory (action) research (Fals-Borda 2001) and cooperative inquiry (Heron 1996). There is thus no single version, to seek one probably suggests a modernist or positivist stance and a quest for a single, objective truth. Instead, different researchers, while trying to meet all of the parts of the first definition, tend to emphasize different parts of it (for many examples, see Reason and Bradbury 2001a).

Bradbury and Reason (2001) suggest researchers should address five quality issues in the new action research: relational praxis, reflexive-practical outcome, plurality of knowing, significant work, and new and enduring consequences/infrastructure. Each of these is explained and discussed below.

2.1 Relational Praxis

The defining characteristic of the emergent worldview of the new action research is participatory (Reason and Bradbury 2001b, p. 6). Our world is not made up of separate things, but of *relationships*, which we cocreate, participate in, and maintain. We cannot stand outside our world, we are necessarily already acting in it as we live and breathe. This worldview combines both positivism and interpretivism. Following positivism, it argues for a *real* reality, which is a state of being in the world, in which we all partake; following interpretivism, it acknowledges that as soon as we try to express this, we enter a maze of human language and socially constructed meanings (Reason and Bradbury 2001b, p. 7). The aim of the new action research is to support and enhance skills for being-in-the-world (Reason and Bradbury 2001b, p. 8). Given that we are *all* acting and being in the world, it also seeks to remove the researcher/subject distinction, aiming instead for a joint inquiry where people who share a problem come together to resolve it.

Heron (1996) argues that there are two complementary kinds of participation in research: *political* participation (concerning the relation between people in the inquiry and the decisions that affect them) and *epistemic* participation (concerning the relation between the knower and the known).

The arguments for political participation are (Heron 1996, p. 21)

- People have a right to participate in decisions about both the method and conclusions in research that seeks to formulate knowledge about them.
- It gives them the opportunity to express their own preferences and values in the research design.
- It empowers them to flourish fully as humans in the study, and be represented as such in its conclusions, rather than being passive subjects of the researchers.
- It avoids their being disempowered, oppressed and misrepresented by the researchers' values that are implicit in any unilateral research design.

The arguments for epistemic participation are (Heron 1996, pp. 20-21)

- Propositions about human experience are of questionable validity if they are not grounded in the researchers' experience.
- The most rigorous way to do this is for researchers to ground the statements directly in their *own* experience as co-subjects.
- Researchers cannot get outside, or try to get outside, the human condition in order to study it. They can only study it through their own embodiment, in joint participation and dialogue with others who are similarly engaged.
- This enables researchers to come to know not only the external forms of worlds and people but also the inner feelings and modes of awareness of these forms.

Heron criticizes quantitative, positivist research *on* people (pp. 25-26). Such research ignores the human right of people to participate in decisions about gaining knowledge of them (i.e., a lack of political participation). It also produces knowledge that is not experientially grounded: the researchers are not involved in the experience examined by the research, and the subjects are not involved in the selection of the constructs that are used to make sense of their experience (i.e., a lack of epistemic participation). Qualitative, interpretive research *about* people is also criticized where the research is designed and interpreted unilaterally by the researcher. However, interpretive researchers do include some participation (in the political sense) if they seek to validate their account with their respondents. Interpretive researchers can also be partially participant (in the epistemic sense) if they do fieldwork involving their own participation in the research setting, rather than being a detached observer. Often, however, decisions about what data to gather and the interpretive models used are not decided jointly with the subjects.

As Brechin writes (1993, p. 73),

> Research tends to be owned and controlled by researchers, or by those who, in turn, own and control the researchers. Those who remain powerless to influence the processes of information gathering, the identification of truth, and the dissemination of findings are usually the subjects of the research, those very people whose interests the research may purport to serve.

Hence qualitative research *about* people is seen as a halfway house between exclusive, controlling research *on* people and fully participatory research *with* people (Heron 1996, pp. 26-30). The new action research calls for research *with, for*, and *by* people (Reason and Bradbury 2001b). This implies that everyone is capable of being a researcher; it is not the preserve of those in laboratories and universities. This can, in turn, lead to ordinary people realizing that

> experts are not the objective, unbiased, disinterested purveyors of truth. Scientists often use "science" to impress or hide political decisions as "scientific." "Science" is not accountable and responsible to the needs of ordinary people but serves the power-holders (Lewis 2001, p. 361).

Recent public debate about the safety of the MMR vaccine or genetically modified crops are evidence of people's increased reluctance to accept scientists' statements unquestioningly.

Participation is often mentioned in the IS literature on action research. For example,

> A major strand of action research is that the practitioners should participate in the analysis, design and implementation processes and contribute at least as much as researchers in any decision making (Avison and Wood-Harper 1990, p. 180).

It is not clear from the IS literature that those practitioners have always been active partners for *all* aspects of an action research project. It is often left unstated, for example, whether they took part in all decisions and reflections about the research methodology and any relevant theories. For example, in the Lancaster school's research into SSM (e.g., Checkland 1981; Checkland and Scholes 1990), it is not clear whether all organizational participants and the postgraduate students involved could choose *not* to use SSM. (For a discussion of aspects of control in action research projects see Avison et al. 2001.)

The particular type of action research known as participatory action research is sometimes cited in the IS literature; for example, the work of Fals-Borda (2001), who has concentrated on liberationist inquiry in underprivileged communities, helping people to understand knowledge as an instrument of power and control, and seeking to raise their consciousness and empower them. But his approach seems to be treated as an interesting alternative to business-based action research, rather than a demonstration of a fundamental part of the worldview of action research. Instead, in IS the interests of an action researcher's co-participants seem often to be conflated with the interests of the organization as, for example, in Kock's definition of action research at the beginning of this section or Baskerville's (1999) definition: "The researcher is *actively involved*, with expected benefit for both *researcher* and *organization.*"

Organizations do not have needs. What are called organizational needs are formulated by powerful groups within the organization (Howcroft and Wilson 2003), possibly to the detriment of others.

There has also been a tradition of participatory design in IS development, developed via action research, especially in Scandinavia (e.g., Bjerknes et al. 1987; Howard 1985). However, participatory approaches too are criticized for not addressing conflict problems from the unequal distribution of power and the irreconcilability of management and worker needs, and for being managerialist in not challenging the power or legitimacy of managers' right to manage (Howcroft and Wilson 2003).

For new action research in business organizations, a goal of improved performance would not be superior to that of human flourishing. This could cause problems for IS action researchers who are based in business schools where managers are seen as the primary clients and research is focused on economic goals.

Explicit attention to the relational praxis issue could enrich IS action research by focusing attention on genuine participation by all affected and studied.

2.2 Reflexive-Practical Outcome

The new action research aims at practical outcomes, as well as conceptual knowledge, asking, for example, "Do people whose reputations and livelihoods are affected act differently as a result of the inquiry?" This applies as much to academic researchers as to other participants, suggesting the need for researchers to provide confessional accounts of their action research.

Drawing on Habermas, Kemmis (2001) (also Carr and Kemmis 1986) distinguishes between three kinds of action research and outcomes: technical, practical, and emancipatory. Much action research (in IS and across the social sciences) is *technical*, concentrating on functional improvements. For example, Kock et al. (1997) discuss an action research study that investigated the thesis that "groupware systems would positively affect productivity and quality of project-related activities." Technical action research would not normally question the goals themselves, nor how the situation in which the action research is being performed has been discursively, socially, and historically constructed (Kemmis 2001, p. 92).

Practical action research is influenced by Schön's (1983) ideas of the reflective practitioner. It has technical aspirations for change, but also aims to inform the practical decision making of the people involved in the research. They aim to improve their functional practice, but also to reflect on and understand how their goals, and the criteria they use for evaluating their practice, are shaped by their own ways of seeing themselves and their context. The research process becomes a form of self-education and the focus is as much on changing themselves as the *subjects* (or authors) of a practice as changing the *outcomes* of their practice (Kemmis 2001, p. 92). Examples can be found in researchers' subjective and confessional accounts of their reflections and personal learning during action research projects (e.g., Mumford 2001).

Emancipatory action research aims to improve technical performance, and the self-understanding of those involved, but also to help them critically evaluate their social or organizational context. They should understand how their functional goals may be limited or inappropriate within a wider view of the situation in which they live or work, and how their self-understandings may be shaped by shared misunderstandings about the nature and consequences of what they do. It aims to "connect the personal and the political in collaborative research and action aimed at transforming situations to overcome felt dissatisfactions, alienation, ideological distortion, and the injustices of oppression and domination" (Kemmis 2001, p. 92). Examples of this type of action research can be found in the women's, civil rights, and land rights movements. In IS, emancipation has been a goal for some researchers, who work within what is summarized as the critical paradigm (Orlikowski and Baroudi 1991). Although Orlikowski and Baroudi found no examples of critical studies and later Walsham (1995) found only a few, recently there has been an increasing number of papers, workshops, and conferences in the critical paradigm (e.g., Adam et al. 2001). However, much of the work in this critical paradigm tends to be analytical, identifying the structures, conventions, and contradictions that prevent human flourishing, without practically empowering those under study to recognize and overcome these barriers (e.g., Howcroft and Wilson 2003). Specifically using action research there have been a few emancipatory projects. For example, Waring (2000, 2002) initially investigated the continuing failure of NHS hospitals to implement integrated information systems but over time focused on emancipation and the role of the systems analyst within IS implementations, adopting a feminist gender lens.

Kemmis (2001; also Carr and Kemmis 1986) places these three types of action research in a hierarchy, with emancipatory action research as the most desirable. However, as Webb (1996) points out, all the words used to promote the ideals of such action research (such as emancipation, autonomy, democracy, consensus, rationality, solidarity, social justice, and community) are themselves problematic and contestable. Power is always present in any group, and cannot be dispersed by rationality; rationality and rational consensus often serve the interests of the powerful; and each of us plays multiple roles of both privilege and oppression. Emancipation is, therefore, neither simple, nor necessarily always desirable.

Explicit attention to the reflexive-practical outcome issue could enrich IS action research by focusing attention on the effect of the research on the researchers, whether technical, practical and emancipatory outcomes of the work are desirable and feasible, and whose interests the research serves.

2.3 Plurality of Knowing

Ways of knowing in new action research are aimed at supporting skills for being-in-the-world (Reason and Bradbury 2001b, p. 8). It is argued that the knowledge emerging from the new action research should have conceptual-theoretical integrity, embrace ways of knowing beyond the intellect, and choose appropriate research methods for finding these multiple ways of knowing.

Action research in IS has also placed strong emphasis on theoretical underpinnings and on outcomes that further contribute to theory. For example,

> The theoretical framework must be present as a premise, otherwise the intervention action is no longer valid as research. The diagnosis document should include explicit theoretical foundations. As the research progresses, the emergence of theory should be recorded carefully in the research notebooks (Baskerville 1999).

For example, soft systems methodology is based on general systems theory, and the action research on SSM led to a reconceptualizing of *system* in an epistemological sense rather than an ontological sense (Checkland and Holwell 1998).

However, the conceptual or intellectualized forms of knowledge as commonly used and produced in academia is not seen as the only form of knowledge outcome in the new action research. The new action research recognizes at least four different types of knowledge (Heron 1996, pp. 52-58; Heron and Reason 2001; Reason 1994a, pp. 42-46).

- *Experiential knowledge*—gained by direct encounter; almost impossible to put into words, being tacit and based on empathy, intuition and feeling
- *Presentational knowledge*—emerges from experiential knowledge; gives the first expression of knowing something through stories, drawings, sculpture, music, dance, etc.
- *Propositional knowledge*—about something in the form of logically organized ideas and theories, as in most academic research
- *Practical knowledge*—evident in knowing how to exercise a skill

These types of knowing require a wider range of methods than commonly found in action research in IS, such as song, dance, and exploring emotions. Since the world view is based on being-in-the-world, it follows that the full range of human sensibilities are appropriate to ways of finding out about fully being and acting in the world. This is likely to cause problems for many researchers who have been trained to remove the subjective and irrational from their research. Compare, for example, Baskerville's rejection of emotion (because it interferes with learning from the research) with Treleaven's welcoming of emotion (because it leads to new knowledge and action).

> The domain [of action research] excludes settings where explicit theoretical frameworks become excluded as the basis for action. A practical implication of this exclusion means that highly emotional social settings where rational

> action planning cannot be shared among the participants, will interfere with the learning from the research (Baskerville 1999).
>
> [Our collaborative inquiry methodology] challenged gendered emphasis placed on rationality in more traditional forms of action research. As participants with emotions and bodies which are themselves often ignored sources of our knowing...we made space in our workplace enquiry to attend to both.... Emotions that accompanied our stories—anger, despair and grief as well as joy with its laughter and well-being—were catalysts to new understanding and acting (Treleaven 2001, p. 262).

Explicit attention to the plurality of knowing issue could enrich IS action research by including the emotional, intuitive, and artistic skills of being in the world.

2.4 Significant Work

It is argued that new action researchers should focus on how they choose where to put their efforts (Bradbury and Reason 2001, p. 452). Worthwhile research will be that which is well-grounded in the everyday concerns of people, and may be seen as increasing in significance as it moves beyond the technical to developing people's capacity to ask fundamental questions about their world. Researchers and readers should ask how the action research helps to call forth a world worthy of human aspiration, so that people might say "that work is inspiring, that work helps me live a better life" (Bradbury and Reason 2001, p. 449).

Conventional IS action researchers who are accustomed to providing knowledge to support organizational objectives (or objectives of the organization's powerful) may find their research challenged by new action researchers who question whether it is worthwhile research. Explicit attention to the significant work issue could enrich IS action research by focusing attention on the kinds of research questions researchers choose to address.

2.5 New and Enduring Consequences or Infrastructure

New action research should also have consequences that endure. These can include new practical and academic knowledge but there are other possibilities too. For example, Bradbury and Reason (2001) ask whether those involved can say, "This work continues to develop and help us." They also ask, "Has the work been seeded so that it can be continued participatively if the initiating researcher moves away?" and, "Does it leave behind new patterns of behavior within a group, or new structures such as centers for action research?" and can others say, "We can use your work to develop our own."

In IS, successful action research is expected to be judged by two criteria: practical achievements in the problem situation and/or learning about the process of problem solving (Checkland and Scholes 1990). Explicit attention to the new and enduring consequences issue could enrich IS action research by recognizing more types of consequences, such as community ties or other forms of social capital, sustainability, and critical awareness.

3 EXAMPLE: A COOPERATIVE INQUIRY INTO METAPHORS

To illustrate the discussion above, this section provides a reflexive account of a study that was based upon the new action research approach using the five quality issues explained above. (For further detail of the study, see Oates 2000, 2002.)

3.1 Background and Motivation

The study involved four participants (myself as a full-time lecturer and part-time Ph.D. student and three systems developers, who were also students). Through cycles of action and reflection, we explored the extent to which conventionally educated information systems developers could adopt a richer model of organizations by using metaphors for organizations, derived in the main from Morgan (1986, 1993) as cognitive structuring devices, and we examined whether the metaphors had relevance for IS development practice.

For this action research, we were developing a theory: the use of organizational metaphors during ISD. We were also developing professional practice— the work of systems developers—and examining whether it could include metaphors to conceptualize their client organizations. Since the project also involved the development of information systems for three organizations, we were concerned with institutional change and development. We were also concerned with our individual learning and development.

My three coresearchers (Alan, Marcus, and Peter) had an average of three years of experience with ISD. They were currently computing degree students in their final year of studies but would return to IS developer roles on completion of their studies. An earlier action research study conducted by a master's student and myself had raised in my mind concerns about the uneven power relationship between lecturers and their research students (Oates 1999, 2000). Did the students' lower status militate against authentic collaboration, and how could academics guard against students reporting outcomes favorable to an academic's favored theory or methodology in the hope of gaining approval and better assessment grades?

I had therefore decided that for any future study involving a student as coresearcher I would work with a *group* of students. A group would at least help to redress the power imbalance; they would have some peer support and outnumber me as their lecturer. I decided to use cooperative inquiry (CI), a type of new action research, as my research strategy, because it is aimed at group research, where knowledge arises through action and joint reflection, with an emphasis on participation by all affected. The most comprehensive guide to CI is Heron (1996). (Additional sources include Heron and Reason 2001; Reason 1988, 1994b, 1994c; Reason and Bradbury 1999, 2001b; Reason and Rowan 1981.)

As recommended by Heron (1996, p. 102), the group agreed that each member of the CI group was free to write an account of the inquiry without submitting a draft to the others, but should make this limitation clear. This account has not been written collaboratively with the group. This is clearly a limitation on any claim that the research findings are based on authentic collaboration.

3.2 Relational Praxis

My coresearchers would return to ISD roles after their studies. Hence they were potential beneficiaries of the research ideas, who should be enabled to participate in its development, evaluation, and dissemination. They had been invited (not coerced) to take part. However, I was concerned about how far the research would be truly participative. My concerns about authentic collaboration included

- whether my academic language, and position of authority over the student coresearchers, might get in the way
- whether doing research initiated to meet my needs would be useful to them
- how to use my expertise in relation to metaphors for organizations and research, and yet do the research collaboratively
- how much I would control what the group did, and how much I could let go, i.e., how collaborative *I* could be

To deal with these concerns, I

- discussed the problem at the first group meeting and explored strategies they could use if I lapsed into lecturer mode
- arranged meetings not in my office but a spare classroom, which was more neutral ground
- ensured all had access to the same data (I circulated my notes on each meeting and deliberately chose not to tape record our meetings, which would have reinforced the idea that I was in charge: setting up the recorder and lending tapes to the coresearchers)
- asked the others what they thought before giving my views, even when questions were directed to me

Eventually I realized I had to accept that a power balance was inevitable, but each of us brought different knowledge and experience to the group. I had more knowledge of research and the use of metaphors, but they had greater expertise of the technical aspects of ISD. I needed their involvement to explore the use of metaphors in practice, but they needed my involvement to help them complete a satisfactory project for their degrees. CI does *not* imply equality, rather, each brings experiences and skills to the group and is willing to share and develop them collaboratively. At the beginning, I had to take the initiative, but through my actions and sharing my thinking, I could help the others take more control.

This issue of achieving authentic collaboration is discussed in many of the CI accounts (e.g., Marshall and McLean 1988; Traylen 1994; Treleaven 1994), and indeed is a significant issue in all non-positivist research (see Lincoln 1998; Lincoln and Denzin

1994). It is an unavoidable challenge where the research was initiated externally by researchers who, obviously, have their own needs or objectives which might not fully coincide with those of the other participants.

Indicators of our successful collaboration and the move from dependence on me as leader include

- increasingly those who arrived first started discussing project issues and did not break off on my arrival
- each of the others led discussions, suggested ideas to the others, and proposed metaphor-based views

It was a concern that I did not participate fully in the experience and action of ISD in the client organizations (epistemic participation). A resolution of this was suggested by Traylen (1994). She helped health visitors explore their hidden agendas in their meetings with clients, but was not a practicing health visitor herself. She realized that she too had hidden agendas in her meetings with the health visitor coresearchers. These could be explored as part of the research, increasing her epistemic participation. Similarly, I realized that I could think of our group as a small organization, and investigate how Morgan's metaphors helped me conceptualize it (see Oates 2000).

I felt there had been authentic collaboration. I also invited the coresearchers to complete a questionnaire anonymously, after the end of our inquiry. Their responses show each was happy with the group process and its findings on metaphors, and no one thought anyone had dominated the discussions.

I feel the CI study did achieve authentic participation among the coresearchers and myself. However, what of others affected by the research? Each of the student coresearchers interacted with people in their client organizations. They were consulted about the information systems being developed, but only in order to develop improved information systems. We did not consider whether a new information system might somehow enhance or prevent their human flourishing. We did agree we should share with people in the client organizations the metaphor ideas. However, each of the coresearchers found it difficult to explain the use of metaphors to others, encountering incomprehension and even derision, so this was not successful. Members of the organizations were not involved in our use of metaphors to understand them. Reflections on the relational praxis issue therefore highlighted this limitation of the study.

3.3 Reflexive-Practical Outcome

Tangible outcomes from the research were working information systems for the three client organizations and student success—each student coresearcher passed the project part of his degree and went on to gain a B.Sc. honors degree, and I ultimately gained my Ph.D. (Oates 2000).

As explained earlier, Kemmis (2001) defines three types of action research outcomes: technical, practical, and emancipatory. This study had a technical focus: developing a systems development methodology that used organizational metaphors to enable a richer view of organizations than that found in conventional systems development. A technical outcome was a prototype methodology (MMM, or multiple metaphor methodology) that we created and refined (Oates 2000).

The study also had practical outcomes in Kemmis' sense of self-education. We each learned abut how we liked to work and research. The student coresearchers all reported that they had enjoyed the group-based research. They felt it was a better way of working than the one-to-one (supervisor-to-student) mode normally used to support student projects because

- group meetings were less intense, stressful, and pressured
- each had the opportunity to sit back and relax, or think things through, while others discussed their projects
- each learned from the others, and there were richer discussions

I too found I enjoyed the cooperative inquiry, and I learned that I *could* reduce my control when working with a student group, gradually abandoning careful plans for each meeting and trusting that learning would emerge from a more unstructured approach where others could take the lead.

Emancipatory outcomes are concerned with those involved critically evaluating their social or organizational context. As discussed above, we did not address this aspect with respect to the employees of our client organizations. We did, however, become critical about our own situations. The student coresearchers criticized the limited education they had received about organizations. Their degree courses (computer science for two, information technology for the third) had concentrated on technical skills and knowledge, yet their future careers would almost certainly entail operating within the complex, dynamic, and fuzzy organizations made up of, and constructed by, people, for which they were ill-equipped. Through reading about, and performing, cooperative inquiry, I became more convinced that we can all be researchers into our own lives and practices, and research should not be restricted to an elite group.

Attention to the reflexive-practical outcomes issue, therefore, highlighted not only tangible, practical outcomes but also self-knowledge and critical awareness outcomes.

3.4 Plurality of Knowing

The work had strong conceptual-theoretical underpinnings. It drew on Morgan's well-established work on organizational analysis via metaphors as well as cognitive psychology theories about how we think via metaphors (e.g., Holyoak and Thagard 1996).

Propositional knowledge was developed as we used metaphors to structure and articulate interpretations of the three client organizations. On the basis of each interpretation, assertions were made about the organizations and/or actions were taken. Qualitative, interpretive accounts were produced, which explored our joint sense-making via metaphors and established the relevance of organizational metaphors to ISD practice in the context of our three client organizations. These can be found in Oates (2000) and the coresearchers' individual project reports (Findlay 1998; Lyons 1998; Thomas 1998).

The other three types of knowledge identified by Heron were also gained. Experiential knowledge was gained in both using metaphors and collaborating in research. We were each changed by having participated in the cooperative inquiry and the systems development projects. Presentational knowledge was produced when we told stories of events we had experienced or observed in organizations. We also used diagrams and other pictures to explore our understanding of the organizations. However, none of us used more artistic forms of expression such as poetry or music. Practical knowledge was gained by each member in learning how to map the metaphors to organizations and link the mapping to ISD issues.

We did allow space for emotions within our inquiry. We frequently laughed as we told stories of our experiences, and we acknowledged when we were feeling tired and had (temporarily) lost interest in the research. We also discussed our anxieties. At the first meeting, we discussed our worries about doing the research and writing reports and a thesis. Later worries we discussed included course demands and uncertainties about future career plans. Hence the group members provided support for each other beyond producing academic, propositional knowledge.

Attention to the plurality of knowing issue provided opportunities to use a wider range of inquiry skills and recognize a variety of knowledge outcomes.

3.5 Significant Work

The study enabled us to practice a method of taking greater account of organizations and the people within them, who were not seen as just cogs in a machine. Through using the metaphors, we were able to articulate richer conceptualizations of the organizations and employees, including their politics and cultures. We were able to identify how these richer conceptualizations did not just lead to greater understanding, but also influenced both our systems development process and product, leading to systems which took greater account of human needs and should mesh better with their social and organizational contexts. As discussed in the outcomes section above, we all also learned something about ourselves and our work and educational situation. I believe the research was worthwhile and significant.

Reflecting on the significant work issue made me consider whether I was justified in involving others in answering my research questions and whether the research aim was really worthy of attention.

3.6 New and Enduring Consequences

The rationale for the work, its conceptual underpinning and its research outcomes are discussed in Oates (2000) and are available for others to try out and evaluate in their own studies. Our process of cooperative inquiry is discussed and evaluated in Oates (2000, 2002) so that others can reconstruct the process and amend it to suit their own group. The student coresearchers have also written accounts of their experience and findings (Findlay 1998; Lyons 1998; Thomas 1998), which other students have consulted. The information systems we developed continue to be used.

The group itself did not endure beyond the end of the coresearchers' studies; each has taken up a position elsewhere in the country. However, each has taken with them an understanding of the use of organizational metaphors and their potential relevance to their systems development practice. One has since reported transferring the approach to a new situation. Follow-up research could find out whether and how they have used the approach in their new jobs.

Through reflecting on cooperative inquiry and the new action research, I also recognized my own preferences about types of research. I now prefer projects within the public, voluntary, and community sectors, where I have found greater openness than in the business sector to ideas of human flourishing.

Reflecting on the enduring consequences issue highlighted for me the study's consequences beyond answering the original research questions.

This section has illustrated the use of the five quality issues and demonstrated some of the possibilities offered by the new action research.

4 SUMMARY AND IMPLICATIONS

This paper has discussed the new action research, and illustrated the discussion by means of a confessional account of an action research study. The paper was structured around five quality issues: relational praxis, reflexive-practical outcome, plurality of knowing, significant work, and new and enduring consequences or infrastructure.

Implications for IS researchers of specifically addressing these five issues are

- increasing attention to human flourishing, addressing the needs of all those involved, not just the powerful
- reflecting on the effects of the research on themselves and considering whether technical, practical, and emancipatory outcomes of their work are desirable and feasible
- recognizing other types of knowledge and knowledge outcomes (e.g., emotional and artistic) as well as propositional and practical knowledge
- reconceptualizing their notions of worthwhile research
- recognizing other potential consequences of their research (e.g., community ties and critical awareness) as well as practical achievements and academic knowledge

The new action research uses new kinds of data and data analysis, inquiry skills, and types of knowledge. A change would be required similar to some researchers' earlier move from quantitative to qualitative data. If the new action research is to be accepted, the IS academic community would need to acknowledge its current bias toward word-based, propositional knowledge and recognize the other types as being of equal value to (or greater value than) propositional knowledge.

The new action research is based on a different worldview, based on action and participation. A paradigm change would be required, similar to some IS researchers' earlier move from positivism to interpretivism. The belief that research into human experience should be carried out by those doing the experiencing implies that everyone is capable of being a researcher; research is not an exclusive preserve of academics. IS researchers would have to ask whether they are willing to let go and share their position as researchers.

Finally, a turn to action would imply more than the use of new types of data and a new worldview. It would raise fundamental questions about the role of IS researchers, why we do research, and for whose benefit.

IS researchers currently choose from a range of strategies and paradigms. The new action research widens the range. It is unlikely that everyone will choose to adopt it, but those that do will be able to explore its relevance to IS research and the appropriateness of the five quality issues discussed here. It may be difficult for it to gain respectability, just as current IS action research is not seen by everyone as acceptable research. IS researchers will need to consider whether their own career objectives are attainable if they choose to do new action research. However, this paper has discussed how IS action research *could* be enriched by taking the turn to action.

REFERENCES

Adam, A., Howcroft, D., Richardson, H., and Robinson, B. (Eds.). *(Re-)Defining Critical Research in Information Systems: Workshop Proceedings*, University of Salford, Salford, UK, July 9-10, 2001.

Avison, D. "The 'Discipline' of Information Systems: Teaching, Research and Practice," in J. Mingers and F. A. Stowell (Eds.), *Information Systems: An Emerging Discipline?*, London: McGraw-Hill, 1997, pp. 113-135.

Avison, D., Baskerville, R., and Myers, M. "Controlling Action Research Projects," *Information Technology and People* (14:1), 2001, pp. 28-45.

Baskerville, R. "Investigating Information Systems with Action Research," *Communications of the Association for Information Systems* (2:19), 1999.

Baskerville, R., and Stage, J. "Controlling Prototype Development through Risk Analysis," *MIS Quarterly* (20:4), 1996, pp. 481-502.

Baskerville, R. L., and Wood-Harper, A. T. "A Critical Perspective on Action Research as a Method for Information Systems Research," *Journal of Information Technology* (11), 1996, pp. 235-246.

Baskerville, R., and Wood-Harper, A. T. "Diversity in Information Systems Action Research Methods," *European Journal of Information Systems* (7), 1998, pp. 90-107.

Bjerknes, G., Ehn, P., and Kyng, M. *Computers and Democracy*, Aldershot, UK: Avebury, 1987.

Bradbury, H., and Reason, P. "Conclusion: Broadening the Bandwidth of Validity: Issues and Choice-Points for Improving the Quality of Action Research," in P. Reason and H. Bradbury (Eds.), *Handbook of Action Research: Participatory Inquiry and Practice*, London: Sage Publications, 2001, pp. 447-455.

Brechin, A. "Sharing," in P. Shakespeare, D. Atkinson, and S. French (Eds.), *Reflecting on Research Practice: Issues in Health and Social Welfare*, Buckingham, UK: Open University Press, 1993, pp. 70-82.

Carr, W., and Kemmis, S. *Becoming Critical: Education, Knowledge and Action Research*, London: Falmer, 1986.

Checkland, P. *Systems Thinking, Systems Practice*, Chichester: Wiley, 1981.

Checkland, P., and Holwell, S. *Information, Systems, and Information Systems: Making Sense of the Field*, Chichester: Wiley, 1998.

Checkland, P. B., and Jenkins, G. M. "Learning by Doing: Systems Education at Lancaster University," *Journal of Systems Engineering* (4:1), 1974, pp. 40-51.

Checkland, P., and Scholes, J. *Soft Systems Methodology in Action*, Chichester: Wiley, 1990.

Fals-Borda, O. "Participatory (Action) Research in Social Theory: Origins and Challenges," in P. Reason and H. Bradbury (Eds.), *Handbook of Action Research: Participatory Inquiry and Practice,* London: Sage Publications, 2001, pp. 27-27.

Findlay, A. "An Information System for Northton Borough Council Structures Department," Unpublished B.Sc. Individual Project Report, School of Computing and Maths, University of Teesside (Pseudonym used), 1998.

Heron, J. *Co-operative Inquiry: Research into the Human Condition*, London: Sage Publications, 1996.

Heron, J., and Reason, P. "The Practice of Co-operative Inquiry: Research 'with' Rather than 'on' People," in P. Reason and H. Bradbury (Eds.), *Handbook of Action Research: Participatory Inquiry and Practice,* London: Sage Publications, 2001, pp. 179-188.

Holyoak, K. J., and Thagard, P. *Mental Leaps: Analogy in Creative Thought*, Cambridge, MA; MIT Press, 1996.

Howard, R. "UTOPIA: Where Workers Craft New Technology," *Technology Review* (14:1), 1985, pp. 43-49

Howcroft, D., and Wilson, M. "Paradoxes of Participatory Practices: The Janus Role of the Systems Developer," *Information and Organization* (13:1), 2003, pp. 1-24.

Kemmis, S. "Exploring the Relevance of Critical Theory for Action Research: Emancipatory Action Research in the Footsteps of Juergen Habermas," in P. Reason and H. Bradbury (Eds.), *Handbook of Action Research: Participatory Inquiry and Practice,* London: Sage Publications, 2001, pp. 91-102.

Kock, N. F. "Myths in Organizational Action Research: Reflections on s Study of Computer-Supported Process Redesign Groups," *Organizations and Society* (4:9), 1997, pp. 65-91.

Kock, N. F., McQueen, R. J., and Scott, J. L. "Can Action Research Be Made More Rigorous in a Positivist Sense? The Contribution of an Iterative Approach," *Journal of Systems and Information Technology* (1:1), 1997, pp. 1-24.

Lau, F. "A Review on the Use of Action Research in Information Systems Studies," in A. Lee, J. Liebenau, and J. I. DeGross (Eds.), *Information Systems and Qualitative Research,* London: Chapman & Hall, 1997, pp. 31-68.

Lewis, H. M. "Participatory Research and Education for Social Change: Highlander Research and Education Centre," in P. Reason and H. Bradbury (Eds.), *Handbook of Action Research: Participatory Inquiry and Practice,* London: Sage Publications, 2001, pp. 356-362.

Lincoln, Y. S. "From the Discourse of 'the Other' to the Other Discourse: Stakeholder-Based Models of Evaluation and the Human Side of Social Welfare," *Scandinavian Journal of Social Welfare* (7:2), 1998, pp. 114-117.

Lincoln, Y. S., and Denzin, N. K. "The Fifth Moment," in N. K. Denzin and Y. S. Lincoln (Eds.), *Handbook of Qualitative Research,* Thousand Oaks, CA: Sage Publications, 1994, pp. 575-586.

Lyons, P. "Internet Retailing Opportunities for Northern DIY," Unpublished B.Sc. Individual Project Report, School of Computing and Maths, University of Teesside (Pseudonym used), 1998.

Marshall, J., and McLean, A. "Reflection in Action: Exploring Organizational Culture," in P. Reason (Ed.), *Human Inquiry in Action: Developments in New Paradigm Research,* London: Sage Publications, 1988, pp. 199-220.

Mathiassen, L. "Reflective Systems Development," *Scandinavian Journal of Information Systems* (10:1&2), 1998, pp. 67-118.

Mingers, J. "The Paucity of Multimethod Research: A Review of the Information Systems Literature," *Information Systems Journal* (13:3), 2003, pp. 233-249.

Morgan, G. *Images of Organization*, Beverly Hills, CA: Sage Publications, 1986.

Morgan, G. *Imaginization: The Art of Creative Management*, Newbury Park, CA: Sage Publications, 1993.

Mumford, E. "Advice for an Action Researcher," *Information Technology and People* (14:1), 2001, pp. 12-27.
Mumford, E. *Designing Human Systems*, Manchester, UK: Manchester Business School, 1983.
Mumford, E. "Job Satisfaction: A Method of Analysis," in K. Legge and E. Mumford (Eds.), *Designing Organizations for Satisfaction and Efficiency*, Teakfield, UK: Gower Press, 1978, pp. 18-35.
Oates, B. J. "Co-operative Inquiry: Reflections on Practice," *Electronic Journal of Business Research Methods* (1:1), 2002, pp. 33-42 (available online at http://www.ejbrm.com).
Oates, B. J. *Metaphors for Organizations During Information Systems Development*, Unpublished Ph.D. Thesis, University of Teesside, Middlesbrough, UK, 2000.
Oates, B. J. "Participation in Information Systems Research," in A. T. Wood-Harper, N. Jayaratna, and J. R. G. Wood (Eds.), *Methodologies for Developing and Managing Emerging Technology Based Systems*, London: Springer Verlag, 1999, pp. 269-279.
Orlikowski, W. J., and Baroudi, J. J. "Studying Information Technology in Organizations: Research Approaches and Assumptions," *Information Systems Research* (2:1), 1991, pp. 1-28.
Rapoport, A. "Three Dilemmas of Action Research," *Human Relations* (23:6), 1970, pp. 499-513.
Reason, P. (Ed.). *Human Inquiry in Action: Developments in New Paradigm Research*, London: Sage Publications, 1988.
Reason, P. "Human Inquiry as Discipline and Practice," in P. Reason (Ed.), *Participation in Human Inquiry*, London: Sage Publications, 1994a, pp. 40-56.
Reason, P. (Ed.). *Participation in Human Inquiry*. Sage, London, 1994b.
Reason, P. "Three Approaches to Participative Inquiry," in N. K. Denzin and Y. S. Lincoln (Eds.), *Handbook of Qualitative Research*, Thousand Oaks, CA: Sage Publications, 1994c, pp. 342-339.
Reason, P., and Bradbury, H. (Eds.). *Handbook of Action Research: Participatory Inquiry and Practice*, London: Sage Publications, 2001a.
Reason, P., and Bradbury, H. "Introduction: Inquiry and Participation in Search of a World Worthy of Human Aspiration," in P. Reason and H. Bradbury (Eds.), *Handbook of Action Research: Participatory Inquiry and Practice*, London: Sage Publications, 2001b, pp. 1-14.
Reason, P., and Heron, J. "A Layperson's Guide to Co-operative Inquiry," Centre for Action Research in Professional Practice, School of Management, University of Bath, 1999.
Reason, P., and Rowan, J. (Eds.). *Human Inquiry: A Sourcebook of New Paradigm Research*, Chichester: Wiley, 1981.
Schön, D. A. *The Reflective Practitioner: How Professionals Think in Action*, New York: Basic Books, 1983.
Thomas, M. An Information System for a Diabetics Unit," Unpublished B.Sc. Individual Project Report, School of Computing and Maths, University of Teesside, 1998.
Torbert, W. R. *The Power of Balance: Transforming Self, Society, and Scientific Inquiry*, Newbury Park, CA: Sage Publications, 1991.
Traylen, H. "Co-operative Inquiry with Health Visitors," in P. Reason (Ed.), *Participation in Human Inquiry*, London: Sage Publications, 1994, pp. 59-81.
Treleaven, L. "Making a Space: A Collaborative Inquiry with Women as Staff Development," in P. Reason (Ed.), *Participation in Human Enquiry*, London: Sage Publications, 1994, pp. 138-162.
Treleaven, L. "The Turn to Action and the Linguistic Turn: Towards an Integrated Methodology," in P. Reason and H. Bradbury (Eds.), *Handbook of Action Research: Participatory Inquiry and Practice*, London: Sage Publications, 2001, pp. 260-272.
Vidgen, R. "Constructing a Web Information System Development Methodology," *Information Systems Journal* (12), 2002, pp. 247-261.

Walsham, G. "The Emergence of Interpretivism in IS Research," *Information Systems Research* (6:4), 1995, pp. 376-394.

Waring, T. "Gender Reflexivity: A Missing Element from Action Research in Information Systems," *Electronic Journal of Business Research Methods* (1:1), November 2002, pp. 50-58 (available online at http://www.ejbrm.com).

Waring, T. S. *The Systems Analyst and Emancipatory Practice: An Exploratory Study in Three NHS Hospitals*, Unpublished Ph.D. Thesis, University of Northumbria, Newcastle, 2000.

Webb, G. "Becoming Critical of Action Research for Development," in O. Zuber-Skerritt (Ed.), *New Directions in Action Research,* London: Falmer Press, 1996, pp. 137-161.

Wood-Harper, A. T. "Research Methods in Information Systems: Using Action Research," in E. Mumford, R. Hirschheim, G. Fitzgerald and A. T. Wood-Harper (Eds.), *Research Methods in Information Systems,* Amsterdam: Elsevier Science Publishers, 1985, pp. 169-191.

ABOUT THE AUTHOR

Briony J. Oates is a senior lecturer in Information Systems in the School of Computing at the University of Teesside, Middlesbrough, UK. She holds a BA (Hons) in German language and literature (1978, Newcastle), an M.Sc. in information technology (1988, Teesside), and a Ph.D. in information systems (2000, Teesside), and has worked as both an academic and a practitioner. Her Ph..D thesis examined the role of organizational metaphors as cognitive structuring devices during information systems development. Her current research interests include the social and organizational aspects of information systems development, development methods for Web-based systems, and e-government. She is also currently preparing a book on research methods, *Researching Information Systems and Computing,* which Sage plans to publish in 2005. Briony can be reached by e-mail at B.J.Oates@tees.ac.uk.

19 THE ROLE OF CONVENTIONAL RESEARCH METHODS IN INFORMATION SYSTEMS ACTION RESEARCH

Matt Germonprez
Case Western Reserve University

Lars Mathiassen
Georgia State University

Abstract Action research has for many years been promoted and practiced as one way to conduct empirical research within the Information Systems discipline. While the approach can lead to highly relevant contributions, researchers are warned against the many risks involved in action research. Based on successful cases of Information Systems action research we explore the role played by conventional research methods in developing and presenting research contributions. The cases suggest that action research lends itself strongly toward multi-method approaches and facilitates the creation of multi-contribution projects. We identify two approaches to mixing action research and conventional research methods—the planned and the emergent approaches—and we argue that action research can be adopted in ways that are no more risky than other conventional approaches to Information Systems research.

Keywords: Action research, IS research methodologies, pluralism

1 INTRODUCTION

In research, as in practice, we constantly balance the best methods to apply to questions and problems. We adopt methods because we have used them in the past, others have used them, and we know how to implement them, only to find out we can not answer all the questions we need to answer. To resolve this, we add additional methods for answering our remaining questions. While accepting findings from the first method, we often need more explanation than the first method could deliver. We therefore constantly run methods sequentially, in parallel, and even at different levels of analysis (Mingers 2001). We have an elaborate understanding of the use of singular research methods, the data they can produce, the questions they can answer, and how they can help develop a high quality contribution to scientific knowledge. We lack, however, a practical understanding of how research methods are mixed and how the use of multi-method approaches supports the researchers in identifying and developing several contributions within the same research project.

This paper explores this issue in the particular context of action research. Action research was chosen because of its support for both positivistic and interpretivistic approaches (Kock 1997b), however, the conclusions can be applied across varying epistemologies, ontologies, and methodologies (Walsham 1995). We build on the exemplar work of John Mingers (2001) on how to mix methods within the Information Systems discipline. We use Mingers' work as a framework to study methodological pluralism in IS and explain how pluralism is used to increase diversity in IS research. Our aim is twofold. First, we want to further our understanding of multi-method approaches to Information Systems research. Second, we want to contribute to a more practical understanding of how action research can be carried out within our discipline. Specifically we address the following research question: What is the role of pluralist methodology in action research within the Information Systems discipline?

Action researchers attempt to develop contributions that are relevant to practice by involving themselves in particular problem situations (Rapoport 1970). While this approach offers important opportunities for exploration and learning, it unfortunately leads to a number of pitfalls. Baskerville and Wood-Harper (1996) have summarized these as (1) lack of impartiality of the researcher; (2) lack of discipline; (3) mistaken for consulting; and (4) context-dependency leading to difficulty of generalizing findings. Many Information Systems researchers hesitate for these reasons to include action research in their repertoire of research methods and younger colleagues are often warned against doing action research because it might slow down or damage their career. We argue in the following that action research lends itself strongly toward multi-method approaches and that a deeper understanding of the multi-method nature of action research can lead us to action research practices that are no more risky than other conventional approaches to Information Systems research.

Mingers proposes mixing methods as a solution for navigating the research process in moving from an original appreciation of a research project to action in implementing the findings. Where one methodology falls short along this path, another can succeed. Action research is sometimes used from the outset as a planned and dominant methodology through which research explanations are provided. Action research emerges in other situations as a useful complimentary method embedded into a larger research process. Drawing upon published cases of successful Information Systems research, we develop and discuss this distinction between *planned* and *emergent* method mixing involving action research. We show in particular how action research invites researchers to engage in additional empirical explorations or develop theoretical thinking based on existing findings. Action research facilitates for these reasons creation of multi-contribution projects.

After having reviewed the literature on action research and Mingers' framework for mixing research methods, the paper presents selected cases of planned and emergent approaches to method mixing involving action research. On that basis, we discuss the implied contributions to research methodology and practice within the Information Systems discipline.

2 ACTION RESEARCH

Information Systems action research is applied research to develop a solution that is of practical value to the people with whom the researchers are working, and at the same time to develop theoretical knowledge of value to a research community (Davison 1998, pp. 3-6). This definition is in line with classical definitions of action research (Rapoport 1970) and it directly addresses concerns by Gustavsen (1993), Levin (1993), and Kock et al. (1997) who state that action research must support both the production of practical outcomes as well as the production of research theory. Baskerville and Wood-Harper (1998) echo the dual outcome perspective of action research. They state that action research is embedded within a practical system it is being used to explain. Within this system, an "unstructured field experiment" takes place where researchers act as change agents to improve practical outcomes and describe a new state of affairs that result from the change (Baskerville and Wood-Harper 1998, p. 91).

Action research links theory and practice through a cyclic and often iterative process (Baskerville and Wood-Harper 1996; Checkland 1991; Hult and Lennung 1980; McKay and Marshall 2001; Susman and Evered 1978). There are many different ways to organize the cyclic process. The model of Susman and Evered (1978) captures well the general approach taken. According to them, action research is executed through one or more cycles of (1) diagnosing (identifying or defining a problem); (2) action planning (considering alternative courses of action for problem solving); (3) action taking (selecting and executing a course of action); (4) evaluating (studying the consequences of the action); and (5) specifying learning (identifying general learning).

Action research allows the researchers to tap directly into practical problem solving through organizational intervention, but it requires constant attention and specific skills to manage the process. There are pitfalls involved (Baskerville and Wood-Harper 1996) and the researchers are confronted with dilemmas related to issues of ethics, goals, and initiative (Rapoport 1970). Researchers need, for each dilemma, to balance a concern for the practical problem and their scientific interests. Leaning too strongly toward practical concerns the process will result in actions that are not theoretically informed and have no cumulative scientific effect. In the other extreme, the process leads away from actions that have relevant implications for the situation at hand. Avison et al. (2001) suggest that action research therefore raises control issues related to the initiation of projects, the authority for action, and the degree of formalization. They offer for each of these managerial concerns a variety of approaches to help structure action research efforts. McKay and Marshall (2001) suggest along the same lines that action researchers should conceive the process as consisting of two parallel, interacting cycles: the problem cycle (focused on the problematic situation) and the research cycle (focused on the scientific goals). This distinction helps researchers separate concerns, adopt appropriate problem solving and research methods, and manage the process toward the dual goals of action research.

We argue in the following that these particular characteristics of action research lend themselves strongly to pluralist methodology and also that a deeper understanding of the multi-method nature of action research can help researchers design and manage their efforts in ways that lead to scientific progress without abandoning practical concerns.

3 MIXING METHODS

Without a strong handle on methodology, research questions cannot be framed, data cannot be analyzed, theory cannot be tested, and science cannot be informed. Specifically, a researcher's choice of method(s) impacts which conclusion can be drawn and it shapes the explanation of results (Kinkaid 1994; van Fraassen 1980; Orlikowski and Baroudi 1991; Salmon 1989). The practical use of methods incorporates researcher beliefs, it is a concrete manifestation of the ontology and epistemology of a field, and it ultimately impacts how research contributes to and shapes scientific paradigms. The question of which research method(s) to use is therefore of primary importance both to the individual researcher and to the continued development of our discipline.

Mingers (2001) offers two fundamental reasons for combining methods in Information Systems research. First, the world we are studying is multidimensional including the material world, the personal world, and the social world (Habermas 1984). To address and understand these different aspects of the world, we need a variety of research methods. Second, research is a process which involves rather different challenges and activities, which will predominate at different times as the process unfolds. To effectively address the challenges involved—including appreciation of the research situation, analysis of data, assessment of explanations, and action to report on or disseminate results (Mingers 2001)—the researcher needs different methods to help develop richer and more reliable results. As researchers move through objective, subjective, and intersubjective perspectives (Habermas 1984) and as they take part in appreciation, analysis, assessment, and action as part of the research process, they are advised to adopt methods across the epistemological perspectives that support the movement from appreciation to action. In doing so, they must balance between relying on tested methods and forging new methodological ground in the examination and explanation of new phenomena. Mingers addresses this thinking through five different types of multi-method research design:

- *Sequential*, where methods are applied in a sequence with results from one method feeding into the next
- *Parallel*, where methods are executed simultaneously with results being transferred between them
- *Dominant*, where one method is adopted as the main approach supplemented by other methods
- *Multi-methodology*, where different methods embodying different paradigms are combined and tailored to a particular project
- *Multi-level*, where the research simultaneously addresses different organizational levels using different methods

Mixing methods requires in each of these cases certain skills to effectively manage the additional complexity within the research process. Researchers must, in Mingers' terms, know how to organize and manage the appreciation, analysis, assessment, and action activities by selecting and mixing a variety of research methods. They need, on a more concrete level, the specific strategies and tactics to practice multi-method research.

4 ACTION RESEARCH CASES

Sometimes researchers have a clear agenda of how they plan on using action research and mixing methods in a research process. Other times, researchers are contributing to a collection of studies, using an emerging set of methods to initiate explorations and provide explanations based on varied epistemological perspectives. We denote the former the *planned approach* where action research becomes an explicit instantiation of Mingers' research process of appreciation, analysis, assessment, and action. We denote the latter the *emergent approach* in which action research is embedded into a broader research context. Mingers' research process represents, in these cases, a meta-level framework that can explain and guide how different methods are selected and mixed in order to inform science and practice. We have explored 12 cases of successful action research within the Information Systems discipline. Three cases are represented in the following two sections to illustrate the planned and emergent approaches of mixing action research. Appendix A presents nine additional cases in summary form.

4.1 The Planned Approach

Researchers often turn to multiple methods in the examination and explanation of research questions. Action research can act as a platform on which explanation can be produced and additional methods can be adopted. Action research is then chosen and articulated as the primary method representing Mingers' research process. There is, from the outset, an overall contractual arrangement between a client (with some problematic situation) and the researchers. It is agreed that the collaboration will serve the double purpose of addressing the client's problematic situation while at the same time advancing scientific knowledge within certain disciplines (Rapoport 1970). The approach is fundamentally based on action research and other, supplementary research methods are adopted as needed.

Mathiassen and his colleagues used the planned approach to study software process improvement (SPI) in four software companies over a three-year period (Mathiassen 2002; Mathiassen et al. 2002). The dual purpose of this project was to improve software practices in four software organizations and at the same time to contribute to knowledge on software engineering and management. Half a dozen researchers and nearly 40 practitioners participated and the project resulted in a number of contributions, including a book (Mathiassen et al. 2002), three Ph.D. dissertations, and more than 30 conference and journal papers. A few, selected publications illustrate how different research methods were adopted and mixed in this planned approach to action research (see Figure 1).

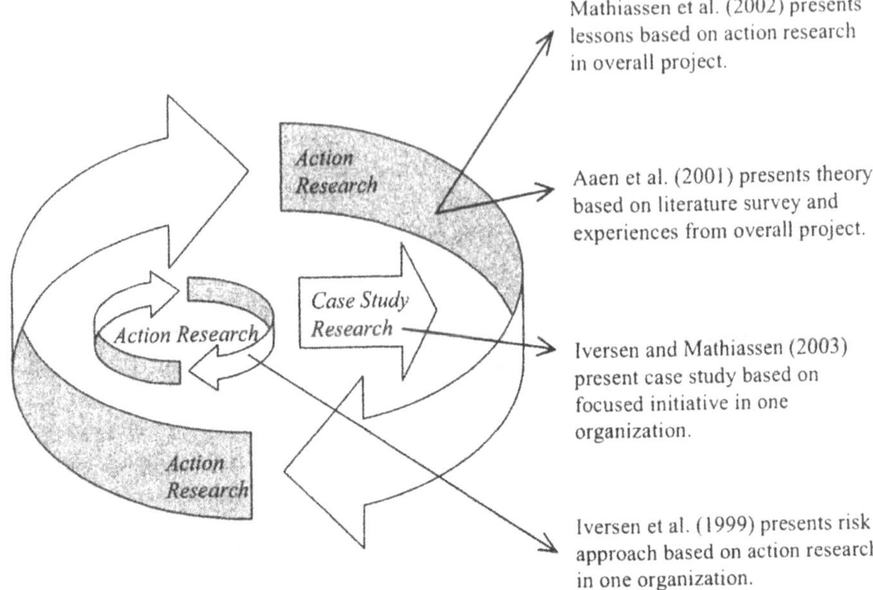
Mathiassen et al. (2002) presents lessons based on action research in overall project.
Action Research
Aaen et al. (2001) presents theory based on literature survey and experiences from overall project.
Case Study Research
Action Research
Iversen and Mathiassen (2003) present case study based on focused initiative in one organization.
Action Research
Iversen et al. (1999) presents risk approach based on action research in one organization.

Figure 1. A Planned Approach to Action Research (Dominant)

Mathiassen et al. (2002) documents the action research in the four organizations. The book includes key lessons from each of the four software organizations together with contributions that focus on specific issues, e.g., assessment of software practices, adoption of knowledge management tactics in SPI, and practical approaches to process implementation. The book presents contributions from the overall project based on action research.

Aaen et al. (2001) offer a conceptual framework for understanding SPI theory and for assessing SPI strategies adopted by specific organizations. The SPI domain is broad with no agreed-upon understanding of the underlying assumptions and key ideas involved in this approach to improving software practices. Aaen et al. provide a conceptual framework based on a systematic survey of the SPI literature and informed by experiences from the four organizations. This publication represents theory development from the overall project based on literature studies and conceptualization of SPI practices in the four organizations.

Other publications were developed based on focused activities within the larger project. These publications emphasize particular issues related to SPI, they draw upon incidents in one or more of the four organizations, and they adopted a variety of theoretical frames and research approaches. Two examples illustrate this. Iversen and Mathiassen (2003) present a traditional case study based on interviews, documents, minutes of meetings, etc., in an attempt to design and implement a software metrics program in one of the four organizations. Action research played no role in the development of this particular publication, but the action research approach to the overall project created the opportunity for the involved researchers to identify and develop this research. Iversen et al. (1999) present an approach to manage risks in SPI projects. The risk management approach was developed in response to specific needs in one of the four software organizations. The opportunity to engage in this particular effort was again created by the overall action research approach. This time the researchers adopted action research for the specific, focused activity.

Action research was adopted as the overall guiding method in the examination of the research across the different contributions. The research process of Mingers (2001), moving from appreciation to action, was enacted through the cycles of action research, i.e., diagnose, plan, act, evaluate, and learn (Susman and Evered 1978). Based on this basic organization of the research process, activities were identified and launched both on the overall project level and within focused initiatives in one or more organization. These activities were based on a mix of research methods including literature surveys, case studies, field experiments, and focused action research efforts (Mathiassen 2002).

Action research served, in this case, as a platform upon which several focused studies were launched through a variety of research methods. Aaen et al. summarize and synthesize findings from a number of focused research activities (see Mathiassen et al. 2002) that were launched and executed in sequence and in parallel (Mingers 2001). The paper is an example of a multi-methodology research design in which literature survey, action research, and theory development are combined to serve the particular needs of this study. The overall SPI project exemplifies multi-level research design. It addresses software engineering and management practices on individual, project, and organizational levels using different methods.

Davison (1998) represents a different example of a planned approach. Action research was used from the outset in combination with case studies to explore how group support systems could be used to improve meeting processes in Hong Kong. Action research was in this case used repeatedly in parallel with case studies as illustrated in Figure 2. This combined approach answered the questions of "what and why?" through case studies and "how to?" through action research. Action research was required as a parallel methodology with case studies to help enact the roles of facilitator, leader, technician, and researcher for the group support systems (Davison 1998).

As with Mathiassen et al. (2002), Davison adopted a planned approach to action research from the very outset of this project. Action research played a dominant role in framing the overall process, but it was used in parallel with case studies and through a sequence of interventions targeting planning, training, and business process re-engineering related to the improved use of group support systems.

4.2 The Emergent Approach

Action research can also be used, not as a planned and dominant methodology from the outset of a research initiative, but as a sequential methodology that emerges and proves helpful as the examination and explanation of the research phenomena unfolds (Mingers 2001). This approach is often evident across several studies embedded within a research discipline and, from the outset, there is no planning of how and when to mix methods across studies.

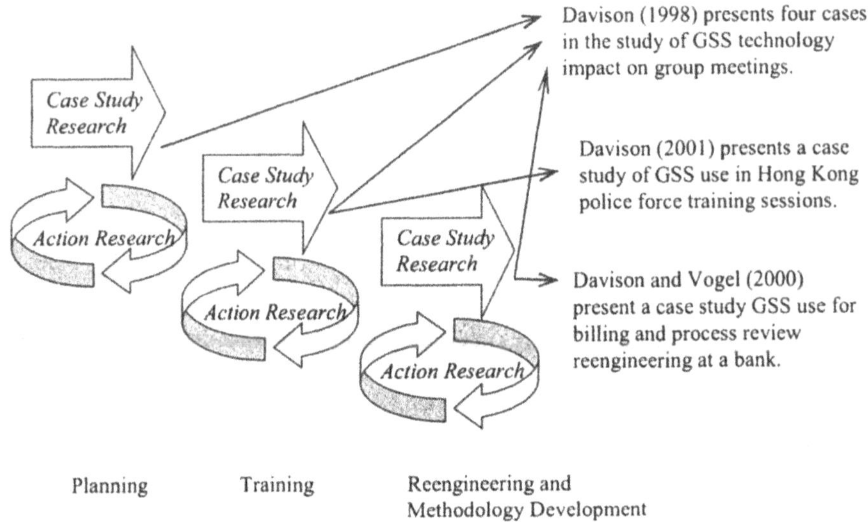

Figure 2. A Planned Approach to Action Research (Parallel and Sequential)

Walls, Widmeyer, and El Sawy (1992) described the need for information system design theories in the context of executive information systems. The design of such systems is comprised of a set of requirements to address a class of problems, a description of artifacts to meet those requirements, kernel theories from the natural and social sciences, and a set of hypotheses that can be used to verify whether the design satisfies the requirements. Markus, Majchrzak, and Gasser (2002) later used action research in the production of such a design theory for emergent knowledge systems. In doing so they followed an iterative process of specifying kernel theory, developing hypotheses, implementing in-use systems, and integrating findings back into the development of new theory. Action research was, in this case, embedded within an overarching research agenda in the development of the TOP Modeler tool by Majchrzak (1997), Majchrzak and Finley (1997), and Majchrzak and Gasser (2000) as illustrated in Figure 3. Kernel theory was modified to the point where a new theory of emergent knowledge systems was provided (Markus et al. 2002). Where Walls et al. developed design theory for executive information systems, Markus et al. developed design theory for systems that support emergent knowledge processes. Where Walls et al. "used empirical evidence from over 20 case studies" (p. 49), Markus et al. "followed the action research strategy" (p. 187).

This research approach provided the involved researchers with a variety of methods (field observation, case study, and action research) to help explain emergent knowledge systems. The use of action research emerged as part of a larger research agenda. It was used in sequence with other methods and not as the planned and dominant methodology from the outset. The resulting design theory can subsequently be further validated, refined, and adopted to other, similar domains. Such explorations can continue the emergent process as well as strengthen and further develop the theory through field experiments, case studies, or renewed action research efforts.

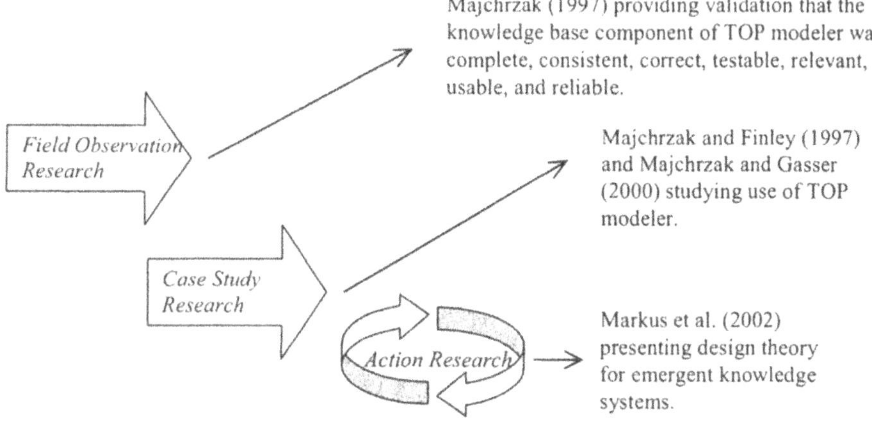
Majchrzak (1997) providing validation that the knowledge base component of TOP modeler was complete, consistent, correct, testable, relevant, usable, and reliable.
Field Observation Research
Majchrzak and Finley (1997) and Majchrzak and Gasser (2000) studying use of TOP modeler.
Case Study Research
Action Research
Markus et al. (2002) presenting design theory for emergent knowledge systems.

Figure 3. An Emergent Approach to Action Research (Sequential)

Iversen et al. (1999) represents a different case of emergent action research (see Figure 1). This study is embedded within a planned and dominant approach to action research (Mathiassen 2002; Mathiassen et al. 2002). The Markus et al. example illustrates how action research can emerge as a useful approach to develop theory based on findings from field observation and case studies. The Iversen et al. example illustrates how a planned and dominant approach to action research leads to a number of events and situations that present the involved researchers with new and unplanned opportunities to engage in additional explorations. Each new initiative can be pursued based on its own mix of methods including new action research efforts. For additional studies representing planned and emergent use of action research, see Appendix A.

5 DISCUSSION

Our studies of successful cases of Information Systems action research offer contributions to scientific knowledge within two areas. First, they further our understanding of multi-method approaches to Information Systems research. Mingers (2001) provides two reasons for adopting pluralist approaches: the multidimensional world and the complexity of research processes. This leads to a two dimensional framework for mapping research methods to practices. The one dimension distinguishes between three different worlds, the material, the personal, and the social world (Habermas 1984), inviting researchers to combine objective, subjective, and intersubjective perspectives. The second dimension distinguishes between four activities, appreciation, analysis, assessment, and action, inviting researchers to combine methods that effectively address the varying concerns and challenges involved as the research process unfolds.

Our research shows that there is a third rationale for pluralist methodology: the potential multiplicity of research contributions related to action research processes. Mingers implicitly assumes a one-to-one relationship between research process and outcome. Each project results in one publication. There is, however, as we have seen, in many cases a one-to-many relationship between the process and its outcome. Each research project can result in several publications and each of these can be based on different research methods or on its own mixture of methods. The multi-publication nature of research practice has two implications. It provides further arguments for adoption of pluralist methodology so a variety and mix of methods can be adopted to serve the different purposes and foci of each contribution. It also highlights the opportunity to use action research to launch additional explorations or to develop theoretical thinking based on existing findings. We suggest consequently that our rationale for and understanding of pluralist methodological opportunities in Information Systems research should be based on three concerns: Which research perspective to adopt? Which research activity to support? Which research contribution to develop?

Second, the presented research contributes to our understanding of how action research can be practiced within the Information Systems discipline. The planned and emergent approaches represent two different, yet complementary ways to consider action research. They are different as the planned approach offers an explicit way to adopt, organize, and manage Mingers' framework for multi-method research. Appreciation, analysis, assessment, and action are easily adopted and enacted through the action research cycle activities of diagnosing, action planning, action taking, evaluating, and specifying learning (Susman and Evered 1978). In the emergent approach, there is no explicit way in which Mingers' activities are enacted. Instead, Mingers' framework can be used directly to reflect on and manage the process as it unfolds. The approaches are complimentary in light of Latour's (1987) two rules of scientific method. The first rule states that scientific facts are carefully constructed from an existing domain of knowledge and that constructed facts are often debatable. Latour's second rule holds that the debate of scientific facts can result in two outcomes: the fact is either relevant or not. If the fact is relevant, it can then be reexamined in light of new methods and tools as well as being subjected to questions of authenticity and research trends. As discussed earlier, Aaen et al. (2001) used action research in a planned approach to propose a framework for examining software process improvement in organizations. From Latour's perspective, Aaen et al. addressed software process improvement as debatable, a relevant issue, and a contribution to an existing domain of knowledge. Based on the relevancy of the Aaen et al. study, subsequent projects ensued, relying on additional methods in an emergent approach to determine the authenticity of software process improvement facts in the particular context of a metrics program and to find out how these facts should be pursued (Iversen and Mathiassen 2003; Iversen et al. 1999). We believe that other lenses could also be used to illustrate how the planned and emergent approaches are used together including Lee's (1991) subjective and objective research approaches and Orlikowski and Baroudi's (1991) description of varied ontologies.

When adopted as research practices, there are important differences between the two approaches to action research. The planned approach is launched from the outset as the dominant approach and it offers a variety of opportunities to explore new issues as they emerge through the process. The emergent approach is embedded into a larger research process and it allows the researchers to explore and further develop earlier findings in practical contexts. When used as theoretical lenses to analyze research activities retrospectively, as we have done in this paper, we can use the two approaches interchangeably. Mathiassen et al. (2002) is presented as an example of the planned approach; but it contains a focused action research initiative (Iversen et al. 1999) that represents an example of emergent action research. Similarly, Markus et al. (2002) is presented as an emergent approach within a wider research agenda, but when viewed in isolation, represents an example of a planned approach to action research.

Our research shows that action research efforts lend themselves to pluralist methodology. First, because multiple methods help create a systematic approach to data collection that can increase the rigor of each individual contribution (Benbasat et al. 1987). Second, because action research typically is involved in multi-contribution projects where each contribution is supported by its own mix of research methods. Appreciating this provides action researchers within our discipline with action strategies that can help them develop more and better contributions from their efforts.

Pluralist methodology helps action researchers address the dilemmas (Rapoport 1970) involved in action research. The ethical dilemma (Rapoport 1970) involves biases when action researchers become too involved in the problematic situation. Such biases can be counteracted by introducing conventional approaches to data collection and analysis that help the researchers triangulate findings and arrive at more reliable results (Benbasat et al. 1987). Goal dilemmas (Rapoport 1970) can be addressed in a similar way by using conventional methods to strengthen the rigor of the research process without abandoning an interest for the problem-solving process (McKay and Marshall 2001). In addressing the initiative dilemma (Rapoport 1970), action researchers are advised to keep an open eye on emerging research opportunities during their involvement in a problematic situation and on that basis identify and execute additional research activities based on a mix of research methods. Such additional studies add to the researchers' scientific results without inducing new initiation costs, and without necessarily interfering with their involvement in practical problem solving. Pluralist strategies will, in this way, guard researchers against the major pitfalls of action research, i.e., lack of impartiality of the researcher, lack of discipline, mistaken for consulting, and difficulty of generalizing (Baskerville and Wood-Harper 1996).

Action research offers unique opportunities for Information Systems researchers to become involved in practical problem solving and to get involved in close collaboration with practitioners from our discipline. These advantages make the approach highly attractive. Action research projects involve, at the same time, complex managerial challenges. There are, however, techniques available for planning, organizing, and controlling action research projects (e.g., Avison et al. 2001; McKay and Marshall 2001) and we have here shown how multi-method thinking and practices further help researchers address the involved dilemmas and pitfalls. We argue on that basis that action research is as attractive as other conventional approaches to Information Systems research, that it offers unique opportunities to engage in research with a strong commitment to relevance, and that it is no more risky, given the right approach, than other conventional approaches within our field.

Further studies are needed to develop more specific tactics that can be adopted in action research efforts. These tactics should combine insights from planning, organizing, and managing projects (e.g., Avison et al. 2001; McKay and Marshall 2001) with pluralist research methodology. Such efforts should also explore the relationship and possible interaction between the professional methods adopted in the problem solving cycle with the mix of research methods that are adopted in the research cycle.

6 SUMMARY

We have used Mingers' framework for combining Information Systems research methods to explore action research practices within our discipline. We have identified two different approaches to action research, the planned and the emergent, and we have demonstrated how the one-to-many relationship between research process and outcome that is typical for action research projects contributes to our general understanding of pluralist methodology. We have also shown that action research is supported well by pluralistic methodology and that it offers important opportunities for additional explorations and for developing theoretical thinking based on existing findings.

This paper is part of an ongoing effort to contribute to increased diversity within Information Systems research (Robey 1996). More action research, especially among the new-comers of our field, will contribute to this diversity by strengthening the position of action research and by increasing the interaction and collaboration between researchers and practitioners within our field. Also, we believe that the planned and emergent approaches exist across different methodologies and, in general, a pluralist methodology can contribute to increased diversity within each individual action research project.

REFERENCES

Aaen, I., Arent, J., Mathiassen, L., and Ngwenyama, O. "A Conceptual MAP of Software Process Improvement," *Scandinavian Journal of Information Systems* (13), 2001, pp. 123-146.

Andersen, I., Borum, F., Kristensen, P.H., and Karnøe, P. *On the Art of Doing Field Studies*, Copenhagen: Copenhagen Business School Press, 1995.

Argyris, C. *Intervention Theory and Method. A Behavioral Science View*, Reading, MA: Pearson Addison Wesley 1970.

Avison, D., Baskerville, R. and Myers, M. "Controlling Action Research Projects," *Information Technology & People* (14:1), 2001, pp. 28-45.

Baskerville, R. L., and Stage, J. "Controlling Prototype Development Through Risk Analysis," *MIS Quarterly* (20:4), 1996, pp. 481-504.

Baskerville, R. L., and Wood-Harper, A. T. "A Critical Perspective on Action Research as a Method for Information Systems Research," *Journal of Information Technology* (11), 1996, pp. 235-246.

Baskerville, R. L., and Wood-Harper, A. T. "Diversity in Information Systems Action Research Methods," *European Journal of Information Systems* (7), 1998, pp. 90-107.

Benbasat, I., Goldstein, D. K., and Mead, M. "The Case Research Strategy in Studies of Information Systems," *MIS Quarterly* (11:3), 1987, pp. 369-386.

Boehm, B. W. "A Spiral Model of Software Development and Enhancement," *Computer* (21:5), 1988, pp. 61-72.

Boehm, B. W. *Software Risk Management*, Washington, DC: IEEE Computer Society Press, 1989.

Borum, F. *Organization, Power, and Organizational Change*, Copenhagen: Handelshøjskolens, Forlag, 1995.

Briggs, R. O., Adkins, M., Mittleman, D., Kruse, J., Miller, S., and Nunamaker, J. F. "A Technology Transition Model Derived from Field Investigation of GSS Use Aboard the U.S.S. CORONADO," *Journal of Management Information Systems* (15:3), 1998/1999, pp. 151-195.

Cassell, C., Fitter, M., Fryer, D., and Smith, L. "The Development of Computer Applications by Non-employed People in Community Settings," *Journal of Occupational Psychology* (61), 1988, pp. 89-102.

Checkland, P. "From Framework Through Experience to Learning: the Essential Nature of Action Research," in H-E. Nissen, H. K. Klein, and R. H. Hirshheim (Eds.), *Information Systems Research: Contemporary Approaches and Emergent Traditions*, Amsterdam: North-Holland, 1991, pp. 397-403.

Davison, R. M. *An Action Research Perspective of Group Support Systems: How to Improve Meetings in Hong Kong*, Unpublished Doctoral Dissertation, 1998, City University of Hong Kong.

Davison, R. M. "GSS and Action Research in the Hong Kong Police Force," Information Technology and People, (14:1), 2001, pp. 60-77.

Davison, R. M. and Vogel, D. R. "Group Support Systems in Hong Kong: An Action Research Project, *Information Systems Journal* (10:1), 2000, pp. 3-20.

DeVreede, G. J. "Collaborative Business Engineering with Animated Electronic Meetings," *Journal of Management Information Systems* (14:3), 1997/1998, pp. 141-164.

Gustavsen, B. "Action Research and the Generation of Knowledge," *Human Relations* (46:11), 1993, pp. 1361-1365.

Habermas, J. *The Theory of Communicative Action, Volume 1: Reason and the Rationalization of Society*, London: Heineman, 1984.

Hult, M., and Lennung, S.-Å. "Towards a Definition of Action Research: A Note and Bibliography," *Journal of Management Studies* (17:2), 1980, pp. 241-250.

Huxham, C., and Vangen, S. "Action Research for Understanding Collaboration Practice: Emerging Research Design Choices," in *Proceedings of the 24th International Congress of Applied Psychology*, San Francisco, August 9-14, 1998a.

Huxham, C., and Vangen, S. "Leadership in the Shaping and Implementation of Collaboration Agendas: How Things Happen in a (Not Quite) Joined Up World," *Academy of Management Journal* (6), 2000, pp. 1159-1175.

Huxham, C., and Vangen, S. "What Makes Practitioners Tick? Understanding Collaboration Practice and Practicing Collaboration Understanding," in *Proceedings of the Workshop on Interorganizational Collaboration and Conflict*, Montreal, April 1998b.

Iversen, J., and Mathiassen, L. "Cultivation and Engineering of a Software Metrics Program," *Information Systems Journal* (13:1), 2003, pp. 3-19.

Iversen, J., Mathiassen, L., and Nielsen, P. A. "Managing Risks in Software Process Improvement: An Action Research Approach," in *Proceedings of the 7th European Conference on Information Systems*, Copenhagen, Denmark, 1999, pp. 370-385.

Kinkaid, H. "Assessing Functional Explanations in the Social Sciences," in M. M. Martin and L. C. McIntyre (Eds.), *Readings in the Philosophy of Social Science*, Cambridge, MA: MIT Press, 1994, pp. 415-428.

Kock, N. F. *The Effects of Asynchronous Groupware on Business Process Improvement*, Unpublished Ph.D. Thesis, University of Waikato, New Zealand, 1997a.

Kock, N. F. "Myths in Organizational Action Research: Reflections on a Study of Computer-Supported Process Redesign Groups," *Organizations & Society* (4:9), 1997b, pp. 65-91.

Kock, N. F., McQueen, R. J., and Scott, J. "Can Action Research be Made More Rigorous in a Positivistic Sense? The Contribution of an Interpretive Approach," *Journal of Systems and Information Technology* (1:1), 1997, pp. 1-24.

Latour, B. *Science in Action: How to Follow Scientists and Engineers Through Society*, Cambridge, MA: Harvard University Press, 1987.

Lau, F. "A Review on the Use of Action Research in Information Systems Studies, in A. S. Lee, J. Liebenau, and J. I. DeGross (Eds.), *Information Systems and Qualitative Research*, London: Chapman & Hall, 1997, pp. 31-68.

Lee, A. "Integrating Positivist Approaches to Organizational Research," *Organization Science* (2:4), 1991, pp. 342-365.

Levin, M. "Creating Networks for Rural Economic Development in Norway," *Human Relations* (46:2), 1993, 193-218.

Majchrzak, A. "Software to Support Socio-Technical Design: The Case of TOP-Integrator," in G. Salvendy, M. Smith and R. Koubek (Eds.), *Design of Computing Systems*, New York: Elsevier, 1997, pp. 229-231.

Majchrzak, A., and Finley, L. "A Practical Theory and Tool for Specifying Socio-Technical Requirements to Achieve Organizational Effectiveness," in J. Benders, J. de Haan, and D. Bennett (Eds.), *The Symbiosis of Work and Technology*, London: Taylor and Francis, 1995, pp. 95-116.

Majchrzak, A., and Gasser, L. "TOP Modeler," *Information, Knowledge, & Systems Management* (2:1), 2000, pp. 95-110.

Markus, M. L., Majchrzak, A., and Gasser, L. "A Design Theory for Systems that Support Emergent Knowledge Processes," *MIS Quarterly* (26:3), 2002, pp. 179-212.

Mathiassen, L. "Collaborative Practice Research," *Information, Technology & People* (15:4), 2002, pp. 321-345.

Mathiassen, L., Borum, F., and Pederson, J. S. "Developing Managerial Skills in IT Organizations: A Case Study Based on Action Learning," *Journal of Strategic Information Systems* (8), 1999, pp. 209-225.

Mathiassen, L., Pries-Heje, J. and Ngwenyama, O. (Eds.). *Improving Software Organizations-From Principles to Practice*, Upper Saddle River, NJ: Addison-Wesley, 2002.

Mathiassen, L., and Stage, J. "The Principle of Limited Reduction in Software Design," *Information, Technology and People* (6:2), 1992, pp. 171-185.

Mathiassen, L., Seewaldt, T., and Stage, J. "Prototyping and Specifying: Principles and Practices of a Mixed Approach," *Scandinavian Journal of Information Systems* (7:1), 1995, 55-72.

Mingers, J. "Combining IS Research Methods: Towards a Pluralist Methodology," *Information Systems Research* (12:3), 2001, pp. 240-259.

McKay, J., and Marshall, P. "The Dual Imperatives of Action Rresearch," *Information Technology & People* (14:1), 2001, pp. 46-59.

Orlikowski, W., and Baroudi, J. "Studying Information Technology in Organizations: Research Approaches and Assumptions," *Information Systems Research* (2:1), 1991, pp. 1-62.

Pava, C. "Designing Managerial and Professional Work for High Performance: A Sociotechnical Approach," *National Productivity Review* (2), 1983a, pp. 126-135.

Pava, C. *Managing New Office Technology: An Organizational Strategy*, New York: Free Press, 1983b.

Pava, C. "Redesigning Sociotechnical Systems Design: Concepts and Methods for the 1990s," *Journal of Applied Behavioral Science* (22:3), 1986, pp. 201-221.

Rapoport, R. N. "Three Dilemmas in Action Research," *Human Relations* (23:4), 1970, pp. 499-513.

Robey, D. "Research Commentary: Diversity in Information Systems Research: Threat, Promise, and Responsibility," *Information Systems Research* (7:4), 1996, pp. 400-408.

Salmon, M. "Explanation in the Social Sciences," in P. Kitcher and W. Salmon (Eds.), *Scientific Explanation; Minnesota Studies in the Philosophy of Science*, Minneapolis, MN: University of Minnesota Press, 1989, pp. 385-409.

Smith, L., Fryer, D. M., and Fritter, M. J. "A Study of Computing Needs of Wageless People in Sheffield," *Medical Research Council/Economic and Social Research Council, and Social and Applied Psychology Memo* 708, University of Sheffield, 1985.

Susman, G. I., and Evered, R. D. "An Assessment of the Scientific Merits of Action Research," *Administrative Science Quarterly* (23), 1978, pp. 582-603.

van Fraassen, B. C. *The Scientific Image*, Oxford University Press, 1980.

VanMannen, J. *Tales from the Field: On Writing Ethnography*, Chicago: University of Chicago Press, 1988.

Vangen, S. *Transferring Insight on Collaboration into Practice*," Unpublished Doctoral Dissertation, University of Strathclyde, Glasgow, 1998.

Vangen, S., and Huxam, C. "Creating a Tip: Issues in the Design of a Process for Transferring Theoretical Insight about Inter-Organizational Collaboration into Practice," *International Journal of Public-Private Partnerships* (1:1), 1998, pp. 19-42.

Walls, J. G., Widmeyer, G. R., and El Sawy, O. A. "Building an Information System Design Theory for Vigilant EIS," *Information Systems Research* (3:1), 1992, pp. 36-59.

Walsham, G. "The Emergence of Interpretivism in IS Research," *Information Systems Research* (6:4), 1995, pp. 376-394.

Ytterstad, P., Akselsen, S., Svendsen, G. and Watson, R. T. "Teledemocracy: Using Information Technology to Enhance Political Work," *MISQ Discovery*, 1996.

ABOUT THE AUTHORS

Matt Germonprez holds a Ph.D. in Information Systems from the University of Colorado at Boulder and is currently an assistant professor at Case Western Reserve University. His research interests focus on computer mediated communication and human-computer interaction. In particular, he is interested in how information technology can be built to support end-user modifications in the context of use. His work draws from the domains of information systems, computer science, and the philosophy of science. Matt can be reached at Germonprez@ CWRU.edu.

Lars Mathiassen holds an M.Sc. in computer science from Århus University, Denmark, a Ph.D. in Informatics from Oslo University, Norway, and a Dr. Techn. in software engineering from Aalborg University, Denmark. He is currently Georgia Research Alliance Eminent Scholar and Professor in Computer Information Systems at Georgia State University. His research interests focus on engineering and management of IT systems. More particularly, he has worked with project management, object-orientation, organizational development, management of IT, and the philosophy of computing. He has co-authored several books, including *Professional Systems Development-Experiences, Ideas and Action*, *Computers in Context: The Philosophy and Practice of Systems Design*, *Object-Oriented Analysis & Design*, and *Improving Software Organizations: From Principles to Practice*. Lars can be reached at Lars.Mathiassen@ eci.gsu.edu.

Appendix A

Additional *Planned* and *Emergent* Cases[1]

Planned Use of Action Research	Illustrative Quote
Kock (1997a): Action research was used to investigate the effects of asynchronous groupware on total quality management and business process reengineering. Action research was a planned approach as iterations through the action research cycle framed subsequent action research.	"The fourth iteration in the AR cycle disseminated all the software applications, previously introduced only locally, throughout the organization. These subsequent iterations generated new hypotheses, reinforced former ones, and also provided ground for refutation of some previous hypotheses" (Kock et al. 1997, p. 15).
DeVreede (1997/1998): Group support systems and animation techniques were used in support of user involvement and organizational change processes. Action research was used as a planned approach within which survey instruments and case study interviews were used to determine perceived session quality and satisfaction.	"During the study, both quantitative and qualitative data from various sources were collected to enable a rich representation of the phenomena under investigation and to permit comparison and contrast of the collected data....Quantitative data sources included system logs of each session and questionnaires completed by the participants after each [action research intervention], and our ongoing observations during each session and during the [case] study as a whole" (p. 144).
Pava (1986): Action research was used in the investigation of concepts in the design of socio-technical information systems for nonlinear decision processes. Action research was a planned approach to frame a traditional case study of socio-technical systems.	"[Embedded in an action research agenda], the social analysis [of the case study] was based on interviews of current and former employees and on longitudinal analysis of human resource data." (p. 212).
Mathiassen, Borum, and Pedersen (1999): The interaction between individual learning and organizational context was studied in realtion to an action learning program. Action research was used in a planned approach in conjunction with a traditional case study in the development, execution, and assessment of an in-house program for educating furture IT managers in a Danish company.	"Our different roles as consultants, trainers, and observers made possible a blend of action research (Argyris 1970; Borum 1995), participant observation (Van Maanen 1988; Andersen et al. 1995), and traditional case study data generation (Yin 1994). This combined approach provides a rich insight into organizational phenomena and allows for validation and triangulation between different types of data." (pg 439).

[1]The cases in this table and those presented in the literature were compiled from Lau (1997) and Michael Myers' Qualitative Research Methods page at http://www.isworld.org, as well as cited papers in published action research studies.

Planned Use of Action Research	Illustrative Quote
Ytterstad, Akselsen, Svendsen, and Watson (1996): Technology was used to address problems surrounding the fact that many politicians do not seek reelection in Norway. Action research was used in a planned approach to determine how politicians used technology to support daily activities and to frame a field trial in the investigation of information technology use in Norwegian politics.	"[The research project] is a cyclic process of investigation that includes the identification and diagnosis of a problem, planning of actions, implementation, and evaluation of results...to analyze political work in terms of content, communication patterns, workload patterns, and advantages and drawbacks as perceived by the politicians. This [analysis] included the following activities: [interviewing, analyzing a questionnaire, analyzing formal documents, analyzing telephone utilization]" (The Project section)
Briggs, Adkins, Mittleman, Kruse, Miller, and Nunamaker (1998/1999): The technology acceptance model was investigated with respect to GSS use. A new model of technology transition is proposed. Action research was used as in a planned approach in parallel with a field study.	"This article presents a 32 month field investigation of an effort to introduce GSS into the daily work processes of the staff of the US Navy's Third Fleet aboard the USS CORONADO. The principles of action research guided the investigation....The project began with [action research] interventions based on the precepts of Davis' [TAM]" (p. 154).
Emergent Use of Action Research	**Illustrative Quote**
Baskerville and Stage (1996): Risk analysis was shown to be an important component in the prototyping of information systems. Action research was used in an emergent approach as it built on the work of Mathiassen and Stage (1992) – (Literature Review/ Observation) and Mathiassen, Seewaldt, and Stage (1995) (Experimental) in the development and evaluation of technology prototyping.	"The third stream of literature [used by Baskerville and Stage 1996] is concerned with general software development frameworks (Boehm 1988, 1989)....The definition of [framework] factors originates from theoretical work (Mathiassen and Stage 1992) that has been further explored in more recent empirical studies (Mathiassen et al. 1995)" (p. 485).
Cassell, Fitter, Fryer, and Smith (1988): Three community groups were examined in the implementation of information technology to unemployed individuals. The SPRITE project [action research] was successful in teaching computing skills that were translated to jobs and the exchange of community information.	"..three community groups consisting of non-employed people were selected as instances of differing autonomous collectives. Non-directive discussions were held in each of the centres on the theme of the relevance of new technology to the group concerned. Discussions were tape-recorded and content-analysed. Analyses were then fed back to each of the groups and a final version negotiated (see Smith, Fryer, and Fritter 1985)" (p. 92).

Emergent Use of Action Research	Illustrative Quote
Pava (1986): New concepts need to be developed in the design of socio-technical information systems. Systems need to support an inherent flexibility, as organizations move away from hierarchical structure in support of nonlinear processes. Action research is used in sequence with prior work on the development of socio-technical systems in the support of nonlinear work.	"To overcome deficiencies of socio-technical system design, alternate approaches must be developed. One has been proposed based on action research projects in nonlinear work systems (Pava 1983a; b). This approach emphasizes new concepts redefining the basic units of social and technical analysis. It also identifies alternatives other than the autonomous work group that can also yield a 'best match' between an organization and its technology" (p. 206).
Huxham and Vangen (2000): Leadership roles play a part in collaborative agenda implementation. Ultimately, collaboration stems from leadership structures, processes, and individuals. Action research was used to contribute to prior empirical research and strengthen the theory of collaboration in social settings.	"This research forms an element of a program that has so far spanned ten years and that is an effort to develop practice-oriented theory on the management of collaboration (Huxham and Vangen 1998a; 1998b; Vangen 1998; Vangen and Huxham 1998). The program is rooted, to a very large extent, in action research" (p. 1161).

20 THEMES, ITERATION, AND RECOVERABILITY IN ACTION RESEARCH

Sue Holwell
Open Systems Research Group
The Open University

Abstract This paper develops three concepts important to the practice of action research—recoverability, research themes, and iteration—by highlighting their applicability beyond single action research studies. The concepts are discussed against a *program* of action research, undertaken by a multidisciplinary research team, with a research focus on local, sector and national levels. This contrasts with the more usual pattern of action research in single situations.

Action research is criticized on the grounds that it lacks generalizability and external validity from one-off studies. Goodness criteria have been derived to address these and other criticisms. The recoverability criterion, less strong than the repeatability of experimentation, is central to these. A second concept, that of research themes, links the recoverability criterion and iteration in action research. Iteration within and between projects and the notion of critical mass, of doing work in more than one setting, address the limitations of single setting studies.

Keywords: Action research, research methodology, multidisciplinary research

1 ACTION RESEARCH IN INFORMATION SYSTEMS

The advocacy and exhortation of the debates about research methodology in the 1980s have been replaced by discussion about the appropriateness of different modes of research, what constitutes good research within a particular methodological stream, and the practicalities of different modes of research (as examples, see Avison et al. 1999; Baskerville et al. 1997; Baskerville and Wood-Harper 1996).

Action research is a case in point. It is now accepted as a relevant approach for IS research, as evidenced by the publication of IS journal issues dedicated to action research (*Information, Technology & People* in 2001 and a forthcoming issue of *MIS Quarterly*).

1.1 Characteristics of Action Research in IS

Several varieties of action research are recognized in IS (Baskerville and Wood-Harper 1998). Other disciplines, notably health and education, where action research is associated with improvement in professional practice, have a different understanding (Hart et al. 1995).

Nevertheless, action research has some generally accepted characteristics: the researcher is immersed in the situation; the work unfolds in response to the situation and not to the researcher's requirements; the questions, problems, and puzzles are taken from the local context; descriptions and theories are built up by iteration within the context and are tested within the situation and there is close collaboration between researchers and actors, (Argyris et al. 1985; Baskerville and Wood-Harper 1996; Burrell and Morgan 1979; Lincoln and Guba 1985; Robson 1993; Whyte 1991).

Any research may be thought of as entailing three elements: (F) some linked framework of ideas and concepts; (M) a way of applying the ideas, and (A) an area of interest in which to apply them (Checkland and Holwell 1998a). For action research, the addition of the research interests being embodied in a set of *themes* is necessary (Figure 1).

Here the researcher, interested in particular themes, declares F and M, then enters a situation in which the themes are relevant and becomes involved as participant and researcher. Work to effect change and improvement follows with the researcher committed to continuous reflection on the collaborative process and its outcomes. This entails trying to make sense of the unfolding experience of A using the declared F and M. This may involve rethinking of earlier phases, and it is the declared F and M that allows this to be done coherently.

In response to criticisms of interpretive research, including action research, in terms of the generalizability of the results from (usually) a single situation and the quality with which it is undertaken, *goodness criteria* have been proposed. These essentially come down to substantiating the approach in the particular case, careful and documented data collection and analysis, iteration, and making all elements explicit at the outset (Baskerville and Wood-Harper 1996; Eden and Huxham 1996; Gummesson 1988; Kock et al. 1997; Lau 1997; Marshall 1990). In particular, if findings are to be taken seriously, then they must be supported by appropriate arguments and/or evidence, i.e., an adequate warrant *against a particular framework* (Checkland 1991; Checkland and Holwell 1998a; Phillips 1992).

2 RECOVERABILITY, THEMES, AND ITERATION

2.1 Recoverability

It is the *recoverability* criterion, that is the crucial one in action research (Checkland and Holwell 1998a). If we imagine a spectrum of knowledge acquisition from experimental natural science at one end to story telling at the other, then along that spectrum

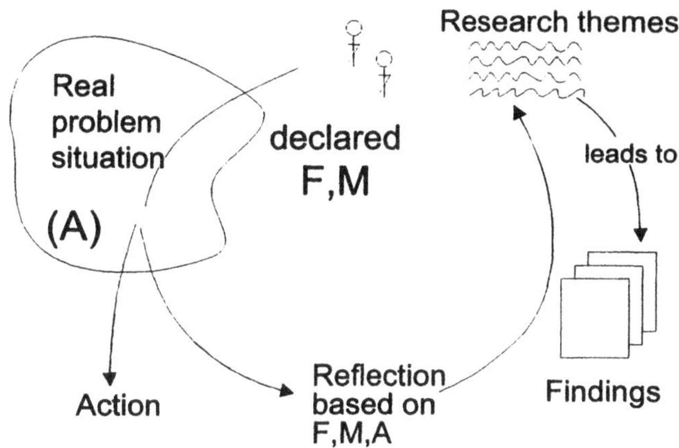

Figure 1. The Cycle of Action Research
(Adapted from Checkland and Holwell 1998b)

will be very different criteria for judging the truth-value of the claims made. Traditional scientific experiments would be at one end and at the other, the weaker criterion that this (research) story is plausible. However, action researchers have to do better than simply settling for plausibility (Checkland and Holwell 1998a).

To do this requires, at the least, that the research process is recoverable by interested outsiders. Therefore, the set of ideas and the process in which they are used methodologically must be stated, because these are the means by which researchers and others make sense of the research. Action researchers should be able to enact a process based on a declared-in-advance methodology that encompasses a particular framework of ideas in such a way that the process is *recoverable* by anyone interested in subjecting the research to critical scrutiny. This principle is almost totally neglected in the action research literature (Checkland and Holwell 1998b) although some practical difficulties that occur if the framework is not declared in advance are noted (Kock 1997).

The points listed as goodness criteria earlier are subsumed under this concept. The *recoverability* criterion is the first of three concepts that are important to the practice of action research.

2.2 Themes

The second concept, that research interests are embodied in a set of *themes,* was referred to in the description of action research in Figure 1. The cycle of action research in Figure 1 has researchers with a set of research themes taking action in a situation relevant to those themes, and exploring it via a declared framework and methodology. Findings may be about any or all of the elements F, M, and A and the themes, and new themes may be added (Checkland and Holwell 1998b). The researcher's interests, embodied in themes are not necessarily derived from a specific context. Rather, they are the longer term, broader set of questions, puzzles, and topics that motivate the researcher. Such research interests are rarely confined to one-off situations.

Given this, and because each new research project is an iteration in a longer term personal research program, then interventions need not be pre-selected (or even negotiated as tightly as Kock, McQueen and Rouse [1996] suggest).

The particular questions, problems, and puzzles through which the themes are explored *do* come from a particular intervention context. Moreover, research themes are unlikely to be completely resolved through a single intervention, and the linking of projects (both forward and backward) via research themes means that iteration can be thought of differently to iteration within and around the action research cycle.

2.3 Iteration

The third concept, *iteration,* is a recognized, and much discussed, characteristic of action research, particularly as a means of addressing criticisms that findings are, first, not generalizable from one-off interventions at single sites (the pattern of classical action research in IS) and, second, that action research lacks rigor (Kock et al. 1997).

However, it can be more complex than repetitions of the cycle through the stages (Checkland 1991; Lewin 1951; Susman and Evered 1978) if thought of in relation to a set of themes explored over time through several different organizational contexts.

3 CONCEPTS IN USE

3.1 The Research Program

The work commenced in the first year as a form of internal market in the UK National Health Service in which purchasers of health services for a local population would contract with providers of health services (general practitioners, hospitals and community units) for the delivery of services(Checkland and Holwell 1998b).

It is a *program* of action research with the prime research objective of understanding the developing nature of the contracting relationship with a view to defining how it could be improved.

It was concerned with sense making at the level of the whole (the UK National Health Service) while maintaining a focus on several layers of detail (organizational levels). It is complex in execution, including several projects overlapping in time, it covers work from different bodies of knowledge, and was undertaken by a seven-member multidisciplinary team with different intellectual traditions and the issues explored cross many organizational boundaries.

The work, done over a 4-year period, followed a three-part design and the overall course of the research is depicted in Figure 2, which is chronological from [1] through [16]. Figures in square brackets hereafter refer to items in Figure 2.

The work involved 30 different city and rural NHS sites, with 60 people taking part in Phase One and 3 to 20 people involved in each of the 10 separate action research projects of Phase Two [9a through 9j].

Phase One consisted of extensive interviewing to gather a range of perceptions of the contracting process as it was being initiated and yielded a richer model [5] than the one used to structure the interviews [1].

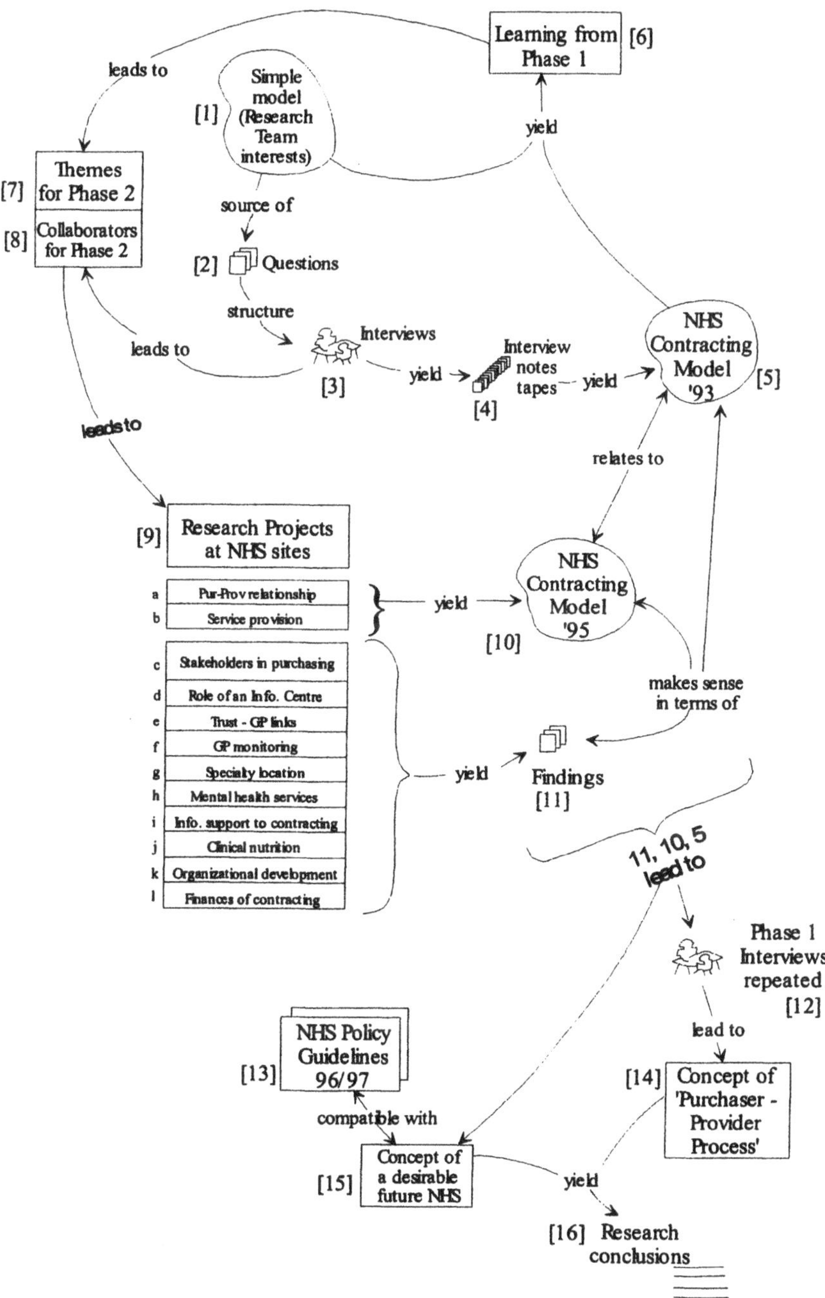

Figure 2. The Overall Arocess of This Research

Phase Two consisted of 12 separate projects [9] carried out with a number of different NHS organizations. Two of these [9a and 9b] produced a still richer model [10] against which the individual findings from the other 10 studies and model [5] also made sense.

Phase Two research outcomes came from the study of specific findings from the research [9a through 9l] and two models, one expressing the aspirations in the NHS at the end of first year [5], the other making sense of the reality two years later [10].

Phase Three consisted of gathering all of the results together to draw general conclusions.

The framework and methodology are declared and both the process and outcomes are clearly linked to them throughout. This addresses three of the four elements suggested for action research in IS (Lau 1997) conceptualization, design, and process.

3.2 Reflection on the Process

This work belongs to the strand referred to in social science as *emerging action research* (Chisholm and Elden 1993; Elden and Chisholm 1993) because it involves multiple levels of organization engaged in the change process and a wide degree of openness in the research process overall (none of the Phase Two projects were known at the beginning). Moreover, because of the number and range of participating organizations it illustrates the concept of critical mass, of doing research in multiple sites, to increase the external validity of action research (Chisholm and Elden 1993).

In doing multidisciplinary research there are difficulties both in observation and in the interpretation of observation, although multidisciplinary perspectives clearly add to the richness of research. There may be less than full conceptual illumination because some team members lack particular intellectual frameworks, and bringing together different intellectual traditions and patterns of work is not always straightforward. Inevitably, there were some arbitrary closures but these were always subject to regular review by the team.

In this work there are instances of single discipline observations on rich data plus a connected multidisciplinary commentary, a middle ground between full theoretical synthesis of all discipline perspectives, and a set of mutually exclusive discipline-based interpretations.

3.2.1 Recoverability

Here the argument that a framework and methodology must be declared in advance is clearly evident, and its evolution in substantive terms traceable. This is not only necessary, as argued earlier, but this program of research, multidisciplinary work with a multilevel research focus, in several research settings over a 4-year period, is simply not possible without a cohering framework of ideas and an overarching methodology.

The research process is clear and recoverable (Figure 2) and the thinking is auditable through the evolution of the models from [1] to [5] in Phase One and [10] at the end of Phase Two. All the findings are mappable onto this latter model. The model of a future vision for the NHS [15] from the end of Phase Three is of a different kind than the others but compatible with them.

3.2.2 Themes

The complexity of the research process, combined with the different interests of the multidisciplinary team, made recognition of the significance of research themes relatively easy. No specific interventions were negotiated to explore any particular themes.

Themes can be thought of as a hierarchy. At the top are the themes that motivate researchers to become involved. Then there are themes more relevant to a particular research program. At the third level, there are themes relevant to particular projects, and finally there are some relevant themes within a particular organizational setting. New themes may be recognized at any time.

Themes are the glue in action research. They make sense of a *program* of research. They give coherence to multiple site, multiple level multidisciplinary research by linking the separate projects and allowing for cross-fertilization between them.

Carrying forward a set of themes makes sense of calls to do several projects on the same topic (Kock et al. 1997) in order to achieve more generalizable outcomes from the evolving set of themes and of undertaking cross-sectional and long-term longitudinal studies.

The similarity in the learning that emerged from separate Phase Two projects is evidence of the thematic linkages within this work (for example, the importance of the changed role for general practitioners).

Finally, taking the concept of themes seriously makes the dilemma of whether to first define your problem and then negotiate settings and collaborators through which to explore it, or whether to find organizations willing to collaborate and then explore their issues (Kock et al. 1996) less of an issue. As long as settings are potentially relevant to research themes, they are appropriate.

3.2.3 Iteration

The pattern of iteration is complex, not least because the research focus was concurrently at several organizational levels (individual purchasers and providers, purchasers and providers as groups, and at the level of the NHS nationally).

Iteration around the stages of a cycle of action research is a recognized characteristic that features in descriptions of action research. Argument that added rigor comes from greater emphasis on iteration generally refers to repetitions of the cycle (diagnose, plan, implement, evaluate, and learn [Susman and Evered 1978]).

The iteration here is similar, but more complex, than that generally described. It is easy to see Figure 2 as being one cycle from diagnosis through to reflection. Phase One of the research design mapping onto the diagnosis stage of Susman and Evered's (1978) five stages of action research, Phase Two mapping onto the action stage, and Phase Three as mapping onto the reflection stage. At one level, this is the case.

However, this is to see the research only at one level, whereas it has multiple layers. There are three kinds of iteration within Phase One. First, iteration was necessary to derive the descriptive concept in the model reflecting the researchers' perspectives [1] used to structure the interviews. Then within Phase One there was diagnosis, action in the form of formal and informal feed back that changed the understanding of the NHS interviewees as they were learning what "contracting" might mean, and reflection in the creation of the model [5]. This sense-making, practice-improving outcome is recognized in action research in social science disciplines (e.g., health) as being the *action*.

Each interview can be seen as an intervention in that the team was a conduit for exchanging perceptions as people in the NHS were developing their own understanding. These, then, are examples of developing ideas from a number of small instances that enables development of a more generalizable model. The interviews were interspersed with formal sessions of reflection. At the end of Phase One the process of building the models leading to the generic model [5] was not only iterative but also included diagnosis, some action, and reflection.

The mapping of Phase Two is more complex again. Each of the action projects in Phase Two fully reflects the iteration that is characteristic of action research. However, this was made more complex because several team members were actually participating in more than one project simultaneously, effectively blurring the boundary between the projects. Again, the importance of carrying a set of *themes* is particularly relevant. Given that the collaborating sites, as purchasers and providers, were also interacting with each other in the course of their daily activity, there was also cross-fertilization of thinking between collaborators from different organizations.

Phase Three can also be seen as a repetition of the full action research cycle. More interviews were conducted, including some at new sites, there was action in the linking of the NHS policy guidelines to the work and the reflection that gave rise to both the concept for the "Future NHS" and the overall conclusions.

4 CONCLUSION

Accounts of action research in IS usually describe single focus, single site studies that are concerned with action as the outcome. However, different kinds of action research are found in other disciplines, such as health and education, where the concern is with the improvement of practice. The research program outlined here is an example. Moreover, its three phases covered a 4-year period, 20 organizations, and included 10 discrete, single site, action research interactions.

Three concepts—recoverability, iteration, and themes—found in various forms in IS action research, and generally discussed under goodness criteria have greater significance in multisite, multidisciplinary research programs than has hitherto been described. Recoverability is exemplified through the research process; complex iteration, between and across separate studies, is described; and attention is drawn to the value of conscious reflecting on a hierarchy of themes that motivate researchers both within a study and over time.

REFERENCES

Argyris, C.; Putnam, R.; and Smith, D. M. *Action Science—Concepts, Methods and Skills for Research and Intervention*, San Francisco: Jossey-Bass, 1985.

Avison, D.; Lau, F.; Myers, M.; and Nielsen, P. A. "Action Research: Making Academic Research Relevant," *Communications of the ACM* (42:1) 1999, pp. 94-97.

Baskerville, R.; Myers, M.; Nielsen, P.A.; and Wood-Harper, A. T. "Panel: The Impact of Action Research on Information Systems," in A. S. Lee, J. Liebenau, and J. I. DeGross

(Eds.), *Information Systems Research: Information Systems and Qualitative Research,* London: Chapman & Hall, 1997, p. 69.

Baskerville, R., and Wood-Harper, A. T. "A Critical Perspective on Action Research as a Method for Information Systems Research," *Journal of Information Technology* (11:2) 1996, pp. 235-246.

Baskerville, R., and Wood-Harper, A. T. "Diversity in Information Systems Action Research Methods," *European Journal of Information Systems* (7) 1998, pp. 90-107.

Burrell, G., and Morgan, G. *Sociological Paradigms and Organizational Analysis*, Aldershot, England: Gower, 1979.

Checkland, P. B. "From Framework Through Experience to Learning: The Essential Nature of Action Research," in H-E. Nissen, H. K. Klein, and R. A. Hirschheim (Eds.), *Information Systems Research: Contemporary Approaches and Emergent Traditions,* Amsterdam: North-Holland, 1991, pp. 397-403.

Checkland, P. B., and Holwell, S. E. "Action Research: Its Nature and Validity," *Systemic Practice and Action Research* (11:1) 1998a, pp. 9-21.

Checkland, P. B., and Holwell, S. E. *Information, Systems and Information Systems: Making Sense of the Field*, Chichester, England: John Wiley & Sons, 1998b.

Chisholm, R. F., and Elden, M. "Features of Emerging Action Research," *Human Relations* (46:2) 1993, pp. 275-297.

Eden, C., and Huxham, C. "Action Research for the Study of Organizations," in S. Clegg, C. Hardy, and W. Nord (Eds.), *The Handbook of Organizational Studies,* Beverley Hills, CA: Sage Publications, 1996, pp. 526-542.

Elden, M., and Chisholm, R. F. "Emerging Varieties of Action Research," *Human Relations* (46:2) 1993, pp. 121-141.

Gummesson, E. *Qualitative Methods in Management Research* Bickley, Bromley, England: Chartwell-Bratt, 1988.

Hart, E., and Bond, M. *Action Research for Health and Social Care: A Guide to Practice*, Buckingham, England: Open University Press, 1995.

Kock, N. F. "Myths in Organizational Action Research: Reflections on a Study of Computer-Supported Process Redesign Groups," *Organizations & Society* (4:9) 1997, pp. 65-91.

Kock, N. F., McQueen, R. J., and Rouse, A. "Negotiation in Information Systems Action Research," in *Proceedings of the First Information Systems Conference of New Zealand,* Los Alamitos, CA: IEEE Computer Society Press, 1996, pp. 164-173.

Kock, N. F., McQueen, R. J., and Scott, J. L. "Can Action Research Be Made More Rigorous in a Positivist Sense? The Contribution of an Iterative Approach," *Journal of Systems and Information Technology* (1:1) 1997, pp. 1-24.

Lau, F. "A Review on the Use of Action Research in Information Systems Studies," in A. S. Lee, J. Liebenau, and J. I. DeGross (Eds.), *Information Systems Research: Information Systems and Qualitative Research,* London: Chapman & Hall, 1997, pp. 31-68.

Lewin, K. *Field Theory in Social Sciences*, New York: Harper, 1951.

Lincoln, Y. S., and Guba, E. G. *Naturalistic Inquiry*, London: Sage Publications, 1985.

Marshall, C. "Goodness Criteria: Are They Objective or Judgement Calls?," in E. G. Guba (Ed.), *The Paradigm Dialog,* Newbury Park, CA: Sage Publications, 1990, pp. 188-197.

Phillips, D. C. *The Social Scientist's Bestiary: A Guide to Fabled Threats to, and Defenses of, Naturalistic Social Science*, Oxford: Pergamon, 1992.

Robson, J. *Real World Research: A Resource for Social Scientists and Practitioner Researchers*, Oxford: Blackwells, 1993.

Susman, G., and Evered, R. D. "An Assessment of the Scientific Merits of Action Research," *Administrative Science Quarterly* (23) 1978, pp. 582-603.

Whyte, W. F. *Participatory Action Research*, Newbury Park, CA: Sage Publications, 1991.

ABOUT THE AUTHOR

Sue Holwell has been a member of the Open Systems Research Group at the Open University since 2002. She teaches postgraduate and undergraduate courses in information systems and systems thinking. Prior to joining the Open University, Sue lectured at Cranfield University and Lancaster University. She has been an active action researcher for many years, collaborating with Peter Checkland, including on this program of research. She is coauthor, with Checkland, of *Information, Systems and Information Systems* and has published about action research, soft systems methodology, and information systems. Before joining academia she worked for 20 years in IS/IT in the Australian Public Service. Sue can be reached at s.e.holwell@open.ac.uk.

Part 5:

Theoretical Perspectives in IS Research

21 THE USE OF SOCIAL THEORIES IN 20 YEARS OF WG 8.2 EMPIRICAL RESEARCH

Donal Flynn
University of Manchester Institute of Science and Technology

Peggy Gregory
University of Central Lancashire

Abstract We study the use of social theories in empirical Information Systems research in the IFIP WG 8.2 conference proceedings since the 1984 Manchester conference. Our results are that interpretivist research and the use of qualitative methods have increased significantly and that only 22 percent of included papers generate theory or concepts according to a narrow definition of theory based on Walsham's classification; the majority of WG 8.2 researchers thus appear reluctant to generalize to theory from their findings, particularly when undertaking interpretivist research. However, using a wide definition of theory that includes researchers' own theory used in their papers, we suggest that additional theory is in fact being generated although in a non-explicit manner. We close by pointing out the benefits of theory generation, inviting WG 8.2 researchers to make their use of theory more explicit and to familiarize themselves with the view that there are forms of generality which are possible within the interpretivist paradigm.

Keywords: Interpretivist research, social theory, generalization, empirical research

1 INTRODUCTION

We discuss trends in the relationship between social theory and empirical research in the papers published by the WG 8.2 community since 1984. The use of theory and its relationship to empirical research in Information Systems has come under increasing attention recently, particularly with regard to research results. Hirschheim and Klein

(2000) state that we need more generalization in our research, and call for papers on historical analysis to build cumulative knowledge and to learn from previous research. Sawyer (2000) discusses the lack of theory development in organizational computing infrastructures and, in Sawyer and Chen (2002), states that there is "almost no proof of concept research in the 8.2 literature." Klein (1999) and Klein and Huynh (1999) point to a lack of theory in interpretive IS research results; they criticize thick descriptions as they "tend to be rather verbose and make it difficult to form a global picture of the social phenomena being researched" (Klein and Huynh 1999, p. 79).

There has undoubtedly been a strong tendency to question the appropriateness, within an intensive research paradigm, of what is often perceived as formal theory (Van Maanen 1995). However, writers such as Silverman (2000) and Eisenhardt (1989) emphasize the ways in which theory may make a contribution within intensive research. In addition, recent articles have appeared that deconstruct the concept of generalizability, pointing out to IS researchers that in fact there are forms of "generality" with which they can feel comfortable, and they need no longer criticize their research for lacking generalizability (Baskerville and Lee 1999). We were thus interested in investigating the extent to which 8.2 researchers were engaged in theory development, and in generalization in particular, from their empirical research.

If theory development can be considered as *output* from the research process, then we were also interested in theory as *input*. We wanted to investigate questions such as were social theories being used by 8.2 researchers (or were studies purely inductive)? Which social theories were being used and were they by famous or by little-known social theorists? How was theory being used (that is, deductively or as sensitizing theory)? Did studies use qualitative, quantitative, or mixed data? Were researchers building on theory and was that theory their own or that of others? To some extent, this part of our investigation was inspired by Jones (2000); however, we went one step further and, rather than regarding citations as an indicator of use of social theory, we wanted to look at those social theories that were deeply integrated into the work of IS researchers. We describe our method in section 2, including selection and validity criteria and the categories used for paper classification, discuss results in section 3, and conclude with a discussion and our conclusions in section 4.

2 METHOD

2.1 Selection Criteria

Our base for data generation was the published IFIP WG.2 conference proceedings since 1984; these are shown in Table 1. Of the 381 papers published in the 17 conference proceedings between 1984 and 2003, 175 papers were included in the study. Keynotes were considered for inclusion but panels were not considered. The criteria for including a paper were that *(1) it reported the conduct of empirical research and (2) social theory was central to the empirical work, either as input, output, or both.* In all but a few cases (five or less) we did not find it difficult to agree on the selection of papers for inclusion, as papers which report on empirical work also deal with some sort of social theory when discussing their results. Papers excluded by these criteria were

Table 1. IFIP WG 8.2 Conferences: 1984-2003. (Adapted and extended from Table 1 in M. Jones, "The Moving Finger: The Use of Social Theory in WG 8.2 Conference Papers, 1975-1999," in R. Baskerville, J. Stage, and J. I. DeGross (Eds.), *Organizational and Social Perspectives on Information Technology,* Boston: Kluwer, 2000, pp. 15-31.)

Date	**Published Proceedings**	**Location**
September 1984	Mumford et al. (1985)	Manchester, UK
August 1986	Bjørn-Andersen and Davis (1988)	Noordwijkerhout, NL
May 1987	Klein and Kumar (1989)	Atlanta, GA, USA
July 1989	Kaiser and Oppeland (1990)	Ithaca, NY, USA
December 1990	Nissen, Klein and Hirschheim (1991)	Copenhagen, DK
June 1992	Kendall, Lyytinen and DeGross (1992)	Minneapolis, MN, USA
May 1993	Avison, Kendall, and DeGross (1993)	Noordwijkerhout, NL
August 1994	Baskerville et al. (1994)	Ann Arbor, MI, USA
December 1995	Orlikowski et al. (1996)	Cambridge, UK
August 1996	Brinkkemper, Lyytinen and Welke (1996)	Atlanta, GA, USA
May-June 1997	Lee, Liebenau and DeGross (1997)	Philadelphia, PA, USA
December 1998	Larsen, Levine and DeGross (1998)	Helsinki, Finland
August 1999	Ngwenyama et al. (1999)	St. Louis, MO, USA
June 2000	Baskerville, Stage and DeGross (2000)	Aalborg, Denmark
July 2001	Russo, Fitzgerald and DeGross (2001)	Boise, ID, USA
December 2002	Wynn et al. (2002)	Barcelona, Spain
June 2003	Korpela, Montealegre and Poulymenakou (2003)	Athens, Greece

literature surveys, thought experiments, and those which exclusively examined texts (less than five); we additionally excluded papers whose focus was research methodology (the great majority of these were theoretical only). Generally the empirical work being reported was undertaken by the paper's authors; however, a few cases were found in which the primary research work was either not undertaken by the authors, or had been completed previously and was being reinterpreted. In these cases, as long as the interpretation or description of the empirical work was a primary feature of the paper it was included. Our definition of *social theory* was

A theory, model, framework, or set of concepts or insights concerning

- *social cognition (e.g., attitudes, values, beliefs)*
- *social behavior (e.g., events, actions taken, structure interacting with action)*

We did not find difficulty in applying this definition as it quickly became obvious that all our included empirical papers dealt with social theory! This is hardly surprising as WG 8.2 focuses on the interaction between IS and the social context. Although we considered formulating a narrower definition, we gave this up not only on the grounds of subjectivity, but also because we wanted to gain an overall picture of social theory use.

2.2 Validity

We approached the need for data validity by initially agreeing on inclusion criteria as well as the categories and their values for paper classification. We read all papers independently and then compared each others' inclusions and categorizations with our own. Where there were differences, they were resolved in early meetings by combining category values (such as case study and ethnography) and in later meetings by revising category definitions, rereading papers, discussion and reaching agreement.

2.3 Paper Classification

We now discuss the categories, and their associated subcategories, that we used to classify each paper and the degree of discussion we required to reach agreement on their meaning and application. We became aware that there were different degrees of subjectivity associated with each of the categories, and that other researchers, with different perspectives to ours, would have classified some papers differently. We defined some subcategories as combinations of other subcategories (for example, field study/survey) and others to be exclusive. Exclusive subcategories occur either as a result of the data (for example, male or female) or where we have decided that a subcategory represents a primary focus. Appendix A details our categorizations for all of the papers.

Research Paradigm. This category is based on the ISWorld description of philosophical perspectives underlying qualitative research (Myers 2004), that identifies three main paradigms: interpretivist, positivist, and critical. However, our interpretation of the positivist paradigm differs, as we decided not to consider theory testing to be a component of this paradigm. The reason for this is that we found that there were many papers that tested theory in an informal way but were interpretivist, in the sense that they were based on an ontological view that reality was socially constructed by peoples' meanings (including those of the researcher), together with a matching epistemology. We decided, therefore, to decouple theory testing from positivism and to regard it in a separate category, discussed below, concerning the relationship of input theory to the research. We took the view that a paper could be based on only one paradigm. There were less than 10 papers where we initially disagreed; this was because the paradigms of those papers were difficult to infer from the information presented.

Research Data. This is based on the three types described in Myers (2004): qualitative data (including data resulting from interviews, documents, questionnaires, participant observation, and researchers' impressions and reactions); quantitative data (data consisting of numbers, with the use of numerical or statistical means to manipulate them in data analysis); mixed qualitative and quantitative data. There were less than five papers where we initially disagreed, mainly on whether a paper was mixed or quantitative.

Research Method. Myers considers (1) qualitative research methods (action research, case study research, ethnography, grounded theory) and quantitative research methods (survey, lab experiment, formal methods, numerical methods). We omitted grounded theory, as it seemed to be more about the relationship of input theory to research (see below). We also found that the distinction between case study and ethnography was not easy to make, as ethnography is characterized by the fact that the researcher "spends a significant amount of time in the field" (Myers 2004). As this information was rarely available from the papers, and as only three papers explicitly claimed to be using ethnographic research, we decided to combine these two methods into one that we termed field study. For quantitative methods, Myers doesn't mention field experiments, so we added these in with lab experiments, using the term *experiment*. We regarded a survey as exclusively collecting data by phone, e-mail, post, or automatic tracing of interaction with the computer. Our final list was action research, field study, survey, experiment, formal methods, numerical methods, and combinations of these methods. There were only a few papers where we initially disagreed, mainly on the combination of methods that papers used. We considered classifying a paper as action research only if the authors stated they were using this method.

Level of social focus. Based on Walsham's (2000) five levels, we added a global level and classified each paper according to six categories that represented the granularity of social focus: (1) personal, (2) group, (3) organization, (4) interorganization, (5) society, and (6) global. Some papers involved more than one level; in these cases, we assigned a primary level of focus to the paper, as well as recording all levels. For perhaps 25 percent of papers, classification required some discussion to reach agreement, involving writing down definitions for each of the levels.

Input theory. Papers were analyzed in terms of the relationship between input theory and their research. Three categories were identified: (1) deductive, (2) sensitizing, and (3) inductive (Bryman 2001). Deductive papers are where a theory is proposed at the beginning and then tested by evidence. Sensitizing papers use theory to organize the empirical research but do not overtly test theory. In inductive papers, empirical phenomena are observed first and inferences drawn from them. Inductive papers almost invariably state that they are adopting this approach with the intention of avoiding prejudgment of the data to be generated. Although these are ideal categories, occupying positions on a spectrum (Sawyer 2000), and although much research is iterative or abductive (Alvesson and Skoldberg 2000), in most cases we categorized the papers by following the main emphasis set by authors. There was some initial disagreement between us, for perhaps 15 percent of papers, mainly concerning whether a paper was deductive or sensitizing; we resolved this by re-reading the paper and reaching agreement.

Own/other theory. The input theory of papers could originate from either (1) the author(s) themselves or (2) others. The first type of paper used a theory generated by the author(s), usually based on the literature or, much more infrequently, on results from previous research by the author(s). The second type used the theory of other authors without modification. There were less than five papers where we initially disagreed on this category.

Output theory. Walsham's (1995) four types of analytic generalization were used to investigate the type of output theory generated by the papers. The types are (1) development of theory, (2) generation of concepts, (3) drawing of specific implications, and (4) contribution of rich insights. Where authors used text only for discussion of their findings, and did not use the term *concept* in their discussion, we (perhaps rather crudely) classified their results as rich insights or specific implications. Where they appeared to be abstracting their results for possible use in contexts other than those presented in the paper, we considered these to be concepts. We are aware that this classification is subjective, but typically, authors were explicit about the fact that they intended to generate a theory. If they had a diagram which appeared to contain concepts shown in relationship to one another, we considered this to be a theory. There were about 10 percent of papers where we initially disagreed on this category.

ISR category. Orlikowski and Iacono (2001) discuss five metacategories of IT—computational, ensemble, nominal, proxy, and tool—based on assumptions about and treatments of IT as an artifact in IS research. We thought that this category would help us relate our results to the wider IS community. This category caused us the most disagreement, concerning interpretation of the meaning and application of the metacategories, as Orlikowski and Iacono do not present the detail of their categorizations of the 10 years of Information Systems research papers they studied. There were about 30 percent of papers where we initially disagreed, and we wrote definitions and application guidance for each metacategory.

Region. We classified papers by the region of the first author, based on the affiliation details given in the papers. The regions are Africa, Asia, Australasia (Australia and New Zealand), North America, South America, and Europe.

Gender. We classified papers by the gender (male or female) of the first author based on the details given in the papers.

3 RESULTS

3.1 Social Theory Papers

Table 2 gives details of the number of included papers (empirical social theory papers) found in each conference. There was only one conference (1984) in which there were no papers that met our criteria. As the theme of this conference was research methods, it is unsurprising that the papers concentrated on theoretical issues rather than the results of empirical studies.

Looking at the data in terms of included empirical social theory papers as a percentage of the total number of papers per year, the high points were in 1994 and 2003 and the low points were in 1984 and 1996. The trend shown by the data is for a gradual increase in the percentage of papers from 1984 to 1994 (with a couple of dips on the way in 1987 and 1990), followed by a sharp decrease between 1995 and 1996, followed by another gradual increase from 1996 to 2003. We conjecture that the dip in the mid-1990s reflects the dichotomy in WG 8.2 between research into IT artifacts and research into social issues, as exemplified by the 1996 method engineering joint WG 8.1/8.2 conference.

We summarize our categorization results in the following sections, mainly in tables but using graphs to indicate interesting trends.

Table 2. Numbers of Papers by Conference

Conference	WG(s)	Total number papers	Empirical Social theory papers	Percent of total
1984	8.2	17	0	0%
1986	8.2	21	7	33%
1987	8.2	15	4	27%
1989	8.2	21	13	62%
1990	8.2	30	8	27%
1992	8.2	17	9	53%
1993	8.2	24	15	63%
1994	8.2	20	15	75%
1995	8.2	21	12	57%
1996	8.1/8.2	19	3	16%
1997	8.2	24	6	25%
1998	8.2/8.6	31	19	61%
1999	8.2	16	6	38%
2000	8.2	27	10	37%
2001	8.2	28	14	50%
2002	8.2	23	13	57%
2003	8.2/9.4	27	21	78%
Total		*381*	*175*	*46%*

3.2 Research Paradigm, Research Data, and Research Method

The papers were first analyzed for their research paradigm, the research data gathered, and the research method used. Table 3 shows a summary of the results for five of the categories we used. The table shows results as a percentage of the included papers in any one year (not as a percentage of all papers).

A clear trend was visible in the data concerning research paradigms, as shown in Figure 1. Over the years the interpretivist paradigm has been the most commonly adopted, with the exception of 1989 in which there were more positivist papers than interpretivist, and the relative number of positivist papers has diminished. In the late 1980s and early 1990s, the mix of positivist and interpretivist papers was fairly even, whereas over the last decade the research emphasis has been far more clearly on the interpretivist side. Only a very small (1 percent) proportion of papers adopt the critical paradigm.

Looking at the research data categories shown in Table 3 there has been a gradual increase in the use of qualitative data over the last 20 years. The number of papers using qualitative data has exceeded all other types of data since 1996, with a marked increase in the percentage of papers using purely qualitative data since 1997. In contrast, during the early period from 1984 until the mid-1990s, research data was more mixed, with the use of quantitative data peaking in 1989. Empirical research in WG 8.2 has been moving more clearly toward being interpretivist and qualitative.

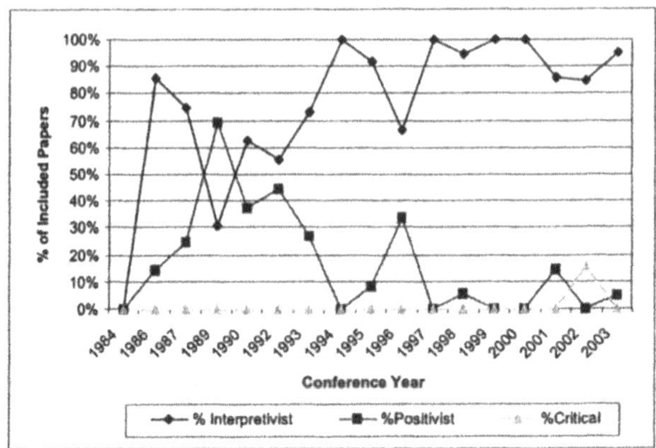

Figure 1. Research in Paradigms

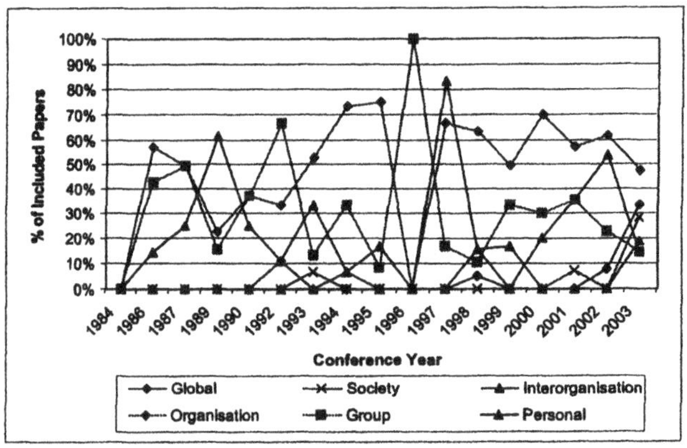

Figure 2. Level of Social Focus

The field study method has been the most widely used research method over all of the years. Surveys and experiments were used more frequently in the late 1980s and early 1990s. Only two papers used numerical methods.

3.3 Level of Social Focus and Conceptualizations of the IT Artifact

Looking at the level of social focus of the papers over all of the years in Figure 2, and considering papers categorized as having more than one level of focus, 54 percent of the papers focus on the organizational level and 26 percent on the personal level. There were very few papers that focused on the global, societal, or interorganizational levels (about 5 percent for each category); the exception to this was the 2003 conference, the theme of which was "Perspectives and Challenges of Organizational Information

Table 3. Summary of Paper Analysis Showing Percentages of Papers in Categories, by Year

	Research Paradigm			**Research Data**			**Research Method**					**Gender**		**Region**					
Year	% Interpretivist	% Positivist	% Critical	% Qualitative	% Quantitative	% Qualitative and Quantitative	% Field study	% Survey	% Field study and Survey	% Action Research	% Experiment	% Male	% Female	% North American	% European	% Australasian	% Asian	% South American	% African
1984	0	0	0	0	0	0	0	0	0	0	0	0	0	0	0	0	0	0	0
1986	86	14	0	43	0	57	86	14	1	14	0	86	14	43	57	0	0	0	0
1987	75	25	0	75	0	25	100	0	0	0	0	75	25	50	50	0	0	0	0
1989	31	69	0	15	54	31	54	46	4	0	31	62	38	85	8	0	8	0	0
1990	63	38	0	63	25	13	50	13	0	0	25	63	38	88	0	13	0	0	0
1992	56	44	0	44	33	22	56	22	0	11	11	44	56	67	22	11	0	0	0
1993	73	27	0	60	20	20	60	33	3	27	0	87	13	13	60	20	0	7	0
1994	100	0	0	93	0	7	100	7	1	0	0	60	40	33	60	7	0	0	0
1995	92	8	0	75	8	17	83	25	1	0	0	83	17	25	67	8	0	0	0
1996	67	33	0	33	0	67	100	0	0	0	0	100	0	0	100	0	0	0	0
1997	100	0	0	83	0	17	100	0	0	0	0	67	33	33	33	33	0	0	0
1998	95	5	0	89	5	5	79	5	0	11	0	58	42	16	74	11	0	0	0
1999	100	0	0	100	0	0	100	0	0	0	0	67	33	33	33	17	0	0	17
2000	100	0	0	100	0	0	90	0	0	10	0	20	80	30	70	0	0	0	0
2001	86	14	0	86	14	0	64	14	0	21	0	86	14	29	71	0	0	0	0
2002	85	0	15	100	0	0	92	0	0	0	8	54	46	23	69	8	0	0	0
2003	95	5	0	76	0	24	95	19	3	0	0	33	67	14	52	19	5	0	10
Total	83	16	1	74	11	15	73	7	7	7	5	62	38	34	53	10	1	1	2

Systems in the Context of Globalization." Given the growing importance of globalization caused by the rapid development of the Internet, this area of research is likely to increase in the future. Empirical studies investigating groups have been represented in each conference over the period, accounting for 26 percent of the total papers.

Results from categorizing the IT focus of the papers, according to the Orlikowski and Iacono (2001) categories for conceptualizing the IT artifact, are shown in Figure 3. Of the papers, 84 percent were in the ensemble category (mostly development and embedded), 10 percent were in the proxy category, 3 percent were in the tool category, 3 percent were in the nominal category, and none were in the computational category. These results are significantly at variance with the review of Sawyer and Chen (2002), who surveyed both Information Systems Research and WG 8.2 papers from 1990 to mid-2001. For the WG 8.2 papers, they found 29 percent were in the ensemble category

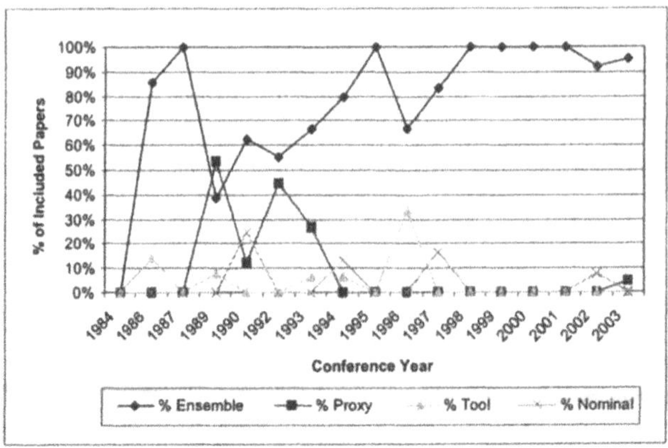

Figure 3. Conceptualizations of the IT Artifact

and 45 percent were in the nominal category. We explain the difference between their results and ours as due to the fact that they included all papers from about 11 years of WG 8.2, in comparison to our inclusion of only empirical papers over a 20-year period. They would, therefore, have included all of the theoretical papers we excluded, many of which might be expected to fall into the nominal category. However, as we had a significant amount of disagreement in our initial categorizations, some of the variance is likely to be explained by differences between interpretations of the categories. This is also suggested by the differences between the results of Sawyer and Chen and by Orlikowski and Iacono on a similar set of Information Systems research papers.

3.4 Gender and Region

From Table 3 it can be seen that there were more papers written by male authors than by female with a ratio of 62:38. This is perhaps not surprising as it reflects the fact that more men than women work in academic departments in universities. However, since 2000, the picture has become more balanced. In both 2000 and 2003 there was a higher percentage of women authors than men authors, and in total over the last five conferences since 1999, there were an even number of papers by male and female first authors (32 each).

Most authors come from either North America or Europe. There is a fair degree of correlation between the venue of the conference and the number of authors coming from that region. Since the conferences have been held either in the United States or Europe it is not surprising that these two groups of authors are the most prominent. Out of 17 conferences, 9 have been held in Europe and 8 have been held in the United States. However the data does not exactly follow the venue of the conference. There is another, broader trend that indicates that whereas in the early years up to 1992 the North American authors were more prominent, since 1993 the European authors have been more prominent. The number of authors from Australasia has been fairly steady since 1990. However the numbers of authors from Asia, Africa, and South America has been very low. The 2003 conference, with its global theme, did attract authors from a wider range of regions; we suggest that WG 8.2 widen its global range of conference venues.

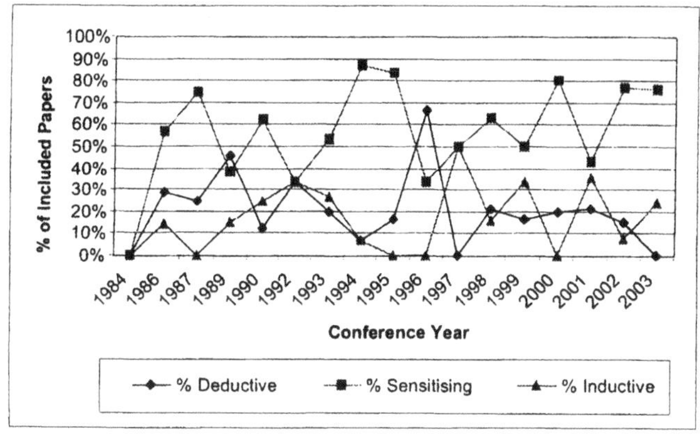

Figure 4. Research Approach: Input Theory Use in Research

3.5 Input Theory: Manner of Use in Research

The results in Figure 4 show that while the number of papers using inductive research approaches has remained fairly steady over the last 20 years, the amount of deductive research being undertaken has gradually diminished and the amount of sensitizing research has increased. For the last few years, the numbers of papers using inductive and deductive approaches has been about the same. However, the trends are not well defined, indicating plurality in the research field.

3.6 Input Theory: Author or Other Source

Input theory can originate either from the author(s) or from others. Figure 5 shows a yearly shifting emphasis but, since 2001, a higher percentage of authors have used the theory of others for their empirical research.

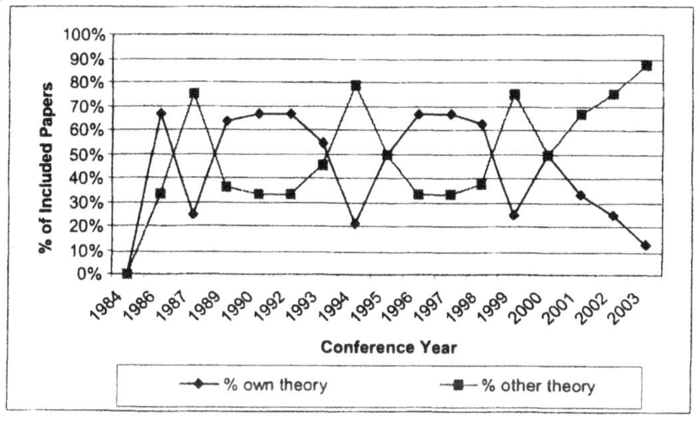

Figure 5. Source of Theory Used

Table 4. Top 10 Social Theories/Theorists

Social Theory/Social Theorist	Number of papers
Actor network theory	9
Structuration theory	5
Foucault	4
Kling (Web model)	3
Orlikowski	3
Activity theory	2
Beck (globalization/identity)	2
Bourdieu	2
Giddens (globalization/identity)	2
Star (boundary objects)	2

Table 4 shows the most frequently used key theories or theorists from the papers. This is a subjective ranking based on our assessment of the key theories and theorists as presented by paper authors. Comparing findings from this survey with Jones' (2000) review of social theorists in WG 8.2 conferences from 1979 through 1999, there are striking similarities between his top 10 list and this one: the actor network theorists, Giddens and Foucault, are still there. Some of the other theorists on Jones's list, such as Burrell and Morgan, Berger, Popper, and Glaser, do not appear because of their focus on research methods. More significantly, the top 10 theories and theorists shown in Table 4 only account for 19 percent of our papers. Therefore, the great majority of papers use less well-known middle range theory, such as Thomas' conflict management model or social presence theory.

3.7 Output Theory: Theory Development

Walsham's (1995) four types of analytic generalization were used to investigate the type of theory generated by the research. Only 22 percent of the papers were found to generate theory or develop concepts; the other 78 percent were found to generate specific implications or contribute rich insights. In Table 5, we show percentages of papers generating theory or concepts. We term this classification *narrow theory output*. We also considered a wider notion of output theory that included all of the papers which had used their own theory as input theory for the empirical work (see section 3.6). The result of this classification is termed *wide theory output* in Table 5. As can be seen from the table, adopting a wide concept of theory results in a much greater proportion of papers (49 percent) that generate theory as a result of their research. We return to this point later.

3.8 Generation of Output Theory

In this section, we investigate the types of papers that generate output theory, i.e., the Walsham (1995) subcategories termed *theory* and *concepts*.

Table 5. Contrasting Narrow and Wide Theory Output

Year of Conference	1984	1986	1987	1989	1990	1992	1993	1994	1995	1996	1997	1998	1999	2000	2001	2002	2003
% Narrow output	0	29	0	15	13	0	27	33	8	0	17	32	50	10	29	31	24
% Wide output	0	57	25	62	63	44	60	40	58	67	50	63	50	50	29	38	29

Research paradigm. We used a narrow view of positivist (excluding papers with a deductive use of theory) and a broad view of positivist (including papers with a deductive use of theory). On a narrow view, Table 6 shows that 25 percent of positivist papers generate theory, compared to 23 percent of interpretivist papers. This was a surprising result as we thought that positivist papers would be more likely to generate theory. As over two-thirds of the interpretivist papers that generate theory are from the last six years, we conclude that there is a trend in interpretivist papers toward theory generation. From a broad view of positivist, 36 percent of positivist papers, compared to only 18 percent of interpretivist (excluding deductive) papers, generate theory, which is a more traditional result.

Research Data. We found that there was little difference (24 percent qualitative, 21 percent quantitative, 19 percent mixed) between types of papers that generated theory. This was surprising as we thought that quantitative papers would be more likely to generate theory.

Research Method. We found that field studies (24 percent) and experiments (25 percent) generate the most theory. In all, 15 percent of survey papers generate theory. This was surprising as we thought that harder methods would be more likely to generate theory. Again, we found it surprising that only 17 percent of action research papers generate theory as one of its hallmarks, differentiating it from consultancy, is held to be its emphasis on the refinement of initial theory (Baskerville and Wood-Harper 1996).

3.9 Gender of First Author

This section investigates the relationship between first author gender and different categories of papers, shown in Tables 7, 8 and 9.

Research paradigm. Male-authored papers adopt a positivist paradigm slightly more than female-authored papers. Using a narrow view of positivist, 18 percent of male-authored papers adopt this paradigm compared to 13 percent of female-authored papers. Using a broad view of positivist, the difference is more marked, as 32 percent of male-authored papers adopt this paradigm compared to 19 percent female-authored papers.

Research Data. Female-authored papers have a slight tendency to use qualitative data more than male-authored papers, as 71 percent of male-authored papers use qualitative data compared to 78 percent female-authored papers. Similarly, male-authored papers have a tendency to use quantitative data more than female-authored papers, as 13 percent of male-authored papers use quantitative data compared to 7 percent female-authored papers.

Table 6. Analysis of Output Theory

		Paradigm					Research Data			Research Method						Level					
Theory-Out		P1	P2	I	IN	C	L	T	Q	A	F	FS	E	N	S	P	G	O	I	S	W
	Total	28	47	145	126	2	129	19	27	12	127	13	8	2	13	27	41	83	8	6	10
Theory (narrow)	*40*	7	17	33	23	0	31	4	5	2	31	2	2	1	2	7	8	21	3	0	1

Key: Paradigm: P1–narrow positivist, P2-broad positivist; I-interpretivist, IN-Interpretivist (not deductive); C-critical. **Research Data:** L-qualitative, T-quantitative, Q-qualitative and quantitative. **Research Method:** A-action research, F-field study, FS-field study and survey, E-experiment, N-numeric. S-survey. **Level:** P-personal, G-group, O-organization, I-inter-organization, S-society, W-global. Cell numbers are numbers of papers.

Table 7. Analysis by Gender of First Author

		Paradigm				Research Data			Research Method					
Gender	*Total*													
		P1	P2	I	C	L	T	Q	A	F	FS	E	N	S
M	*108*	18	35	89	0	77	14	17	11	71	9	5	1	11
F	*67*	9	13	56	2	52	5	10	1	56	3	3	1	3

Key: Paradigm: P1–narrow positivist, P2-broad positivist; I-interpretivist; C-critical. **Research Data:** L-qualitative, T-quantitative, Q-qualitative and quantitative. **Research Method:** A-action research, F-field study, FS-field study and survey, E-experiment, N-numeric, S-survey. Cell numbers are numbers of Male- or Female-authored papers.

Table 8. Analysis by Gender of First Author

Gender	*Total*	Level 1						Level 2							
		P	G	O	I	S	W	P	G	O	I	S	W	SL	ML
M	*108*	16	21	59	5	2	5	28	25	66	6	2	5	85	23
F	67	11	20	24	3	4	5	18	21	29	3	6	5	53	14

Key: Level: P-personal, G-group, O-organization, I-inter-organization, S-society, W-global; SL-single level, ML-multiple level. Cell numbers are numbers of Male- or Female-authored papers.

Research Method. We found it interesting that male-authored papers (10 percent) use the action research method much more than female-authored (1 percent) papers. This might be due to the fact that most action research takes place in organizations, typically run by men, and that female researchers may find that this culture is not conducive to such research. Female-authored papers use the survey method (3 percent) less than male-authored papers (11 percent). The great majority (84 percent) of female-authored papers use the field study method compared to male papers (66 percent).

Level. The level 1 figures in Table 8 refer to those papers which focus on only one level or, for papers that focus on more than one level, where we have made a decision as to the primary level on which they focus. The level 2 figures refer to all of the levels on which papers focus.

Female-authored papers (43 percent) focus on the organizational level much less than male-authored papers (61 percent), but focus more on the group level (31 percent female-authored papers compared to 23 percent male-authored papers), and the society level (9 percent female-authored papers compared to 2 percent male-authored papers). About the same proportion (26 percent) of male-authored and female-authored papers focus on the personal level. With regard to focus on single or multiple levels, 21 percent of both male-authored and female-authored papers focus on multiple levels, with the personal and organization levels being the level of focus most frequent in multiple level papers.

Input theory. Female-authored papers have a different approach to the use of theory on input, as 13 percent use theory deductively with 67 percent using theory sensitizingly. In comparison, for male-authored papers, 22 percent use theory deductively with 60 percent using theory sensitizingly. Male-authored papers (40 percent), compared to female-authored papers (30 percent), have more of a tendency to use their own input theory. Of these male-authored papers, 47 percent test their theory, while of these female-authored papers, 35 percent test their theory.

Output theory. In all, 27 percent of male-authored papers and 16 percent of female-authored papers generate theory, according to our narrow definition in section 3.7 above. If we consider the wider definition, the difference is more marked, with 67 percent of male-authored papers and 46 percent of female-authored papers generating wide theory. Thus male-authored papers tend to generate their own theory more, compared to female-authored papers.

Region. Of the Australasian papers, there are more female-authored papers (59 percent) than male-authored, with 41 percent of papers from North America and 33 percent from Europe female-authored papers.

ISR category. We found that female-authored papers (43 percent) fall into the embedded ensemble category more than male-authored papers (35 percent); male-authored papers also fall into the nominal, proxy, and tool categories more than female-authored papers.

3.10 Region of First Author

This section investigates the relationship between first author region and different categories of papers, shown in Tables 10, 11, and 12.

Table 9. Analysis by Gender of First Author

Gender	*Total*	Theory-In					Theory-Out		Region							ISR Category						
		D	S	I	OT	DOT	T1	T2	AA	AF	AN	AS	AU	EU	ED	EE	EP	ES	N	P	T
M	*108*	24	65	19	43	20	29	72	1	2	35	1	7	62	44	38	3	2	5	11	5
F	*67*	9	45	13	20	7	11	31	1	1	24	0	10	31	25	29	6	0	1	6	0

Key: Theory-In: D-deductive, S-sensitising, I-inductive; OT-own theory, DOT-deductive own theory. **Theory-out:** T1-narrow theory, T2-wide theory. **Region:** AA-Asia, AF-Africa, AN-North America/Canada, AS-South America, AU-Australasia, EU-Europe. **ISR Category:** ED-ensemble development. EE-ensemble embedded, EP-ensemble production network, ES- ensemble structure; N-nominal; P-proxy; T-tool. Cell numbers are numbers of Male- or Female-authored papers.

Table 10. Analysis by Region of First Author

		Paradigm				Research Data			Research Method					
Region	*Total*	P1	P2	I	C	L	T	Q	A	F	FS	E	N	S
AN	*59*	19	23	39	1	36	12	11	2	40	4	5	7	1
AU	*17*	0	3	17	0	14	0	3	1	13	2	0	1	0
EU	*93*	7	18	85	1	75	5	13	9	70	7	6	0	1

Key: Paradigm: P1–narrow positivist, P2-broad positivist; I-interpretivist; C-critical. **Research Data:** L-qualitative, T-quantitative, Q-qualitative and quantitative. **Research Method:** A-action research, F-field study, FS-field study and survey, E-experiment, N-numeric, S-survey. Cell numbers are numbers of North American (AN), Australasian (AU) or European (EU) papers.

Table 11. Analysis by Region of First Author

Region	*Total*	Level 1						Level 2							
		P	G	O	I	S	W	P	G	O	I	S	W	SL	ML
AN	*59*	14	17	24	1	3	0	19	18	25	1	3	0	50	9
AU	*17*	3	6	5	1	0	2	3	6	7	1	1	2	14	3
EU	*93*	8	17	53	5	2	8	21	20	60	6	3	9	70	23

Key: Level: P-personal, G-group, O-organization, I-inter-organization, S-society, W-global; SL-single level, ML-multiple level. Cell numbers are numbers of North American (AN), Australasian (AU) or European (EU) papers.

Table 12. Analysis by Region of First Author

Region	*Total*	Theory-In					Theory-Out		ISR Category							
		D	S	I	OT	DOT	T1	T2	ED	EE	EP	ES	N	P	T
AN	*59*	15	33	11	26	12	16	38	18	21	3	0	3	2	12
AU	*17*	3	9	5	3	2	3	5	8	6	1	0	1	0	1
EU	*93*	14	67	12	34	12	18	39	41	38	5	2	2	2	3

Key: Theory-In: D-deductive, S-sensitising, I-inductive; OT-own theory, DOT-deductive own theory. **Theory-Out:** T1-narrow theory, T2-wide theory. **ISR Category:** ED-ensemble development. EE-ensemble embedded, EP-ensemble production network, ES- ensemble structure; N-nominal; P-proxy; T-tool. Cell numbers are numbers of North American (AN), Australasian (AU) or European (EU) papers.

Research paradigm. In all, 32 percent of North American papers adopt a narrow positivist paradigm. This proportion is significantly more than the 8 percent of European papers that are positivist. No Australasian papers are positivist from this view. On the broad view of positivist (including papers with a deductive use of theory), North American papers are again in the majority, increasing to 39 percent, European papers increase to 19 percent, while Australasian papers increase to 18 percent.

However, these results do not necessarily support the perception that North America continues to favor the positivist paradigm, as the majority of its positivist papers date from the late 1980s and early 1990s. In the last five years, the proportions of positivist papers (on both narrow and broad views) from North American and Europe are approximately equal.

Data. A higher proportion (81 percent) of European and Australasian papers use exclusively qualitative data, compared to 61 percent of North American papers. In contrast, 20 percent of North American papers use exclusively quantitative data, while Australasian papers do not use quantitative data exclusively at all. About the same proportion of all papers use mixed qualitative and quantitative data. However, following the trend of the positivist paradigm, although not to the same extent, more of the recent North American papers use qualitative data exclusively.

Research Method. In all, 75 percent of European and Australasian papers and 68 percent of North American papers use the field study method exclusively. A total of 10 percent of European papers use action research, compared to 3 percent and 6 percent respectively for North American or Australasian papers. Surveys are not used exclusively in Australasian papers. A total of 12 percent of North American papers use the experimental method, compared to no European papers.

Level. In all, 25 percent of European papers focus on more than one (typically two) levels; the most popular of these are organization-personal, group-personal and global-organization. In comparison, 18 percent and 15 percent of Australasian and North American papers respectively focus on more than one level. The primary focus of European (57 percent) and North American (41 percent) papers is on the organizational level. In contrast, the primary focus of Australasian papers is the group level. North American papers focus on the personal level more than other regions. The interorganizational, society, and global levels together account for only 13 percent of papers.

Input theory. Sensitizing use of input theory is most popular for European papers. In contrast, a greater proportion of North American and Australasian papers use theory deductively and inductively. North American papers use theory deductively (25 percent) more than other regions, whereas Australasian papers use theory Inductively (29 percent) more than other regions.

The authors of North American papers (44 percent), compared to the authors of Australasian papers (18 percent), are more inclined to use their own input theory. Of the North American papers, 20 percent test their theory, while of the Australasian and European papers, 13 percent test their theory.

Output theory. There are some major differences between regions: measured widely, 64 percent of North American papers generate theory, compared to 42 percent of European and 29 percent of Australasian papers.

ISR category. In all, 72 percent of North American papers fall within the ISR embedded category, with 20 percent of papers in the proxy category. This is less than European and Australasian papers, where 92 percent and 88 percent respectively fit the embedded category.

4 DISCUSSION AND CONCLUSIONS

This review of the past 20 years of IFIP WG 8.2 conferences has found that empirical research has changed over the years in response to the dialogue within the community. One of the clearest changes over the period has been the move toward use of the interpretivist research paradigm and qualitative data. The debate over research paradigms that started at the 1984 conference was not a call to drop positivist research altogether but a call for a wider diversity in research. There were some doubts expressed at the 1984 conference about whether the community would allow this to happen. The need for legitimation of interpretivist research discussed by many authors (King and Applegate 1997) thus appears to have been achieved.

Relating our study to similar work, we found that our results were significantly at variance with those of Sawyer and Chen (2002). For the WG 8.2 papers surveyed, they found 29 percent of papers were in the ensemble category and 45 percent of papers were in the nominal category, compared to our results where 84 percent of the papers were in the ensemble category and 3 percent were in the nominal category. Although some of this variance can be accounted for by the fact that Sawyer and Chen included all papers from an 11-year period, whereas we included only about half from a 20-year period, the differences between the Sawyer and Chen and the Orlikowski and Iacono surveys indicate that there are clearly large differences between categorization of individual papers, possibly reflecting different views on the position of the socio-technical boundary between IT as an artifact and its context.

Highlights of our results are

- In all, 83 percent of the papers adopt the interpretivist paradigm, 74 percent use qualitative data and 73 percent use field studies.
- The organization is the main level of focus (54 percent), followed by group (26 percent) and personal (26 percent).

- In all, 63 percent of the papers use sensitizing theory, 19 percent deductive and 18 percent inductive.
- In all, 36 percent of the papers use authors' own input theory.
- In all, 22 percent of the papers generate theory.
- Contrary to expectation, we did not find that positivist papers, papers that used quantitative data, and survey papers were more likely to generate theory.
- Male-authored papers are more likely to be positivist, to generate theory and to focus on the organizational level.
- North American papers are more likely to generate theory and less likely to be qualitative.

We found that the top 10 social theories were only used by about a fifth of the papers, perhaps indicating that authors are casting their net widely when looking for theories to help them make sense of research settings. We were intrigued to find that our top 10 theories were similar to Jones' (2000) findings. This implies that researchers do not necessarily "spray" fashionable citations about in their work; on the contrary, if a theory is cited, it generally means it is deeply integrated into the research.

Jones concluded his paper by stating, "The question for IS researchers, therefore, is not whether they should engage with social theory, but how to do so." As we have found, on the narrow definition of theory, only 22 percent of papers generate theory or concepts, according to Walsham's (1995) classification. The impressions of the commentators cited in the introduction appear to be borne out by the evidence. Grunow (1995), quoted in Sawyer (2000), finds 82 percent of organizational research papers did not contribute to theory development.

However, in section 3.7 above, we introduced a wide definition of output theory that includes authors' own theory where it is used as input theory. Our reason for taking this view is that, in the 36 percent of papers that use authors' own theory as input theory, there are very few citations to previously published work for the theory. In this way, it appears that authors may in fact be generating theory, but in a non-explicit manner, by using it as sensitizing or deductive theory. On this wide definition, 49 percent of papers generate theory.

The majority of WG 8.2 researchers appear to be reluctant to generalize to theory or concepts from their findings, particularly when undertaking interpretivist research. However, Baskerville and Lee (1999) point out that IS researchers are unnecessarily handicapping themselves with well-meaning but scientifically inaccurate conceptions of the conditions under which generalization may be claimed, and proceed to give clear examples of how progress in this area may be made. Taken together with our finding that, according to our wide definition of theory, theory is being generated in papers in a non-explicit manner, there thus appears to be room in the future for WG 8.2 researchers to familiarize themselves with the view that there are forms of generality that are possible within the interpretivist paradigm. If they can be convinced of its possibilities, they can make theory more explicit and can further the theory development aspect of their work toward generalization, with its benefits of portability of results between research studies, relevance for practitioners, and possible contribution from the IS discipline to other disciplines.

Concerning the process of paper categorization in which we have been engaged, our experience is that this has been a learning process whereby the meaning and application of our categories has iteratively emerged from several readings of the papers, our studies of IS research methodology literature, discussions, conceptualizations, and (re)definitions. Our understanding of some of the categories changed over the duration of this process. From the categorization discussion earlier we indicated that, for some categories and subcategory boundaries, we had difficulty in constructing definitions about which we could agree. From our experience in writing this paper, we recommend paper categorization of a substantial body of IS research as a good approach to questioning and clarifying our basic concepts.

REFERENCES

Alvesson, M., and Skoldberg, K. *Reflexive Methodology*, London: Sage Publications, 2000.

Avison, D.; Kendall, J. E.; and DeGross, J. I. (Eds.). *Human, Organizational, and Social Dimensions of Information Systems Development*, Amsterdam: North-Holland, 1993.

Baskerville, R., and Lee, A. S. "Distinctions Among Different Types of Generalizing in Information Systems Research," in O. Ngwenyama, L. D. Introna, M. D. Myers, and J. I. DeGross (Eds.), *New Information Technologies in Organizational Processes: Field Studies and Theoretical Reflections on the Future of Work*, Boston: Kluwer Academic Publishers, 1999, pp. 49-65.

Baskerville, R.; Ngwenyama, O.; Smithson, S.; and DeGross, J. I. (Eds.). *Transforming Organizations with Information Technology*, Amsterdam: North-Holland, 1994.

Baskerville, R.; Stage, J.; and DeGross, J. I. (Eds.). *Organizational and Social Perspectives on Information Technology*, Boston: Kluwer Academic Publishers, 2000.

Baskerville, R., and Wood-Harper, A. T. "A Critical Perspective on Action Research as a Method for Information Systems Research," *Journal of Information Technology* (11), 1996, pp. 235-246.

Bjørn-Andersen, N., and Davis, G. B. (Eds.). *Information Systems Assessment: Issues and Challenges*, Amsterdam: North-Holland, 1988.

Brinkkemper, S.; Lyytinen, K.; and Welke, R. J. (Eds.). *Method Engineering: Principles of Method Construction and Tool Support*, London: Chapman & Hall, 1996.

Bryman, A. *Social Research Methods*, Oxford: Oxford University Press, 2001.

Eisenhardt, K. M. "Building Theories from Case Study Research," *Academy of Management Review* (14:4), 1989, pp. 532-550.

Grunow, D. "The Research Design in Organization Studies," *Organization Science* (6:1), 1995, pp. 93-103.

Hirschheim, R., and Klein, H. K. "Information Systems Research at the Crossroads: External Versus Internal Views," in R. Baskerville, J. Stage, and J. I. DeGross (Eds.), *Organizational and Social Perspectives on Information Technology*, Boston: Kluwer Academic Publishers, 2000, pp. 233-254.

Jones, M. "The Moving Finger: The Use of Social Theory in WG 8.2 Conference Papers, 1975-1999," in R. Baskerville, J. Stage, and J. I. DeGross (Eds.). *Organizational and Social Perspectives on Information Technology*, Boston: Kluwer Academic Publishers, 2000, pp. 15-32.

Kaiser, K. M., and Oppeland, H. J. (Eds.). *Desktop Information Technology*, Amsterdam: North Holland, 1990.

Kendall, K. E.; Lyytinen, K.; and DeGross, J. I. (Eds.). *The Impact of Computer Supported Technologies on Information Systems Development*, Amsterdam: North-Holland, 1992.

King, J. L., and Applegate, L. M. "Crisis in the Case Study Crisis: Marginal Diminishing Returns to Scale in the Quantitative-Qualitative Research Debate," in A. S. Lee, J. Liebenau, and J. I. DeGross (Eds.), Information Systems and Qualitative Research. London: Chapman & Hall, 1997, pp. 28-30.

Klein, H. K. "Knowledge and Methods in IS Research: from Beginnings to the Future," in O. Ngwenyama, L. D. Introna, M. D. Myers, and J. I. DeGross (Eds.), *New Information Technologies in Organizational Processes: Field Studies and Theoretical Reflections on the Future of Work*, Boston: Kluwer Academic Publishers, 1999, pp. 13-25.

Klein, H. K., and Huynh, M. Q. "The Potential of the Language Action Perspective in Ethnographic Analysis," in O. Ngwenyama, L. D. Introna, M. D. Myers, and J. I. DeGross (Eds.), *New Information Technologies in Organizational Processes: Field Studies and Theoretical Reflections on the Future of Work*, Boston: Kluwer Academic Publishers, 1999, pp. 79-95.

Klein, H. K., and Kumar, K. (Eds.). *Systems Development for Human Progress*, Amsterdam: North-Holland, 1989.

Korpela, M., Montealegre, R., and Poulymenakou, A. (Eds.). *Organizational Information Systems in the Context of Globalization*, Boston: Kluwer Academic Publishers, 2003.

Larsen, T. J., Levine, L., and DeGross, J. I. (Eds.). *Information Systems: Current Issues and Future Challenges*, Laxenburg, Austria: IFIP, 1998.

Lee, A. S., Liebenau, J, and DeGross, J. I. (Eds.). *Information Systems and Qualitative Research*, London: Chapman & Hall, 1997.

Mumford, E., Hirschheim, R., Fitzgerald, G., and Wood-Harper, T. (Eds.). *Research Methods in Information Systems*, Amsterdam: North-Holland, 1985.

Myers, M. D. "ISWorld: Qualitative Research in Information Systems" (available online at www.qual.auckland.ac.nz/; last accessed 24 January 2004).

Ngwenyama, O., Introna, L. D., Myers, M. D., and DeGross, J. I. (eds). *New Information Technologies in Organizational Processes: Field Studies and Theoretical Reflections on the Future of Work*, Boston: Kluwer Academic Publishers, Kluwer, 1999.

Nissen, H-E., Klein, H. K., and Hirschheim, R. (Eds.). *Information Systems Research: Contemporary Approaches and Emergent Traditions*, Amsterdam: North-Holland, 1991.

Orlikowski, W. J., and Iacono, C. S. "Research Commentary: Desperately Seeking the 'IT' in IT Research—A Call to Theorizing the IT Artifact," *Information Systems Research* (12:2), June 2001, pp. 121-134.

Orlikowski, W. J., Walsham, G., Jones, M. R., and DeGross, J. I. (Eds.). *Information Technology and Changes in Organizational Work*, London: Chapman & Hall, 1996.

Russo, N. L., Fitzgerald, B., and DeGross, J. I. (Eds.). *Realigning Research and Practice in Information Systems Development: The Social and Organizational Perspective*, Boston: Kluwer Academic Publishers, 2001.

Sawyer, S. "Studying Organizational Computing Infrastructures: Multi-Method Approaches," in R. Baskerville, J. Stage, and J. I. DeGross (Eds.), *Organizational and Social Perspectives on Information Technology*, Boston: Kluwer Academic Publishers, 2000, pp. 213-231.

Sawyer, S., and Chen, T. T. "Conceptualizing Information Technology in the Study of Information Systems," in E. H. Wynn, E. A. Whitley, M. D. Myers, and J. I. DeGross (Eds.). *Global and Organizational Discourse about Information Technology*, Boston: Kluwer Academic Publishers, 2002, pp. 109-131.

Silverman, D. *Doing Qualitative Research: A Practical Handbook*, London: Sage Publications, 2000.

Van Maanen, J. "Fear and Loathing in Organizational Studies," *Organization Science* (6:6), 1995, pp. 687-692.

Walsham, G. "Globalization and IT: Agenda for Research," in R. Baskerville, J. Stage, and J. I. DeGross (Eds.), *Organizational and Social Perspectives on Information Technology*, Boston: Kluwer Academic Publishers, 2000, pp. 195-210.

Walsham, G. "Interpretive Case Studies in IS Research: Nature and Method," *European Journal of Information Systems* (4:2), June 1995, pp. 74-81.

Wynn, E. H., Whitley, E. A., Myers, M. D., and DeGross, J. I. (Eds.). *Global and Organizational Discourse about Information Technology*, Boston: Kluwer Academic Publishers, 2002.

ABOUT THE AUTHORS

Donal Flynn has 11 years experience in the UK, Netherlands, and Belgium as programmer, database designer, project consultant and project manager. He has degrees in Chemical Physics (B.Sc., University of Kent), Sociology and Psychology (M.Sc., Imperial College, University of London) and Conceptual Modeling (Ph.D., University of East Anglia). His research interests are in the socio-organizational processes that underlie the interaction between IT systems and their human and organizational contexts. He is currently a senior lecturer in Information Systems at the Department of Computation, University of Manchester Institute of Science and Technology. He can be reached at donal.flynn@umist.ac.uk.

Peggy Gregory is currently a senior lecturer in Computing at the University of Central Lancashire, UK. She has been working in academic computing for the past 10 years, having previously worked in the business sector as a programmer and IT Manager. She has degrees in Theology (B.A., University of Bristol) and Software Development (M.Sc., University of Huddersfield). Her research interests are in the development and use of information systems in their organizational context.

Appendix A

Papers and Their Categorizations

Year	Categorized Papers
1986	Davis/Srinivasan, M, I, QL, F, O, 1, A, T, EE, AN; Robey, M, I, QQ, F, G, 2, A, S, ED, AN; Sandstrom, F, I, QQ, F, G, 2, B, R, EE, EU; Davis/Hamann, M, I, QQ, A, O, 2, A, R, EE, AN; Etzerodt, M, I, QL, F, PG, 3, , R, EE, EU; Blackler, M, I, QL, F, O, 2, B, S, EE, EU; Lange, M, P, QQ, FS, O, 1, A, T, TS, EU
1987	Ciborra, M, I, QL, F, O, 2, A, R, ED, EU; Gurstein, M, I, QL, F, PO, 2, B, S, ED, AN; Kendall, M, P, QQ, F, G, 2, B, C, ED, AN; Hellman, F, I, QL, F, G, 1, B, S, ED, EU
1989	Kling, M, I, QQ, FS, G, 2, A, S, EE, AN; Seror, F, I, QQ, F, P, 2, B, R, EE, AN; Gogan, F, I, QL, F, O, 3, , T, PP, AN; Brown, F, P, QQ, FS, O, 1, B, R, PD, AN; Yap, F, P, QT, S, P, 3, , S, PP, AA; Klepper, M, P, QQ, FS, O, 1, A, R, PD, AN; Olfman, M, P, QT, E, P, 1, A, T, PP, AN; George, M, P, QT, FS, G, 2, B, R, EE, AN; Sein, M, P, QT, E, P, 2, A, S, PP, AN; Frank, M, P, QT, S, P, 1, A, S, PP, AN; Carlsson, M, I, QL, F, P, 2, A, S, EE, EU; Webster, F, P, QT, E, P, 1, A, S, EE, AN; Lin, M, P, QT, E, P, 1, B, S, TS, AN
1990	Elam, F, P, QQ, F, G, 2, B, R, ED, AN; Cooper, M, P, QT, N, O, 1, A, S, PC, AN; Banville, M, I, QL, F, O, 2, B, R, EE, AN; Davies, F, I, QL, F, O, 2, A, R, N, AU; Calloway, F, I, QL, E, G, 3, , R, ED, AN; Toraskar, M, I, QL, F, P, 3, , T, EE, AN; Lee/Goldstein, M, P, QT, S, G, 2, A, S, N, AN; Baskerville, M, I, QL, E, P, 2, A, R, ED, AN

Year	Categorized Papers
1992	Wynekoop, F, P, QT, S, P, 1, A, R, PP, AN; Sumner, F, P, QQ, F, O, 3, , R, PD, AN; Aaen, M, I, QT, S, OW, 3, , R, PD, EU; Becker, F, P, QT, E, G, 2, A, R, ED, AN; Wanninger, M, I, QL, F, G, 1, A, S, ED, AN; Vician, F, P, QQ, F, G, 1, A, R, PP, AN; Davies, F, I, QL, F, G, 3, , R, ED, AU; Jones, M, I, QL, F, G, 2, B, R, ED, EU; Baskerville, M, I, QL, A, OG, 2, B, R, ED, AN
1993	Willcocks, M, P, QT, S, O, 3, , S, P, EU; Millett, M, P, QQ, SF, O, 2, A, S, P, EU; Kaasboll, M, IP, QQ, SF, P, 2, A, S, ED, EU; Frietas, M, P, QT, S, P, 2, B, T, TI, AS; Goldkuhl, M, I, QL, A, G, 2, A, S, ED, EU; Morley, F, I, QL, A, P, 3, , T, ED, EU; Guimares, M, P, QT, F, P, 1, A, T, PP, AN; Trauth, F, I, QL, F, S, 2, A, R, EP, AN; Fitzgerald, M, I, QQ, SF, O, 3, , R, ED, EU; Parkin, M, I, QL, F, O, 1, B, T, PD, AU; Heiskanen, M, I, QL, A, O, 2, B, R, ED, EU; Fischer, M, I, QL, F, O, 2, B, S, EE, EU; Little, M, I, QL, F, O, 3, , S, ED, AU; Ledington, M, I, QL, A, G, 1, A, S, ED, AU; Jones, M, I, QL, F, PO, 2, B, R, ES, EU
1994	Applegate, F, I, QL, F, O, 2, A, T, EE, AN; Ciborra, M, I, QL, F, O, 2, B, C, N, EU; Willcocks, M, I, QQ, SF, O, 2, B, T, EE, EU; Janson, M, I, QL, F, O, 1, B, R, N, AN; Hales, M, I, QL, F, OG, 2, A, T, ED, EU; Korpela, F, I, QL, F, OG, 2, B, S, ED, EU; Davies, F, I, QL, F, G, 2, B, S, ED, AU; Douzou, F, I, QL, F, G, 2, B, R, EE, AN; Mumford, F, I, QL, F, G, 2, A, R, ED, EU; Wilson, M, I, QL, F, O, 2, B, R, EE, EU; Jones, M, I, QL, F, O, 2, B, R, EE, EU; Bjorn-Andersen, M, I, QL, F, O, 2, B, C, EE, EU; Qureshi, F, I, QL, F, I, 2, B, S, EE, EU; Zimmerman, M, I, QL, F, PO, 3, R, EE, AN; Nance, M, I, QL, F, O, 2, B, R, T, AN
1995	Louw, F, I, QL, F, O, 2, A, R, ED, EU; Grint, M, I, QQ, FS, O, 2, B, R, EE, EU; Hamilton, M, I, QL, F, O, 2, B, R, ED, AU; Gallivan, M, I, QL, F, G, 2, B, T, ED, AN; Guimaraes, M, P, QT, S, P, 1, A, S, EE, AN; Scarbrough, M, I, QL, F, O, 2, A, S, ED, EU; Gasson, F, I, QQ, S, O, 1, A, S, ED, EU; Vidgen, M, I, QL, F, O, 2, B, R, EP, EU; Monteiro, M, I, QL, F, O, 2, B, R, EE, EU; Bowker, M, I, QL, F, O, 2, A, R, ED, AN; Bloomfield, M, I, QL, F, P, 2, B, R, EE, EU; Ciborra, M, I, QL, F, O, 2, A, R, EE, EU
1996	Van Slooten, M, I, QQ, F, G, 1, A, R, ED, EU; Mathiassen, M, I, QL, F, G, 2, B, R, ED, EU; Peters, M, P, QQ, F, G, 1, A, R, TI, EU
1997	Urquhart, F, I, QL, F, P, 3, , T, ED, AU; Janson, M, I, QL, F, PO, 3, , R, N, AN; Mantelaers, M, I, QL, F, PG, 3, , R, ED, EU; Sawyer, M, I, QL, F, PO, 2, A, R, EE, AN; Romm, F, I, QQ, F, PO, 2, B, R, EE, AU; Silva, M, I, QL, F, O, 2, A, R, EE, EU
1998	Howcroft, F, I, QL, F, P, 2, A, T, EE, EU; Riva, M, I, QL, F, O, 1, A, T, ED, EU; Aaen, M, I, QL, A, O, 1, A, T, ED, EU; Iversen, M, I, QL, A, O, 2, B, S, ED, EU; Harvey, F, I, QL, F, O, 2, B, R, EE, AU; Baskerville, M, I, QL, F, O, 3, , T, ED, AN; Seppanen, M, I, QL, F, I, 2, A, S, ED, EU; Monteiro, M, I, QL, F, O, 2, B, R, EE, EU; Gasson, F, I, QL, F, G, 2, B, T, ED, AN; Kaplan, F, I, QL, F, O, 2, B, R, EE, AN; Urquhart, F, I, QL, F, P, 3, , S, ED, AU; Levy, F, I, QL, F, O, 2, A, R, EE, EU; Butler, M, I, QL, F, O, 2, B, R, EE, EU; Damsgaard, M, I, QL, F, IW, 2, A, R, EE, EU; Leist, F, P, QT, N, O, 1, A, T, ED, EU; Heiskanen, M, I, QL, F, I, 2, A, R, ED, EU; Brooke, F, I, QL, F, G, 1, A, R, ED, EU; Kirveennummi, M, I, QL, F, O, 2, A, R, EE, EU; Kautz, M, I, QQ, S, PO, 3, , S, EE, EU
1999	Sauer, M, I, QL, F, I, 3, , S, EE, AU; O'Donovan, M, I, QL, F, O, 1, A, T, EE, AF; Scheepers, M, I, QL, F, O, 2, B, R, ED, EU; Sarker, M, I, QL, F, O, 3, , T, ED, AN; Karsten, F, I, QL, F, G, 2, B, R, ED, EU; Spitler, F, I, QL, F, G, 2, B, T, EE, AN

Year	Categorized Papers
2000	Braa, F, I, QL, F, O, 2, A, C, ED, EU; Crowston, M, I, QL, F, PO, 2, A, S, ED, AN; Lamb, F, I, QL, F, O, 2, B, R, EE, AN; Pouloudi, F, I, QL, F, O, 2, B, R, ED, EU; Aanestad, F, I, QL, F, O, 2, B, R, EE, EU; Mark, F, I, QL, F, G, 2, A, R, EE, AN; Grunden, F, I, QL, F, O, 1, A, S, ED, EU; Karsten, F, I, QL, F, G, 2, A, S, EE, EU; Wilson, F, I, QL, F, PO, 2, B, S, EE, EU; Eriksen, M, I, QL, A, G, 1, B, R, ED, EU
2001	Baskerville/Stage, M, I, QL, F, G, 1, A, T, ED, AN; Baskerville/Pries-Heje, M, I, QL, F, OG, 3, , T, ED, AN; Hedstrom, F, I, QL, F, P, 3, , S, ED, EU; Lang, M, P, QT, S, O, 3, , S, ED, EU; Moreton, M, I, QL, F, O, 1, A, T, ED, EU; Nandhakumar, M, I, QL, F, POG, 2, B, R, EP, EU; Allen, M, I, QL, F, O, 3, , R, EE, EU; Eschenfelder, F, I, QL, F, O, 2, B, R, ED, AN; Henfridsson, M, I, QL, A, O, 2, B, R, ED, EU; Brooks, M, P, QT, S, P, 3, , S, EE, EU; Rehbinder, M, I, QL, F, PG, 1, A, T, ED, EU; Nielsen, M, I, QL, A, O, 2, B, S, ED, EU; Rose, M, I, QL, A, G, 2, B, S, ED, EU; Bonner, M, I, QL, F, PS, 2, B, R, EP, AN
2002	Lamb, F, I, QL, F, O, 3, , C, EE, AN; Edenius, M, I, QL, F, P, 2, B, R, EE, EU; Wastell, M, I, QL, F, O, 2, B, R, EE, EU; Wilson, F, C, QL, F, PO, 2, B, R, EE, EU; Monod, M, I, QL, F, O, 2, B, R, N, EU; Kvasny, F, C, QL, F, PG, 2, A, R, EE, AN; Metcalfe , M, I, QL, E, G, 1, A, T, ED, AU; Atkinson, M, I, QL, F, G, 1, A, T, ED, EU; Abbot, F, I, QL, F, OW, 2, B, R, EP, EU; Wagner, F, I, QL, F, P, 2, B, R, EE, AN; Rose, M, I, QL, F, PO, 2, B, T, ES, EU; Ellingsen, M, I, QL, F, PO, 2, B, R, EE, EU; McGrath, F, I, QL, F, PO, 2, B, R, EE, EU
2003	Firth, F, I, QL, F, G, 2, B, S, EE, AU; Hasan, F, I, QL, F, G, 2, B, R, EE, AU; Jutla, F, I, QL, F, O, 2, A, R, EE, AN; Okunoye, M, I, QQ, F, OW, 2, A, T, EE, EU; Samiotis, M, I, QL, F, O, 2, B, T, EE, EU; Chilundo, M, I, QL, F, SO, 3, , R, EE, AF; Joham, F, I, QQ, SF, SW, 3, , S, EP, AU; Kopanaki , F, I, QL, F, I, 2, B, T, EE, EU; Seror, F, I, QL, F, S, 3, , R, EP, AN; Stein, M, I, QQ, SF, OW, 2, B, R, EE, AU; Abbot, F, I, QL, F, OW, 2, B, R, EP, EU; D'Mello, F, I, QL, F, POW, 2, B, R, EP, EU; Kiely, F, I, QL, F, O, 3, , R, ED, EU; Mursu, F, I, QQ, SF, SO, 2, B, R, ED, EU; Nicholson, M, I, QL, F, OW, 2, B, R, ED, EU; Sharma, M, I, QL, F, I, 3, , T, ED, AA; Byrne, F, I, QL, F, PG, 2, B, S, ED, AF; Madon, F, I, QL, F, PS, 2, B, R, EE, EU; Prasopoulou, F, I, QL, F, I, 2, B, R, EE, EU; Jaghoub, F, I, QL, F, SW, 2, B, R, EE, EU; Meso, M, P, QQ, S, I, 2, B, T, PD, AN

Data categories listed in order after first author name: ***Gender*** - M = Male, F = Female; ***Research Paradigm*** - I = Interpretivist, P = Positivist, C = Critical; ***Research Data*** - QL = Qualitative, QT = Quantitative, QQ = Mixed Qualitative and Quantitative; ***Research Method*** - F = Field study, S = Survey, FS = Field study & Survey, E = Experiment, A = Action Research, N = Numerical methods; ***Level*** - W = Global, S = Social, I = Interorganisational, O = Organizational, G = Group, P = Personal; ***Input Theory*** - 1 = Deductive, 2 = Sensitizing, 3 = Inductive; ***Own/Other Theory*** - A = Author's own, B = Another author; ***Output Theory*** - T = Theory, C = Concepts, R = Rich Insights, S = Specific Implications; ***ISR Category*** - T= Tool (TL = labor substitution, TP = productivity, TI = information processing, TS = social relations), P = Proxy (PP = perception, PD = diffusion, PC = capital), E = Ensemble (ED = development project, EP = production network, EE = embedded system, ES = structure), N = Nominal; ***Region*** - AA = Asia, AF = Africa, AN = North America, AS = South America, AU = Australasia, EU = Europe

22 STRUCTURANTION IN RESEARCH AND PRACTICE: Representing Actor Networks, Their Structurated Orders and Translations

Laurence Brooks
Chris Atkinson
Brunel University

Abstract This paper sets out to describe how the StructurANTion theoretical framework and tools derived from it can be used to represent translations of humanchine actor networks and their structurated orders, to underpin both research and practice in integrating IS and organizational transformation.

Keywords: Structuration theory, actor network theory, humanchine, translations, patient-centered information system

1 INTRODUCTION

This paper sets out to illustrate how, within the StructurANTion framework, an existing humanchine actor network, the actors within it, its information system component, and its associated structurated order are translated from one network into another. To do this a number of instruments that capture the translation of a network are introduced and illustrated.

The proposition is that such translations occur when an emergent focal actor (or actors) within a humanchine actor network invoke the *emancipatory* structure inherent within all structurated actor networks. This extends the concept (Atkinson and Brooks 2003) of actor networks as being structurated *hybrid societies* of humans and non-humans. As such it is part of an ongoing project to deploy social and socio-technical theories as *mutable* conceptual actors for use within the field of organizational and social information systems (IS) as a research discipline and concrete real-world practices. We wish to contribute to, but in no way fully address, the provocative question posed for this conference as to whether theory can *really* inform IS practice. Therefore, our approach is to explore the role of social theory in IS and the contribution of the IS field toward social research. This is accomplished by integrating two of the strong social theory candidates already deployed in IS research: structuration theory (ST) and actor network theory (ANT). Our illustration draws on a case study based on a project undertaken within the breast surgery unit of a UK National Health Service (NHS) hospital.

2 STRUCTURATION AND ACTOR NETWORK THEORY

Given the constraints of a conference paper and the conceptual complexity of both ST and ANT, it is not possible here to give a full account of StructurANTion's two constituent theories. It will therefore be assumed that the reader has some familiarity with structuration theory (Giddens 1979, 1984a) and actor network theory (ANT) (Callon 1986; Latour 1999). Particularly as they have not gone unnoticed or under-used within the IS research community. The former is cited by Jones and Karsten (2003) as having been deployed in 225 IS related papers/studies over the past 15 years, the latter as having been increasingly internationally deployed and cited within IS research journals and conference papers by those undertaking interpretive analyses of IS-related interventions (Tatnall and Gilding 1999; Walsham 1997). We argue that both are playing an important role in the emergence of IS as a significant academic discipline.

A more detailed description of StructurANTion theory and how the two constituent theories relate to this proposed theoretical hybrid is provided by Atkinson and Brooks (2003).

The main feature of ST drawn on in this paper is the recursive relationship between a person's psychologically located structures of domination, legitimation, and signification (Giddens 1979) that (mediated through their cognitive modalities) they draw on for their agency and interactions with others. It is the aggregations and combinations of these structures within the actions of millions of people that make up social systems and their institutions. The nature of such structurally mediated agency is highly routinized, resulting in stable social systems and interpersonal relationships. As a result, these social systems recursively persist over time and space. However, they do not always stay the same, as a consequence of an individual's capability for reflexive monitoring of their own and others' actions. Combining this with the unintended consequences of such actions, these social systems gradually change over time and space.

In contrast, the salient aspect of ANT is the manner in which humans and machines (or nonhumans) respond to a problem and are abruptly translated, through the machinations of a focal actor, into hybrid humanchine actor networks. Actor network translation with its moments of problematization, intéressemment, enrolment, and mobilization will be central to this paper. The manner in which focal actors seek to inscribe their interests into the network, to address a problem they have identified as salient, is important. Through successive translations, actor networks (humanchines) display increasingly concerted agency, robust organization, and identity. They become increasingly black-boxed. This is often achieved through the coordinating role of technologies. Such networks and their actors are, however, always under threat of being themselves translated by more powerful focal actors into other networks. Networks are also networks of other networks (Latour 2001). They are often in contention with each other, seeking constantly to translate each other. It is these features of network translation that will be the central theme of this paper.

3 CONSTRUCTING THE STRUCTURANTION HYBRID

This section discusses the arguments behind the bringing together of StructurANTion's two constituent theories of ST and ANT. It identifies StructurANTion theory as a mutable conceptual device, not only capable of forming alliances with people to interpret the world but, through such alliances, one potentially capable of affecting agency.

Atkinson and Brooks (2003) have already discussed the procedural grounds (the *how* rather than substantial *why*) that support the melding of ST and ANT to form StructurANTion. These are founded in the discussions of the nature and role of social theory by Giddens with respect to ST and by Law and Urry (2002) with respect to ANT. Giddens observes that is possible to use theory "not only to understand the world, but as a *cognitive device* with which to change it" (Giddens 1979; emphasis added). Giddens conceives of this as the workings of the double hermeneutic. People interpret theory and mutate it, using it as a cognitive device, a conceptual mechanism, a conceptual actor, to inform human understanding of the world and to underpin their agency. Law and Urry also acknowledge that theory is a conceptual artifact, one that is capable of being enrolled and translated into a real world humanchine network and used as a mutable *theoretical actor* "that can help to bring into being what it supposedly discovers."

What is advocated by these authors is no less than the social shaping of theory or given a device like the nature of theory the social shaping of a cognitive technology. When translated into the real world, ST and ANT can be seen by their progenitors and these author's as *mutable mobiles* (Law and Mol 2000). Mutable mobiles are actors (artifacts) capable of being translated; metamorphosed, melded, and mobilized in response to prevailing circumstances. Through an exercise of the double hermeneutic, in alliance with human actors, they are not only a means of understanding the world but of changing it.

How has StructurANTion come about? StructurANTion has come into being as a theoretical actor through the mutation and conjoining of its two progenitor theories by this paper's authors. They have pursued this by an alliance with, and application of, the double hermeneutic actor to ANT and ST. StructurANTion, like all other theories, is seen by the paper's authors, as itself a mutable mobile.

As to the substantive *why* grounds for melding the two theories into the StructurANTion theory hybrid: This is a deliberate exercise of the double hermeneutic. It is an attempt to gain both interpretive and practical leverage from being able to understand how socio-technical networks, societies of humans and nonhumans, come into being, persist, and change. The latter occurs in either a slow evolutionary, gradual, manner or (as will be explored later) in an abrupt transformational way. The double hermeneutic adopted here is, in ANT terms, itself a cognitive actor, a device, a stratagem of theory translation by the authors.

Together with people, in this case the authors, who enact it, the nonhuman technique that is the double hermeneutic makes up an interpretative humanchine (where a machine is the material manifestation of some form of technique) network whose intention is to mutate theory and demonstrate its use. This actor network, in turn, contributes to this paper artifact and its content of discursive actors. In turn, this network along with all its other actors in the academic paper production process, such as the laptop, the Internet, the English language, the traces of other authors and their papers, reviewers, editors, the review committee, the IFIP 8.2 authoring Web site, etc., all constitute the conference paper producing humanchine actor network. Such a humanchine actor network is itself a medium, or a machination (Latour 1987), convened for translating papers of an appropriate quality and participants into the conference's wider humanchine actor network, whose annual recurrence in turn makes up the collective actor network that is the IFIP 8.2 conference.

A humanchine within StructurANTion theory, from this example, is therefore the prime actor, a human and nonhuman duality that perpetrates agency. As Latour (1999) says, "Purposeful action and intentionality may not be properties of objects, but they are also not properties of humans either. They are properties of institutions [collectives of humans and nonhumans], apparatuses, or what Foucault called dispositifs." This we have designated here as humanchines. He also says, "It is not aircraft that fly people, but airlines," and more chillingly, "It is not guns or people that kill people, but people *and* guns that kill people," i.e., humanchines.

The neologism of the humanchine, as described here, is intended to convey and emphasize the entangling, even the interpenetration, of the human and nonhuman as the entity that acts. In the integration of the human with the machine it is the human/nonhuman artifact duality that constitutes the agency perpetrating actor. It is difficult to identify any instances where humans exhibit agency and socially interact, and where nonhumans, artifacts (in some form or other), are not present. Language alone is an all-pervasive artifact. In its oral, visually scripted, and signed manifestations it is the most fundamental and omnipresent of artifacts, mediated in many instances by information technologies. Along with people, language perpetrates agency and facilitates social intercourse.

The concept of the humanchine has been adopted to capture the duality of people and artifacts that constitute a collective, whose agency and trajectory is more than the sum of its constituent actors (Latour 1999). There may be only one human and one machine in a humanchine networked duality as in the case of the Latour's "gun-man" or there may be thousands, as in the case of an airline. No matter how many are involved, it is this all pervasive humanchine duality that is at the center of StructurANTion theory and that perpetrates action. They also constitute information systems.

Such collectives not only endure over space and time, but also change, usually imperceptibly, although sometimes dramatically, as new humanchine organizational networks emerge. The airline that a conference participant translates into by becoming a passenger in order to fly to a conference is not the same as the one first encountered five weeks prior when the flight was booked, never mind five years later. Yet, like IFIP 8.2 itself, it lives on as a complex humanchine networked actor.

Such humanchine networks, however, do not sustain themselves solely through the, often Machiavellian, stratagems of the focal actor who brought such organizational networks, societies of people and artifacts, into being in the first place. In the case of the airline, this was most probably a now deceased aviator entrepreneur or a corporate buyout. There have to be mechanisms inherent in the routinized relationships and agency perpetrated by the animate and non-animate actors that constitute humanchines, whether they are an airline or a conference, which secures these networks' existence, over time, yet enables them also to change. This raises the questions, by what processes do complex collective humanchines, people, and technologies persist yet change gradually? This, in turn, prompts the obverse question. What are the mechanisms that come into play when such networks are transformed dramatically?

To explore this phenomenon and to address these questions, the authors, in league with the conceptual device that is Giddens' double-hermeneutic have engaged in an exercise of Law and Urry's theory mutation. The StructurANTion theoretical hybrid is the outcome of this exercise.

4 THE STRUCTURANTION THEORY HYBRID

In this section we explore which specific features of the progenitor theoretical actors of structuration theory and actor network theory have been drawn on and why (see Figure 1), as well as what has been co-opted from elsewhere. In a later section we shall explore how such hybrid humanchine networks, including their information systems, are translated from one into another.

As examples, the airline and IFIP 8.2, cited above, are two ends of the network spectrum. One is highly ordered, socially stratified, functionally complex, and physically located, while the other is very fluid in form, functionally simple, socially much less stratified and non-located, almost virtual. Yet both are constituted out of people and artifacts that interact together and show combined agency, persist, and change. What then does a theoretical hybrid need to extract from ANT and ST to provide an ontology that is a cogent theoretical description of both organizational networks?

ANT shows us how humans and the nonhuman come together to form organizational actor networks that exhibit agency, whereas ST does not; it only sanctions the existence of human actors. Conversely, ANT has no concept of a network of humans and nonhumans as having any form of intrinsic inherent structures binding it together, enabling through reflexivity actor agency and interactions.

Within ANT, the network's actor's communications, sanctions, and power are not mediated by commonly shared modalities and structures; networks are relationally flat. A network is bound together (if by anything) by a controlling focal actor who aligns all the other actors' interests with theirs through the moments of translation and on whose behalf it acts. ST does address networks, through the recursive cycle of agency and, modally mediated, cognitive structures, but only with respect to humans. If we combine the two theories together we get a structurated, persistent human/machine network that exhibits agency. There is a problem with this as nonhumans do not have minds and therefore cannot host ST structures.

Structures, according to Giddens, only exist in peoples' minds. They autonomically facilitate human agency and therefore change intrinsically as that human agency changes,

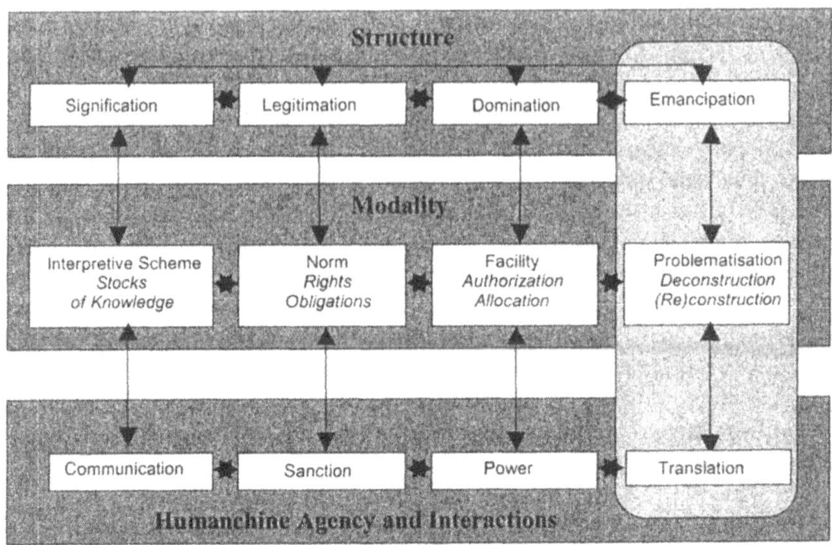

Figure 1. The Structures, Modalities, and Agency of Actors Within the StructurANTion Framework

in response to external events and the actor's recursive and reflexive monitoring of them. They are therefore directly inaccessible. Social structures within ST change as a result of a human's autonomic, unconsciously mediated reflexivity, not by conscious reflection. Conversely, modalities are far more accessible as they manifest themselves in the individual or collective human actions. People can recount, through *discursive consciousness* (Giddens 1984b) how they and their fellow actors deploy their actions, their stocks of knowledge, their rights and obligations, as well as their capacity to authorize others actions and allocate resources.

By observing modalities as they manifest themselves in their collective agency and other actor's responses, we can infer what they are. They also manifest themselves through mediating artifacts, in documentation, and in formalizations of action, for example clinical protocols, as well as in the formal and informal rules of behavior, in particular social and organizational settings. They can also be accessed through human actors' discursive consciousness, getting people within a network to talk about their work and the manner in which they interact. Through such approaches we can, to some degree, surface these modalities as requirements that any prospective technology has to accommodate and hopefully enhance. These can then be built into machines, including information technologies. If adopted and effective, they will replicate, expand, and augment existent human, agency, and modalities as well as other machine functionalities. They may also provide opportunities for new forms of humanchine agency. Humans may be persuaded through interaction with the technology to change their agency and hence effect change in their modalities and structures (e.g., mobile phones). Jones (1999) calls this process "the double mangle."

Within StructurANTion, to be functionally effective and persist, humanchine networks are integrated within organizations at the point of agency through an alignment of their human and machine modalities. As a result, they replicate the organizational network's existing structures of signification, legitimation, and domination. Within organizations these constitute structurated orders, with modalities shared by actors. Each act by an actor in the network reinforces the existing structurated order's forms of signification, legitimation and domination. While the network's actor's behaviors may change their structural relationships, their structurated order remains the same. This begs the question: How then do such networks fundamentally change their structurated order?

Structuration relies on change being an emergent property of widespread individual agency/structure relationships and the unintended consequences of their actions. In doing so it reinforces the existing structurated order or possibly changes it incrementally. ANT, on the other hand, relies on some form of focal actor, either human or nonhuman, to drive change and the actors in the network together, subsequent to its translation through an alignment of actor interests. This necessitates human and nonhuman actors traversing one or a series of (in ANT terms) obligatory passage points. Such change is often dramatic, a step change.

As an alternative to replicating existing structures through the modalities within technologies, we can construct machines to be *machinations*. Such machine machinations have the focal actor's desired modalities and anticipated forms of future humanchine agency "built" into them. Instead of melding with the existing structurated order, the IS acts as a fulcrum around which change is levered. In alliance with machine machinations, the focal actor seeks to perpetrate changes through exerting power and political agency in order to bring agency, modalities, and ultimately structures of the human and other technologies in line with those built into their technology. Latour (1987) reflects,

> The simplest means of transforming the juxtaposed set of allies into a whole that acts as one is to tie the assembled forces to one another, that is to build a *machine*. A machine as its name implies, is first of all, a machination, a stratagem, a kind of cunning, where borrowed forces keep one another in check so that none can fly apart from that group.

This has already been identified and argued to involve evoking what has been termed the *emancipatory structure* (Atkinson and Brooks 2003), which is inherent within the structurated order of every network (see Figure 1). This structure, unlike the others, is drawn on by an actor who seeks to be a focal actor (either as an individual or a collective of human and nonhuman actors) of a new network, one that addresses the issues they have identified as arising from the existing network's behaviors as founded on what they problematize as the unacceptable modalities of the existing structurated order. It originates when an actor problematizes not just network behaviors but its underpinning structurated order and seeks to replace both with a new network.

The concept of emancipation has in the last 200 years been taken to refer to freedom from slavery or Marx's capitalist oppression and class struggle. However, in its more general sense, as we are using here, it is defined as "the setting free from legal, social, or political restraint" (Newton 2003). So the translation of a humanchine network entails setting the human and nonhuman actors free from their current structurated orders and (through traversing the moments of translation by which a network comes into being) reestablishing these actors within it. This applies to existing actors. New actors, such as an IS, could be developed, *in situ*, within the network or brought in through procurement processes.

In the next section we explore the translation of the structurated order of a clinical decision-making network, encompassing information systems along with people.

5 TRANSLATING STRUCTURATED ORDERS

In order to better understand the ideas being presented, this section discusses the translation of a humanchine actor network, the agency of actors from which it is constituted, and the structurated order that the actors in the network recursively reproduce. The following example draws on action research undertaken within a UK hospital (Atkinson 1997). That research studied the entire breast cancer surgical service network and the prospective role of a to-be-procured, hospital-wide clinical IS. This was undertaken on behalf of the UK NHS Executive (Atkinson 1997; Atkinson and Peel 1998) as part of a project exploring the nature and scope of the electronic patient record. Initially the focus is the structurated order of a clinician-led breast cancer decision-taking actor network, captured in Figure 2 and explored in Table 1.

The existing structurated order of this network had been a physician-centered breast cancer service and was being challenged, problematized, and transformed by sympathetic clinicians and patient representations. Power over treatment decisions had resided with oncologists and surgeons. However, enlightened clinicians and patient advocates were challenging this situation. As part of a major hospital-wide IS procurement, they wanted to inscribe a patient- centered approach to care into the new application.

What was happening within the hospital was both a social and technical change leading to a new structurated order. Their scoping and procurement of a new patient-centered information system (PCIS), with patient-centered clinical protocols, was intended to support and reinforce a new women (and male) patient-centered service. They were pursuing this, we would argue, by their drawing on the emancipatory structure latent within the structurated orders of all networks, as set out within the StructurANTion theory (Atkinson and Brooks 2003). The narrative of this transformation from a physician to patient-centered service, the actors, and their behaviors, is illustrated in Table 2. This sets out the activities of the actors as they initially problematize and subsequently deconstruct the existing network with its clinician-centered structurated order. This subsequently leads to the (re)construction of a new patient-centered actor network and the emergence of a patient-centered breast cancer clinical-decision making process and structurated order (see Figure 3 and Table 3).

The following section details the steps by which a transformation might be seen to be taking place. As with all theory, the search is for a way in which to make it useful/useable. While these are broad stages, they provide an understanding of the change pathways, with the intention that future work will provide tools and support for effecting or at least affecting these transformations.

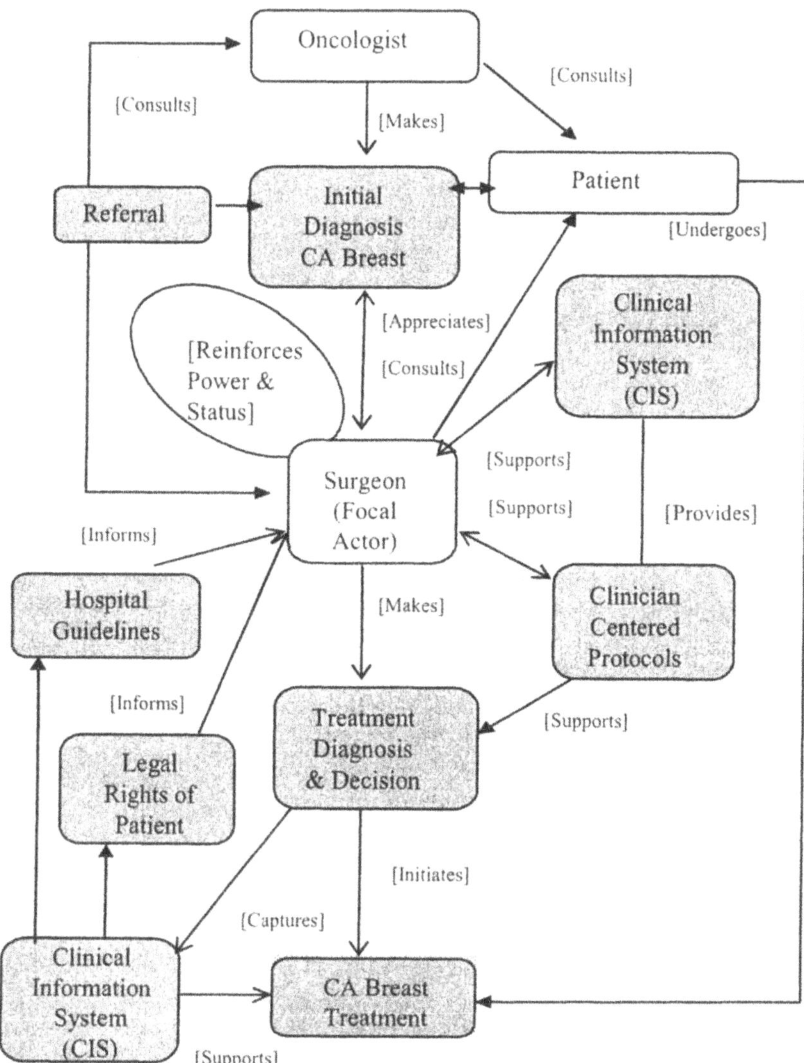

Figure 2. Clinician-Humanchine Breast Cancer Treatment Decision-Making Actor Network (human actors are shown not shaded)

Table 1. Structurated Order of Clinician-Centered Humanchine Breast Cancer Treatment *(Human actor = not shaded. Artifact actor = shaded)*

Structure	Signification	Legitimation	Domination
Modality / Actor	Interpretive Scheme *Stocks of Knowledge*	Norm *Rights & Obligations*	Facility *Authorization Allocation*
GP Referral	GP referral in clinical terminology drawing clinical knowledge and of the patient.	Initiate process of care;be appropriate referral.	Authorize and allocate clinicians and resources.
Oncologist	Clinical knowledge diagnostic expertise skills expressed in clinical terms and language.	Make an appropriate diagnosis on behalf of patient. To assess/add patient information. Ensure that diagnosis is in line with current practice and protocols.	Authorize resources to support diagnosis. Allocate clinical expertise to support diagnosis. Allocate patient to a surgeon for diagnosis/ treatment.
Patient	Lay knowledge and experience of disease and its consequences.	Receive and comply with clinically appropriate diagnosis and treatment. Entitled to a second opinion.	No capacity to allocate or authorize resources. Clinical personnel allocate resources on their behalf.
Surgeon (Focal Actor)	Clinical stocks of knowledge to underpin diagnosis and treatments.	Make a diagnosis on behalf of the patient and identify a clinically appropriate treatment.	Authorize and allocate cognitive and diagnostic resources to identify, make diagnosis, and treat their patient.
Clinical Information System (CIS)	Clinical and patient information available in clinical and managerial forms.	Provide and capture patient information. Provide clinician access to CA breast diagnosis protocols and capture diagnosis and treatment.	Allocate clinical information and protocols and capture compliance (or variant) behaviors by other actors.
Diagnosis of CA Breast	Expressed in clinical terms.	To be addressed via clinically appropriate treatment.	Authorize, allocate clinical resources to capture and address diagnosis.
Clinician-Centered Protocols	Expressed in clinical terms.	To have compliance demonstrated. To provide clinicians with appropriate diagnostic procedures. o have compliance demonstrated	In league with oncologist and surgeon formulate diagnosis and treatment decision and capture compliance.

Structure	Signification	Legitimation	Domination
Modality / Actor	Interpretive Scheme *Stocks of Knowledge*	Norm *Rights & Obligations*	Facility *Authorization Allocation*
Hospital Guidelines	Inscribed in clinical and administrative terms.	To be complied with and support hospital in avoiding litigation.	Mitigate patient litigation by authorizing clinician compliance.
Legal Rights of Patient	Inscribed in legal/ legislative languages.	Provide patient with information on their legal rights/obligations.	Allocate to the patient certain rights with respect to their hospital treatment.
Treatment Decision	Expressed in clinical terminology.	Right to address obligated to be appropriate.	Authorization of clinical personnel and allocation.
Treatment	Draws on clinical knowledge and language within network actors.	To provide the most clinically appropriate treatment against diagnosis and decision.	Allocate resources and allocate clinicians necessary to patient treatment.

6 BREAST CANCER TREATMENT DECISION MAKING ILLUSTRATION

The following illustrates a typical StructurANTional analysis of a translational change from a physician-centered to a patient-centered decision making context.

This happened in a hospital where the focal actor changed from being the clinical expert, accompanied by their protocols, IS, and technologies, to the patient. With help from sympathetic clinicians and managers, they deconstructed the existing network (see Figure 2), translated it, and then reconstructed it (see Figure 3). Employing a StructurANTional analysis of a network and its evolutionary changes, we have identified five broad stages.

Stage 1: Delineate how prior to translation the human and nonhuman actors in the existing network behave individually and in their relationships toward each other (Figure 2).

Stage 2: Identify the modalities of the structures of signification, legitimation, and domination of each actor in the network (Table 1).

Stage 3: Explore the emancipatory structure at work in the problematization of the existent actor network structurated order by a self appointed focal actor. This is manifest in the deconstruction and subsequent (re)constructional translation through the mobilization of actors and the emancipatory structure inherent within the network (Table 2).

Stage 4: Delineate how the actors in the newly emergent network behave individually and collectively and make manifest the new structurated order.

Stage 5: Identify the humans and machine artifacts of the actor network and their modalities within the newly emergent network and the structurated order they (re)produce (Figure 3).

Table 2. The Emancipatory Structure in Action

	Emancipatory Structure and Humanchine Translational Agency
Start	*Existing Actor Network – Physician Centered System*
Modality Problematization Existing Network—see Figure 2 and Table 1	• Recognition and acknowledgment by **patient group** and **sympathetic clinicians** that today current **surgeon/consultant** dominated **breast cancer (BC) decision** making and delivery is untenable, unethical, and disempowering to those who are in need of or currently undergoing **surgical (and other forms treatment)** as well as **maintenance and remedial treatment**. • Share this perception with other **patients** and with clinicians in **the department** and **hospital** that would be or had the potential to be sympathetic to the idea of **patient-centered clinical decision making and treatment**. • Explore this situation and various perceptions of it as untenable with other **patient's** and **clinicians** and other **interested parties**. • Confirm this perception with other **patient's** and **parties** who would be potential allies within the **clinical body**. • Identify research sources of information on current practices in **BC decision making** and the introduction **of patient-centered services** as a basis for creating a vision of **future services.** • Establish a case for confronting and addressing the current BC decision making within the **hospital department**(s) for **BC services** and their delivery. • Enroll politically other **patients, managers, and clinicians** into this point of view. • Explore how **patient notes, clinical information system**, **protocols**, and **guide-lines** support clinical decision making and exclude the **patient** through limiting **access** and the **arcane clinical languages** used
Modality Deconstruction Intéressemment Enrolment	• Seek out powerful **allies and stakeholders** within and outside the organization that share the same opinion with respect to current BC decision making practice and who are willing to support and become involved in changing current practices. • Put together a case and seek out **allies and resources.** • Challenge **existing diagnostic and treatment practices** covertly in the process of care and overtly within the **hospital** practices and through the **local/national media**. • Identify those **clinicians** that will support the case for change within the **hospital** those who will consider it and those who resist it at all costs. • Explore and seek to persuade **nursing practitioners and managers** to become involved in the processes of change. • Explore how **current practices, clinical protocols, and guidelines** within and accesses through the current **clinical record information system** (**CRIS**) disenfranchise the **patient** from taking their own decision making. • Offer alternatives to these **protocols and guidelines** based on **research and services provided elsewhere** in patient-centered **BC hospital** and **community services.** • Identify and offer **different models of clinical decision making**, research findings on the nature and role of practices and **protocols** that will empower the **patient decision taker**.

	Emancipatory Structure and Humanchine Translational Agency
Modality (Re)Construction Mobilization Obligatory Passage Point Intéressemment	• Engage in those **internal political** and **wider public** processes to raise awareness and change attitudes within the clinical body currently providing breast care services. • Persuade **leaders** within the **clinical body and executive management** to take seriously and allocate the funds necessary to create a internal an fiscal climate in which patient-centered BC can be realized. • Lobby widely in **appropriate arenas** raise the profile of issues within current BC services and how they need to be addressed. • Gain support and commitment within the **clinical body** including **consultants** and **junior staff** to delivering patient-centered care. • Confront and change **clinician attitudes** to the point where they accept that it is the person's right to take control of her/his **decision** making for **diagnosis and treatment of breast cancer**. • Fight and engage in politics with **resisters** in hospital necessary to the emergence of the new services and offer images and demonstrations of the efficacy of them and their appropriateness to the modern situation. Gain initial buy-in from **major actors**. • Develop **patient-centered protocols and guidelines** accessible and in a language and format that is comprehensible to the **patient** and are clinically effective and acceptable to the **clinical body**. • Provide these via a **patient clinical information system** that will support them in their decision making and subsequent chosen treatment regimens. Such a **PCIS** would be based on successful applications developed and used elsewhere —either in-house or by an appropriate **application/service provider** aware of the sensitivities and issues of such an application and its use. • Identify and commission a **PCIS application developer** who could offer and/or develop a **PCIS** that would be patient-centered. • Work participatively with the **application provider, patients, clinicians, and managers** to develop a **PCIS** that is patient-focused, one that **the patient** has access to, that is in a language and format that is readily accessible, and that genuinely empowers them in their decision making and ongoing treatment. Deploy **participative prototyping** approach to facilitate this process, on an ongoing basis over **application** life time. • Ensure **PCIS** integrates with **hospital and primary care IS**. • Reform the **nursing role** for some to **breast cancer patient advocates and counselors**, empowering them in this role. • Establish within the current **BC care services** a **patient-centered service** accessible and available to all, some patient may wish to delegate it to their **chosen physician**. • Carry out a **patient-centered research and development program** to continuously improve these **patient-centered breast cancer services** and demonstrate their efficacy.
End	*New Actor Network–Humanchine Patient-Centered System (see Figure 3 and Table 3)*

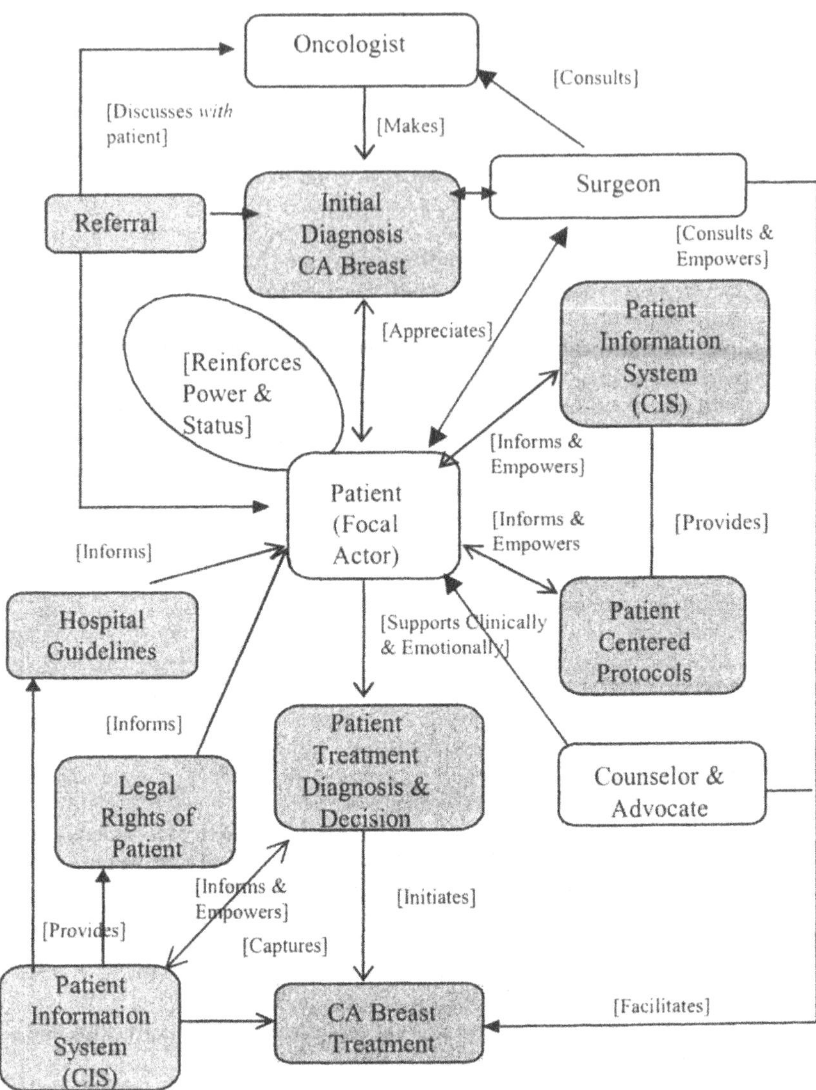

Figure 3. Patient-Centered Humanchine Breast Cancer Treatment Decision-Making Network (human actors are shown not shaded)

Table 3. The New Structurated Order of Patient-Centered Humanchine Breast Cancer Treatment Decision-Making Actor Network *(Human actor = not shaded. Artifact actor = shaded)*

Structure	Signification	Legitimation	Domination
Modality / Actor	Interpretive Scheme *Stocks of Knowledge*	Norm *Rights Obligations*	Facility *Authorization Allocation*
Referral	Patient/GP referral in patient and clinical terms.	Initiate process of care by appropriate referral.	Authorize and allocate clinicians and resources.
Oncologist	Clinical knowledge linked with diagnostic expertise skills expressed in clinician's and patient's language.	Support the patient with diagnosis, aiding them in their decision making. Access/add patient information with the patient in the PCIS.	Authorize surgeon and allocate theater plus post-operative resources in line with diagnosis and treatment decision.
Patient (Focal Actor)	Expressions of their insight into their body, fears, concerns, and needs using their own language and insights.	Patient right to take the decision on treatment as to what happens to their body. Right to effective treatment in line with their decision. Self obligated to seek decision and resources or abrogate them to clinician.	Right to authorize and allocate professional and technical resources in line with their decision making on what is most appropriate treatment for them.
Surgeon	Clinical knowledge linked with diagnostic and surgical expertise and skills expressed in clinicians and patient's language.	Obligated to support the patient in her clinical decision making. Right to disagree and withdraw from caring for patient, while offering alternatives.	Capacity to ensure clinical human, material, and informational resources are available to the patient and the procedures undertaken.
Diagnosis CA Breast	Expressed in a way that is understandable to the patient and in line with current clinical terminology.	Medically cogent diagnosis that is also commensurate with the patient's expressed requirements.	Allocates appropriate clinical resources. Authorizes clinical personnel to make diagnosis.
Patient Clinical Information System (PCIS)	Provides clinical information to patient and clinician in appropriate languages.	Provides information to patient and clinician in format and content appropriate to both.	Allocate access to clinical information and clinical artifacts to the patient and clinicians. Authorize communications between patient and clinicians.

Structure	Signification	Legitimation	Domination
Modality / Actor	Interpretive Scheme *Stocks of Knowledge*	Norm *Rights Obligations*	Facility *Authorization Allocation*
Patient's Treatment Decision	Signifies, in lay and clinical language, what is needed to be undertaken to address patient's diagnosis.	Give the patient the right to make a decision on treatment to meet her diagnosis.	Allocates to patient the resources and information necessary to address her treatment decision; also authorizes access to clinical expertise necessary.
Patient-Centered Protocols	Provides a road map for patient use in support of her clinical decision making on treatment, commensurate with effective practice.	Enshrines the rights and obligations of patient and clinician necessary to undertake a diagnosis and proceed to treatment.	Allocates resources and authorizes the clinical personnel necessary to realizing the patient's decision on breast cancer treatment.
Hospital Guidelines	Provides, in lay and technical language, those clinician practices a patient can expect from a clinician.	Set out the rights and obligations of both patient and clinician with respect to treatment and services. Hospital legitimized.	Provides a framework to guide the patient in the allocation of resources and authorization of personnel.
Legal Rights of Patient	Sets, in lay and technical language, the rights of the patient about diagnosis and treatment.	Identifies what the patient and clinician can/cannot expect of their legal rights within clinical practice. Enshrined in the constitution.	Sets out the resources, personnel, and artifacts that the patient and the clinician can draw on in protecting their legal rights.
Treatment	Provides a description in lay and clinical professional terms of the treatment decided upon and how it will be carried out.	Sets out the rights and obligations of the humans and artifacts undertaking the treatment to address the patient's diagnosis.	Allocates the physical and authorizes the human and professional resources necessary to materialize patient treatment decision.

We recognize that the template for analysis set out here is tentative, naïve, and overly linear, given the complexity of real-world translations. We offer it as a point of departure, acknowledging that it would change through the demands of, and reflections on, future research and development using StructurANTion. The diagrammatic translations are themselves part of the cognitive development of ideas about the changes and, therefore, are an essential component in our understanding of that change. We now consider each stage in turn.

6.1 Stage 1: The Existent Network its Actors, Structures, and Modalities

Within the proposed StructurANTional analysis, the first task is to represent the structurated order existent within the initial, and problematized, organizational actor network(s). This entails delineating the actors and the complimentary modalities of their common structures that are drawn on in the pursuit of individual and collective agency. For this network, these are shown in Table 2, with the nonhuman elements shaded. From this table, it could be inferred that machines and artifacts have structures. We will argue, in subsequent papers, that this is not the case. Rather, it is at the level of modality that humans and machines are conjoined (or not) within the network. For example, the clinical information system hosts diagnostic protocols whose significating knowledge base and language are commensurate with those of the surgeon and oncologist, but not the patient. However, because the humanchine duality has a human component within it and they do have structures, then the humanchine also has structures that are recursively reproduced through its agency; as per Giddens' (1984a) structuration theory. The question for future research is how these structures are mobilized (or not) through effectively integrated human and nonhuman modalities via the double mangle of Jones (1999).

Figure 2 identifies the network's actors and their relationships. Table 1 presents the modalities of this network's structures inherent within humans and inscribed within is nonhuman artifacts and technologies.

The power of the network's mutually reinforcing structurated order within the modalities of both humans and nonhumans is epitomized here by the clinical protocol. Such protocols embody all three modalities and are enacted by humans and nonhumans. When enacted by the physician, they sanction the allocation of clinical resources, and authorization of clinical practitioners' behaviors; when followed, they legitimate clinical decision making and obligations, they embed clinical knowledge, and they are expressed in clinical language. They also allocate the patient to a passive role of the recipient of the decision. The protocols also reside within the modalities of the clinical information system and the physician's mind and are made manifest in their joint agency when the humans and technologies in the network enact it to arrive at a clinical decision.

Figure 2 illustrates how the actors within a network are identified and their relationships are represented. In Table 1, the modalities of the structurated order of the existent clinician-dominated actor network are captured and represented. Table 2 then sets out the process of translation of the actors from within the clinician-dominated actor's network to one in which a patient-centered structurated order presides.

6.2 Stage 2: Actor Behavior and Relationships

The next task of the analysis is to explore the relationships and interactions of the actors as they draw upon these modalities to constitute the network and the collective agency that, in this case, arrives at a clinician-made treatment decision for the patient (see Figure 2). Note how in Figure 2 all of the human and machine actors convene around the focal actor, the surgeon, to address their problematization of the need for a treatment diagnosis for their patient. The clinical information system, along with the clinical protocols, are particularly influential in sustaining this structurated order as they have the surgeon's and other clinician's interests inscribed within their modalities and functional behavior. This is manifest not only in their exclusive access to the protocols and the passive role of the patient within them but the artifact of clinical terminology in which they are expressed. The relationships they have with each other all reflect the position of power the surgeon has within the network. They are all geared up to proclaim the surgeon as the center of the network. In doing so, this subordinates the patient to that of passive recipient of its treatment decision.

6.3 Stage 3: Translation of the Existing Structurated Order

This stage of the analysis (see Table 2) focuses on what transpires as a result of the problematization and deconstruction of the network's existent structurated order of the surgeon focal actor and the emergence of a new one, the patient-focused actor network.

In this case, the stage consists of an exploration of the problematization of the current clinician-centered actor network by a newly emergent focal actor: here the patient along with her clinician and technological allies. It is a struggle for power within the network. The actors are shown in bold in Table 2. This results in the current clinician-centered network's deconstruction and its subsequent (re)constructional translation into a patient-centered clinical decision making actor network, all of which (see Table 2) is achieved through the mobilization of a multiplicity of other actors facilitated by the emancipatory structure inherent within the existent network. They form a problematization network, temporarily convened to by the focal actor. When mobilized they effect change in the existing network. Some, but not all of them, then are further translated to constitute the patient-centered humachine breast cancer treatment decision-making actor network.

The demonstrational analysis in this paper would suggest that translation is inevitable and mono-directional. In reality, this is far from the case. From an initial problematization, the trajectory of translation is potentially poly-directional, multi-dimensional, and its outcome unclear. The emergent structurated order could be as easily at odds with the initiating problematization as in concert. Both research and development should anticipate the uncertainty of translation within real-world situations.

6.4 Stage 4: Delineate the Behaviors of Actors in the Newly Emergent Network

Having identified the moments of translation, the next stage is to capture or anticipate its outcome. The emergent actor network (see Figure 3) in this example centers on supporting the patient as the focal actor, with new roles for the surgeon in enabling them to make the decision and the nurses acting as counselor and advocate. The IS provides patients with information access; providing them with graphical tools and protocols in a language that is both accessible to support their own decisions, yet is also clinically cogent. This newly emergent patient-centered structurated order is delineated in Table 3.

6.5 Stage 5: Identify the Humans and Machine Artifacts of the Actor Network and Their Modalities Within the Newly Emergent Network

Having, either through research or design, identified the newly translated actor network and its agency, the next stage is to identify the modalities of the humans and artifacts within the network. This is done in order to define the structurated order of a newly emergent network. For research, they provide insights into the translations of actor networks, and their orders of agency and structure. If, however, the StructurANTion framework is used to underpin practice, then this analysis could support a focal actor in addressing their chosen problematization. These are set out in Table 3, which reveals how the modalities of the new patient centered network have changed and how they manifest themselves in how the humans behave and also, of relevance here, how the PCIS changes in its functionality and modalities.

7 CONCLUSION

This paper set out to argue for and illustrate the concept of the humanchine duality of the actor network as being at the focus of IS research within the context of organizational development. Information technologies are, we assert, themselves nonhuman actors. They convene together with the human beings that use them to constitute humanchine IS. Such systems are themselves hybrid actors within a wider humanchine network that collectively exhibits agency (for example, breast cancer decision-making). Such humanchine networks persist and change over time and space. The neologistic hybrid, the humanchine, has been adopted to capture the essential duality of the human and nonhuman actors who as networks exhibit a collective agency. This analysis is based on the authors' StructurANTion framework, itself a hybrid theory whose own neologism reflects the humanchine duality of its core concept.

The discussion here has focused on the manner in which such networks change dramatically. In StructurANTion terms, they are translated by a focal actor invoking a network's emancipatory structure. The result is a new network possessed of a structurated order that is markedly different from its network progenitor. The illustration used was the movement from a clinician-centered breast cancer decision-making actor network to one centered on the woman patient as the focal actor. The ramifications for the technology actor and its human users are that structural modalities (embedded within their minds) and its functionality have to change. Drawing on the StructurANTion framework, IS as a practice needs to both understand the its own role within the existing network and to work with those who are affecting the network's translation, to create a new technological application whose modalities and functionality are aligned with the emergent network. To achieve this in practice, our next step is another, greater challenge. To paraphrase Machiavelli, from the early explorations of ANT, "It must be considered that there is nothing more difficult to carry out nor more doubtful of success nor more dangerous to handle than to initiate a new [structurated] order of things."

REFERENCES

Atkinson, C. J. *A Project to Support the Combined Development of the Integrated Patient Management System (IPMS) and Clinical Practice in XXX NHS Trust*, Report to the NHS Executive Information Management Group and Chief Executive of the XXX NHS Trust, 1997.

Atkinson, C. J., and Brooks, L. S. "StructurANTion: A Theoretical Framework for Integrating Human and IS Research and Development," in J. Ross and D. Galletta (Eds.), *Proceedings of the Ninth Americas Conference on Information Systems*, Tampa, FL, 2003, pp. 2895-2902.

Atkinson, C. J., and Peel, V. J. "Transforming a Hospital through Growing, Not Building, an Electronic Patient Record System," *Methods of Information in Medicine* (37), 1998, pp. 206-310.

Callon, M. "Some Elements of a Sociology of Translation: Domestication of the Scallops and the Fishermen," in J. Law (Ed.), *Power, Action and Belief: A New Sociology of Knowledge*, London: Routledge and KeeganPaul, 1986.

Giddens, A. *Central Problems in Social Theory*, Basingstoke, UK: Macmillan, 1979.

Giddens, A. *The Constitution of Society*, Cambridge, UK: Polity Press, 1984a .

Giddens, A. *The Constitution of Society: Extracts and Annotations*, Berkeley: University of California Press, 1984b.

Jones, M. "Information Systems and the Double Mangle: Steering a Course Through the Scylla of Embedded Structure and the Charybdis of Strong Symmetry," in T. J. Larsen, L. Levine, and J. I. DeGross (Eds.), *Information Systems: Current Issues and Future Changes*, Laxenburg, Austria: IFIP, 1999, pp. 287-302.

Jones, M., and Karsten, H. "Review: Structuration Theory and Information Systems Research," Research Paper No. 2003/11, Judge Institute of Management Studies, Cambridge University, 2003.

Latour, B. "Gabriel Tarde and the End of the Social," in P. Joyce (Ed.), *The Social and its Problems*, London: Routledge, 2001.

Latour, B. *Pandora's Hope: Essays on the Reality of Science Studies*, Boston: Harvard University Press, 1999.

Latour, B. *Science in Action*, Boston: Harvard University Press, 1987/

Law, J., and Mol, A. "Situating Technoscience: An Inquiry into Spatialities," unpublished working paper, Centre for Science Studies and the Department of Sociology, Lancaster University, and the Department of Philosophy, the University of Twente, 2000.

Law, J., and Urry, J. "Enacting the Social," Working Paper, Centre for Science Studies and Sociology Department, Lancaster University, 2002.

Newton, H. *Newton's Telecom Dictionary: Covering Telecommunications, Networking, the Internet, Computing, and Information Technology*, Gilroy, CA: CMP Books, 2003.

Tatnall, A., and Gilding, A. "Actor-Network Theory and Information Systems Research," in *Proceedings of the 10th Australasian Conference on Information Systems*, Victoria University of Wellington, Wellington, New Zealand, 1999, pp. 955-966.

Walsham, G. "Actor-Network Theory and IS Research: Current Status and Future Prospects," in A. S. Lee, J. Liebenau, and J. I. DeGross (Eds.), *Information Systems and Quality Research,* London: Chapman & Hall, 1997, pp. 466-480.

ABOUT THE AUTHORS

Laurence Brooks is a lecturer in the Department of Information Systems and Computing at Brunel University. He previously was a lecturer in the Department of Computer Science at the

University of York and held a research position in the Judge Institute of Management Studies, Cambridge University. He gained a Ph.D. in Industrial Management from the University of Liverpool and a B.Sc. in Psychology from the University of Bristol. His current research interests are in the role that social theory might play in contributing to our understanding of information systems in areas such as health information systems and collaborative work support systems. He is a member of the UK Academy for Information Systems (UKAIS) national board. Laurence can be reached at laurence.boroks@brunel.ac.uk.

Chris Atkinson is a senior lecturer in Information Systems in the UMIST Department of Computation. Until recently he was with Brunel University's Department of Information Systems and Computing Science. Originally a civil engineer, he undertook an M.Sc. and Ph.D. at Lancaster University in soft systems with a particular focus on systemic metaphor and its role in organizational problem solving. He has worked as both an academic and practitioner, focusing on how to integrate information systems development and organizational change, especially within healthcare settings. To that end, he has evolved and extensively deployed the Soft Information Systems and Technologies Methodology (SISTeM). Actor-network theory has recently emerged as an important framework for research, practice and methodological development. Its integration with structuration theory has proved fruitful as a further area for research and development. His field of study and practice has centered on working with multi-professional teams, clinicians, managers, and information systems practitioners in affecting integrated organizational development. Chris may be contacted via christopher.atkinson@brunel.ac.uk.

23 SOCIO-TECHNICAL STRUCTURE: An Experiment in Integrative Theory Building

Jeremy Rose
University of Aalborg

Rikard Lindgren
Viktoria Institute

Ola Henfridsson
Viktoria Institute

Abstract When it comes to investigating the relationship between the social and the technical, the Information Systems (IS) discipline has been a net importer of theories. These theories often carry differing interpretations of central concepts, which then become both confusing and difficult to integrate. In response to calls for IS to become a reference discipline in its own right (in other words, a theory exporter), this paper offers an example of integrative theory development. Instead of adapting a theory from another discipline or building a theory from empirical data, we examine the structure concept in some of its various theoretical adaptations in IS and try to integrate them to produce theory focusing on IS concerns while resolving some of the major areas of contention. Both social and technological versions of structure are investigated through three theoretical IS perspectives drawn from different reference disciplines. The first perspective relates to social theories (principally structuration theory), the second to linguistic theories (principally the structural linguistics of Chomsky), and the last to science studies (principally actor-network theory). The objective is to study areas of agreement and contention around the structure concept. Areas of agreement can be incorporated into integrative theory development, whereas areas of contention must be resolved (a far more difficult task). The resulting theoretical model is illustrated with a case study involving competence management systems design and use at Volvo Information Technology in Göteborg, Sweden.

Keywords: Integrative theory development, socio-technical structure, theoretical IS perspectives

1 INTRODUCTION

Primarily drawing its theory from reference disciplines, the Information Systems (IS) discipline can be described as an applied research field. This is often associated with discipline immaturity and sometimes prompts calls for IS to become a reference discipline in its own right (Baskerville and Myers 2002), usually followed by articles claiming that it now is (Nambisan 2003). Since the definition of *applied* is here *theory importing*, IS logically needs to become a theory exporter. This implies that there would be a body of reasonable integrated and consistent theory that related disciplines would find attractive, work with, and publish in their own journals. In this ambition, adaptive theory making (theory making which adapts a theory from another discipline) has two major drawbacks. The first is that it is not easy to re-export. The best that normally happens is that it provides a minor contribution to the original reference discipline. The second disadvantage is that it imports conceptual schemas carrying their own meanings, which conflict with the meanings of imports from other reference disciplines. For example, *network* in IS means rather different things depending on whether the reference discipline is communications, economics, mathematics, marketing, psychology, or science studies. The result is that IS theoreticians have rather few generally shared concepts to work with in the process of developing consistent theory.

In this paper, we develop an example of integrative theory development. Here we compare contributions in the IS literature from which to exploit different reference disciplines and try to discover what they have in common that is central to the IS tradition. In defining the IS tradition, we follow Lee (2001): it "examines more than just the technological system, or just the social system, or even the two side by side; in addition, it investigates the phenomena that emerge when the two interact." Thus, socio-technical phenomena emerging in the context of designing, implementing, and using information systems are the platform for theory advances.

Starting with three examples of adaptive theory making (structurational, linguistic, and infrastructural accounts), we develop a theoretical model for capturing what is essential to IS, rather than what was essential to the contributing disciplines (Lee 2001). We do not start with an existing theory because this would normally be dependant on a reference discipline. Instead, we start with an empirical situation (the design and use of competence management systems at Volvo Information Technology, hereafter Volvo IT, in Göteborg, Sweden), and a much used (and abused) concept: structure. Structure is a term that is used in many different ways in the IS literature, for example, to mean a formal organizational power distribution, or the relation of the different parts of a computer program. However, we here examine structure primarily from the standpoint of a particular European tradition of thinking that has been influential in IS, where structure refers to the common enduring pattern both in social interaction, and in the linguistic and semiotic discourse in which those interactions primarily take place. In IS (unlike the original disciplines), discussions of structure should be related to computer systems, which are simultaneously systems of ideas expressed in the logic of software, and material (physical) systems. The objective of this paper is to present an integrative model of structure useful for both academics and practitioners in improving IS practice. The model is illustrated with a case study of competence system design and use at Volvo IT.

2 STRUCTURE AND TECHNOLOGY

According to the *Oxford English Dictionary*, structure is "1, the way something is constructed or organized. 2, a supporting framework or the essential parts of a thing. 3, a constructed thing; a complex whole; a building." In this section, we examine some of the prevalent theoretical uses of the term in the IS literature as it relates to technology and computer systems.

2.1 The Dimensions of Structure

Some features of the discussion are common across different theorizations of structure, such as the ordering of components, persistence (endurance or stability over time and space), influences on related events or understandings, where structure is understood to be situated. These common features are characterized as the dimensions of structure (see Table 1).

These dimensions also form a partial definition of our understanding of the structure concept—that is to say, if some theoretical concept is examined that can be placed on all or most of these scales, it is likely that we are discussing something related to structure, even if the theoretical concept has another name (such as network in actor-network theory [Monteiro 2000] or formative context [Ciborra and Lanzara 1994]).

In the following sections, we examine three examples of adaptive theory focusing on different aspects of structure in IS. In each case, we examine both the theory and (briefly) the theoretical context or perspective. The purpose of the discussion is to identify theoretical understandings relating to the dimensions of structure outlined above. The theory analysis compares the three perspectives in relation to the structure dimensions.

2.2 Adaptive Theory Example 1: Structurational Theories of IS (Social Theory Perspective)

One way of framing the relationship between technical and social, or more precisely the relationship between a computer-based artifact and its social context, is to discuss the place of computer artifacts in social structure. This discussion is exemplified by recent contributions to the IS literature which have sought to adapt structuration theory (Giddens 1979, 1984; Giddens and Turner 1987) to theorize the IS field. A particular form of virtual social structure (and its emergent relationship with human agency) is of course central to structuration theory; IS adaptations try to establish the role of computerized technologies within than relationship, since this is not a project that Giddens himself undertook. Giddens (1984) thought of structure as

> rules and resources recursively implicated in social reproduction, thus institutionalized features of social systems have structural properties in the sense that relationships are stabilized across time and space....[Structure] exist[s] only as memory traces, the organic basis of human knowledgeability, and is instantiated in action.

Table 1. Dimensions of Structure

Dimensions of Structure	Description
Formation	How a structure comes into existence. Is it preexisting as a law of nature, socially constructed—perhaps emergent from social actions—or built as an artifact?
Ordering	How the underlying pattern, framework, foundation, or logic of structure can be understood.
Location	Whether structure is virtual (i.e., social) or material (physical) or both. Can structure therefore be located in technology, making it acceptable to talk of technology structures (and consequently technology effects), or is it outside technology?
Scope	How widespread a pattern or framework need be to be thought of as structure. Should a social structure be society-wide, for instance, or can it be local to small groups of people or individual situations?
Influence	How determinant a structure is in enabling or constraining related events and human actions. Should it be thought of as a rule system, which is determinant of external events and actions, or as influential (but not determinant) at some conscious level which people can understand and at least partly resist, perhaps working at an unconscious level to influence or control our actions? Does it constrain, or enable, or both?
Evolution	The extent to which structure is enduring or persistent or stable over time, whether and how it can alter or be altered.

IS academics had some difficulty with the virtuality of Giddens' structure concept and in some structurational technology models social structure is seen to be *inscribed* (Akrich 1994) into an information system. Typically this happens during the process of construction, with the finished product later influencing the behavior of its users: "designers incorporate...structures into technology...once complete the technology presents an array of social structures" (DeSanctis and Poole 1994). Once inscribed in technology, structure is no longer virtual, but material, and thus displays different kinds of endurance and influence which are not really compatible with Giddens' duality of structure and agency.

Orlikowski's earlier contributions (e.g., the duality of technology model, Orlikowski 1992), adopted a much moderated and qualified version of structure-in-technology: technology is both constituted by human agency and helps constitute institutional practice. Thus "information technology facilitates and constrains human action through the provision of interpretive schemes, facilities and norms" (Orlikowski and Robey 1991). However, in her most recent contribution, Orlikowski (2000) moves back to Giddens' position and locates structure entirely outside technological artifacts. By separating the artifact from its use and focusing on technologies-in-practice (the recurrent interaction that users have with technologies) Orlikowski relocates structure in the minds of users. Orlikowski (2000) asserts that technology structures are not external or independent of human agency; they are not out there, embodied in technologies simply waiting to be appropriated. Rather they are virtual, emerging from peoples repeated and situated interaction with particular technologies. In order to develop the virtual technology structure position, she has to rather underplay or ignore the effect of the technology designer in constituting the artifact according to a certain set of social norms (structure).

While Orlikowski's later position is more consistent with Giddens' view of structure (as rules and resources existing only in memory traces) than other earlier structurational IS positions, it is not entirely typical of thinking in other IS literatures with a social theory background. For example, it is commonplace among actor-network theorists to stress both the way that the social is inscribed into technological objects (see below). A compromise position is adopted by the social shaping school (Bijker et al. 1987; MacKenzie and Wajcman 1985). Grint and Woolgar (1997) argue that

> inasmuch as technology embodies social aspects it is not a stable and determinate object (albeit one with political preferences inscribed into it), but an unstable and indeterminate artifact whose precise significance is negotiated and interpreted but never settled.

According to this account, social structure can be located in technological artifacts, but cannot be programmatically inscribed and later read off; social actors retain interpretive flexibility and the role of technology in social structure is fluid and emergent.

2.2 Adaptive Theory Example 2: Deep and Surface Structures (Linguistic Perspective)

Drawing on Chomsky's distinction between deep and surface structures in language, Wand and Weber (1995) distinguish an external view of phenomena around a computer system from the internal view represented in the black box of the system itself. The internal view consists of three types of structure (see Table 2).

Reflecting the dominant positivism of the American literature base, the success or failure of an information system's design is related to its correspondence to an external organizational reality or real world system, which is not obviously apparent (surface), but sometimes hidden (deep): the "unwritten rules of organizational functioning... [which] act to guide behavior by determining and defining appropriate and inappropriate behavior" (Leifer et al. 1994). Leifer et al. note that "deep structure...consists of the values, beliefs and unwritten rules in an organization...failure to identify this is one of the reasons why information systems fail." In a modified, realist formulation, Wand and Weber argue that "if information systems are to fulfil the requirements established for them, they must correctly embed the meaning of someone or some group's perception of the real-world system." The claim is that deep structures are both more *true* and more enduring than surface structures (which are of course easier to capture in the form of requirements): "good deep structures (in the information system) provide inherent stability to information systems in the face of change" (Wand and Weber 1995).

Table 2. Wand and Weber's (1995) Three Types of Structure

Structure Type	Description
Surface structure	The way the system represents itself in the form of interface, inputs, and outputs to the user.
Deep structure	Reflecting the meaning or underlying rule set of the real-world system the computer system is intended to model.
Physical structure	The technological implementation of the computer system.

Truex and Baskerville (1998) investigate the fidelity of such characterizations of deep structure to Chomsky's original theory and find that "the deep structure concept is intentionally used in the IS literature in a fairly loose way...as metaphorical and inspirational." Whereas Chomsky refers to the structure of language and its universal underlying grammar, the IS contributions view language as the window to the social structures that govern (in their view) organizational life. These are expressed in terms which resemble the expression of social structure discussed above. It is assumed that "surface structures are observable [via the medium of language] and that deep structures may be uncovered through them" (Truex and Baskerville 1998).

Although the IS adaptation of deep structure is discussed here because it is both well-known and explicitly relevant to our theme, it should be recognized that the Chomskian version of structure in linguistics is neither typical of the mainstream of IS interest in linguistic and semiotic theories, or much followed up in more recent literature. Most theories of language like to treat structure as enduring underlying patterns, and as rule sets, which either relate to or govern speech acts. These structures both constitute and are constituted by daily speech acts, much in the manner that Giddens relates social behavior (action) to social structure. Chomsky, however, goes further than most in tying these structures to underlying physiological traits of human brains.

The idea that a social order could be inscribed into a computer system and later decoded has its roots in an analogy with the writing and reading of texts (see Grint and Woolgar 1997). Understanding software as text (Lutje 2000) makes it hard to disentangle computer systems from the web of social and linguistic structures that are involved in its production and use. Computers become another media (like film or television) deeply embedded in the transmission of culture. Lutje (2000) further suggests that software as text can be understood as a network or web of semiotic signs. Semiotic understandings of the world can be used in the design process (Andersen 2001; Andersen and Mathiassen 2002; Liu 2000; Stamper 1996; Stamper et al. 2000), or in the understanding of user interpretation and the cultural effects of software (Stamper 1988). Although the materiality of software is a little elusive, computer systems are normally thought of as technological artefacts, and material artefacts can also be understood as signs, both reflecting and projecting heavily mediated interpretations of their social surroundings. Dahlbom and Mathiassen (1995) argue that this relationship means that changing the linguistic and social structures (culture) of a situation also normally involves changing the material structures (such as computer systems).

Linguistic and semiotic theorizations of IS overlap considerably with those that come from social theory. Stamper's (1973) semiotic framework, for example, contains a layer referring to the social world (beliefs, expectations, functions, commitments, contracts, law, culture). Conversely Orlikowski adopts Giddens (1984) duality of structure model, which incorporates signification, interpretive schemes, and communication. Hill (1988) neatly captures this overlap by speaking of technology as a *cultural text* whereby "experience of reality is mediated according to the values, assumptions, and focused capabilities that are sedimented into technological systems."

In what Hill describes as the "tragedy of technology," humans experience the world as "the remorseless working of things" wherein technology appears to command culture, and thus the trajectory of society. The script can be altered; however, the power of technological enframing makes us unaware of this possibility. Here Hill points to the invisibility of our socio-linguistic structures—we normally speak and act without conscious reference to them. Moreover, there are limits to our ability to understand and interpret the structures that we enact, and further limits to our individual and collective ability to respond to them. Interpretive flexibility in the face of technology exists only in so much as technology can be understood and responded to. If technologies appear inflexible to us, then we can only act in their structural shadows, despite the social constructionists' insistence upon interpretive flexibility.

2.4 Adaptive Theory Example 3: IS Infrastructure (Social Study of Technology Perspective)

Ciborra (2000) offers a fairly wide-ranging discussion of information systems as infrastructure. Here the materiality of computer systems as structure simply cannot be ignored. Dahlbom (2000) suggests four essential features of infrastructure which reflect the dimensions of structure identified earlier: it is a societal foundation, a stable structure, a common resource, and a common standard. Ciborra and his colleagues use Star and Ruhleder's (1996) eight dimensions of information infrastructures: embeddedness, transparency, reach or scope, learned as part of membership, links with conventions of practice, embodiment of standards, built on an installed base, and visible upon breakdown.

These dimensions highlight the relationship between social practice and infrastructure, where even the apparently technical details are socially constructed. As Monteiro and Hanseth (1996) note, "standards are neither ready made nor neutral: they inscribe organizational behavior deeply within their 'technical' details." Infrastructure is normally no more present in our consciousness than the socio-linguistic structures discussed earlier. Infrastructure is also emergent: "one important characteristic of infrastructure is to be 'recursive': it feeds upon existing infrastructures and represents the platform for future infrastructures" (Ciborra 2000).

However, the Star and Ruhleder's infrastructure dimensions also point to a central feature of the information infrastructure discussion: the understanding of structure as social and material structures enmeshed. The need to theorize both material and social elements explains the adoption of actor-network theory by information infrastructure theorists (Hanseth 2000; Hanseth and Braa 2000; Holmström and Stalder 2001; Monteiro 1998; Monteiro and Hanseth 1996). Part of the project of actor-network theorists is to overcome "the divide between material infrastructure and social superstructure" (Latour 1991). Thus ANT adopts the principle of symmetry, in which both the human and nonhuman can be analyzed according to the same principles (for a critique of this resolution, see Rose et al. (2003). In the information infrastructure literature, cables, computers, developers, users, and standards can all function as actors in the formation of a actor-network. ANT theorists assume the incorporation of socio-linguistic structures, so-called inscriptions, into technical objects (Akrich 1994). As argued by Callon (1991) and Monteiro and Hanseth (1996), inscriptions represent human interests embodied in socio-economic-technical networks that later define and distribute roles to human and nonhuman actors.

Although carefully distancing himself from the structure/agency cliché, which he disdainfully refers to as the *pont aux ânes*, Latour (1999) regretfully acknowledges the structural connotation of the modern, internet related, use of the word *network.* Although ANT is primarily concerned with the active formation of networks, many structural features of networks are also apparent, such as stability and influence (Callon 1991). Stressing the role of the material in stabilizing networks, Grint and Woolgar (1997) note that "there is a constant need to establish and re-produce the network. In part this can be achieved through material embodiment. Indeed networks based solely on human relations tend to be weak."

Irreversibility in networks, echoing Hill's (1988) argument, helps form the trajectory of society. Where the socio-linguistic structures of society are inscribed into the material technology text, technology becomes "society made durable" (Latour 1991) and retreat becomes difficult. Thus, for the information infrastructure theorists, the material and the social are inextricably entwined in the formation and evolution of new, enduring network/structures, their ordering, and influence.

2.5 Theory Analysis

In this section, each of the six dimensions of structure is examined through the perspective of the three adaptive theory examples. The objective is to outline the major areas of agreement and points of contention in the treatment of structure. The analysis is summarized in Table 3.

Formation: Structure is largely thought of as recursive; that is, emergent from previous structure and emerging from the present structure. Structuration theorists see it as entirely socially constructed, whereas infrastructure theorists see it as emergent from both social and technological structures, where the technological components are both socially shaped, but also influential in their own right. The language theorists are more concerned with preexisting rule sets, but only those influenced by Chomskian linguistics relate those to natural biological laws.

Ordering: This dimension reflected some diversity; the structurational view focuses on social structure and human agency in a technology context, the linguistic view on a relation between different levels of structure, and the infrastructural perspective on networks of human and nonhuman components. The perspectives are rather more complementary then contradictory, however, in as much as they could to some extent be combined.

Location: This dimension is the subject of contention. Orlikowski's (2000) structure-entirely-outside-technology position is a minority position. Wand and Weber

Table 3. Dimensions of Structure Against Different Theoretical Perspectives

Dimensions of Structure	Description	Social Structure (Structuration Theory)	Linguistic/Semiotic Structure (Deep, Surface Structure)	Infrastructure (Actor-Network Theory)
Formation	How a structure comes into existence.	Emergent through action.	Deep structure as law of nature (Chomsky); otherwise emergent through communication.	Socio-technically emergent.
Ordering	How the foundation of structure can be understood.	Structure in human memory traces, technology use.	Structure inscribed in technology.	Structure in technology and social practice.
Location	Whether structure is virtual (i.e., social) or material (physical) or both.	Structure enables and constrains action, humans retain interpretive flexibility.	Structure influential over communication (linked culture and technology text create illusion of determined society trajectory [Hill]).	Infrastructure enables or disables upon breakdown.
Scope	How widespread a pattern or framework need be, to be thought of as structure.	Micro and macro.	Micro and macro.	Organizational and societal.
Influence	How determinant a structure is in enabling or constraining related events and human actions.	Social structure in relation to human agency.	Mirroring of social structure in technology (Wand and Weber); linguistic structure in relation to communicative acts.	Socio-technical systems or networks.
Evolution	The extent to which structure is enduring over time.	Humans effect changes in a social context.	Both material and social structure implicated in change.	Emergent socio-technical change with technology as actor.

(1995) understand structure as internal to a computer system, the semiotics-based theorists assume that material (technological) structure mirrors and reinforces socio-linguistic structure, and the infrastructure theorists assume a socio-technical location for structure. However, in each case, structure is only partly visible.

Scope: This dimension was less contentious. Although some contributions were more targeted at an organizational scope and some at a society level, there was a fairly general assumption that structure could operate at a micro and a macro level simultaneously. The infrastructuralists were the exception: here structure had to be a least organization-wide.

Influence: Influence is another slightly contentious dimension. The Chomskian theorists were the only ones to speak of determination: deep structures as rule systems determining organizational behavior. Otherwise structures were thought of as influential, but the degree of influence was disputed. The orthodox social constructivist position insists on humans' interpretive flexibility in the face of technology, whereas Hill's technology-as-text position suggests a much more powerfully influential role for technology. This notion of structure as beyond everyday human understanding and beyond humans' capacity to resist is also partly captured in the adaptation of the deep structure concept. The infrastructuralists had a slightly different position: structure as enabler (and consequent disabler upon breakdown).

Evolution: All perspectives regarded structures as relatively enduring. There is disagreement, however, on how structures can be changed, with highly knowledgeable humans as the driving force in the structurational accounts, but technologies and non-human actants more significant in other accounts.

In sum, we conclude from this analysis that the most contentious point about structure in the IS literature is its location. Briefly put, are the computer systems a part of our conception of structure or are they not? Accordingly, our integrative theory model concentrates on the relation of material structure to virtual structures (see section 4).

3 RESEARCH METHODOLOGY

3.1 Research Approach

The research presented in this paper can be classified as interpretive IS research (Walsham 1995a) in three ways.

First, following Markus and Robey's (1988) distinction between variance and process theories in IS research, the integrative model proposed can be regarded as a process theory meaning that "outcomes are not conceived as variables that can take on a range of values, but rather as discrete or discontinuous phenomena." Rather than providing a framework that explains structure in IS (e.g., consisting of a set of elements or variables with causal linkages), our intent is to understand the deeper structure of a phenomenon, which is believed can then be used to inform other settings" (Orlikowski and Baroudi 1991). We view the proposed integrative model as an attempt to provide an understanding of structure useful for both academics and practitioners to improve IS practice.

Second, our literature review is concentrated on more or less interpretive approaches to structure. Organizational scientists have long discussed the relationship between structure and technology. Various schools of organization theory such as Taylorism, the socio-technical school, and contingency theory (Galbraith 1973; Lawrence and Lorsch 1967) have explored this issue in order to understand how these elements interact with other organizational elements such as goals, participants, and environment (cf. Leavitt 1965). Recently, however, the IS field has adopted approaches to structure (and technology) that focus more on social structure and information technology, rather than on formal structure and production technologies. Recent adaptations of structuration theory (DeSanctis and Poole 1994; Orlikowski 1992, 2000; Orlikowski and Robey 1991), linguistic theory (Truex and Baskerville 1998), and actor-network theory (Hanseth and Monteiro 1997; Monteiro 2000) are all examples of such approaches. We acknowledge this development in the field of IS by focusing on approaches to IS and structure that reflect this trend.

Third, the case used for developing and illustrating the theoretical model was conducted as part of an action research project on design and use of competence systems. The study draws on qualitative data collected through data sources such as focus groups, interviews, and observations.

3.2 Research Design

The research design is presented in Figure 1. We selected three approaches to structure in IS that all have been developed as theoretical adaptations of the reference disciplines in which they are grounded. The structurational, language, and infrastructural approaches display both agreements and disagreements over dimensions of structure such as formation, ordering, location, scope, influence, and evolution. On the basis of these agreements and disagreements, we present the integrative model of structure as a way of reconciling some difficulties with the theorization of structure found in current IS literature. The empirical data was also used to help develop the model, but in the paper serves to illustrate the explicative relevance of the integrative model in the context of competence systems design and use at Volvo IT.

3.3 Data Collection and Analysis

The case study at Volvo IT (for a description of the research site, see subsection 5.1) was conducted between November 1999 and December 2001. The study used multiple data sources including document reviews, focus group sessions, interventions, interviews, participant observation, testing of systems, and workshops.

Semi-structured interviews covering topics such as competence development, competence management, and work practice were conducted with 26 respondents in different job positions (account managers, business area managers, CEO, competence development managers, HR personnel, management consultants, project managers, sales managers, and programmers). The interviews lasted approximately 45 minutes and were later transcribed to facilitate the data analysis.

The data analysis was conducted applying the principles of interpretive field studies (Klein and Myers 1999). Consistent with these principles, we analyzed the data in an open-minded manner in order not to downplay potential new perspectives and ideas emerging in the iterative process of going back and forth between the theoretical conceptions that guided the research design and the story which the data tell (Klein and Myers 1999; Walsham 1995b). Concurring with the open coding technique originating in grounded theory (Strauss and Corbin 1997), the empirical data was analyzed in an iterative way, where initial categories generated were revised and refined until they adequately explained the data material.

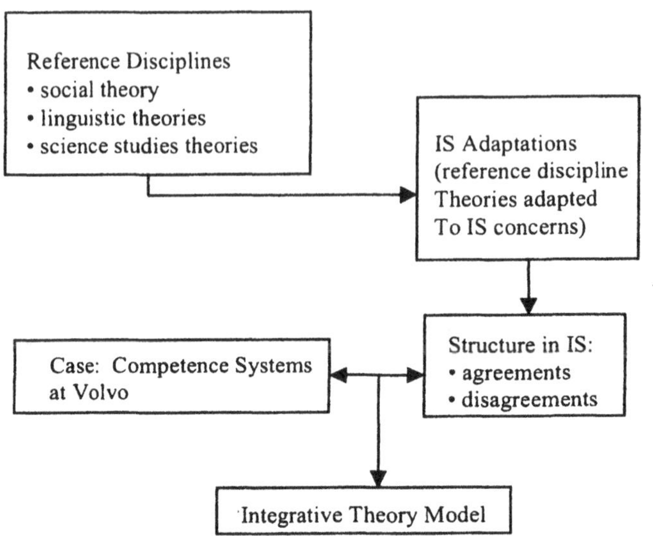

Figure 1. Research Design

4 INTEGRATIVE THEORY MODEL: DEEP, SURFACE, MATERIAL STRUCTURES

The integrative theory model (Figure 2) reflects some relative consensus about structure (see section 2.5 on theory analysis). It is recursive (formation); that is structure is seen as the preexisting frame for both speech acts and practical acts, which then reproduce and transform their formative structures. In this sense, it evolves out of itself. Structures represent shared patterns (ordering) of thought and belief which also become represented in material patterns (there are many makes and designs of computer, but they share many common characteristics). Structures operate at a local level (scope) to influence how we speak and act, but the cumulation of such influences results in society-wide structure. Structures influence, rather than determine, our actions, but we do not necessarily have full understanding of them, so they can influence our behavior in ways about which we are not conscious. Where we are conscious of them we do not necessarily always have the ability to avoid their influence. Structures endure (evolution) but not necessarily unchanged—they can evolve slowly or fast. A more contentious issue is location (is a computer system itself a material structure?), and this is explicitly addressed by the model. Here we distinguish between socio-linguistic structures and material structures, where the type of material structure we are primarily interested in is computer systems, a computing infrastructure, or a computing technology.

Socio-linguistic structures represent the virtual patterns of thought and behavior of groups, organizations, and societies, which are also represented in the discourse and patterns of actions of their members. Socio-linguistic structures may be thought of as schemes of signification, legitimation, and domination, as values, norms, cultural associations, and shared cultural histories and meanings which both provide the formative context within which all communicative and non-communicative actions are

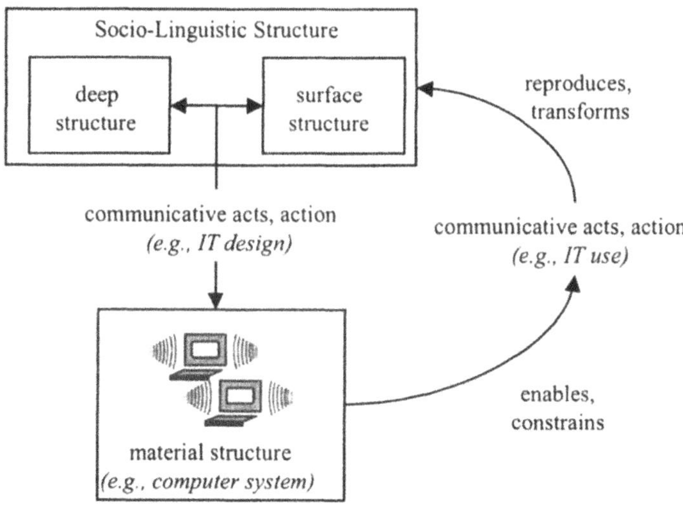

Figure 2. Deep, Surface, Material Structures

taken and, at the same time, the reference system against which they are interpreted and evaluated. In any social situation, discourse, or domain, some socio-linguistic structures are more in focus and more articulated than others. The structures which are ready at hand, easily comprehensible, and accessible we refer to as surface structures. For example, much of the business information systems literature focuses on task structure. However, many other structures may also be important features of the situation, although less apparent. They may be out of focus (i.e., simply not much thought or talked about in the situation), difficult to articulate (perhaps for political reasons), hard to recognize (such as widely held cultural prejudices), or unconscious. These are the deep structures of the situation. To some extent, the deeper structures can be revealed by discussion and analysis, but they are never fully knowable. There may be much interplay between deep and surface structures, but they should not be thought of as a rule system in which they determine each other, or determine communicative acts and actions.

Socio-linguistic structure is the context for human actions of all kinds. We distinguish two kinds of actions: those which are primarily speech-based (or communicative), and those which are primarily behavioral or task oriented—many actions are both. Structure can be less or more influential, but is never absent. One particular form of human action is the design, manufacture, and installation of material objects, including technological artefacts and computer systems. Some surface structures are normally consciously inscribed into material objects, but they may also carry the unconscious imprint of deep structures. Material structure (e.g., computing infrastructure) thus normally reflects socio-linguistic structure. Where the material objects are shared across groups or more widely, they can also be structural in the sense that they influence communicative and practical acts.

Material structure influences human action to a greater or to a lesser degree. Social structure is the reference system against which material structure is interpreted; however, material structure has physical characteristics that somewhat delimit interpretation. If the computer system locks the building at four o'clock, you cannot simply walk through the locked door. Socio-linguistic structures always refer to existing material structures (put another way, our current situation always contains both socio-linguistic and material structures). Therefore, socio-linguistic structures and material structures together make the context in which action is taken.

Material structures, together with their interpretations in the light of social structures, influence the actions of their users. A technology or computer system user has some, but not absolute, freedom to interpret the inscriptions that the designers consciously or unconsciously wrote into the technology object. Human actors have interpretive flexibility, the ability to interpret and respond to material structures in different ways, but this flexibility is heavily bounded by our cognitive limits. In using the technology objects, the users take actions, which often serve to reproduce the social structures inscribed into the objects, but may also change (transform) them.

In the following section, the theoretical model is used to analyze and explain competence management systems development at Volvo IT.

5 COMPETENCE MANAGEMENT SYSTEMS DEVELOPMENT AT VOLVO IT

5.1 Competence Management at Volvo IT

Volvo IT is the Volvo Group's resource and expertise center for IT systems. At the time of the research project (April-June, 2000), Volvo IT had approximately 2,500 employees. Approximately 1,400 of those worked in Sweden and roughly 900 in the Göteborg area where the head office is located. Analogous to many large organizations, the problem of knowing who within the organization knows what has become part of Volvo IT's surface structure or discourse. In 1999, when Volvo IT was formed, this problem became particularly evident. Consolidating the old Volvo data with the systems developers and other IT personnel from the product companies, Volvo IT expanded from 900 to 2,400 employees. The prevalent approach of using, for example, Excel spreadsheets as instruments (material structure) for project configuration and competence management became unmanageable. Volvo IT had only a vague overall picture (i.e., an undeveloped surface structure) of existing competence within the organization and could not conduct goal-directed competence management on neither the organizational nor the individual level. Resulting from this, Volvo IT decided to strengthen their competence management process by initiating a number of activities and projects.

An important initial activity was the attempt to establish a common understanding/shared discourse of the notion of competence in the organization. In this work, the project management group used a published corporate report where competence was defined in terms of five aspects: skills, knowledge, experiences, relationships, and values. In addition to these aspects, motivation was identified as an inner source of energy required for activating competence (AB Volvo 1987). While most project members seemed to appreciate such an understanding of competence, some actors clearly hesitated. In other words, it was unclear how far the surface structure or discourse was accepted, indicating a potential conflict with deep structures. The typical argument of those who questioned this competence view was that qualitative aspects like relationships, values, and motivation are mere complicating factors in that they are too difficult to codify and make explicit. The competence perspective suggested by the project management group (attempt to establish a surface structure) was obviously at odds with some wider understanding (deep structures) and the task of establishing a common appreciation or shared discourse/surface structure of competence in the organization turned out to be more complex than anticipated. These contrasting perspectives on competence resonate well with the underlying rationales of job-based and skill-based approaches to competence management.

5.2 Job-Based and Skilled-Based Approaches to Competence Management

Drawing on an underlying conception or deep structure of work as individuals matched to job roles (Fombrum et al. 1984; Ghorpade and Atchinson 1980), most existing approaches to competence management build on job descriptions (Lawler and Ledford 1992). Viewing jobs as relatively fixed positions requiring specific skills, the job-based approach is focused on filling jobs with individuals with the requisite skills to perform them. In this discourse, competence is a set of properties required for a specific job (Spencer and Spencer 1993). The focus is not skill per se, but rather what type of skills are needed to perform a certain work task (McClelland, 1973). Based on formalized explicit descriptions covering work tasks and required competence, employees' competencies are made visible and measurable. This perspective (underlying way of thinking or deeper structure) of competence management is associated with conventional managerial ideas of organization and working such as command and control, division of thinking and doing, hierarchical structures, machine bureaucracies, planning, and rational analysis.

Contrary to the job-based approach to competence management, the skilled-based approach is concentrated on the individual (Lawler and Ledford 1992). This approach/discourse argues that competence management activities need to be aligned with the development of individuals so that the organization ends up with the right competence profile for each individual employee. Organizations should develop person descriptions that describe what competencies an individual has to develop to be effective in a specific work area. In order to support individual employees' ambitions to develop new competencies and take on new responsibilities, there is no permanent assignment of work activities. Instead, self-managing work teams are assigned responsibility for the performance of a particular work process (Lawler and Ledford 1992). This way of thinking about competence management can be described as an interest- and development-based view, where highly skilled and motivated independent workers take control of their own competence development, forming communities of interests, actively developing their own knowledge and skills, and directing the development of their careers. In this context, the deeper or underlying structure of management is about coordination and facilitation, decentralization of power, and democracy.

5.3 Project 1

In an attempt to take a firmer grip on its competence management, Volvo IT initiated a pilot installation of Tieto Persona/Human Resource (TP/HR) in late 1999. Designed to support mapping, categorization, and visualization of an organization's competencies, TP/HR was based upon a preestablished competence classification where competencies were defined as functional skills (practical work tasks) and technical skills (methods or techniques used to perform the tasks). At the outset of the project, managers of the pilot group clearly articulated that TP/HR should be adapted and configured to handle both qualitative and quantitative aspects of competence: skills, knowledge, experiences, relationships, values, and motivation (an apparently agreed surface structure was thus developed). This could be described as an inclusive understanding of competence, reflecting both job-based and skill-based approaches. Although relationships, values, and motivation obviously are difficult to codify and make explicit, these qualitative aspects of competence were considered obvious parameters to include in the system. As the pilot project advanced this changed, however, and it was decided that these parameters were not to be covered by TP/HR's formalized competence classification (meaning that this apparently common surface structure was in trouble). Instead, the employees themselves were supposed to handle information about relationships, values, and motivation. In line with the rationale of the job-based approach to competence management, the pilot group thus concentrated exclusively on skills, knowledge, and experiences. Despite agreed communicative actions, people actually acted differently.

Volvo IT's idea was to use the representations of competencies as provided by TP/HR to match tasks with qualified people or to get an expert's view of a particular problem. It was assumed that the required competence resides somewhere in the organization and the TP/HR system's role was to support the identification of that particular competence in a rationalistic and effective way. This logic builds on the assumption that tasks are recurrent and competencies are largely stable over time and therefore reusable. The TP/HR system was primarily a management tool, including features for measuring the status of employees' competencies and gap analyses (here material structure reflected a deeper structure related to management command and control thinking). It was assumed that employees would regularly feed the system with competence information, although they did not get much in return. On the basis of managerial structure, it was assumed (wrongly) that employees do what they are instructed. This producer/consumer dilemma undermined employees' motivation to use the TP/HR system.

Since relationships, values, and motivation were not included in the TP/HR's formalized competence classification, some employees used the free text area function to indicate personal experiences, competencies outside one's direct line of work, hobbies, or remarks about future plans (i.e., contrasting the designers' intentions as inscribed into TP/HR, users used their interpretive flexibility to use the system to reflect a different structure of competence). Although this area was indexed and thus searchable, the information could not be aggregated and it was not in any way related to the formal competence structure of the system. While pilot group managers initially communicated the need for a competence management system conveying a multifaceted appreciation of competence, the result of the TP/HR project was in fact a system substantiating and reinforcing traditional job-based principles of competence management.

The experience from the TP/HR project offered an opportunity for our research team to introduce and evaluate a technology, which, by being based on interest-driven actions instead of formalized representations, contrasted the basic tenet of TP/HR. More specifically, the first project inspired us to develop a competence management system (material structure) that inscribes a different surface structure (and maybe unconsciously the deeper individual choice structure) and investigate how such a system would influence wider organizational structures and practices.

5.4 Project 2

Building on the experience from the TP/HR project, our research team initiated a second project on the design and use of competence management systems. In line with the rationale of action research, the intention of this project was partly to make the organizational members aware of and appreciative of a broader understanding of competence (i.e., change the organizational surface structure in respect to competence), including the skill-based approach, partly to inform the design of competence management systems capable of embracing this new conception (i.e., inscribe structure into a computer system).

Designed and implemented as a recommender system (RS), the Volvo Information Portal (VIP) prototype was intended to provide the employees with targeted and relevant intranet information. In addition to the standard RS function, we added a *find competence* feature. This feature enabled the VIP users to enter a natural-language text describing a specific interest, e.g., database administration on an Oracle system. VIP would then list all users with matching agents, i.e., all users who had agents actively searching for information related to the specified interest. Obviously, the VIP prototype did not locate people with formalized competence but people with an interest in the subject area. To label this feature find competence was a deliberate provocation intended to cause the organizational members to reflect upon the relation between interest and competence. Reflecting a far more skill-based approach to the understanding of competence, VIP was designed to reinforce that understanding among the people who used it.

VIP was released on Volvo's intranet in January 2000 and 50 or so individuals were notified of its existence. No formal training was offered but an introduction e-mail was submitted to all interested parties and the prototype had built-in help files. Although the prototype was not explicitly announced or promoted to the larger audience, it was generally available to all Volvo employees. When we conducted our research, during the period April to June 2000, approximately 20 users had active agents. On the basis of how organizational members used this prototype to find information in which they were interested, our research team was able to inquire into how personal interest, embodied in information seeking activities, could be a means for identifying competence.

On the basis of our involvement in Volvo IT's competence management activities and experiences from TP/HR and VIP, we could derive three qualitatively different perceptions of competence: (1) competence as formalized description, (2) interest as competence, and (3) interest beyond competence. Although all three perceptions could be found in dialogues with Volvo IT employees, the first category clearly represents the dominating unreflected perspective, or deep structure. The introduction of the VIP system, however, made people question their assumptions and brought about a debate. To our surprise, we found many users ready to testify to the importance of an interest- and action-driven (skill-based) competence view. A tangible outcome of our research is that Volvo IT has applied some of the results produced. Currently, Volvo IT is conducting a project aimed at improving the organization's competence management worldwide. Based on lessons learned from TP/HR and the VIP system, Volvo IT has decided that personal interest profiles should be included in the organization's competence descriptions. As a result of the research, there has been an organization-wide change, even among managers, in the surface structure related to competence. What was before passive and un-reflected (deep structure) is in the present situation more surface.

6 CONCLUSION

In this experiment in integrative theory building, we studied three related understandings of structure, as expressed in the IS literature. All of the perspectives are heavily dependent on theories from other disciplines. There is much overlap in the understanding of structure between the adaptations of linguistic and social theories. In both cases, structure is viewed as important bedrock and context in the development and use of computer systems. However, the linguists start with structure as the regularities and rule sets of language and its use (which is then expanded to encompass human behavior), whereas the social theorist start with the study of social practice (which cannot be understood without including the study of meaning expressed in language). The comparison with the infrastructure debate, however, exposes the main point of contention, which is, not surprisingly, the computer systems themselves. Should these be considered structural or not? If so, how can the processes by which they become structural and their later influence be explained?

We chose to follow the convention of thinking of a computer system as a material artifact (though there is much that is not particularly material about software) and explain the relation of structure and actions in terms of deep structure, surface structure, and material structure. In doing this, we adopt Chomsky's terminology, as Wand and Weber did, but without adopting his specific theoretical stance. We simply observe that a great deal of the effort IS researchers expend in adopting and adapting European theories of social behavior and meaning is devoted to trying to understand patterns of meaning and behavior in IS development and use that go beyond functional work-based actions of users and simple denotational uses of language. Both surface and deep structures can be seen as influential in the design and use of competencies management systems at Volvo IT. Underlying structural perceptions of competence (as job- or skill-based), in relation to other structural understandings of how a company should be managed relate to the choice and development of computer systems intended to assist with those activities. The computer systems themselves used in practice serve to help reproduce or to change the underlying structures. Understanding how the deep and surface structures of social practice and discourse are consciously and unconsciously embedded in material computer systems, by the actions of designers, and subsequently interpreted by other stakeholders and reincorporated in the production and reproduction of linguistic and social structures continues to be one of the most challenging areas of inquiry in the discipline.

We developed, in this paper, an example of integrative theory building. This form of theory building is contrasted with the more usual adaptive theory building, which borrows a theory from another discipline and adapts it to IS phenomena. Integrative theory building sets out to synthesize existing IS theories by examining their similarities and resolving their differences. Both forms of theory building can be combined with the collection and analysis of empirical data, but are not primarily reliant on it. Integrative theory building can be seen to have some advantages in comparison to the other forms of theory building. In relation to building theory based on empirics, it draws strength from serious consideration of the historical traditions of thought represented in theory. In relation to adaptive theory building, it is well-focused on IS concerns, because the theories integrated have already been adapted to the IS domain. It can also be seen as a cumulative form of IS theory making, in that it builds on earlier contributions (the American positivist tradition also seeks to be cumulative in adding to scientific knowledge). A potential disadvantage is that it is never likely to include the latest developments in related disciplines, which may be highly relevant. However, in relation to the development of the field, it has the advantage of trying to assimilate the understandings of previous researchers instead of minimizing them in the search for something more relevant or trendy—the next killer theory.

Nevertheless, there seem to be some natural limits to what can be achieved with integrative theory building. Theories to be integrated must be chosen with care for their relative similarities. The occasional positivist theory in this study jarred rather with the generally socially constructionist tone. It seems to be necessary to include some background to individual contributions to relate them both to their background discipline and to the wider use of that discipline within IS, but that makes the theory hard to understand and to present coherently—book form is possibly more appropriate than article form. It is further rather doubtful whether the resulting theory can be accepted by other researchers wedded to their background disciplines—integrative theory is always bound to be inconsistent or trivial in relation to any specific background discipline that it tries to incorporate. Despite these doubts, we suggest that integrative theory development (which bridges disciplines and builds on existing insights to define what is central to IS) is a viable way to build truly exportable IS theory, although it is likely to take some time and hard work.

REFERENCES

AB Volvo. "På rätt spår" [On the Right Track], Corporate Report, 1987.

Akrich, M. "The De-Scription of Technical Objects," in J. Law (Ed.), *Shaping Technology/ Building Society: Studies in Sociotechnical Change*, Cambridge MA: The MIT Press: 1994, pp. 205-224.

Andersen, P. B. "What Semiotics Can and Cannot Do for HCI," *Knowledge-Based Systems* (14:8), 2001, pp. 419-424.

Andersen, P. B., and Mathiassen, L. "Semiotics in Information Systems Engineering," *Semiotica* (142:1/4), 2002, pp. 381-395.

Baskerville, R. L., and Myers, M. D. "Information Systems as a Reference Discipline," *MIS Quarterly* (26:1), 2002, pp. 1-14.

Bijker, W. E., Hughes, T. P., and Pinch, T. (Eds.). *The Social Construction of Technological Systems: New Directions in the Sociology and History of Technology*, Cambridge, MA: MIT Press, 1987.

Callon, M. "Techno-Economic Networks and Irreversibility," in J. Law (Ed.), *A Sociology of Monsters: Essays on Power, Technology and Domination*, London: Routledge, 1991, pp. 132-161.

Ciborra, C. (Ed.). *From Control to Drift*, Oxford: Oxford University Press, 2000.

Ciborra, C. U., and Lanzara, G. F. "Formative Contexts and Information Technology: Understanding the Dynamics of Innovation in Organizations," *Accounting, Management and Information Technologies* (4:2), 1994, pp. 68-93.

Dahlbom, B. "Postface: From Infrastructure to Networking," in C. Ciborra (Ed.), *From Control to Drift*, Oxford: Oxford University Press, 2000, pp. 212-226.

Dahlbom, B., and Mathiassen, L. *Computers in Context*, Oxford: Blackwell, 1995.

DeSanctis, G., and Poole, M. S. "Capturing the Complexity in Advanced Technology Use: Adaptive Structuration Theory," *Organization Science* (5:2), 1994, pp. 121-147.

Fombrum, C., Tichy, N. M. et al. *Strategic Human Resource Management*, New York: John Wiley & Sons, 1984.

Galbraith, J. *Designing Complex Organizations*, Reading, MA: Addison-Wesley, 1973.

Ghorpade, J., and Atchinson, T. J. "The Concept of Job Analysis: A Review and Some Suggestions," *Public Personnel Management* (9), 1980, pp. 134-144.

Giddens, A. *Central Problems in Social Theory*, London: Macmillan Press, 1979.

Giddens, A. *The Constitution of Society*, Cambridge, MA: Polity Press, 1984.

Giddens, A., and Turner, J. H. *Social Theory Today*, Cambridge, MA: Polity Press, 1987.

Grint, K., and Woolgar, S. *The Machine at Work*, Cambridge, MA: Polity Press, 1997,

Hanseth, O. "Actor-Network Theory and Information Infrastructures," in C. Ciborra (Ed.), *From Control to Drift*, Oxford: Oxford University Press, 2000.

Hanseth, O., and Braa, K. "Who's in Control: Designers, Managers or Technology," in C. Ciborra (Ed.), *From Control to Drift*, Oxford: Oxford University Press, 2000.

Hanseth, O., and Monteiro, E. "Inscribing Behavior in Information Infrastructure Standards," *Accounting, Management & Information Technology* (7:4), 1997, pp. 183-211.

Hill, S. *The Tragedy of Technology*, London: Pluto Press, 1988.

Holmström, J., and Stalder, F. "Drifting Technologies and Multi-Purpose Networks: The Case of the Swedish Cashcard," *Information and Organization* (11), 2001, pp. 187-206.

Klein, H., and Myers, M. "A Set of Principles for Conducting and Evaluating Interpretive Field Studies in Information Systems," MIS Quarterly (23:1), 1999, pp. 67-93.

Latour, B. "Technology is Society Made Durable," in J. Law (Ed.), *A Sociology of Monsters: Essays on Power, Technology and Domination*, London: Routledge, 1991, pp. 103-131.

Latour, B. "On Recalling ANT," in J. Law and J. Hassard (Eds.), *Actor Network Theory and After*, Oxford: Blackwell Publishers, 1999, pp. 15-25.

Lawler, E. E., and Ledford, G. "A Skill-Based Approach to Human Resource Management," *European Management Journal* (10:4), 1992, pp. 383-391.

Lawrence, P. R., and Lorsch, J. W. *Organization and Environment: Managing Differentiation and Integration*, Boston: Harvard University Press, 1967.

Leavitt, H. J. "Applied Organizational Change in Industry: Structural, Technological and Humanistic Approaches," in J. March (Ed.), *Handbook of Organizations*, Chicago: Rand McNally, 1965, pp. 1144-70.

Lee, A. "Editor's Comments: *MIS Quarterly's* Editorial Policies and Practices," *MIS Quarterly* (25:1), 2001, pp. iii-vii.

Leifer, R., Lee, S., and Durgee, J. "Deep Structures: Real Information Requirements Determination," *Information & Management* (27:5), November 1994, pp. 1-27.

Liu, K. *Semiotics in Information Systems Engineering,* Cambridge, England: Cambridge University Press, 2000.

Lutje, I. *Software som Tekst*, Aalborg, Denmark: Aalborg Universitetsforlag, 2000.

MacKenzie, D., and Wajcman, J. (Eds.). *The Social Shaping of Technology*, Milton Keynes, England: Open University Press, 1985.

Markus, M. L., and Robey, D. "Information Technology and Organizational Change: Causal Structure in Theory and Research," *Management Science* (24:5), 1988, pp. 583-598.

McClelland, D. C. "Testing for Competence Rather than for 'Intelligence,'" *American Psychologist* (28), 1973, pp. 1-14.

Monteiro, E. "Actor-Network Theory and Information Infrastructure," in C. U. Ciborra (Ed.), *From Control to Drift*, Oxford: Oxford University Press, 2000, pp. 71-86.

Monteiro, E. "Scaling Information Infrastructure: The Case of Next-Generation IP in the Internet," *Information Society* (14:3), 1998, pp. 229-245.

Monteiro, E., and Hanseth, O. "Social Shaping of Information Infrastructure," in W. J. Orlikowski, G. Walsham, M. Jones, and J. I. DeGross (Eds.), *Information Technology and Changes in Organizational Work*, London: Chapman & Hall, 1996.

Nambisan, S. "Information Systems as a Reference Discipline for New Product Development," *MIS Quarterly* (27:1), 2003, pp. 1-18.

Orlikowski, W. J. "The Duality of Technology: Rethinking the Concept of Technology in Organizations," *Organization Science* (3:3), 1992, pp. 398-429.

Orlikowski, J. "Using Technology and Constituting Structures: A Practice Lens for Studying Technology in Organizations," *Organization Science* (11:4), July-August 2000, pp. 404-428.

Orlikowski, W. J., and Baroudi, J. J. "Studying IT in Organizations: Research Approaches and Assumptions," *Information Systems Research* (2:1), 1991, pp. 1-28.

Orlikowski, W. J., and Robey, D. "IT and the Structuring of Organizations," *Information Systems Research* (2:2), 1991, pp. 143-169.

Rose, J., Jones, M., and Truex, D. "The Problem of Agency: How Humans Act, How Machines Act," in *ALOIS Workshop: Action in Language, Organizations and Information Systems*, Linköping University, Linköping, Sweden, 2003, pp. 91-106.

Spencer, L. M., and Spencer, S. M. *Competence at Work: Models for Superior Performance*, Chichester, England: John Wiley & Sons, 1993.

Stamper, R. "Analyzing the Cultural-Impact of a System," *International Journal of Information Management* (8:2), 1988, pp. 107-122.

Stamper, R. *Information in Business and Administrative Systems*, New York: John Wiley and Sons, 1973.

Stamper, R. "Signs, Information, Norms and Systems," in B. Holmqvist, P. B. Andersen, H. Klein, and R. Posner (Eds.), *Signs of Work*, Berlin: Gruyte. 1996.

Stamper, R., Liu, K. C., Hafkamp, M., and Ades, Y. "Understanding the Roles of Signs and Norms in Organizations: A Semiotic Approach to Information Systems Design," *Behavior & Information Technology* (19:1), 2000, pp. 15-27.

Star, S. L., and Ruhleder, K. "Steps Toward an Ecology of Infrastructure: Design and Access for Large Information Spaces," *Information Systems Research* (7:1), 1996, pp. 111-34.

Strauss, A., and Corbin, J. *Grounded Theory in Practice*, London: Sage Publicatoins, 1997.

Truex, D., and Baskerville, R. "Deep Structure or Emergence Theory: Contrasting Theoretical Foundations for Information Systems Development," *Information Systems Journal* (8:2), 1998, pp. 99-118.

Walsham, G. "The Emergence of Interpretivism in IS Research," *Information Systems Research* (6:4), 1995a, pp. 376-394.

Walsham, G. "Interpretive Case Studies in IS Research: Nature and Method," *European Journal of Information Systems* (4), 1995b, pp. 74-81.

Wand, Y., and Weber, R. "On the Deep Structure of Information Systems," *Information Systems Journal* (5:3), July 1995, pp. 203-223.

ABOUT THE AUTHORS

Ola Henfridsson is the manager of the Telematics Group at the Viktoria Institute, Göteborg, Sweden. Dr. Henfridsson is also an assistant professor in Informatics at the School of Information Science, Computer and Electrical Engineering, Halmstad University. He holds a Ph.D. degree in Informatics from Umeå University, Sweden, and is a member of the editorial board of the *Scandinavian Journal of Information Systems*. Dr. Henfridsson has published his research in journals such as *Accounting, Management & Information Technologies*, *Database*, *The Information Systems Journal*, *The Journal of Information & Knowledge Management*, and *The Scandinavian Journal of Information Systems*. Dr Henfridsson can be reached at ola.henfridsson@viktoria.se.

Rikard Lindgren is a member of the Telematics Group, Viktoria Institute, Göteborg, Sweden. Dr. Lindgren is also an assistant professor in Informatics at School of Economics and Commercial Law, Göteborg University, and at the School of Information Science, Computer and Electrical Engineering, Halmstad University. He holds a Ph.D. degree in Informatics from Göteborg University, Sweden. Dr. Lindgren has published his research in publication outlets such as *The European Journal of Information Systems*, *The Journal of Information & Knowledge Management*, *The Journal of Knowledge Management Practice*, and *The Scandinavian Journal of Information Systems*. Dr. Lindgren can be reached at rikard.lindgren@viktoria.se.

Jeremy Rose is an associate professor at the Department of Computing Science, Aalborg University, Denmark. He is currently a member of the PITNIT and SPV research projects in Denmark, and active as a member of the IFIP WG8.2 community. His research interests are principally concerned with IT and organizational change, the management of IT, and IS development. The research approach combines empirical insights with theoretical perspectives to produce practically useful knowledge around various contemporary technologies in different types of organizational settings. He has published in management, systems and IS forums. Further details and some publications are available at http://www.cs.auc.dk/~jeremy/. He can be reached at jeremy@cs.auc.dk.

24 EXPOSING BEST PRACTICES THROUGH NARRATIVE: The ERP Example

Erica L. Wagner
Cornell University

Robert D. Galliers
Bentley College

Susan V. Scott
London School of Economics

Abstract The phrase *best practice* has entered into common parlance in contemporary business discourse, yet recent research has shown that the construction of industry standards and their inscription into software packages is not straightforward. Organizations increasingly find they are bound to accept project outcomes that have emerged as a consequence of negotiations between an installed base of consultancy or software vendor solutions and local context. We adopt a narrative approach to analyze the negotiation of a best practice design during the implementation of an ERP system. Having adopted the position that the IT artifact is part of an ensemble of networked agencies that shift over time, we then use an actor-network perspective to trace the different sources, agencies, and affects of inscription during the ERP project. Doing so highlights the politics involved in localizing an IT artifact and the issues raised when software vendors and sector specific partners collaborate with the intention of manufacturing a commercially viable ERP package intended to represent the embodiment of best practice. The paper contributes to IS research discourse by demonstrating the application of narrative analysis in longitudinal interpretive field studies.

Keywords: Best practice, ERP systems, appropriation, implementation, actor network theory, narrative analysis, interpretive research, longitudinal research.

1 INTRODUCTION

Enterprise resource planning (ERP) systems aim to standardize business processes and to integrate functions, data, and organizational structures through software reference models (Kumar and van Hillegersberg 2000). Vendors frequently claim that ERP systems are inscribed with state-of-the-art best practices for a particular industry, saving organizations from reinventing the wheel and perpetuating local business practice anomalies (Cortada 1998; Davenport 2000). Recent research has critically evaluated the notion of so-called best practices, raising awareness that these benchmarks are driven by consultants and software vendors seeking economies of scale for their products and services (Newell et al. 2000; Sawyer 2001; Walsham 2001). Swan et al. (1999, p. 284) argue that IS best practices are "illusory and potentially disruptive because of the different interests between technology suppliers and IS adopters."

In this paper we build upon this perspective using a distinctive approach that combines actor-network theory (Callon 1986; Latour 1987) with narrative research methods (Wagner 2003) in order to analyze the inscription of priorities during the configuration of best practice software packages as part of a large-scale IS implementation. Using empirical material gathered during a longitudinal study we illustrate the tensions that are generated between local working patterns and the technological constraints of the system (Hanseth and Braa 1998; Walsham and Sahay 1999). We develop this further by examining the way in which the ERP system is made to work despite significant mismatches between local context and best practice.

This research extends and develops existing critiques that have focused upon the way in which best practice relates to business process reengineering (Newell et al. 2000), interfirm knowledge transfer (Timbrell et al. 2001), benchmarking IT practices (Cragg 2002), and evidence-based information systems (Atkins and Louw 2000). Our particular contribution centers upon the use of ANT and narrative approaches to trace the ebb and flow of different sources, agencies, and affects of inscription during a multiyear ERP project. In other words, during the negotiations that surround the efforts to construct an industry gold standard, whose interests are left behind and whose go forward? What can we learn about the status, nature, and plausibility of best practice software designs by studying a rich ERP case study?

The case study provides particular interest as it documents the attempt by a leading United States university to design, in collaboration with an international software vendor, an ERP product intended to represent the embodiment of best practice in academic administration. The empirical material presented provides a window into the local processes involved in constructing products that are subsequently sold to an international market as a global standard. Told over time as the phases of change unfolded, these narratives give insights into why organizations buy into the best practice mentality, how different coalitions form to manage the implications of modernization, and the effort required to make best practice ERP software packages work.

2 METHODOLOGY AND CONCEPTUAL SCHEME

We use a broadly interpretive research approach (Walsham 1993) in which multiple perspectives and articulations of interests are sought out as the basis of analysis. A longitudinal case study was conducted from June 1999 to August 2000 with an Ivy League university (Ivy) and their ERP development partner, a multinational software vendor (Vision). Five field site visits were made to Ivy (each lasting an average of eight weeks) in order to conduct interviews with project members and Ivy employees. These visits resulted in 137 interviews[1] with 53 project stakeholders (see Table 1).

Preparatory background visits to Ivy during research design inspired the selection of a narrative approach. The lead researcher was struck by the insistence of those she met that the ERP project had been "talked up" so much it would not be allowed to fail. This intriguing insight led her to craft a methodology that captured events and opinions as they unfolded over time. It was imperative to document these in a timely fashion, charting their progress and analyzing the process by which they are effectively buried.

In narrative interviews, participants are asked to describe current experiences, with the researcher listening for convincing explanations of "why things are the way they are" (Bruner 1990). The researcher is not expecting to access the truth; instead, the narrative is treated as a performance in which events are retold in a particular order and for particular effect. The narratives gathered are interpreted as "artifactual representations of the intentions, actions or goals of actors situated in time" (Czarniawska 1998). The task is to interpret these artifacts in order to come to a finer reading of the situation and provide plausible theoretical explanations (Bruner 1990).

When handling narrative data it is important to recognize that narratives move on and interviewees reframe stories in subsequent accounts, black- boxing issues that used to be open controversies, repositioning themselves, redefining priorities. Rather than seeing this as a problem, changing narratives can form a point of analytical leverage. Here, the narrative approach enabled us to reveal insights into processes of negotiation over time.

Table 1. Interviews Conducted at Ivy

Actors Interviewed	Organizational Role	Number of Interviews			
		Summer 99	**Winter 99**	**Spring 00**	**Summer 00**
6	Central administrative leader	4	2	7	3
13	End-user	6	8	12	5
8	Faculty	0	1	7	3
6	Project management	5	5	7	5
20	Project team member	19	11	15	12
	Total interviews by phase	34	27	48	28
53	← Actors	Totals		Interviews →	137

[1]See Wagner (2003) for a detailed discussion of narrative research and methodological approaches.

After the first round of interviews, it became apparent that the issues and actors referred to in the narrative accounts clustered around the negotiations taking place about ERP project priorities. These, in turn, highlighted multiple interpretations of best practice reflecting the embeddedness of actors in context and vested interests. Listening to interview tapes, and combing transcripts for controversial actors as well as organizational allies, helped in following the action. While individual accounts of change do not in themselves create change, woven into these individual narratives are connections and politics that highlight the basis for coordination (Bruner 1990). By tracing the path of these connections, insights emerge as the constitution of issues and processes of negotiation. This approach to research can only ever by partial; however, a fieldwork journal and color coding of charts were used in order to enhance rigor. The process is rarely neat, since participants often interweave their personal narratives with references to collective, institutional, international, or cultural issues.

The approach challenges the convention of selecting a discrete unit of analysis. Traditional sociological analysis has a tendency to separate macro, meso, and micro levels of focus. Instead, a narrative approach allows scope for global, collective, and local negotiations to be considered simultaneously. In order to analyze such data it is necessary to find a conceptual scheme that supports this multilevel approach.

It is for this reason that we maintain narrative and ANT work sympathetically together. Latour (1999) has argued that the foundations of ANT are based upon tracing relationships between actors rather than on their relative size. For example, when studying business practice within a prestigious university, actor narratives should determine the action and transport the research focus. United States accounting trends become actors only when and if interviewees define them as such. Such trends are not interpreted as the macro-level structure that influences local action but rather as a resource employed by delegate actors to translate interests. The criteria for their inclusion became a reference in narrative data concerning their involvement in Ivy's ERP initiative.

Through attention to language, the strength of ties within and between networks becomes apparent. Narrative is particularly helpful for studying the constitution of agency and the production of networks because individual stories of negotiation speak on behalf of a network of interests. An individual's account of change when viewed from an actor-network perspective is interpreted as a delegate or spokesperson for a particular set of interests.

ANT provides yet another point of analytical purchase for narrative data. When IT is involved in a situation to which the narrative refers, it is given status as an actor. A stream of actor-network studies exist that focus on conceptualizing ERP as an organizational actor with interests that actively influence project negotiations (cf. Hanseth and Braa 1998, 1999; Scott and Wagner 2003). ANT does not define agency as a solely human characteristic, instead asking that researchers follow the action as it unfolds without specifying in advance who or what participates in negotiations. Although controversial (cf. Collins and Yearley 1992), ANT's broad constitution of agency can provide an opportunity to consider on whose behalf delegate actors work and what interests they inscribe (cf. Law 1999).

Since a major aim of IS research is to theorize the agency and role of IT within contemporary society (Orlikowski and Iacono 2001), our goal was to "give voice" to nonhuman actors. These narratives reflect particular values and politics and represent an epistemological ordering (Scott 2000) that can influence the future of work practices. In this study, when a reference was made to a group, or a cause or action was attributed to nonhumans such as the ERP system itself, a delegate was interviewed and technical documentation reviewed.

The combination of ANT and narrative approaches creates a theoretical context in which to question how some stories become more accepted than others. Why do particular actors, issues, or events fade away, no longer making significant appearances in the narrative data gathered, while others continue to dominate? Latour (1999, p. 307) calls this latter state of achievement a "matter of fact—something so ingratiated within a community...that its presence is indisputable and obvious." By analyzing narrative data over time using ANT, we come to appreciate the effort that this achievement takes in a community. ANT is a theoretical apparatus for understanding change and order, highlighting the processes of negotiation through which interests become inscribed and translated. For a particular storyline to rise to prominence, rival sets of interests have to be dismantled.

Inscription of interests implies a prior process of negotiation about "what will be carried forward...and what will be left behind to make room for the new" (Latour 1999, p. 71). This provides insights into what it means to be connected or disconnected from a socio-technical ensemble. The successive inscription of interests into material form provides us with a partial history of what circulated within an ontological network. However, if we follow the processes of inscription, we can capture the properties that are gained as well as those that are lost in each stage of transformation (ibid). Inscription marks the achievement of order in the translation of interests and heralds the formation of a new network configuration where the present is similar but different from the past. These translations are made more permanent and tangible over time as they are inscribed into a more material form (Akrich and Latour 1992). In the context of an ERP system, this can create an installed base (Hanseth 2000) that may resist change and contribute to the shaping of an organizational future that favors a particular worldview.

3 ANALYZING THE NEGOTIATION OF BEST PRACTICE AT IVY

3.1 Negotiating Buy-In: Forming the Best Practice Coalition

To set the scene for this empirical material we evoke pressing concerns characterizing use of IT in higher education institutions in the 1990s. Y2K inspired the replacement of many legacy systems and higher education management were aware of the need to show compliance with increasingly keen risk management regulations. These conditions created a market for ERP and a race among IT vendors for association with a brand claim to best practice. Vision was interested in developing a strategic partnership with Ivy to create a standard ERP package that would be sold internationally. To achieve this Vision acquired contextual expertise that would help define the patterns of use associated with university best practices. These patterns would be designed into ERP modules and then connected to Vision's preexisting government package to create another industry-specific product.

But, Ivy end-users expressed resistance to the system. Their stories reveal both the negotiations that preceded the design of best practices and how such decisions excluded alternative practices.

> When this project started they told us that it was going to make our lives easier. At least we expected [it] to have the grant accounting system that has worked for many years...why break something that [isn't] broken?....[The vice president] let the bean counters run wild so the next thing you know they've designed grant accounting based on some textbook budgeting trend. They take nearly 30 percent of my grant money in indirect costs so they can administer my money and then they have the balls to create a system that doesn't give me any of the answers I need to manage my projects. (Faculty Member)

Faculty felt the project team were purposefully designing their preferences *out of* the ERP system in order to assert central control over faculty research accounting. The negotiations that preceded this can be illuminated through the power brokering of Ivy's budget director who had the ear of the vice president for finance—a respected leader who championed the Vision partnership. The director used this relationship to legitimize his corporate accounting perspectives. He argued that faculty should be required to manage their research projects as professional businesses.

> Our value here is the creation and transmission of knowledge. When I ran my own business, my value was *making the most money*...I don't find the tools to get to those values very different...I would say that the mentality that we've had for managing is primitive to say the least....We want to move people toward a [time-phased] management model.

The notion that grant accounting practices were antiquated successfully enrolled project team members who agreed to design the ERP system around perceived corporate best practices. The budget director's narrative is interpreted as a delegate or spokesperson for corporate accounting interests. His vision was repeated almost verbatim within the narratives of project team actors. In this way the team are expressing their enrolment within the director's network, connecting themselves to a powerful group of corporate accounting interests. Further, the director's network gained momentum through the inscription of interests into material objects such as time-phased budgeting and the financial management module. These nonhuman actors are delegates that work on the director's behalf in order to translate conflicting network interests, thereby perpetuating the stability of a corporate accounting vision.

When the ERP system went live, the Ivy community was presented with a concept that changed previous methods, processes, and motivations for grant accounting in ways that Ivy staff found difficult to manage. Despite these tensions, this product design became represented by Vision as the embodiment of university best practices. Yet as a faculty advocate said, "no university...we know of is using the [time-phased] method as implemented in the [Ivy-Vision] ERP."

Faculty interpreted the Ivy-Vision design as representing more than an administrative change.

> By making a decision to go with [Vision]...senior management...was making it impossible for [Ivy] to continue doing business in *fragmented silos*. Like it or not, you've got to work with a new way of accounting. *It's integrated—it's slower, it's a pain in the ass*. And the faculty who used to do it the old way for years decide *it's absolutely terrible*—they don't want to do it *[because] it's not [Ivy's] way*....You don't like it? You're out of the...picture. If you're more inclined to accept the changes...you're in the narrow universe of people we'll work to have consensus with.

This quote implies that the practices inscribed into the ERP system by a small group of powerful actors surfaced as the best way of working—an indisputable organizational reality. Potential Vision clients see the Ivy-Vision system as the gold standard for university administration while Ivy is still radically customizing its business processes, organizational structure, and ERP technology because of the recalcitrance of end-users to accept Vision's best practice design. We now turn our focus to these challenges.

3.2 Translating the Best Practice ERP System

Faculty members and their staff were frustrated that a business-oriented rationale had been prescribed by the project team without consultation. They felt insulted that their grant accounting functionality was not prioritized in the best practice design.

The best practice product aimed to be an achievement of order. However, the emergence of Excel-based shadow systems illustrated that creating a standard ERP package is not the same as achieving user acceptance of that standard. Supporting the appropriation of an ERP system for use by Ivy actors required more and more resources. The project team was forced to reintroduce aspects of the legacy systems. The post-installation customization efforts helped to temper the rigid best practice design.

In response to faculty pressure, the project team agreed on three courses of action. First, they would leave the mainframe legacy accounting system running until ERP functionality was created. Second, they would meet the faculty's requirements by designing legacy "financial commitments" into the system. Third, two interim transaction support centers (TSC) would be staffed with clerical workers acting as boundary objects (Star and Griesemer 1989) between the ERP system and Distributed managers.

> A customized legacy-ERP application was "bolted" onto the ERP accounting module. The project team presented this as a user-friendly solution with a Web-based interface and promptly turned off the mainframe legacy system. The bolt-on was meant to further enroll recalcitrant faculty by keeping administrative work activities within the boundaries of the system. Instead, business managers used the bolt-on to shift boundaries away from the project team toward their own needs. The bolt-on converted raw data into familiar financial commitments through ERP-based software code thus freeing users

> from the need to become familiar with the ERP. Staff would then import these data into Excel to create their "old-world" faculty reports on commitments.

This translation process was more convoluted and time-consuming than the legacy environment, but was preferred by administrators who needed control of financial commitments. It required them to flip back and forth between different definitions of best practice in the management of faculty grants. The bolt-on fused two design rationales (best practice and legacy), by acting as a point of translation for the administrators. Notwithstanding, the administrators themselves were required to act as "marginal people" (Bowker and Star 1999) who sat at the crossroads between faculty expectations and managerial imperatives.

As the administrators began to develop trust in the bolt-on and the support centers, they began asking TSC staff to do their reconciliations. This was not well received by the project team for three reasons. First, central managers wanted to discourage detailed monthly reconciliations by clerical staff in favor of a conceptual shift toward time-phased budgeting. Second, they wanted the centers and the bolt-on to be viewed as temporary, capable of being disbanded in favor of full migration to the ERP environment. However, faculty and administrators commended the transaction support centers, and their responsibilities are increasing. As one TSC manager explains their role,

> we're supposed to be gone soon but we've been able to do a lot of trouble shooting...we get a lot of phone calls. We help people who're still using... Excel...and the bolt-on but also have to understand the [Vision] system. In the beginning it was extremely stressful [because] all these administrators worked very hard and they [weren't] getting anywhere...we bailed them out...people were...stressed to the point of tears....So we let them keep their models and we say don't worry, give it to us, we'll get it into [the system].

Administrators were becoming further entrenched in their legacy accounting practices because they were able to maintain their preferred work practices when communicating with the TSC. Center staff would then be responsible for translating this language into the ERP system. This was troublesome for the project team because departmental users were expressing a backlash against their attempt to shift the translation effort. Third, the drive to rationalize administrative paperwork and accompanying work processes was diminished because each administrator could interact in their own way with center staff, based on the personal relationships that were developing. Together, these three concerns indicate the power of the supposedly silenced to challenge the achievement of order within Ivy. The centers still exist at the time of writing, over three years after the installation of the best practice ERP design. By insisting on and negotiating flexibility in the creation of these translation points, the faculty were able to gain concessions by shaping the process through which they would interact with the best practice system. The following, from a project leader, recounts a story signifying the complexity of enrolling faculty into the time-phased approach.

> We're trying to make things as simple and straightforward as possible and we've failed miserably so far.....The grant reports are the best example of the

> difficulties we have....The Economics professor that used to be Provost...called the [current] Provost really angry because he couldn't read his grant report. The [financial controller] sat down with him and every concept he was asking for was on that report. But he couldn't see it and his business manager couldn't explain it so she's been making him Excel reports....This guy's smart! He knows what he's doing and he can't even read the report, and I thought that was pretty telling. So now faculty aren't using the ERP and what we have...is a very expensive data repository, and still a lot of silos of micro-computing.

By obscuring familiar, valued legacy concepts, the design failed to translate faculty interests. The system that was being naturalized within Ivy was a hybrid whose design fused together best practices with the values and politics of multiple networks and blurred the boundaries of administrative practice.

Ivy's locally accepted interpretation of Visions' best practice standard will almost certainly include a potentially incompatible set of constituent elements including Excel software, the bolt-on, interim support centers, and accompanying business processes. We consider the implications of this hybrid design next.

3.3 Redefining Best Practice and Achieving Order

So, Ivy worked to achieve order through coalition building and translating between espoused best practice and the working needs of those hosting it. For example, the multiple translations that occurred during the grant accounting controversy illustrate how compromise tends to provide diverse groups enough resources to make the system work by repairing conflict and building a good enough solution to translate the interests of previously disparate networks. The moments of enrolment and translation emphasize the inclusion of heterogeneous concepts of work within artifacts expected to be used in daily administrative routines (Scott and Wagner 2003).

The ERP system's reliance on the bolt-on reveals that multiple, often conflicting perspectives can come to reside within a single technological artifact, thereby integrating diverse organizational groups. The system that becomes appropriated by Ivy will result from negotiations between different groups. Similarly, it may be surmised that this system will continue to evolve over time as interests change and organizational goals shift.

Ivy actors updated the University's grand narrative to incorporate the uncertainty of contemporary administrative practice. At the same time this contemporary narrative brings forward time-honored traditions for which actors negotiated a place within Ivy's future operating environment. In so doing, far more actors must be enrolled into the story because the narrative is sympathetic to multiple perspectives.

> Starting over was not an option apparently so we still don't have either system fully in place....[Administrators] clearly understand that the faculty are their customers, and try to provide...services they most need and value. Why has Central management become invested in changing...from a tried-and-true system...to one that hasn't been tested...and is less efficient and useful? My guess is that there's too much water under the bridge to go back—which is why

> we're building our own. This is the most fascinating part of the whole story! I wish I knew how it turns out, or even how to turn it around! (Medical School Manager)

As Latour (1999) notes, being able to sustain order over time requires a great deal of resources. The more diffuse the network ensemble (from project team to University community), the more difficult it is to retain order because heterogeneity reigns (faculty interests will differ from those administrative leaders) and there will always be "a bombardment of offers" where alternative ontological networks are vying for dominance.

Although an orthodox ERP system might never be the best practice at Ivy, the evidence suggests that the design of multiple perspectives into systems is a resilient and ingenious act of will. As expressed by the newly appointed director of Integrated Administrative Systems.

> So what's the good stuff?...One is that I find that the people who were... ***actually involved*** in project implementation—in spite of all the arguments...are a pretty cohesive and dedicated group. They understand how things work and they try very hard...everyone wants to do their best...but what we think is best differs. We'll make it work but... there's a tendency to...fall back to the old way of thinking...So it's...a balancing act... you can't keep...all these independently minded ducks in a row, especially when those ducks are Nobel Prize winning faculty.

Above all, Ivy's project narrative has been one of making it work: prioritizing workable solutions over ideology and purported best practice. Unlike previous studies that have depicted powerful ERP systems forcing change on a victimized organization, our analysis reveals greater potential symmetry in the negotiation between human and nonhuman actors. Vision's strategy of selling Ivy's practices as a solution for their higher education market assumes a high-degree of homogeneity between research universities and community colleges. This has implications not only for North American institutions whose activities are quite diverse, but also for universities world-wide where the "commercialization of education" (Noble 1998) is a far less familiar concept. Notwithstanding, the Ivy-Vision alliance helped create the vendor's higher education industry solution, which is being marketed internationally as best practice— despite Ivy's adaptation of it.

3.4 Reflections

The findings open a Pandora's box illustrating the politics of creating such standards. First, the notion of inscribed best practices can impact local acceptance and use of standard software products. Second, the term best practice does not imply majority rule. IT-based best practices are constructed by those actors who are able to materialize preferred patterns of use. Within Ivy, a small group of actors controlled this process and drove the design decisions related to what was classified as best. Third, while literature exists on best practices, the process of arriving at them is not considered to any extent. Instead, these actions are black-boxed and assumed. We argue that this is storing up trouble for software companies and the reputation of packaged software (ERP)/ outsourcing strategies. If Ivy was unable to use the best practice solution it helped design, how can other institutions be expected to find success by implementing the higher education solution propagated on the Vision web site?

Fourth, there is a relationship between best practices and the bandwagon effect (Fujimura 1996), where the latter refers to the extent to which one is influenced by what others have done. Perpetuating the ERP bandwagon (Kremers and van Dissel 2000; Kumar and van Hillegersberg 2000) into the 21st century is in part about selling the best practice ideal—something that seems to require the collusion of client organizations, management consultants, and software vendors. For example, Ivy currently finds itself between a rock and a hard place in terms of communicating their ERP experiences. Any discussion related to the recalcitrance of faculty and end-users could influence the sales of the Vision product. Ivy fears that low sales would limit Vision's future investment in the product. Since Ivy is contractually bound to periodic upgrades and needs in-house Vision programming expertise, they have chosen to publicly perpetuate the myth of their ERP system as the higher education gold standard.

> We've a vested interest in getting [Vision] to increase investment in the higher education ERP but we're also interested in an honest dialogue with universities. So how do we encourage other universities to go with [Vision] when the process has been so difficult? We think that by continuing to be development partners...we'll have a lot of influence over how the product does develop in the long run and it will get easier for other universities. Ultimately, the bigger [Vision's] higher education business, the better the applications will end up being...because they will invest more in our [sector]. (Project Team Leader)

We argue that inscription of work practices into ERP technology necessarily involves leaving behind historically valued activities judged by a small group of stakeholders to be less than best for common industry needs. The irony is that Ivy was itself unable to enroll its end-user community into adopting the standard package. Instead, obtaining local acceptance and use of the package has required the project team to revisit the pre-ERP environment to create a more inclusive definition of *local* practices. This does not simply mean that more legacy practices are incorporated into the ERP system, but that in creating a working system, Ivy wove together historically valued practices with the ERP design and created a system that is best for their local context. This leads us to reflect upon the hollowness of best practice claims and to consider what we mean when we talk about efficiency gains in reusing software designs.

4 DISCUSSION

This study has shown that the challenge of software appropriation is to both understand the negotiations involved in making a system work and to enroll enough actors at the right time(s) to create a critical mass of users who accept the system and will then mobilize others to do so. Particular attention has been paid to the translation of different interests into technology-based practices. Through an actor-network perspective we understand the temporary and partial nature of these inscribed standards. Critical reflection of the political agendas that inform and often drive design decisions help produce a historically situated account of how things come to be defined as best. The challenge for IS scholars is to reassess our understanding of when to involve actors in making IT work, and focus research attention on the use of technology in practice.

In this section we begin by discussing the notion of involvement during large software projects. We argue that making a system work requires *appropriation* which can only result from knowing who and what to involve in the initiative, and also when to leverage that knowledge.

4.1 Creating "Good Enough" Practice

In order to move the stalled project forward, the project team had to make a choice between creating a system that would work for Ivy, and sticking to their so-called gold standard. The team chose the former and began to appropriate a local solution from best practice rhetoric. The notion of IS appropriation, previously considered as the relatively static final phase of development (Collins 1993), is seen as being more dynamic in nature (Pan et al. 2003; Robertson et al. 1996). This study supports the latter interpretation by illustrating appropriation of a system, not as a one-time event of achieving order, but rather as an iterative process where order is achieved, controversy accommodated, and relative stability regained over time.

The value of extending the analysis beyond the installation of the best practice system is that we focus on *achievement* itself as a unit of analysis. Our study of successive inscriptions of interests into the Ivy-Vision ERP development provided a partial history of what circulated within the ontological network of project members. This provided insights into what it meant to be connected or disconnected from Ivy's socio-technical ensemble. However, by following the processes of achieving such inscriptions, we are able to report on the properties that were gained as well as those that were lost in each phase of transformation. We saw that the creation of an installed base contributed to shaping Ivy's organizational future; the ERP environment is hard for users to reject completely, but the achievement of order that has emerged from the ERP project is temporary and partial. The functionality that had been lost in the design of the installed base was later found to be a key actor in maintaining the enrolment of Ivy actors in the ERP-enabled environment. The longitudinal nature of this study enabled us to analyze how Ivy managed to move forward in spite of the complications that accompanied the project team's appropriation of best practice.

Ivy's appropriation of the ERP system was achieved through compromise. The investment made it difficult to reject the software outright. However, when the project team attempted to align the wider Ivy community with the system, they aggravated many socio-technical arrangements that had constituted Ivy's very existence. In seeking to align faculty with a business storyline, the team were faced with a further round of negotiations, enrolments, and translations of interest, which ironically produced, not alignment, but another reconfiguration of best practice.

Debates surrounding legacy accounting versus corporate best practice dominated shaping Ivy's future. The question to be answered was not just about whose stories would be recognized but also whose perspectives would inform and guide the financial management of faculty dollars. By following the *processes of inscription*, we learn about the system properties that are gained as well as those that are lost in each phase of transformation. This provides insights into what it means to be connected or disconnected from a socio-technical ensemble. An actor-network analysis provided insight into Ivy's struggle to answer these questions. Our findings not only illuminate the process by which the relative value of potential best practices are debated, but also the way specific actor-networks can shape negotiation. We recall the quotation by a technical team leader who proposed that Ivy actors without plausible and forward-focused stories were excluded from the "narrow universe of people included in the consensus picture." However, we saw actors being included in negotiations without wholly subscribing to the dominant vision.

The creation of design modifications and TSC illustrate how appropriation can be achieved despite consensus. Actors can coordinate without always having to share the same epistemological position. However, actors whose narratives speak only of the past are trying to combat the irreversible arrow of time (Adam 1995) that pushes us forward. As such, these actors do not provide a plausible narrative that can be translated into the future. Their stories cannot be considered equal to others. The effect is that they become silenced because of their irrelevance to the negotiations occurring within the change initiative. We were able to access some of these stories but we found that their interests failed to be designed into IT artifacts because they were unconnected to the network alliances shaping Ivy's future.

4.2 Theorizing the IT Artifact

The choice of a narrative research approach, used to apply ANT, arose from our interest in analyzing the adoption and use of best practices over time. Our findings show that these black-boxed practices get deconstructed through use. They are then reconstructed and take on a hybrid form; the prescribed, generic processes become infused with local value. The study of technology-in-use is a timely topic because it directly contributes to a broad discussion about the lack of research within IS that theorizes the IT artifact (Benbasat and Zmud 2003; Orlikowski and Iocono 2001; Weber 2003). This gap is especially problematic for IFIP WG 8.2 because of participant interest in the "relationships and interactions between information systems, information technology, organizations and society" (Fitzgerald et al. 1985).

Our challenge is to cultivate research that adopts an "ensemble view of technology focused on the interplay between people and technology" (Orlikowski and Iacono 2001, p. 126). The current ensemble view literature is categorized as either concentrating on how technologies come to ***be*** in the historical style of actor-network theoreticians, or on how technologies come to ***be used*** (ibid). Here, we build on these ideas by not separating how technologies come to be or be used. Instead, we interpret the process of technology development and its subsequent appropriation as continuous and situated in time. The narrative application of ANT allows researchers to follow the continuous evolution of technologies as they come to be defined as working IS—accepted and used. This is in keeping with Benbasat and Zmud's (2003) call for research that focuses on the relationships between the IT artifact, its use, and its impact on practices and capabilities.

As more organizations either partner with vendors to modify software in an attempt to create best practice products for their industry sector, or adopt such software based on these claims, it is important for us to analyze, theorize, and translate to practitioners the implications of such initiatives. The socio-technical ensemble view of technology is one that is gaining legitimacy within the business community (Schrage 2003; Schweisberger and Chatterjee 2002). However, the notion is black-boxed itself, providing a convenient way of talking about complex systems. We concentrate here on ways that we can understand and guide negotiations, and the inscription of interests into artifacts. It is not just that new interests are designed into software, rather it is that new and old comingle, that history informs the present and can shape the future. We argue that the combination of narrative and ANT provides distinctive insights into this process, and promising tools for studying the interplay between humans and nonhumans in particular contexts.

4.3 Reflections on Method

We suggest that one way to respond to the need to theorize the IT artifact is to echo Pettigrew's (1985) call for longitudinal research, in this instance focusing on phased field visits in order to yield insights on how technologies are introduced, shaped, and reshaped over time, and with what organizational consequences. This might be accomplished by focusing on the dynamics of IT-enabled change processes, rather than making hard and fast statements based on findings from a particular phase of the project. Robertson et al. (1996) use a model of different IT project phases to understand the influence of suppliers and organizational stakeholders in the diffusion and appropriation of software products sold as best practice solutions. Their analysis emphasizes the influence that these phases had on preceding and subsequent negotiations involved in appropriating software. The field work design of this study was aimed at analyzing the dynamics of IT-enabled change itself rather than a particular outcome.

> narrative...as the central organizing principal of the research, draws on multiple stories clustered around negotiation to help the researcher follow emergent action and study the interconnections between people and technology over time (Wagner 2003).

Such methodological tools are employed to remind us that the primary goal is to understand a

> longitudinal sequence of drama that allows varying readings to be taken of the development of the organizational information system, of the impact of one drama on successive and even consequent dramas, and of the kinds of mechanisms that lead to, accentuate, and regulate the impact of each drama (Pettigrew 1990, p. 275).

In recognition of the multiple influences revealed in this study, its context is organized around the transformation of an Ivy community that is continually being reformed through the stories, negotiations, and actions of its actors. Further, Ivy is made durable through the design and modification of its material objects, software, work procedures, and structures (Latour 1991).

4.4 Limitations and Contributions

We suggest that creating an IS that is accepted as a matter of fact by constituent actors is more likely when its design takes into account valued legacy practices and incorporates the past into best practice designs. We argue that the likelihood of a system being appropriated within an organization increases when there is continuity between the legacy environment and proposed future activities. This is fundamentally at odds with the best practice agenda. Vision's system failed to be accepted as a matter of fact because its accounting module ignored the grand narrative that had informed Ivy practice over many years. Vision's best practice accounting functionality was informed by a corporate business logic that ignored budgeting and grants management activities in universities. Appropriating the accounting functionality was difficult to achieve because administrative tasks involved multiple interest groups whose views differed dramatically. This suggests that integration cannot always be achieved by mandating a standard view to which all actors must subscribe.

The analysis of the post-installation customization of Ivy's best practice system suggests that integration is better understood as a process of selectively negotiating in order to create a good enough solution that meets multiple and interpenetrating needs. The analysis of boundary objects shows the relevance of local practice in IS design. It would appear that a future vision should enroll actors, but so too should negotiators consider what works for the organization, based on past activities and current imperatives.

Process-oriented researchers will know how difficult it is to take background assumptions of ANT and narrative and apply it in the field. We provide four key points that helped us study IT change in an ANT-narrative mode. The first relates to extending the notion of agency to both human and nonhuman actors. This is in order to follow action as it unfolds without specifying in advance who or what participates in negotiations. Next, it is important to recognize that we begin our research of IT-enabled change at a particular moment in the organization's history where the future is being planned for and negotiated. A network is formed that connects the interests of multiple actors. This network is constituted by the relationships between actors and indicates that other groups are being dismantled as a result of this enrolment. Third, we do not set *a priori* boundaries on the scope of the research context. Finally, our notion of IT-enabled change as an achievement of order is fundamental to an ANT-narrative approach (Monteiro 2000). Conceptualizing order as a temporary state emphasizes multiple agencies of change and order, providing a nonlinear account of negotiation.

4.5 Future Research and Unanswered Questions

We argue that over time the best practice trend will be diluted as organizations implement inflexible software-based standards and find them problematic. We question the notion of best practices as synonymous with relinquishment of valued legacy work practices. We consider what stories concerning the ERP system were brought forward from the best practice design only to be later overwritten by customization efforts. One area for future research would be to consider the extent of a potential backlash to the ERP trend as a result of its best business practice design.

The Ivy-Vision ERP package is currently being sold as inscribing best practices for universities globally. As people try to apply these practices to local situations, and as circumstances change, we argue that they will begin to evaluate the extent to which local practices are of greater value than those mandated by best practice. Such a choice was adopted by Ivy but not without a price. It would be interesting to consider the extent to which the best practice concept is storing up trouble for software providers, and to critically evaluate outsourcing strategies and application service provision agreements.

REFERENCES

Adam, B. *Timewatch: The Social Analysis of Time,* Cambridge: Polity Press Ltd, 1995.

Akrich, M., and Latour, B. "A Summary of Convenient Vocabulary for the Semiotics of Human and Nonhuman Assemblies," in W. E. Bijker and J. Law (Eds.), *Shaping Technology/ Building Society*, Cambridge, MA: MIT Press, 1992, pp. 259-264.

Atkins, C., and Louw, G. "Reclaiming Knowledge: A Case for Evidence-Based Information Systems," in H. Hansen, M. Bichler, and H. Mahrer (Eds.), *Proceedings of the 8th European Conference on Information Systems*, Vienna, Austria, 2000, pp. 39-45.

Benbasat, I., and Zmud, R. W. "The Identity Crisis Wwithin the IS Discipline: Defining and Communicating the Discipline's Core Properties," *MIS Quarterly* (27:2), 2003, pp. 183-194.

Bowker, G., and Star, S. L. *Sorting Things Out: Classification and its Consequences*, London: MIT Press, 1999.

Bruner, J. S. *Acts of Meaning*, Cambridge, MA: Harvard University Press, 1990.

Callon, M. "Some Elements of a Sociology of Translation: Domestication of the Scallops and the Fishermen of St. Brieuc Bay," in J. Law (Ed.), *Power, Action and Belief: A New Sociology of Knowledge?*, London: Routledge & Kegan Paul, 1986, pp. 196-231.

Collins, H. M. "The Structure of Knowledge," *Social Research* (60), 1993, pp. 95-116.

Collins, H. M., and Yearley, S. "Journey into Space," in A. Pickering (Ed.), *Science as Practice and Culture*, Chicago: University of Chicago Press, 1992.

Cortada, J. W. *Best Practices in Information Technology : How Corporations Get the Most Value from Exploiting Their Digital Investments*, Upper Saddle River, NJ: Prentice Hall, 1998.

Cragg, P. "Benchmarking Information Technology Practices in Small Firms," *European Journal of Information Systems* (11), 2002, pp. 267-282.

Czarniawska, B. *A Narrative Approach to Organization Studies. Volume 43 Qualitative Research Methods*, London: Sage Publications Inc., 1998.

Davenport, T. H. *Mission Critical: Realizing the Promise of Enterprise Systems*, Boston: Harvard Business School Press, 2000.

Fitzgerald, G.; Hirschheim, R. A.; Mumford, E.; and Wood-Harper, A. T. "Information Systems Research Methodology: An Introduction to the Debate," in E. Mumford, R. Hirschheim, G. Fitzgerald, and A. T. Wood-Harper (Eds.), *Research Methods in Information Systems*, Amsterdam: North-Holland, 1985.

Fujimura, J. H. "Crafting Science: Standardized Packages, Boundary Objects, 'Translation'," in A. Pickering (Ed.), *Science as Practice and Culture*, Chicago: Chicago University Press, 1992, pp. 168-211.

Hanseth, O. "The Economics of Standards," in C. Ciborra and Associates (Eds.), *From Control to Drift: The Dynamics of Corporate Information Infrastructures*, Oxford: Oxford University Press, 2000, pp. 56-70.

Hanseth, O., and Braa, K. "Hunting for the Treasure at the End of the Rainbow: Standardizing Corporate IT Infrastructure," in O. Ngwenyama, L. Introna, M. D. Myers, and J. I. DeGross (Eds.), *New Information Technologies in Organizational Processes: Field Studies and Theoretical Reflections on the Future of Work*, London: Chapman & Hall, 1999, 121-140.

Hanseth, O., and Braa, K. "Technology as Traitor: SAP Infrastructure in Global Organizations," in R. Hirschheim, M. Newman, and J. I. DeGross (Eds.), *Proceedings of the 19th International Conference on Information Systems*, Helsinki, Finland, 1998,pp. 188-196.

Kremers, M., and van Dissel, H. "ERP System Migrations," *Communications of the ACM* (43:4), 2000, pp. 53-56.

Kumar, K., and Van Hillegersberg, J. "ERP Experiences and Evolution," *Communications of the ACM* (43:4), 2000, pp. 23-26.

Latour, B. *Pandora's Hope: Essays on the Reality of Science Studies*, Cambridge, MA: Harvard University Press, 1999.

Latour, B. *Science in Action: How to Follow Scientists and Engineers through Society*, Cambridge, MA: Harvard University Press, 1987.

Latour, B. "Technology is Society Made Durable," in J. Law (Ed.), *A Sociology of Monsters: Essays on Power, Technology and Domination*, London: Routledge, 1991, pp. 103-131.

Law, J. "After ANT: Complexity, Naming and Topology," in J. Law and J. Hassard (Eds.), *Actor Network Theory and After*, Oxford: Blackwell Publishers, 1999, pp. 15-25.

Monteiro, E. "Actor-Network Theory," in C. Ciborra and Associates (Eds.), *From Control to Drift: The Dynamics of Corporate Information Infrastructures*, Oxford: Oxford University Press, 2000.

Newell, S., Swan, J., and Galliers, R. D. "A Knowledge-Focused Perspective on the Diffusion and Adoption of Complex Information Technologies: The BPR Example," *Information Systems Journal* (10), 2000, pp. 239-259.

Noble, D. F. "Digital Diploma Mills: The Automation of Higher Education," *First Monday* (3:1), 1998.

Orlikowski, W. J., and Iacono, C. S. "Research Commentary: Desperately Seeking the 'IT' in IT Research—A Call to Theorizing the IT Artifact," *Information Systems Research* (12:2), Number 2, 2001, pp. 121-134.

Pan, S. L.; Newell, S.; Galliers, R. D.; and Huang, J. "Overcoming Knowledge Integration Challenges During ERP Adoption: The Role of Social Capital," under review, 2003.

Pettigrew, A. M. "Contextualist Research: A Natural Way to Link Theory and Practice," in E. Lawler (Ed.), *Doing Research That Is Useful in Theory and Practice*, San Francisco: Jossey-Bass, 1985, pp. 222-249.

Pettigrew, A. M. "Longitudinal Field Research on Change: Theory and Practice," *Organization Science* (1:3), 1990, pp. 267-291.

Robertson, M.; Swan, J.; and Newell, S. "The Role of Networks in the Diffusion of Technological Innovation," *Journal of Management Studies* (33:3), 1996, pp. 333-359.

Sawyer, S. "A Market-Based Perspective on Information Systems Development," *Communications of the ACM* (44:11), 2001, pp. 97-102.

Schrage, M. "Why IT Really Does Matter," *CIO Magazine*, August 1, 2003 (available online at http://www.cio.com/archive/080103/work.html).

Schweisberger, J., and Chatterjee, A. "Effective CRM Implementations," *Hospitality Upgrade*, 2002 (available online at www.hospitalityupgrade.com/).

Scott, S. V. "IT-Enabled Credit Risk Modernization: A Revolution Under the Cloak of Normality," *Accounting, Management and Information Technologies* (10:3), 2000, pp. 221-255.

Scott, S. V., and Wagner, E. L. "Networks, Negotiations, and New Times: The Implementation of Enterprise Resource Planning into an Academic Administration," *Information and Organization* (13:4), 2003, pp. 285-313.

Star, S. L., and Griesemer, J. R. "Institutional Ecology, 'Translations' and Boundary Objects: Amateurs and Professionals in Berkeley's Museum of Vertebrate Zoology, 1909-39," *Social Studies of Science* (19), 1989, pp. 387-420.

Swan, J.; Newell, S.; and Robertson, M. "The Illusion of 'Best Practice' in Information Systems for Operations Management," *European Journal of Information Systems* (8), 1999, pp. 284-293.

Timbrell, G.; Andrews, N.; and Gable, G. "Impediments to Inter-Firm Transfer of Best Practice: In an Enterprise Systems Context," *Australian Journal of Information Systems*, Special Edition, 2001, pp. 116-125.

Wagner, E. L. "Interconnecting Information Systems Narrative Research: Current Status and Future Opportunities for Process-Oriented Field Studies," in E. Wynn, E. A. Whitley, M. Myers, and J. I. DeGross (Edsl;), *Global and Organizational Discourse About Information Technology*, Boston: Kluwer Academic Publishers, 2003, pp. 419-436.

Walsham, G. *Interpreting Information Systems in Organizations*, Chichester, England: Wiley, 1993.

Walsham, G. *Making a World of Difference: IT in a Global Context*, Chichester: John Wiley and Sons, 2001.

Walsham, G., and Sahay, S. "GIS for District-Level Administration in India: Problems and Opportunities," *MIS Quarterly* (23:1), 1999, pp. 39-65.

Weber, R. "Editor's Comments: Still Desperately Seeking the IT Artifact," *MIS Quarterly* (27:2), 2003, pp. iii-xi.

ABOUT THE AUTHORS

Erica L. Wagner is an assistant professor of Information Systems at Cornell University's School of Hotel Administration. She is trained as an accountant and social scientist, and recently completing her Ph.D. (2002) at the London School of Economics, University of London. Her research centers on the ways in which information and communication technologies (ICT) are made to work within different organizational contexts. As a qualitative field researcher, she attempts to emphasize how negotiating with technology necessarily implies the reordering of organizational reality through design, implementation, and customisation activities. Erica can be reached by email at elw32@cornell.edu.

Robert D Galliers is the Provost of Bentley College and was previously Professor of Information Systems and Research Director in the Department of Information Systems at the London School of Economics (LSE). Before joining LSE, he served as Professor and Dean of Warwick Business School in the UK, and earlier as Professor and Head of the School of Information Systems at Curtin University in Australia. He has a Ph.D. from the LSE and an Honorary D.Sc. from Turku School of Economics & Business Administration, Finland. A leader in the field of management information systems, he is editor-in-chief of the *Journal of Strategic Information Systems*, and a fellow of the Association for Information Systems (AIS), the British Computer Society, and the Royal Society of Arts. He is past president of the AIS and was program cochair of the 2002 International Conference on Information Systems. He has held visiting professorships at INSEAD, France, University of St Gallen, Switzerland, City University of Hong Kong, the Institute for Advanced Management Studies, Belgium, National University of Singapore, Hong Kong Polytechnic University, and Bond University, Australia. He has published widely in leading international IS journals and has also authored/edited a number of books, the most recent being the third edition of the best seller *Strategic Information Management* (Butterworth-

Heinemann, 2003), with Dorothy Leidner; *Rethinking Management Information Systems* (Oxford University Press, 1999) with Wendy Currie, and *IT and Organizational Transformation* (Wiley, 1998) with Walter Baets. Bob can be reached at galliers@LNMTA.bentley.edu.

Susan Scott is a lecturer in the Information Systems Department at the London School of Economics where she pursues research focusing of the role of information systems in the transformation of financial services. She is Academic Director of the Moving Markets research project and teaches courses on IS in Business. Her background includes degrees in African History and Politics (SOAS), Analysis, Design and Management of Information Systems (LSE), and a Ph.D. in Management (Cambridge). Susan can be reached at S.V.Scott@lse.ac.uk.

25 INFORMATION SYSTEMS RESEARCH AND DEVELOPMENT BY ACTIVITY ANALYSIS AND DEVELOPMENT: Dead Horse or the Next Wave?

Mikko Korpela
University of Kuopio

Anja Mursu
University of Kuopio

Abimbola Soriyan
Obafemi Awolowo University

Anne Eerola
University of Kuopio

Heidi Häkkinen
University of Kuopio

Marika Toivanen
University of Kuopio

Abstract We argue that the currently dominant methods in Information Systems are not satisfactory for emancipatory research and development whose starting point is work. Activity theory was proposed as such an emancipatory research-cum-development approach in IS a decade ago. However, the potential identified in the theory has not fully materialized. As our own contribution toward making activity theory more operational in IS, we present an elaborated framework, ActAD, and review our experience in applying it to descriptive research, practical analysis, and constructive research. We claim that in order to fully unleash the potential of activity theory, activity-based methods should be developed further for IS requirements analysis projects and IS implementation projects, as well as for facilitating software development. The most appropriate way of developing such applied methods is through collaborative action research in real-life information systems work—the information systems practitioners developing their own work through activity analysis and development, with researcher participation.

Keywords: Activity theory, emancipatory research, work development, information systems development, methodology

1 INTRODUCTION: WORK IS THE KEY

Ever since the Manchester conference on Information Systems research methods in 1984, the international research community has recognized that a diversity of research approaches or philosophies exists within IS. According to Orlikowski and Baroudi (1991), three broad categories are commonly identified: positivist, interpretive, and critical. In this paper, we take the last mentioned, emancipatory or developmental standpoint (i.e., we are not only interested in understanding information systems within organizations, but also in developing "better" information systems). What, then, does better mean? What criteria should good information systems meet?

Different theorists in different times have named different factors as the most important, distinctive aspects of information systems. The very term information system implies that *information* is what it all comes down to. Those who share this view delve into information flows and entity-relationship models, regardless of the technology and purpose of information processing. Others underline *technology*, usually equating it with computers, and particularly equating information systems with software systems. Those with cybernetic backgrounds emphasize the term *system*, searching only for systemic entities and ignoring how bits and pieces of information and communication technology (ICT) are used in organizations. Researchers applying Habermas stress the *communication* aspects of information systems, viewing them as language games. In the very title of IFIP WG 8.2, *social and organizational* aspects of information systems are emphasized, with the view that the term information system refers to the organizational processes and resources of information management. Finally, the *human or individual* actors of information systems are pointed out as the starting point much too seldom.

In our mind, all these aspects—information, technology, system, communication, organization and the individual—are important factors, but still only elements. None of these viewpoints sufficiently explains the *purpose* for which information systems exist.

Our starting point is that purposeful *work* is the proper holistic viewpoint that binds the elements together. Individuals in organizations need information, use technology, and communicate, in more or less systemic ways, to jointly produce some services or products (use-values). An improved information system in this viewpoint means better facilitation to the workers to do their work. This can be achieved by means of better information, by means of more efficient technology, by means of better organization, by means of more humane work conditions, etc.

What kind of methodologies are available in current literature for IS research and practice whose starting point is work with an emancipatory and developmental approach? In this conference, the pioneering and foundations laying role of the socio-technical approach, soft systems methodology and the so-called Scandinavian school since the Manchester 1984 conference is self-evident. Considering the recent IFIP WG 8.2 conferences, however, two theoretical frameworks dominate: actor-network theory (Walsham 1997) and structuration theory (Walsham and Han 1991). The former is commonly regarded as tedious, the latter as over-general, and neither of them pays much attention to work practice. Both of them are only research methodologies, and not very easily transformed into practical information systems development methodologies.

Ethnomethodology and other ethnographic research methods in IS (Myers 1999) do pay attention to work practice, but they are better suited to descriptive analysis than developmental or design purposes. Moreover, ethnomethodology does not provide a theoretical basis for understanding work.

Alter (1999, 2001, 2002) has suggested the *work system method* as an approach for understanding and analyzing systems in organizations whether or not information technology (IT) plays an essential role. In Alter's words,

> a work system is a system in which human participants and/or machines perform business processes using information, technologies, and other resources to produce products and/or services for internal or external customers.

While Alter's approach fits well with our objectives by starting from purposeful work as the systemic unit of contextualized analysis, his framework is not based on established social theory of work.

Regarding methods for information systems practice, participatory design (Kensing and Blomberg 1998) provides diverse experiences and methods for emancipatory information system development, but with little or no theoretical foundation. In software engineering, Robertson and Robertson (1999) provide a requirements analysis and design methodology with a rare, explicit starting point in work analysis, but narrow down its scope only to process chains and software design.

Something more operational than actor-network theory and structuration theory but more theoretically founded than participatory design, ethnomethodology, or work system method is needed. In this paper we study the applicability of activity theory as a methodology for work-oriented IS research and development, and present experiences in applying an elaborated activity-theoretical framework.

2 ACTIVITY THEORY IN INFORMATION SYSTEMS

Activity theory has a broad and long research tradition, which emphasizes that human activity is culturally and historically formed, mediated, and defined by its object. It is beyond the scope of this paper to provide a tutorial on the evolution of activity theory since the 1920s; Hedegaard et al. (1999) provide a concise introduction. There are several widely different traditions within activity theory, but the activity-theoretical framework most commonly applied in Information Systems is developmental work research (DWR) developed by Engeström (1987, 1999).

Activity theory was first presented as an IS research approach in the IFIP WG 8.2 conference in Copenhagen that followed the first Manchester conference on research methods. Kuutti (1991) suggested, in line with the standpoint of this paper, that the very object of analysis in IS should be *activity system* instead of *information system*. Similarly, Bødker (1991) discussed the potential of activity theory as an analytical framework in understanding computer-based artefacts as instruments for work activities and materials for systems design.

During the decade after the Copenhagen conference, activity theory has been applied to research in the fields of human-computer interaction, computer-supported cooperative work, and information systems in a few studies (Bardram 2000; Bertelsen and Bødker 2000; Bødker 1997; Mwanza 2002; Redmiles 2002; Vrazalic and Gould 2001). Some research tradition can be found in Aarhus, Denmark, in Ronneby, Sweden, in Oulu, Finland, as well as in smaller groups in the United States, Australia, and New Zealand.

In Finland, dozens of developmental work research projects have been conducted, and many of them have involved information technology and information systems in some way. However, IS researchers have not been involved in these projects, so the methods have not been developed toward improved applicability to IS. Furthermore, very little has been published in English on the methodological development in these projects. There is no coherent textbook or methodological guide available for the international audience on DWR (e.g., regarding the formalized heavy work development method called Change Laboratory). Thus, the DWR experience in Finland has not enriched the IS research and development methods.

Activity theory in general and Engeström's developmental work research in particular emphasize that activity is a collective phenomenon involving several actors. The studies reviewed above, however, tend to reduce activity into a set of actions by an individual—the doctor's activity, the patient's activity, etc. The analytical model does not guide the researcher clearly enough into studying the individual and collective aspects of work activities within the same framework. This leads to a relatively weak support of designers of cooperative systems.

Activity theory has not been applied in IS as much as was expected in the early 1990s (Iivari and Lyytinen 1998). Is it a dead horse, gone with the wind, or can it be revived within IS?

In our mind, an analytical framework for emancipatory, work-oriented IS research and practice should meet the following requirements:

1. The starting point must be work activity as a systemic entity.
2. Technology, including computer-based technology, must be seen as a tool to facilitate work, embedded in the work system.
3. Both collective and individual aspects of work need to be taken into account.
4. Work systems need to be studied in their organizational context.
5. The analytical framework must be based on a sound theoretical basis.
6. The analytical framework must be applicable to both descriptive studies and practical development.
7. The analytical framework must be applicable both to technological development by software and IS professionals and to the developing of the work practice itself by the workers.

Activity theory provides a good starting point for developing a framework that satisfies these requirements and for methodologies that make use of the framework.

Activity theory and DWR are, however, very generic approaches that need to be adapted before they can be feasibly used by IS researchers and practitioners. As long as such adaptations of activity theory are not available, it will not spread in IS research and practice.

Our research group has for several years elaborated on an activity analysis and development framework, ActAD, that could be adapted to information systems research and practice. While accepting all of the fundamental aspects of Engeström's model, we aimed at making it more operational.

3 ACTIVITY ANALYSIS AND DEVELOPMENT

In this section we present a brief summary of the ActAD framework (originally published in Mursu et al., 2003; for more detailed accounts, see Korpela et al. 2000, 2002a, 2002b). In section 4 we present experiences in using the framework as a tool in analyzing and developing information systems.

Let us first illustrate our concepts with an example from intensive care in a hospital. In order to bring a seriously ill baby back to normal life, several healthcare professionals obviously need to engage in a care *process* and take action within it (Figure 1). The poor health of the baby is the *object* of the activity in this case, and the improvement of her health condition is the intended *outcome*, which is the purpose the of activity.

Looking closer into the process of the activity, in Figure 2 we have two of the *actors* (nurses) required in the care process engaged in *actions*, making use of different *means of action* (both physical and mental). In order to achieve the intended outcome, it is important that the individual actions are coordinated toward the shared goal. Various *means of coordination and communication* are employed to that end; in this case, a vital signs monitor helps the nurses to coordinate their actions.

When the health condition of the baby has improved sufficiently, the intensive care activity is completed and the responsibility of her further care is handed over to an ordinary ward (Figure 3). The outcome of the intensive care activity transforms into the object of the ward activity. A third type of means, *means of networking*, is required to mediate the relation between the activities. In this case, patient documents are such a means.

The concepts introduced in the intensive care case in an informal manner are now presented as an abstract framework (Figure 4). The framework starts from the elements of a *mediated action by an individual person* (Figure 4, broken line); the subject or actor, the object of the action, the instruments or means (both mental and physical) needed for the action, as well as the goal (Vygotsky 1978). For instance, a carpenter uses a hammer and applies his skills to some planks and nails in order to construct a scaffolding for a building.

In practice it almost always takes several actions by several individuals to produce any useful service or product (Figure 4, lower half); for instance, a number of carpenters, brick-layers, and electrical workers are needed to build a house. In activity theory, such a set of mediated actions on a shared object by a number of actors, directed by a (more or less consciously) jointly aspired outcome, is called an *activity* (Leontiev, 1978). It is important to notice that individual hūman actions can only be understood through the collective activity of which they are a part. Instead of bunches of uncoordinated actions, work, in practice, consists of systemic activities subordinating the actions in a purposeful way.

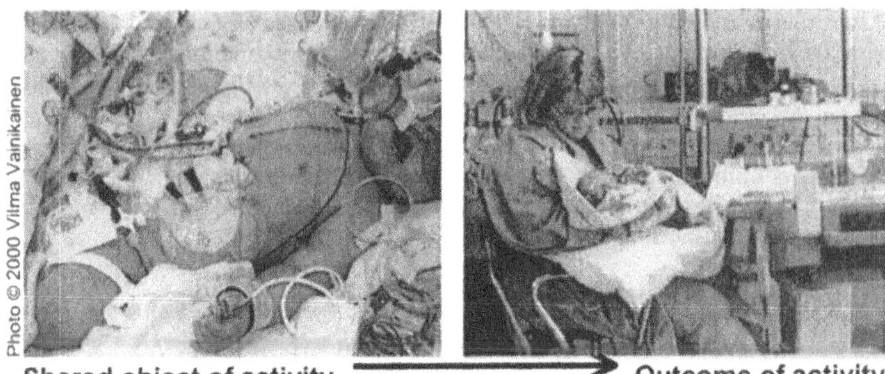

Figure 1. The Object and Intended Outcome of an Activity; Neonatal Intensive Care as the Case

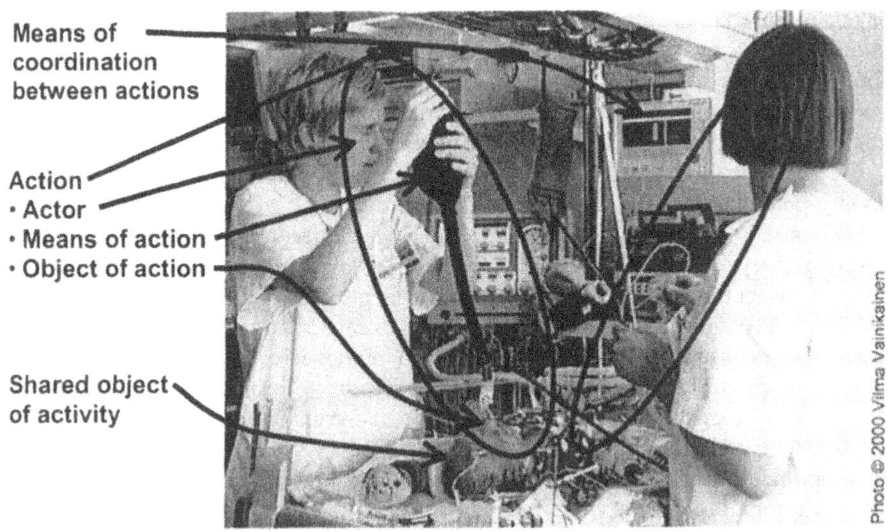

Figure 2. Two Actors Performing Actions That Are Part of the Activity, with Two Types of Means

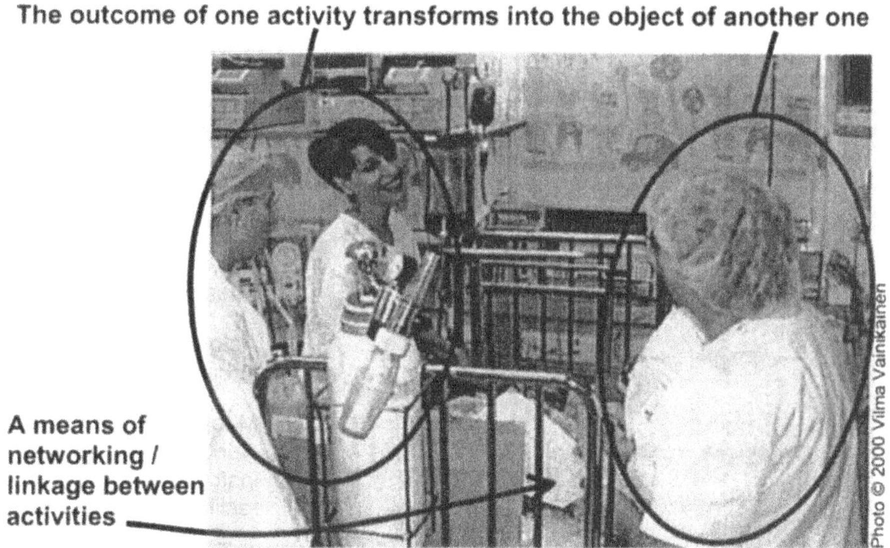
The outcome of one activity transforms into the object of another one
A means of networking / linkage between activities
Photo © 2000 Vilma Vainikainen

Figure 3. Intensive Care Activity Handing over to a Ward Activity

In addition to the instruments or means of the individual actions, other kinds of mediating instruments—social infrastructure—are also needed within an activity, as emphasized by Engeström (1987). The actions need to be oriented by *means of coordination and communication* (Figure 4, upper half); a blueprint, division of labor, meetings, and rules, for instance, among a construction team.

According to Engeström (1987), work activity as a real-world phenomenon is *systemic* by nature. That is, there must be a relative fit between the elements of a work activity, a *mode of operation* (Figure 4, large oval). When an activity evolves over time, it moves from one relative fit to another, from one mode to another, in historical phases. Today's house-building activity is quite different from what it was 20 or 100 years ago, both in terms of the elements (actors, object, and means) and in terms of the systemic mode of the activity. *Contradictions*, imbalance within and between various elements and the mode, are the force driving the activity to transform.

Finally, activities do not stand alone. The elements of one activity are produced by other activities, and the outcome of one activity is usually needed in one or more other activities (Figure 4, smaller ovals; Engeström, 1987). Construction workers do not cut and saw trees into planks on a building site any more, but buy planks sawn elsewhere in a sawmill activity. Mediation is also needed between the activities, and this is achieved by *means of networking* (Korpela *et al.*, 2000). For instance, advertisement, contracts, and transportation are needed to link sawmill and house construction activities.

Networks of activities constitute the "metabolism of use-values" in society; *i.e.*, activities are the "organs" that produce, exchange, and consume use-values in the "body" of society. The networks are split by organizational boundaries and accompanied by other kinds of societal relations (e.g., financial ones dealing with exchange values). Social theories other than activity theory should be applied to the organizational, financial, and wider societal contexts of activities (Korpela et al. 2001a).

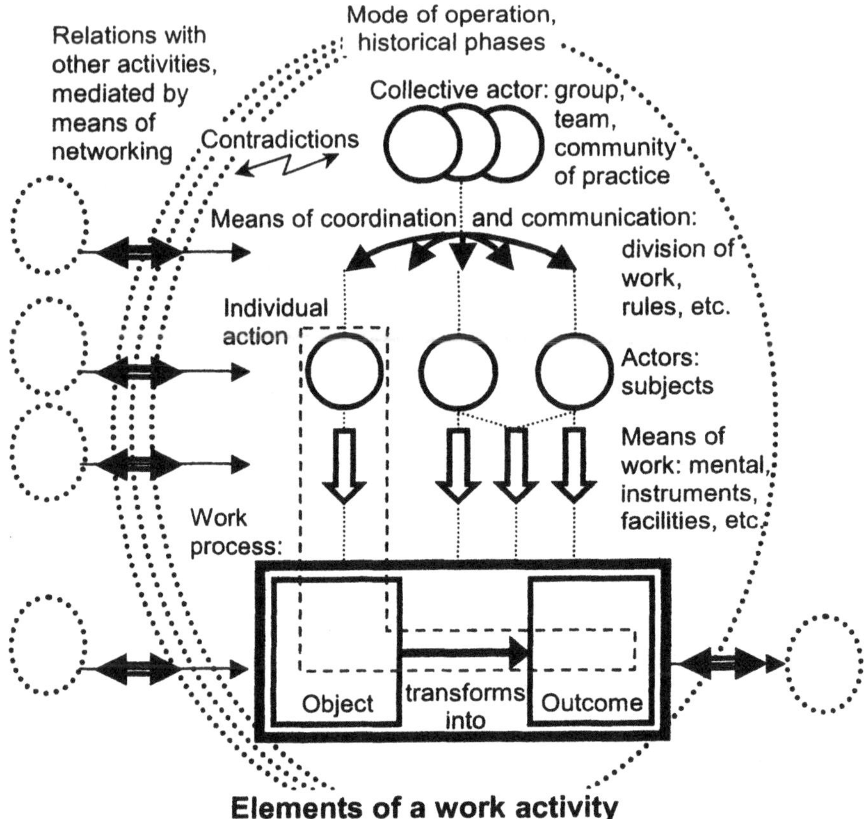

Figure 4. Three Different Types of Means in Activity Networks (Mursu et al. 2003; Adapted from Korpela et al. 2000).

Compared with Engeström's original model, ActAD includes both individual actions and collective activity intertwined in the same model, emphasizes the systemic mode, generalizes the means of coordination and communication, and introduces the means of networking.

According to the model, ICT can be used as a means in three different ways: as a means of work in individual actions, as a means of coordination and communication between actions in an activity, and as a means of networking between activities. However, the scopes overlap. From the viewpoint of supporting work activities by ICT, all three types of means are needed.

Echoing Kuutti's voice from 1991, we maintain that the systemic unit that IS researchers and practitioners should consider first is (work) activity in all its aspects and dynamics. It is not very relevant whether or not the information management facilities and processes within an activity or a network form an information *system*; the point is whether the use of ICT facilitates the objectives of the activity.

However, while ICT is becoming increasingly ubiquitous, it is one of the most important types of means in and between many information-laden activities. The need for new information-technological means is one of the most common sources for change in work, and information systems projects are one of the most common forms of change (Korpela et al. 2002a, 2002b).

4 APPLICATION EXPERIENCES

The proof of the pudding is in the eating. We have applied ActAD as an analytical framework in descriptive research, as lenses that enable students and laymen to better look at their own work, and as an explorative requirements analysis method in constructive research.

4.1 Experiences in Descriptive Research

In a joint Finnish-Nigerian project, where the main objective was to produce empirical evidence and understanding of the practice and problems of IS development in Nigeria, the software and information systems activities in three Nigerian software companies were analyzed using the ActAD framework (Mursu et al. 2002).

The framework, first illustrated by an imaginary IS project using cartoon-like figures, was used as an agenda in group discussions and interviews with Nigerian software practitioners. Afterward, the results of the interviews were presented to the informants using the same frameworks but the informants' own data. The interviewees grasped the model without any effort. The discussions remained mostly in a descriptive mood, but in a few situations the software professionals started to identify inadequacies in their own activities and to discuss potential remedies.

Due to the highly constrained circumstances and timescales, the discussions usually did not proceed to the prior historical phases of the activities. However, the ActAD framework provided a practicable tool for studying previously unknown activities, and the result was a rich picture—if only descriptive—of the information systems development activity in Nigeria (Mursu et al. 2002).

In the same project, Soriyan (2004) applied the framework to the analysis of 14 years of experience of a hospital software development project in an academic environment in Nigeria. The analysis dealt with the activity networks, organizational and international settings, as well as historical phases of the project. The analysis highlighted the role of different organizations' management activities and top managers, particularly when the project was to transform itself from in-house development to product development.

The descriptive studies of the Nigerian software development and information systems development also resulted in some preliminary normative guidelines, providing practitioners with methods for taking issues of sustainability and social impact into account in information systems implementation. These methods lean on the analysis of activity networks. The methods are, however, currently a work in progress.

The ActAD framework was tested and further developed during the Finnish-Nigerian project, in order to be suitable for IS research purposes. The interviewees grasped the framework without effort, and during the discussions they considered it as inspirational.

4.2 Experiences with Analytical Lenses

Good analytical theory should be like lenses that make us see our world brighter. In several small cases, we have provided the ActAD framework as a lens—as a checklist of key issues to look at—for laymen for a rapid analysis of their work, service, or activity chain.

The first experiment took place in Nigeria, where nurses and a general practitioner in a local health center were introduced to an early form of the ActAD framework by a couple of activity diagrams illustrated by cartoon-like figures (Korpela et al., 2000). They were then asked to analyze their own healthcare activities and activity networks using a list of questions based on the framework. The idea was to start identifying needs for improvement in the manual information system in their health center. The experiment worked surprisingly well; it was actually easier for the healthcare professionals to analyze their work than for local software academics to analyze the activity network around the manual information system.

The other experiences are from nine continuing education and Master's courses in Finland on information systems in healthcare and social services, starting from 1998. The students were nurses, social workers, medical doctors, and software practitioners. In some courses, the students were introduced to the ActAD framework during a one-hour lecture. They were then asked to select a recent case when they had been involved in a patient-care activity, either as an actor or as an object (patient). The students were instructed to identify the object and the intended outcome, the actors required, and the different types of means. The task was to specifically discuss the role of the IT-based means and the possible needs for improvement in the information-technological facilitation of the activity. Especially revealing were the cases where the student had been an object of care; it was much more difficult for students in a healthcare professional's position to recognize what other actors actually did in the process and how the activity chain continued from one organization to another.

Multi-professional groups have used simple ActAD-based methods in four continuing education courses in conducting a developmental feasibility study. The methods applied are similar to the rapid assessment method described in the following section.

4.3 Experiences in Constructive Research

ActAD has been used for explorative analysis by two teams in a large software integration research and development project in Finland.

4.3.1 Rapid Participatory Assessment Method for Integration Needs

The first team developed a *rapid participatory assessment method for integration needs*, adapted from the training courses described above, using maternity care as the case setting (Häkkinen 2003). First, activities, networks, and formal organizations in the target domain are roughly identified. Major stakeholders within the activity network are then invited into two discussion sessions, no more than a couple of hours each. In the first session (critique workshop), the stakeholders elaborate on the details of the tentative description of the activity network and identify bottlenecks or problems. In the second session (abridged future workshop) the focus is on suggesting solutions—not only software integration but all types of solutions.

4.3.2 Explorative Requirements Analysis Method

The other team was given the task of developing methods for grasping a previously unexplored grey area of inter- and extra-organizational activities, in search of requirements for possible new software products or interfaces (Toivanen et al. 2003). Home care was selected as the case domain. Again, the *explorative requirements analysis method* developed by the team begins with developing a rough overview of the activities, networks, and organizations involved in order to identify major stakeholders. The number of social and healthcare professionals and other parties involved in providing home care services in this case study was astonishing.

Representatives of key stakeholder groups were then interviewed, using a semi-structured interview guide that followed the activity framework but utilized terms specific to the home care domain instead of abstract terms such as *actors* or *means*. On the basis of the interviews, preliminary descriptions of the findings were collected and structured into documents and rich pictures on a wall. The findings were presented, considered, and processed forward with the stakeholders in workshops.

After the workshops, revised descriptions of the information needs in the *network of activities* were constructed. The outcome was depicted as a systematic description of the activity chains in the domain. Core activities were identified and described down to the level of actions, where the "empty spots" (lack of software) and requirements for new software and their integration with legacy systems can be identified. Using the acquired understanding, the purpose and the advantages of the would-be software and the first sketch for its architecture can be derived. The team has reached this stage by the time of writing this paper.

Software design today usually starts by describing use cases (*i.e.*, the communication between human beings and software). In this case this was not possible because of the grey area; the inter- and extra-organizational activities were not sufficiently understood. Similarly to Robertson and Robertson (1999), we found it important to understand the work of the organizations and their personnel before proceeding to use case design. Furthermore, understanding the cooperation and communication between and inside organizations proved important.

Thus far the experiences have been promising. This method is significantly more laborious than the rapid assessment method, but correspondingly leads into a more detailed requirements description. The method gave us a lot of information needed for developing the work practices and the technological solutions, including "humanware," hardware, and software. However, in order to produce software specifications, our method needs further development. Next we will proceed from the descriptions of the key activities to describing the actions in which computer-based systems are required, and further to a software architecture and unified modeling language (UML; Fowler 2000) models.

In conclusion, the explorative method proceeds from activity networks to software engineering.

1. Work is modeled as a *network of activities* within and between organizational boundaries, embracing all of the aspects of the ActAD framework presented in section 3 and Figure 4. In addition, clusters of information are emphasized as *concepts of work* and the relationships between them. This step is divided into three parts: gathering, structuring, and describing the information needs.
2. Using the results of the first step, more accurate functional and quality requirements of the intended software are designed. The key activities are described in more detail. The idea is that the process of an activity (how the object is transformed into the outcome through actions) is described in terms of the actions of various actors over time (see the generic sketch in Figure 5). A combination of the activity framework and UML's activity diagrams and use case diagrams are considered for this task (the term *activity* as used in UML has a completely different meaning than in activity theory). The purpose is to explore the *information needs* in more detail, and to identify empty spots for potential software; for instance, the needs for currently lacking computer-based means of action, coordination, or networking (applications or components). Simultaneously, the *information system architecture* and the first draft of the *software architecture* are defined. The software architecture specifies the software and its structure (*i.e.*, components), relationships between new software and legacy systems, and its relationships with the environment. Finally, the software architecture is evaluated against the functional and quality requirements. If some further architectural evolution is required, step 2 is repeated accordingly.

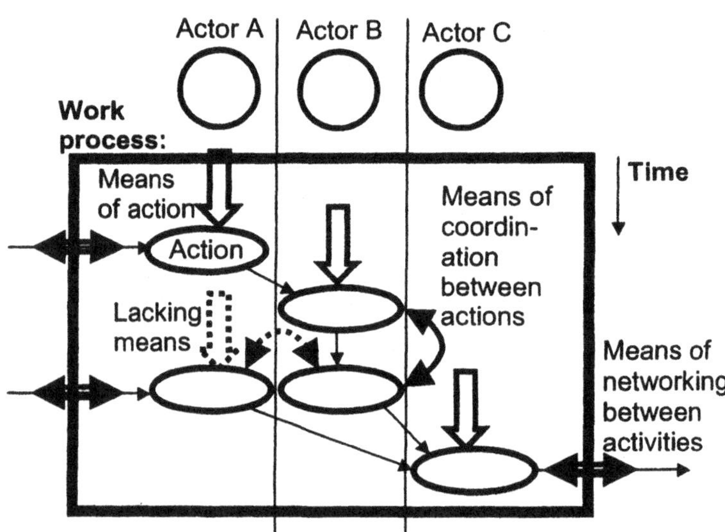

Figure 5. Zooming in from ActAD to UML

3. The *requirements specifications* of each piece of software identified in the previous step are then generated utilizing the architecture and action descriptions. Since the specifications must be understandable to software engineers, we propose the use of UML diagrams (Fowler 2000) in this step. Thus far we have emphasized component diagrams, activity diagrams, and use case diagrams (including scenarios), which we use in our method in the implementation, testing, and introduction phases. The derivation of class diagrams from the ActAD framework needs further research.

5 DISCUSSION: CHALLENGES AHEAD

In this section we will first assess activity theory in general and the ActAD framework in particular against the objectives we set at the end of section 2. We will then evaluate the application experience thus far. At the end we shall discuss what remains to be done if activity theory is to fulfil the promises that were put forward in 1991.

5.1 Assessment of the Analytical Framework

Comparing the ActAD framework against the seven requirements for an improved analytical framework stated at the end of section 2, some of the requirements are met by the very theoretical approach behind it. That is, activity theory is based on the position that activities are real-life entities that abound in work practice and other domains of life, and this position is based on a century-long tradition of theoretical work, not just speculation (requirement number 5). The systemic nature of activity and its substructure of actions are specifically emphasized in the Leontievian branch of the activity theory and further developed by Engeström (requirement number 1). However, Engeström's (1987) argumentation about the origins of the structure of activity is not backed up by evidence, and we have replaced it by a more straightforward linkage to Vygotsky's and Leontiev's work.

Engeström's original framework binds the individual, collective, and technological aspects of work together (requirement number 2 and number 3), and we have further elaborated on this by first clarifying the presence of several individual actors and actions within a collective activity and second by identifying three different types of mediating artefacts (means of work/action, means of coordination and communication, means of networking). Unlike Engeström, we have emphasized that wider societal contexts of activities should be studied by other social theories. In practice, we have included organizational boundaries and financial relations in the analysis when relevant (requirement number 4). We regard these modifications to Engeström's developmental work research framework as our main theoretical and methodological contribution thus far.

Developmental work research has proved to be a particularly suitable approach to holistic development of work by its actors, in addition to the approach's descriptive power (requirement number 6). Several dozen successful work development projects have been undertaken, particularly in Finland, using different variants of the methodology. ActAD does not add major new methodological innovation to this body of experience, but is a more readily comprehensible representation for nonexperts than the original triangles.

Finally, requirement number 7 states that IS frameworks should be applicable to developing technologies in the same way as to developing the other aspects of activities. In this respect, ActAD takes some steps forward from other activity-theoretical approaches by identifying the three different classes of mediating artefacts, as well as by indicating how the process of "object-transforms-into-outcome" is made up of actions by individuals. We have suggested that actions that utilize IT can be equated with *use cases*, which are requirements descriptions understood by software engineers (Korpela et al. 2001a). However, this aspect of the framework is still not sufficiently developed for information systems practitioners to use it as an everyday requirements specification method. The explorative method described in section 4.3 is the first practical step in this direction.

The three types of IT-based means identified in ActAD are studied by three different design-oriented subdisciplines. As Grudin (1994) has pointed out, the means of work in individual actions is the focus of *human-computer interaction* (HCI) studies, while the means of coordination and communication between actors in an activity are the focus of *computer-supported cooperative work* (CSCW), and the means of networking between activities are mainly studied in *information systems* (IS/MIS). However, the scopes overlap. From the viewpoint of supporting work activities by IT, all three types of means are needed and therefore HCI, CSCW, and IS viewpoints should be considered in most developmental endeavors. Each viewpoint will require its own type of further development of the activity-theoretical analytical methods into methods for practical design.

5.2 Assessment of the Application Experience

There are a number of ways in which a genuinely practical theory can be applied to the IS domain (Korpela et al. 2000).

1. People participating in any activity can *analyze their own activity* and identify requirements for new *IT-based means that facilitate their work* (a shopping list for new software).
2. An IS professional can *analyze any other person's work* and identify requirements for new *IT-based products* that can be developed to facilitate a class of activities at a number of customer sites (a blueprint for a software product).
3. IS professionals, like any other people, can *analyze their own IS development and maintenance activities* and identify requirements for new *methods and practices (means) for IS development* (a shopping list for new methods).
4. IS researchers can *analyze IS practitioners' work activities* and identify requirements for more appropriate *theories, methods, and education* to facilitate the IS practitioners' work in general (a blueprint for relevant education).
5. IS researchers, like any other people, can *analyze their own activities of IS research and education* and identify requirements for better *theoretical and methodological means for IS* (a shopping list for new theories, frameworks and research methods).

In the cases described in section 4, the emphasis has been on the first (section 4.2), second (section 4.3), and fourth (section 4.1) types of application. As mentioned in the previous section, the use of activity theory by information systems practitioners (application type 2) has mainly been descriptive in nature, without a clear transition to a constructive mode by software engineers.

The most challenging way of applying Activity theory would be a combination of types 3 and 4: collaborative action research by information systems practitioners and researchers on what are the most pressing tensions, imbalances, or contradictions in every-day IS practice today, and what kind of new means might be required to address these developmental challenges. This type of action research has been conducted far too seldom, the Danish MARS and MUST projects and the software process improvement tradition being the main exceptions. We argue that activity theory and the developmental work research methods would be exceptionally helpful for information systems practitioners; they could reflect on their own work development needs, if they have first tried the same methods for analyzing other people's needs. The activity concepts would provide for a common language for practitioners and action researchers. The next major step on our research agenda is to proceed to such collaborative IS activity analysis and development. However, few software or consulting companies are ready to invest their resources in developing their work practices.

We have seen no signs of IS researchers' serious analysis of their own work activities (type 5). Maybe it is too bizarre an idea to study oneself by one's own frameworks.

5.3 What Needs to Be Done

The discussion above leads to the conclusion that activity theory in general and the ActAD framework in particular have a great methodological potential for IS research and practice, but this potential is still largely not realized; activity theory in IS is not a dead horse, but maybe a sleeping one. What should be done to bridge the gap between the promise and the reality?

Since we are discussing an emancipatory, developmental approach, the crucial issue is whether the methodology is suitable for *practical application in everyday situations*, and whether it proceeds *from descriptive to constructive uses* (developing new organizational information systems and possibly new software artefacts). If the activity-theoretical methodologies in IS will not meet these criteria, they will remain marginal or die away. However, even excellent methodologies do not spread by themselves but need to be accompanied by educational efforts.

We propose that the following tasks are necessary to make activity theory deliver in IS as a constructive method and not only as a descriptive method:

1. The generic concepts, frameworks, and methodologies of activity theory need to be better *applied and operationalized to IS*. That is, versions of activity analysis frameworks are needed which are specifically tailored to information systems analysis and development, paying attention to information flows and IT-based means, without losing the holistic view. We need activity-based methods for IS requirements analysis projects and for IS implementation projects. Furthermore, we need activity-based methods that bridge IS analysis with software development—from activity analysis and development through action case modeling to use

cases and conceptual models. The methods and sample case descriptions need to be distributed as a "cookbook on activity theory in IS" to make them easily available to practitioners, and in order for them to be used in IS education. Besides the experience based on ActAD, various other approaches including developmental work research, change laboratory, work system method, participatory design, and ethnomethodology should be reviewed for the cookbook.
2. The most appropriate way of developing such applied methods is through collaborative action research in real-life information systems work: IS practitioners developing their own work through activity analysis and development with researcher participation. This ensures that the methods to be developed will be truly practical, something emerging from pressing needs within information systems work.
3. The wider societal and organizational contexts of activities need to be better incorporated into the analytical framework, especially as far as issues like sustainability, affordability, and socio-economic impact of information systems are concerned.
4. Finally, true theoretical discussion is required to strengthen the socio-scientific basis of the frameworks and methodologies. It is not sufficient to simply apply Engeström's triangles or our ovals as such.

6 CONCLUSION

The conference's call for papers challenged authors to address issues such as "Is theory irrelevant? Can theory inform practice?" In this paper we argued that the currently dominant methods in IS are not satisfactory for emancipatory research and development whose starting point is *work*. We reviewed the experience in applying activity theory in the field of IS, and concluded that the potential identified in it already a decade ago has not fully materialized. As our own contribution toward making activity theory more operational in IS, we presented the ActAD framework, distinguished its differences from Engeström's original model, and reviewed our experience in applying the framework to descriptive research, practical analysis, and constructive research. We claimed that in order to fully unleash the potential of activity theory, activity-based methods should be further developed for IS requirements analysis projects and IS implementation projects, as well as for facilitating software development.

We regard that activity theory has the necessary elements for becoming a relevant theory that can inform practice in IS. The task ahead is to transform the generic frameworks into genuinely practical research and development methodologies. ActAD is one step forward and so far our experiences are encouraging. However, further steps are needed. While the strong theoretical foundation provides rigor, the degree of relevance can be increased by borrowing from participatory design, ethnomethodology, and software engineering.

ACKNOWLEDGEMENTS

The paper is based on a research funded by the Academy of Finland through the INDEHELA-Methods project no. 39187 in 1998-2001 and INDEHELA-Context no. 201397 in

2003, as well as on a research funded by the National Technology Agency Tekes through the PlugIT project no. 40246/02 and 90/03 in 2001-2004.

REFERENCES

Alter, S. "A General, Yet Useful Theory of Information Systems," *Communications of the AIS* (1:13), 1999, pp. 1-69.

Alter, S. "Which Life Cycle: Work System, Information System, or Software?," *Communications of the AIS* (7:17), 2001, pp. 1-53.

Alter, S. "The Work System Method for Understanding Information Systems and Information System Research," *Communications of the AIS* (9), 2002, pp. 90-104.

Bardram, J. "Temporal Coordination: Of Time and Collaborative Activities at a Surgical Department," *Computer Supported Cooperative Work*, (9:2), 2000, pp. 157-187.

Bertelsen, O. W., and Bødker, S. "Introduction: Information Technology in Human Activity," *Scandinavian Journal of Information Systems* (12:1), 2000, pp. 3-14.

Bødker, S. "Activity Theory as a Challenge to Systems Design," in H-E. Nissen, H. K. Klein and R. Hirscheim (Eds.), *Information Systems Research: Contemporary Approaches and Emergent Traditions*, Amsterdam: North-Holland, Amsterdam, 1991, pp. 551-564.

Bødker, S. "Computers in Mediated Human Activity," *Mind, Culture, and Activity* (4:3), 1997, pp. 149-158.

Engeström, Y. "Activity Theory and Individual and Social Transformation," in Y. Engeström, R. Miettinen and R. Punamäki (Eds.), *Perspectives on Activity Theory*, Cambridge, England: Cambridge University Press, UK, 1999, pp. 19-38.

Engeström, Y. *Learning by Expanding: An Activity-Theoretical Approach to Developmental Research*, Helsinki: Orienta-Konsultit, 1987.

Fowler, M. *UML Distilled: A Brief Guide to the Standard Object Modeling Language*, Reading, MA: Addison-Wesley, 2000.

Grudin, J. "Groupware and Social Dynamics: Eight Challenges for Developers," *Communications of the ACM* (37:1), 1994, pp. 92-105.

Hedegaard, M.; Chaiklin, S.; and Jensen, U. J. "Activity Theory and Social Practice: An Introduction," in S. Chaiklin, M. Hedegaard and U. J. Jensen (Eds.), *Activity Theory and Social Practice: Cultural-Historical Approaches*, Aarhus, Denmark: Aarhus University Press, 1999, pp. 12–30.

Häkkinen, H. "Rapid Method for Integration Requirements Assessment: Case Maternity Care," in S. Laukkanen and S. Sarpola (Eds.), *Electronic Proceedings of the 26th Information Systems Research Seminar in Scandinavia*, Helsinki School of Economics, Helsinki, 2003.

Iivari, J., and Lyytinen, K. "Research on Information Systems Development in Scandinavia: Unity in Plurality," *Scandinavian Journal of Information Systems* (10:1/2), 1998, pp. 135-185.

Kensing, F., and Blomberg, J. "Participatory Design: Issues and Concerns," *Computer Supported Cooperative Work* (7:3/4), 1998, pp. 167–185.

Korpela, M.; Eerola, A.; Mursu, A.; and Soriyan, H. A. "Use Cases as Actions Within Activities: Bridging the Gap Between Information Systems Development and Software Engineering," abstract, in *Proceedings of the Second Nordic-Baltic Conference on Activity Theory and Sociocultural Research*, Ronneby, Sweden, 2001a, p. 51.

Korpela, M.; Mursu, A.; and Soriyan, H. A. "Two Times Four Integrative Levels of Analysis: A Framework," in N. L. Russo, B. Fitzgerald and J. I. DeGross (Eds.), *Realigning Research and Practice in Information Systems Development: The Social and Organizational Perspective*, Boston: Kluwer Academic, 2001b, pp. 367-377.

Korpela, M.; Mursu, A.; Soriyan, H. A.; and Eerola, A. "Information Systems Research and Information Systems Practice in a Network of Activities," in Y. Dittrich , C. Floyd, and R. Klischewski (Eds.), *Social Thinking—Software Practice*, Cambridge, MA: MIT Press, 2002a, pp. 287-308.

Korpela, M.; Mursu, A.; Soriyan, H. A.; and Olufokunbi, K. C. "Information Systems Development as an Activity," *Computer Supported Cooperative Work* (11:1/.2), 2002b, pp. 111-128.

Korpela, M.; Soriyan, H. A.; and Olufokunbi, K. C. "Activity Analysis as a Method for Information Systems Development: General Introduction and Experiments from Nigeria and Finland," *Scandinavian Journal of Information Systems* (12:1), 2000, pp. 191-210.

Kuutti, K. "Activity Theory and its Applications to Information Systems Research and Development," in H-E. Nissen, H. K. Klein and R. Hirschheim (Eds.), *Information Systems Research: Contemporary Approaches and Emergent Traditions*, Amsterdam: North-Holland, 1991, pp. 529-549.

Leontiev, A. N. *Activity, Consciousness and Personality*, Englewood Cliffs, NJ: Prentice-Hall, 1978.

Mursu, A.; Soriyan, A.; and Korpela, M. "ICT for Development: Sustainable Systems for Local Needs," in-progress research paper presented at the IFIP TC8 & TC9/WG8.2 & WG9.4 Joint Working Conference on Information Systems Perspectives and Challenges in the Context of Globalization, Athens University of Economics and Business, Athens, 2003.

Mursu, A.; Soriyan, H. A.; and Korpela, M. "Risky Business: A Case Study on Information Systems Development in Nigeria," in S. Krishna and S. Madon (Eds.), *Information and Communication Technologies and Development: New Opportunities, Perspectives and Challenges*, Bangalore: Indian Institute of Management, 2002, pp. 385-401.

Mwanza, D. *Towards an Activity-Oriented Design Method for HCI Research and Practice*, Unpublished Ph.D. Thesis, Open University, Milton Keynes, UK, 2002 (available online at http://iet.open.ac.uk/pp/d.mwanza/Phd.htm).

Myers, M. D. "Investigating Information Systems with Ethnographic Research," *Communication of the AIS* (2:23), 1999, pp. 1-20.

Orlikowski, W. J., and Baroudi, J. J. "Studying Information Technology in Organizations: Research Approaches and Assumptions," *Information Systems Research* (2), 1991, pp. 1-28.

Redmiles, D. "Introduction," *Computer Supported Cooperative Work: Special Issue on Activity Theory* (11:1/2), 2002, pp. 1-11.

Robertson, S., and Robertson, R. *Mastering the Requirements Process*, Reading, MA: Addison-Wesley, 1999.

Soriyan, H. A. *A Conceptual Framework for Improved Information Systems Development Methodology*, Unpublished Ph.D. Thesis, Obafemi Awolowo University, Ile-Ife, Nigeria, 2004.

Toivanen, M.; Eerola, A.; and Korpela, M. "From Information Systems Requirements to Software Components: Home Care Case," in S. Laukkanen and S. Sarpola (Eds.), *Electronic Proceedings of the 26th Information Systems Research Seminar in Scandinavia*, Helsinki School of Economics, Helsinki, 2003.

Vrazalic, L., and Gould, E. "Towards an Activity-Based Usability Evaluation Methodology," in M. Anders, A. Opdahl, and S. Bjørnstad (Eds.) *Electronic Proceedings of 24th Information Systems Research Seminar in Scandinavia* , 2001.

Vygotsky, L. S. *Mind in Society: The Development of Higher Psychological Processes*, M. Cole. V. John-Steiner, and S. Schribner (Eds.), Boston: Harvard University Press, 1978.

Walsham, G. "Actor-Network Theory and IS Research: Current Status and Future Prospects," in A. S. Lee, J. Liebenau and J. I. DeGross (Eds.), *Information Systems and Qualitative Research*, London: Chapman & Hall, 1997, pp. 466-480.

Walsham, G., and Han, C-K. "Structuration Theory and Information Systems Research," *Journal of Applied Systems* (17), 1991, pp. 77-85.

ABOUT THE AUTHORS

Mikko Korpela is Research Director of the Healthcare Information Systems R&D Unit, University of Kuopio, Finland, and a docent at the Department of Computer Science at the same university. He achieved his D.Tech. degree in Information Systems at the Helsinki University of Technology in 1994. He is a member of IFIP WG 8.2 and 9.4. His research interests include West African political history, activity development, and information systems development particularly in healthcare. Mikko can be reached by e-mail at mikko.korpela@uku.fi.

Anja Mursu is an acting professor of Software Engineering at the Department of Computer Science, University of Kuopio, Finland. She achieved her Ph.D. in Information Systems in 2002 at the University of Jyväskylä, Finland. Her research interests include information systems development, especially in developing countries, risk assessment, sustainability of information systems, activity theory, and recently also usability of information systems. Anja can be reached by e-mail at anja.mursu@uku.fi.

Abimbola Soriyan is a lecturer at the Computer Science and Engineering Department, Obafemi Awolowo University, Ile-Ife, Nigeria. She achieved her Ph.D. on information systems development in Nigerian software industry at the same university in 2004. Her research interests include information systems development with emphasis on healthcare. In particular, she is interested in why many developing countries have not embraced IT despite the success stories in the West. Abimbola can be reached by e-mail at hasoriyan@yahoo.com.

Anne Eerola is a professor in Software Engineering at the Department of Computer Science, University of Kuopio, Finland. She achieved her Ph.D. degree in 1994 at the same university on the analysis, design and maintenance of object oriented systems. She has worked for 10 years in the software industry and another 10 years as a teacher in the university. Her research emphasizes requirements modeling, software architectures and components, design patterns, and software testing. Anne can be reached by e-mail at anne.eerola@cs.uku.fi.

Heidi Häkkinen, R.N., is a Health Informatics student at the Department of Health Policy and Management, University of Kuopio, Finland. She has been working in the field of clinical information systems for the past six years. Currently she works for a research project in health care systems integration and can be reached at heidi.hakkinen@uku.fi.

Marika Toivanen is a Ph.D. student at the Department of Computer Science, University of Kuopio, Finland. She did her Master's thesis about human factors in software processes in 2002. Since then, she has been studying information systems development methods. Currently she focuses on eliciting methods of health care systems requirements. Marika can be reached at marika.toivanen@uku.fi.

26 MAKING SENSE OF TECHNOLOGICAL FRAMES: Promise, Progress, and Potential

Elizabeth Davidson
David Pai
University of Hawaii, Manoa

Abstract In a seminal paper, Orlikowski and Gash (1994) articulated a conceptual framework for technological frames of reference (TFR) to lay the groundwork for a systematic approach to socio-cognitive research on information technology. This work is widely cited as a justification for social and socio-cognitive analysis of IT, but a limited number of studies utilizing and further developing the frames concept have been published in the ensuing decade. In this paper, we review the promise of the technological frame concept, assess theoretical and methodological progress evident in TFR publications, and consider how potential contributions of TFR may be realized in future research. In doing so, we consider how limitations in TFR research to-date might be addressed, including the feasibility of a rapprochement with quantitative research methods, of TFR analysis at the industry level of analysis, and of action research approaches.

Keywords: Technological frames of reference, qualitative research methods, social cognition, social construction of technology, information systems research

1 INTRODUCTION AND MOTIVATION

In 1994, Wanda Orlikowski and Debra Gash introduced the concept of technological frames of references (TFR) to the Information Systems community in their paper entitled "Technological Frames: Making Sense of Information Technology in Organizations." The influence of the TFR concept in the IS and other academic communities is evident in several ways. A citation search on the ISI Social Science Citation Index yielded a list of 52 citations, ranging from traditional IS research journals to engineering and computer science outlets (e.g., *IEEE Transactions on Engineering Management, Human-Computer Interaction*, and *Journal of the Operations Research Society*) and beyond IT to technology journals (e.g., *Science, Technology and Human Values* and *Technology Analysis and Strategic Management, Social Studies of Science*). Using an Internet search engine (Google) to target this paper, we found hundreds of "hits," ranging from Ph.D. seminar reading lists to citations in unpublished white papers and research descriptions posted to Internet Web sites.

Given the apparent wide appeal and exposure of the technological frames concept, we were surprised to find relatively few published reports of research that actually conducted a TFR analysis or further developed the theoretical framework. Instead, publications citing Orlikowski and Gash's paper typically referred to the notion of frames in passing, or more generally to socio-cognitive approaches. A few papers indicated that the frames concept would be actively engaged, then failed to do so, pursuing other theoretical approaches instead. Intrigued by this seeming contradiction between the intuitive appeal of the TFR concept and its actual use in IS research practice, we examine the promise for theoretical and practical contributions outlined in Orlikowski and Gash's framework and the theoretical, methodological, and practical progress evident in the eight journal and conference publications we did locate. From this analysis, we consider limitations in TFR research and suggest several ways in which its potential might be realized in future work.

The paper is organized as follows. We first review key aspects of the technological frames concept outlined by Orlikowski and Gash, and its intellectual roots in the social construction of technology and organizational cognition literatures. We then review publications that have utilized the frames concept and, guided by Orlikowski and Gash's blueprint for future research, assess how these studies have contributed to development of the TFR concept. In our discussion, we consider aspects of TFR research that may constrain its use in IS research and highlight various approaches to help researchers interested in socio-cognitive investigations to realize the potential contributions of this framework. In doing so, we consider whether rapprochement with quantitative research methods is feasible and what the theoretical and practical implications might be. We note that action research is a particularly promising approach for TFR studies that could heighten contributions to practice. Finally, we suggest that TFR analysis could be used to better understand frame formation around emerging technologies within organizational fields, and that such studies could inform intra-organizational analysis of frame formation and change as well as interorganizational analysis.

2 CHARACTERIZING TECHNOLOGICAL FRAMES

Orlikowski and Gash (1994) based their argument for technological frames as a theoretical lens in IS research on the premise that people act on their interpretations of the world (Berger and Luckmann 1967; Weick 1979), including their interpretations regarding technologies in organizations. A variety of IS researchers had previously addressed socio-cognitive and interpretive issues, focusing particularly on the ways in which technology designers' cognitive understanding influences the design of technical artifacts (cf. Boland 1978, 1979; Bostrom and Heinen 1977; Dagwell and Weber 1983; Markus and Bjørn-Andersen 1987) and on the alignment of users' expectations with the intended design (cf. Ginsberg 1981). To lay the groundwork for a socio-cognitive perspective in IT research focused on frames of reference, Orlikowski and Gash defined technological frames as "that subset of members' organizational frames that concern the assumptions, expectations, and knowledge they use to understand technology in organizations. This includes not only the nature and role of the technology itself, but the specific conditions, applications, and consequences of that technology in particular contexts" (p. 178). They posited that frames exist at the individual level, but individuals within relevant social groups tend to develop shared frames of reference that guide their interactions around technology and their understanding and use of technology. The groups they identified in an empirical study included technologists who implemented organizational IT systems, users of these systems, and managers who decided which technologies to adopt or champion.

Orlikowski and Gash's major interest was sensemaking around information technologies and organizational changes related to technology, particularly the difficulties that arise when the frames of relevant social groups differ. They defined *congruence* as the alignment of TFRs along key elements, that is, similarity in structure (common categories) and content (common values) and posited that when there is incongruence in the frames of key organizational stakeholder groups, problems such as misaligned expectations, contradictory actions, resistance, skepticism, and poor appropriation of IT may result (p.180). In their empirical study, Orlikowski and Gash identified incongruence between the interpretations of a new IT of managers, users, and technologists and their expectations for its uses, which contributed to underutilization of the system.

To better understand the promise and limitations of the TFR concept in IS research, we examined the intellectual roots of Orlikowski and Gash's framework in two literatures: research on the social construction of technology (SCOT) (cf. Bijker 1995; Bijker et al. 1987) and organizational studies of social cognition (cf. Bartunek 1984; Gioia 1986; Weick 1979; for a comprehensive literature review, see Walsh 1995). Here, we reflect on how Orlikowski and Gash selectively drew from these literatures in their framework; later we will consider how further developing these intellectual roots might benefit future TFR research.

2.1 Technological Frames and SCOT

Orlikowski and Gash derived their notions of *interpretive flexibility of technology* (also see Orlikowski 1992), *relevant social groups*, and *shared frames* primarily from SCOT theories. Orlikowski and Gash's articulation of the TFR concept has similarities with Bijker's (1995) definition of a technological frame: some frame elements Bijker defined—goals, key problems, users' practices, and so on—are evident in the frame dimensions Orlikowski and Gash identified (e.g., technological strategy motivation, criteria for success, ease-of-use, training); other Bijker elements, such as tacit knowledge, are implicit in Orlikowski and Gash's definition of TFR as taken-for-granted assumptions. A substantive difference is that Orlikowski and Gash's frame concept is strictly socio-cognitive and does not include the technology artifact, whereas Bijker includes cognitive, social, and material elements (particularly, the technology artifact) within his definition of a frame (p. 126).

There are other notable differences in the TFR concept and SCOT analysis. Orlikowski and Gash defined TFRs as individual, socio-cognitive structures that may be shared among members of a relevant social group—a definition akin to group-level analysis in social cognitive research. Bijker defined technological frames as social, rather than socio-cognitive, structures, noting "a technological frame structures the interactions among the actors of a relevant social group. Thus it is not an individual's characteristic, nor a characteristic of systems or institutions; technological frames are located between actors, not in actors or above actors" (p. 123). Orlikowski and Gash's research interests are consistent with SCOT goals of understanding how frames influence the design and interpretations of technology artifacts. However, their use of TFR analysis varied from SCOT research in level of analysis and emphasis. Orlikowski and Gash's framework directs attention to the interpretive influences of *existing* TFRs *within* organizations, particularly in situations in which incongruence exists, and of frame changes resulting from organizational IT adoption. SCOT studies such as Bijker's consider technological frames at the societal level and examine processes through which frames take shape and stabilize or change. Orlikowski and Gash characterized TFRs as a process theory, but they did not explicitly incorporate SCOT process concepts such as *closure* (acceptance of a dominant technology interpretation across relevant social groups) or *stabilization* (strengthening of a TFR within a social group).

2.2 Technological Frames and Social Cognitive Studies

To highlight the influence of the organizational social cognitive literature on the TFR concept, we draw on Walsh's (1995) comprehensive synthesis and review, which incorporated studies and findings cited by Orlikowski and Gash. Walsh coined the term *knowledge structure* to represent the myriad constructs that have been used in social cognitive research (i.e., frames of reference, schema, interpretive schemes, scripts and so on), defining them as "a mental template that individuals impose on an information environment to give it form and meaning" (p. 281). Knowledge structures are individually held, although much research has been conducted at the group (cf. Gioia 1986; Isabella 1990; Walsh and Farley 1986), organizational (cf. Bartunek and Moch 1987; Daft and Weick 1984) and even industry (cf. Porac et.al 1989) levels of analysis. Orlikowski and Gash grounded their TFR concept at the individual level but addressed group level frames by defining them as shared aspects of individual frames. In doing so, they avoided the debate about whether higher-level socio-cognitive structures exist, independent of individual structures. Walsh cautioned, however, that we should "view organizational cognition as much more than some kind of aggregation or even congregation of individual cognitive processes" (p. 304).

Organizational research on social cognition generally examines frame content (domains of knowledge) or structure (organization and integration of knowledge), with studies of content far outnumbering studies of structure (Walsh 1995). By examining sensemaking related to information technology in organizations, Orlikowski and Gash identified frame content in a new context for social cognitive research; their empirical study identified three domains relevant to the adoption of a groupware technology (*nature of technology*, *technology strategy*, *technology-in-use*). They cautioned that these frames are not all-inclusive nor will they necessarily apply across contexts; instead, frames are context-sensitive and should not be defined *a priori*.

Integration and differentiation are two aspects of frame structure typically considered in organizational research, reflecting cognitive theories of information processing that underlie much social cognitive research (Walsh 1995). Structural studies at the group level of have relied primarily on cause mapping (Weick and Bourgon 1986), network analysis (Krackhardt 1987), or multidimensional scaling (MDS) techniques (Walsh et al. 1988) to derive knowledge structures (Walsh 1995, p. 301). Orlikowski and Gash were primarily concerned with the social rather than the cognitive influences of frames and thus gave little attention to these structural dimensions. Structure was briefly discussed in terms of domains of knowledge and the implications for frame congruence.

> We define the notion of congruence in technological frames as referring to the alignment of frames on key elements or categories. By congruent, we do not mean identical, but related in structure (i.e., common categories of frames) and content (i.e., similar values on the common categories)" (Orlikowski and Gash 1994, p. 180).

Organizational researchers have examined how frames develop, how they represent the information environment, and, particularly, how they are used in managerial sensemaking, decision-making, and action (Walsh 1995). Similar to many organizational social cognitive studies (cf. Bartunek 1984; Bartunek and Moch,1987, Daft and Weick 1984; Gioia 1986), Orlikwoski and Gash's central interest was the *use* of technological frames in sensemaking and interpretation related to IT and organizational change, particularly the consequences of frame use where frame incongruence exists. They suggested that recognition of incongruence and interventions aimed at overcoming it might result in frame change (i.e., frame alignment). Issues related to frame development and frame representation of an informational environment were not explicitly addressed.

3 DEVELOPMENTS IN TFR RESEARCH

We conducted a literature search to assess how Orlikowski and Gash's (1994) TFR framework has influenced IS research practice. To identify published reports employing TFR analysis, we conducted a citation search on ISI Social Science citation index. This search (conducted in July, 2003) produced 52 citations to Orlikowski and Gash's paper, but on further examination, only five publications actually conducted some form of TFR analysis. We also searched electronic journal index databases EBSCO and ABI/ INFORM (ProQuest) on the terms "technology frame" and "technological frame." Additionally, we searched for the same terms on the ACM Digital Library and eMISQ (via the AIS website www.aisnet.org). These searches produced three additional studies, all conference papers, which utilized TFR analysis, for a total of eight published research reports. We also searched the Internet but found, despite numerous "hits," only white papers or research descriptions. Since we could find no evidence these papers had been published in a refereed academic journal or conference proceedings, we did not include them in our analysis, but we do recognize them as evidence of the growing interest in TFR research. Table 1 lists the eight published studies we identified and summarizes key aspects of each; Orlikwoski and Gash's study is included for comparison.

Orlikowski and Gash's paper appears to have established a genre for TFR studies, which other researchers have utilized in a variety of organizational and technological settings. All of the TFR studies we identified involved in-depth case studies at one or more organizations, utilized interviews as a key source of data collection, and relied (the majority, exclusively) on qualitative data analysis methods. Of these publications, five reported on studies of technology development or implementation, whereas one used TFR analysis to assess training related to technology, one to assess policy formation associated with a new technology, and one to examine operational support. Similar to Orlikowski and Gash's study, six studies examined frames at a point in time (snapshot) although some historical reconstructions were attempted from interview data or written documents; one longitudinal study examined shifts in frames during an extended requirements study and another examined framing processes in a manner akin to SCOT analysis.[1] Finally, six of the eight studies were primarily concerned with frame incongruence, as was Orlikowski and Gash' study, while two examined frame change processes and consequences.

To assess how the TFR framework has been used and extended in these studies, we revisited Orlikowski and Gash's suggestions for future research and categorized the areas they identified as primarily theoretical development, methodological enhancement, or contributions to practice. We organize our discussion of subsequent TFR studies around each area.

3.1 Theoretical Developments

Defining additional frame domains and content: Orlikowski and Gash identified three general domains of TFRs, which were relevant in their empirical study (nature of technology, technology strategy, technology in use), but they cautioned that other frame domains might be relevant in different contexts. At first look, the TFR studies we reviewed suggest no common or general frame domains have emerged, that frame domains are in fact context-specific. Each paper used unique titles for frame domains,

[1]Our literature search revealed a number of studies utilizing the SCOT technological frame approach, or mentioning both SCOT and TFR, but pursuing SCOT analysis. Because we wanted to focus on Orlikowski and Gash's TFR framework, we did not include these studies. The McLoughlin et al. (2000) study incorporated sufficient aspects of both SCOT/TFR that we included it here

Table 1. Academic Publications Reporting on a TFR Study and Analysis

Research	Study Context	Frame Domains	User Groups
Orlikowski and Gash (1994)	Lotus notes implementation in large consulting firm • Early, ongoing implementation • Snapshot of frames	Nature of technology Technology strategy Technology-in-use	Technologists Users
Barrett (1999)	EDI service in the London insurance market • Longitudinal study of development, introduction, system use • Snapshot of frames derived from cultural assumptions	Nature of technological change Nature of business transactions Importance of market institutions	IT professionals Senior managers Brokers and underwriters (Users)
Davidson (2002)	Sales information system at an insurance company • Longitudinal study of requirements and pilot implementation • Frame change over time	IT delivery strategies IT capabilities and design Business value of IT IT-enabled work practices	System developers System constituents Executive
Gallivan (2001)	Reskilling IT professionals for client/ server technologies at a telecommunications company • Implementation • Snapshot of frames	Vision of reskilling/ type of change	Change managers IT professionals Miscellaneous others
Khoo (2001)	Design of peer review policies for the Digital Library for Earth System Education (DLESE) • Design, implementation • Snapshot of frames	N/A (mentions library models and peer review models)	"Communities of perception" Library that is digital Digital artifact as library
Lin and Cornford (2000)	Replacement of an e-mail system in a financial institution • Pre-implementation • Snapshot of frames	The nature of problems Requirements for the system Images of implementation Issues around use	Office IS group User group Management group
McLoughlin et al. (2000)	Implementation of team-based cellular manufacturing technology in three companies • Longitudinal during implementation • Process of frame closure and stabilization	Treated as unidimensional (a "frame" in the SCOT sense)	Various stakeholders

Research	Study Context	Frame Domains	User Groups
Sahay, et al. (1994)	Implementation of GIS technology • Implementation • Snapshot of frames	N/A (identified a detailed list of issues/problems with GIS technology)	Experts (IT) Users
Shaw et al. (1997)	Effectiveness of computer systems support at elevator company • Operations, ongoing support • Snapshot of frames	Technology in use Technology strategy Ownership of technology Nature of technology	MIS staff Management End-users

identified different numbers of domains, and described domain content in contextually bound terms. However, a close examination of the frame domains described in these research reports suggests some similarities across settings. For example, Barrett's (1999) "nature of technological change" and Davidson's (2002) "business value of IT" overlapped with Orlikowski and Gash's "technology strategy" category. Lin and Cornford's (2000) "requirement for the system" and Davidson's "IT capabilities and design" bore resemblance to the "nature of technology" domain. Aspects of the "technology in use" category were evident in "nature of business transactions" (Barrett 1999), "IT-enabled work practices" (Davidson 2002) and "issues around use" (Lin and Cornford 2000). This variation in naming, number, and precise definition of frame domains is consistent with the idiographic nature of TFR analysis.

Several studies introduced new frame domains. Barrett identified "importance of market institutions" within the organizational field he examined: "these cultural assumptions consider not only patterns of meaning or meaning structures held by these groups, but also how they are embedded in relations of power and norms of the organization" (pp. 3-4). By incorporating analysis of cultural elements, he addressed institutional and structural influences of TFRs on actors' interpretations and actions around IT adoption. Orlikowski and Gash briefly discussed the institutional implications of frames but did not develop this topic in their paper. Davidson highlighted "IT delivery strategies" as a salient domain during requirements determination and demonstrated possible conflicts between this domain and the "business value of IT" domain. Gallivan (2001) applied TFR analysis to re-skilling of IT workers. Linking TFRs to the type of organizational change envisioned, he discussed "visions of re-skilling" as first, second or third order changes.

Understanding of incongruence, frame convergence, and divergence: Orlikowski and Gash suggested the concept of incongruence would require further development to better understand where incongruence occurs, the degree of incongruence that might be tolerable, and the processes through which incongruence might increase or decrease. The studies in Table 1 offered a variety of insights on incongruence. Sahay et al. (1994) proposed an analytic method (multidimensional scaling or MDS) for quantitatively assessing the degree of incongruence in frames, which we will discuss shortly. However, Davidson's analysis illustrated that incongruence may vary across frame domains, relevant social groups, and over time as the salience of domains shift in response to change triggers. This socio-cognitive instability could frustrate methodological attempts to quantify and capture frame incongruence within domains across groups. Lin and Cornford drew on ANT theory to address incongruence, suggesting that frame alignment could be managed by one group's social translation of other group's frames toward their own frame.

Khoo (2001) posited that in some settings, the degree of frame difference might exceed incongruence and defined an incommensurate level of interaction as

> one in which the concepts of one frame cannot be understood in terms of the concepts of the other frame...while incongruence allows for the same data to be interpreted in different ways, incommensurability stresses that the data are in themselves differently constituted (p. 158).

Khoo drew parallels to Kuhnian paradigms, suggesting relevant social groups are defined by their technological frames and their interpretations of a technology; this approach is akin to SCOT analysis, although he does not reference SCOT theorists.

Frame change over time, change triggers: Similar to Orlikowski and Gash, six of the eight studies we identified reported the TFR analysis primarily as a snapshot of frames, or treated frames as static and unchanging over time. Thus, they did not address frame change or change triggers. Without longitudinal data, change processes are difficult to detect. Past frames can be reconstructed using retrospective interviews and written project archives, as Lin and Cornford apparently did to examine frame changes through social translation processes. Reconstructing past frames from interview data can be subject to retrospective sensemaking, that is, of reconstructing past frames inaccurately through the lens of current frames, and lacks the detailed access to the oral discourse that real- time field data collection allows.

Davidson's combination of retrospective interviews, project document reviews, interviews, and longitudinal on-site observations to reconstruct earlier frames and to trace current frames illustrated how various data collection approaches, in combination, allowed a longer timeframe for analysis of frame change than typically would be feasible for an on-site field study. In Davidson's study, frame change was evident in shifts in frame domain salience, triggered by reorganizations, environmental stimuli, or technology changes, rather than in changes in domain content. For example, the "IT-enabled work practices" domain was not salient to developers or users during design or installation, but when technology was actually implemented and used, this domain became highly salient to users, and latent incongruence between technologists' and users' frames emerged. Salience shifts were evident in the oral discourse as changes in project and design metaphors and narratives; these subtle discourse changes would be difficult to detect without ongoing field observation.

Frame tolerance or rigidity: Orlikowski and Gash's TFR framework did not explicitly consider structural aspects of frames that have been addressed in the organizational social cognition literature. Nor did the framework incorporate process concepts such as stabilization and closure from SCOT theory. The authors suggested further research would be needed to investigate frame tolerance and rigidity and their association with incongruence and its consequences. Because the majority of publications we examined produced a static snapshot of frames, these issues have not yet been thoroughly explored in TFR research. However, two studies, employing longitudinal designs, did consider frame structuring processes. McLoughlin et al. (2000) examined evidence of frame closure and stabilization in the process of adopting and adapting to a cell manufacturing technology and linked these frame development stages to political processes in the firms studied. Davidson's analysis demonstrated how lack of stability in frames and in the framing process destabilized the requirements determination process. She suggested that excessive stability (or fluidity) in frames could contribute to unwarranted escalation (or de-escalation) of commitment to an IT project despite contradictory contextual cues (Keil and Robey 1999). Neither study examined frame structure explicitly, leaving this area of the TFR framework relatively unexplored and undefined.

Complementing political analysis: Orlikowski and Gash maintained that the TFR concept

> allows us to explain and anticipate outcomes that are not captured by other perspectives, such as political or structural contingency models. For example, while a political perspective may explain particular outcomes that are due to the loss or gain of power by a group, it cannot explain contradictory outcomes due to different interpretations of a technology (p. 199).

Orlikowski and Gash's concern with interventions to resolve incongruence implies that some groups will change their frames to align with other groups' frames. While frame change may occur through education and experience, power and influence are likely to come into play. In fact, although the TFR studies we examined did highlight interpretive processes, political processes were tightly interwoven in the analysis, suggesting that it is difficult to isolate socio-cognitive from power/political processes (Markus and Bjørn-Andersen 1987). Davidson observed, based on her analysis, that "Interpretive power is brought to bear when dominant frames form the basis from which others develop their understanding of technology." Lin and Cornford drew on actor-network to explain how one group manipulated another's frames related to the new technology to gain support for their own technology selection. Barrett's TFR analysis at the organizational field level highlighted the power relations embedded in cultural norms and interpretations that form the basis of frames. And McLoughlin et al. explicitly linked their TFR analysis to political processes that were brought to bear to establish a dominant frame and to extend its influence.

3.2 Methodological Refinements

TFR analysis requires eliciting deeply held assumptions, expectations, and knowledge and assessing incongruence and inconsistencies in frames thus described. Orlikowski and Gash utilized interview data, which they analyzed using qualitative content analysis methods. These methods are well suited to the interpretive assumptions underlying the research framework but are labor intensive, difficult to specify, and hard to replicate. The eight publications we identified nonetheless all used similar, qualitative methods, albeit with some variations. In keeping with Orlikowski and Gash's call to attend to language and metaphors, Davidson examined the influence of metaphors and stories. Metaphors were cast as symbolic indicators of TFRs, as well as sensemaking devices that aided in interpretation. They also played a communicative role in sharing meaning among organization members. Organizational stories had similar functions. Shaw et al. (1997) used a qualitative approach to analyzing frames with quantitative measures to assess end-user satisfaction with support services. However, the authors' causal hypothesis that frame incongruence resulted in dissatisfaction was not analytically supported.

Moving further into the quantitative realm for TFR analysis, Sahay et al. used interviews to collect data and content analysis to identify a list of implementation problems with a geographical information system technology. They then used a dual-coding technique to collect quantitative rating data to compare how each user group viewed these problems. With this data, they quantitatively assessed inconsistencies, or incongruence, between groups over each of the specific problems. Their use of dual-coding integrated qualitative and quantitative approaches for identifying frame domains and content, providing a systematic method to quantify congruence/incongruence across frame domains and user groups. The goal of this paper was to illustrate the multidimensional scaling method in TFR analysis; the quantitative analysis was not tied to measurements of outcomes such as dissatisfaction with the system. Thus, the analytic advantages of quantitative measurement of TFR incongruence were not fully demonstrated.

3.3 Contributions to Practice

Orlikowski and Gash suggested three primary ways in which TFR analysis could benefit practice. First, frames could be identified and articulated early in technology projects to determine if incongruence exists among relevant social groups, allowing an opportunity for groups to reconcile differences. Lin and Cornford's study illustrated how organizational groups might recognize and adjust expectations in early project stages. Second, changes in frames could be tracked over time to determine if incongruence is developing or if problems due to interpretation are likely. Davidson and McLoughlin et al. demonstrated how longitudinal frame analysis might be conducted. Finally, frame analysis could be directed toward technologists, particularly when external consultants are involved in implementation efforts. Although technologists' TFRs were examined in most of the studies we reviewed, intra-group frame inconsistency or incongruence was not the focus of the research.

These studies illustrate the potential for application of TFR analysis in practice but they do not give direct evidence that this approach has actually affected practice. If it has, we should see its influence in two ways. One is if organization members, or external consultants who act as change agents or project managers, utilize TFR analysis in project intervention. We searched the Internet to seek out business press reports, but we found only academic research references. This does not mean that TFR analysis is not used in practice, only that there are no business press reports. A second avenue for influencing practice is through action research projects conducted jointly by academic researchers and practitioners. One of the studies we identified, the McLoughlin et. al. study of cell manufacturing technologies, was conducted as an action research project.

The other studies may have reported results to research sites, a practice advocated by Orlikowski and Gash, and in this way had a direct influence on practice at the research sites, but if so, the results are not reported in these papers.

4 REALIZING THE POTENTIAL OF TECHNOLOGICAL FRAMES

Technological frames of reference provide a flexible approach to explore interpretive issues in information technology design, implementation, and use. Our analysis outlines the contributions Orlikowski and Gash (1994) hoped their framework would make in IS research and highlights further developments in TFR analysis in subsequent studies. However, despite its apparent intuitive appeal, the TFR framework has not generated the level of research activity that theories such as the technology acceptance model (Davis 1989) have spawned. Why, if technological frames provide a necessary and useful perspective to study persistent issues with technology design, adoption, and organizational change, have few actual studies been performed and the results published? And how could interest in TFR analysis be energized? Building on our earlier analysis, we now consider three areas that pose significant challenges in TFR studies and suggest possible ways in which these challenges might be managed.

4.1 Is There a Better Way to Identify and Analyze TFRs?

Methodologically, TFR researchers have relied on organizational case studies and qualitative research methods (open-ended interview techniques, content analysis, grounded theory techniques) to elicit frame domains. These are rigorous methods to explain what has happened in a given set of circumstances, but they are not easily scaled for multiple case studies or survey research. Collecting and analyzing data on technological frames is labor intensive and time-consuming. Moreover, the researcher's efforts to define frames in one research setting cannot be easily amortized in subsequent studies in other research sites, due to idiographic definition of frames. Thus, it is difficult to build a comparative body of empirical findings, as has been done with positivist, quantitative research streams, for example, TAM studies using standard instruments for *ease of use* and *usefulness* constructs.

Could a quantitative method to measure and compare frames at individual and group levels, and to more precisely measure frame incongruence, increase research interest in TFR analysis? Would such an approach be consistent with epistemological assumptions about technological frames? Idiographic, qualitative approaches have been used in organizational social cognitive studies, but a number of organizational researchers have also used quantitative methods for frame assessment (for a summary, see Walsh 1995). This suggests that alternatives to qualitative content analysis methods may be possible.

Sahay et al. (1994) illustrated a mixed qualitative/quantitative method for TFR research—multidimensional scaling (MDS). They first used interviews to solicit dimensions from study subjects, then used content coding to develop the frame dimensions. Subjects were asked to assess and order each dimension using an MDS instrument; the results were subjected to statistical analysis to analyze frame similarities and differences.

This methodological approach was not inconsistent with the idiographic assumptions of TFR analysis: frames were defined within the research context. Furthermore, subjects were prompted to provide frame domains in their own words, although some structuring by researchers happened during content analysis. Used in this way, MDS can be consistent with interpretive assumptions underlying the TFR framework.

Repertory grid techniques are another set of quantitative methods that could be employed "to better understand how various information system stakeholders (users, managers, information system professionals) think about IT in their organizations" (Tan and Hunter 2002, pp. 39-40). At the core of these techniques is identification of elements, constructs, and links between them. Elements and constructs may be supplied to the research subjects, or elicited from them. An individual's construct system can then be subjected to a variety of analysis, from a basic content analysis to more sophisticated statistical techniques to quantify individual structural elements such as element distance, construct centrality, and element preference, as well as comparative structural measures such as cognitive differentiation, complexity, and integration (Tan and Hunter 2002, p. 49). This cognitive mapping tool could be used to elicit individuals' technological frames (cognitive maps), evaluate differences between frames, and monitor changes in frames over time. Tan and Hunter (2002) suggested that using the repertory grid technique would not only aid researchers studying cognitive processes, but would inform the practitioners who participate in exercises, helping them to recognize and possibly adjust tacit assumptions about IT.

We are cautiously optimistic that techniques such as MDS and repertory grid could make the task of TFR data analysis more manageable. By providing techniques to assess and compare frame structure, these techniques could facilitate theory development related to the structural aspects of frame incongruence, an area that has received little attention in TFR research thus far. On the other hand, while these techniques can be applied in ways that are consistent with the interpretive underpinnings of TFRs, wide-scale adoption of their use might be the first step down the slippery slope to a positivist straightjacket. The putative objectivity and scientific precision of quantitative, statistical methods would likely require similarly precise measurement methods for outcome variables, promote nomological rather than idiographic definitions of frame domains to promote cross-study comparisons, and inevitably lead to generalized cause-effect hypothesis about independent and dependent theoretical constructs. If such a transformation of TFR assumptions and methods were to occur, could the move from organizational field studies to controlled laboratory experiments be far behind?

We paint this scenario with tongue-in-cheek, and in the spirit of the 2004 IFIP 8.2 Manchester conference, to highlight differences in positivist and interpretive research assumptions. While Orlikowski and Gash did not rule out quantitative approaches to studying technological frames, their intent was to advance process theories of IT in organizations, in which humans' interpretations of IT play an analytic and theoretic role. Quantitative data collection and analysis are typically used to measure constructs and test variance theories hypotheses. Use of these methods might make the technological frames concept more appealing to positivist researchers and promote its wider use, but the studies would be quite different from the interpretive studies that have thus far characterized TFR research. Whether quantitative TFR studies (positivist or interpretive) would be more informative, or even as informative, to practitioners as well-organized qualitative feedback remains to be demonstrated.

4.2 Do We Really Understand Frame Incongruence?

Identifying frame incongruence and its consequences was a major contribution Orlikowski and Gash posited from TFR analysis. Subsequent TFR studies have focused almost exclusively on these issues. While incongruence is important, merely noting that different groups think differently about information technologies, and that differences can cause problems, is not very satisfying in the long run.[2] The TFR framework does not address frame structure or structural aspects of incongruence, and subsequent studies have skirted these questions. Moreover, the assumption that incongruence is necessarily detrimental has not been critically examined. Organizational social cognitive research provides some evidence that heterogeneity in frames may be inconsequential or even beneficial in some circumstances (Walsh et al. 1988).

How could the conceptual definition and implications of incongruence be refined in TFR research? Given the prevalence of problematic IS implementation projects, there should be no difficulty finding research sites where incongruence abounds. Cognitive mapping and analysis methods such as MDS or repertory grid techniques could be helpful to develop measures of and analyze incongruence. Longitudinal case studies would make it easier to monitor emerging incongruence, particularly if incongruence shifts over time, as Davidson (2002) suggests. However, longitudinal field research is a risky prospect, particularly since the phenomena of research interest may not develop in a particular case or during the timeframe of a field study. Short-term case studies reduce the time investment and allow some retrospective TFR analysis, but they are subject to the limitations we outlined earlier. This makes it difficult to assess frame change or change incongruence.

Perhaps more importantly, *post hoc* explanations of what went wrong are of limited value to practitioners dealing with the consequences of failed projects. One solution to this dilemma, which could help researchers assess how critical a role and what types of roles incongruence actually plays in organizational IS/IT activities, is action research. We will not attempt to review the benefits and limitations of action research here but merely point out that action research approaches are gaining legitimacy within IS academic circles and can enhance the practical relevance of IS research (Avison et al. 1999). In an action research project, TFR researchers would work with practitioners during IS/IT project activities to assess TFRs and to plan, monitor, and adjust interventions, such as tackling frame incongruence. By observing, taking action, and observing outcomes, researchers could elicit the phenomena they are interested in for research purposes and observe unfolding cause/effect processes. In these ways, they would be able to assess the effectiveness of TFR analysis for improving organizational outcomes. As Tan and Hunter (2002) suggested, practitioners who participate in these projects could benefit directly.

---

[2]Thanks to Sussan Gasson for this insight.

4.3 Is There More to Technological Frames than Inconguence?

While much theoretical work remains to be done on the concept of incongruence, we suggest the importance of frame incongruence in organizational IS/IT activities has been established in the IS literature by Orlikowski and Gash (1994) and subsequent studies, and that it is time to move on to new topics of theoretical interest. Walsh (1995) commented that when socio-cognitive structures matter *in use*, it becomes increasingly important to understand how they *develop* and *change* at individual, group and organizational levels. SCOT theory similarly highlights the importance of frame formation and change, particularly *stabilization* and *closure*. The TFR framework outlined by Orlikowski and Gash (1994) addressed these questions at a high level, noting only that individuals' frames develop and change through experience, education, and sometimes through planned interventions. The small body of TFR research we reviewed provided some additional insights, for example, how individuals' frames arise from cultural assumptions (Barrett 1999), are communicated or diffused though stories and metaphors (Davidson 2002), and change through social translation (Lin and Cornford 2000) and political processes (McLoughlin et al. 2000).

Sociologists like Meyer and Rowan (1977) and DiMaggio and Powell (1983) have emphasized the influence of external belief structures and cultural frames on actors. Powell and DiMaggio (1991) defined an *organizational field* as "those organizations that, in the aggregate, constitute a recognized area of institutional life: key suppliers, resources and product consumers, regulatory agencies, and other organizations that produce similar services or products" (pp. 64-65). Within this institutional domain, organizations share common meaning systems and interact primarily with each other. Scott commented that studies at the organizational field level of analysis are critical to understanding *institutional logics*—those widely held beliefs and socio-cultural structures that inform practice.

> Individuals do construct and continuously negotiate social reality in everyday life, but they do so within the context of wider, pre-existing cultural systems: symbolic frameworks, perceived to be both objective and external, that provide orientation and guidance (e.g., Goffman 1974) (Scott 2001, p. 41).

Technological frames of actors—individuals, groups, and organizations— within an organizational field are shaped by the field's institutional logics.

We suggest it could be interesting and valuable to examine frame development and change at the organizational field levels of analysis, as Barrett (1999) did with his study of the London insurance market. Information systems and technologies are increasingly utilized in interfirm relationships and transactions; individual and organizational sensemaking related to IT innovation happens not only within firm boundaries, but also among firms. At this level of TFR analysis, IT vendors, customers/user organizations, academics, consultants, and other stakeholder groups become relevant social groups, whose sensemaking, actions, and interactions shape technological artifacts (Bijker

1995). These frames (with regard to specific technological innovations) constitute the external belief structures that influence actors' frames within organizations. Understanding how frames develop, diffuse, and change within organizational fields could inform our understanding of technology-related change within industrial sectors as well as within particular organizations. Swanson and Ramiller (1997) have called for research at this level of analysis, using the term *organizing vision* to describe socially constructed interpretations of new technologies, the rationale for IT adoption, and shared expectations for how a new IT innovation will affect practices. This type of broad study of technological frames would move away from social cognitive research in the direction of SCOT research, where socio-technical theories such as actor-network theory (Latour 1987) could provide insights on frame change processes.

5 CONCLUDING THOUGHTS AND REFLECTIONS

Technological frames of reference research represents the type of IS endeavor that the 1984 Manchester IFIP 8.2 conference advocated: socially informed, organizationally based, methodologically rich, and relevant to practice. Although we did not find a wealth of TFR publications stemming from Orlikowski and Gash's (1994) seminal paper, we did find widespread interest in the TFR concept and a handful of significant studies, published in respected IS journals and outlets, conducted by a variety of IS researchers, many with ties to IFIP Working Group 8.2.

Despite its intuitive appeal and potential explanatory power, we may never see the level of research activity centered on the technological frames of reference concept as we have, for example, around the technology acceptance model. TFR studies as a genre are based on interpretive assumptions, labor-intensive qualitative research methods, organizational field studies, and, ideally, longitudinal research designs. Furthermore, TFR research is ideographic, making it difficult to leverage the researcher's investment in one study across multiple future studies. Much IS research is conducted from a positivist perspective, using controlled, quantitative data collection methods (laboratory studies, surveys) and statistical analysis techniques. "Publish or perish" tenure pressures in academia tend to favor these research approaches and productivity measured in number of publications. Given these institutional realities, there may be a limited market for the TFR concept as a basis of IT research. Adoption of mixed qualitative/quantitative methods such as multidimensional scaling or repertory grid analysis may increase the appeal of TFR analysis somewhat, among the broad audience of IT researchers.

Without the disciplinary innovations inspired and carried out by the IFIP 8.2 community over 20 years, notably greater acceptance of qualitative methods and interpretive approaches (Walsham 1995), technological frame research may never have taken root at all. Like the larger body of research carried out by IFIP 8.2 researchers and kindred spirits, TFR research has established a beachhead, set up camp, and conducted successful forays into the territory of the dominant paradigm. Much theoretical work remains to be done. In our analysis and discussion, we highlighted areas where TFR concepts and methods require further development and suggested three areas that we find promising for future study. We are hopeful that a similar analysis in another decade will reveal that the TFR research has continued to grow and to thrive.

REFERENCES

Avison, D.; Lau, F.; Myers, M.; and Nielsen, P. "Action Research," *Communications of the ACM* (42:1), 1999, pp. 94-98.

Barrett, M. I. "Challenges of EDI Adoption for Electronic Trading in the London Insurance Market," *European Journal of Information Systems* (8:1), 1999, pp. 1-15.

Bartunek, J. "Changing Interpretive Schemes and Organizational Restructuring: The Example of a Religious Order," *Administrative Science Quarterly* (29), 1984, pp. 355-372.

Bartunek, J., and Moch, M. "First Order, Second Order, and Third Order Change and Organization Development Interventions: A Cognitive Approach," *Journal of Applied Behavioral Science* (23), 1987, pp. 483-500.

Berger, P. L., and Luckmann, T. *The Social Construction of Reality*, New York: Anchor, 1967.

Bijker, W. *Of Bicycles, Bakelites, and Bulbs: Towards a Theory of Sociotechnical Change*, Cambridge, MA: MIT Press, 1995.

Bijker, W.; Hughes, T.; and Pinch, T. *The Social Construction of Technological Systems*, Cambridge, MA: MIT Press, 1987.

Boland, Jr., R. "Control, Causality and Information Systems Requirements," *Accounting, Organizations and Society* (4:4), 1979, pp. 259-272.

Boland, Jr., R. "The Process and Product of Systems Design," *Management Science* (24), 1978, pp. 887-898.

Bostrom, R., and Heinen, J. "MIS Problems and Failures: A Socio-Technical Perspective, Part I—The Causes," *MIS Quarterly* (1:1), 1977, pp. 17-32.

Daft, R., and Weick, K. "Toward a Model of Organizations as Interpretative Systems," *Academy of Management Review* (9:2), 1984, pp. 284-295.

Dagwell, R., and Weber, R. "System Designers' User Models: A Comparative Study and Methodological Critique," *Communications of the ACM* (26:11), 1983, pp. 987-997.

Davidson, E. "Technology Frames and Framing: A Socio-Cognitive Investigation of Requirements Determination" *MIS Quarterly* (26:4), 2002, pp. 329-358.

Davis, F. "Perceived Usefulness, Perceived Ease of Use, and User Acceptance of Information Technology," *MIS Quarterly* (13:3), 1989, pp. 318-340.

DiMaggio, P., and Powell, W. "The Iron Cage Revisited: Institutional Isomorphism and Collective Rationality in Organizational Fields," *American Sociological Review* (48), 1993, pp. 147-160.

Gallivan, M. "Meaning to Change: How Diverse Stakeholders Interpret Organizational Communication About Change Initiatives," *IEEE Transactions on Professional Communication* (44:4), 2001, pp. 243-266.

Ginzberg, M. "Early Diagnosis of MIS Implementation Failure: Promising Results and Unanswered Questions," *Management Science* (27:4), 1981, pp. 459-478.

Gioia, D. "Symbols, Scripts, and Sensemaking: Creating Meaning in the Organizational Experience," in H. Sims Jr., D. Gioia, and Associates (Eds.), *The Thinking Organization*, San Francisco: Jossey-Bass, 1986, pp. 49-74.

Goffman, I. *Frame Analysis*, New York: Harper, 1974.

Isabella, L. "Evolving Interpretations as a Change Unfolds: How Managers Construe Key Organizational Events," *Academy of Management Journal* (33:1), 1990, pp. 7-41.

Keil, M., and Robey, D. "Turning Around Troubled Software Projects: An Exploratory Study of the De-escalation of Commitment to Failing Courses of Action," *Journal of Management Information Systems* (15:4), 1999, pp. 63-88.

Khoo, M. "Community Design of DLESE's Collections Review Policy: A Technological Frames Analysis," in *Proceedings of the first ACM/IEEE-CS Joint Conference on Digital Libraries*, Roanoke, VI, June 24-28, 2001.

Krackhardt, D. "Cognitive Social Structures," *Social Networks* (9), 1987, pp. 109-134.

Latour, B. *Science in Action*, Boston: Harvard University Press, 1987.

Lin, A., and Cornford, T. "Framing Implementation Management," in W. J. Orlikowski, S. Ang, P. Weill, H. C. Krcmar, and J. I. DeGross (Eds.), *Proceedings of the 21st International Conference on Information Systems*, Brisbane, Australia, 2001, pp. 197-205.

Markus, M. L., and Bjørn-Andersen, N. "Power Over Users: Its Exercise by System Professionals," *Communications of the ACM* (30:6), 1987, pp. 498-504.

McLoughlin, I.; Badham, R.; and Couchman, P. "Rethinking A-Political Process in Technological Change: Socio-Technical Configurations and Frames," *Technology Analysis & Strategic Management* (12:1), 2000, pp. , 17-37.

Meyer, J., and Rowan B. "Institutional Organizations: Formal Structure as Myth and Ceremony," *The American Journal of Sociology* (83), 1977, pp. 340-363.

Orlikowski, W. "The Duality of Technology: Rethinking the Concept of Technology in Organizations," *Organization Science* (3:2), 1992, pp. 398-427.

Orlikowski, W., and Gash, D. "Technology Frames: Making Sense of Information Technology in Organizations," *ACM Transactions on Information Systems* (12:2), 1994, pp. 174-207.

Porac, J.; Thomas, H.; and Baden-Fuller, C. "Competitive Groups as Cognitive Communities: The Case of Scottish Knitwear Manufacturers," *Journal of Management Studies* (26:4), 1989, pp. 397-416.

Powell, W., and Dimaggio, P. J. *The New Institutionalism in Organizational Analysis*, Chicago: University of Chicago Press, 1991.

Sahay, S.; Palit, M.; and Robey, D. "A Relativist Approach to Studying the Social Construction of Information Technology," *European Journal of Information Systems* (3:4), 1994, pp. 248-258.

Scott, W. *Institutions and Organizations* (2nd ed.), Thousands Oaks, CA: Sage Publications, 2001.

Shaw, N.; Lee-Partridge, J.; and Ang, J. S. K. "Understanding End-User Computing through Technological Frames," in K. Kumar and J. I. DeGross (Eds.), *Proceedings of the 18th International Conference on Information Systems*, Atlanta, GA, 1997, pp. 453-459.

Swanson, E., and Ramiller, N. "The Organizing Vision in Information Systems Innovation," *Organization Science* (8:5), 1997, pp. 458-474.

Tan, F. and Hunter, M. "The Repertory Grid Technique: A Method for the Study of Cognition in Information Systems," *MIS Quarterly* (26:1), 2002, pp. 39-57.

Walsh, J. "Managerial and Organizational Cognition: Notes from a Trip Down Memory Lane," *Organization Science* (6:3), 1995, pp. 280-321.

Walsh, J., and Farley, L. "The Role of Negotiated Belief Structures in Strategy Making," *Journal of Management* (31), 1986, pp. 873-896.

Walsh, J.; Henderson, C.; and Deighton, J. "Negotiated Belief Structures and Decision Performance: An Empirical Investigation," *Organizational Behavior and Human Decision Processes* (42), 1988, pp. 194-216.

Walsham, G. "The Emergence of Interpretivism in IS Research," *Information Systems Research* (6), 1995, pp. 376-394.

Weick, K. *The Social Psychology of Organizing*, Reading, MA: Addison-Wesley, 1979.

Weick, K., and Bougon, M. "Organizations as Cognitive Maps: Charting Ways to Success and Failure," in H. Sims Jr., D. Gioia, and Associates (Eds.), *The Thinking Organization*, San Francisco: Jossey-Bass, 1986, pp. 102-135.

ABOUT THE AUTHORS

Elizabeth J. Davidson is an associate professor in the Department of Information Technology Management at the University of Hawaii. In her research, she has examined how organization members conceptualize opportunities for using information technology and negotiate IT development initiatives. She is currently studying teamwork in global IT support and the use of information technologies in scientific research. Elizabeth has extensive experience as a project manager and business analyst in the information technology industry in the United States. She can be reached at davidson@cba.hawaii.edu.

David Pai is a doctoral candidate in the Communications and Information Sciences program at the University of Hawaii. His research interests include adoption, implementation and use of information technologies, cognitive aspects of human-computer interaction, and symbiotic nature of IS research paradigms. David has a Master's degree in MIS, and prior to entering academia, he worked as lead engineer developing rocket engine systems.

27 REFLECTION ON DEVELOPMENT TECHNIQUES USING THE PSYCHOLOGY LITERATURE: Over Two Decades of Bias and Conceptual Blocks

Carl Adams
University of Portsmouth

David E. Avison
ESSEC Business School

Abstract Analysis and development techniques have played an important role in information systems, providing support for developers in structuring and directing tasks. They also provide cognitive support in collecting, collating, analyzing, and representing information about system requirements and attributes. However, by developing previous work further, in particular by classifying techniques into six generic types and transcribing these onto problem/solution space diagrams, this paper argues that by directing tasks and dictating what and how information is collected and represented, techniques can bias developers' understanding of system requirements and attributes.

The 1984 IFIPWG8.2 conference showed how IS can be informed by literature in our foundation disciplines. By drawing on the psychology literature to develop a classification of techniques, this paper shows some of the potential biases inherent in techniques. The classification is applied to many techniques which have contributed to development activity. Through an understanding of the conceptual blocks embedded in them, the paper hopes to inform practice about the selection and mixing of development techniques. More generally, the paper suggests a reexamination of our assumptions when undertaking IS development.

The techniques that limit problem and solution scope most and also provide the most cognitive and conceptual biases are the more formal, objective ones, and we recommend that less formal techniques are also used in practice. Parallels can be drawn with the movement encouraging the use of qualitative research approaches in IS research inspired by the 1984 conference. Research methods and techniques also provide support in the form of directed tasks, activities, guidance on data collection, analysis, and representation. Drawing on this parallel, it could be argued that quantitative, formal research approaches may introduce more conceptual biases than less formal qualitative, approaches, and that the latter should be used instead of, or at least alongside, quantitative approaches.

1 INTRODUCTION

Coincidently, around 20 years before the 1984 conference, another landmark event covering the use of computer applications took place in Manchester: the National Computing Centre (NCC) in the UK was formed with a mission to provide UK businesses and organizations with an understanding of how "to enable effective use of information technology" (NCC 2001). The NCC started as a UK government organization to collect and disseminate best practice in computing among its members and businesses. Part of this best practice covered the use of development methods and techniques, in particular by advocating the systems development life cycle or the waterfall model.

The best practice techniques included flow charts, decision tables, structure charts, functional charts, flow diagrams, physical layout charts, grid charts, string diagrams, relationship charts, and a variety of lists and formal documents and forms that systems analysts could use to help capture and represent system attributes. These techniques were the mainstay of traditional systems analysis as covered in many works at that time (Condon and 1974; Couger and Knapp 1974; Daniels and Yeates 1971; FitzGerald 1981; Gane 1979; Kindred 1980; Lee 1979, 1984; Lott 1971; Senn 1989; Wetherbe 1979; Yeates 1973). Many of these development techniques had some diagrammatic attributes as "it is usually easier and more intelligible to record identified procedures in diagrammatic rather than in narrative form" (NCC 1978, p. 116).

Techniques have evolved, and this evolution has similarities with other evolutionary processes. Many of the newer techniques were based on the earlier traditional techniques, bringing out the good practices which seemed most appropriate to address the current development environment of the time. However, sometimes newer techniques are little more than renaming of older reincarnations. Other techniques have evolved into something substantially different as the environment focus changes and best practice develops to address that. Some even claim to be completely new varieties. However, even new breeds usually have characteristics that share some visual or process attributes. For instance, the early flow charts and flow diagrams identified as good practice by the NCC have influenced the design of data flow diagrams.

These development techniques, and their reincarnations, have been very useful to analysts and developers over the decades: they provide structure, direct tasks, and provide cognitive tools with which systems analysts and developers can collect, collate, analyze, and represent information about system requirements and attributes. These techniques are usually seen in a positive light.

There is significant published research about the potential negative aspects of using methodologies and tools (along with that discussing their potential

> benefits). Techniques, on the other hand, are seen largely as benign, very often as simple aids to help carry out a task and are used in many methodologies. They might be seen as supporting the collection, collation, analysis, representation or communication of information about system requirements and attributes (or a combination of these) (Adams and Avison 2003, p. 203).

Techniques are generally seen as providing neutral support for developers, enabling them to learn about system requirements and deal with the many challenges of development. However, as Adams and Avison (2003) show, the leading processes and framing attributes of techniques can blinker systems analysts' understanding of system requirements and attributes: "Techniques may provide barriers to problem cognition rather than enlighten, and visual and linguistic influences may blinker perception in one direction" (p. 217).

Each development technique has its own unique set of characteristics distinguishing it from other techniques. Adams and Avison examined about 80 techniques and produced a two-dimensional classification based on visual/ language and paradigm/process influences.

> By classifying the characteristics of techniques, [we] indicate how different types of technique are likely to influence problem cognition, and in doing so [have] tried to map the framing effect of techniques. In truth, the discussion of these framing effects has shown that they are complex and interwoven. However, there are two continua that are striking. These are the degree of openness of the approach and the degree to which the technique is rule based. Our major classification places techniques, therefore, into one of four quadrants: prescriptive/ closed; prescriptive/open; non-prescriptive/closed; and non-prescriptive/open....It should be a concern to information systems developers that in our classification of over 80 techniques, over three-quarters are found in the closed rather than open paradigm, suggesting that the vast majority of techniques used in systems work have limited vision. Further, the majority are also prescriptive, limiting perception by restricting the user to a set of rules that should not be transgressed. When they are supported by tools, these rules may be enforced on their users. Even when no tools are in use, these rules are likely to be enforced as "standards" and transgressing them seen as not acceptable (Adams and Avison 2003, p. 220).

One implication from the earlier work is that techniques may limit the scope of understanding system requirements and attributes. The present paper develops this concept further, drawing on works from psychology, in particular by looking at conceptual blocks. To examine the scope of understanding, a classification of techniques is developed based on attributes that affect the problem/solution space.

The rest of this paper is structured as follows: First, works on conceptual blocks are examined. Then relevant attributes of techniques are identified and used to classify techniques. These groupings are considered in terms of problem and solution spaces. The classification is then considered in terms of conceptual block and biases. The paper concludes with a discussion about how this might inform practice.

Table 1. Four Main Areas of Conceptual Blocks (Adapted from Adams 1987)

Perceptual Blocks	**Cultural and Environmental Blocks**
• Seeing what you expect to see— stereotyping • Difficulty in isolating the problem • Tendency to delimit the problem area too closely (that is, imposing too many constraints on the problem) • Inability to see the problem from various viewpoints • Saturation (for example, disregarding seemingly unimportant or less visible aspects) • Failure to utilize all sensory inputs	• Cultural blocks could include: – Taboos – Seeing fantasy and reflection as a waste of time – Seeing reasons, logic, numbers, utility, practicality as *good*; and feeling, intuition, qualitative judgments as *bad* • Regarding tradition as preferable to change – Environmental blocks could include: – Lack of cooperation and trust among colleagues – Having an autocratic boss – Distractions
Emotional Blocks	**Intellectual and Expressive Blocks**
• Fear of taking risks • No appetite for chaos • Judging rather than generating ideas • Inability to incubate ideas • Lack of challenge and excessive zeal • Lack of imagination	• Use of appropriate cognitive tools and problem solving language

2 CONCEPTUAL BLOCKS

One of the key works on cognitive blocks is by Adams (1987), who identified four main types, which are represented in Table 1.

Each of these potential blocks can be applied to technique attributes. For instance, perceptual blocks may occur where a technique dictates activities that delimit the problem area, impose constraints, or represent a problem from a particular viewpoint. Emotional blocks may occur where a technique is prescriptive and limits the generation of new ideas. Cultural and environmental blocks may occur where there are strong paradigms or where there are strong traditional approaches to activity. Intellectual and expressive blocks may occur where the technique uses inappropriate tools to capture or represent system attributes.

Groth and Peters (1999), focusing on creative, innovative, and lateral thinking perspectives, examined barriers to creative problem solving among managers. They identify a long list of perceived barriers to creativity including fear of failure, lack of confidence, environmental factors, fear of success and its consequences, fear of challenge, routines, habits, paradigms, preconceived notions, rules, standards, tunnel sight, internal barriers, structure, socialization, external barriers, money, rebellion, health and energy, mood, attitudes, desire, and time. They grouped the perceived barriers into self-imposed, professional environment, and environmentally imposed categories. Fear of some sort seems to be a predominant barrier. This seems to imply that there will be a natural tendency to use familiar techniques, not to move outside the scope of the technique or the expected practice of the environment.

3 CHARACTERISTICS OF TECHNIQUES

To examine further how techniques may introduce conceptual blocks, we need to examine the different attributes of techniques in greater detail. Adams and Avison (2003) collated a list of technique attributes that may contribute to framing system requirements, including

- Visual attributes, for example, visual representation and structure of technique output
- Linguistic attributes, for example, terminology and language used—not just English language, but also others such as mathematical and diagrammatical
- Genealogy attributes, for example, history of techniques, related techniques
- Process and procedure attributes, for example, description and order of tasks
- People attributes, for example, roles of people involved in tasks
- Goal attributes, for example, aims and focus of techniques
- Paradigm attributes, for example, discourse, taken–for-granted elements, cultural elements
- Biases, for example, particular emphasis, items to consider, items not considered
- Technique or application-specific attributes

Some of these characteristics of techniques are explicit, for instance where a particular visual representation is prescribed. Other characteristics might be less obvious, such as the underlying paradigm. Many of the characteristics are likely to be interwoven, for instance the visual and linguistic attributes might be closely aligned with the genealogy of a technique.

To bring together particular attributes of techniques we draw upon the approach used by Waddington (1977) who discusses natural attributes for grouping items. Waddington describes our basic, or natural, methods of ordering complex systems, the most basic of which relies on identifying simple relationships, hierarchies, patterns, and similarities of characteristics. Humans are good at looking for patterns and structure to understand and interpret complex systems: Humans automatically group attributes of a complex system into simpler constructs. The grouping process itself is a little subjective and different people may naturally group techniques differently. For examining the complex array of differing techniques, the focus was on attributes that may impact problem understanding. Initially, the visual representation, process, and linguistic attributes seemed the most appropriate for examining possible blocks to problem understanding (i.e., by defining problem representation, defining processes and language to describe the problem arena). The grouping process started with examining the visual attributes (i.e., presentation) of techniques bringing together techniques with similar presentation characteristics. Then techniques with similar process attributes were collated. Four main groups emerged, two of them based on visual presentation (diagrams or lists/matrices), and two of them based on process and language, one using problem reducing terms and the other using brainstorming terms. The remaining techniques fell into two further groups one looking at different future scenarios and another dealing specifically with conflict situations. The resulting classifications are

(1) brainstorming approaches, (2) relationship approaches, (3) scenario approaches, (4) reductionist approaches, (5) matrix approaches, and (6) conflict approaches. The grouping process had a few iterations; for instance, initially the relationship approach was split into two groups, one using more formal presentations than the other. An earlier iteration of classification can be seen in Adams and Avison (2002). This grouping process provides a starting point for examining underlying structures in techniques and attributes that may affect problem and solution scoping for a technique.

4 REPRESENTING PROBLEM AND SOLUTION SPACES ADDRESSED BY TECHNIQUES

"The future, it has been suggested, is a combination of the known and the unknowable. The proportion of the latter tends to rise as the time-scale extends" (Rosenhead 1992, p. 194). Techniques may be used to identify likely or desirable futures and reduce these, increasing future possibilities to a manageable level. To represent the scoping of how problems are addressed by a technique we use problem and solution space diagrams, or "trumpet of uncertainty" diagrams (Rosenhead 1992, p. 200). The problem and solution space diagrams look like a trumpet or cone where at the smaller end is the current situation (problem, requirements, or our knowledge of those), and as time progresses the total number of all possible situations expands exponentially, represented by a wide bell shape at the other end. Applying a technique is effectively trying to predict the future situation out of all possible future situations. This is represented in Figure 1. Problem and solution diagrams are then applied to each of the six groups.

The brainstorming approaches (Figure 2) perform similar brainstorming and lateral thinking activities. Representations for brainstorming techniques vary, some containing lists and others diagrams (e.g., a mind map). Brainstorming is often associated with Edward de Bono (1970, 1977), who covered it as one of a set of lateral thinking techniques, although there are earlier claims (e.g., Clark 1958). The problem or solution scope is typically expanded to a wide range of areas, but only considered briefly and sometimes without focus, and is sometimes seen as a scattergun approach. There is usually some further activity in selecting some of these areas for further investigation or analysis. A possible representation of the problem solution space for brainstorming approaches is given in Figure 2.

Relationship approaches (Figure 3) are techniques where diagrams are central. Relationship approaches seem to have defined structures and representational relationships between component parts. Included in this grouping are network diagrams (for example, Bicheno 1994) and cognitive mapping (Eden 1992). The problem or solution scope seems to be limited to that defined by the diagram structures.

Scenario approaches (Figure 4) also have linguistic attributes based around future scenarios. There may be some brainstorming or similar activity to identify what the scenarios could be, then there is a deeper investigation of those. In others examples there are given scenarios to work through. Representation in these approaches can vary from lists of items to diagrams. The problem or solution scope is selectively widened; however, the widened areas are examined in depth.

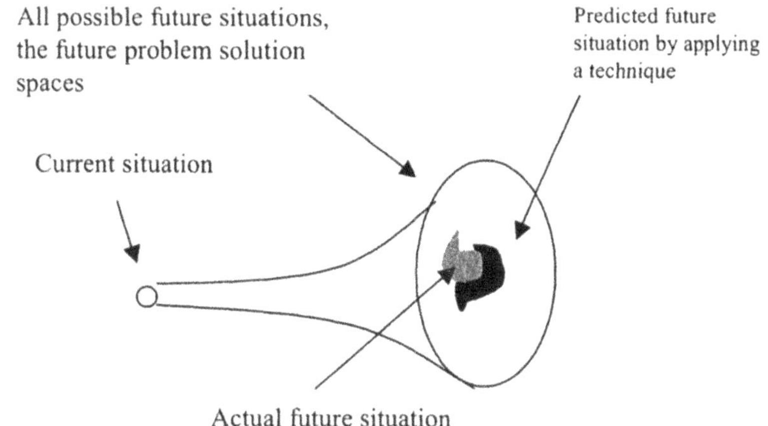

Figure 1. Representation of Problem and Solution Spaces

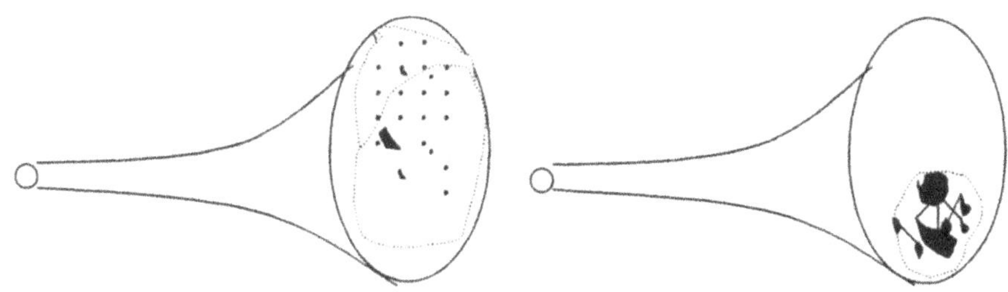

Figure 2. Problem and Solution Spaces for Brainstorming Group

Figure 3. Problem and Solution Spaces for Relationship Group

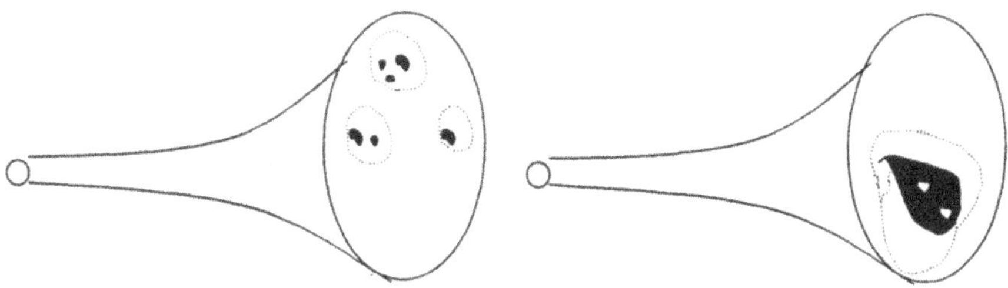

Figure 4. Problem and Solution Spaces for Scenario group

Figure 5. Problem and Solution Spaces for Reductionist group

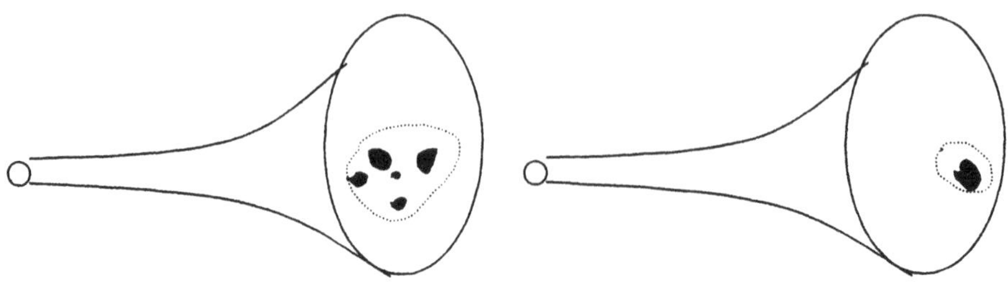

Figure 6. Problem and Solution Spaces for Matrix Group

Figure 7. Problem and Solution Spaces for Conflict Group

Reductionist approaches (Figure 5) are techniques that share similar linguistic and terminology attributes, in particular those reducing the problem area into smaller component parts. They also seem to have similar visual attributes based on well-defined structures. The problem or solution scope of these techniques may be limited and directed by the underlying reductionist paradigms (e.g., the solution will materialize by going through the processes defined by the technique).

Matrix approaches (Figure 6) are techniques that present information in lists, matrices or tables. There may be some activity involving the identification of items to include in the lists and activity in identifying attributes of those items. For instance, a list of requirements may be compared or analyzed against a list of factors or constraints. The problem or solution scope may be constrained by the items included within the list or matrix. It will, therefore, be difficult to consider items or issues not on the list.

Conflict approaches (Figure 7) have linguistic and terminology attributes revolving around conflict and conflict resolution. There is usually some activity to identify and represent attributes of the conflict from the main stakeholders' perspective. This would then be used to develop appropriate resolution strategies. The problem solution scope is likely to be constrained to the perceived problem vista from each of the main stakeholders.

The allocation of techniques to particular groups involves a certain amount of subjectivity, and others may put some techniques in another class. In addition, a technique may have attributes placing it in more than one group (e.g., a scenario technique may have brainstorming attributes and a brainstorming technique may have some relationship attributes). In addition, if a technique can be classed in more than one grouping, this will give us more information about the technique. Indeed, some techniques may be a collection of other techniques. The grouping portrays a message or underlying structure about the scoping of a problem and the proposed solution that a technique considers. The problem and solution space figures provide a vista on the scoping of techniques. This provides a basis on which to consider wider cognitive and conceptual blocks. To understand these potential blocks further we will draw on additional literature, again much of it from psychology.

Perceptual blocks may be embedded in techniques that *dictate* activities and define a particular scope. This would be relevant to relationship and matrix approaches on a visual perspective or language and paradigm perspective for reductionist approaches. Emotional blocks may be embedded in techniques which limit scope for innovation, changes to processes and rules, as well as limiting scope for representation. Again, relationship and reductionist approaches fit here. However, on a wider interpretation of the emotional blocks covered by Adams (1987) and closer to the internal barriers, attitudes, habits, and self-imposed attributes discussed by Groth and Peters (1999), people are likely to prefer some techniques and tasks more than others. Some people may like the prescriptive support of reductionist techniques, while others may be more at home with less prescriptive brainstorming activity. Conversely, people will be less at home with less preferred techniques as well as new, untried techniques. From this perspective, emotional blocks will be relevant to all of the categories of techniques. Cultural and environmental blocks may be embedded in techniques that have strong paradigms and rules, such as reductionist approaches. Intellectual and expressive blocks may be embedded in techniques where the tools are inappropriate for capturing, representing, or analyzing system attributes. This could be appropriate to all of the categories of techniques depending on the task context.

For further understanding of potential blocks to understanding, we need to examine other works. The following potential cognitive and conceptual blocks are slightly arbitrarily grouped into three areas: visual representational attributes, language and discourse attributes, and individual orientation attributes.

5 VISUAL REPRESENTATIONAL ATTRIBUTES

5.1 Framing Influences

We are initially drawn to the framing effect (Tversky and Kahneman 1973, 1974, 1981), which states that different representations of essentially the same situation will result in different decisions, choices, or understanding. Some of the earliest and most influential movements of cognitive psychology, that of the *Gestalt psychologists* (Gillam 1992; Honderich 1995; Wetheimer 1923), contributed to our understanding of the framing effect.

> In Gestalt theory, problem representation rests at the heart of problem solving—the way you look at the problem can affect the way you solve the problem....The Gestalt approach to problem solving has fostered numerous attempts to improve creative problem solving by helping people represent problems in useful ways (Mayer 1996).

The key element here is that the way in which a problem is represented will affect the understanding of the problem. The implication for techniques is that the visual, linguistic, and other representation imposed by a technique will impact understanding on system requirements. The Gestalt movement in cognitive psychology has a (comparatively) long history and has had a big impact on the understanding of problem solving. The movement has spawned various strands of techniques such as lateral thinking and some other creative techniques. Gillam (1992) gives a more current examination of Gestalt theorists and works, particularly in the area of perceptual grouping, that is, how people understand and group items. Gillam shows that perceptual coherence (that is, grouping) is not the outcome of a single process (as originally proposed by Gestalt theory) but may be best regarded as a domain of perception (that is, the grouping process is likely to be more complex, influenced by context and other aspects).

5.2 Category Inclusion

A prescriptive structure is likely to exert influence on problem cognition. For instance, hierarchy and tree structures are likely to exert some influence on problem cognition by binding attributes together (for example, on the same part of a tree structure) and limiting items to the confines of the imposed structure. In cognitive psychology this is known as *category inclusion* (Anderson and Bower 1973). "One enduring principle of rational inference is category inclusion: categories inherit the properties of their superordinates" (Sloman 1998). The implication is that techniques dictating hierarchical structures will force a (self-perpetuating) category inclusion bias. The reverse is also likely to be true: an element in one branch of a hierarchical structure will automatically have different properties to an element in another branch of the hierarchical structure. For instance, take a functional breakdown of an organization (such as that described in Yourdon and Constantine 1979). One might conclude from category inclusion that a task in an accounting department will always be different to a task in a personnel department, which clearly may not be the case as both departments will have some similar tasks, such as ordering the stationary.

However, this category inclusion is not universally the case. As Sloman (1998) shows us, the process is likely to be more complex. The initial premise surrounding a situation is likely to be related to the underlying paradigm. Dictating a hierarchical structure in itself may not result in category inclusion biases. However, coupled with an underlying closed paradigm, it may more likely result in category inclusion biases. Along the same theme are proximity influences and biases. The understanding of items can be influenced by the characteristics of other items represented in close proximity.

5.3 Order Influences

Perceptual processing is profoundly influenced by the sequence in which the information is presented and the relational constructs of information (Mulligan 1999). The order and number of items in a list will influence how people understand (and recall) items and how people categorize them. The initial placing of information items, either in lists or diagrams, will affect understanding of the importance, location and relationship of those items.

6 LANGUAGE AND DISCOURSE ATTRIBUTES

6.1 Discourse Influences

The discourse and language used to describe a problem is likely to play a role in problem understanding. Adams (1987) discusses various types of languages of thought used in problem representing and solving. People can view problems using mathematical symbols and notation, drawings, charts, pictures, and a variety of natural verbal language constructs such as analogies and scenarios. Further, people switch consciously and unconsciously between different modes of thought using the different languages of thought. The information systems development environment is awash with technical jargon and language constructs. In addition, different application areas have their own set of jargon and specific language. Individual techniques have their own peculiar discourse consisting of particular language, jargon, and taken-for-granted constructs, all of which may exert influence. For instance, the initial discourse used affects understanding of a problem situation, particularly in resolving ambiguities (Martin et al. 1999) by setting the context with which to consider the situation.

The initial discourse surrounding requirement identification may be providing leading questions and processes. In addition, cognitive psychology literature indicates that there will be a different weight attached to *normative* than to *descriptive* representations and results of techniques. The basis for this is the understanding/ acceptance principle, which states that "the deeper the understanding of a normative principle, the greater the tendency to respond in accordance with it" (Stanovich and West 1999).

Language aspects highlight another set of possible influences, that of communication between different groups of people (such as between analysts and users). Differences of perspective between different groups of people in the development process has been discussed within the IS field under the heading of softer aspects or as the organizational or people issues (for example, Checkland 1981; Lederer and Nath 1991; Sauer 1993). Identifying differences and inconsistencies can be classed as a useful task in identifying and dealing with requirements (Gabbay and Hunter 1991). From cognitive psychology there are also other considerations. Teigen's (1988) work on the language of uncertainty shows that there is often more than the literal meaning implied in the use of a term, such as contextual and relational information or some underlying other message. The use of language is very complex. The implications are that even if a technique prescribes unambiguous language and constructs, there may well be considerable ambiguity when it is used.

6.2 Goal Influences

Goal or aim aspects also profoundly influence problem understanding by providing direction and focus for knowledge compilation (Anderson 1987). Goals influence the strategies that people undertake to acquire information and solve problems. Further, when there is a lack of clear goals, people are likely to take support from a particular learning strategy, which will typically be prescribed by the technique.

> The role of general methods in learning varies with both the specificity of the problem solver's goal and the systematicity of the strategies used for testing hypothesis about rules. In the absence of a specific goal people are more likely to use a rule-induced learning strategy, whereas provision of a specific goal fosters use of difference reduction, which tends to be a non-rule-induction strategy (Vollmeyer et al. 1996).

The implications are that techniques with clear task goals will impact the focus and form of information collection (for example, what information is required and where it comes from, along with what information is not deemed relevant) and how the information is to be processed. Further, if there are no clear goals, then people are likely to rely more heavily on the learning method prescribed by the technique.

7 INDIVIDUAL ORIENTATION ATTRIBUTES

Some of the individual orientation attributes are similar to the emotional blocks identified earlier.

7.1 Preference Influences

There are also likely to be individual preferences, and corresponding biases, for some techniques or specific tasks within techniques. As Puccio (1999) suggests, "the creative problem solving process involves a series of distinct mental operations," that is, in collecting information, defining problems, generating ideas, and developing solutions, people will express different degrees of preference for these various operations. In the information systems domain, Couger (1995) has noted similar preferences: "It is not surprising that technical people are predisposed towards the use of analytical techniques and behaviorally orientated people towards the intuitive techniques."

In addition, there may be some biases between group and individual tasks, a point taken up by Poole (1990), who notes that group interaction on such tasks is likely to be complex with many influences. The theme is also taken up by Kerr et al. (1996), who investigated whether individual activities are better than group activities (that is, have less errors or less bias), but their findings are inconclusive.

> the relative magnitude of individual and group bias depends upon several factors, including group size, initial individual judgement, the magnitude of bias among individuals, the type of bias, and most of all, the group-judgment process....It is concluded that there can be no simple answer to the question "which are more biased, individuals or groups?" (Kerr et al. 1996).

To address the potential individual or group biases, many authors suggesting techniques recommend some consideration of the make-up of the different groups using them (for example, Bicheno 1994; Couger et al. 1993), although they give limited practical guidance on doing so.

7.2 Functional Fixedness

The Gestalt psychologists also indicate a potentially strong influence on problem understanding, that of functional fixedness "prior experience can have negative effects in certain new problem-solving situations...the idea that the reproductive application of past habits inhibits problem solving" (Mayer 1996). The implication is that habits learned using previous techniques would bias the application of new techniques. This is a particular concern in view of the evolution of techniques over the last few decades, and implies that new techniques will have embedded biases as will analysts using a new technique, who will be biased in some way by the habits learned in previously used techniques.

8 APPLICATION

As the discussion shows, attributes influencing cognition are likely to be complex and involved. There is much potential for biases based on language, ordering of items, leading processes and questions, and a variety of other attributes of techniques. In addition, the earlier discussion on the evolution, use, and adaptation of techniques indicates that there is likely to be considerable variation in applying a technique. However, we can collate the identified potential cognitive blocks and apply them to the different groups of techniques. This provides a picture of likely biases and cognitive and conceptual blocks for each group. For instance, reductionist techniques are likely to dictate many aspects of the development environment including the processes, representation, and even underlying paradigm. Consequently, such techniques are likely to exert cognitive and conceptual blocks in representation, language, and culture or environment. In addition, the lack of flexibility may also increase emotional block. These potential blocks are summarized for each grouping in Table 2.

In the appendix, we categorize around 70 techniques into the six types listed in Table 2. It is suggested that developers might ensure that they use techniques from different categories, even for a similar task, to reduce the risk of cognitive and conceptual biases due to framing and scoping influences. In addition, there are likely to be further influences, such as individual biases toward different types of techniques (or tasks within them), negative versus positive framing, and a range of perceptual blocks. It is interesting to note that reductionist approaches, with prescriptive processes and paradigms, seem to have the highest potential for introducing cognitive blocks and biases, particularly limiting scope within the mindset of the paradigm. The matrix and relationship approaches seem have high potential to limit scope by the lists and diagrams used. The brainstorming and lateral thinking approaches seemed to have the least potential for introducing biases; however, they also provided the least amount of support in collating and representing system attributes. It is interesting to note that the largest groups of techniques in the appendix are the reductionist, relationship, and matrix approaches, which seems to indicate that biases and cognitive blocking is firmly embedded in the analyst tool box.

9 CONCLUSION

Regarding implications for research, there is clearly a need to investigate the phenomenon of blocks and biases in further understanding. Potential fruitful areas could be in mixing techniques to reduce bias and examining biases in practice.

In addition it is interesting to note that there are parallels between *research* methods and techniques and *development* techniques: Research methods and techniques also provide researchers with guidance and support to do their research by directing tasks, dicta-

Table 2. Potential for Cognitive and Conceptual Blocks and Biases

	Potential Cognitive and Conceptual Blocks						
Group	**Perceptual Blocks**	**Emotional Blocks**	**Cultural and Environmental Blocks**	**Intellectual and Expressive Blocks**	**Visual Representational Attributes**	**Language and Discourse Attributes**	**Individual Orientation Attributes**
Brain-storming	Low	Low/ Medium *	Low	Low/ Medium/ High #	Low/ Medium	Low/ Medium ##	Low/ Medium/ High **
Relation-ship	High	High/ Medium *	High	Low/ Medium/ High #	High	Medium/ High	Low/ Medium/ High **
Scenario	Low-Medium	Medium *	Medium	Low/ Medium/ High #	Medium	Medium	Low/ Medium/ High **
Reduc-tionist	High	High (lack of task flexibility) *	Very High	Low/ Medium/ High #	High	High	Low/ Medium/ High **
Matrix	High	High/ Medium *	High	Low/ Medium/ High #	High	Medium	Low/ Medium/ High **
Conflict	Low-Medium	Medium *	Medium	Low/ Medium/ High #	Medium	Medium/ High	Low/ Medium/ High **

Table key:
*Or Low/Medium/High depending on individual's orientation to tasks
**Depends on individual's orientation to tasks and previous experience
#Applying the most suitable cognitive tools for a given task is context dependent
##Could be high if there is a lack of clear goals

ting what information to collect, how to collate and analyze it, and how to represent it. These are likely to have a similar set of biases to the development techniques covered in this paper. There has been a predominance of quantitative objective research approaches just as there is a predominance of objective type development techniques. Our research suggests that just as the objective techniques provide the greatest scope for cognitive and conceptual biases and blinkering, so too do the objective research approaches.

For individual research projects, several questions are raised on what biases are introduced when using particular research techniques. Does a particular research technique blinker the researcher's perception of the phenomenon being investigated, and if so how? Should the selection of research techniques also include consideration of reducing possible biases in understanding the phenomenon being investigated? For example, in this research, development techniques were initially categorizing under a natural approach indicated by Waddington (1977), however, a different approach to classifying techniques may have highlighted other technique attributes and consequently other possible types of bias.

This paper has contended that techniques influence problem understanding during information systems development. The influences can be considered under certain representational attributes. The psychology literature indicates how these attributes are likely to affect problem understanding. These biases become more prominent when one considers that the results of a technique (that is, diagrams, tables, etc.) may be used by different groups of people than the group that produced them (for example, analysts may produce some charts and tables which will be used by designers) and this is likely to perpetuate such biases throughout the development process.

Techniques have evolved with best practice being developed to address the needs of the business and development environment. However, this paper has shown that techniques are not neutral. They have the potential to introduce bias and cognitive and conceptual blocks, blinkering the analyst's and developer's view of system requirements and attributes. These non-neutral attributes of techniques have also evolved along with the best practice elements.

One of the results of this study is that there is a predominance of techniques with prescriptive representational structures (i.e., matrix approaches and relationship approaches) and/or prescriptive reductionist paradigms, supporting the findings of Adams and Avison (2003), who found a predominance of techniques in the closed objective paradigm. The objective techniques seem to have the biggest potential for reducing problem and solution scope and introducing bias.

There are implications for research, both in topic areas requiring further investigation and in possible biases inherent in research techniques. This is of concern given the predominance of quantitative objective research approaches, just as there is a predominance of objective type development techniques, which seem to offer the greatest scope for cognitive and conceptual biases.

REFERENCES

Adams, C., and Avison, D. E. "Dangers Inherent in the Use of Techniques: Identifying Framing Influences," *IT and People* (16:2), 2003, pp. 203-234.

Adams, C., and Avison, D. E. "Macro Analysis of Techniques to Deal with Uncertainty in Information Systems Development: Mapping Representational Framing Influences," in D. Bustard, W. Liu, and R. Sterritt (Eds.), *Soft-Ware 2002: Computing in an Imperfect World*, Berlin: Springer-Verlag, 2002, pp. 280-299.

Adams, J. Conceptual Blockbusting: A Guide to Better Ideas, Harmondsworth, England: Penguin Books, 1987.

Anderson, J. R. "Skill Acquisition: Compilation of Weak-Method Problem Solutions," *Psychological Review* (94), 1987, pp. 192-210.

Anderson, J. R., and Bower, G. H. *Human Associative Memory*, Washington, DC: Winston, 1973.

Andrew, J., and Moss, T. *Reliability and Risk Assessment*, Harlow, England: Longman Scientific and Technical, 1993.

Avison, D. E., and Fitzgerald, G. *Information Systems Development: Methodologies, Techniques and Tools* (3rd ed.), London: McGraw-Hill, 2003.

Batiste, J., and Jung, J. "Requirements, Needs, and Priorities: A Structured Approach for Determining MIS Project Definition," *MIS Quarterly* (8:4), 1984, pp. 215-227.

Beard, R. *Risk Theory*, London: Methuen, 1969.

Bennet, P.; Cropper, S.; and Huxham, C. "Modeling Interactive Decision: The Hypergame Focus," in J. Rosenhead (Ed.), *Rational Analysis for a Problematic World: Problem Structuring Methods for Complexity, Uncertainty and Conflict*, Chichester, England: John Wiley and Sons, 1992, pp. 283-314.

Bicheno, J. *The Quality 50: A Guide to Gurus, Tools, Wastes, Techniques and Systems*, Buckingham, England: PICSIE Books, 1994.

Camden. *Guidelines to the Establishment of Hazards Analysis and Critical Control Points*, Technical Manual 19, Camden Food and Drink Association, Glouchestershire, UK, 1987.

Carley, M. *Rational Techniques in Policy Analysis*, Aldershot, England: Gower Publishing, 1980.

Chapman, C. B. "Model and Situation Specific OR Methods: Risk Engineering Reliability Analysis of LNG Facility," *Journal of Operational Research* (35:1), 1984, page 27.

Chapman, C. B. "Risk Management," *Project Management* (18), 1990, p. 5.

Checkland, P. *Systems Thinking, Systems Practice*, Chichester, England: John Wiley, 1981.

Clark, C. *Brainstorming: The Dynamic New Way to Create Successful Ideas*, Garden City, NY: Doubleday and Co, Inc., 1958.

Condon, R. J. *Data Processing Systems Analysis and Design*, Reston, VA: Reston Publishing Company, 1974.

Couger, D. *Creative Problem Solving and Opportunity*, Danvers, MA: Boyd and Fraser Publishing, 1995.

Couger, D.; Higgins, L.; and McIntyre, S. "(Un)Structured Creativity in Information Systems Organizations," *MIS Quarterly* (18:4), December 1993, pp. 375-397.

Couger, J. D., and Knapp R. W. (Eds.). *System Analysis Techniques*, New York: John Wiley & Sons, 1974.

Daniels, A., and Yeates, D. *Basic Training in Systems Analysis* (2nd ed.), London: Pitman Publishing, 1971.

de Bono, E. *Lateral Thinking: A Textbook of Creativity*, Harmondsworth, England: Penguin Books Ltd., 1977.

de Bono, E. *The Mechanism of Mind*, Harmondsworth, England: Penguin Books Ltd., 1969.

de Bono, E. *The Use of Lateral Thinking*, Harmondsworth, England: Penguin Books Ltd., 1970.

Eden, C. "Using Cognitive Mapping for Strategic Options Development and Analysis (SODA)," in J. Rosenhead (Ed.), *Rational Analysis for a Problematic World: Problem Structuring Methods for Complexity, Uncertainty and Conflict*, Chichester, England: John Wiley and Sons, 1992, pp. 21-42.

Fagan, M. "Design and Code Inspections to Reduce Errors in Program Development," *IBM Systems Journal* (15:3), 1976, pp. 182-211.

FitzGerald, J. *Fundamentals of Systems Analysis* (2nd ed.), Chichester, England: John Wiley and Sons, 1981.

Flynn, D. J. *Information Systems Requirements: Determination and Analysis*, London: McGraw-Hill, 1992.

Fortune, J., and Peters, G. *Learning from Failure: The Systems Approach*, Chichester England: John Wiley and Wiley, 1995.

Friend, J. K., and Hickling, A. *Planning Under Pressure: The Strategic Choice Approach*, Oxford: Butterworth-Heinemann, 1997.

Gabbay, D., and Hunter, A. "Making Inconsistency Respectable: A Logical Framework for Inconsistency Reasoning," *Lecture Notes in Artificial Intelligence* (535), 1991, London: Imperial College, London, 1991, pp. 19-32.

Gane, C. P. *Structured Systems Analysis: Tools and Techniques*, London: Prentice-Hall, 1979.

Gajpal, P. P.; Ganesh, L. S.; and Rajendran, C. "Criticality Analysis of Spare Parts Using the Analytic Hierarchy Process," *International Journal of Production Economics* (35), 1994, pp. 293-297.

Geschka, H. "Creativity Techniques in Germany," *Creativity and Innovation Management* (5:2), 1996, pp. 87-92.

Gillam, B "The Status of Perceptual Grouping: 70 Years after Wertheimer," *Australian Journal of Psychology* (44:3), 1992, pp. 157-162.

Grey, S. *Practical Risk Assessment for Project Management*, Chichester, England: John Wiley and Sons, 1995.

Groth, J., and Peters, J. "What Blocks Creativity? A Managerial Perspective," *Creativity and Innovation Management* (8:3), 1999, pp. 179-187.

Honderich, T. (Ed.). *The Oxford Companion to Philosophy*, Oxford: Oxford University Press, 1995.

Howard, N. "Metagame Analysis," in J. Rosenhead (Ed.), *Rational Analysis for a Problematic World: Problem Structuring Methods for Complexity, Uncertainty and Conflict*. Chichester, England: John Wiley and Sons, 1992, pp. 249-265.

Jantsch, E. *Technological Forecasting in Perspective*, Report for the Organization for Economic Co-operation and Development (OECD), Paris, 1967.

Kerr, N. L.; MacCoun, R. J.; and Kramer, G. P. "Bias in Judgement: Comparing Individuals and Groups," *Psychological Review* (103:4), 1996, pp. 687-719.

Kindred, A. R. *Data Systems and Management: An Introduction to Systems Analysis and Design* (2nd ed.), London: Prentice-Hall, 1980.

Land, F. "Adapting to Changing User Requirements," *Information and Management* (5), 1982, pp. 59-75.

Lederer, A., and Nath, R. "Managing Organizational Issues in Information System Development," *Journal of Systems Management* (42:11), 1991, pp. 23-39

Lee, B. *Basic Systems Analysis* (2nd ed.), London: Hutchinson, 1984.

Lee, B. (Ed.). *Introducing Systems Analysis and Design*, Manchester, England: NCC Publications, 1979.

Lott, R. W. *Basic Systems Analysis*, London: Canfield Press, 1971.

Martin, C.; Vu, H.; Kellas, G.; and Metcalf, K. "Strength of Discourse Context as a Determinant of the Subordinate Bias Effect," *The Quarterly Journal of Experimental Psychology* (52A:4), 1999, pp. 813-839.

Mayer, R. E. *Thinking, Problem Solving, Cognition* (2nd ed.), New York: W. H. Freeman and Company, 1996.

Mulligan, N. W. "The Effects of Perceptual Inference at Encoding on Organization and Order: Investigating the Roles of Item-Specific and Relational Information," *Journal of Experimental Psychology* (25:1), 1999, pp. 54-69.

NCC. *From the White Heat of Technology to the Knowledge E-conomy, 1966-2001*, National Computer Centre: Manchester, England: NCC Publications, 2001 (available online at http://www.ncc.co.uk/ aboutncc/pdfs/ncc351.pdf).

NCC. *Introducing Systems Analysis and Design*, Volume 1, National Computing Centre, Manchester, England: NCC Publications, 1978.

Obolensky, N. *Practical Business Re-engineering: Tools and Techniques for Achieving Effective Change*, London: Kogan Page, 1995.

Pinto, J., and Slevin, D. "Critical Success Factors in Successful Project Implementation," *IEEE Transactions on Engineering Management* (EM 34:1), 1987, pp. 22-27.

Poole, M .S. "Do We Have Any Theories of Group Communication?," *Communication Studies* (41:3), 1990, pp. 237-247.

Puccio, G. "Creative Problem Solving Preferences: Their Identification and Implications," *Creativity and Innovation Management* (8:3), 1999, pp. 171-178.

Rosenhead, J. "Robustness Analysis: Keeping your Options Open," in J. Rosenhead (Ed.), *Rational Analysis for a Problematic World: Problem Structuring Methods for Complexity, Uncertainty and Conflict,* Chichester, England: John Wiley and Sons, 1992, pp. 193-218.

Rosenhead, J., and Mingers, J. (Eds.). *Rational Analysis for a Problematic World Revisited: Problem Structuring Methods for Complexity and Conflict* (2nd ed.), Chichester, England: John Wiley and Sons, 2001.

Sauer, C. *Why Information Systems Fail: A Case Study Approach*, Henley, England: Alfred Waller, 1993.

Senn, J. A. *Analysis and Design of Information Systems* (2nd ed.), New York: McGraw-Hill, 1989.

Sloman, S. A. "Categorical Inference Is Not a Tree: The Myth of Inheritance Hierarchies," *Cognitive Psychology* (35), 1998, pp. 1-33

Stanovich, K. E., and West, R. F. "Discrepancies between Normative and Descriptive Models of Decision-Making and the Understanding/Acceptance Principle," *Cognitive Psychology* (38), 1999, pp. 349-385.

Teigen, K. H. "The Language of Uncertainty," *Acta Psychologica* (68), 1988, pp. 27-38.

Tversky, A., and Kahneman, D. "The Framing of Decision and the Rationality of Choice," *Science* (221), 1981, pp. 453-458.

Tversky, A., and Kahneman, D. "Judgement Under Uncertainty: Heuristics and Biases," *Science* (185), 1974, pp. 1124-1131.

Vesely, S. "A Time-Dependent Methodology for Fault Tree Evaluation," *Nuclear Engineering and Design* (13), 1970, pp. 337-360.

Vollmeyer, R.; Burns, B. D.; and Holyoak, K. J. "The Impact of Goal Specificity on Strategy Use and the Acquisition of Problem Structure," *Cognitive Science* (20), 1996, pp. 75-100.

Waddington, C. H. *Tools for Thought*, St. Albans, UK: Paladin, Frogmore, 1977.

Wertheimer, M. "Untersuchunngen zur lehre von der Gestalt," *Psychologischo Forshung* (4), 1923, pp. 301-350.

Wetherbe, J. C. *Systems Analysis for Computer-Based Information Systems*, St. Paul, MN: West Publishing Company, 1979.

Yourdon, E., and Constantine, T. *Structure Design: Fundamentals of a Discipline of Computer Program and Systems Design*, Englewood Cliffs, NJ: Prentice-Hall, 1979.

Yeates, D. *An Introduction to Systems Analysis and Design*, Milton Keynes, England: Open University Press, 1973.

ABOUT THE AUTHORS

Carl Adams is a principle lecturer in Information Systems and e-commerce at the University of Portsmouth, UK. His research interests include IS development practice, mobile applications and development, trust, electronic commerce, and cultural and social implications of technology. He can be contacted by e-mail at carl.adams@port.ac.uk.

David Avison is Professor of Information Systems at ESSEC Business School, Paris, France after nine years at the School of Management at Southampton University for nine years. He is also Professor of Information Systems at Brunel University in England and visiting professor at University Technology, Sydney, Australia. He is joint editor of Blackwell Science's *Information Systems Journal* now in its 14th volume. A past chair of WG 8.2, David is currently vice chair of IFIP Technical Committee 8 and represents France on that committee.

Appendix

(1) Brainstorming approaches	
Affinity diagram	Bicheno (1994)
Association/images technique	Couger, Higgins, and McIntyre (1993)
Brainstorming	Clark (1958), De Bono (1977), Waddington (1977)
Brainwriting—shared enhancements variation	Couger, Higgins, and McIntyre (1993), Geschka (1996)
Bug List	Couger, Higgins, and McIntyre (1993)
Delphi	Carley (1980), Waddington (1977)
Fagan Reviews	Fagan (1976)
Five Ws and the H	Couger, Higgins, and McIntyre (1993)
Force field analysis	Couger, Higgins, and McIntyre (1993)
Lateral thinking techniques	De Bono (1969, 1970, 1977)
Nominal group technique (NGT)	Couger, Higgins, and McIntyre (1993)
SIL—suggested integration of problem elements	Couger, Higgins, and McIntyre (1993)
Wildest idea	Couger, Higgins, and McIntyre (1993)
(2) Relationship approaches	
Analytic hierarchy process	Gajpal, Ganesh, and Rajendran (1994)
Boundary examination	Couger, Higgins, and McIntyre (1993)
Cognitive mapping	Eden (1992)
Critical path analysis (CPA), method (CPM)	Jantsch (1967)
Decision trees	Avison and Fitzgerald (2003)
Fault tree analysis	Andrew and Moss (1993), Vesely (1970)
Influence diagrams, interrelationship diagrams	Bicheno (1994)
McKinsey 7 S Framework	Obolensky (1995)
Network techniques	Jantsch (1967)
Planning assistance through technical evaluation of relevance numbers (PATTERN)	Jantsch (1967)
Precedence diagramming method (PDM) network	Obolensky (1995)
Program evaluation and review technique (PERT)	Jantsch (1967)
Reliability networks	Andrew and Moss (1993)
Strategic choice	Friend and Hickling (1997), Rosenhead (1992)
Strategic options development and analysis (SODA)	Eden (1992)
Tree analysis	Andrew and Moss (1993)
Value chain analysis	Obolensky (1995)
(3) Scenario approaches	
Future analysis	Land (1982)
Robustness analysis	Rosenhead (1992), Rosenhead and Mingers (2001)
Scenario planning/writing/analysis	Carley (1980)

(4) Reductionist approaches	
Common cause failures (CCFs)	Andrew and Moss (1993)
Hazard and operability studies (HAZOP)	Andrew and Moss (1993)
Hazards analysis and critical control points (HACCP)	Camden (1987)
Maintainability analysis	Andrew and Moss (1993)
Morphological approaches	Couger, Higgins, and McIntyre (1993), Geschka (1996), Jantsch (1967),
Preliminary hazard analysis	Andrews and Moss (1993)
Risk assessment/engineering/management	Andrew and Moss (1993), Beard (1969), Chapman (1984, 1990), Grey (1995)
Synergistic contingency evaluation and review technique (SCERT)	Chapman (1984)
Systems failure method (SFM)	Fortune and Peters (1995)
(5) Matrix approaches	
Attribute association	Couger, Higgins, and McIntyre (1993)
Bug list	Couger, Higgins, and McIntyre (1993)
Critical success factors (CSF)	Flynn (1992), Pinto and Slevin (1987)
Cross-impact matrices	Waddington (1977)
Decision matrices	Jantsch (1967)
Decomposable matrices	Couger, Higgins, and McIntyre (1993)
Dimensional analysis	Couger, Higgins, and McIntyre (1993)
External dependencies	Oblensky (1995)
Failure Modes and Effect Analysis (FMEA)	Andrew and Moss (1993), Bicheno (1994), Fortune and Peters (1995)
Five "Cs" and "Ps"	Obolensky (1995)
Five whys	Bicheno (1994)
Force field analysis	Couger, Higgins, and McIntyre (1993)
Johari window of knowledge	Obolensky (1995)
Markov chains, Markov analysis	Andrew and Moss (1993)
Matrix techniques, matrix analysis	Bicheno (1994), Geschka (1996), Jantsch (1967)
Opposition-support map	Obolensky (1995)
Options matrix	Obolensky (1995)
Rapid ranking	Andrew and Moss (1993)
Requirements, needs and priorities (RNP)	Batiste and Jung (1984)
Shareholder value analysis (SVA)	Obolensky (1995)
Stakeholder analysis	Obolensky (1995)
SWAT analysis (strengths, weaknesses, opportunities and threats)	Obolensky (1995)
Value chain analysis	Obolensky (1995)
Value engineering/management	Obolensky (1995)
(6) Conflict approaches	
Gaming, game theory	Jantsch (1967)
Hypergames	Bennet, Cropper, and Huxham (1992)
Metagames	Howard (1992)
RBO - rational bargaining overlaps	Obolensky (1995)
Simulation	Andrew and Moss (1993), Carley (1980)
Systems failure method (SFM)	Fortune and Peters (1995)

Part 6:

Systems Development: Methods, Politics, and Users

28 ENTERPRISE SYSTEM AS AN ORCHESTRATOR OF DYNAMIC CAPABILITY DEVELOPMENT: A Case Study of the IRAS and TechCo

Chee Wee Tan
National University of Singapore

Eric T. K. Lim
Nanyang Technological University

Shan Ling Pan
National University of Singapore

Calvin M. L. Chan
National University of Singapore

Abstract Corporations are perpetually hunting for ways to develop exclusive, sustainable, and competitive advantages that will enable them to leapfrog ahead of their industrial adversaries. Notably, the debut of enterprise systems (ES) during the recent decade has given rise to frequent talk of the utilization of integrative, IT-inspired business mechanisms to achieve the much sought-after but elusive competitive edge. Others, however, have argued that the search for sustainable competitiveness should instead be anchored in organizational efforts to cultivate and build up firm-specific dynamic capabilities. Cognizant of the various perspectives, this paper takes a holistic approach in proposing the achievement of sustainable competitive advantages by examining the manner in which ES adoption can contribute to the forging of dynamic capabilities. In particular, Montealegre's (2002) process model of capability development is adopted as the analytical framework to explore the strategization of ES development in two different organizations, with the main distinction being that one of them subscribes to commercially available SAP applications while the other chooses to develop its ES in-house. Through comparing and contrasting evidence from both cases, this study attempts to decipher how ES adoption can be strategized to develop strategic capabilities and understand the implications between off-the-shelf and bespoke ES in affecting the process of dynamic capability development.

Keywords: Management information systems, enterprise systems, dynamic capability development, dynamic capabilities perspective

1 INTRODUCTION

Prized as the holy grail of modern competition, information integration has steered management information systems (MIS) design toward an integrative architecture (Kumar and Hillegersberg 2000). Particularly, the emergence of enterprise resource planning (ERP) systems as configurable software applications that integrate the complete range of business activities within an organization (Howcroft and Light 2002) has accelerated the rate of MIS diffusion across businesses (Adam and O'Doherty 2000).

Around the period when ERP systems were gaining prominence as replacements for obsolete legacy systems (Holland and Light 1999), Davenport (1998) revived the concept of enterprise systems (ES) as a more generic representation of an expanding MIS spectrum which provides real-time operating information to support rapid "sense-and-response" business models (Davenport 2000a; Rosemann and Watson 2002). Collectively, ES present a comprehensive and synchronous definition of integrated MIS and their advancements, including innovations such as supply chain management (Papazoglou et al. 2000), customer relationship management (Goodhue et al. 2002), and knowledge management (Alavi and Leidner 2001) systems.

Embedded within this broad range of ES are exemplary business paradigms devised for increasingly sophisticated business information requirements (Markus 2000), with Davenport (2000a) and Hayman (2000) predicting that future generations of ES would craft new capabilities for corporations to enhance their competitiveness through acute sensitivity to fluctuating market conditions (Davenport 1998, 2000b).

Contemporary management literature has popularized the perception of the firm's ability to manipulate distinctive internal competencies relative to environmental dynamics as the key determinant of its competitive sustainability (Ginsberg and Venkatraman 1985; Lieberman and Montgomery 1988). Specifically, Barney (1997) and Miyazaki (1995) advised companies to be proactive in developing distinct inherent abilities that are inimitable.

Based on the above, this article postulates a potential convergence between ES adoption and capability development, i.e., the holistic commitment associated with ES projects may compel enterprises to perform a thorough, systemic review of every aspect of business operations, which in turn bestows organizations with the perfect opportunity to sculpt their strategic capabilities. To analyze the contribution of ES development in forging dynamic capabilities, we adopt Montealegre's (2002) process model of capability development. This framework is applied to two different settings: a governmental institution and a commercial establishment. The main distinction between these two enterprises hinges on the fact that the public agency designed and developed its ES in-house whereas the profit-oriented corporation relied on the off-the-shelf SAP system. Through comparing and contrasting evidence from these cases, this paper endeavors to address the following research question: How can ES adoption be strategized for the purpose of dynamic capability development and what are the differences, if any, between in-house and off-the-shelf ES?

2 THEORETICAL FOUNDATION

The strategization of information technology (IT) to attain a competitive edge is ritualistic among private institutions (Ives and Learmonth 1984; McFarlan 1984) and contributes to an extensive list of classical business applications (Clemons 1991). Notwithstanding these testimonial cases of successful MIS, the feasibility of IT-based competitive sustainability remains debatable within academia (see Mata et al. 1995; Mykytyn et al. 2002). Citing reasons such as the prevailing adoption of IT as a *strategic necessity* (Clemons 1986) and the possibility of generating even deadlier reactions from rivals through creative duplication (Kettinger et al. 1994; Vitale 1986), many have contested the viability of IT-derived competitiveness and emphasized that research in this domain should focus on "describing how, rather than systematically why" IT can deliver strategic benefits (Reich and Benbasat 1990, p. 326).

Unsurprisingly, in light of their copious organizational influence and the substantial implementation investments they require, ES are readily conceived by scholars as the next logical candidate for the reimbursement of competitive value (Ross and Vitale 2000). As IT-based business solutions, ES are touted as configurable software packages that purportedly enable the collation of transaction-oriented data and functional processes into a singular infrastructure (Lee and Lee 2000; Markus et al. 2000a; Markus and Tanis 2000). Nevertheless, despite the projected benefits of prepackaged ES (Markus et al. 2000b), there remain unresolved adoption hurdles.

Implicit within ES packages are business principles that emulate industry best practices (Everdingen et al. 2000). These posited business paradigms, as predefined by the vendor, serve as convenient templates for corporations to mirror competitive praxis, although in many instances the projected benefits of the implemented ES do not materialize (Markus and Tanis 2000). The failures have been attributed to a blend of socio-technical constraints surrounding ES, such as their complexities, their customization difficulties, and the presence of cultural misfits underlying their inherent business process assumptions and those of the adopting organization (Howcroft and Light 2002; Lee and Lee 2000; Soh et al. 2000). While we do not underestimate the aforementioned technological and organizational challenges of ES implementation, the purpose of this paper is to shed light on how competitive benefits can be manifested through ES adoption, rather than the reason why they can or cannot be realized.

As conceived by Rosemann (1999), the fundamental notion of ES is analogous to the developmental objective in mapping the entire array of enterprise business processes into an integrated infostructure. From this perspective, ES are predominantly operational commodities that double up as "the key element of an infrastructure" which conveys a holistic business solution to adopters (Rosemann and Watson 2002, p. 201). Yet, despite the consensus among researchers of the strategic significance of ES, their exact business potential has not been exploited beyond the extrapolative predictions of existing MIS trends (Davenport 2000a; Hayman 2000; Markus et al. 2000b). Consequently, the question of how ES can deliver competitive qualities continues to evade answering in strategic MIS research and, specifically, ES literature.

One particular response to this theoretical and empirical challenge has approached the management of ES from the stance of strategic capability development (Davenport 2000a; Hayman 2000). Drawing from the resource-based view (Penrose 1959), where firms are presumed to possess specific, time-independent resource differences (Amit and Schoemaker 1993; Mahone and Pandian 1992), managerial theorists have stressed the need for businesses to exploit their unique attributes in shaping market positions (Priem and Butler 2000). Their call also resonates in the MIS arena where Kettinger et al. (1994) hinted at a normative relationship between IT and core competencies by suggesting that "the sustainability of competitive advantage may be achieved by leveraging unique firm attributes with information technology to realize long-term performance gains" (p. 31).

Developing this idea further, our research subscribes to the dynamic capabilities perspective (DCP) as a theoretical basis to evaluate the relationship between ES adoption and strategic capability development. Dynamic capabilities, as defined by Teece et al. (1997), reflect a "firm's ability to integrate, build and reconfigure internal and external competencies to address rapidly changing environments" (p. 516), i.e., they drive the creation, evolution, and assimilation of resources to formulate novel value-creating strategies (Grant 1996).

Following Eisenhardt and Martin (2000), this study understands dynamic capabilities to be the "firm's processes that use resources—specifically the processes to integrate, reconfigure, gain and release resources—to match and even create market change...[or] the organizational and strategic routines by which firms achieve new resource configurations" (p. 1107). We also take into account Wheeler's (2002) justification in a study on net enablement that DCP captures the essence of the "dynamic process of recreating and executing innovative options to gain and sustain competitive advantage" within organizations competing in digitally networked environments (p. 127). In short, DCP presumes a continuum of business competencies renewal to achieve congruence to ever-changing competitive circumstances, thus making it palatable to our proposed research objective.

Specifically, this study adopts the process model of capability development proposed by Montealegre (2002) for data analysis. In the model, the acquisition of dynamic capabilities is presented as an incremental sequence of phases that intuitively depict the patterned and evolutionary nature of their development. In addition, for each stage, the essential organizational capability is highlighted together with a prescribed course of action (see Figure 1).

Establishing strategic direction is imperative for any company to generate exclusive insights into future scenarios that are obscure to competitors (Amit and Schoemaker 1993). Therefore, the *capability to strategize* based on the in-depth appreciation of internal capacities versus external surroundings is decisive of a firm's competitiveness

Phase 1 Establishing Direction	Phase 2 Focusing on Strategy Development	Phase 3 Institutionalizing the Strategy
Key Capabilities Developed at Each Phase:		
Capability to Strategize	Capability to be Flexible	Capability to Integrate and Engender Trust
Global benchmarking and training	Integrating resources into core activities	Gaining internal commitment
Learning from past experience and history	Experimenting	Investing in complementary infrastructure
Absorbing knowledge as a unified group at the top of the organization	Investing in, leveraging and co-opting resources	Strengthening external relationships

Figure 1. Process Model of Capability Development (Montealegre 2002)

(Hamel and Prahalad 1994). To gain such abilities, Montealegre advocated that managers should (1) initiate global benchmarking and training exercises to expand corporate resources (Oliver 1997), (2) learn from past experiences and history in dealing with recurring problems (Maidique and Zirger 1985), and (3) ensure that the top management absorb knowledge as a unified group in order to consolidate knowledge capital (Alavi and Leidner 2001).

Once the strategic course is charted, Montealegre believed, the *capability to be flexible* is critical for the development and execution of supportive strategies that are highly adaptable to unanticipated disruptions (Jarvenpaa and Leidner 1998). To accomplish this, priorities must be assigned to (1) integrating resources into core activities to ensure that they are effectively channeled to support the strategic routines of the organization, (2) experimenting to optimize task performance (Dosi and Marengo 1992), and (3) investing in, leveraging, and co-opting resources to maximize the potential within imported assets and tacit competencies in fulfilling corporate objectives (Chakravarthy 1997).

Finally, a firm's business mission is dependent on the commitment and cooperation across a diversity of internal and external stakeholders (Freeman 1984). To shape a collaborative community, companies must possess the *capability to integrate and engender trust* (Teece et al. 1997) among different stakeholder categories. To serve this purpose, Montealegre expounded on the participation of corporations in (1) gaining internal commitment such that newly acquired abilities can be infused into the collective skill sets of employees or within special routines embedded in the organization's operations and knowledge base (Nelson and Winter 1982), (2) investing in complementary infrastructure which supplements the corporate strategy, and (3) strengthening external relationships in order to boost responsiveness to market variations.

Reflecting on previous discourses, it is not difficult to discern a casual link between the aptitudes of ES and the managerial issues to be tackled for dynamic capability development. In effect, the portrayed business activities demand a cross-functional platform be forged through the integration of enterprise-wide processes. By virtue of their systemic properties, ES may perhaps be the integrative backbone necessary to orchestrate such an undertaking.

3 METHODOLOGY

This study adopts an in-depth case research method. According to Yin (1984), case research is "an empirical inquiry that investigates a contemporary phenomenon within its real-life context, when the boundaries between phenomenon and context are not clearly evident and in which multiple sources of evidence are used" (p. 23). It is appropriate in scenarios where the research question is exploratory (Yin 1984) and exists within a broader sociological context, necessitating rich descriptions of the social environment (Strauss and Corbin 1990). Moreover, case study offers an opportunity to engage in theory building for topics where there is relatively little prior knowledge (Eisenhardt 1991).

For this investigation, two case studies were conducted at two separate sites, each over a six-month period using several methods of data collection (Benbasat et al. 1987). At each location, focused interviews (Merton et al. 1990) were conducted with organizational members associated with the conceptualization, implementation, and post-implementation phases of the ES development life cycle (Ross and Vitale 2000). By soliciting information from representatives involved throughout these stages, the investigators were allowed a feel of the entire ES adoption experience. The gathered data was further triangulated with conversational evidence from external stakeholders (Orlikowski 1993). Through these exercises, a qualitative in-depth collation of data points was made within the study environments (Lacity and Janson 1994) that focused specifically on developmental and managerial issues pertaining to the ES with lesser emphasis on their technicalities (Eisenhardt 1991).

Given the prescribed methodology, objective data compilation was impossible as interactions between researchers and human subjects participating in the enquiry would inevitably alter the perceptions of both parties (Walsham 1995). This research thus accepts a more interpretivistic perspective (Walsham 1995) of the data gathered, i.e., that the contextual understanding of the investigators provides supplementary background information that influences evidence interpretation (Lacity and Janson 1994). Also, as part of data analysis, preliminary themes isolated in the earlier half of the study were incorporated into subsequent interview sessions to stimulate a deeper scrutiny of the emerging issues. Data collection concluded when the themes became repetitive and the information appeared to reach saturation (Glaser and Strauss 1967).

4 CASE DESCRIPTION

For this study, the Inland Revenue Authority of Singapore (IRAS) and TechCo were selected as appropriate venues to examine the functionality of ES in forging strategic capabilities.

4.1 IRAS: The Case of the Inland Revenue Integrated System (IRIS)

IRAS was inaugurated in 1992 to reengineer an outdated tax administration system marred with sluggish manual tax processing procedures and the excessive accumulation of paper tax-documents. IRAS was experiencing shortfalls in human resources that led to unproductive and lengthy tax cycles with an estimated discrepancy of 300 returns being left uncollected annually. Such red tape proved to be fatal to customer relations as taxpayers making enquiries became frustrated with the long delays at the tax agency in locating their personal tax folders.

To address such administrative backlogs, the Inland Revenue Integrated System (IRIS) was developed in-house to integrate tax processing functions into a unified infostructure. Designed as a modular system, IRIS comprises application components catering to specific tax processes

> *IRIS comes with different modules—or you can call them components: the pipeline to process tax returns, the enforcement module, the case management module, and then there is the payment module, data module, and another very specific module to handle property accounts....Basically, it is the integration of all these modules that makes up IRIS.* (Director, IRIS Design)

Besides, according to IRAS statistical approximation, 80 percent of tax returns are deemed to be normal and do not require intensive verification by tax officers. It is thus a waste of manpower to validate every case physically. Being implicitly fashioned, IRIS is equipped with inbuilt predefined evaluation criteria to process standard tax returns, with the remainder being routed to the respective tax officers with the appropriate domain knowledge. This routing is fully automated with the Workflow Management System (WMS).

Together with the launch of an Internet-based income declaration interface for taxpayers, IRIS was seamlessly assimilated into the range of tax processing activities to create the Electronic Filing (e-Filing) system, which has to date attracted 924,014 e-filers or approximately 50 percent of the entire taxpaying population. With the e-Filing system, the bureaucratic reputation of IRAS was reversed. In a recent 2001 survey, 94.1 percent of individual taxpayers, 89.6 percent of corporate taxpayers and 94.6 percent of goods and services taxpayers expressed satisfaction with the agency's services, which were found to be convenient, as well as competently and courteously provided.

From the description, IRIS represents an exemplification of the concept of ES and exhibits many characteristics of its commercial counterparts. Therefore, the case of IRAS presents a unique occasion to uncover how internally crafted ES can contribute to strategic capability development.

4.2 TechCo: The Case of the SAP System

In contrast to IRAS' governmental setting, TechCo resembles a textbook example of ES adoption. Being an international leader in electrical engineering and electronics,

TechCo employs a total of 447,000 people in over 190 countries. Given its geographically dispersed organizational structure, TechCo reckoned that it would benefit from fostering an accommodating environment for multi-regional collaborations. Moreover, the general proliferation of highly customizable products and services had contributed to the dissemination of specialized knowledge resources.

TechCo had also completed several acquisitions to expand its businesses which resulted in inconsistent IT infrastructures and incompatible business operations. In some extreme cases, the subsidiaries had progressed independently of corporate direction, thus prompting an urgent need to reconnect these disjointed business divisions. Taking stock of these restraints to company performance, TechCo decided to adopt ES as its core business strategy to manage the dynamism of its fast-expanding business empire. It selected the SAP/R3 package, which was based on the SAP AG system with the rollout being estimated at U.S. $32 billion.

TechCo's SAP implementation aimed to establish a shared service hub that would support cross-functional information sharing measures to complement the company's business competencies in the virtual sphere. Through exploiting SAP's technological faculties, TechCo hoped to incorporate its vast network of decentralized business systems into a homogeneous IT architecture that would enforce information transmission standards to promote collaboration among its business entities. Specifically, these reengineering efforts were tuned toward stimulating intra-organizational cooperation and manifested as restructured business processes coupled with revised policies for information standardization.

In sum, TechCo acknowledged the assembly of information and knowledge resources through the SAP system as being vital to the framing of a shared mission to manage its increasingly complex business functions. The case of TechCo, therefore, offers the alternative of examining the research topic from a commercialized angle.

5 CASE ANALYSIS AND FINDINGS

Based on Montealegre's (2002) framework, the findings from this study are analyzed as a chain of events corresponding to the model of dynamic capability development.

5.1 Phase 1: Establishing Direction

The articulation of a corporate vision is widely documented in the strategic management literature as the means for organizations to mitigate uncertainties in facing business challenges (Amit and Schoemaker 1993). To do that, managers must possess a tight grasp of the organization's internalized capacities with respect to environmental inconsistencies (Alchian and Demsetz 1972; Hamel and Prahalad 1994). This *capability to strategize* requires the fostering of "conditions that will let management capture and exploit the knowledge that already exists throughout the organization" (Montealegre 2002).

The installation of ES in both IRAS and TechCo led to business renewal aimed at revitalizing core competencies. At IRAS, the stigma of a governmental tax collector coupled with administrative inefficiencies had stifled positive taxpayer relationships. To repair its image, IRAS' management resolved to revamp the taxation experience through implementing IRIS to integrate the various tax processing functions.

> *We visualized a very efficient tax [processing] system and we started examining our existing business processes. We concluded that a majority of the tax returns were processed by tax officers who only required a short duration of approximately 20 minutes to complete each of them. We believed that this process could be automated....We would get the machine to perform 80 percent and leave 20 percent to be handled by tax officers whenever necessary and without compromising accuracy as in the manual system. It is our objective to ease all these jams and bottlenecks in the rear [back-end tax processing] by moving the tasks into the system.* (CIO, IRAS)

Similarly, TechCo faced a situation where the disconnected IT infrastructures and corporate plans prompted the need to reunify the autonomous business subsidiaries under a single umbrella. To prepare for SAP migration, TechCo established a cross-cultural implementation team to engineer a network of social linkages to encourage communication among divisional managers.

> *The familiarization effort provided the team with better knowledge and appreciation of the organization's culture, practices and systems, which proved, later, to be very useful in the implementation of the SAP system.* (Manager, TechCo)

It is plausible, from the preceding examples, that the knowledge momentum accompanying ES conceptualization was the prime motivator for these organizations to perform rigorous assessments of their hidden proficiencies. Through cross-functional knowledge-sharing arrangements, the managers in both cases demonstrated improved operational awareness, derived from greater transparency in the business routines which spanned functional boundaries. In other words, ES adoption enhanced the capability of both organizations to strategize by providing a premeditated environment for managers to thoroughly reflect upon ritualistic procedures in order to identify and extract core competencies that would uphold the corporate mission. Particularly, this auxiliary power of ES can be considered in relation to the three managerial issues raised by Montealegre in conjunction with the development of strategizing abilities.

Global Benchmarking and Training: Global benchmarking is a handy instrument for gauging an organization's competitive status and the effectiveness of its resource configuration (Oliver 1997). Unfortunately, it is usually tedious to embark on such an extensive evaluation campaign. ES, however, conveniently overcome this predicament with the inclusion of the best business practices predominant in the industry (Soh et al. 2000). TechCo, for instance, employed the SAP system as the key element of its business strategy to simulate an intra-organizational knowledge-sharing network on par with the prevailing standards.

Despite the convenience, commercialized ES do occasionally suffer from incompatibilities with specialized business needs. Indeed, the evidence suggests that ES should not be blindly accepted as definitive yardsticks, but rather assessed objectively based on corporate requirements. For instance, the configuration of preexisting taxation systems was not a viable option for IRAS.

> *There are actually systems being implemented in the States, but when we brought over such a system, we discovered that 80 percent [of it] was different from our functions. It would require wholesale customization for our purpose—basically, rewriting.* (IRIS System Manager)

Learning from Past Experiences and History: One common issue in ES adoption for developing dynamic capabilities is the preservation of the knowledge accumulated from specialization in current work processes and legacy systems (Pan et al. 2001). These tacit knowledge assets are frequently entrenched within operating procedures that are seldom stand-alone (Blacker 1995). Often, such firm-specific expertise is invaluable in keeping enterprises afloat in competitive tides (Grant 1996). For TechCo, the SAP system served as the beacon directing knowledge extraction, and the task at hand became one of soliciting tacit wisdom through relationship building across divisional borders. This contributed to the convergence of organizational knowledge resources (Baskerville et al. 2000).

> *The documentation and participative discussion between the team and various business units have helped us understand the different work practices and identify processes which the business units have in common.... With so many [business] processes being supported and developed through the [SAP] system, it is important to know the level of impact on cross-functional operations.* (Respondent, TechCo)

Conversely, for the bespoke ES, without the restrictions of an embedded knowledge base, the immediate concern in knowledge retention is to ensure a fair user representation during system conceptualization.

> *Before it [the development of IRIS] began, each user branch would have already identified their core users or experts. They were very experienced users, so they represented their functional group and any decision with regard to the system would be made by these people.* (IRIS System Engineer)

To summarize, tacit knowledge was archived in both institutions as an indispensable record of an organization's historical proficiencies. This indicates that the development of ES grants the ideal opportunity for firms to ascertain their dispersed knowledge competencies for fusion into the system.

Absorbing Knowledge as a Unified Group at the Top of the Organization: Top management endorsement has always been a deterministic priority in ES adoption, due to the systems' immense appetite for corporate resources and their prerequisite of cross-functional participative commitment (Martin et al. 1999; Ross 1999). Nonetheless, for ES adoption to induce dynamic capability development, managerial enthusiasm must go beyond political and budgetary sponsorship to proactively redefine the firm's underlying business philosophy.

In both cases, the senior management displayed keen interest and passion in knowledge sharing during ES conceptualization. This strategic involvement offered valuable insights that might otherwise have been imperceptible at the operational level.

> *I think the management played a big part in deciding policy issues.... There were certain things which might infringe on certain policies, so whether to have them or not, the management had to decide.* (IRIS System Engineer)

5.2 Phase 2: Focusing on Strategy Development

Chakravarthy (1997) stressed that the development of a strategy to operationalize business objectives should conform to principles of flexibility for institutions to remain vigilant and responsive toward unpredictable circumstances (Jarvenpaa and Leidner 1998). This management philosophy, as maintained by Montealegre, pivots on the extent of managerial control versus the changeability of the organization (Eisenhardt and Martin 2000).

The assimilation and manipulation of information flow through ES adoption assisted IRAS and TechCo to not only enforce better control of all aspects of their business activities, but also reconstruct their knowledge distribution channels for prospective growth. The implementation of IRIS, for example, enabled IRAS to maintain a profile of every taxpayer, and to associate each transaction with a specific customer for efficient tracking. Moreover, the chronological classification and consolidation of transaction data into individual digital folders facilitated sharing between tax officers on knowledge such as the most effective manner in which to serve a particular taxpayer.

> *Whenever a taxpayer approaches us, the first thing we do is to retrieve his record. Using IRIS, a click on their identity card number will get us a record of their last conversation with us, and whom they have spoken to; everything is there. In the past, we wouldn't know who the last person to handle the case was, unless the taxpayer himself had taken down the name of the officer he had contacted.* (IRIS System Engineer)

The situation was identical in TechCo, where a Shared Service Center (SSC) was initiated to coordinate the ES initiative and manage centralized SAP services. This included administering information and knowledge connections among multiple business units.

> *The SAP control forces business units to transfer information management to the SSC.... Most of the business units want the SAP to integrate their business processes. They depend on the SSC to manage the information exchanges and know-how for cross-functional integration.* (Manager, TechCo)

From the evidence, it is clear that ES allowed each of the organizations to tackle problems as a cohesive entity through dynamic and systematic information management. ES can thus be seen as an integrative platform for enterprises to incorporate and reconfigure resources flexibly as a metamorphic architectural response to external stimuli, i.e., an ES package is analogous to a jigsaw puzzle in which companies piece together their competitive strategies. The exploitation of the integration technology of ES thus amplifies a firm's capability to be flexible by crafting an adaptable infrastructure for continuous organizational renewal. Specifically, this correlation can be examined in terms of the three managerial components identified by Montealegre.

Integrating Resources into Core Activities: The introduction of ES redefines the resource configuration of an organization through seamless integration across core operational activities. However, in coping with prepackaged ES integration, the prime concern is the probability of a mismatch between the business processes and resource requirements of the application and those of the adopter (Howcroft and Light 2002).

> *There are certain logistic practices being adopted in the SAP system that differ from our legacy logistic system. The sales and distribution module in the SAP system has made database implementation different from what we used to have in our distribution channels....Since the SAP system is new, we are unsure if it can adapt to changes.* (Distribution Manager, TechCo)

Conversely, integration poses less of a hassle for in-house developed ES; they are closer to the organization's projected techno-structure.

> *When we designed IRIS, we didn't even think about what the current process was. We did everything from scratch; it didn't even match the existing work processes. Basically, it's a total revamp.* (IRAS Interviewee)

Experimenting: Experimentation is a familiar managerial tool for stimulating progressive improvements in operational routines (Dosi and Marengo 1992). With respect to ES adoption, experimentation was evident in both companies as they struggled to find ways to merge resources into a sophisticated modular architecture for optimal flexibility. For both cases, this was achieved through a phased approach to implementation. This allowed each organization to evaluate the system at consecutive stages so as to fine-tune it for cross-functional performance.

> *Due to the complexity of IRIS, we decided to divide its implementation into phases. This gave us a chance to look at how each module was performing and make changes to accommodate the users, if necessary.* (IRIS System Engineer)

Investing in, Leveraging and Co-opting Resources: In both IRAS and TechCo, the implementation of ES was only a preliminary step toward an integrated value chain, and was complemented by expenditure on technological extensions. As described earlier, to achieve full automation of the taxation system, IRAS virtually expanded IRIS' competencies by connecting the system to a Web-based tax declaration interface for direct data entry. This satisfies Hayman's (2000) vision of ES as a bridge between front-end transactions and back-end processes. Similarly, by relying on the SAP system to facilitate information integration, TechCo was able to encourage its various subunits to cooperate in knowledge sharing. In sum, ES provided the necessary socio-technical foundation for both organizations to leverage and co-opt resources in fulfilling their strategic goals.

5.3 Phase 3: Institutionalizing the Strategy

It is common to express the achievement of business missions within corporations as a product of underlying commitment and cooperation across the multiplicity of stakeholders (Freeman 1984). Therefore, companies must build a united community to integrate knowledge (Nonaka 1994) and engender trust (Teece et al. 1997) among diverse stakeholders.

Naturally, with ES implementation as a vehicle for organizational reinvention, the remaining challenge lies in the institutionalization of these systems as parallel-running complements to the corporate strategy, especially in light of the resulting revolution in functional paradigms concomitant with reengineered business processes. For instance, the adoption of IRIS altered conventional thinking on tax processing by placing the onus of information authentication on the taxpayer. Initially, many tax officers were against these modifications.

> *The principal concept was that we must accept the new tax filing model [80/20 rule] contained in IRIS and, therefore, there must be a change in mindset. Of course, there were many obstacles....A number of tax officers would argue: "No, this [manual tax return verification process] is the right way. We must still check..." and things like that.* (CIO, IRAS)

Similar resistance was apparent in TechCo where most users were reluctant to take on new responsibilities or contribute during conferences for fear of being ostracized. Compounding this problem, some of the employees were apprehensive about the SAP system.

> *By the structuring of SAP in the department, it seems like there will be less work to handle. And what I am concerned with is that I will be made redundant since I have been relieved of certain duties. I feel that I am being replaced.* (System User, TechCo)

It is vital for stakeholders to identify with the renewed business vision encapsulated within the implemented ES package that serves as the cornerstone to an organization's competitive strategy (Pan et al. 2001). In both IRAS and TechCo, it is clear that this alignment was hampered by technical and psychological barriers, which hindered the efforts of both organizations to cultivate trust with their stakeholders and to integrate their knowledge.

To eliminate the impediments, change management was carried out to mollify physical and mental resistance. Moreover, the evidence suggested that the ES in the two establishments were serving as the centerpiece to an extended enterprise solution. Through process integration across corporate hierarchies, the ES in IRAS and TechCo acted as a technological platform for establishing symmetrical communication with external associates to formalize strategic partnerships that would invariably boost the capability of the organization to integrate stakeholders and engender trust with them.

Gaining Internal Commitment: From the above, it may be perceived that both technical and sociological reservations against the ES being implemented had obstructed the reception of the ES within both organizations (Pan et al. 2001). In response, IRAS created a change management team to palliate the restructuring process through tailored courses and continuous training for system users. With repeated efforts in familiarizing tax officers with IRIS' functions, the resistance was eventually broken down.

> *Some are common applications such as the workflow imaging system, which everybody uses. However, there are certain applications which only affect specific user modules....The change management team had to tailor-make training sessions for these other users....Moreover, it's not just one-time training; it's retraining, retraining and retraining.* (Tax Officer, IRAS)

At TechCo, a participative policy was initiated to nurture understanding and trust among system users.

> *After the participative policy was introduced, I noticed that personal responsibility was emphasized, employees were not blamed when things went wrong, but rather, they were often offered help. Mistakes made were taken as "lessons learnt" and viewed as opportunities for learning. Most people [now] take it upon themselves to solve problems and ensure things are smooth flowing, even if it [the problem] does not arise within their sphere of work. The ownership and responsibility for failures have been reduced to a minimum.* (System Engineer, TechCo)

The case findings suggest that ES adoption is a sophisticated socio-technical ordeal, with resistance being rooted in both systemic functionalities and psychological dispositions. Our study of the two cases also shows that change management may proceed with a two-pronged approach: developing customized training programs to inject functional familiarity into users and introducing reinforcing policies to empower stakeholders through relationship building to eradicate anxieties toward the refined business model (Ross 1999). Although the complexities of ES adoption have often been accredited with the creation of overwhelming internal resistance, the evidence from our study indicates that redefined business processes together with proper change management may actually bridge the mental distance among employees and promote congenial intra-organizational cooperation.

Investing in Complementary Infrastructure: The acquisition of complementary resources helps in directing the evolutionary path of corporate proficiency (Teece et al. 1997), especially at the stage of capability development where a logical extension could translate into the integration of knowledge-sharing cum diffusion mechanisms into the fabric of the ES to allow coworkers to share experiences on the business processes reengineered to cut across divisional perimeters. As remarked by an SAP system user,

> *We depended on individual colleagues when problems shifted out of our job scope. There were organizational changes within the functionality of the SAP system, which made it difficult to know sufficiently to support all aspects of the system.*

In TechCo, the knowledge management gizmo was the newly-established e-Business Center that catered to the creation and sales of knowledge products synthesized through cross-functional collaborations.

Strengthening External Relationships: With their respective cohesive infostructures, IRAS and TechCo set out to consolidate their relationships with external stakeholders. IRAS managed to link up many employers to have the relevant tax information of individual workers transferred directly into IRIS during each tax cycle. To its credit, IRAS has to date clinched data transfer agreements under the auto-inclusion scheme for approximately 46 percent of all employees in the country.

The direct transmission of tax information relieves IRAS of intensive data capturing efforts and permits controls to be incorporated into the system to validate data accuracy.

> *The question of concern when we adopted the 80/20 rule was that only 20 percent of all the data being pumped through would be checked. Where was the control? How would we know we were getting all the [necessary] information? This gave birth to another idea of getting the information directly from the employer and automatically into the system. [Now] we have no worries about whether income is understated because we source it [the information] directly from a third party. Moreover, having auto-inclusion means we can reduce our tremendous efforts in data capturing.* (CIO, IRAS)

The case of IRAS demonstrates how ES can facilitate extended enterprise solutions (Markus et al. 2000b). The increased interorganizational connectivity underscores the competitive value of recognizing shared partnerships as a dynamic competency. In fact, the adoption of ES superimposes technological extendibility on collaborative ideology to tender business options that are otherwise impermissible.

Based on the above analysis, the interdependencies between ES and dynamic capability development may be mapped into a process model for ES adoption (see Figure 2).

6 CONCLUSION

Inevitably, given the popularity of ES, much has been deliberated on their worthiness as competitive weapons (Markus et al. 2000b; Ross and Vitale 2000). Critics, for example, have queried the absence of the concept of *enterprise* in ES (Davenport 1998; Hayman 2000). Through a pilot study of how ES adoption can contribute to dynamic capability development, this paper attempts to provide a prefatory answer to this theoretical and empirical question.

Based on a comparison of two cases, we conclude that ES can be strategic compatriots in the capability development process of organizations. The process model of ES adoption proposed in this paper captures the essence of this relationship. Through the model, we argue that the conceptualization of an ES initiative increases a firm's capability to strategize by providing a premeditated structure that aids the identification and extraction of knowledge assets embedded within business routines. This is succeeded by the implementation of the ES as an integrative backbone to assimilate and reconfigure isolated core competencies for optimal flexibility, effectively enhancing the

Phase 1 Establishing Direction **Conceptualization**	Phase 2 Focusing on Strategy Development **Implementation**	Phase 3 Institutionalizing the Strategy **Strategization**
Key Capabilities Developed at Each Phase:		
Capability to Strategize	Capability to be Flexible	Capability to Integrate and Engender Trust
Provides a premeditated structure for the identification and extraction of core competencies	Provides the integrative technological platform for resource configuration	Provides the centerpiece of an extended enterprise solution
Deciding the best approach to system development	Customizing the system for optimal resource adaptability	Training and empowerment to remove technical and psychological barriers within stakeholders
Retaining and codifying the tacit knowledge embedded within legacy systems	Phased approach to system implementation	Expanding knowledge management capabilities on sharing and diffusion.
Proactive participation by the top management during system design	Technological extensions supported by an integrated infostructure	Leveraging on a homogeneous infostructure to virtually integrate stakeholders

Figure 2. Process Model of ES Adoption for Dynamic Capability Development

firm's capability to be flexible. Finally, the strategization of an ES investment as the centerpiece of an extended enterprise solution becomes an essential maestro to a firm's capability to integrate and engender trust among its stakeholder groupings. Also included within the proposed framework are pragmatic suggestions for practitioners to consider in adopting ES. These propositions correspond to Montealegre's (2002) prescription of managerial actions to be undertaken in developing dynamic capabilities. The model thus offers practitioners a methodical approach for the strategization of ES investments to recover competitive value.

Furthermore, the contrasting approaches to ES adoption in IRAS and TechCo offer unique insights into the management of ES-inspired dynamic capabilities development. Strikingly, the integration of prior knowledge is a more tedious process with commercial ES packages than systems developed in-house because the underlying business models embedded within off-the-shelf ES may be incompatible with the existing operational procedures of an organization. Resistance also poses more of a hassle for off-the-shelf ES as in-house development allows users to formulate the perimeters of the ES, making them more accepting of the systems.

Apart from practical implications, the framework raises a number of questions which may be considered as future research directions. First, generalization is a definite liability of case studies, and as such, follow-up studies should be conducted to validate the preliminary findings of this study. Second, the contrast between in-house and off-the-shelf ES opens up unexplored territory for further investigation. Third, the discrepancies in knowledge retention, resource integration, and stakeholder management point to significant variations within the broad notion of ES, which in turn warrant further investigations to establish their implications for capability development.

In sum, the role of ES adoption is tantamount to that of an *orchestrator* in deciding the rhythm of capability development, i.e., ES serve as nerve centers that coordinate the operational manifestation of the capability development process.

REFERENCES

Adam, F., and O'Doherty, P. "Investigating the Reality of ERP Implementations," *Journal of Information Technology* (15:4), 2000, pp. 305-16.

Alavi, M., and Leidner, D. E. "*Review:* Knowledge Management and Knowledge Management Systems: Conceptual Foundations and Research Issues," *MIS Quarterly* (25:1), 2001, pp. 107-36.

Alchian, A. A., and Demsetz, H. "Production, Information Costs and Economic Organization," *American Economic Review* (62), 1972, pp. 777-94.

Amit, R. H., and Schoemaker, P. J. H. "Strategic Assets and Organization Rent," *Strategic Management Journal* (14), 1993, pp. 33-46.

Barney, J. B. *Gaining and Sustaining Competitive Advantage*, Reading, MA: Addison-Wesley, 1997.

Baskerville, R., Pawlowski, S., and McLean, E. "Enterprise Resource Planning and Organizational Knowledge: Patterns of Convergence and Divergence," in W. J. Orlikowski, S. Ang, P. Weill, H. C. Krcmar, and J. I. DeGross (Eds.), *Proceeding for the 21st International Conference on Information Systems*, Brisbane, Australia, 2000, pp. 396-406.

Benbasat, I., Goldstein, D. K., and Mead, M. "The Case Research Strategy in Studies of Information Systems," *MIS Quarterly* (11:3), 1987, pp. 372-86.

Blackler, F. "Knowledge and the Theory of Organizations: Organizations as Activity Systems and the Reframing of Management," *Journal of Management Studies* (30:6), 1995, pp. 863-85.

Chakravarthy, B. "A New Strategy for Coping with Turbulence," *Sloan Management Review* (38:2), 1997, pp. 69-82.

Clemons, E. K. "Information Systems for Sustainable Competitive Advantage," *Information & Management* (11:3), 1986, pp. 131-136.

Clemons, E. K. "Investment in Information Technology," *Communications of the ACM* (34:1), 1991, pp. 22-36.

Davenport, T. H. "The Future of Enterprise System-Enabled Organizations," *Information Systems Frontiers* (2:2), 2000a, pp. 163-80.

Davenport, T. H. *Mission Critical: Realizing the Promise of Enterprise Systems*, Boston: Harvard Business School Press, 2000b.

Davenport, T. H. "Putting the Enterprise into the Enterprise System," *Harvard Business Review* July-August, 1998, pp. 121-31.

Dosi, G., and Marengo, L. "Some Elements of Evolution Theory of Organizational Competences," in R. W. England (Ed.), *Proceedings of the 10th World Congress of the International Economic Association*, Moscow, Russia, 1992, pp. 157-179.

Eisenhardt, K. M. "Better Stories and Better Constructs," *Academy of Management Review* (16:3), 1991, pp. 620-27.

Eisenhart, K. M., and Martin, J. A. "Dynamic Capabilities: What Are They?," *Strategic Management Journal* (21), 2000, pp. 1105-1121.

Everdingen, Y., Hillegersberg, J., and Waarts, E. "ERP Adoption by European Midsize Companies," *Communications of the ACM* (43:4), 2000, pp. 27-31.

Freeman, R.E. *Strategic Management: A Stakeholder Approach*, Boston: Pitman, 1984.

Ginsberg, A., and Venkatraman, N. "Contingency Perspective of Organization Strategy: A Critical Review of the Empirical Research," *Academy of Management Review* (9:3), 1985, pp. 421-434.

Glaser, B., and Strauss, A. *The Discovery of Grounded Theory*, Chicago: Aldine, 1967.

Goodhue, D. L., Barbara, H.W., and Watson, H. J. "Realizing Business Benefits Through CRM: Hitting the Right Target in the Right Way," *MIS Quarterly Executive* (1:2), 2002, pp. 79-94.

Grant, R. M. "Prospering in Dynamically-Competitive Environments: Organizational Capability as Knowledge Integration," *Organization Science* (7:4), 1996, pp. 375-387.

Hamel, G., and Prahalad, C. K. *Competing for the Future*, Boston: Harvard Business School Press, 1994.

Hayman, L. "ERP in the Internet Economy," *Information Systems Frontiers* (2:2), 2000, pp. 137-139.

Holland, C. P., and Light, B. "A Critical Success Factors Model for Enterprise Resource Planning (ERP) Implementation," *IEEE Software* (16:3), 1999, pp. 30-35.

Howcroft, D., and Light, B. "A Study of User Involvement in Packaged Software Selection," in L. Applegate, R. Galliers, and J. I. DeGross (Eds.), *Proceedings of the 23rd International Conference on Information Systems*, Barcelona, Spain, 2002, pp. 69-77.

Ives, B., and Learmonth, G. P. "The Information System as a Competitive Weapon," *Communications of the ACM* (27:12), 1984, pp. 1193-201.

Jarvenpaa, S., and Leidner, D. "An Information Company in Mexico: Extending the Resource Based View of the Firm to a Developing Country Context," *Information Systems Research* (9:4), 1998, pp. 342-361.

Kettinger, W. J., Grover, V., Guha, S., and Segars, A. H. "Strategic Information Systems Revisited: A Study in Sustainability and Performance," *MIS Quarterly* (18:1), 1994, pp. 31-57.

Kumar, K., and Hillegersberg, J. "ERP Experiences and Evolution," *Communications of the ACM* (43:4), 2000, pp. 23-26.

Lacity, M., and Janson, M. A. "Understanding Qualitative Data: A Framework of Text Analysis Methods," *Journal of Management Information System* (11:2), 1994, pp. 137-155.

Lee, Z., and Lee, J. "An ERP Implementation Case Study from a Knowledge Transfer Perspective," *Journal of Information Technology* (15:2), 2000, pp. 281-288.

Lieberman, M., and Montomery, D. "First Mover Advantages," *Strategic Management Journal* (9:1), 1988, pp. 41-58.

Mahone, J. T., and Pandian J. R. "The Resource-Based View Within the Conversation of Strategic Management," *Strategic Management Journal* (13:5), 1992, pp. 363-80.

Maidique, M. A., and Zirger, B. J. "The New Product Learning Cycle," *Research Policy* (14:6), 1985, pp. 299-309.

Markus, M. L. "Paradigm Shifts—E-Business and Business/Systems Integration," *Communications of the AIS*, (4:10), 2000, pp. 1-44.

Markus, M. L., and Tanis, C. "The Enterprise System Experience—From Adoption to Success," Chapter 10 in R. W. Zmud (Ed.), *Framing the Domains of Information Technology Management*, Cincinnati, OH: Pinnaflex, 2000, pp. 173–207.

Markus, M. L., Axline, S., Petrie, D., and Tanis, C. "Learning from Adopters' Experiences with ERP: Problems Encountered and Success Achieved," *Journal of Information Technology* (15:2), 2000a, pp. 245-65.

Markus, M. L., Petrie, D., and Axline, S. "Bucking the Trends: What the Future May Hold for ERP Packages," *Information Systems Frontiers* (2:2), 2000b, pp. 181-93.

Martin, E. W., Brown, C. V., DeHayes, D. W., Hoffer, J. A., and Perkins, W. C. *Managing Information Technology: What Managers Need to Know*, Upper Saddle River, NJ: Prentice Hall, 1999.

Mata, F. J., Fuerest, W. L., and Barney, J. B. "Information Technology and Sustained Competitive Advantage: A Resource-Based Analysis," *MIS Quarterly* (19:4), 1995, pp. 487-505.

McFlaran, F. W. "Information Technology Changes the Way You Compete," *Harvard Business Review* (62:3), 1984, pp. 98-103.

Merton, R. K., Fiske, M., and Kendall, P. L. *The Focused Interview: A Manual of Problems and Procedures* (2nd ed.), New York: Free Press, 1990.

Miyazaki, K. *Building Competencies in the Firm: Lessons from Japanese and European Optoelectronics*, New York: St. Martin's Press, 1995.

Montealegre, R. "A Process Model of Capability Development: Lessons from the Electronic Commerce Strategy at Bolsa de Valores de Guayaquil," *Organization Science* (13:5), 2002, pp. 514-531.

Mykytyn, K., Mykytyn Jr, P. P., Bijoy, B., McKinney, V., and Bandyopadhyay, K. "The Role of Software Patents in Sustaining IT-Enabled Competitive Advantage: A Call for Research" *Journal of Strategic Information Systems* (11), 2002, pp. 59-82.

Nelson, R. R., and Winter, S. J. *An Evolutionary Theory of Economic Change*, Cambridge, MA: Harvard University Press, 1982.

Nonaka, I. "A Dynamic Theory of Organizational Knowledge Creation," *Organization Science* (5:1), 1994, pp. 14-37.

Oliver, C. "Sustainable Competitive Advantage: Combining Institutional and Resource-Based Views," *Strategic Management Journal* (18:9), 1997, pp. 697-713.

Orlikowski, W. "Case Told as Organizational Change: Investigating Incremental and Radical Changes in System Development," *MIS Quarterly* (7:3), 1993, pp. 309-340.

Pan, S. L., Newell, S., Huang, J., and Wan, K. C. "Knowledge Integration as a Key Problem in an ERP Implementation," in V. Storey, S. Sarkar, and J. I. DeGross (Eds.), *Proceeding for the 22nd International Conference on Information Systems*, New Orleans, 2001, pp. 321-28.

Papazoglou, M. P., Ribbers, P., and Tsalgatidou, A. "Integrated Value Chains and Their Implications from a Business and Technology Standpoint," *Decision Support Systems* (29:4), 2000, pp. 323-342.

Penrose E. T. *The Theory of the Growth of the Firm*, New York: Wiley, 1959.

Priem, R. L., and Butler, J. E. "Is the Resource-Based 'View' a Useful Perspective for Strategic Management of Research?," *Academy of Management Review* (26:1), 2000, pp. 22-40.

Reich, B. H., and Benbasat, I. "An Empirical Investigation of Factors Influencing the Success of Customer-Oriented Strategic Systems," *Information System Research* (1:3), 1990, pp. 325-347.

Rosemann, M. "ERP Software: Characteristics and Consequences," in J. Preis-Heje, C. Ciborra, and K. Kautz (Eds.), *Proceedings of the 7th European Conference on Information Systems*, Copenhagen, Denmark, 1999, pp. 1038-1043.

Rosemann, M., and Watson, E. "Integrating Enterprise Systems in the University Curriculum," *Communications for the AIS* (8), 2002, pp. 200-18.

Ross, J. W. "Dow Corning Corporation: Business Process and Information Technology," *Journal of Information Technology* (14), 1999, pp. 253-266.

Ross, J. W., and Vitale, M. R. "The ERP Revolution: Surviving vs. Thriving," *Information Systems Frontier* (2:2), 2000, pp. 233-241.

Soh, C., Sia, S. K., and Tay-Yap, J. "Cultural Fits and Misfits: Is ERP a Universal Solution?," *Communications of the ACM*, (43:4), 2000, pp. 47-51.

Sprott, D. "Componentizing the Enterprise Application Packages," *Communications of the ACM* (43:4), 2000, pp. 63-69.

Strauss, A. L., and Corbin, J. *Basics of Qualitative Research: Grounded Theory Procedures and Techniques*, London: Sage Publications, 1990.

Teece, D. J., Pisano, G., and Shuen, A. "Dynamic Capabilities and Strategic Management," *Strategic Management Journal* (18:7), 1997, pp. 509-533.

Vitale, M. R. "The Growing Risks of Information Systems Success," *MIS Quarterly* (10:4), 1986, pp. 327-334.

Walsham, G. "The Emergence of Interpretivism in IS Research," *Information Systems Research* (6:4), 1995, pp. 376-394.

Wheeler, B. C. "NEBIC: A Dynamic Capabilities Theory for Assessing Net-Enablement," *Information Systems Research* (13:2), 2002, pp. 125-146.

Yin, R. *Case Study Research: Design and Methods*, London: Sage Publications, 1984.

ABOUT THE AUTHORS

Chee Wee Tan is a recent graduate from the M.Sc. Program in the Department of Information Systems, School of Computing, at the National University of Singapore and holds a B.Sc. (Honors) from the same university. He is currently employed as a teaching assistant in the School where he tutors undergraduate courses on Information Systems. His research interests reside in the areas of enterprise systems, e-government, and knowledge management. His works have been published in *Journal of the American Society for Information Science and Technology*, *European Journal of Information Systems*, and a number of international conferences on Information Systems. He is also the author of two textbooks for the undergraduate modules he has taught. Chee Wee Tan can be reached at tancw@comp.nus.edu.sg.

Eric T. K. Lim holds a B.A. (Honors) from the School of Communication and Information at the Nanyang Technological University in Singapore. He is currently pursuing his interest in research and collaborating with scholars from the National University of Singapore on several academic projects. Some of these projects have resulted in publications that have appeared in *Journal of the American Society for Information Science and Technology* and the proceedings of the European Conference on Information Systems. He has also been involved in the coordination of government-funded national surveys as well as an international online survey initiated by MTV Asia. Eric Lim can be reached at emani@singnet.com.sg.

Shan Ling Pan is an assistant professor and Coordinator of the Knowledge Management Laboratory in the Department of Information Systems of School of Computing at the National University of Singapore. Dr. Pan's primary research focuses on the recursive interaction of organizations and information technology (enterprise systems), with particular emphasis on issues related to work practices, cultures, and structures from a knowledge perspective. Specifically, he is interested in understanding the complex issues related to the adoption, implementation, and use of enterprise systems in organizations. Some of these enterprise systems include knowledge management systems, enterprise resource planning systems, and customer relationship management systems. Presently, he is the principal investigator of two research projects, "E-Commerce and Global Strategies" and "Awareness Information for Teamwork in Organization," both funded by the National University of Singapore. Dr. Pan's research work has been published in *IEEE Transaction on Engineering Management*, *Journal of the American Society for Information Science and Technology*, *Communications of ACM*, *Information and Organization*, *Journal of Strategic Information Systems*, *Journal of Organizational Computing and Electronic Commerce*, *European Journal of Information Systems*, *and Decision Support Systems*, as well as Academy of Management meetings and the International Conference on Information Systems. Shan Ling Pan can be reached at pansl@comp.nus.edu.sg.

Calvin M. L. Chan is currently a Ph.D. candidate and a teaching assistant in the Department of Information Systems, School of Computing, at the National University of Singapore. He holds a B.Sc. (Honors) from the University of Warwick as well as a Diploma from the Singapore Polytechnic. Prior to his Ph.D. studies, he has worked for a number of years as an IT project consultant in the public sector. He is also an executive committee member of the Information & Knowledge Management Society and has publications in the *Journal of the American Society for Information Science and Technology*, the proceedings of the Hawaii International Conference on System Sciences, and the proceedings of the Americas Conference on Information Systems. Calvin Chan can be reached at cchan@comp.nus.edu.sg.

29 ON TRANSFERRING A METHOD INTO A USAGE SITUATION

Brian Lings
University of Exeter

Björn Lundell
University of Skövde

Abstract Many things can militate against the successful transfer of IS methods from research to commercial environments. In this paper we synthesize a framework for reasoning about method transfer. Four main themes emerge from analysis of the relevant literature: the importance of a clear conceptual framework for a method; support for learning; usability within a defined context; and acceptability to stakeholders. These themes are elaborated in the paper, and also illuminated, by reference to Langefors' infological equation and from experience gained in four case studies of method transfer. We claim that there is an onus on both method developers and those responsible for method adoption to consider all identified aspects, in an attempt to minimize inherent tensions between methods in concept and methods in action.

Keywords: Method transfer, infological equation, experience of transfer, framework

1 INTRODUCTION

We consider here the issue of transferring a method from a method development environment to method users within defined usage contexts within commercial **organizations**. In particular, we present a framework for reasoning about method transfer. Such a framework highlights the key issues to be considered when reasoning about methods and their potential adoption, and is intended as a guide to method users (what should be considered in evaluating a method for possible adoption?), and method developers and promoters (what should be addressed when developing and describing a method?).

Much effort has been expended in the development of methods within the IS community, but the value and nature of methods have been subjects of ongoing debate.

The position taken here is that there remains a broad consensus on the need for methods at least as a guide to assist thinking and acting within IS development.

In considering IS research, Moody (2000) notes the fundamental importance of influencing and being influenced by practice, claiming that "the fundamental problem that needs to be solved in IS research is not a methodological problem nor a theoretical problem, but a knowledge transfer problem" (p. 359).

In this sense, method development without method transfer into a real usage situation is of little value. This sentiment is well represented in the literature (see, for example, Avgerou and Cornford 1993; Bubenko 1986; Fitzgerald 1996; Goldkuhl 1994; Kitchenham et al. 1997; Lyytinen 1987; Moody 2002; Shanks 1996; Wynenkoop and Russo 1995). Amplifying on this, Fitzgerald (1996), based on Ward (1992), adds the prerequisite that such proof "requires the method or technique to have been successfully applied to a non-trivial problem situation" (p. 12).

Method transfer must acknowledge the situated nature of IS development practices—that is, the concept of "method in action" (Fitzgerald et al. 2002)—an issue well addressed in the literature (e.g., Bansler and Bødker 1993; Baskerville et al. 1992; Fitzgerald 1997, 1998; Fitzgerald et al. 2003; Floyd 1986; Grant and Ngwenyama 2003; Hughes and Wood-Harper 2000; Introna and Whitley 1997; Jayaratna et al. 1999; Mathiassen and Purao 2002; Nandhakumar and Avison 1999; Ramesh et al. 2002; Russo et al. 1995; Stolterman 1991; Truex et al. 2000; Wastell 1996).

In the light of the above, the successful transfer of a given method must be interpreted as developing, within a given context and based on the method, a successful method in action.

The framework presented specifically focuses on what affects the success of any method transfer activity. In essence, success must be judged by (1) ease of deployment, (2) success of process (in terms of acceptability to stakeholders and management), and (3) effectiveness of outputs (in terms of a sense of ownership by stakeholders and value judgement by managers and other stakeholders). We use these ideas to reflect on our experience of transfer of the *2G* method into a number of company contexts. This qualitative method was developed for use in the evaluation of IT products, and is intended to aid in reasoning about the implications of adopting a product (Lundell and Lings 2003).

Other frameworks related to methods can be found in the literature, but have differing perspectives and foci. The NIMSAD framework (Jayaratna 1994), while addressing some method transfer issues, focuses on inherent properties of methods themselves rather than method transfer. The framework of Fitzgerald et al. (2002) focuses on ISD method use and is centered on method-in-action. It raises the issues of both rational and political aspects of methods and emphasizes the complexity and dynamic nature of the context in which a method is used.

2 ON METHODS

The terms *method* and *methodology* have been used in a variety of ways in the literature (see, for example, Wynekoop and Russo 1995), and considerable confusion can be generated by intermixing their usage. In order to avoid confusion, we use the term method to refer to documented ways of working in IS development and IT evaluation. In this sense, it is appropriate to refer to SSADM as an IS development method and the

2G method as an evaluation method. We use the term methodology only in the context of research methodology, namely when describing our approach to developing the framework proposed.

In this section, we clarify what we mean by method, which we term method-in-concept to distinguish it from method-in-action. Loosely speaking, method-in-concept refers to a method as understood by its stakeholders, and in this sense it is a social construction: a shared set of values and assumptions identified with a method within a professional community of method developers and users. For example, Hesse (2001) states that the rational unified process (RUP) is based on a phase-oriented software life-cycle model. If this is a shared view of the professional users of RUP, then it can be said to be a part of its underlying philosophy. However, the real claim is that there are no examples of RUP-in-action that are non-phase oriented, and there is reason to believe that this situation will persist.

Different authors have offered different definitions of method. It is clear that the tension between method-in-concept and method-in-action is reflected in the definitions offered, which can be viewed as taking different perspectives on the same phenomenon.

For example, in discussing the scope of object-oriented methods, Rumbaugh et al. (1991) and Henderson-Sellers *et al.* (2001) both take a method-in-concept perspective and identify a number of components (or constituents) for a method. This emphasizes the technical perspective of development.

From a more socio-technical perspective of development, Hirschheim et al. (1995), with a slight method-developer emphasis, define method-in-concept to be "an organized collection of concepts, methods,[1] beliefs, values and normative principles supported by material resources" (p. 22).

By way of contrast, Fitzgerald et al. (2002) offer a more method-user emphasis in their definition of a method as a

> coherent and systematic *approach*, based on a particular *philosophy* of systems development, which will guide developers on *what* steps to take, *how* these steps should be performed and *why* these steps are important in the development of an information system (p. 5).

A more sociological perspective, distinguishing between methods as perceived by method users and as perceived by method developers, is offered by Floyd (1986), who states,

> We consider methods not so much as static, well-defined objects, but as dynamic sources of ideas to be tailored to a given situation and transformed by use...there is a subtle interplay between the system development process as it is (in our view), as it should be (in our view), and as it should be (according to the method's view) (pp. 30-31).

[1]The term method is here used to denote "a well-defined description" of "a way of accomplishing a task" (p. 11). Hirschheim et al. use the term methodology to refer to method as used here.

If method-in-concept is to be transferred into method-in-action, then both social and technical issues must be addressed. This raises issues both with respect to method development and method deployment. These issues are further complicated when a method is to be supported in a tool. A method as implemented in a tool (method-in-tool) is likely to differ from its method-in-concept and any associated method-in-action. There are many reports of the tensions resulting from these differences (for further discussion, see Lundell and Lings 2004). We focus here on transfer of method-in-concept into method-in-action, but there are clearly also implications for transfer to method-in-tool.

For us, Avgerou and Cornford (1993) sum up the importance of method transfer when they claim that a "theoretically sound [method] which cannot be successfully communicated to and adopted by ordinary organizations and businesses is of little practical value" (p. 280).

While focusing on method transfer in this paper, we acknowledge that it is an aspect of broader issues that have received attention in the literature, including method adoption and diffusion (e.g., Beynon-Davies and Williams 2003) and technology transfer (e.g., McMaster et al. 1997).

3 RESEARCH METHODOLOGY

Our primary goal has been to establish the major factors affecting method transfer. A further goal has been to use this knowledge to plan transfer of the *2G* method into real company contexts, and to analyze our experience of transfer. To this end we have used literature analysis followed by a number of case studies. Our literature analysis emphasized coverage of relevant literature on the topic, and so was "not confined to one research methodology, one set of journals, or one geographic region" (Webster and Watson 2002, pp. xv-xvi). In fact method transfer emerged as one of the most significant themes within a broader literature study related to method validation.

Initially, we used scientific databases for searching high quality papers and conducted an extensive search of relevant journals and conference proceedings. Besides our use of keywords and citations (both forward and backward) for identification of sources, the search also involved systematic browsing.

The study started with a broad search for themes, and in particular factors affecting the effectiveness of methods.[2] The rich body of literature available for analysis includes sources from several (partly overlapping) areas which could be broadly categorized as IS assessment and evaluation (e.g., Etzerodt and Madsen 1988; Hirschheim and Smithson 1988; Kitchenham et al. 1997), IS methods (e.g., Avison and Fitzgerald 2003; Fitzgerald et al. 2002; Veryard and Macdonald 1994), and research methodology (e.g., Marshall and Rossman 1999; Maxwell 1996). None of our own papers was used in the review.

A set of themes emerged which was then used to direct further searches. Extensive note files were maintained throughout, and text relevant to themes annotated accordingly. As themes evolved, annotations evolved also. Occasionally, theme changes

---

[2]It should be noted that we were motivated to consider these specific areas by the development effort behind the *2G* method.

were such that significant reinterpretation of the set of note files was necessary. A framework for considering method transfer was thus iteratively evolved through a methodology that borrowed from qualitative coding techniques. A small proportion of the sources consulted is considered strongly relevant for informing a theory of method transfer.

The framework was used to plan and analyze four studies of transfer of the *2G* method, each conducted in a different IS development company. Each case study had a method user new to the method and based in a company for the equivalent of four months full time (see, for example, Rehbinder et al. 2001). All studies used individual open interviews; two studies supplemented these with group interviews. Method user experience of transfer was considered across the four case studies. For two of the case studies, the field notes taken by the method user were supplemented by one of the method developers through *post facto* stakeholder open interviews. Stakeholder feedback was also obtained through respective company seminars, involving (at least) the method user, a method developer (who took notes), and stakeholders involved in the case study. For the purpose of dissemination, other company representatives (including managers) also participated in these seminars, and interestingly also proffered feedback on the method application as well as its deliverables.

4 METHOD TRANSFER: AN INFOLOGICAL PERSPECTIVE

To understand method transfer it is enlightening to consider information transfer between human actors. Langefors' early work (1973; see also Langefors 1995) constitutes a foundation for "the Scandinavian approach" in the field of Information Systems. As part of this work, Langefors (1973) defined the infological equation in an attempt to explain the relationship between data and information, and so theorize about information transfer.

In a sense, the equation is a response and reaction to the early work in computing, which did not discriminate between data and information, in that Langefors' definition of information is dependent on a human aspect (which he refers to as *pre-knowledge*). The equation has influenced a number of thinkers (see, for example, Nissen 1995). In a recent elaboration,[3] Langefors himself has stressed the importance of shared pre-knowledge for all forms of communication, including oral, and that this continues to have major relevance in IS.

The equation (Langefors 1973, p. 248; 1987, p. 90; 1995, p. 144) is based on the assumption that information I is obtained through an interpretation i that is dependent not only on data D, but also on the pre-knowledge S of the observer and the available time t for interpretation by the observer:

$$I = i(D, S, t)$$

[3]Invited talk, University of Skövde, April 24, 2003: "ADB i ett historiskt perspektiv."

For example, given a certain available time *t* for observation, the resulting information *I* will be limited if *D* is extensive but *S* is limited; and similarly if *D* is limited even if *S* is extensive. Of course, *S* will evolve over time, as it constantly changes with experience.

In fact, to Langefors, *S* is interpreted as complete life experience, and of course no two people can have the same *S*. For information transfer to take place, two people only require "sufficiently similar" pre-knowledge (Langefors 1987, p. 90). Langefors himself acknowledges that not all of an actor's pre-knowledge is relevant to interpretation of data. Here we, perhaps controversially, interpret sufficient similarity to be also context dependent, in that not all of an actor's pre-knowledge is relevant to any *given* interpretation. To characterize this context dependency, we will refer to congruence of pre-knowledge between communicators to distinguish it from Langefors' notion of sufficient similarity.

Considering method transfer from this perspective, we can intuit that a method user's facility with a method will be a function of the available method description *D*, the user's relevant pre-knowledge *S*, and the time *t* for interpretation. It should be noted that *D* represents communicable data, which will include method documentation but may include any form of communication. Hidding (1997), in a study of methods and the use of their descriptions by practitioners, makes a number of relevant observations which we interpret in the light of the equation. For a method user, *D* will include the full documentation of the method which, according to Hidding, is usually "voluminous and detailed in different sets of folders and binders" (p. 104). However, many practitioners "had internalized the [method] to the point that it had become subconscious" (p. 105), in fact to the level that some even claimed that they did not use any method. This is indicative of very high relevant pre-knowledge *S* ("practitioners no longer 'interpreted' methodology, as they had 'compiled' it" [p. 105]).

It should also be noted that if *D* is large, then practitioners for various reasons will not be willing to invest large amounts of time *t* and so high pre-knowledge *S* with ready access to relevant parts of *D* will be demanded. Such behavior was noted by Hidding: practitioners demand ready access to method documentation but will typically be prepared to read only around 30 pages as required (p. 106).

While not directly influencing the framework for method transfer presented in this paper, we use Langefors' equation in interpreting many of the concepts that emerged within the framework.

5 THE FRAMEWORK FOR METHOD TRANSFER

Through our literature analysis and our early experience of method transfer (Lundell and Lings 1999), the framework for reasoning about method transfer represented in Figure 1 has been evolved.

5.1 Clear Conceptual Framework

All methods have an underlying philosophy, whether implicit or explicit, and offer at least some guidance for a way of working. The extent to which they are prescriptive

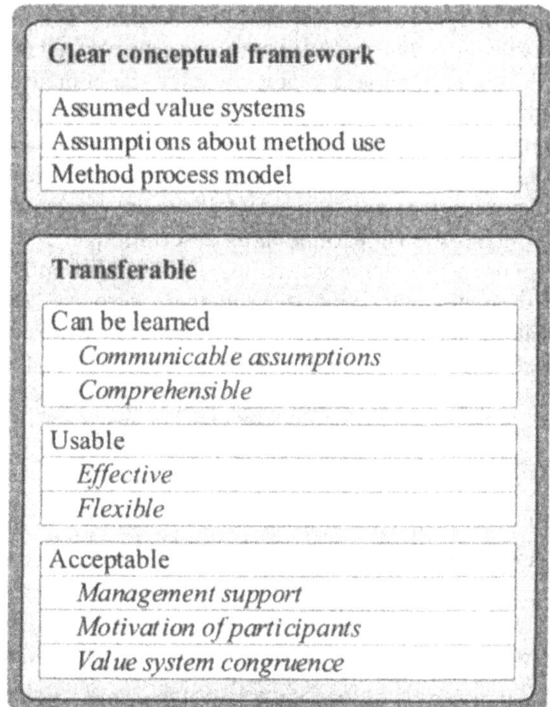

Figure 1. A Framework for Reasoning about Method Transfer

in the latter varies, as does the consequent detail to be conveyed to any potential method user (*D* in Langefors' equation). The (often tacit) conceptual framework on which a method draws can usefully be considered as contributing to an understanding of the pre-knowledge of the method developer—which we will indicate by S_m. Of course, this is an inexact analogy. A method may be developed by many developers, each with their own pre-knowl edge, and therefore each with their own method-in-concept. This may mean that inconsistent interpretations exist of the conceptual framework of a method and thereby what is good practice with respect to the method. We acknowledge this complexity, but for the purposes of exposition will assume a coherent S_m.

It is clear that a user's pre-knowledge is fundamental in influencing the ease with which a method may be transferred, and thereby influences the perception of its success. In other words, efficient transfer requires congruence for the method between S_m and the user's *S*.

5.1.1 Assumed Value Systems

The traditional view of IS development and evaluation as a technical process has been questioned. For example, Myers and Young (1997) note the focus of much information systems research on the "social, political and organizational aspects of IS development" (p. 224), and Bennetts et al. (2000) claim that it is "clear that one key issue in ISD is organizational politics" (p. 194). Furthermore, as software development is a social process (Pflegger 1999, p. 34), or at least partly a social process (Sawyer and Guinan 1998, p. 552), it is important to acknowledge the complexity involved in the social and political issues of human communication. Therefore, it seems that a consideration of human issues with respect to methods and method usage is of fundamental importance. Interestingly, as noted by Beynon-Davies and Williams (2003), the dynamic systems development method (DSDM) does have some acknowledgment of cultural issues and organizational learning in its description.

However, methods vary widely in their underlying philosophies and a mismatch with the culture of the deployment context could be disastrous (Carroll 2003). Indeed, Avison and Fitzgerald (2003) identify underlying philosophical assumptions of methods as "perhaps the most important aspect" when comparing methods (p. 55), but note that these are frequently not made explicit, making them difficult to assess.

5.1.2 Assumptions about Method Use

Assumptions may be made about the intended usage of a method. Some method providers demand that their users are certified (as with SSADM). Assumptions may also be made about the nature of the objectives of method use. For example, a method may be directed at the production of safety critical software (such as VDM); such a method would be unlikely to suit a research environment targeting the production of prototype software systems. Further, some may specifically cater for scaling, but many may not (Laitinen et al. 2000).

Method developers must also be clear about the intended context for use of a method. For example, it may be assumed that a method will be applied in a production environment (such as may be the case for COTS). This will be detached from any specific context in which the product will be deployed. Alternatively, a method may be designed for a high level of stakeholder participation (for example, SSM), within the context of the intended software usage situation. It may also be that different assumptions are made about different stages within a method.

5.1.3 Method Process Model

By method process model we mean any implied way of working within a method. Certainly there are different views on appropriate models of such processes. For example, according to Nandhakumar and Avison (1996), "it has been widely acknowledged that information systems development in practice is not a sequential process" (p. 210), and Kruchten (1999) claims benefits for RUP in its iterative processes. In fact in Multiview2, there is "no implied precedence in the four components of the methodology...since all four are co-present" (Avison et al. 1998, p. 130). Some methods will emphasize contingency, perhaps offering support for method tailoring, while others, such as SSADM, will emphasize a "clear prescriptive structure, in which every step in the development process is precisely delineated" (Wastell 1996, p. 26). For the purpose of method transfer, it is important that any inherent assumptions about process be conveyed to a method user.

5.2 Transferable

We will use the infological equation as an analytic device to elaborate on important factors for method transfer. In particular, we make the following observations.

First, efficient transfer requires the intended method user to have sufficient pre-knowledge (S) to interpret the data available about a method (D). In other words, a good transfer strategy will acknowledge the need to align S with the pre-knowledge of the method developers (S_m).

Second, if a large volume of documentation is assumed necessary for detailing a method (i.e., there is a large D), then even with S reasonably congruent with S_m, the time necessary for transfer may be large. For example, a proficient user of the rational unified process (RUP) is likely, on moving to object-oriented process, environment, and notation (OPEN), to have closely congruent pre-knowledge (S with S_m for OPEN), but may still take a long time to fully interpret the new method.

Of course, documentation cannot be complete as aspects of a method will always be implicit. Efficient transfer, therefore, requires high congruence between S and S_m or, more realistically, a means of increasing the congruence between S and S_m. This may be facilitated, for example, by supplementing the published material about a method (D) through communication with other method users (i.e., by opening other communication channels; see, for example, Hughes and Wood-Harper 2000, p. 401). The cost of each of these new communications must be acknowledged as effectively extending the time for interpretation of D, and for a given situation it may be too costly to achieve the required congruence between S and S_m.

With these issues in mind, we consider each transfer factor in turn.

5.2.1 Communicable Assumptions

In essence, *communicable assumptions* refers to information interchange concerning the conceptual framework of a method. The reason for communication is to assist in aligning the potential method user's pre-knowledge with the assumptions behind the method (i.e., increasing the congruence between S and S_m), and for this to happen these assumptions must be made available.

Of course, as pointed out by Russo and Stolterman (2000), whether design knowledge can be communicated to practitioners is an open question, but if such knowledge is not communicable then it is not possible to change the way practicing designers view the design process, and method transfer becomes an impossibility. We would characterize this scenario by observing that it would reflect very low relevant pre-knowledge (S) of the method receiver, with no realistic way of increasing S with the resources available.

5.2.2 Comprehensible

Method developers must consider how to communicate underlying method assumptions, and those involved in method transfer must adopt strategies for efficient communication of them to potential method users. This is unlikely to be achieved simply by increasing *D*. In fact, Introna and Whitley (1997) argue that this is not possible. Hence, alternative strategies need to be devised. For example, Hidding's (1997) reference to "voluminous and detailed" documentation was to point out that it was inaccessible at the point of need, and so did not assist comprehension in the way intended. This may be helped by mechanisms to assist the retrieval of relevant documentation, but during transfer it is difficult for a putative method user to know what may be relevant.

Hence, a further implication of transfer is the need for access to human expertise. This can take the form of interaction with colleagues, training, real-time support, and mentoring. The key factor is access to a human actor knowledgeable about the method and the context of its deployment—in other words, with pre-knowledge that is congruent with both S and S_m. For example, a commercial method consultant will be strongly congruent with S_m but not necessarily with S, whereas a method user within an organization may well have pre-knowledge less congruent with S_m but may be more effective in support of a colleague because of congruence of local knowledge.

5.2.3 Effective

As noted by Wynekoop and Russo (1997), there is no single way of assessing effectiveness. We consider effectiveness to be the extent to which use of a method is seen to contribute to achieving organizational goals. However, perceived ineffectiveness may reflect ineffectiveness of transfer rather than inappropriateness of the method for the particular usage situation.

Effectiveness of transfer can be considered to be the extent to which S has evolved toward congruence with S_m, thus allowing high information flow, and therefore rapid progression toward method expertise. Method users who have deeply understood and internalized the principles behind a method are more likely to be more innovative in their ways of working with the method. If this level is not reached, there is a danger that following a method can stifle creativity (Wastell 1996).

5.2.4 Flexible

Flexibility includes "the ability of the [method] to be adapted and improved" (Veryard and Macdonald 1994, p. 270). In particular, it is essential to establish whether the particular assumptions behind the application of the method are too particular to an envisaged usage context. There is also the related issue of whether the method scales for use in different usage situations within a context. As noted earlier, it is unclear whether many methods lend themselves to being scaled down (Laitinen et al. 2000). Furthermore, it may also be important to be able to adapt methods to local work practices (Fitzgerald et al. 2003).

In terms of the infological equation, adapting a method to a context is effectively a process of adjusting a method in order to increase congruence for the method between the method developer and the particular method users. Hence, the method developer's pre-knowledge, S_m, becomes more closely congruent for the method, within the usage situation, with the S of each method user.

5.2.5 Management Support

Lyttinen (1987) observes that "One reason for the abundance of IS design approaches is that it is quite easy to develop a method, but difficult to get it accepted" (p. 4). It is clearly the case that management support is important in influencing such acceptance (see, for example, Kozar 1989).

The investment required in adopting a method is likely to be high, and no matter how efficient the transfer, t is unlikely to be short enough for managers. Thought must, therefore, be given to how this can be ameliorated. For example, it may be possible for methods to offer managers useful early indications of success. If benefits can be expected from method use early in method transfer, then investment in adoption is more likely.

5.2.6 Motivation of Participants

According to Beynon-Davies and Williams (2003), methods are "the vehicles by which practitioners...introduce changes to development practices" (p. 30). Taken alongside the claim of Viller and Somerville (2000) that methods "are unlikely to be adopted in industry unless they can be integrated with existing practice" (p. 169), a potential tension can be observed. We see this as supporting the view of Grant and Ngwenyama (2003), who identify the background knowledge (i.e., S) and motivation of those applying a method as an important factor in affecting the outcome of the application of a method.

5.2.7 Value System Congruence

It is clear that information exchange is maximized if method user pre-knowledge (S) is congruent for the method with the method developer's pre-knowledge (S_m). It is important to note, however, that methods "implicitly or explicitly demonstrate the value sets of their creators" (Jayaratna 1996, p. 26). That is, the value systems of the creators of a method are embodied in S_m.

At the same time, "Methodology users' values play a significant role in terms of the choice of methodology. It is natural to adopt or use a methodology that is congruent with their value systems" (Jayaratna et al. 1999, p. 33). That is, a method is more likely to be adopted if there is congruence for the method between the value systems implicit in S and S_m.

6 OBSERVATIONS FROM CASE STUDIES WITH THE *2G* METHOD

Our own model of case studies for method transfer involves a number of roles, including method developer, method user, mentor, and participant. In the studies reported here, a method developer acted as mentor for each method user. Each method user was based in a different company, with each study taking place over a period of from four to six months. Method users were given access to senior developers—the participants—who made time available for open interviews.

In the rest of this section, we use our experience from four case studies on method transfer, here referred to as alpha, beta, gamma, and delta, to illuminate the main themes of the method transfer framework.

6.1 On the Conceptual Framework of the *2G* Method

There is an underlying assumption behind the *2G* method that evaluation is a socio-technical activity, and that evaluation framework development is a key early phase in any evaluation activity. The method differs in two main respects from other systematic methods for developing an evaluation framework. First, it does not use concepts that have been defined *a priori*. Instead, the definition of concepts evolves during analysis. Second, it does not use an *a priori* structure for interrelating these concepts. Instead, interrelationships emerge during analysis. Therefore, we would characterize its approach as primarily data driven: it is a qualitative method, informed by grounded theory, and requires a user to be sympathetic to qualitative techniques.

The method grounds data both from an organizational and a technological perspective, using an iterative, two-phase process. The focus is on organizational need in the first phase, but shifts in the second phase to how needs might be met through current technology.[4] The method is not prescriptive, but gives clear guidelines concerning the use of different kinds of data source in the two phases, and how frameworks may be evolved. It is important that both phases of the method application take place in the organizational setting in which the technology under investigation would be used. The method is intended as a general method, scalable according to context.

The development of an evaluation framework is an evolutionary process involving data collection, analysis, and coding. These activities are not inherently sequential; each can affect (and trigger) the others so that, in essence, all activities are going on together. This characteristic is inherited from grounded theory. In practice, the method uses a process model of iteration between its two phases, but there is no rigid assumption about the speed of this iteration. For example, phase change may occur several times within an interview session, or each session may be devoted to a single phase.

6.2 On the Transfer of the *2G* Method: Can it Be Learned?

None of the users had prior exposure to the *2G* method, or to qualitative techniques in general. There was variation in the level of prior work experience, and in specialism, but all were sympathetic to qualitative ideas and had outgoing personalities. It is an acknowledged weakness of the studies that only in one case (delta) was the method user employed in the respective company prior to undertaking the study, and so had been genuinely a part of the context of the *2G* method application (for one year).

Initially, selected readings were offered, both of documentation of the method and of specific aspects of qualitative techniques, such as conducting open interviews and

[4]Interpreted broadly as "any method, technique, tool, procedure or paradigm" (Pfleeger 1999, p. 111) used in IS development or maintenance.

coding. Tutoring in the underlying assumptions behind the method was continued through mentoring, which took place outside the organizational context. Method users were also required to gain familiarity with the phenomenon under evaluation, which was different for each case study, but this is not considered specific to the *2G* method.

In the alpha study, the method user had a systems analysis background; in the other studies, the method users had a more technical background. However, all adapted readily to conducting open interviews. Coding was in general more problematic, requiring mentor assistance in the early stages of each method application.

The nature of the context of the delta study led to initial misunderstanding of the exact role of the second phase of the method, requiring further input from the mentor. In the other studies, the nature of the two phases was more clearly understood, and only the usual problems of field work using elite interviews, namely ready access to senior developers, was experienced.

6.3 On the Transfer of the *2G* Method: Usable?

In each case study, the framework produced was presented to participating stakeholders for comment after completion. In the alpha study, organizational need was felt to be very well reflected in the framework but technical detail was not fully developed. There were two contributory factors here. First, a very broad scope for the interviews demanded the analysis of large amounts of rich data. Second, the method user had restricted pre-knowledge of the technology under consideration, which made coding more challenging than anticipated. The beta study also produced a rich framework. It was initially difficult to delineate the desired scope of the study, which caused the framework to be broader than eventually required. This again inhibited its refinement with an appropriate level of technical detail.

The beta study required the method user to apply the *2G* method in a novel way, namely in a post-usage situation. This meant that open interviews were more challenging. Interviewees had already compiled the knowledge gained from their experience, and accessing information about their experience at the time of tool usage was difficult. This contributed to the initial difficulty in delineating the appropriate scope of the study.

The gamma and delta studies were conducted in smaller companies and with a narrower focus. The *2G* method was applied in a lightweight fashion, using group interviews and coding with limited interrelationships. In these studies, the level of technical detail in the resulting frameworks was considered by the stakeholders to be good. The delta study also applied a novel technique in phase 2, using prototypes to stimulate discussion.

6.4 On the Transfer of the *2G* Method: Acceptable?

In all studies care was taken to explain the nature of the method to be used, requirements in terms of stakeholder commitments, and expected organizational benefits, both in terms of the delivered output (the evaluation framework) and organizational learning. Initial contact was made with all participants, and method users were based in the companies throughout the studies. It was felt important to build trust, and therefore to explain exactly how the method user would handle confidentiality—an important aspect politically, as testified in post-study interviews.

In all of the case studies, a sizeable majority of stakeholders have been enthusiastic participants, which has contributed significantly to the level of success. In the alpha study, feedback from *post facto* analysis suggested that the method was highly transparent, and appeared to be a natural way of working. From the beta study, one stakeholder was committed to following the application in detail: an initial skepticism of the way of working as a method user changed over the course of the study so that, by the end, there was management commitment to using the method internally. In the delta study the resulting evaluation framework was adopted by the company as the basis on which to select a product for adoption, but the method itself was again seen as transparent. In all of the studies there was feedback that participation was an informative experience, even where a delivered framework was considered to require further development.

7 CONCLUSIONS

In this paper we have reasoned about method transfer as a special case of knowledge transfer. We have presented a framework for considering the issue of method transfer, which we claim has implications for method development. We have also applied the framework to reasoning about our own experience of method development and transfer.

The infological equation has helped us to reason about method-in-concept and method-in-action. In particular, it is clear that there can be no realistic assumption that method-in-action will ever reflect a method-in-concept as understood by a method developer. This would require full congruence, for the method, of the method user's and method developer's pre-knowledge. Given that in practice there is likely to be more than one method user, such congruence is even more unlikely.

One reaction to this situation may be to attempt to engineer a method for a specific context, effectively tailoring a method in full appreciation of the given pre-knowledge of the intended method users. However, this again is unrealistic when viewed in terms of the infological equation. Not only is user pre-knowledge difficult to access, but it will vary between users and change over time. Hence, an engineered method will still be a method-in-concept, with all the implications that that implies. Further, the engineering itself suggests a requirement for documentation of the engineered method and its rationale. This implies added investment, which may not be forthcoming.

One final position is to look to IS developers to evolve effective practice in their own context, without requiring adherence to any method-in-concept. This can be said to acknowledge real practice, and can be seen as offering freedom for creativity among developers. This treats methods-in-concept as merely exemplars, to be used to develop further the pre-knowledge of developers. This removes the problem of congruence with an external method developer's pre-knowledge, although the different pre-knowledge of the individual users may remain a problem. From a management perspective, acceptability may be lowered because of the need to control projects. The fact that working practices will, to a large extent, be embodied rather than documented can be a two-edged sword for managers. Lack of prescribed procedures may be attractive for recruitment, but may militate against quick and effective utilization of new staff.

Behind all method use is the question of method support, and in particular the role that tools can play in IS development. It may be that IS development requires some form of tool support, but it is unclear what form this should take. This is a particular question for those who advocate direct support for methods in tools, such as in CASE.

We believe that the role of IS researchers is to attempt to minimize the inherent tensions between method-in-tool, method-in-concept, and method-in-action. For method proponents, this means explicitly addressing the issues raised: concerning a clear conceptual framework, highlighting early benefits of adoption, and explicitly addressing scalability. Further research is required into how methods can be presented to better support comprehension, evaluation for contextual relevance, and adaptation. Open questions also remain concerning evaluation, adaptation and diffusion within defined contexts. More situated research is required to aid in understanding the complex dynamics of method adaptation and diffusion within real organizations, and more support must be offered for organizations wishing to tailor methods for their own environments. Finally, there is a need for more research into the nature of effective tool support for methods. As a minimum, tools must be non-prescriptive with respect to method; at best we need to know how to build tools that transparently support ISD in all its variety.

We claim that the major factors underlying these issues are those identified in the method transfer framework presented here. They relate to congruence of users' pre-knowledge with a method's underlying principles, and effective ongoing support for the method.

ACKNOWLEDGMENT

The authors are grateful to Anders Malmsjö for discussions about interpreting the infological equation.

REFERENCES

Avgerou, C., and Cornford, T. "A Review of the Methodologies Movement," *Journal of Information Technology* (8:5), 1993, pp. 277-286.

Avison, D., and Fitzgerald, G. *Information Systems Development: Methodologies, Techniques and Tools* (3rd ed.), Maidenhead, England: McGraw-Hill, 2003.

Avison, D. E.; Wood-Harper, A. T.; Vidgen, R. T.; and Wood, J .R .G. "A Further Exploration into Information Systems Development: The Evolution of Multiview2," *Information Technology and People* (11:2), 1998, pp. 124-139.

Bansler, J. P., and Bødker, K. "A Reappraisal of Structured Analysis: Design in an Organizational Context," *ACM Transactions on Information Systems* (11:2), 1993, pp. 165-193.

Baskerville, R.; Travis, J.; and Truex, D. "Systems Without Method: The Impact of New Technologies on Information Systems Development Projects," in K. E. Kendall, K. Lyytinen, and J. I. DeGross (Eds.), *The Impact of Computer Supported Technologies on Information Systems Development*, Amsterdam: North-Holland, 1992, pp. 241-269.

Bennetts, P. D. C.; Wood-Harper, A. T.; and Mills, S. "An Holistic Approach to the Management of Information Systems Development: A View Using a Soft Systems Approach and Multiple Viewpoints," *Systemic Practice and Action Research* (13:2), 2000, pp. 189-205.

Beynon-Davies, P., and Williams, M. D. "The Diffusion of Information Systems Development Methods," *The Journal of Strategic Information Systems* (12:1), 2003, pp. 29-46.

Bubenko Jr., J. A. "Information System Methodologies: A Research View," in T. W. Olle, H. G. Sol, and A. A. Verrijn-Stuart (Eds.), *Information Systems Design Methodologies: Improving the Practice*, Amsterdam: North-Holland, 1986, pp. 289-318.

Carroll, J. "The Process of ISD Methodology Selection and Use: A Case Study," in *Proceedings of the 11th European Conference on Information Systems,* Naples, June 16-21, 2003.

Etzerodt, P., and Madsen, K. H. "Information Systems Assessment as a Learning Process," in N. Bjørn-Andersen and G. B. Davis (Eds.), *Information Systems Assessment: Issues and Challenges,* Amsterdam: North-Holland, 1988, pp. 333-345.

Fitzgerald, B. "An Empirical Investigation into the Adoption of Systems Development Methodologies," *Information & Management* (34:6), 1998, pp. 317-328.

Fitzgerald, B. "Formalized Systems Development Methodologies: A Critical Perspective," *Information Systems Journal* (6:1), 1996, pp. 3-23.

Fitzgerald, B. "The Use of Systems Development Methodologies in Practice: A Field Study," *Information Systems Journal* (7:3), 1997, pp. 201-212.

Fitzgerald, B.; Russo, N. L.; and Stolterman, E. *Information Systems Development: Methods in Action*, London: McGraw-Hill, 2002.

Fitzgerald, B.; Russo, N. L.; and O'Kane, T. "Software Development Method Tailoring at Motorola," *Communications of the ACM* (46:4), 2003, pp. 65-70.

Floyd, C. "A Comparative Evaluation of System Development Methods," in T. W. Olle, H. G. Sol, and A. A. Verrijn-Stuart (Eds.), *Information Systems Design Methodologies: Improving the Practice*, Amsterdam: North-Holland, 1986, pp. 19-54.

Grant, D., and Ngwenyama, O. "A Report on the Use of Action Research to Evaluate a Manufacturing Information Systems Development Methodology in a Company," *Information Systems Journal* (13:1), 2003, pp. 21-35.

Goldkuhl, G. "Välgrundad Metodutveckling," Research Report LiTH-IDA-R-94-04, Department of Computer and Information Science, Linköping University, Linköping, Sweden, January, 1994 (in Swedish).

Henderson-Sellers, B.; Collins, G.; Due, R.; and Graham, I. "A Qualitative Comparison of Two Processes for Object-Oriented Software Development," *Information and Software Technology* (43:12), 2001, pp. 705-724.

Hesse, W. "Dinosaur Meets Archaeopteryx? Seven Thesis on Rational's Unified Process (RUP)," in J. Krogstie, K. Siau, and T. Halpin (Eds.), *Proceedings of the Sixth CAiSE/IFIP8.1 International Workshop on Evaluation of Modeling Methods in Systems Analysis and Design (EMMSAD'01)*, Interlaken, Switzerland, 2001, pp. VII-1–VII-8.

Hidding, G. J. "Reinventing Methodology: Who Reads It and Why?," *Communications of the ACM* (40:11), 1997, pp. 102-109.

Hirschheim, R.; Klein, H. K.; and Lyytinen, K. *Information Systems Development and Data Modeling: Conceptual and Philosophical Foundations*, Cambridge, England: Cambridge University Press, 1995.

Hirschheim, R., and Smithson, S. "A Critical Analysis of Information Systems," in N. Bjørn-Andersen and G. B. Davis (Eds.), *Information Systems Assessment: Issues and Challenges*, Amsterdam: North-Holland, 1988, pp. 17-37.

Hughes, J., and Wood-Harper, A. T. "An Empirical Model of the Information Systems Development Process: A Case Study of an Automotive Manufacturer," *Accounting Forum* (24:4), 2000, pp. 391-406.

Introna, L. D., and Whitley, E. A. "Against Method-*ism*: Exploring the Limits of Method," *Information Technology & People* (10:1), 1997, pp. 31-45.

Jayaratna, N. "Choice of a Methodology for Information Systems Development!," in N. Jayaratna and B. Fitzgerald (Eds.), *Lessons Learned from the Use of Methodologies: Information*

Systems Methodologies—Fourth Conference on Information Systems Methodologies, British Computer Society: Information Systems Methodologies Specialist Group, 1996, pp. 23-28.

Jayaratna, N. *Understanding and Evaluating Methodologies—NIMSAD: A Systemic Framework*, London: McGraw-Hill, 1994.

Jayaratna, N.; Holt, P.; and Wood-Harper, A. T. "Criteria for Methodology Choice in Information Systems Development," *The Journal of Contemporary Issues in Business and Government* (5:2), 1999, pp. 30-34.

Kitchenham, B.; Linkman, S.; and Law, D. "DESMET: A Methodology for Evaluating Software Engineering Methods and Tools," *Computing & Control Engineering Journal* (8:3), 1997, pp. 120-126.

Kozar, K. A. "Adopting Systems Development Methods: An Exploratory Study," *Journal of Management Information Systems* (5:4), 1989, pp. 73-86.

Kruchten, P. *The Rational Unified Process*, Reading, MA: Addison-Wesley, 1999.

Laitinen, M.; Fayad, M. E.; and Ward, R. P. "Thinking Objectively: The Problem with Scalability," *Communications of the ACM* (43:9), 2000, pp. 105-107.

Langefors, B. "Distinction between Data and Information/Knowledge," *Information Age* (9:2), 1987, pp. 89-91.

Langefors, B. *Essays on Infology: Summing Up and Planning for the Future*, Lund, Sweden: Studentlitteratur, 1995.

Langefors, B. *Theoretical Analysis of Information Systems* (4th ed.), Lund, Sweden: Studentlitteratur, 1973.

Lundell, B., and Lings, B. "Method in Action and Method in Tool: A Stakeholder Perspective," *Journal of Information Technology*, 2004 (forthcoming).

Lundell, B., and Lings, B. "The *2G* Method for Doubly Grounding Evaluation Frameworks," *Information Systems Journal* (13:4), 2003, pp. 375-398.

Lundell, B., and Lings, B. "Validating Transfer of a Method for the Development of Evaluation Frameworks," in A. Brown, and D. Remenyi (Eds.), *Proceedings of the Sixth European Conference on the Evaluation of Information Technology*, Brunel University, Uxbridge, U.K., November 4-5, 1999, pp. 255-263.

Lyytinen, K. "A Taxonomic Perspective of Information Systems Development: Theoretical Constructs and Recommendations," in R. J. Boland Jr. and R. A. Hirschheim (Eds.), *Critical Issues in Information Systems Research*, Chichester, England: John Wiley & Sons, 1987, pp. 3-41.

Marshall, C., and Rossman, G. B. *Designing Qualitative Research* (3rd ed.), Thousand Oaks, CA: Sage Publications, 1999.

Mathiassen, L., and Purao, S. "Educating Reflective Systems Developers, *Information Systems Journal* (12:2), 2002, pp. 81-102.

Maxwell, J. A. *Qualitative Research Design: An Interactive Approach*, Thousand Oaks, CA: Sage Publications, California, 1996.

McMaster, T.; Vidgen, R. T.; and Wastell, D G. "Technology Transfer: Diffusion or Translation?," in T. McMaster, E. Mumford, E. B. Swanson, B. Warboys, and D. Wastell (Eds.), *Facilitating Technology Transfer Through Partnership: Learning from Practice and Research*, London: Chapman & Hall, 1997, pp. 64-75.

Moody, D. "Building Links between IS Research and Professional Practice: Improving the Relevance and Impact of IS Research," in W. J. Orlikowski, S. Ang, P. Weill, H. C. Krcmar, and J. I. DeGross (Eds.), *Proceedings of the 21st international conference on Information systems*, Brisbane, Australia, 2000, pp. 351-360.

Moody, D. L. "Validation of a Method for Representing Large Entity Relationship Models: An Action Research Study," in S. Wryzca (Ed.), *Proceedings of the 10th European Conference on Information Systems*, 6-8 June, Gdansk, Poland, 2002, pp. 391-405.

Myers, M. D., and Young, L. W. "Hidden Agendas, Power and Managerial Assumptions in Information Systems Development: An Ethnographic Study," *Information Technology & People* (10:3), 1997, pp. 224-240.

Nandhakumar, J., and Avison, D. E. "The Fiction of Methodological Development: A Field Study of Information Systems Development," *Information Technology & People* (12:2), 1999, pp. 176-191.

Nandhakumar, J., and Avison, D. "Information Systems Development Methodologies in Use: An Empirical Study," in N. Jayaratna and B. Fitzgerald (Eds.), *Lessons Learned from the Use of Methodologies: Information Systems Methodologies—Fourth Conference on Information Systems Methodologies,* British Computer Society: Information Systems Methodologies Specialist Group, 1996, pp. 205-214.

Nissen, H.-E. "The Infological Equation Opening Two Perspectives on Information Systems," in B. Dahlbom (Ed.), *The Infological Equation: Essays in Honor of Börje Langefors*, Gothenburg Studies in Information Systems, Report 6, 1995, pp. 11-46.

Pflegger, S. L. "Understanding and Improving Technology Transfer in Software Engineering," *Journal of Systems and Software* (47:2-3), 1999, pp. 111-124.

Ramesh, B.; Pries-Heije, J.; and Baskerville, J. "Internet Software Engineering: A Different Class of Processes," *Annals of Software Engineering* (14), 2002, pp. 169-195.

Rehbinder, A.; Lings, B.; Lundell, B.; Burman, R.; and Nilsson, A. "Observations from a Field Study on Developing a Framework for Pre-Usage Evaluation of CASE Tools," in N. L. Russo, B. Fitzgerald and J. I. DeGross (Eds.), *Realigning Research and Practice in Information Systems Development*, Boston: Kluwer Academic Publishers, 2001, pp. 211-220.

Rumbaugh, J.; Blaha, M.; Premerlani, W.; Eddy, W.; and Lorensen, W. *Object-Oriented Modeling and Design*, Englewood Cliffs, NJ: Prentice-Hall International, 1991.

Russo, N. L., and Stolterman, E. "Exploring the Assumptions Underlying Systems Development Methodologies: Their Impact on Past, Present and Future Research," *Information Technology & People* (13:4) 2000, pp. 313-327.

Russo, N. L., Wynekoop, J. L., and Walz, D. B. "The Use and Adaptation of System Development Methodologies," in M. Khosrowpour (Ed.), *Proceedings of the 1995 International Resources Management Association International Conference*, Atlanta, Georgia, May 21-24, 1995, p. 162.

Sawyer, S., and Guinan, P. J. "Software Development: Processes and Performance," *IBM Systems Journal* (37:4), 1998, pp. 552-569.

Shanks, G. *Building and Using Corporate Data Models*, Unpublished Ph.D. Thesis, Monash University, Melbourne, 1996.

Stolterman, E. "About Design and Methods: Some Reflections Based on an Interview Study," *Scandinavian Journal of Information Systems* (4), 1992, pp. 137-150.

Truex, D.; Baskerville, R.; and Travis, J. "A Methodical Systems Development: The Deferred Meaning of Systems Development Methods," *Accounting, Management & Information Technology* (10), 2000, pp. 53-79.

Veryard, R., and Macdonald, I. G. "EMM/ODP: A Methodology for Federated and Distributed Systems," in A. A. Verrijn-Stuart and T. W. Olle (Eds.), *Methods and Associated Tools for the Information Systems Life Cycle,* Amsterdam: Elsevier, 1994, pp. 241-273.

Viller, S., and Sommerville, I. "Ethnographicaly Informed Analysis for Software Engineers," *International Journal of Human-Computer Studies* (53), 2000, pp. 169-196.

Ward, P. T. "The Evolution of Structured Analysis, Part II: Maturity and its Problems," *American Programmer* (5:4), 1992, pp. 18-29.

Wastell, D. G. "The Fetish of Technique: Methodology as a Social Defense," *Information Systems Journal* (6:1), 1996, pp. 25-40.

Webster, J., and Watson, R. T. "Guest Editorial: Analyzing the Past to Prepare for the Future: Writing a Literature Review," *MIS Quarterly* (26:2), June 2002, pp. xiii-xxiii.

Wynekoop, J. L., and Russo, N. L. "Studying System Development Methodologies: An Examination of Research Methods," *Information Systems Journal* (7:1), 1997, pp. 47-65.

Wynekoop, J. L., and Russo, N. L. "Systems Development Methodologies: Unanswered Questions," *Journal of Information Technology* (10:2), 1995, 65-73.

ABOUT THE AUTHORS

Brian Lings was awarded a doctorate in Computer Science from the University of East Anglia in 1975 and holds an academic medal from the awards committee of the Australian Computer Society. After a number of years at the University of Queensland, Australia, he joined the Department of Computer Science, now within the School of Engineering, Computer Science and Mathematics at the University of Exeter, UK. He chairs the steering committee of the British Network for Cooperation on Databases (BNCOD). His recent publications center on the areas of tool evaluation and development, and on issues of consistency maintenance and interchange in multi-model environments. He is a codeveloper of the *2G* method, and continues to be active in applications of the method, and in issues related to its theoretical basis and transfer. Much of his work is conducted in collaboration with colleagues at the University of Skövde, Sweden. He can be reached at B.J.Lings@exeter.ac.uk.

Björn Lundell has been a staff member at the University of Skövde, Sweden, since 1984. He received his M.Sc. (1991) in Computer Science from the University of Skövde, Sweden, and his Ph.D. (2001) from the University of Exeter. He has been active in international standardization (within the Swedish working group directly corresponding to ISO JTC1/SC7, i.e., software engineering), on drafts to ISO 14102, as well as on the published standard ISO/IEC 14102. He is a codeveloper of the *2G* method, and has acted as method user and mentor in a variety of company contexts. His research is published in a variety of international conferences and journals. He has a general interest in qualitative methods and his research centers on the issues of applications of grounded theory in technical domains, database modeling, technology evaluation, method development, and theoretical and practical aspects of method transfer into real organizational usage. To this end he has established active links with a number of Swedish companies. Björn can be reached at bjorn@ida.his.se.

30 FROM CRITICAL THEORY INTO INFORMATION SYSTEMS PRACTICE: A Case Study of a Payroll-Personnel System

Teresa Waring
University of Newcastle upon Tyne

Abstract Modern organizations both in the private and public sector are seen to be increasingly reliant, in terms of achieving improvements and service targets, on the efficient provision of information to enable administrative and managerial decision making. A key barrier to effective ICT introduction and integration of information systems has been identified as the complex social, organizational, and political issues endemic within organizations, preventing true discourse. This paper describes how an approach based upon some of the emancipatory principles of Jurgen Habermas may be used to develop an innovative approach to participative process and information flow modeling. This approach was used within a UK Hospital Trust in the North East of England to facilitate the integration of two departments and the procurement of a computerized payroll-personnel system. The results of the action research project are described and conclusions drawn as to the success of the approach and the role of the systems analyst within this type of project.

Keywords: Information systems, participatory action research, emancipation, critical social theory

1 INTRODUCTION

The main aim of this paper is to explore what it means to conduct critical research in the area of integrated information systems (IS) implementation. The work that is described here was initially inspired by Hirschheim and Klein (1989), Kendall and Avison (1993), and Alvesson and Willmott (1992) where the concept of *emancipation* features highly. According to Alvesson and Willmott (1992, pp. 432-435),

> emancipation describes the process through which individuals and groups become freed from repressive social and ideological conditions, in particular those that place socially unnecessary restrictions upon the development and articulation of the human consciousness. The intent of Critical Social Theory (CST) is to facilitate clarification of the meaning of human need and expansion of autonomy in personal and social life....Emancipation necessarily involves an active process (or struggle) for individual and collective self-determination....Any substantial and lasting form of emancipatory change must involve a process of critical self-reflection and associated self-transformation.

The work described in this paper has been to critically investigate potential emancipatory principles for systems analysis, design, and development, synthesized from the wider literature, and then to translate these principles into practice within the context of IS implementations. Fundamentally, this has been through an exploration of the changing role of the systems analyst to enhance participant communication and discourse during the implementation process. The research took place over a five year period and included four major integrated systems implementations within three hospitals in the North East of England.

The paper focuses on one of these implementations where the hospital (Hospital Z) was having difficulties with its payroll service as well as its personnel department. The suggested way forward was an integrated payroll-personnel system. The study was further complicated by the merger of Hospital Z with another large local community healthcare provider (Community Trust) and the on-going difficulties that ensued. The research approach taken for this particular study was based on Habermas's (1974) three-stage methodology, with insight derived from other authors writing in the area of CST.

The first section of the paper provides an overview of the theory that has informed the study. The second section comprises the research methodology and fieldwork undertaken within the hospital. The third section takes a reflexive approach to the discussion of the empirical material. The fourth section draws some conclusions on carrying out critical research for the participants and researcher.

2 CRITICAL SOCIAL THEORY AND ITS RELATION TO INFORMATION SYSTEMS IMPLEMENTATION RESEARCH

Critical social theory (CST) is the name given to a school of thought which originated in the 1930s from certain scholars associated with the Institute of Social Research at the University of Frankfurt—Horkheimer, Adorno, Fromm, and Marcuse—and more recently Habermas (Lyytinen and Klein 1985). Lyytinen and Klein state that "CST has as a fundamental concept the belief that any dynamic social theory must view society and its parts as highly dynamic— it can be changed by its members"(p. 220).

CST is relatively well established within management studies and increasingly is becoming known in the Operational Research and Information Systems fields (Alvesson and Willmott 1996; Flood and Jackson 1991; Flood and Romm 1996; Hirschheim and

Klein 1989, 1994; Lyytinen 1992; Lyytinen and Klein 1985 Mingers 1992; Warren and Adman 1999). In the IS field, the critical social theory of Habermas has been developed from two perspectives, both of which express a common connection between critical social theory and IS research (Brooke 2002; Lyytinen 1992).

1. A critique of scientism and relationships between theory and practice (Habermas 1972)
2. The nature of social action and the type of knowledge it is based upon (Habermas 1984, 1987)

2.1 The Theory of Communicative Action

In the area of IS development and implementation, it is this second perspective that has been the focus of a number of researchers. Habermas (1972) has expressed concern about how technical knowledge interest (where a desire to control outcomes is preferred to more discursive communication leading to an ideal situation where people are freed from domination and control) has come to dominate society through technocracy (Alvesson and Skoldberg 2000, p. 115). Habermas (1972) argues that there is a need to restore man's ability to engage in critical reasoning and not be steered by ideas and values that have not been subjected to scrutiny. To do this within the context of an information systems implementation, there is a need to develop communicative competence or action (Habermas 1984).

Habermas (1984) challenges us to examine critically all forms of communication and in particular where agreements are made. Very often there is a political and a power dimension that is overlooked. He encourages a move toward more informed approaches to discursive action. Discursive action is oriented toward the cooperative search for truth, the clarification of unclear message content, the analysis of the intended use of the messages, and so forth toward the attainment of more *ideal speech situations*. The four criteria below define the validity of communications in a complex, social, and political environment (Habermas 1984).

1. *Clarity* (can what is being said be understood by the receiver?)
2. *Truthfulness* (is what is being said truthful?)
2. *Sincerity* (is what is being communicated done with good intent?)
4. *Social acceptability* (is the communication in keeping with the values and norms of the receiver?)

Discursive action is aimed at justifying any or all of the four claims should one become the subject of doubt. This requires that all actors respect certain ground rules when claims are made for and against raised claims in the pursuit of rational justification (Habermas 1979). This would require organizations to be transformed towards a structure where all actors have a chance to express opinions, to enter or leave the discourse, and to honor what Habermas (1979) calls the "force of better argument." Habermas recognizes that the conditions of the ideal speech situation represents an ideal but this does not of itself undermine its significance. He argues that each time a theoretical or practical argument is pursued with the intention of reaching a rational agreement, an ideal speech situation is presupposed (Held 1980). From a Habermasian perspective, the analysis of the ideal speech situation shows it to involve assumptions about the institutional context of interaction and the end point of this argument is that the structure of speech is held to involve *the anticipation of a form of life in which truth, freedom and justice are possible.*

Habermas' theory of communicative action has appealed to IS researchers in the area of development and implementation. Hirschheim and Klein (1989) identify the systems analyst as being integral to the implementation of information systems. These individuals could be involved in many phases of systems acquisition from carrying out feasibility studies to training users on the new system. They are at the interface between both the old and the new system as well as being the main information/communication conduit for the project. Successful communication between stakeholders within the context of an integrated systems implementation is perhaps the major critical factor leading to the eventual outcome of the project.

Much of the work of a critical nature that has emerged in relation to IS implementation has been largely hypothetical and in most cases untried (Lyytinen 1992). This includes the work of Hirschheim and Klein (1989) and of Ngwenyama (1991).

2.2 Toward a Theory of Emancipatory Practice for the Systems Analyst

The emancipatory philosophy of Habermas and its attack on positivism (scientific and technical domination of society) has been attractive to a number of academics. Yet this work is highly theoretical and has been criticized because of its lack of engagement with the practical (Held 1980). Habermas has provided little guidance for those wanting to advance emancipatory action and change or indeed on how to conduct critical research (Forester 1993; Johnson 1999). Nevertheless, Mingers (1992) constructively suggests that the way forward is to be guided by Habermas's (1974) key principles. These principles involve three stages.

Stage 1: The development of critical theories about the nature of the social situation in terms of the position and true interests of the actors within a social structure.

Stage 2: Use these theories to enlighten concerned actors as to their position. This may lead to authentic insights and changed attitudes. Mingers argues that it is only success at this stage that provides the validation of the theories.

Stage3: The enlightened social group chooses tactics and strategy to be adopted in the actual political struggle.

Laughlin (1987) used these stages to guide the approach he adopted in exploring accounting systems. First of all, the researchers examined in detail how the systems arose, their historical context, technical and other pertinent issues. They discussed this with the "researched" and then explored with the group how changes could be made.

The three stage methodology (Habermas 1974) and the principles of CST have the potential to be translated into a more socially informed approach to systems analysis within the context of highly political and socially complex integrated systems projects with pluralistic goals. Nevertheless, Alvesson and Willmott (1996) believe that it is important to consider other areas of management and organizational theory to support the move from theory into practice. They commend the work of Forester (1992) in the field of organizational studies as being appropriate and also suggest that the work of Payne (1991, 1992) and Ulrich (1987) in relation to Habermas's theory of communicative competence have much to recommend them.

2.2.1 Change Agents and Emancipatory Practice

Systems analysts have also been referred to as agents of change (Walsham 1993b) as they assist in bringing new information systems into organizations. Bradshaw-Camball (1990) as cited by Payne (1992) sees the role of the CST change agent as an individual who tries to help articulate alternative possibilities that may lead to opposing or reconstructing the dominant organizational ideology. An alternative reality is generated through a dialectic between the consultant and the organization members that

> traditional power bases that come from being the "expert" quickly disappear.... Being open to public debate and criticism is an essential skill for the change agent committed to the radical humanist paradigm. This is threatening at times but if modeled for all organizational participants will set a norm of self reflection which will strengthen and enable the change process to continue (Bradshaw-Camball 1990, p. 255, as cited by Payne 1992, p. 246).

Carrying out this approach informed by CST does have its difficulties. Gray (1989) describes encounters which may arise where strong conflicts surface "over race, gender, organizational mission and ideology, distribution of power and concepts of leadership" (Gray 1989, p. 394).

Payne (1992) argues that change agents working toward emancipatory practice need to develop or have other skills besides systems-related knowledge. He believes that without a high level of creativity and a high level of communications skills they will not be able to operate in a "variety of situations where they will be faced with many difficult personal challenges" (p. 247).

Taking this further, McKay and Romm (1992) and Romm (1995) see the critical change agent as someone who

- Critically educates staff
- Develops skills within an organization to level the playing field
- Widens participation to include all stakeholders

Payne (1996) does have some misgivings about such prescriptive forms of action and believes that the micro-emancipation ideals of Alvesson and Willmott (1996) that advocate incremental change could be a way forward. He offers to the critical change agent the opportunity to reflect on their use of language. For example, do they use language that alienates the organizational leaders? Alternatives are suggested in Table 1.

Table 1. Alternative Phrases (Payne 1996)

Emancipatory Language	Alternative Phrases
Oppression	Ethics development
Emancipation	Fairness, integrity
Empowerment	Social responsibility

2.2.2 Operational Research and Emancipatory Practice

In terms of developing the practical intent of CST, there are two main areas of research in Operational Research which have bearing on the practice of the systems analyst. The first is total systems intervention (Flood and Jackson 1991) and the second is critical systems heuristics (CSH) (Ulrich 1987). It is this second areathat is examined here.

In CSH, there is an attempt at a form of practical discourse which provides the necessary mediation between reason as an *intellectual ideal* and practice as an *experiential reality*. Underpinning Ulrich's practice is Habermas's concept of the ideal speech situation and his theory of communicative competence. Ulrich explores the role of the expert and their use of justification break-offs that tend to occur when people claim expertise in decision making and do not provide any rationale or defense of their claims. Ulrich argues that by applying polemic, the non-experts can impose the burden of proof or justification on those who try to dominate and control the situation.

2.2.3 Information Systems Research and Emancipatory Practice

The application of radical humanist principles in information systems development and implementation has been developed further by Klein and Hirschheim (1993) by considering various theoretical approaches which might lead to better understanding although not necessarily emancipation. They analyze four projects which they believe exhibit emancipatory potential.

- Kerola's reconceptualized SDLC (Kerola 1985)
- MARS project (Lanzara and Mathiassen 1984)
- UTOPIA project (Ehn 1988)
- SAMPO project (Lehtinen and Lyytinen 1983)

Klein and Hirschheim (1993) conclude that even though there are no emancipatory IS methodologies and possibly with the weight of evidence mitigating against ISD methodologies, then the quest should be for guiding radical humanist principles: "Just as democracy thrives through living traditions and wise practices, so Radical Humanist principles could flourish through emergent traditions and enlightened practices of ISD" (Klein and Hirschheim 1993, p. 275).

Some of these principles have also emerged in the work of other authors.

The systems analyst as a moral agent: Walsham (1993a) offers a modification to the *analyst as emancipator* role and uses the term *moral agent*, which places an emphasis on the analyst's own actions rather than the emancipation of others.

Walsham (1993a, p. 283) argues that "a focus on self-reflection and understanding will normally be related to changed action involving others." This could involve questioning the approaches to systems development or implementation or the specific goals of the system. By reflection, the systems analyst could begin to take action on issues immediately. The introduction of an ethical dimension to the role of the systems analyst would be a departure from the traditional *systems expert* role where ethical issues have yet to become pertinent.

Emancipatory systems development: Although they discount the use of systems development methodologies, Hirschheim and Klein (1994) choose to pursue the belief that some actually have some emancipatory principles and can be *reformulated* to achieve these emancipatory ideals more comprehensively. They see the ETHICS (Effective Technical and Human Implementation of Computer Systems) as developed by Mumford (1983) as having the potential to be reformulated.

Hirschheim and Klein (1994) believe that the main issue in ISD is the nature of participation. Recognizing the chief concerns of the critical schools of management and the nature of participation, they suggest that the principal weakness of participation in ISD is the failure to acknowledge a political dimension. They argue that organizations are historically constituted and may not have a tradition or a structure that facilitates participation in the radical humanist form.

Wilson (1997) is concerned about advocating the adoption of a variant of ETHICS as some of the suggestions in the Hirschheim and Klein (1994) work appear to be anti-emancipation. For example, they describe the use of facilitators to ensure "that everyone contributes and is listened to" (p. 93) and the "emancipatory methodology" will be used to overcome "wilful unresponsiveness by an individual." These menacing overtones are at odds with Habermas's ideals of a future unalienated and uncorrupted society and betray the authors historicity, which all IS researchers face when trying "to place their own project on a footing that is different" (Wilson 1997, p. 202).

The ethical systems analyst: In terms of emancipatory practice and the increasing use of IT to deliver systems which have the potential to impact heavily on the societal organization of work, Wood-Harper et al. (1996) urge the systems analyst to have a greater awareness of ethical theory. They argue for an explicit analysis of the implications of design decisions using a basic understanding of this theory.

Wood-Harper et al. believe that this approach could be very successful in the National Health Service (NHS) to assist systems analysts to define requirements where there may be conflicting ethics between groups such as government, hospitals, business managers, nurses, doctors, and patients.

The theory linked to practice which has been examined in this section can be interpreted in a manner that could provide a framework for action on the part of the systems analyst/researcher to explore further and is represented in Figure 1.

Alternatively, the framework may be used as a heuristic in order to inform practice and be improved upon in response to new and better understanding.

The journey into emancipatory praxis from the perspective of the researcher/researched systems analyst is done against a background of functionalist theory and practice and the critical social theory of Habermas that has not engaged practice. The framework in Figure 1 has as its starting point Habermas's critical methodology that guides all action. However, Habermas did not move his theory into practice and, there-

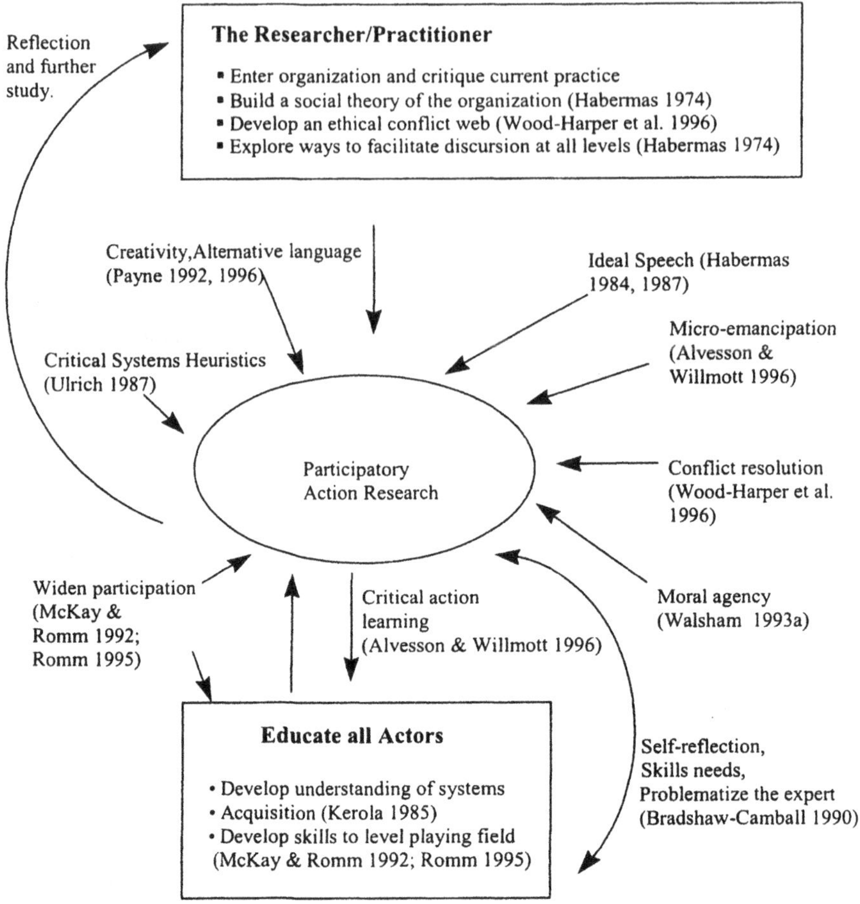

Figure 1. A Framework for Emancipatory Practice

fore, the theories of other authors considered appropriate are integrated into the framework. At no point is the framework intended to be prescriptive and, in reality, the various contributions may depend upon the actors (including the researcher) involved and whether they want to move the project into action. The intention of the researcher would be to develop practice that would incrementally move toward *micro-emancipation* (Alvesson and Willmott 1996) for all concerned.

The literature has indicated that emancipation begins by developing reflexive skills within an organization from which a social theory about that particular organization will emerge through analysis of the collected data and personal experience. From this, it should be possible to establish where there is conflict and there needs to be an exploration of the nature of the conflict and why it has occurred. Opportunities for discursion must also be investigated prior to any intervention. This approach formed the basis of research conducted at a hospital in the North East of the UK, **Hospital Z**.

3 RESEARCH METHODOLOGY

The research described in this paper was part of a larger critical study into the changing role of the systems analyst acting in a more overt emancipatory role within IS projects (Waring 2000; Waring and Wainwright 2002). The methodology used in the project was participatory action research (PAR) informed by principles derived from critical social theory (Stringer 1999; Udas 1998). Stringer suggests that an authentic, socially responsive methodology must enable participation, acknowledge people's equality of worth, provide freedom from oppressive debilitating conditions and enable the expression of people's full potential.

Udas proposes that participatory action research (PAR) must be underpinned by some fundamentals. The first is that PAR questions the nature of knowledge, research, and methods. Second, the nature of knowledge in PAR is for improvement of practice, not for the construction of an abstract theory-base. The PAR assumption of the nature of knowledge is that it is created by local practitioners, environments, and historical factors. Third, the findings and value of research are retained locally. Fourth, the researcher must be prepared to be flexible and creative.

Udas continues by outlining certain methodological principles that apply to PAR.

- It is participant centered and non-alienating.
- Researchers/facilitators enter a project clear about their own theory of social change and can share this with participants in a democratic way.
- The research methods are based on mutual respect and trust and facilitate collaborative inquiry, potential benefits, and acceptance of each party's responsibilities.

Stringer (p. 35) states that participation is most effective when it enables significant levels of active involvement and people to perform substantial tasks. It must provide support for people as they learn to act for themselves and encourages plans and activities that people are able to accomplish themselves. Finally, it deals personally with people rather than with their representatives or agents.

PAR is inherently political. Udas writes that "PAR is predicated on the democratic notion that oppressed and marginalized people can transform their social realities through education, research and action while forwarding their own value system" (p. 606).

PAR must be aimed toward social justice, involve critical reflection on practice, question assumptions on which practice is based, and promote collaborative collective action (Udas 1998). It is a continuing cycle of research activities involving active participation of practitioners. It is anti-positivist. It is not problem solving although it may help to do so. It is a process having value itself. It is a means of self-examination, improvement, and emancipation, not an instrument to recreate a status quo.

Within the overall framework of the PAR approach, five primary methods of data collection were utilized.

Research Diary: This diary recorded project team meetings, meetings with vendors of the potential integrated systems, training sessions for staff on the modeling tool, and my own reflections on the experience and the research.

Document analysis: NHS policy documents, departmental plans, procedure manuals, historical data, tender documents, PRINCE project management documents, finance and personnel data and statistics, departmental structures, and job descriptions.

Participant and non-participant observation: As a member of the project team for the duration of the procurement, I was actively involved in all discussions and developments. These meetings were duly recorded. I was also allowed access to departments during normal working periods to observe how each department operated.

Participant workflow/information flow modeling: As part of the emancipatory methodology, all project team members and some of the departmental staff were trained in the use of a graphical process modeling tool and went on to produce models of work processes and information flows as they were and then how they would want them to be. These models were collected and analyzed.

Semi-structured interviews with key staff: The project manager, personnel officer, the assistant director of finance, and a member of the vendor's team were interviewed. These interviews lasted between one and two hours, were audio-taped, transcribed, and fed back to the respondents for comments and amendments.

The analysis of the data collected was then related to each of the three stages of Habermas's (1974) methodology as shown in Table 2.

3.1 Findings of the Research

The findings from the fieldwork can be structured using the three-stage methodology of Habermas (1974). However, the data collected was vast and it is difficult to communicate its richness within the confines of the paper.

Stage 1: This section describes the highly stressful and intensely political climate that surrounded the proposed implementation of an integrated payroll-personnel system.

Table 2. The Analytical Framework

Description	Research Approach
Stage 1: The development of critical theories about the nature of the social situation in terms of the position and true interests of the actors within a social structure.	Diary, observation, semi-structured interviews, document analysis, workflow/information modeling exploring history, politics, power, relationships, and information flows.
Stage 2: Use these theories to enlighten concerned actors as to their position. This may lead to authentic insights and changed attitudes.	Feedback of the analysis of Stage 1 by all participants to the project team. CST-informed discussion.
Stage3: The enlightened social group chooses tactics and strategy to be adopted in the actual political struggle.	Discussion of strategy to move project forward to procurement of new integrated system. Further modeling of "to be" system and incorporation of models into contract with vendor. Interview with vendor team member.

Hospital Z serves a mainly inner-city population. Similar to other UK hospitals, it has experienced problems meeting successive government demands of ever-increasing information. A particularly problematic area was that of payroll and personnel. Hospital Z was unable to provide regular and reliable statistics on sickness, absence and staff turnover—essential under the new UK Labour government. There were also internal management and control problems and difficulties relating to the facilities management company that managed the payroll. A decision was taken by the executive board to purchase and implement an integrated payroll and personnel system. To complicate matters further, in April 1997 it was understood that a merger was being planned between Hospital Z and Community Trust, based about three miles away. Community Trust had its own separate human resource management (HRM) department, but their payroll was managed by Hospital Z.

No expertise in systems procurement: Following the decision to procure a new integrated payroll/personnel system, a project team was set up. This consisted of staff from HRM and from finance within Hospital Z and was headed by a project manager who previously had been employed by NHS supplies and had some experience of the NHS procurement process but had never led such a project. None of the project team members had ever been involved in a procurement. There was a real concern about how the procurement decision might be made as Hospital Z had a history of failed systems and poor relations with vendors.

Culture clashes and dominant personalities: In terms of departmental cultures and approaches to work in Hospital Z, the finance department (who ran the payroll) had a closed gate-keeping culture: digital locks on all doors, staff needing to account for their movements. They were lead by Pat, a strong-willed finance director. The HRM department had open access, where doors were ajar and staff were very laid back. Their manager was Vicki, who had a more consultative approach to management. Although staff from both departments vaguely knew each other, they did not work closely together. In the early stages of the project the team met once a week to build up a rapport. However, throughout these early meetings, proceedings were dominated by the director of finance who insisted that whatever system was procured, it had to meet all of her requirements.

Jobs on the line: Before Hospital Z had started to progress the system procurement further there was a merger and the staff from Community Trust joined Hospital Z. A new project team had to be established. However, the logistics of this were not easy. The Community Trust management block was located three miles away. Their HRM department was organized on a more formal basis than in Hospital Z. Community Trust staff normally had very little contact with Hospital Z during the course of their day-to-day business.

The first meeting of the newly constituted project team was described as "*tense.... Staff were eyeing one another suspiciously. Two HRM departments in one Trust. Were there going to be job losses? Who would be doing what when the new system arrived?*" (Senior HRM Officer, Hospital Z). There was a great deal of frustration and suspicion as staff began to worry about job losses and role changes.

Total breakdown of communications: As relationships deteriorated within the project team, various political agendas began to emerge and threatened the success of the project. The project manager wrote the specification document for the procurement without apparent consultation and in a technical language that the project team did not fully understand. The team were unhappy: Who had he consulted? How did he know what was required? From the hospital's perspective, it was vital to get the project underway so that staff would get paid when the arrangement with the facilities management company ran out. A robust personnel system was also required that addressed many of the needs of Hospital Z and newly merged Community Trust. This impasse needed to be addressed if the procurement of the integrated system was to continue. It was essential that staff started communicating with one another and that important organizational issues were addressed.

It was at this point that the author began her intervention when she was invited into the Trust in order to try and facilitate the discussions that needed to take place regarding the implementation. The social situation as described above was recognized by staff and emerged through interviews with individuals. However, it needed to be explored in more depth to determine fact from fiction and bring clarity to what was rapidly becoming a very complex political situation. It required staff to be civil and establish some degree of sincerity in the discussions (Habermas 1984).

The approach that was to be explored involved participation by all staff within the project and it could not take place without their agreement. I did not want to be seen as an expert entering the situation and becoming a dominant force but as someone working with staff in this challenging situation (Payne 1992).

The finance staff were the most antagonistic and thought that they should decide on the outcome of the procurement process. However, after lengthy debate where expressions such as "*giving staff a fair chance*" and the "*need to be responsible*" (Payne 1996) were heard, eventually, with the agreement of all of the project team, the research was allowed to progress.

It was apparent that project team members had little understanding of the business processes in other departments and this understanding was necessary if everyone was to have full participation in the system procurement discourse. We needed to critically educate staff and develop skills to level the playing field (Romm 1995). Establishing a common language to facilitate project communications was required and discussed with staff. They recognized that jargon used by IT professionals was not appropriate and intimidated them. Yet it was important to them that they knew and understood what was about to happen.

Acting as a moral agent(Walsham 1993a), I trained project team staff in a systems modeling technique that allowed them to go back to their departments and graphically model their information flows and business processes. Over a period of two weeks, HRM and the finance staff developed models of their systems and explored a better understanding of their information needs regarding a new system. The modeling tool used was a PC-based software package that allowed simple process modeling techniques representing activities, information flow inputs, outputs, controls, and mechanisms. This software had been selected from a previous iteration of the action research project (Waring 2002). This software was flexible, easy to use, and could professionally produce diagrams that were understandable by all levels of employee. The aim was to create a simple communicative, shared and common language to enable equal participation and discourse to take place among key stakeholders. Information flows, work tasks, and processes could be quickly and easily represented, stored, and disseminated in a graphical and understandable format that would serve as the focus for discussion and debate.

Stage 2: The main objective in Stage 2 was an attempt at a form of practical discourse to enlighten all project team members and through them staff in their departments as to the findings in Stage 1. This focus on exploring the social situation represented a departure from traditional systems analysis. It was recognized that goals were contestable and that managerial preconceptions of problems and their solutions should not be accepted before conditions for effective discourse (ideal speech situation) had been created (Ulrich 1987).

It was agreed that each project team member (non-experts) would do a short presentation of the business processes and information flows and requirements in their particular area. Many of the models, as shown in Figure 2, were not very sophisticated but they provided staff with a better understanding of what was going on in areas of HRM and finance of which they knew very little.

Figure 3 illustrates how, by decomposing the "Process request" in Figure 2, a further level of detail is revealed that may expose inappropriate or inefficient practices.

The models made overt what staff did in their departments and their information needs. The director of finance found it difficult to use justification break-offs and dominate discussions. Other staff could question her assertions and had confidence when asking for explanations. Everyone in the project team began to develop a better understanding of how their systems worked.

There were many issues relating to the new system that had yet to be discussed. How would the system be administered? How would the data get into the new system. Who would be in charge overall? Finally, over a period of four weeks, the author assisted the team (primarily facilitating discussions through the use of process modeling tools) in outlining in detail how staff from all of the participating departments would be reorganized and how the processing would take place. Discussions became very heated as staff began to realize the implications of integration. Many ethical issues emerged and needed to be addressed (Wood-Harper et al. 1996). Further meetings had to be arranged to discuss other political issues that had arisen from the modeling exercise including the requirement for new sickness and absence reporting from within the hospital and the need for time sheets. One issue then became apparent as the project proceeded. Should the departments of payroll and personnel be merged?

Stage 3: Although Stage 2 was traumatic, it did bring to the surface many issues that would have been buried until after the new system had been installed. Eventually, over the course of a month of lengthy discussion, the project team agreed as to how they would progress and were then ready to meet the potential vendors. They had decided that if a vendor could not provide their requirements as they had discussed, they would not purchase an integrated system. The senior management of Hospital Z, who had followed the progress of the project, approved their decision.

At that time, there were limited choices for hospital payroll-personnel systems and only one supplier came through the NHS's tender process. Through discussions with other sites, the project team had learned about the tactics used by vendors to sell their systems and to minimize their involvement in the actual implementation. The project team spent some time rehearsing how discussions with the vendor might go and how all of their needs might be taken into consideration (McKay and Romm 1992; Romm 1995).

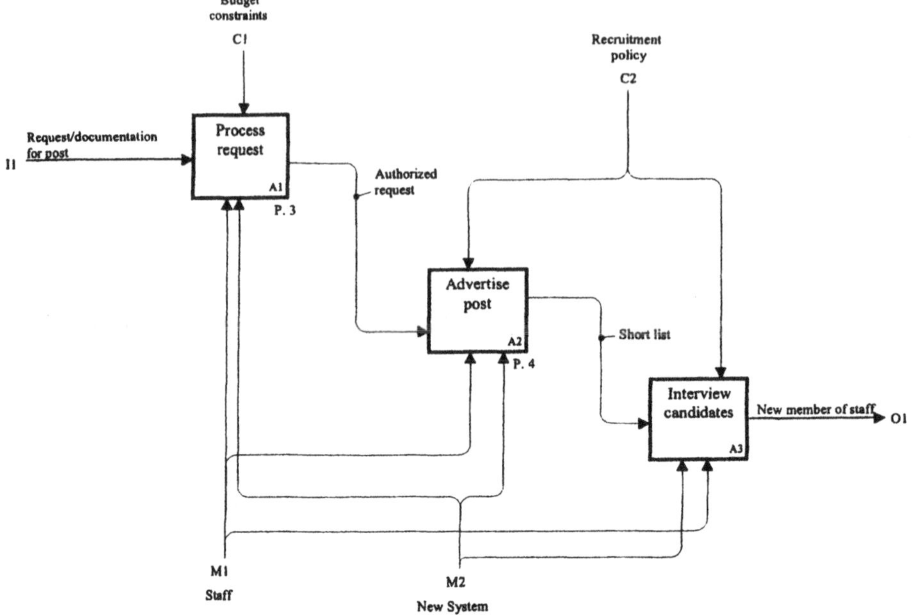

Figure 2. Part of the Modeling Done by HRM Staff

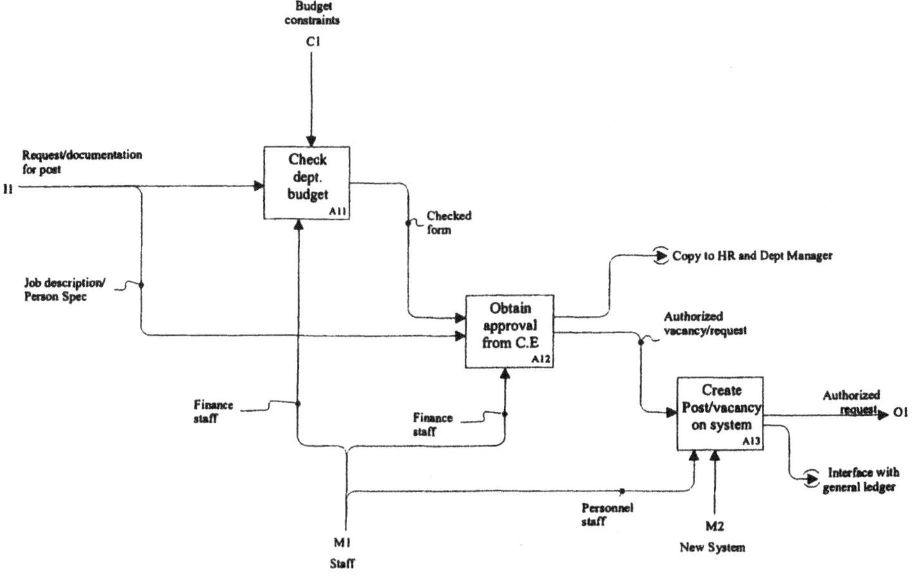

Figure 3. Decomposition of Process Request

When contract negotiation began, the vendor was surprised to meet a full project team and astonished at one so fully appraised of their requirements (interview with the vendor's lead negotiator, 1998). At the final two-day meeting with the vendors of the integrated system, Integrated PP, prior to signing the procurement contract, the process modeling diagrams were presented and the project team talked through with the vendors how they would like the system to operate. A number of the organizational issues, which had been raised during the modeling, had yet to be successfully addressed. However, all were aware of what they were and negotiations continued. In order to take the project team's work into consideration the models of the "to-be" system were incorporated into the procurement contract. This included a new module to be developed by the vendor to facilitate the administration of car leasing. The contract was signed in July 1997 with installation and training commencing in September 1997. Personnel went live in January 1998 and payroll in April 1998. After this, difficulties remained to be ironed out but staff continued to work on it.

4 DISCUSSION

The essential aim of this section is to draw together the variety of primary and secondary source material referred to thus far and synthesize some key issues that have emerged from the research. The emphasis is on evaluation of, and reflection on, the different strands of the research and how they collectively contribute to an understanding of emancipatory practice for the systems analyst within the context of an integrated systems implementation.

The section is intended to be reflexive in that it includes philosophical reflection and the problematization of the researcher's assumptions, interpretations, and interactions with the empirical material (Alvesson and Skoldberg 2000). It is important that the research is scrutinized carefully for evidence of emancipation because, rather than looking for a change in thinking only, a critical approach also looks for changes in actions. These actions could be those of the researcher or of the participants in the research.

Reflexivity and the Framework for Emancipatory Practice: Critical research needs to comprehend the empowerment of the individual. How do the knowledge and critical reflections the participants gained through the research process assist them in freeing themselves from repressive social and ideological conditions (Johnson 1999)? Here, I was the researched subject who wanted to become emancipated in order to develop socially responsible systems analysis theory and practice. I was also the researcher who wanted to investigate how this theory and practice might affect others within an integrated systems implementation and change their behavior and actions. Thus, the framework for emancipatory practice (Figure 1) reflects these dimensions in a fairly simplistic manner and engages them through action research. As has already been stated in section 1, the framework was never intended to be a mechanistic model developed within a functionalist paradigm but a heuristic that allowed the researcher to explore and learn within an action environment. The framework has also developed as further theory has emerged during the course of the project and this reflects the developmental and exploratory nature of the work.

Reflexivity and the Three Stage Model of Habermas: Traditional functionalist texts on systems analysis (e.g., Ashworth and Goodland 1990) have not explicitly encouraged the systems analyst to develop an understanding of the social situation of the users within the organization. However, it must be stated that interpretivist texts do view this as important (Checkland 1981; Walsham 1993b).The systems analyst operating in the functionalist paradigm has had a job to do and this usually began with investigation of the current information system exclusive of its political and social dimension. Yet, by not addressing the political nature of implementation, the systems analyst has ignored the medium through which the new system will be negotiated.

Becoming a critical researcher is developmental and takes time for reflection as well as practice. In **Stage 1**, Johnson (1999) argues that there is no one social theory; each is dependent upon the particular organization. It should show the historical development of social conditions within an organization, the organizational culture, structures, and actions that may shape participants' views and constrain their actions. This can only be done by presenting empirical findings and theories that show "the historicity and constructedness of social conditions" (Johnson 1999, p. 7). From the perspective of an integrated information systems implementation, it must also consider the history of IT within the organization at the micro and macro levels for further insight.

In a similar manner to Laughlin (1987), the approach chosen for this research was to study in depth the history of information management and technology within the NHS as a whole to develop an understanding of how government policies were affecting the local hospitals. This historical pre-understanding of the macro organizational environment is very important to the critical systems analyst before examining the local situation as it gives many insights into how and why actions are taken in a particular manner.

When joining the payroll/personnel project in Hospital Z, I was thrown immediately into the charged atmosphere of the joint Trust project team meeting. Here, staff *appeared* to have agreed what was needed but were in paralysis as to how to move on. The suppressed conflict was being managed by the senior directors of the hospital by their reference to "no loss of jobs" and "you don't need to purchase an integrated system if you cannot agree on one." Through interviews with staff on the project team it was apparent they did not believe this. Open critique of the situation and practice was initially very limited and had to be facilitated. It was at the project team meetings that the participating staff members needed clarity of purpose and truthfulness of intention. Whether the participants were sincere and what they stated within meetings was socially acceptable emerged as the process modeling was undertaken (Habermas 1984). Payne (1992) argues that the researcher needs to be creative in these types of situations and possess good communication skills themselves. I had my interpretation of the social situation, but was it a shared view.

- The NHS has a poor record of IS implementation success.
- Hospital Z with an acknowledged history of failed projects, was proposing to allow three departments with no experience of systems acquisition to undertake this procurement.
- Not only were they undergoing a major merger where staff sensitivities about job losses and redeployment were at the fore but these warring parties were also expected to sit down and act rationally about a new information system that would radically change their working practices.

- Departmental cultures were totally different.
- The project manager found it difficult to move the procurement forward as he readily acknowledged that he lacked the patience and the political skills to deal with the situation.
- The technical specification for the system and the business case had been written by the project manager in collusion with the director of finance without the specific needs of the project team or their respective departments.

However, I felt that all participants in the project needed to clearly understand this situation and articulate it in a way that allowed them to develop their own critical social theories of their situation. Offering the project team the opportunity to be educated in the language of process modeling was my approach to creativity. It gave them a chance to explore their own departments in a way they had never done and to determine how they might like to work in the future. This is not to say that it could not be done in other ways. They adopted the modeling approach because they could not envisage how the implementation could be progressed and they actually liked the professional look of the models. We established a set of ground rules for the models and I acted in a consultant role if they had problems in developing their models. I tried to ensure that the models belonged to them and were their interpretation of what took place. Although the average size of the project team was 12 (depending on agenda issues), the participation in the modeling was extended to staff in the various departments in a cascade manner

There is obvious overlap between stages and moving into **Stage 2** appeared to happen when the initial modeling of the "as-is" situation had been completed. How do they demonstrate that they have been *enlightened* and have changed attitudes? This is not easy and happens over a period of time at different rates for individuals (possibly never for some). The project team was encouraged to explore what happened in all three departments and share that knowledge. They devoted a series of team meetings to present their findings and discuss problems and issues. Hard copies of all process models were distributed prior to the meetings and individuals took it in turn to present theirs. After each presentation, staff had the opportunity to ask questions about departmental activity and why it was undertaken in a particular manner. These sessions provided a controlled and safe environment for staff as they learned about each other's work and the implications of integration. They gained confidence, over this time period, in their ability to probe about issues that were coming to the fore of consciousness—role changes, redundancies, new working practices.

Moving into **Stage 3** and agreeing on a way forward in this implementation is not as simple as deciding on the new system to be purchased. Choosing to do nothing was not an option for reasons explained earlier. The team were united in a goal of procuring a new system but they needed to decide on the tactics and strategy that would allow them an opportunity to get one that would satisfy all of their needs and make the work easier. Before meeting any potential vendor, they needed to be sure that they knew what they wanted in their language and not in the language of the technical specification written by the project manager. They chose to model the "to-be" system and incorporate aspects that they felt would enhance their work. The enemy became the vendors who, in the organizational history of Hospital Z, had not provided appropriate systems and had been instrumental in their failure.

The hospital did procure the system and, before leaving the project, the participating staff appeared relatively happy about the future. Nevertheless one cannot be sure about everyone and their experience of the process.

5 CONCLUSION

The project described here was one of four and provided an opportunity to engage in a critical approach to systems analysis. I deliberately set out with the intention of engaging in and exploring the politics of organizational life through the medium of integrated information systems projects. I invited the reader of this research to engage with and challenge the interpretations presented here. I would now like to draw the following conclusions.

Organizational participants in critical research can be dramatically changed by the process: This has implications for sponsors of this type of critical IS research and for the researchers themselves. Allowing staff to question accepted practice within an organization and challenge the dominant ideology requires a leap of faith that many would not be prepared to allow (Reynolds 1999). In the NHS, there is an adherence to functional methodologies that pervade all information management and technology projects—SSADM, POISE, PRINCE. It can be almost impossible to exclude them from information management and technology projects as funding from central bodies is predicated upon these methodologies being used. Users, too, who take a critical stance on an IS project may reflect on the consequences of this stance and its impact on their working lives outside the project.

As Brookfield (1994) comments,

> critical learners perceive that if they take a critical questioning of conventional assumptions, justifications, structures and actions too far they will risk being excluded from the cultures that have defined and sustained them up to that point in their lives (p. 208).

Users of IS must balance the risks of being marginalized with the need to be emancipated and this is a struggle, not only within themselves, but with their fellow workers and managers. If the culture has been one of control and domination, then those working therein, both management and staff, will have great difficulty accepting a critical approach to IS acquisition and consequently an organization with an appropriate climate is necessary.

There is a personal cost of CST research: Payne (1992) states that emancipatory practice can be difficult from a personal perspective. Staff may become critical and threaten the established norms of the organization, they may become stressed by the process, the outcomes may not be to the liking of the management. Further development of the research may be blocked in that particular organization. Emancipatory intent also has had other consequences that IS researchers and other practitioners might find disquieting. Once a critical perspective is developed, it can spill over into all other activities and can be detrimental to social relationships, work, and career. The need to expose injustice can be overwhelming at times and may lead to the researcher being labeled as awkward, dysfunctional, or not a team player, and if this is in the work environment it can act as a barrier to promotion. Therefore, it is important that researchers who become involved in critical research develop coping strategies. This could involve seeking out and networking with like-minded people, leaving the organization and joining one sympathetic to critical theory. It could also involve a need for mentoring for researchers who choose to approach research in this way.

REFERENCES

Alvesson, M., and Skoldberg, K. *Reflexive Methodology*, London: Sage Publications, 2000.

Alvesson M., and Willmott, H. C. "On the Idea of Emancipation in Management and Organization Studies," *Academy of Management Review* (17:3), 1992, pp. 432-464.

Alvesson, M., and Willmott, H. *Making Sense of Management: A Ccritical Introduction*, London: Sage Publications, 1996.

Ashworth, C., and Goodland, M. *SSADM—A Practical Approach*, New York: McGraw-Hill, 1990.

Bradshaw-Camball, P. "Organizational Development and the Radical Humanist Paradigm: Exploring the Implication," in L. R. Jauch and J. L. Wall (Eds.), *Best Papers Proceedings of the Academy of Management*, 1990, pp. 253 -257.

Brooke, C. "What Does it Mean to be 'Critical' in IS Research?," *Journal of Information Technology* (17), 2002, pp. 49-57.

Brookefield, S. D. "Tales from the Dark Sside: A Phenomenography of Adult Critical Reflection," *International Journal of Lifelong Education* (13:3), 1994, pp. 203-216.

Checkland, P. *Systems Thinking, Systems Practice*, Chichester, England: Wiley, 1981.

Ehn, P. *Work-Oriented Design of Computer Artifacts*, Stockholm: Arbetslivscentrum, 1988.

Flood, R. L., and Jackson, M. C. "Total Systems Intervention: A Practical Face to Critical Systems Thinking," in R. L. Flood and M. C. Jackson (Eds.), *Systems Practice, Volume 4 in Critical Systems Thinking: Directed Readings*, New York: Wiley, 1991, pp. 321-337.

Flood, R. L., and Romm, N. R. A. "Emancipatory Practice: Some Contributions from Social Theory and Practice," *Systems Practice* (9:2), 1996, pp. 113-128.

Forester, J. "Fieldwork in a Habermasian Way," in M. Alvesson and H. Willmott (Eds.), *Critical Management Studies*, London: Sage Publications, 1992, pp. 46-65.

Forester, J. "Critical Theory and Organizational Analysis," in G. Morgan (Ed.), *Beyond Method*, London: Sage Publications, 1993, pp. 234-246.

Gray, B. "The Pathways of My Research: A Journey of Personal Engagement and Change," *Journal of Applied Behavioral Science* (25), 1989, pp. 383-398.

Habermas, J. *Communication and the Evolution of Society*, London: Heinemann, 1979.

Habermas, J. *Knowledge and Human Interests*, London: Heinemann, 1972.

Habermas, J. *The Theory of Communicative Action. Volume 1: Reason and the Rationalization of Society*, London: Heinemann, 1984.

Habermas, J. *The Theory of Communicative Action. Volume 2: Lifeworld and Systems: A Critique of Functionalist Reason*, London: Heinemann, 1987.

Habermas, J. *Theory and Practice*, London: Heinemann, 1974.

Held, D. *Introduction to Critical Theory*, Berkeley: University of California Press, 1980.

Hirschheim, R. A., and Klein, H. K. "Four Paradigms of Information Systems Development," *Communications of the ACM* (32:10), 1989, pp. 1199-1216

Hirschheim, R. A., and Klein, H. K. "Realizing Emancipatory Principles in Information Systems Development: The Case for ETHICS," *MIS Quarterly* (18:1), March 1994, pp. 83-109.

Johnson, S. "Doing Critical Organizational Research: An examination of Methodology," in C. H. J. Gilson, I. Grugulis, and H. Willmott (Eds.), *Proceedings of the First International Conference on Critical Management Studies*, University of Manchester, July 1999.

Jonsson, S. "Action Research," in H-E. Nissen, H. K. Klein, and R. Hirschheim (Eds.). *Information Systems Research: Contemporary Approaches and Emergent Traditions*, Amsterdam: North-Holland, 1991, pp. 371-396.

Kendall, J. E., and Avison, D. E. "Emancipatory Research Themes in Information Systems Development: Human, Organizational and Social Aspects," in D. E. Avison, J. E. Kendall, and J. I. DeGross (Eds.), *Human, Organizational and Social Dimensions of Information Systems Development*, Amsterdam: North-Holland, 1993, pp. 1-12.

Kerola, P. "On the Fundamentals of a Human-Centered Theory for Information Systems Development," *Report of the 8th Scandinavian Seminar on Systemeering*, Aarhus, Denmark, August 14-16, 1985, pp. 192-210.

Klein H. K., and Hirschheim, R. "The Application of Neohumanist Principles in Information Systems Development," in D. E. Avison, J. E. Kendall, and J. I. DeGross (Eds.), *Human, Organizational, and Social Dimensions of Information Systems Development*, Amsterdam: North Holland, 1993, pp. 263-280.

Lanzara, G., and Mathiassen, L. *Mapping Situations Within a Systems Development Project*, DIAMI PB-179, MARS Report 6, Department of Computer Science, Aarhus University, 1984.

Laughlin, R. "Accounting Systems in Organizational Contexts: A Case for Critical Theory," *Accounting, Organizations and Society* (12:5), 1987, pp. 479-502

Lehtinen, E., and Lyytinen, K. *The SAMPO Project: A Speech-Act Based Information Analysis Methodology with Computer Based Tools*" Report WP-2, Department of Computer Science, Jyväskylä University, 1983.

Lyytinen, K. "Information Systems and Critical Theory," in M. Alvesson and H. C. Wilmott (Eds.), *Critical Management Studies*, London: Sage Publications, 1992, pp. 159-180.

Lyytinen, K., and Klein, H. "The Critical Social Theory of Jurgan Habermas as a Basis for a Theory of Information Systems," in E. Mumford, R. Hirschheim, G. Fitzgerald, and A. T. Wood-Harper (Eds.), *Research Methods in Information Systems*, Amsterdam: North-Holland, 1985, pp. 219-236

McKay, V. I., and Romm N. R. *People's Education in Theoretical Perspective*, Cape Town, South Africa: Maskew Miller Longman, 1992.

Mingers, J. "Technical, Practical and Critical or Past, Present and Future?," in M. Alvesson and H. C. Wilmott (Eds.), *Critical Management Studies*, London: Sage Publications, 1992, pp. 90-112.

Mumford, E. *Designing Human Systems: The ETHICS Method* 1983, Manchester, England: Manchester Business School, 1983.

Ngwenyama, O. "The Critical Ssocial Ttheory Approach to Information Systems: Problems and Challenges," in H-E. Nissen, H. K. Klein, and R. Hirschheim (Eds.), *Information Systems Research: Contemporary Approaches and Emergent Traditions*, Amsterdam: North-Holland, 1991, pp. 267-280.

Payne, S. L. "Critical Systems Thinking: A Challenge or Dilemma in its Practice?," *Systems Practice* (5:3), 1992, pp. 237-249.

Payne, S. L. "Ethical Skill Development as an Imperative for Emancipatory Practice," *Systems Practice* (9:4), 1996, pp. 307-317.

Payne, S. L. "A Proposal for Corporate Ethical Reform," *Business and Professional Ethics Journal* (10:1), 1991, pp. 67-88.

Reynolds, M. "Grasping the Nettle: Possibilities and Pitfalls of a Critical Management Pedagogy," *British Journal of Management* (9), 1999, pp. 171-184.

Romm, N. R. "Knowing as Intervention," *Systems Practice* (8), 1995, pp. 137-167.

Stringer, E. T. *Action Research* (2nd Ed.), London: Sage Publications, 1999.

Udas, K. "Participatory Action Research as Critical Pedagogy," *Systemic Practice and Action Research* (11:6), 1998, pp. 599-628.

Ulrich, W. "Critical Heuristics of Social Systems Design," *European Journal of Operational Research* (31), 1987, reprinted in R. Flood and M. C. Jackson (Eds.), *Critical Systems Thinking: Directed Readings*, New York: Wiley, 1991, pp. 103-116..

Walsham, G. "Ethical Issues in Information Systems Development: The Analyst as Moral Agent," in D. E. Avison, J. E. Kendall, and J. I. DeGross (Eds.), *Human, Organizational, and Social Dimensions of Information Systems Development*, Amsterdam: North Holland, 1993b, pp. 281-294.

Walsham, G. *Interpreting Information Systems in Organisations*, Chichester, England: John Wiley and Sons, 1993b.

Warren, L., and Adman, P. "The Use of Critical Systems Thinking in Designing a System for a University Information Systems Support Service," *Information Systems Journal* (9), 1999, pp. 223-242.

Waring T. S *The Systems Analyst and Emancipatory Practice: An Exploratory Study in Three NHS Hospitals*, Unpublished Ph.D. Thesis, University of Northumbria at Newcastle, 2000.

Waring T. S., and Wainwright D. W. "Enhancing Clinical and Management Discourse in ICT Implementation," *Journal of Management in Medicine* (16:2/3), 2002, pp. 133-149.

Wilson, F. A. "The Truth Is Out There: The Search for Emancipatory Principles in Information Systems Design," *Information Technology and People* (10:3), 1997, pp. 187-204.

Wood-Harper, A. T., Corder, S., Wood, J. R. G., and Watson, H. "How We Profess: The Ethical Systems Analyst," *Communications of the ACM* (39:3), March 1996, pp. 69-77.

ABOUT THE AUTHOR

Teresa Waring is a senior lecturer in Information Systems in the Business School at Newcastle University, UK. She has a Ph.D. and an M.Sc. in computer-based information systems and her main research area of interest is the implementation of information systems. Although Teresa has spent most of her career in education, she has acted as a consultant to a large number of organisations both in the public and private sectors and retains close links with many of them. She has presented research papers at many conferences both in the UK and Europe and has a number of internationally refereed journal articles. In 2002, she won the best paper award at the Business Innovation in the Knowledge Economy Conference at IBM Warwick, UK, and was highly commended for her paper in the *International Journal of Operations and Production Management*. She can be reached at t.s.waring@ncl.ac.uk.

31 RESISTANCE OR DEVIANCE? A High-Tech Workplace During the Bursting of the Dot-Com Bubble

Andrea Hoplight Tapia
Pennsylvania State University

Abstract Under certain circumstances, a critical orientation to the study of workplace deviance/resistance is necessary to understand ICT-enabled workplace culture and employee behavior. The critical orientation to workplace deviance characterizes acts in opposition to an organization with the potential to do harm as semi-organized, group resistance to organizational authority. The questions that drive this research are, does technology enable deviance? When does an act of social deviance become an act of resistance against domination? The answers depend on the perspective of the labeler. To discuss these, I offer the example of a case study of a small software development company called Ebiz.com. For the first few years of the existence of Ebiz.com, the social control exerted on the employees increased yet there were no observable or discussed acts of employee retaliation. I argue that the social environment of the dot-com bubble allowed several myths to propagate widely and affect human behavior. As the market began to fail, and dot-coms began to close, the employees seemed to recognize their situation and enact deviant behavior, or resist. Most importantly, what I have learned from this work is that ICT work may lead to increased deviant or resistant behaviors and that ICT work may also provide a means to do increased deviant or resistant behavior.

Keywords: Dot-com, deviance, resistance, critical theory, organizations, workplace

1 INTRODUCTION

Does technology enable deviance? When does an act of social deviance become an act of resistance against domination? The answer depends on the perspective of the labeler. Workplace deviance has been defined for the most part as if there were some objective standard by which to determine what behavior is potentially harmful or whether or not it violates organizational norms. Most lay people would claim that there is a moral standard widely accepted by society of which behaviors are right and wrong. However, the judgment of whether some behaviors are norm-violating or whether they are potentially harmful can be very subjective. The determination of what is and is not deviant workplace behavior depends on who is asked to make that assessment.

Much of the wider research on the information and communications technology (ICT) enabled workplace has taken a normative, managerialist, or essentialist (Avgerou 2002) orientation. Using this orientation, employee actions that run contrary to organizational norms and values and may potentially cause harm to the organization are labeled as employee deviance (Bennett and Robinson 2000, 2003; Keen 1981; Marakas and Hornik 1996; Markus 1983). These actions are framed as costing the organization time, resources, and money. The deviant is portrayed as receiving legitimate social stigma, punishment, and banishment from the organizational home. In almost all cases, the deviant is portrayed as a low-level, individual employee with unfounded gripes against the organization or an unstable personality. This normative orientation results in support of the status quo and sustenance of the control of organizational elite.

Perhaps a more useful way to look at the these antiorganizational acts in the workplace, especially in the ICT-enabled workplace, is to adopt a critical orientation in which these acts are characterized as semi-organized, group resistance to organizational authority. The rationality and benevolence of organizational leadership is questioned in the following ways: Behavior is rational, efficient, and effective for whom? Whose goals are being pursued? What interests are being served? Who benefits?

Essential to understanding the critical orientation toward acts contrary to organizational norms and values with the potential to cause harm to the organization are the following six elements.

1. They are rarely committed by a solitary individual. Groups of employees who occupy similar organizational roles/space plan and enact them together.
2. They are infused with emotional qualities, such as anger, frustration, jealousy, and resentment, but are rarely directed at another single employee holding the same or lower organizational status.
3. They are rarely committed by mentally unstable employees.
4. They are committed by employees who occupy all levels of an organization, not only the lowest-level, shop-floor, blue-collar employees.
5. They are sporadic in nature. These acts are not committed continuously throughout the entire life cycle of an organization. The acts are tied to particular organizational and managerial policies, changes, and acts.
6. They are committed in information and communication technology rich environments in which the presence and use of the technology may allow for these acts to take different forms, reach wider audiences, and have more intense effects.

Several examples of acts contrary to organizational norms and values with the potential to cause harm to the organization include a system administrator distributes his root password outside the organization; a software developer installs a back door to her program so that she can access it at a later time without detection and permission; a Web-designer intentionally writes highly esoteric and complicated code so that it cannot be shared with other employees; aprogrammer writes a worm that deletes company files and destroys company back-ups; a contractor inserts a software worm into each company's information system while he works for that company; a Web master engages in credit card fraud by obtaining and selling credit information she obtained from her employment; a network manager creates and distributes electronic counterfeit coupons and sweepstakes giveaways from her employer to her friends. All of these behaviors are intentional acts, initiated by organizational members that violate norms of the organization and have the potential to harm the organization.

My purposes in this paper are to clearly explain and support the argument for applying a critical orientation to the study of workplace deviance/resistance. In order to accomplish this, I present some of the relevant arguments in the literature on workplace deviance, ICT-enabled workplaces, and critical theory. I then ground these using a case study of a small software development company.

2 WORKPLACE CHANGE

"With few exceptions, research has proposed that changes in communication technologies are tightly linked with changes in organizations" (Fulk and DeSanctis 1995). Since the early 1990s, "we are now seeing a new type of postindustrial, post bureaucratic, post-Fordist workplace"(Burris and Daday 2001). While most authors agree that a change has taken place, they are divided as to the nature of the change.

Studies of ICTs in workplace organizations can be seen to have taken three general orientations: normative, analytical, and critical (Sawyer and Tapia 2003; Sawyer et al. 2004). The normative orientation refers to research whose aim is to recommend alternatives for professionals who design, implement, use, or make policy about ICTs. An analytical orientation refers to studies that develop theories about ICTs in institutional and cultural contexts, or to empirical studies that are organized to contribute to such theorizing. The critical orientation refers to examining ICTs from perspectives that do not automatically and uncritically accept the goals and beliefs of the groups that commission, design, or implement specific ICTs (e.g., Wastell 2002).

On the normative side, this new workplace has been characterized by social scientists as having a decentralized locus of control, a reduction of hierarchy, an upskilling of work, a centrality of educated knowledge workers, and more flexible democratic forms of work environment (Adler 1992; Block 1990; Clegg 1990; Hirshhorn, 1984; Piore and Sabel 1984; Smith 1990, 1997, 1998). Hammer and Champy (1993, p. 4) state that "the real power of technology is not what can make the old processes work better, but that it enables organizations to break old rules and create new ways of working—that is, to reengineer."

A scholar of the normative or managerialist orientation would look at workplace deviance and discuss it in terms of the harm it causes to the work organization, usually in terms of costs. For example, annual cost estimates range from $4.2 billion for violence (Bensimon 1997), to $200 billion for theft (Buss 1993), to $7.1 billion for corporate security against computer/information attacks, and in less direct costs such as increased insurance premiums (Allen et al. 1996; Bensimon 1997; Slora et al. 1991).

On the critical side, scholars hold the belief that although workplaces have changed, they have remained highly centralized and have adopted new forms of managerial control, including new forms of peer-driven and self-driven control, along with a polarized workplace involving expert and nonexpert sectors bringing up strong issues for gender and race (Burris 1998; Burris and Daday 2001; Hodson 1995, 1996, 1997, 1999; Prechel 1994; Vallas 1999; Vallas and Beck 1996).

If the relationship between ICT and organizational culture is seen as mediated by an exercise of power—a system of authority and domination that asserts the primacy of one understanding of the physical world and one prescription for social organization over others—then the choice of technology represents an opportunity to affect not only the performance at work but also the status, influence, and self-concept of those promoting change. New technology may be far less attractive for what it does than for what it says symbolically about its creators, and users.

3 SOCIAL DEVAINCE AND THE WORKPLACE

Deviance is defined as

> Behavior and characteristics that some people in society find offensive or reprehensible and that generates—or would generate if discovered— in those people disapproval, punishment of condemnation of, or hostility toward the actor or possessor (Goode 1997, p. 37).

Much of the work that has focused on deviants selected "nuts, sluts, and perverts" (Liazos 1994) as the subjects of study for several reasons: they were easily identifiable as deviants, they were located at the bottom of the socio-economic ladder, access to them was easily obtained, and they provided catchy titles and sensational articles that caught the public eye. There was no clear study of workplace deviance since all deviance could happen at any place at any time as long as the definition of the deviance and the deviant fit the parameters above.

During the past century, psychologists studied workplace deviance most often and social psychologists focused on behaviors such as theft, work-slowing, and sabotage among blue-collar, lower-level employees. The types of behaviors tended to be oriented toward plant floor behaviors rather than actions typical of the boardroom such as fraud, harassment, or embezzlement. Almost all conceptualizations of workplace deviance were limited to the actions of individuals rather than the deviant actions of groups, whole organizations, or even industries.

In the ICT-enabled workplace, to a passerby, a programmer engaged in code writing could be creating legitimate or non-legitimate code. This legitimate and illegitimate behavior could also be comingled throughout the day and this passerby could never discern the illegitimate, deviant behavior. An average ICT employee may have many windows open on his or her desktop at the same time and may shift between them as part of regular, legitimate employment, as well as aspects of deviant behavior. In order to detect the illegitimate behavior, the detector needs to be as complex, sophisticated, and technically knowledgeable as the deviant him- or herself.

4 THE ICT-ENABLED WORKPLACE

Only very recently has the construct of social deviance expanded to encompass the office, laboratory, and boardroom and the deviants expanded to encompass the managers, technicians, accountants, and other diverse employees. It is not possible to discuss the modern workplace without talking about the role and place of information and communication technologies (ICTs). It is essential to view the modern workplace as the social environment into which ICTs are embedded (Mackenzie and Wajcman 1999; Orlikowski and Iacono 2001). This social constructivist model sees ICT as embedded in a web of meaning encompassing the organizational structure, functions, norms, values, and patterns of behavior. It is impossible to treat work, technology, and the people doing the work independently.

Research into forms of deviance and resistance in the computerized work environment lags far behind its prevalence in today's workplace (Colclough, and Tolbert 1992; Hollinger 1986; Hollinger and Clark 1982; Oakes and Cooper 1998; Raelin 1986; Sewell 1998; Sewell and Wilkinson 1992; Vardi and Wiener 1996; Wiseman and Bromiley 1996). ICT workplace deviance may include sabotaging computer programs, stealing proprietary information, executing viruses and hacking into private computer space. Not surprisingly, organizations spend billions annually to offset cyber attacks. While the incidence of computer crime has risen, there has been little or no movement on computer deviance/resistance committed at the workplace against the work organization or fellow employees.

The exception to this is in the area of *cyberloafing*. Technological changes have at once revolutionized the way we do work and, at the same time, multiplied the opportunities employees have to be *un*productive at work. Computer misuse or cyberloafing in the workplace is something with which employers are, or should be, increasingly concerned (Lim et al. 2001; Mastrangelo et al. 2001). Lim et al. defined cyberloafing as the act of employees using the company's Internet access during work hours to surf non-work related Websites and to send personal e-mail.

Perhaps the most notable exception to this is Wilson and Howcroft's (2000) work on the resistance to a new information system among the female nursing staff at a hospital. In this case, Wilson and Howcroft illustrate the deliberate acts of resistance, social deviance (my words), committed by the nurses when they found the information system to be incompatible with their organizational mission and role as caregivers. In this case, the authors clearly reject the normative, managerialist orientation, which they state pervades the field of Information Systems research, and select a critical orientation. This orientation allows them to see the nurses as asserting their ability to define their role within the organization through acts of resistance to what they perceived as organizational domination.

5 CRITICAL THEORY, DEVIANCE AND RESISTANCE

This critical orientation can be seen to have its roots in the critical theory of the Frankfurt School (Heidegger 1977; Horkheimer and Adorno 1972; Marcuse 1982). Critical theory, in general, can be characterized to be explicitly concerned with critiquing domination with an orientation toward praxis focused against domination. If there is one central concept running throughout the literature of critical theory, it is domination. Critical theory is also oriented toward helping people understand why and how they are dominated, and then empowering people to do something to ameliorate their misery.

The theoretical standpoint taken in this paper is that ICT has been institutionalized as a multifaceted force of industries, techniques for carrying out tasks in organizations, and principles for organizing that is closely associated with a particular form of business management. The narrowness of the managerialist perceptions and normative knowledge that has been prevalent in much of the information systems literature and practice has been subject to a great deal of critical debate. For the most part, the normative orientation can be seen as the managerialist orientation, supporting the status quo, seeking to further the interests, through increased efficiency, effectiveness, and product output, of the managerial class. Examples of this can be seen in Keen (1981), Markus (1983), and Marakas and Hornik (1996), all of whom view resistance (to the implementation of IT in their cases) as a message that something is wrong rather than as a barrier to overcome. The analytical orientation, in contrast, can be seen as a scientific, hands-off orientation in which information is gathered and categorized but rarely used by theorists themselves to enact any sort of social change.

Essential to understanding how critical theory has been applied to ICTs and organizations is the belief that ICTs are not neutral and embody the values of a particular industrial civilization and especially of its elites, which rest their claims to hegemony on technical mastery. Shields (1997) states that

> newer frameworks view technological change as a process whereby competing groups of technical experts and entrepreneurs bring technical, political, professional, economic and other values and interests to bear in trying to frame and resolve contested technological designs in their favor (p. 198).

He contends that technologies are not value neutral instruments. They are self-consciously fashioned by social groups who intentionally promote their values and interests while intentionally undermining others. Feenberg (1991) finds that the modern industrialized world has brought new forms of oppression, and he suggests that society has the ability to select the forms of technology that it will adopt, thus granting it agency in the face of oppression.

As discussed above, deviance is defined as causing harm, or the threat or potential to cause harm, to one's organization. The very definition of deviance reflects a normative, managerial orientation. The interests of the organization's management are those that are most often discussed as the victim of employee deviance. When workplace deviance is characterized, it is usually in terms of the extreme costs to institutions and organizations. Even when psychologists have attempted to find the causes of workplace deviance, they have attributed it to two principal causes: deviance as a reaction to experiences and deviance as a reflection of one's personality. In other words, workplace deviance is seen as a result of a reaction to perceived frustration and injustices, or seen as a personality flaw such as lack of control and aggressive tendencies. In almost all cases, deviance is framed as an individual issue, not a social issue

On the other hand, critical theorists would see deviant behavior as inherently social, an act of a group, and as a conscious act of rebellion or resistance to real subjugation by the dominant administrative coalition. Critical theory may form the basis for explaining what appears to be an irrational response to ICT-enabled organization to the managerialist scholar. A critical theorist would not see these responses as irrational or deviant. They would characterize them as acts of resistance or acts of self-empowerment of the dominated class. Several authors have recently applied a critical orientation to workplace deviance. Dehler and Welsh (1998), for example, assert that the current normative definitions of workplace deviance are social constructions that support the status quo and sustain the control of the organizational elite. Critical theory is proposed as a better lens through which to view behavior that violates norms of the organization.

6 A CASE STUDY

This research began as a larger study in which three small software development companies were examined at various points during their life cycles. The goals of a larger study were to understand the organizational culture and structure of the small software development company and its relationship to technology during the dot-com bubble. I present some of this data from one of the three cases, which, in many ways, is representative of the other two.

I chose to focus on small software development companies during the dot-com bubble because of the rapid boom to bust cycle in which they existed. These types of transitions are excellent phenomena to study the relationships among organizational culture, structure, and information technology. The connections and effects among these are forming, raw and visible to those trained to observe such phenomena. The boom can be described as comprising entrepreneurs who enthusiastically set up Web-based enterprises selling everything from infrastructure, services, domain names, advertising, toys, graphics, and anything else. In 1999, the NASDAQ gained 128 percent. In 1999, there were 546 IPOs that raised over $69 billion. The average first-day gains of IPOs in 1999 were 68 percent compared to 23 percent in 1998.

However, during the years 2000 and 2001, this bubble burst. Stock prices plunged, investors lost confidence, and Web-based businesses started closing down. Evidence of the bust includes the 4,854 Internet companies acquired or shut down (3,892 acquisitions and 962 shutdowns) during the first quarter of 2000. Moreover, the first 16 months of the bust saw 44 shutdowns per month. The bust was a wide-ranging occurrence; failed companies included Internet content providers, infrastructure companies, Internet-services providers, and providers of Internet access (Webmergers.com 2002). This is a case study of a single, small software development company that was born and died during the years between 1996 and 2001, the era that has come to be known as the dot-com bubble. Ebiz.com was a small but rapidly growing firm that wrote business-to-business software and constructed Websites specifically geared to large-scale e-commerce. Their product was custom software, tailored to the user, with long-term service contracts

Charting the company's time-line reveals a life span typical of other failed small software development companies during the period of the dot-com bubble and subsequent collapse. The company was founded in 1996 by two veteran software developers. By the beginning of 2000, Ebiz.com was up to 90 employees. However, for various reasons discussed below, 65 employees were laid off during the fall of 2000. In January 2001, the remaining employees were laid off and the doors were closed. Another company purchased the software and hired 15 of the original Ebiz.com employees. This second company also had failed by August 2001.

The question that arises is how typical was Ebiz.com in terms of its organization culture during the dot-com bubble? While there are some significant efforts currently being made to chronicle, archive, and preserve what remains of dot-com era organizational materials (Webmergers.com 2002), organizational culture is nearly impossible to capture once a company is defunct and employees have scattered. Written materials such as business plans, handbooks, and organizational charts cannot begin to capture the richness of the culture of these now defunct dot-coms. The words and thoughts of the individuals involved in the situation most completely convey the cultural underpinnings of the dot-coms. It is important to note here that although I cannot know how pervasive the cultural traits discussed here were during this time period, I have seen similar behavioral and cultural patterns among the three small software development companies that were included in my larger study.

6.1 Increases in Social Control Efforts

The facts are that, during the course of the year that I spent with Ebiz.com, the management increased social control efforts over the employees. In some cases, ICT choices were used to increase the level of social control over employees. It is also clear that the employees acquiesced to all of the social control efforts enacted by the managers while blind dot-com optimism was still the flavor of the news across the country. As the market began to fail, and dot-coms began to close, the employees seemed to recognize their situation and enact deviant behavior or, as some would say, resist.

There is strong evidence that the owners and managers used several techniques to increase their control over their employees. They created an organizational culture that included the following elements.

(1) The manipulation of operating systems and programming languages to maximize owner control over workers and products.
(2) The dissolution of the boundaries between home and work life. Employees' physical and social needs were met by the workplace.
(3) The creation of a culture based on crisis that rewarded heroic behavior.
(4) The creation of self-policing, co-programming teams that developed systems of concerted control over each other.

The dot-com boom and bust resemble other episodes in history of unreasonable individual and corporate speculation. The dot-com bubble is characterized as a period of rapid economic growth with individuals and corporations taking risks they might not have taken and traditional business practices and social values being ignored.

The dot-com bubble is described as a period of enormous contagion of optimism, constantly changing opportunity, *ad hoc* organizational structures, very rapid growth, highly mobile workers, massive early investment that exerts enormous pressure to produce the goods quickly in order to turn cash-flow positive, fast and often unpredictable rate of change, and a loss of traditional human resources programs and regulations. I argue that the social environment of the dot-com bubble allowed several myths to propagate widely and affect human behavior.

- **The myth of Silicon Valley**. This myth was the belief that during the dot-com bubble, any intelligent, hard-working individual could become a millionaire before the age of 25 working in the IT industry.
- **The myth of circumventing rules**. This myth was the belief that during the dot-com bubble any intelligent, hard-working individual need not follow traditional pathways to wealth. The rules were gone. The legitimate, socially acceptable means to wealth including education, 20 years of steady employment, 20 years of smart investing, and long-term real estate ownership, for example, were seen as circumventable. A culture of the get-rich-quick mentality was created.
- **The myth of the future downtime**. This myth allowed any intelligent, hard-working individual to believe that the dot-com bubble was a short-term phenomenon in which one had to seize the opportunity while the opportunity was there. Work as hard as possible for a short time, and the rest and relaxation would come later. Hard work now guaranteed huge payoffs in the near future.
- **The myth of engineer managers.** This myth was the belief that during the dot-com bubble any intelligent, hard-working software engineer could do the job of owner and manager of any new business.

The owners and managers of Ebiz.com used these myths to increase their control over the workers. They were aware of them, manipulated them, and took advantage of them. The employees gambled that the high cost to them at the time would pay off in the future. They were speculating that their backbreaking labor would fill their metaphorical pans with gold. They acquiesced to the owner's and manager's demands because they believed that they would become millionaires soon, they believed that they could become a millionaire by unconventional means, they believed that once they made it they could rest, and they believed that the managers and owners knew what they were doing.

In the following sections, I will detail several examples of social control enacted by the management of Ebiz.com and the resulting acts of resistance (deviance) as the myths began to fail.

6.2 Software Change: Social Change

Within Ebiz.com, several technological changes transpired during the investigation period. The biggest change was the programming language in which the company created its products. The company began programming in PHP, a language considered open-source, free, and uncontrolled. The small group of original programmers who were responsible for most of the initial products was a tightly knit group who exerted significant control over the business. In a surprising move, the owners decided to change the language from PHP to JAVA. The employees of Ebiz.com characterized JAVA as unstable, complicated, more tightly controlled, and inferior to PHP. They moved from their own proprietary architecture to an off-the-shelf, more standardized architecture.

The employees believed that they had made the switch because the owners wanted to please a very large client and that JAVA, as an object-oriented language, was easier to sell to the nontechnical managers of their clients. At the time of the change, the owners fired the original PHP programmers and physically moved around the other employees. When asked why, the owner stated, "I don't like the way these guys [the PHP programmers] were operating, all isolated from the rest. I especially didn't like their issues around marking their territory...I can't have any little fiefdoms here."

The owners then asked all of the remaining programmers to rewrite all of the existing code in JAVA. Most employees were not skilled in JAVA, so the slow learners were fired; the quick learners were forced up a steep learning curve. To fill the organizational holes a few new, hot-shot Java programmers were hired. The employees that were kept on were expected to spend 80 to 100 hours at the office rewriting products in JAVA while spending their free time learning the intricacies of the language itself. The owners instituted a testing policy in which all programmers had to pass the on-line Brainbench JAVA certification test.

The older employees, who were previously seen as experts in PHP, were thus placed on a level playing field with all other employees. The hierarchy was destabilized, restructured, and competition was fostered between employees to see who would learn and adapt the fastest. This was measured by the date on which they took their Brainbench test as well as their score and ranking. Effectively, the owners' choice of moving to a new programming technology allowed them to reassert control over the workplace, the employees, and the relationship with the clients. In the case of Ebiz.com, technological change was the result of a struggle for power in which the owners gained control by eliminating the need for difficult-to-control-experts and replacing them with new technology and new employees who were seen as easier to control.

This was only possible in an era in which the wider culture was infused with the belief that software programmers were disposable, short-term employees. In this case, the PHP programmers easily moved on to other work and new JAVA employees were easily hired. The dot-com bubble was a time of excess, not scarcity. The owners/managers of Ebiz.com had little fear of a lack of workers and had every hope of enticing new employees to come work for them.

6.3 Boundary Loss: Control Gain

Many small software companies existed because of a single initial technological innovation and were forced to continue to produce innovations if they were to continue to exist. Managers concluded that not only did they have to hire creative people but also they had to foster creativity daily.

Playful work environments that foster exploration appear to help drive the innovation that defines the high-tech sector. The owners of Ebiz.com were aware of this management trend and used it to create their own organizational culture. They recreated the elements of a playroom in the workplace including filling the environment with toys, colors, and music. They removed most of the internal walls of the building, covered all remaining walls with erasable whiteboards, and provided all of the employees with colored markers. They organized game times each day where all of the technical teams played together. A few times a week, meal times were organized where the employees ate together on company property. Constant free, cool and fun junk food was provided to all employees.

One owner said,

> All the game playing, well…its stress relief, first off. It helps the employees avoid burnout longer. If I encourage them to play everyday they'll go back to work with a fresh start afterwards, and work harder and better because of it. After they let off some steam killing their virtual coworkers, they can attack a problem with a clean slate and maybe come up with something they hadn't thought of before.

I suggest that Ebiz.com used this play room management style to create an atmosphere in which owners demanded increasing inputs in hours and effort from their employees, increased employee competition, increased self- and peer-generated control systems, increased hierarchies within technical and nontechnical employees, and do this all in an atmosphere where the dominant ideology is that the employees have more autonomy and fun at work.

Perlow (1998, 1999, 2001) states that organizational culture assumes a crucial control function in knowledge-based organizations where the work preformed is creative, open-ended, individually styled, and highly demanding. Attempts are made to elicit and direct the required efforts of members by controlling the experiences, thoughts, and feelings that guide their actions. The intent is for the workers to be driven by internal commitment, strong identification with company goals, and intrinsic satisfaction from work. It compels employees not only to do what is expected at work but also to conform to norms that determine how they lead their lives outside their work environment.

Ebiz.com does this by erasing the boundary between home and work. Ebiz.com identified the activities that employees would do at home, such as play computer games, watch TV, lounge on the couch or bean bag chairs, each lots of junk food, and hang out with friends, and incorporated all of these elements into the Ebiz.com work environment. The owners created a work environment that was so much like home that it became a second home for many employees. They provided higher quality technology, food, games and atmosphere than many of the employees could hope to afford on their own, further inducing them to stay a little longer. One Ebiz.com owner said, "I tried to make work as much like home as possible so that it would be easier for the guys to spend a lot of time here." The employees felt that work was so fun and comfortable that staying there for 12 to 15 hours a day was less painful, and perhaps even desirable. A member of the development team said,

> I used to go out for coffee and lunch almost every day, but now what's the point? They make coffee at work and there are always some food things in the fridge downstairs. I just heat up a burrito or somebody orders pizza and I don't have to even leave my desk.

6.4 Teams and Crisis: Concerted Social Control

Management by teams has been said to grant employees autonomy and a more democratic work environment through determining their own work organization, communicating horizontally with the organization instead of up a hierarchy, building close relationships with suppliers, and sharing information. (Hodson, 1995, 1996, 1997, 1999; Smith 1990, 1997, 1998). However, the critical view of team-based management states that supervision, responsibility, and discipline are often shifted from managers to peers without any compensation or security. Workers are asked to do more without any increase in pay. There has been a shift from traditional bureaucratic control to concerted control in that workers collaborate to develop the means of their own control. They control their behaviors through a complex system of values, norms, and rules. Increased production pressures and intensification of work have been found to be legitimated by the peer relationships among the teams and as the team encouraged workers to push themselves to the limit for the good of the work group (Barker 1993; Smith 1990).

The culture of software developers celebrates and rewards workers' intensity and total devotion to work. The culture develops a system based on constant crisis and a reward system based on individual heroics, which results in workers doing whatever it takes to solve the crisis of the moment. The managers and peers model the desired behavior themselves, also putting in long hours.

One of the managers of Ebiz.com stated,

> I put in very long hours. I try to get here every day at about 7:00 a.m. and I never get out before 8:00 at night. Sometimes I stay even later. I've pulled a few all-nighters here and I come in at least for a few hours every weekend. Once I get home I also get a little done there. I've got a pretty good computer and connection there too. When the deadlines are getting close, you just gotta get it done no matter how much time it takes. I know it's a lot to ask, but I'd never ask anybody to do anything that I wasn't willing to do myself.

Ebiz.com also developed a culture of time one-up-manship, in which employees challenged each other to stay for longer and longer hours. For example, one of the employees said to another, "I was here last night until 10:00 finishing up the clean-up on that code." In response, the other team member stated, "Oh yeah, well I've been here until at least 10:00 every night this week. I'll probably have to pull an all-nighter tonight or tomorrow just to get it all cleaned up by the deadline."

One of the owners stated, "I really didn't realize that the guys were pulling that one-up-manship stuff until recently. Had I known that was going on during the big crunch time, I would've been cheering. That's exactly what I wanted to hear." From this statement, it is obvious that the owners desired the employees to goad each other into working longer and longer hours, releasing the owners from the position of having to ask for additional time directly from the employees.

Evidence was found among the teams of Ebiz.com to support the shift to concerted control. Individual members of the various teams stated multiple times that they felt that they had to work long hours for the good of the team. A member of the development team said, when asked what he was working on,

> I'm trying to get this script hammered out. We've got a deadline in a few days and I don't want to be the slacker here. Its bad if [another developer] has to bail me out and pick up my slack just to get the stuff done on time. I'm just trying to hold my own, you know, pull my own weight.

On several occasions while observing, it was noted how two members of the development team were always working to get the others to work faster. They would play loud thrash music to motivate. They would yell over the din to the others to get going, to write clean code and write it fast.

One owner said,

> Putting these guys into teams was one of the best things I ever did. They get each other to work harder and better. If one of the members of their team is not doing their work, the other guys come down on him pretty hard. They do my job for me. Plus, I think they like getting criticism from one of the guys rather than from me. Its not so top down that way.

6.5 Resistance

In 2001, the economy began to falter. Stock prices plunged, investors lost confidence, and Web-based businesses started closing down. Further evidence that the organizational culture that developed at Ebiz.com was tied to the dot-com bubble are the changes that occurred after the bubble began to burst.

Ebiz.com laid off half its employees and its most lucrative account was cancelled. The employees of Ebiz.com began to doubt whether they would become millionaires as they had hoped to be, or even if they would have jobs in 6 months. They started to doubt the expertise of the managers' ability to run the company and lead the workers. They seemed to develop a collective consciousness of the number of hours they had been spending at work and the little they had to show for it. They began to complain.

As with all forms of resistance, these strategies were never direct, tended to be more for the benefit of the other exploited workers, and had a high cost attached to them if they were recognized as resistance and the resistor was singled out.

The most dramatic and damaging form of resistance that was enacted was coincidentally around the time the company switched from PHP to JAVA and began to have financial problems. The employees became aware of the financial problems despite the managers efforts to keep it secret. The old employees that had been repositioned in the hierarchy because of the move to JAVA began to put out very buggy code. These were intelligent, accomplished programmers who did not make these errors randomly. Only after several months did the managers hire an expert JAVA programmer who recognized the extent of the damaged code. Since the code had been worked on by the entire development staff, it was impossible to determine who had caused the major bugs. The project was scrapped and started over.

As discussed earlier, every wall was covered with erasable whiteboard material to encourage creativity. Several drawings appeared on the white boards. One depicted a development team member being sexually assaulted by a member of management from behind. Another drawing portrayed the development team waving from a boat deck labeled "The Titanic" with the management as its captain. A third image portrayed the cartoon robot from the TV show *Futurama* demanding that next time he wrote an interface it would be with hookers and blackjack. There were clearly several different artists for each of the drawings. The owner responded with very visual anger in which he called an impromptu meeting, screamed at the developers, and asked them who had done it. No one volunteered any names. When later asked, the developers only smirked and refused to talk about the authors or artists. Nearly all of the technical team members mentioned the drawings on the whiteboards during interviews with the research team.

Another form of resistance was also the computer games. Ironically, the games were seen as integral to tying the playroom culture together by the management. When the dot-com bubble burst, the carrot disappeared and a more traditional managerial stick appeared. The employees began to complain about the hours and the lack of economic compensation. The owners began to complain about the gaming getting in the way of getting real work done. The owners tried to take more control of the gaming and joined in the games themselves. The games soon developed into an adversarial system of developers against owners and managers. The developers organized a point and ranking system to depict just how badly they had thrashed the managers.

Soon the Sega Dreamcast system was taken away from the workers altogether. The employees then began to play Unreal tournament with each other from their desks. If someone were to see them without seeing the fronts of their monitors, they would appear to be working diligently on coding. However, the percentage of the workday spent playing games rather than coding went up dramatically after the Sega Dreamcast equipment was taken away.

As the company began to falter financially and the first layoffs were announced, the company kitchen was also closed. The managers stated that they could no longer afford to stock the kitchen with food for everyone. They also could no longer afford to rent the additional office space needed for the lounge, game room, and kitchen, so the rooms and the equipment would be unavailable to the employees from that point forward. The employees reacted by taking breakfast, lunch, and coffee breaks together as the entire development staff for several hours a day. They would leave the office together around 9:00 a.m. for Starbucks and return around 9:45 a.m. to begin work. They would then leave again around 12:30 p.m. for lunch and return around 2:00 p.m. Finally they would take a mid-afternoon break en masse around 4:00 p.m. that might last until the end of the day.

In response, the managers instituted a whole-company meeting every morning at 8:30 a.m. At first all of the developers were present and mostly engaged. However, after one week, fewer and fewer developers arrived before 9:00 a.m. After the second week of this new policy, no developers came to the morning meeting at all.

7 CONCLUSIONS

The central argument in this work is that under certain circumstances a critical orientation to the study of workplace deviance/resistance is necessary to understand ICT-enabled workplace culture and employee behavior. I began this paper with two questions. Does technology enable deviance? When does an act of social deviance become an act of resistance against domination? The answer depends on the perspective of the labeler.

In the case of Ebiz.com, we can see that it is clear that not all acts of social control result in acts of workplace deviance or resistance. There is strong evidence that the owners and managers used several techniques to increase their control over their employees, including the manipulation of operating systems and programming languages to maximize owner control over workers and products, the dissolution of the boundaries between home and work life, the creation of a culture based on crisis that rewarded heroic behavior, and the creation of self-policing, co-programming teams that developed systems of concerted control. For the first few years of the existence of Ebiz.com, the social control exerted on the employees increased yet there were no observable or discussed acts of employee retaliation. I argue that the social environment of the dot-com bubble allowed several myths to propagate widely and affect human behavior. The clear answer is to be found in the power of the myths employed by the managers of Ebiz.com. Those myths, the myth of the Silicon Valley, the myth of circumventing rules, the myth of future downtime, and the myth of engineer managers, were presented earlier.

As the market began to fail, and dot-coms began to close, the employees seemed to recognize their situation and enact deviant behavior, or resist. Employees at Ebiz.com committed several acts that can be construed as deviance or resistance. The employees intentionally

- produced error filled code
- publicly graphically depicted themselves being assaulted by the managers
- publicly graphically depicted the company as failing
- publicly graphically depicted the product as poor quality
- dramatically increased game playing time at work
- dramatically increased off-site breaks from work
- dramatically decreased hours spent at work.
- directly disobeyed managers when told not to engage in game playing
- directly disobeyed managers' requests for morning meetings

The question that must be asked is, do these acts fit the criteria for resistance and can they be analyzed in terms of the critical orientation? Essential to understanding the critical orientation toward these acts are the following six key elements.

- Acts are contrary to organizational norms and values with the potential to cause harm to the organization.
- Acts are rarely committed by a solitary individual. Groups of employees who occupy similar organizational roles/space plan and enact them together.

In all cases the acts of deviance committed by the employees of Ebiz.com were known to all other employees and in most cases done together. These acts were not solitary and secret from other employees.

- Acts are infused with emotional qualities, such as anger, frustration, jealousy and resentment, but are rarely directed at another single employee holding the same or lower organizational status.

The employees of Ebiz.com began to develop a sense of fear that they would lose their jobs, not find another one, or be forced to go back to school. They were angry with the managers for poorly managing Ebiz.com and letting it fail. They were very disappointed that their dreams of becoming young millionaires would not come true. There is very clear emotional content to all of their comments at this stage in the business; however, I have no direct evidence that the acts committed were charged with these emotions.

- Acts are rarely committed by mentally unstable employees.

I have no evidence that any of the employees at Ebiz.com were mentally unstable. It would be highly unlikely that they all were unstable.

- Acts are committed by employees who occupy all levels of an organization, not only the lowest-level, shop-floor, blue-collar employees.

At Ebiz.com, the hierarchical ladder approached being flat. At the developer level, there were only two types of employees: team leaders and team members. Both forms of employees participated in the acts of resistance. None of the employees considered themselves to be blue-collar.

- Acts are sporadic in nature. These acts are not committed continuously throughout the entire life cycle of an organization. The acts are tied to particular organizational and managerial policies, changes, and acts.

It is clear that the acts committed by the employees began when the company, and the dot-com myths, began to fail. There was a decided lack of acts in opposition to the organization during the first few years of the company's existence, despite increasing efforts to control the employees.

- Acts are committed in information and communication technology rich environments in which the presence and use of the technology may allow for these acts to take different forms, reach wider audiences, and have more intense effects.

Several acts in the case of Ebiz.com would not have been possible without certain forms of ICT available to the employees. The acts of code sabotage took place in a completely virtual environment. The form of sharing the coding work, co-programming, reviewing each other's work, and building on each other's code allowed for a large amount of anonymity on the part of the saboteurs. The act of switching game playing platforms from a console game in a separate room to PC-based games using the company LAN, used the technology available to the employees to take back control of their time.

Most importantly what I have learned from this work is that ICT work may lead to increased deviant or resistant behaviors and that ICT work may also provide a means to do increased deviant or resistant behavior. ICT work may provide both the setting and the vehicle for increased workplace deviance/resistance. In the case of Ebiz.com, some of the acts committed by the employees in opposition to the organization could be characterized as traditional office workplace deviance, such as extended lunch breaks. However, other behaviors, such as product sabotage of code, extended game playing, and cyberloafing, were enabled and defined by the ICT nature of the workplace. In the case of the extended game playing, even after the managers thought they had removed the vehicle for deviant behavior—the Sega Dreamcast system—the employees found a new, and perhaps better, method to engage each other in deviant behavior. The Internet and the workplace LAN provided the setting and the game Unreal Tournament provided the vehicle for increased deviance.

The determination of what is and is not deviant workplace behavior depends on who is asked to make that assessment. Much of the wider research on the information and communications technology-enabled workplace has taken a normative or managerialist orientation. Using this orientation, employee actions that run contrary to organizational norms and values and may potentially cause harm to the organization are labeled as employee deviance. These actions are framed as costing the organization time, resources, and money. In almost all cases, the deviant is portrayed as a low-level, individual employee with unfounded gripes against the organization or an unstable personality. The critical orientation to workplace deviance characterizes these same acts as semi-organized, group resistance to organizational authority.

Social context matters. These small business failures were different than other business failures. They happened within a society that held some overarching beliefs that inspired certain organizational style behaviors. The pervasive dot-com, get-rich-quick mentality acted as a catalyst that urged new, small software development companies to be created, to take certain business forms, and to act in certain ways. Employees within the software development teams seemed to negotiate a fine line between acceptance and resistance of the workplace norms espoused by the owners. At the beginning of this study, the employees were clearly in support of the workplace norms governing their time and effort; they accepted the long hours and the lack of remuneration for working overtime. Upon hire, the casual clothing, the games and playtime, the relaxed atmosphere, the nontraditional and non-bureaucratic environment dazzled them. Along with this environment came the feeling that the employees were climbing aboard a ship that was sailing toward incredible success. In the case of Ebiz.com, work was constructed to be more and more like play in order to attract the highly productive computer mavericks who could write the killer application overnight. This situation was time-bounded by the fact that the dot-com bubble burst by the end of 2001 and the get-rich quick mentality along with it.

REFERENCES

Adler, P. *Technology and the Future of Work*, New York: Oxford Press, 1992.

Allen, R. E., and Lucero, M. A. "Beyond Resentment: Exploring Organizationally Targeted Insider Murder," *Journal of Management Inquiry* (5), 1996, pp. 86-103.

Avgerou, C. *Information Systems and Global Diversity*, Oxford: Oxford University Press, 2002.

Barker, J. R. "Tightening the Iron Cage: Concertive Control in Self-Managing Teams," *Administrative Science Quarterly* (38), 1993, pp. 408-37.

Bennett, R. J., and Robinson, S. L. "The Development of a Measure of Workplace Deviance," *Journal of Applied Psychology* (85), 2000, pp. 349-360.

Bennett, R. J., and Robinson, S. L. "The Past Present and Future of Workplace Deviance Research," in J. Greenberg (ed.), *Organizational Behavior: The State of the Science*, Mahwah, NJ: Lawrence Erlbaum Associates, 2003.

Bensimon, H. "What to Do about Anger in the Workplace," *Training and Development*, September 1997, pp. 28-32.

Block, F. *Postindustrial Possibilities*, Berkeley, CA: University of California Press, 1990.

Burris, B. "Computerization of the Workplace," *Annual Review of Sociology* (28), 1998, pp. 141-157.

Burris, B., and Daday, G. "Technocratic Teamwork: Mitigating Polarization and Cultural Marginalization in an Engineering Firm," in S. Vallas (Ed.), *The Transformation of Work*, New York: Elsevier, 2001, pp. 241-262.

Buss, D. "Ways to Curtail Employee Theft," *Nation's Business*, April 1993, pp. 36-38.

Clegg, S. R. *Modern Organizations*, Newbury Park, CA: Sage Publications, 1990.

Colclough, G., and Tolbert III, C. M. *Work in the Fast Lane: Flexibility, Divisions of Labor, and Inequality in High-Tech Industries*, Albany, NY: State University of New York Press, 1992.

Dehler, G. E., and Welsh, M. A. "Problematizing Deviance in Contemporary Organizations: A Critical Perspective," in R. W. Griffin, A. O'Leary-Kelly, and J. M. Collins (Eds.), *Dysfunctional Behavior in Organizations: Violent and Deviant Behavior*, Stamford, CT: JAI Press, 1998, pp. 241-269.

Feenberg, A. *Critical Theory of Technology*, New York: Oxford University Press, 1991.

Fulk, J., and DeSanctis, G. "Electronic Communication and Changing Organizational Forms," *Organization Science* (6:4), 1995, pp. 337-349.

Goode, E. *Deviant Behavior*, Upper Saddle River, NJ: Prentice Hall, 1997.

Hammer, M., and Champy, J. *Reengineering the Cooperation: A Manifesto for Business Revolution,* New York: Harper Collins, 1993.

Heidegger, M. *The Question Concerning Technology*, New York: Harper and Row, 1977.

Hirschhorn, L. *Beyond Mechanization*, Cambridge, MA: MIT Press, 1984.

Hodson, R. "Dignity in the Workplace Under Participative Management: Alienation and Freedom Revisited," *American Sociological Review* (61:5), 1996, pp. 719-738.

Hodson, R. "Group Relations at Work: Co-worker Solidarity, Conflict, and Relations with Management," *Work and Occupations* (24:4), 1997, pp. 426-452.

Hodson, R. "Organizational Anomie and Worker Consent," *Work and Occupations* (26:3), 1999, pp. 292-323.

Hodson, R. "Worker Resistance: An Underdeveloped Concept in the Sociology of Work," *Economic and Industrial Democracy* (16), 1995, pp. 79-110.

Hollinger, R. "Acts Against the Workplace: Social Bonding and Employee Deviance," *Deviant Behavior* (7), 1986, pp. 53-75.

Hollinger, R., and Clark., J. "Employee Deviance: A Response to the Perceived Quality of the Workplace," *Work and Occupations* (10), 1982, pp. 97-114.

Horkheimer, M., and Adorno, T. "The Culture Industry: Enlightenment as Mass Deception," in *Dialectic of Enlightenment*, New York: Herder and Herder, 1972, pp. 120-167.

Keen, P. "Information Systems and Organizational Change," *Communications of the ACM* (24:1), 1981, pp. 24-33.

Liazos, A. "The Poverty of the Sociology of Deviance: Nuts, Sluts, and 'Perverts,'" in C. J. Curran and C. M. Renzettti (Eds.), *Theories of Crime*, Needham Heights, MA: Allyn & Bacon, 1994, pp. 372-395.

Lim, V., Loo, G., and Teo, T. *Perceived Injustice, Neutralization and Cyberloafing at the Workplace*, Washington, DC: Academy of Management, 2001.

Mackenzie, D., and Wajcman, J. *The Social Shaping of Technology*, Philadelphia: Open University Press, 1999.

Marakas, G. M., and Hornik, S. "Passive Resistance Misuse: Overt Support and Covert Recalcitrate in IS Implementation," *European Journal of Information Systems* (5:3), 1996, pp. 208-220.

Marcuse, H. "Some Social Implications of Modern Technology," in A. Arato and E. Gebhardt (Eds.), *The Essential Frankfurt School Reader*, New York: Continuum, 1982.

Markus, M. L. "Power, Politics, and MIS Implementation," *Communications of the ACM* (26:6), 1983, pp. 430-444.

Mastrangelo, P., Everton, W., and Jolton, J. "Computer Misuse in the Workplace," unpublished manuscript, University of Baltimore, 2001.

Oakes, L., and Cooper, D. "Business Planning as Pedagogy: Language and Control in a Changing Institutional Field," *Administrative Science Quarterly* (43:2), 1998, pp. 257-292.

Orlikowski, W., and Iacono, C. S. "Research Commentary: Desperately Seeking the 'IT' in IT Research—A Call to Theorizing the IT Artifact," *Information Systems Research* (12:2), 2001, pp. 1 121-134.

Perlow, L. "Boundary Control: The Social Ordering of Work and Family Time in a High Tech Corporation," *Administrative Science Quarterly* (43), 1998, pp. 328-357.

Perlow, L. A. "Time to Coordinate: Toward an Understanding of Work-Time Standards and Norms," *Work and Occupations* (28:1), 2001, pp. 91-111.

Perlow, L. "Time Famine: Toward a Sociology of Work Time," *Administrative Science Quarterly* (44), 1999, pp. 57-81.

Piore, M., and Sabel, C. *The Second Industrial Divide*, New York: Basic Books, 1984.

Prechel, H. "Economic Crisis and the Centralization of Control Over the Managerial Process," *American Sociological Review* (59), 1994, pp. 723-745.

Raelin, J. A. "An Analysis of Professional Deviance Within Organizations," *Human Relations* (39), 1986, pp. 1103-1130.

Sawyer, S., and Tapia, A. "The Computerization of Work: A Social Informatics Perspective," in J. George (Ed.), *Social Issues of Computing*, Upper Saddle River, NJ: Prentice Hall, 2003, pp. 93-109.

Sawyer, S.; Tapia, A.; Pesheck, L.; and Davenport, J. "Mobility and the First Responder," *Communications of the ACM* (47:3), March 2004, pp. 62-65.

Sewell, G. "The Discipline of Teams: The Control of Team-Based Industrial Work Through Electronic and Peer Surveillance," *Administrative Science Quarterly* (43:2), 1998, pp. 397-428.

Sewell, G., and Wilkinson, B. "Someone to Watch Over Me: Surveillance Discipline and Just-in-Time Labor Process," *Sociology* (26), 1992, pp. 271-289.

Shields, M. A. "Reinventing Technology in Social Theory," *Current Perspectives in Social Theory* (17), 1997, pp. 187-216.

Slora, K., Joy, D., and Terris, W. "Personnel Selection to Control Employee Violence," *Journal of Business and Psychology* (5:3), 1991, pp. 417-426.

Smith, V. "The Fractured World of the Temporary Worker: Power, Participation, and Fragmentation in the Contemporary Workplace," *Social Problems* (45:4), 1998, pp. 411-430.

Smith, V. *Managing in the Corporate Interest: Control and Resistance at an American Bank*, Berkeley, CA: University of California Press, 1990.

Smith, V. "New Forms of Work Organization," *Annual Review of Sociology* (23), 1997, pp. 315-339.

Vallas, S. P. "Rethinking Post-Fordism: The Meaning of Workplace Flexibility," *Sociological Theory* (17:1), 1999, pp. 68-101.

Vallas S. P., and Beck, J. P. "The Transformation of Work Revisited: The Limits of Flexibility in American Manufacturing," *Social Problems* (43:3), 1996, pp. 339-361.

Vardi, Y., and Wiener, Y. "Misbehavior in Organizations: A Motivational Framework," *Organization Science* (7:2), 1996, pp. 152-165.

Wastell, D. "Organizational Discourse as a Social Defense: Taming the Tiger of Electronic Government," in E. Wynn, E. A. Whitley, M. D. Myers, and J. I. DeGross (Eds.), *Global and Organizational Discourse about Information Technology*, 2002, pp. 179-195.

Webmergers.com. "Researchers Launch Site to Collect Dot-Com Era ''Blueprints,''' Webmergers.com, June 26, 2002 (available online at http://www.webmergers.com/data/article.php?id=59).

Wilson, M., and Howcroft, D. "The Role of Gender in User Resistance and IS Failure," in R. Baskerville, J. Stage, and J. I. DeGross (Eds.), *Organizational and Social Perspectives on Information Technology*, Boston: Kluwer Academic Publishers, 2000, pp. 453-471.
Wiseman, R., and Bromiley, P. "Toward a Model of Risk in Declining Organizations: An Empirical Examination of Risk, Performance and Decline," *Organization Science* (7:5), 1996, pp. 524-543.

ABOUT THE AUTHOR

Andrea Hoplight Tapia is an assistant professor of Information Sciences and Technology at the Pennsylvania State University. Andrea completed a National Science Foundation funded post-doctoral fellowship at the University of Arizona entitled "Universities in the Information Age." Her Ph.D., from the University of New Mexico, is in the area of Sociology and focuses on the study of technology, culture, and workplace organizations. Her most recent work examines the nature of computer-centered, high-tech industry. She is particularly interested in the how the workplace and employer-employee relations change when in a high-tech environment. At the core of her research is her interest in the social values attributed to technology and the power structures that arise within organizations due to the manipulation and use of those techno-values, in other words, techno-social capital. Andrea can be reached by email at atapia@ist.psu.edu.

32 THE POLITICS OF KNOWLEDGE IN USING GIS FOR LAND MANAGEMENT IN INDIA

S. K. Puri
Sundeep Sahay
University of Oslo

Abstract This paper focuses on understanding the knowledge politics that inhibit effective use of geographic information systems (GIS) for managing the land degradation problem in India. It is argued that the issues of power and politics of knowledge are ubiquitously embedded in representation of the problem domain and the technology itself. Addressing these issues is an inseparable part of the challenges to information systems design and implementation. Theoretical perspective is first developed around political considerations involved in the co-construction and use of knowledge domains relevant to the design of GIS applications to address land degradation. This theoretical framework is drawn upon to analyze the politics of representation, the politics of invisible work, and the politics of institutions observed in the case of a GIS implementation in rural India. The analysis also demonstrates how the insidious impacts of such politics may be somewhat mitigated by creating socio-material networks to cultivate communicative action that leads to better design and technology acceptance by the end users.

Keywords: GIS, rural development, India, politics of knowledge, Habermas, participation, socio-material networks, communicative action.

1 INTRODUCTION

Various attempts to use geographic information systems (GIS[1]) for land manage-

---

[1]The spatial data on which GIS operates, which distinguishes it from other IS applications, is linked to locations which implies that each data element has a unique set of relations to all other data (Reeve and Petch 1999, p. 99).

ment in many developing countries, including India, have not yielded very effective results (Sahay and Walsham 1997). While various social and organizational reasons have been emphasized as contributing to this relative lack of success, we argue that the power and politics of knowledge that surround the issue of land degradation and the technology itself are crucial aspects to understanding the design and implementation challenges.

Politics of knowledge around GIS applications in the land management domain in India arise because of the various interest groups including international agencies, government administrators, scientific institutions, and community beneficiaries. Contributing to the challenge is the nature of the land degradation phenomenon that has been shaped by various historical and institutional forces. Dealing with land degradation, which implies the loss or attenuation of the vigor and productivity of land (Sehgal and Abrol 1992), is an important aspect of the government's agenda because agriculture contributes to nearly 29 percent of the country's GDP (as compared to 2 percent, for example, in the U.S.) (Singh 2002). The Indian agriculture sector employs 69 percent of its workforce, as against 2 percent in the U.S. (ibid.), prompting it to be described as a biomass-based civilization (Gadgil 1993). India also contributes to a significant proportion of the World's biodiversity (WCMC 1996). Consequently, the heightening land degradation problem has attracted global attention in venues such as in the 1992 Rio Earth Summit. Arising from the national prioritization and international attention, considerable financial resources are being invested to improve land use practices, including better information management through the use of ICTs like GIS and remote sensing.

Political meanings have been historically inscribed into the land degradation problem, being shaped by the interests of those in power. The historically existing colonial interests of the British in exploiting forest and other land-based resources (Gadgil and Guha 1995) were further reinforced by the government's post-independence policies of state control over much of the common property land resources (Kumar et al. 2000). Land development programs, by and large, were designed and executed without the involvement of the local people and communities. These policies represented a "relic of the colonial administrative practices" (Haeuber 1993, p. 492). In recent years, there have been some signs of shifts taking place in these policies arising from the inculcation of more integrated watershed-based resource management models, adoption of participatory practices, and large-scale attempts to introduce GIS technology. The use of this rather complex technology, however, brings in its own particular political dynamics arising from questions of who is trying to introduce it, who owns it, and what kinds of knowledge get used or excluded in the process.

Drucker (1988) describes the organizational structure adopted by the British during their 200-year rule of the Indian subcontinent as "totally flat" (p. 49). The subcontinent was divided into nine administrative provinces. Each district in a province was headed by an Indian Civil Service (ICS) officer, called district collector (referred to as district officer by Drucker) who reported directly to the provincial secretary. The principal tasks of a district collector (DC) were maximizing revenue collection and maintaining law and order. The entire subcontinent was thus managed with a relatively lean bureaucracy. The independent India adopted a federal democratic polity, in which both the central and state governments are involved in development management and administration of the about 500 districts. The flat organizational structure described by Drucker has since given way to a system of top-down, hierarchical governance. In the context of use of GIS for land development, the typical organizational structure is shown in Figure 1. The role of the DC, which now subsumes socio-economic development of the district, however, continues to be crucial in the implementation of development related policies and plans.

The aim of this paper is to develop a perspective on the politics of knowledge that elaborates on the nature of knowledge and the dynamics around it that arise in the context of GIS applications for land management in India. Using this political perspective, we draw upon Habermas to develop a critical approach on how these political considerations can be addressed more meaningfully to guide GIS implementation efforts. In the next section, we develop a theoretical perspective on the politics of knowledge in relation to GIS for land management. In section 3, the research approach is presented, followed by the case study description in section 4. Section 5 presents the case analysis, drawing upon the theoretical framework presented in section 2. In the final section, some ideas are presented on how to meaningfully engage with the political challenges.

2 THE POLITICS OF KNOWLEDGE IN GIS APPLICATIONS FOR LAND MANAGEMENT

This section focuses on developing a theoretical perspective around the politics of knowledge, specifically as it relates to the application of GIS technology for land management, arising from the needs and interests of the various competing groups (Floyd 2002, p. 203). In the context of land degradation, three domains of knowledge that are relevant are technical, scientific, and indigenous. These are discussed.

2.1 Technical Knowledge

Technical knowledge relates to GIS and remote sensing technologies, and the epistemological basis on which it is situated. GIS, which is an information system that is designed to work with data referenced by spatial or geographic coordinates (Star and Estes 1990), has its roots in the scientific principles of cartography and mathematics, situated in a positivist epistemology, which are employed in map-making within standard scientific representation of knowledge and cognition (Harley 1992). Sahay (1998, p. 184) has argued that the positivist epistemology in which GIS is grounded was contributed to by the quantitative revolution of the 1980s within geography departments in U.S. universities, which emphasized the "conceptualization of reality in spatial, map-based terms," and promoted a rational conceptualization of the world in which space, assumed to be objective, "can be dominated and restructured very efficiently and rapidly through the use of GIS." Harvey (1989) argued that GIS technology furthered the objective image of maps, based on the mathematical rigor of their preparation, and the assumptions that the correspondence between maps and the real world was unproblematic and context-free. This naive empiricism (Taylor 1990) is grounded in the belief that the technical and the social realms exist in mutual exclusivity (Sahay 1998), and represents a form of power that can be presented in the "guise of scientific disinterestedness" (Poster 1982, quoted in Harley 1988, p. 279).

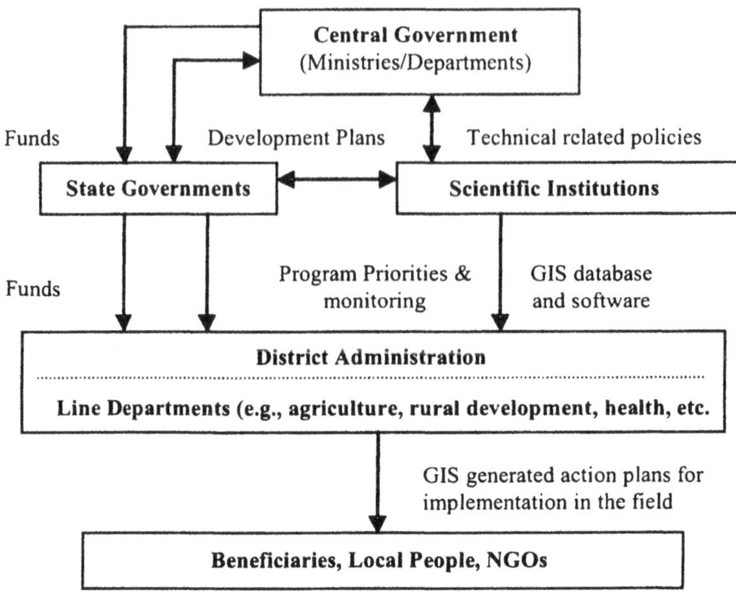

Figure 1. Interplay between Concerned Agencies

The political nature of construction and use of cartographic knowledge is typically exemplified by the detailed cartographic survey and mapping of India undertaken during the British rules. It proved indispensable for "the rationalization of the extraction of surplus, administrative strategies and techniques of control" (Baber 2001, p. 44).

2.2 Scientific Knowledge Around Land Management

Scientific knowledge concerns the application domain that derives from the scientific parameters relating to, for example, soil types, slope of the land, vegetation patterns, runoff, etc. Two relevant domains of scientific knowledge relate to selecting spatial themes for analyses and modeling techniques employed.

The major relevant spatial themes concern land use and land cover data, geomorphology, drainage, groundwater potential, etc. (NRSA 2002). Typically, maps of scales 1:50,000 or smaller are used in India in GIS applications for land management, which are not sufficient to design micro-level interventions. Another important concern is the availability of maps and, citing reasons of security and military concerns, many governments, especially in developing countries, are reluctant to share the maps with the public and sometimes also with other government departments (Fox 1991). This reluctance tends to ensure that most GIS development work continues to be largely confined to select scientific departments.

A key aspect of addressing land degradation is the use of GIS modeling. Approaches to GIS modeling, which are largely shaped within a positivist epistemology, tend to rely on remotely sensed data. As a result, these models are limited in depicting the complex reasons of land degradation arising from socio-cultural aspects such as anthropogenic pressures, and other deep-rooted political considerations. Sahay and Walsham (1996) provide a vivid example of this with respect to the criteria adopted for building the models. They argued that scientists tend to use rationalistic criteria of profit maximization in the models to specify land use strategies, while the farmers when asked indicated their preference for the criteria of risk minimization.

Hoeschele (2000) described the organizational politics in Kerala state of India which led to the GIS model overestimating the extent of wastelands by deliberately misrepresenting the current land use data. As a consequence, the local bureaucrats argued that farmers were not able to manage community lands effectively and thus required dominant government intervention. Inappropriate use of apparently neutral technology can lead to further marginalization of the poor and the powerless, and also underscores how modeling may be misleading if the relevant social data are not properly taken into account.

As a result of the politics around scientific knowledge, land degradation, despite being a complex social-historical phenomenon, tends to be modeled primarily on a scientific basis. This leads to a separation between sites of technological production and locations of its use. This separation is reflected in how GIS systems are designed and produced by scientists in laboratories for subsequent transfer to the districts. The divide arising from a specialization of technology contributes to a "design from nowhere" (Suchman 1994, p. 27).

2.3 Indigenous Knowledge

Indigenous knowledge is held by local communities and evolves over time through being field-tested for its suitability to local needs, conditions, and ethos (Mundy and Compton 1995). It may be defined as the "systematic body of knowledge acquired by local people through the accumulation of experiences, informal experiments, and intimate understanding of the environment in a given culture" (Rajasekaran et al. 1994, p. 26). Such knowledge has historically been excluded and made invisible from the westernized scientific models of land and water management on the assumption that it was inferior, unscientific, and static (Howard and Widdowson 1996, 1997). As a result, this context-specific knowledge, which is embedded in the practice of community members (Banuri and Marglin 1993), has remained largely unarticulated to external domains like those of the state and scientists.

This condition of marginalization of indigenous knowledge is, however, gradually being revised due to various reasons such as the changes in politics of development aid, examples of the breakdown of technology-driven applications, cases of success of community-driven efforts, and a high level of political advocacy by non-governmental organizations (NGOs), activist groups, and international conventions. For example, the 1992 Rio Earth Summit formally recognized the importance of indigenous knowledge in achieving sustainable development, and the issue found mention in 17 of the 40 chapters of Agenda 21 (Mathias 1994). Indigenous knowledge is not static, as assumed in scientific thinking, and undergoes changes with learning as a response to new external stimuli, and communities adapt and integrate new technologies into their knowledge domain over time (ibid.). Agrawal (1995) perceived the divide between scientific and indigenous knowledge based on methodological considerations rather than substantive grounds as superficial, arguing that any form of knowledge is embedded within a specific social context, which influences the processes around knowledge creation and use.

In summary, the theoretical discussion above has tried to describe the disparate knowledge systems that are in play in the context of GIS for land management, the political interests that shape how these systems are manifested, and the challenges in their synthesis. These conceptual ideas provide the basis to analyze the case in section 5, after the presentation of the research approach and the case in sections 3 and 4 respectively.

3 RESEARCH APPROACH

The research strategy adopted for the present study, overview of the research setting, approach used for data collection, and analysis are presented.

3.1 Background

Efforts to use GIS for land management in Anantapur district were first taken up in 1995 under a large-scale technology initiative of the Department of Space in India, in which a natural resource spatial database for the district was developed by the Andhra Pradesh Remote Sensing Application Centre (APSARAC) mainly using satellite remote sensing data. The application software was developed by other leading scientific institutions like NRSA and the Space Application Centre. Subsequently, during 2001-02, a local GIS team redesigned the database for which both spatial and non-spatial data were collected mainly through physical surveys conducted in association with the local people and NGOs using a global positioning system (GPS). It constituted a major departure from the earlier top-down methods in which scientific institutions assumed ownership of the design mostly without consulting end-users in the concerned district departments and the communities who have a livelihood stake in how land use is planned. The motivation to take up this study in Anantapur arose from the desire to understand the impact of this locally inspired, bottom-up approach on technology diffusion and its use in context of the problem domain.

3.2 Research Setting

Anantapur, a poorly developed district of Andhra Pradesh, is situated in the rain-shadow zone of peninsular India, with a low annual precipitation of 521 mm. A total of 31 percent of its land area is heavily degraded.[2] The district has historically been drought-prone, the adverse impact of droughts having been further exacerbated in recent times due to massive deforestation, excessive withdrawal of ground water, increasing soil salinity, etc. (Rao et al. 1993).

---

[2]Source: *Anantapur District—A Profile*, dated December 20, 2002, complied by district administration.

3.3 Research Method: Data Collection and Analysis

The case study described in this paper is based on field work carried out in Anantapur district during 2002-2003. In all, we conducted 81 semi-structured interviews with officials of the concerned district/line departments, concerned scientists, bureaucrats in the state and central governments' rural development departments, academics from a local university, and NGOs. We also participated in meetings with villagers.

Individual interpretations of interviews were transcribed to identify relevant issues and also to prepare summaries. These were subsequently discussed between the authors and respondents. Examination of the secondary data (maps, guidelines, etc.) provided insights into the efforts made by the district administration to involve communities and NGOs both in development and GIS work.

The research method adopted in this study was the interpretive case study approach, which proceeds on the assumption that "the social world is essentially relativistic and can only be understood from the point of view of individuals who are directly involved in the activities which are to be studied" (Burrell and Morgan 1979, p. 5). The interpretive paradigm seeks to understand the fundamental nature of the social world, as it is, at the level of subjective experience (ibid., p. 31), how people assign meaning to those experiences (Devine 1995, p. 138), and how intersubjectivity is constructed. The focus is not on establishing truth claims but on understanding the processes through which intersubjectivity is reached. Interpretive approaches in IS research take the stance that knowledge of reality is socially constructed by human actors (Walsham 1995) through shared meaning (Klein and Myers 1999). The case study is a useful approach to adopt when the phenomenon of study is difficult to disengage from its context, for example, the introduction of a new technology like GIS in rural settings of India.

4 THE CASE STUDY

We discuss the case under four main themes, viz. (1) how GIS-based recommendations provided earlier by scientific institutions fared in the district, (2) how institutional structures were redefined to facilitate design and implementation of the development agenda set by people, (3) design and use of a locally identified GIS database and applications to support development, and (4) the trajectory of participatory processes, and implicating knowledge of people both in development and GIS work.

4.1 Applying GIS-Based Recommendations of Scientific Institutions

The methodology adopted by scientific institutions comprised the use of remotely sensed satellite data along with other resource and topographic maps, and socio-economic data obtained from secondary sources to develop locale-specific action plans for land and water management at the micro-watershed level (about 500 hectare area). These plans suggested land use and water augmentation measures based on GIS analysis and modeling of the above data vis-à-vis a set of decision rules based on criteria of soil classification, ground water potential, and slope (NRSA 2002).

The action plans for Anantapur were accordingly generated by APSRAC in 1995. We asked the project director (responsible for land/water development programs as well as ICT/GIS-related work in the district) how the above database and the accompanying software had been put to use. She explained that these action plans were based on 1991 satellite data. The land degradation situation had since changed due to natural phenomenon like recurring drought, reclamation efforts made in the intervening years, etc. Therefore, action plans and concerned maps needed continuing review and update. APSRAC's recommendations were consequently being used as one parameter for prioritization of areas to initiate current developmental activities. The scale (1:50,000) of these maps was also too small to be of much practical use in the field.

We were keen to find out how participation of end-users was viewed and taken up by scientists. A senior scientist, referring to practices adopted until recently, stated

> There exists a cultural gap between scientists working in so-called elite institutions vis-à-vis the prospective users in district line departments and village communities. Therefore, effective communication amongst these different groups is problematic. A useful dialogue requires the two parties to be genuinely convinced of its relevance, which requires a positive change in mindsets in the context of your question. What is really happening is that action plans are sent to districts, some training is provided to district staff, they might make suggestions knowing from past experience that scientific institutions would not change their style of functioning, while scientists are equally convinced that they are doing their "job" while being aware that the outputs will not be seriously considered. To tell you the truth, it is a political game being played all the time.

The view of middle-level officials in line departments was that scientists and the concerned institutions took considerable time to respond to their needs and queries, they were not available for discussion when required, not being colocated, and the district had no administrative or functional control over them. In summary, the tension among scientists and users implicated in GIS work at districts was evident.

4.2 Redefining Institutional Structures

The watershed-based rural development guidelines promulgated by the government of India in 1995 envisage that such development would no longer be carried out solely under the government umbrella (Government of India 1995), but with active cooperation and involvement of the local people. The recommended procedure is to assign development design and implementation to local Watershed Development Teams (WDTs) comprising officials of concerned line departments and elected individuals from the watershed villages, to be chaired by a non-official. WDTs have also been financially empowered.

The project director explained that in Anantapur, this decentralized model was taken a step closer to empowerment of people by making provisions that while the WDTs would be established as above, the actual design of village level development activities be decided by the communities in *gram sabha.*[3] Implementation of works corresponding to each identified activity was also taken up by people themselves through user-teams elected in *gram sabha*, with the power to incur expenditure. It was realized by the district administration that if ICTs including GIS were to be effectively used to support development, then these facilities had to be locally available so that the database and application design was carried out as per local vision and needs. Also, availability of trained GIS professionals on-the-spot and within the local administrative control would greatly assist in developing new applications as the needs arose, as well as to update the databases to ensure their currency and topicality. Accordingly, IT/GIS infrastructure had been locally instituted.

4.3 Designing the GIS Database and Applications Locally

We met the local GIS scientists several times to understand how the new database had been generated. It was explained that data collection had been carried out by field teams set up locally for this purpose. Each team comprised a civil engineer, one person from the village concerned, one NGO representative, and a scientist from the local GIS unit. During 2001-02, these teams surveyed villages to mark the latitude/longitude values of various relevant features such as wells, check dams, agriculture parcels, etc., on cadastral maps[4] (1:80,00 scale) using a GPS. The present condition and use of these features, and the relevant attribute data, was also noted alongside. These maps (with the manually recorded data) were sent to NRSA for preparation of a digital database for GIS use, which had since been completed (Rao et al. 2003). It was now routinely used locally for multifarious purposes. For example, a recent water audit in Anantapur district had been carried out by utilizing this database in partnership with the British Department for International Development (ibid.).

We witnessed several demonstrations of the GIS work being done in Anantapur, and the use of the locally created database. One striking example was how survey and recording of all water harvesting structures, and subsequent analysis of this data, led to the identification of 29 redundant structures out of a total of 176 built under various government programs in the past years. Besides incurring wasteful expenditure, such redundant constructions potentially had a negative impact on downstream water availability and recharge of ground water. The GIS helped to make visible the inefficiencies associated in past projects, and provided the impetus for change. As a result, the district administration has formally decided to revive the traditional water harvesting structures, and future proposals for new constructions would need to be closely justified.

---

[3]*Gram sabha* is the village council, the body constituted of all adult members of the village.

[4]Land records with ownership are shown on cadastral maps in India. This mapping has generally not been based on any systematic scale or datum.

4.4 Enabling Participation and Implicating Indigenous Knowledge

The facilitation of a decentralized approach to development, and local improvisation to strengthen it further, came about because of the personal belief of the DC in the positive role of people's empowerment in achieving meaningful development. During a meeting with villagers, we asked their opinion about the changes brought about by the administration in development procedures. An elder responded that

> Things are different now. Our voice is listened to and we implement the activities approved in the *gram sabha*. There is not much interference from the government people. Their people check accounts and also monitor work progress. All this has happened because of wise political leadership in the state and of the DC here in Anantapur. He visited this village three times in two years to see how things were progressing. I have never seen anything like this in all my life.

We asked them whether they were overawed by the presence of officials and scientists during meetings of *gram sabhas*. The response was emphatically in the negative although it was admitted that villagers were somewhat sceptical of the new openness in the beginning. We asked them how it was that the scientists and engineers now listened to them. The villagers said that it was mainly because of the interest taken by the DC in ensuring that development activities and their locations were decided in *gram sabhas*.

In the meetings of *gram sabhas*, one of which we witnessed, development plans for the area were discussed and finalized. The local understanding of the people about land, water, and vegetative resources and their perceptions on how these should be developed and used were jointly noted by administrators and GIS scientists. People explained their perspective through participatory mapping. Resource maps were drawn by them on the ground (not to scale) to depict the location of various existing resources, and the proposed location of mooted development activities. The scientists acknowledged that some of the elders had an astute sense of the local topography and the drainage pattern including how traditional water harvesting structures had been used beneficially in the past. These markings by the community members on the ground maps were subsequently incorporated into the GIS database. The government officials and the GIS scientists present also had an equal opportunity in these meetings to put forward their points of view, and the final plan for local development was evolved through negotiations in these meetings.

5 CASE ANALYSIS

Through the case description, we have identified several issues including those relating to participatory processes and how these were enabled or not, how knowledge of the community was elicited and integrated with the GIS approach, and how the institutional structures were redefined to facilitate more effective GIS efforts. The case analysis presented relates to three issues concerning the politics of representation, the politics of invisible work, and the politics of institutions. Following this, the manner in which the integration of various disparate knowledge systems was attempted, given the existing politics, is described.

5.1 The Politics of Representation

In the case described, the GIS technology was being used to, first, represent the land degradation problem, second, to model intervention strategies, and third, to depict the changes that came about as a result of the interventions. Various issues become relevant in examining the effectiveness of these representation processes. For example, the small scale of maps used (1:50,000) is not effective to represent crucial aspects of land degradation such as micro-level dynamics of how soil erosion, an important determinant of land degradation, is taking place. But by drawing upon context-specific inputs from the community members, richer resource perspectives can be introduced and the reasons for land degradation more effectively incorporated. There are similar issues relating to modeling, for example, relating to its more selective use primarily for prioritizing the extent of degradation in different areas, and based on these selecting particular geographical areas for action. The more specific intervention strategies were not based on GIS models but through drawing upon the understandings of what the community members considered appropriate.

These examples suggest issues around the politics of representation and how they were dealt with in the particular context. The politics of representation has been an important issue for discussion among researchers especially from within the science and technology studies (STS) and feminist domains (Verran 1995). In the context of land degradation, the politics of representation arises from at least two issues. The first concerns the kind of maps that are used, and the associated issues of who owns them and how they are used. This leads to the historical problem of how land degradation has been represented and problematized primarily based on the interests and visions of remote sensing scientists in the Indian context (Sahay and Walsham 1997). The second issue concerns modeling. Historically, attempts to construct land intervention strategies have been primarily within a natural science epistemology, with attempts to apply similar principles to systems development (Klein and Lyytinen 1992). As emphasized through the case, information system design, especially in the context of land degradation, requires a different approach to modeling than that practiced in natural sciences due to the added complexities of the socio-cultural aspects.

5.2 The Politics Around Making Certain Forms of Knowledge Invisible

An important insight from this case relates to the manner in which knowledge that the community members, especially the elders, have about the location of traditional resources such as waterbodies was elicited through various forms of dialogue. Such knowledge which has traditionally remained invisible in the context of designing information systems, has parallels with Bowker and Star's (1999, p. 230) description of "invisible work" in the context of nurses' roles in hospitals. These researchers described nurses' work to include the interplay between professional practice, information systems, classification, and organizational change. They describe how important work carried out by nurses as part of overall patient care in hospitals tended to be marginalized at the expense of the more visible work of doctors. This marginalization leads to an incomplete representation of work and poorly designed information systems. Recent feminist writings have also emphasized the importance of taking into equal consideration such invisible work (Star and Strauss 1999). Such work is often based on unarticulated knowledge and embedded in work practices developed over time. The setting of the workplace for nurses' roles is different from that of the community in the context of land management, and thus brings out different challenges to its articulation.

5.3 The Institutional Politics Involved Around GIS Use

Analysts like Pfeffer (1992) have argued that two key aspects contributing to institutional politics are the struggle over centralization and decentralization, and the ambiguity around what goals are in question. Questions of centralization/decentralization are associated with issues of power asymmetries and how they are negotiated through the introduction of new information systems. In the context of land management, the district administrators sometime see scientists' efforts to embed technology into their workplace as an intrusion into their domain. Similarly, scientists view the inclination of some administrators to locally control the GIS as something not required or desirable. Decentralization of control over the use of GIS and access to resources is thus a contentious issue, and strongly negotiated. The ambiguity of goals around GIS-based land management, interventions stems from a number of issues. The first concerns the long gestation periods of these projects compounded by the problem of government officials being frequently transferred, which makes accountability extremely diffused and ambiguous. The second concerns the multi-departmental nature of land management which makes it problematic to pin down responsibility to particular people or positions. A third issue relates to the fact that GIS helps to develop land use strategies that have direct budget implications, bringing in various interests such as contractors and politicians who are not directly accountable. The resulting ambiguity around the aims and objectives of the land management and GIS efforts contribute to create the potential for institutional politics to be exercised.

The analysis emphasizes the multiplicity and situated forms of knowledge, its dynamic nature, and it being intricately linked to the practices and political interests of different relevant groups. Knowledge around GIS is not something objective and "out there" but, as the STS researchers have argued over the years, knowledge is a construction of its social, political, and historical contexts (Berger and Luckmann 1967). This political perspective around the analysis helps to emphasize the dynamics of representation, invisible work, institutional context, and their interconnections.

5.4 Integrating Disparate Knowledge Systems

Some strategies for developing integration of technical and organization knowledge have been developed in the context of GIS in Western organizations, for example, through the adoption of socio-technical design approaches which emphasize identification of user needs, seek to develop commitment to participation, and propose strategies to help absorb technological change (Campbell and Masser 1995). These researchers also pointed out that organizations that successfully adopted GIS had focused on (1) implementing simple applications albeit of fundamental importance to the users, to start with, and (2) recognizing and keeping in view limitations of the organization (particularly, accepted practices and availability of resources). While these findings emerged in Western workplace settings, they provide some useful insights, especially to emphasize the multiplicity and situated nature of knowledge forms, and how they should be considered in an integrated manner.

User participation in the design of IS has evinced varying trajectories, methodologies, and aspirations in the Western contexts (Asaro 2000), a basic motivation being "people who are affected by a decision or event should have an opportunity to influence it" (Schuler and Namioka 1993, p. xii). One strand of this approach, viz. ETHICS (Mumford 1995) consciously takes into account the cultural and other socially relevant value systems of the people who would use and be impacted by the proposed system. Through participatory methods, the designers and users are expected to arrive at consensus about shared system goals, and bring in the open the conflicts among stakeholders. Such thinking is also evident in the work of Checkland and Scholes (1990) in their articulation of the soft systems methodology. However, these approaches arose primarily within Western organizational settings and their socio-economic and political contexts. These do not adequately emphasize how the broader interorganizational structures can be addressed in developing-country contexts, given the historically existing attitudes of the bureaucrats and scientists and the different community politics, and largely an absence of a tradition of participatory approaches in development programs.

A starting point in the process of creating a framework for dialogue is the need for scientists to acknowledge the importance of indigenous knowledge, which in itself is a critical change, as it challenges the existing and deep-rooted assumptions of the superiority of scientific knowledge. The challenge is to try and develop greater intersubjective understanding as a basis to create relevant sociomaterial networks while acknowledging the structural conditions, especially around power and politics, within which it is constructed. While the rhetoric of indigenous knowledge and empowerment of marginalized group is a recurrent theme in current debates around development thinking, its advocates rarely "emphasize that significant shifts in existing power relations are crucial to development" (Agrawal 1995, p. 416).

One source for the development of such a critical perspective comes from Habermas, whose ideas have to a limited extent been drawn upon by IS researchers. Habermas posits "science and its project of improving the human conditions as a collaborative effort where people (scientists and non-scientists) work together to achieve its ends" (Ngwenyama 1991, p. 271). Creating such intersubjective understanding underscores the importance of free and fair communication. Habermas' theory of communicative action (1984) seeks to unravel the set of norms which constitute the basis of all social action by human actors. Instrumental action is aimed at the object (agent) to act as per sender's (actor's) dictates and needs, agents being assumed as passive or inanimate recipients. The top-down, externally driven mode of participatory development, and decision-oriented IS models which shaped the prevalent "managerial ideology by strengthening the instrumental rationalization of organizations" (Lyytinen 1992, p. 166), conform to this type of action. Strategic action aims at similarly controlling another rational actor's response but provides some opportunity to the recipient actor to act otherwise or differently. Thus, the outcome is rendered uncertain, being determined by the actions of the originating actor. Participatory development approaches in which the beneficiaries play a role in defining the agenda of development arise from strategic action. In communicative action, the objective is to achieve mutual consensus and understanding based upon a shared pool of implicit assumptions and beliefs, through means of formal or informal dialogues. Development of such communicative action helps to redress the dominance of instrumental rationality guided by scientific thinking, and engage with the problem of alienation and colonization of the life-world (ibid.) experienced by the users. Participatory development approaches in which the agenda is jointly set by those in authority and the local communities represent forms of discursive action. Examples of such action, at least in the IS domain, are rather rare.

The Habermasian ideal speech situation (IDS) (Habermas 1984) seeks to outline conditions that favor the development of communicative action in which the pros and cons of an action can be debated by the groups on the criteria of clarity, veracity, sincerity, and social responsibility (Lyytinen 1992). Such dialogue, it is argued, can help provide the basis for creating effective sociomaterial networks in which scientists locate themselves in wider set of accountabilities. These ideas of Habermas were drawn upon by Hirschheim and Klein (1994) to try and provide a linkage between a critical perspective and emancipatory goals of IS design.

The concept of IDS has attracted several critiques, particularly from feminists like Braaten (1995) who point to Habermas' purely procedural form of consensus, and the focus on the conditions for discourse which ignore emotions and feelings linked to the cognitive and intellectual maturity of the subject. Nevertheless, the IDS provides useful insights on how enabling conditions should be established to facilitate more meaningful dialogue between relevant groups.

The procedure used in the case described to develop dialogue between the scientists and community focused on demystifying the technology by making selective use of the participatory modeling approach. The scientists used the GIS only to prioritize the areas for land degradation, and then sought input from community members to identify the traditionally existing waterbodies on paper-based maps. These inputs were then incorporated by the scientists into the GIS database.

This approach helped to deal with the politics of representation by providing a mechanism by which people could try and agree upon what was being represented and whether it was adequate or not. The resource maps served as useful boundary objects (Star and Griesemer 1989) for scientists to understand the local environmental conditions and people's needs and aspirations for development. These maps allowed for increased opportunities for analysis and reflection, to identify stakeholder interests more closely, and to provide a spatial structure to discussion on water harvesting approaches. The maps were simple to understand although they inscribed complex social and resource perspectives, and provided a starting point for establishing a dialogue between scientists and the local communities.

This participatory development approach, where the agenda was jointly set by those in authority and the local communities, also helped to engage with the politics of invisible work by "providing a channel and environment for symbolic interaction" (Lyytinen 1992, p. 166). Through communicative action, the knowledge of the communities, which was historically rendered invisible in and through the use of technical and scientific methodologies, could be freely expressed and embedded in the GIS design. This understanding could be articulated and then inscribed more meaningfully with the analytical strength provided by GIS.

Putting both scientific and indigenous knowledge to collective use acknowledges the contextual and situated nature of reality, and provides mechanisms to deal with the politics of institutions. Practicing disparate knowledge systems together is a political process, since it concedes contesting cognitive authorities (for example, of scientists and communities in the land degradation scenario), while seeking to locate work responsibility within their diverse and contesting authority structures, according them equal consideration (Verran 1998). This cultural approach explicitly acknowledged that there are arenas of political and social domination of the majority "by the most powerful within society" (Campbell and Masser 1995, p. 19), and that participation on its own cannot redress power and political asymmetries (Beck 2002). An implication of this is to understand how approaches to participatory design can be developed that do not magnify power asymmetries. The case presented provides an example of one such approach by trying to demystify the technology so that dialogue between scientists and community members could take place on relatively more equal terms.

6 CONCLUSION

The efficacious role of both end-user participation and integration of relevant knowledge for improving development efforts and GIS-based interventions to address land degradation has been emphasized in the above theoretical discussion. It has been argued that local knowledge of the environment can provide context-specific understandings that can complement scientific knowledge. The link between participation and knowledge for meaningful use of GIS is crucial. Participation under conditions approaching Habermasian IDS can potentially lead to the "crossing of boundaries" (Suchman 2002, p. 93) and the creation of procedures in which the land managers and scientists are able to interact. It then provides the bridge between technical and indigenous knowledge to facilitate development of more meaningful IS. Suchman (1994) has argued for reframing of objectivity to develop alternative systems of technology production and use. Her argument emphasizes that the objective epistemology of scientific and technical knowledge needs to be rearticulated to "multiple, located and partial perspectives that find their objective character through ongoing processes of debate" (Suchman 2002, p. 92) to accommodate lived experience. IS designs based on such critical perspectives then seek to foster mediation between stakeholders and provides the bridge to cross the boundaries.

ACKNOWLEDGMENTS

We thank the two anonymous reviewers for making several insightful suggestions to improve upon an earlier draft of this paper.

REFERENCES

Agrawal, A. "Dismantling the Divide between Indigenous and Scientific Knowledge," *Development and Change* (26:3), 1995, pp. 413-439.

Asaro, P. M. "Transforming Society by Transforming Technology: The Science and Politics of Participatory Design," *Accounting, Management and Information Technologies* (10), 2000, pp. 257-290.

Baber, Z. "Colonizing Nature: Scientific Knowledge, Colonial Power and the Incorporation of India into the Modern World-System," *British Journal of Sociology* (52:1), 2001, pp. 37-58.

Banuri, T., and Marglin, F. A. "A System-of-Knowledge Analysis of Deforestation, Participation and Management," in T. Banuri and F. A. Marglin, (Eds.), *Who Will Save the Forests? Knowledge, Power and Environmental Destruction*, London: Zed Books, 1993, pp. 1-23.

Beck, E. "P for Political: Participation Is Not Enough," *Scandinavian Journal of Information Systems* (14), 2002, pp. 77-92.

Berger, P. L., and Luckmann, T. *The Social Construction of Reality*, London: Allen Lane, The Penguin Press, 1967.

Bowker, G. C., and Star, S. L. *Sorting Things Out: Classification and its Consequences*, Cambridge, MA: MIT Press, 1999.

Braaten, J. "From Communicative Rationality to Communicative Thinking: A Basis for Feminist Theory and Practice," in J. Meehan (Ed.), *Feminists Read Habermas: Gendering the Subject of Discourse*, London: Routledge, 1995, pp. 139-161.

Burrell, G., and Morgan, G. *Sociological Paradigms and Organisational Analysis: Elements of the Sociology of Corporate Life*, London: Heinemann, 1979.

Campbell, H., and Masser, J. *GIS and organizations: How Effective Are GIS in Practice?* London: Taylor & Francis, 1995.

Checkland, P., and Scholes, J. *Soft System Methodology in Action*, Chichester, England: John Wiley, 1990.

Devine, F. "Qualitative Methods," in D. Marsh and G. Stoker (Eds.), *Theory and Methods in Political Science*, London: Macmillan Press, 1995, pp. 137-153.

Drucker, P. F. "The Coming of the New Organization," *Harvard Business Review*, January-February 1988, pp. 45-53.

Floyd, C. "Towards Knowledge Co-construction," in C. Floyd, G. Kelkar, S. Klein-Franke, C. Kramarae, and C. Limpangog (Eds.), *Feminist Challenges in the Information Age—Information as a Social Resource*, Opladen, Germany: Leske + Budrich, 2002, pp. 203-222.

Fox, J. M. "Spatial Information for Resource Managers in Asia: A Review of Institutional Issues, *International Journal of Geographical Information Systems* (5:1), 1991, pp. 59-72.

Gadgil, M. "Biodiversity and India's Degraded Lands," *Ambio* (22:2-3), 1993, pp. 167-172.

Gadgil, M., and Guha, R. *Ecology and Equity: The Use and Abuse of Nature in Contemporary India*, London: Routledge, 1995.

Government of India. *Guidelines for Watershed Development*, Department of Land Resources, Ministry of Rural Development, Government of India, New Delhi, 1995.

Habermas J. *The Theory of Communicative Action: Reason and Rationalization of Society* (Volume I), Boston: Beacon Press, 1984.

Haeuber, R. "Development and Deforestation: Indian Forestry in Perspective," *The Journal of Developing Areas* (27), 1993, pp. 485-514.

Harley, J. B. "Deconstructing the Map," in T. J. Barnes and J. S. Duncan (Eds.), *Writing Worlds*, London: Routledge, 1992, pp. 231-247.

Harley, J. B. "Map, Knowledge and Power," in D. Cosgrove and S. Daniels (Eds.), *The Iconography of Landscape: Essays on the Symbolic Representation, Design and Use of Past Environments*, Cambridge, UK: Cambridge University Press, 1988, pp. 277-312.

Harvey, D. *The Condition of Postmodernity*, Oxford, UK: Basil Blackwell, 1989.

Hirschheim, R., and Klein, H. "Realizing Emancipatory Principles in Information Systems Development: The Case for ETHICS," *MIS Quarterly* (18:1), 1994, pp. 83-109.

Hoeschele, W. "Geographic Information Engineering and Social Ground Truth in Attappadi, Kerala State, India," *Annals of the Association of American Geographers* (90:2), 2000, pp. 293-321.

Howard, A., and Widdowson, F. "Traditional Knowledge Advocates Weave a Tangled Web," *Policy Options/Options Politiques*, April 1997, pp. 46-48.

Howard, A., and Widdowson, F. "Traditional Knowledge Threatens Environmental Assessment," *Policy Options/Options Politiques*, November 1996, pp. 34-36.

Klein, H. K., and Myers, M. D. "A Set of Principles for Conducting and Evaluating Interpretive Field Studies in Information Systems," *MIS Quarterly* (23:1), 1999, pp. 67-93.

Klein, H. K., and Lyytinen, K. "Towards a New Understanding of Data Modelling," in C. Floyd, H. Zullighoven, R. Budde, and R. Keil-Slawick (Eds.), *Software Development and Reality Construction*, Berlin: Springer-Verlag, 1992, pp. 203-219.

Kumar, A., Bren, L., and Ferguson, I. "The Use and Management of Common Lands of the Aravalli, India," *The International Forestry Review* (2:2), 2000, pp. 97-104.

Lyytinen, K. "Information Systems and Critical Theory," in *Critical Management Studies,* M. Alvesson, and H. Willmott (eds.), Sage Publications, London, 1992, pp. 159-180.

Mathias, E. "Indigenous Knowledge and Sustainable Development," Working Paper No. 53, International Institute of Rural Reconstruction, Y.C. James Centre, Silang, Cavite 4118, Philippines, April 1994.

Mumford, E. *Effective Systems Design and Requirements Analysis: The ETHICS Approach*, London: MacMillan, 1995.

Mundy, P., and Compton J. L. "Indigenous Communication and Indigenous Knowledge" in D. M. Warren, D. Brokensha, and L. J. Slikkerveer (Eds), *The Cultural Dimension of Development: Indigenous Knowledge Systems*, London: Intermediate Technology, 1995, pp. 112-123.

NRSA. *Integrated Mission for Sustainable Development,* National Remote Sensing Agency, Department of Space, Government of India, Hyderabad, 2002.

Ngwenyama, O. K. "The Critical Social Theory Approach to Information Systems: Problems and Challenges," in H-E. Nissen, H. K. Klein, and R. Hirschheim (Eds.), *Information Systems Research: Contemporary Approaches and Emergent Traditions*, Amsterdam: North-Holland, 1991, pp. 267-280.

Pfeffer, J. *Managing With Power: Politics and Influence in Organizations*, Boston: Harvard Business School Press, 1992.

Rao Rama Mohan, M. S., Batchelor, C. H., James, A. J., Nagaraja, R., Seeley, J. and Butterworth, J. A. *APRLP Water Audit*, Andhra Pradesh Rural Livelihood Project, Department of Rural Development, Hyderabad, 2003.

Rao, R. S., Venkataswamy, C., Rao, C. M., and Rama Krishna, G. V. A. (1993) "Identification of Overdeveloped Zones of Ground Water and the Location of Rainwater Harvesting Structures Using an Integrated Remote Sensing Based Approach—A Case Study in Part of the Anantapur District, Andhra Pradesh, India," *International Journal of Remote Sensing* (14:17), 1993, pp. 3231-3237.

Rajasekaran, B., Martin, R. A., and Warren, M. "A Framework for Incorporating Indigenous Knowledge Systems into Agricultural Extension Organizations for Sustainable Agricultural Development in India," *Journal of International Agricultural and Extension Education*, Spring 1994, pp. 25-31.

Reeve, D., and Petch, J. *GIS Organisations and People: A Socio-Technical Approach*, London: Taylor & Francis, 1999.

Sahay, S. "Implementing GIS technology in India: Some Issues of Time and Space," *Accounting, Management and Information Technologies* (8:2-3), 1998, pp. 147-188.

Sahay, S., and Walsham, G. "Implementation of GIS in India: Organizational Issue and Implications," *International Journal of Geographic Information Systems* (10:4), 1996, pp. 385-404.

Sahay, S., and Walsham, G. "Social Structure and Managerial Agency in India," *Organization Studies* (18:3), 1997, pp. 415-444.

Schuler, D., and Namioka, A. "Preface," in D. Schuler and A. Namioka (Eds.), *Participatory Design: Principles and Practice*, Mahwah, NJ: Lawrence Erlbaum, 1993, pp. xi-xiii.

Sehgal, J. L., and Abrol, I. P. "Land Degradation Status: India," *UNEP Desertification Bulletin* (21), 1992, pp. 24-31.

Singh, P. "Globalisation Prospects: Realising an Agricultural Dream," *The Hindu Survey of Indian Agriculture 2002*, The Hindu Group of Publications, Chennai, India, 2002, pp. 15-21.

Star, J., and Estes, J. *Geographic Information Systems: An Introduction*, Englewood Cliffs, NJ: Prentice-Hall, 1990.

Star, S. L., and Griesemer, J. R. "Institutional Ecology, 'Translations,' and Boundary Objects: Amateurs and Professionals in Berkeley's Museum of Vertebrate Zoology, 1907-39," *Social Studies of Science* (19), 1989, pp. 387-420.

Star, S. L., and Strauss, A. "Layers of Silence, Arenas of Voice: The Ecology of Visible and Invisible Work," *Computer Supported Cooperative Work* (8), 1999, pp. 9-30.

Suchman, L. "Located Accountabilities in Technology Production," *Scandinavian Journal of Information Systems* (14:2), 2002, pp. 91-106.

Suchman L. "Working Relations of Technology Production and Use," *Computer Supported Cooperative Work* (2), 1994, pp. 21-39.

Taylor, P. J. "Editorial Comment: GKS," *Political Geography Quarterly* (9), 1990, pp. 211-12.

Verran, H. "Imagining Ownership—Working Disparate Knowledge Traditions Together," *Republica* (3), 1995, pp. 99-108.

Verran, H. "Re-imagining Land Ownership in Australia," *Postcolonial Studies* (1:2), 1998, pp. 237-254.

Walsham, G. "Interpretive Case Studies in IS Research: Nature and Method," *European Journal of Information Systems* (4), 1995, pp. 74-81.

WCMC. (World Conservation Monitoring Centre) *Guide to Information Management in the Context of the Convention on Biological Diversity*, United Nations Environment Programme, Nairobi, Kenya, 1996.

ABOUT THE AUTHORS

Satish Puri has an M. Tech. in computer science from the Indian Institute of Technology (IIT), Bombay, and a Ph.D. from the University of Oslo. He has worked on design of information systems for forestry sector, mapping of wastelands using remotely sensed satellite data, and applications of GIS for wasteland management. He may be reached at puri_sk@hotmail.com.

Sundeep Sahay is a professor in the Department of Informatics at the University of Oslo, Norway. Dr. Sahay has been actively involved in researching social, organizational, and managerial implications of GIS systems, both in developed and developing countries. His current research interests include health information systems and issues relating to global software organizations. He has published extensively both in IS and organization studies. He may be contacted at sundeeps@ifi.uio.no.

33 SYSTEMS DEVELOPMENT IN THE WILD: User-Led Exploration and Transformation of Organizing Visions

Margunn Aanestad
University of Oslo

Dixi Louise Henriksen
IT University of Copenhagen

Jens Kaaber Pors
Roskilde University

Abstract This paper addresses an increasingly significant category of IT use: that of user-led deployment of generic technologies in organizational settings. Three case studies of such deployment are presented: a Web-based collaboration application developed in-house and deployed in distributed work of a multinational pharmaceutical company, a commercial groupware application deployed in the merger of a Northern European financial company, and a communication infrastructure for multimedia telemedicine in a Norwegian hospital. The activities studied were not fully organized in formal development projects, but were to a large extent initiatives "in the wild" where users influenced directions and outcomes of the process. In all three cases, we found a slow transformation of the initial organizing visions and intentions, a successive addition and adjustment of various technological components, and gradual alterations in work practice. In this paper, we classify this work as development in order to emphasize the importance of such redesigns, tailoring, and adaptations of the technologies that take place in use settings. In closing, the paper discusses the implications for the position and contribution of the IS professional in informing this process.

Keywords: Users, IS professional's role, generic technologies, configuration development

1 A DIFFERENT REALITY?

An emphasis on methodology has been central to the field of IS research since its inception (Avison and Fitzgerald 2003; Truex et al. 2000). This emphasis and the specific methodologies arose in a context where information systems were typically custom developed for the first time as stand-alone systems, designed for specific purposes in a relatively stable organizational context. This is not the case today. Computer networks are pervasive, most organizations host several information systems, and generic software, such as commercial off-the-shelf packages, is widely used. The demands on the in-house IT personnel are increasingly related to negotiations with vendors on purchasing and updating software components, and to tailoring and integrating these software components into the enterprise's portfolio of information systems. The issue of legacy systems together with a turbulent business environment, where frequent mergers and acquisitions induce a demand to link and integrate very diverse and heterogeneous information systems, makes integration technologies (like middleware, warehouses, and portals) a hot issue. In summary, information systems in organizations are no longer developed from scratch, but tend to emerge as a complex layering of components where practical interoperability as well as standardization trends have become new crucial issues.

This shift in what developing an information system means today is to some degree reflected in the IS research field. Truex, Baskerville, and Klein (1999), for example, discuss IT for new emergent organizations and state that several of the commonly accepted goals of the IS field are obsolete, namely "the idea that systems should support organizational stability and structure, should be low maintenance, and should strive for high degrees of user acceptance" (p. 123). They propose an alternative view that advocates ways in which systems should be under constant development rather than fully specified, thus allowing for constant adjustments and adaptations, or emergence. Other works reevaluate the inherited concepts, methodologies, and understandings, for example, in the renewed emphasis on development and implementation processes characterized by drift and unpredictability (e.g., Aanestad and Hanseth 2000; Ciborra 2000) and increased attention to the socially situated and context-dependent nature of information systems development (e.g., Dittrich et al. 2002; Fitzgerald et al. 2002).

In this paper, we take these shifts in practice and recent research as a point of departure. The paper offers new insights from three case studies of actual deployment processes of generic technologies in geographically distributed organizations. The application areas of the technologies in all three cases are characterized by employment in a voluntary manner for noncritical processes. This implies more leeway than normally found with mandatory use of organization-wide information systems for vital and strategic purposes, such as accounting and planning. This exposition aims to unpack empirically the abstract notions of emergence, drift, and context-dependency of technologies by scrutinizing in particular the role of *organizing visions* (Bloomfield and Vurdubakis 1997; Klecun-Dabrowska and Cornford 2002; Swanson and Ramiller 1997) and *users as configuration developers* (Bansler and Havn 1994). Initial organizing visions were in all cases found to be active in the discourses of managers and developers initiating the deployment process. These were vague yet significant in shaping and directing the employment of the specific technology. An evolving idea or guiding concept was to some extent built in either as material affordances or functionalities as well as a set of more open-ended possibilities and facilities to be actualized in practice. We found that initial organizing visions multiplied and were transformed through user-led exploration of these possibilities. Users were thus found to be central actors, in particular in taking up the role of a configuration developer adjusting both technological components and related work practices. These findings suggests new ways in which IS professionals can assess organizing visions and assist users to explore and realize the potential of generic technology in relation to particular and shifting settings.

2 THREE CASES OF REALIZATION

The cases presented in the following are three intensive and longitudinal field studies all employing an empirical approach and methods inspired from ethnography and science, technology and society studies (STS). In all three case studies, prior studies in the actor-network tradition (ANT) within STS have provided the main analytical resources as well as a methodological commitment toward always attending to local practice and specificity. ANT, for example, takes the production of scientific facts and technologies as a topic of study (e.g., Latour 1999) and abandons searching for drivers, determinants, or explanations of the phenomenon under study. Instead of seeking universal explanations for what guides and shapes the development process, how it may be controlled or made more predictable, the research approach in all three case studies has been to work empirically by analyzing the particular places and situations in which an information system is realized in practice. We have focused on the processes, practices, and specific circumstances through which the technologies and work practices become integrated and co-emerge. The three cases have been the main case studies for the three authors' Ph.D. work (Aanestad 2002; Henriksen 2003; Pors 2004). These studies have progressed in parallel and along the way we have compared and discussed our findings. This paper presents the outcome of these discussions presenting selected aspects of the cases, all of which have been made anonymous.

The first case concerns a Web-based information system developed in-house in the multinational pharmaceutical company, PharmaCo, and the development company, PharmaCoIT (Henriksen 2003). The system is called PharmoWeb and has been developed and deployed to support work in long-term and globally distributed pharmaceutical projects. The technology consists of a project home page displaying project news and events to pharmaceutical project members, and a document-sharing section where project members can store, organize, and easily retrieve relevant document files and meeting summaries. The second case concerns a commercial generic groupware application employed in a large financial company located in Northern Europe, BankCo (Pors 2004). Following the merger of several financial companies in the year 2000, the newly merged company-wide Communications Department decided to purchase the groupware application Lotus QuickPlace to support the geographically distributed project groups and organizational sections. The third case describes the user-managed, gradual, and emergent development of a local infrastructure for multimedia telemedicine in a Norwegian hospital (Aanestad 2002). Part of this process occurred in relation to a telemedicine project where live audio and video from surgical procedures was transmitted between hospitals.

A reoccurring similarity found in these three cases has been the lack of methodology or planning (mentioned earlier as a traditional concern in the IS field). Rather, these cases present examples of how generic technologies were deployed and realized through ongoing exploration, additions, and adjustments. In this paper, we focus on the organizing visions expressing the motivation behind the activities involving IT and the reasons why they started. We discuss the process to realize this vision through gradual transformations of initial intentions, through additions and adjustments to the technological components, and continuous alterations of the work practices involved.

2.1 First Case: Ongoing Development from In-House to the Generic

PharmoWeb was first introduced in 1998 and has since been redeveloped in four versions. The first version of PharmoWeb was developed by a librarian closely connected to the particularities of pharmaceutical work in a research department. It was later redeveloped by an internal IT department for wider use in other departments and across varying phases of pharmaceutical work ranging from research to production to marketing. Besides use in research and development projects, PharmoWeb was also more recently set up as a home page portal for other departments and was also sold and implemented in other companies such as a large cleaning services firm and a public transport corporation. As such, PharmoWeb evolved as an integrated part of differing work practices and became increasingly generic. In the most recent version, parts could be sold, combined, and customized according to the needs of diverse user settings, for example, purchasing the home page and news components, the document section, but omitting a calendar function or a progress-reporting module. This development history spans a range of differing organizing visions attached by the various actors involved. For example, PharmoWeb as the main tool for joining together geographically distributed pharmaceutical projects.

2.1.1 Adjusting the PharmoWeb Vision to the Particularities of Pharmaceutical Project Work

Along with the organizational restructuring of PharmaCo, a new management department was established as the central place from which pharmaceutical development projects were to be planned, managed, and monitored. This new management department thus became the official site for coordinating projects, facilitating communication, and ensuring progress in relation to project plans, budgeting, and competition from products developed in other pharmaceutical companies. As part of the management department's strategy of supporting communication and coordination within development projects, each project was to "own" a PharmoWeb. A management guideline stated that PharmoWeb should be the development project's main tool for sharing documents, that all project members should have reading access, and that at least core group members should have uploading rights. Besides this very loose vision and guideline, no strategy was defined as to formalize or establish the use of PharmoWeb in development projects. Over time, PharmoWeb came to form a part of projects in a variety of ways, creating both a central project site for broadcasting general project news as well as a set of private workspaces used for collaboration in subgroups within pharmaceutical projects. In most projects, the project assistant, the right hand of the project manager, became responsible for setting up and maintaining PharmoWeb. Also, the department decided to cofinance the development of a second version, and several of the project assistants were actively involved in stating project needs and demands based on their previous experiences, testing of prototypes, and continually pinpointing problems for new versions.

Project assistants maintained PharmoWeb for use on a daily basis. Regular activities of the project assistants contributed in the form of news and pictures to the home page section, or as meeting minutes documents that could keep track of project discussions and decisions. The project assistants also put together and uploaded monthly status reports and various formal documents and charts describing the particular project and work organization. They regularly maintained an overall plan where one could find upcoming deadlines and follow recent project accomplishments. Project assistants also ensured that all project members are entered as users, continually maintained yellow pages information and user groups by adjusting members and delegating appropriate access rights, such as read only, upload to some sections and not others, etc. They adjusted menu categories and subdirectories and suggested conventions for file naming and the use of restricted areas. One also posted a work-around on the news page listing how to avoid uploading glitches that came up now and then with large files.

A head project assistant started surveying all PharmoWebs in use and occasionally provided individual project assistants with support and suggestions on how to reorganize menu categories or adjust the layout on a page. She later wrote a PharmoWeb manual for the project assistants and started to create her own extensions to PharmoWeb such as active server pages and a calendar functionality incorporated in PharmoWeb. For example, the calendar was made upon demand from the project assistants to satisfy the need for the coordination of the numerous meetings and events taking place within the large development projects. She built the calendar system by tying excel spreadsheets together with an automatic generation of html pages. These skills, a formal position, and this new work of maintenance (that in practice are fully integrated with a range of other tasks the project assistants carry out) are alterations of existing work practices that have emerged alongside the deployment of a new, partly generic technology. This ongoing maintenance can be seen as over time shaping PharmoWeb as useful in relation to the specificities of pharmaceutical development projects in PharmaCo.

2.1.2 The Resulting Configuration(s)

In the study of PharmoWeb's deployment in the different parts of these dispersed pharmaceutical projects, an array of metaphors were used to envision PharmoWeb and the varied role it plays in pharmaceutical project work. Some spoke of a shared and accumulative archive or repository for project relevant information. Others used PharmoWeb as a dynamic workspace or place for collaborating on working documents across geographical distances and obstacles of time zone differences. Also, some mentioned that because projects are large and dispersed and different members are involved or active at different times, or perhaps working in several projects at once, PharmoWeb provides a way of seeing and following project progress, gaining an overview of current status, and identifying active project members. As deployed in pharmaceutical work, PharmoWeb was enveloped in many drifting visions and intentions and was set up in variety of ways within pharmaceutical projects. The deployment of PharmoWeb can be thought of as, on the one hand, partly achieving the librarian's initial vision of developing a new tool that may tie pharmaceutical projects together and, on the other hand, user-led emergence of a number of different versions of PharmoWeb-in-use: for example, ways in which access rights are delegated to entire projects creating a broadcasting medium or by limiting space to subgroups encouraging private workspaces.

The next case looks at the kinds of exploration that take place with the implementation and use of a fixed standard package technology, QuickPlace, and argues that the possibilities and potentials of generic technology are explored, tried out, and experimented with in ways that likewise are not centrally managed, not foreseen, nor necessarily corresponding to initial organizing visions.

2.2 Second Case: User-Led Deployment of a Commercial Groupware Package

The commercial off-the-shelf groupware application Lotus QuickPlace (now branded IBM Lotus Team Workplace) was deployed to facilitate the merger forming the major Northern European financial company BankCo. The justification and initial organizing vision was to support distributed work groups by providing them with a virtual workspace in the same way as these groups had previously been able to share files and working documents via local area networks (LANs) in their national settings. The groupware was acquired by the Communications Department responsible for the corporate intranet to mediate transnational communication in the merged organizational units. The initial organizing vision of QuickPlace was as an interim substitute for the intranet of each of the premerger companies. The investment in licensing the software along with resources for hosting it internally was considered sufficient for the subsequent use. Only scant central support of the groupware employment, therefore, existed in the form of training, guidelines, or efforts to align it with established organizational practices such as workflow and project models. For the integration of the virtual workspace with the work practice, the users were left with the tutorial provided by the manufacturer of the software and local experiences within BankCo of the existing infrastructures for collaboration, most importantly e-mail and LAN drives.

Despite this rudimentary support, the spread of the groupware application was rapid and uncontrolled. Within two years, the number of virtual workspaces grew to more than 100 with a total of 3,000 users and more than 20 Gb of uploaded documents. These workspaces were deployed for differing long-term and short-term tasks such as coordination and documentation of the activities in development projects. The only formal criteria for setting up a virtual workspace was that the requesting manager should send an e-mail to the department of IT Operations, stating that it was intended for employment within a geographically distributed setting involving two or more of the national headquarters of BankCo. The systems administrator of the QuickPlace server in IT Operations would then set up an instantiation of the groupware application as a virtual workspace and return the password to the requesting person granting him or her managing rights along with instructions of how to invite other members. As a consequence, it was left entirely to the local manager to integrate the groupware and make the organizing vision concrete in a workable way. The most successful examples of integration were found in situations where one or several persons took up the responsibility of realizing a certain mode of employment regarding the groupware application. The translation of documents such as financial reports described below is an example of this kind of integration.

2.2.1 Local Facilitators

The efforts of local managers and facilitators played a central part in binding elements of the specific distributed work practice and the virtual workspace together. In terms of explorations, adjustments, and additions to the technology, only a limited amount of technical configuration took place locally. In the few situations identified in the study with an achievement of integration, the major part of the work of configuration was carried out to align the groupware with other infrastructures, their mode of employment, and the established social protocols of the group members. Despite the extended functionalities of the groupware application provided in the virtual workspaces, such as shared calendars, discussion forums, and support for workflow integration, the changes were made mostly to accommodate the constraints of the basic functionality of the technology, namely uploading, searching for, and downloading documents.

The decentralized spreading and uneven employment of QuickPlace in BankCo with dispersed islands of successful deployment can in part be ascribed to the vagueness of the organizing vision. The fact that the groupware technology was introduced in terms of a LAN substitute until the intranet becomes established for the merged company did not encourage the situated exploration of possibilities. The resources for trying out and experimenting with the groupware in practice were typically not present. The ways it became integrated in certain settings were local appropriations of the initial organizing vision. Here local facilitators took it upon themselves to carry out the configuration development and align the groupware with other infrastructures of the work practice and the social protocols of the collaboration.

2.2.2 The Resulting Configuration(s)

Such a successful integration in a local, specific configuration was found in the translation section of the Communications Department. This section managed to integrate QuickPlace in their recurrent task of translating quarterly financial reports. The translators were geographically spread throughout the company and, due to the confidentiality and security issues imposed to avoid insider trading, the drafts for the final report could not be circulated by e-mail. The manager of the translation section requested a QuickPlace and made the translators use this infrastructure for the distribution of texts by designating specific folders in the virtual workspace for the uploading of translated documents. As an effect of this coordination mediated by QuickPlace, an overview of the collaboration was visible in a novel way via the page-views generated by the QuickPlace application showing the contents of folders. After some trial runs and experimentation with the technology, the groupware was also employed for the critical translation of the annual financial report where the translators travel to one headquarters and are collocated in one building. In this situation, the local area network drive could just as well be employed from a safety point of view, but the superiority of the integration of groupware with the work practice caused the translation team continue to use QuickPlace for exchanging documents. This practice was since transferred to the translation of other publications where the security issues were not prominent, but rather the improved overview motivated the spread of the employment of groupware. In this way, the deployment of QuickPlace involved both utilization of new functionalities and successive adjustments to the ordering of the contents in the virtual workspaces as well as alterations in the distributed work practices.

2.3 Third Case: An Emerging Infrastructure for Telemedicine

The third case is from a research and development department in a Norwegian hospital that developed image-guided technologies and procedures. A lot of visitors were coming to the department and a local analog transmission facility was set up between the operating theaters and an external room, in order to avoid disturbing the operating theater team more than necessary. Images from overview cameras in the operating theaters and from several other image sources (e.g., x-ray, videoscope, ultrasound equipment) were transmitted and displayed in the external room. Two-way audio through microphones and loudspeakers facilitated conversation between the operating team and the guests. After the cameras and microphones were installed in the operating theater, the workers expressed uneasiness over the fact that people present in the external room could watch and listen to the activities in the operating theater without themselves being seen or heard. To address this issue, a key was installed beside the wall-mounted camera in the operation theater. This key had to be turned on by the operation theater personnel in order for the transmission of video signals to be possible, and when the key was turned, a red "on air" lamp was lighted. In this way, the personnel would be informed about the onlookers and were able to control the transmission.

Later it was realized that this transmission facility offered a possibility for telemedicine. Cooperation with surgeons at a nearby hospital resulted in the establishment of a research project, involving connection to a digital broadband network. The aim was to assess the technology and see whether it suited the demands of surgeons, including the image quality. The early visions in the project included the establishment of educational services for surgeons under training (i.e., demonstrations of novel or complex surgical procedures), as well as consultations with other experts around specific patient cases.

When the local transmission facility was hooked up to the external telemedicine network, new challenges arose related to the surveillance issues mentioned above. During the first test transmissions in the project, some receivers (at the other side) would come and go according to their local duties. The operation team did not know how many or exactly who the receivers were, or whether they where present or not at any given moment. If there had been quiet time periods without interaction, the operating surgeon might ask explicitly who was watching. As the technology provided a two-way audio- and video-connection, an image was actually transmitted from the receivers' site, but this image was (in the beginning of the project) only being displayed in the external room. When this discussion arose, the image signal was forwarded into the operating theater and displayed on a free monitor. The key and the red light were complemented by the image, and the whole team could see who the receivers were.

2.3.1 Realizing the Network through Diverse Usage

Early in the project period, many technical tests and proof-of-concept transmissions were carried out. When proper usage was supposed to start, the project encountered difficulties with summoning the intended receivers, i.e., the surgeons, at given points of time. It was difficult for the surgeons at the receiving site to change work schedules to accommodate the transmission, in particular since it was not included in any formal training program and did not give credit points. Thus, in order to utilize the network access and video digitizing equipment fully, the scope of the project's activities was expanded. Several sessions were arranged for other hospital departments, including several other medical specialties, ranging from whole-day regional seminars to half-hour lunch meetings. Nurses and other groups also took up use of the facility.

Planning and coordination work became more important. During the planning sessions, the specific configurations of devices would be laid out according to the information needs and content of the transmission. A presenter might want to use a PowerPoint slide show, to show a VHS, or to project digital x-ray images during a presentation. These activities demanded detailed technical planning in order to be able to transmit them to the receiver site, and a variable number of video converters, routers, VHS players, etc., would need to be connected. These diverse transmissions significantly impacted the development of the internal infrastructure in terms of technical characteristics. In order to be able to perform a given task, specific pieces of equipment had to be borrowed or purchased for these purposes. A client department, for which a seminar was arranged, might buy minor devices that were needed for their transmission, e.g., an extra microphone, which afterwards were donated to the project. The resulting portfolio or collage of equipment was thus shaped by these activities, and also deter-mined which use areas were possible to serve, i.e., what kind of transmission could be handled.

2.3.2 The Resulting Telemedicine Infrastructure

The technicians working with this telemedicine infrastructure were conscientious objectors, doing 14 months service at the hospital as an alternative to army service. Most of them had an engineering or computer science background, in addition to specific personal skills (e.g., one was a musician and knew microphone and loudspeaker technology well; others knew digital video editing). They had a central role during telemedicine transmissions. Prior to any transmission, they had to establish the connection to the other side, verify proper image transfer both ways, and do sound checking. The required images from the operating theater had to be selected, the sound quality adjusted, and the transmission monitored during its whole extent. The technicians were equally central in both long-term and short-term development of the local infrastructure. The interesting point is that their work was not supposed to revolve around the telemedicine project from the beginning. Due to scarcity of office space, they were allocated the external room as their working place, and were an available and cheap resource when the project activities required support personnel. This work was increasingly being recognized as important, and after the project ended a permanent position for telemedicine support was established at the hospital. The external room continued to function as the technological hub for external and internal audio/video transmissions, not just for this department but for the whole hospital. Several departments were using the facilities frequently, in particular for connecting to the operation rooms during professional meetings or during student teaching sessions.

In this case, we see that unforeseen problems arise and are dealt with through additions and adjustments of equipment and practices. What would be necessary for making the technology work and be useful became obvious through actual use. However, the telemedicine project failed in establishing a market for broadband-based educational services for surgeons. These visions were too grand and required establishing new structures (e.g., educational programs) that were beyond the scope of the actors and the project. Rather than realizing the organizing visions that initially were circulating among the surgeons and managers, the project resulted in realizing changes in existing work and communication practices, new competencies, and a new control room well equipped for a variety of transmissions.

3 ALLOWING FOR THE USER-LED TRANSFORMATION OF ORGANIZING VISIONS

These cases illustrate the unpredictable and situated nature of the deployment of generic technologies; how adjustments and alterations of both technology and work practice emerge in sync. The cases show that the realization of information systems in organizations through exploration, additions, and adjustments was slow, incremental, and proceeded in unexpected directions. This was the case of PharmoWeb's different set-ups as broadcasting medium and private workspaces, the emergent role of QuickPlace in the work of language translation, and the adjustments of the telemedicine technology in practice.

The processes we describe on the level of the local and situated work practice were not strictly planned projects and in some cases not even formally defined as projects, but rather "initiatives in the wild" of exploring possibilities and continuously assessing the efforts to integrate technology and work practice. This implies that the activities were to some degree voluntary, since this was not about strategic or critical systems. These activities happened at the margins of ordinary work.

Even if the degree of exploration was high, there still was rationality behind deploying these technologies, a vision that influenced the subsequent process. It is this vision, or the underlying rationality, that we call the organizing vision. In all three cases, the technology was selected and implemented because some people somewhere intended for it to do something. In the BankCo and telemedicine cases, management were the carriers of the initial vision, while in the PharmaCo, case the interest from management was significant during the deployment process. What was the role of these initial visions? On the one hand, an organizing vision remains to some extent imaginary and abstract, for example, descriptions of what the system can do when touted by manufacturers, vendors, and developers. Yet these visions also functioned as a resource for apprehending future use. It enabled others to act by mobilizing an organizing vision and, by complying with this vision a distinct aspect of what constitutes a generic technology was shaped. Thus the organizing vision was significant in both starting and shaping the process, e.g., through defining the participants and the initial use areas. However, its abstract and vague character is also important. In an actor-network terminology, an organizing vision might thus be thought of as an emergent actor that gains agency through diverse and shifting relations to practices and people, their commitment, and sense of ownership. An organizing vision is something that can gain strength and have influence not because it is black-boxed, strong, and rigid, but precisely because it is adaptive and transformative (de Laet and Mol 2000).

In all three cases, this vision was somewhat vague and did not fully determine the process. During exploration and practical assessment of the technology, the organizing visions were translated, transformed, adjusted, and multiplied. In the first and second case, PharmaCo and BankCo, several visions and configurations coexisted. For example, PharmoWeb was imagined and deployed successfully as a centralized portal in one project and as a set of private subgroup workspaces in another. In this case, as well as the other two cases, exact outcomes and benefits of the process were unknown and unpredictable at the outset. From our cases, we may suggest that the outcomes (what we call the resulting configurations) tend to be more modest in their results than the initial organizing visions but are made concrete through this transformation. We conclude that the vagueness of organizing visions might actually be seen as productive for the successful realization of generic technologies or, as Truex et al. write, "usefully ambiguous" (1999, p. 121).

Second, all three cases bring forth the important role of active users and use mediators in the deployment processes that are carried out as configuration development with generic standard software packages. In all three cases, particular groups of individuals (project assistants, translators, technicians) gained a central role in realizing instantiations of generic information systems in local circumstances. These findings challenge the traditional temporal sequencing of phases in terms of analysis–design–construction–implementation and suggests that, in the case of generic technologies, the loci of development work may not reside with professional systems developers alone. All three case studies employing generic technology thus exhibit side-effects of the division of labor in the development process of generic systems, where the standard packages developed and sold by software companies are developed further by the consuming organizations (Grudin 1991).

We suggest that it is through such user-led configuration development that organizing visions become realized slowly as adjustment of technology and work practices through a joint negotiation. The word negotiation should not be taken to imply formalized discussion; on the contrary, most of the material from our cases shows that this is usually informal and ad hoc. For example, in the financial company, BankCo, the amount of configuration development carried out by IT Operations was limited to the initial setup of the welcome page and a custom logo, leaving the rest of the configuration development to the local managers acting as facilitators to integrate the groupware with local practices. In the hospital setting, doctors, nurses, and the recently employed technicians took part in such configuration development. In relation to all three cases, a transformation of organizing visions, configuration development, and the notion of user-led deployment was productive for conceptualizing the realization of a generic technology in organizational settings. In all three cases, we suggest that making more resources available to these configuration developers could have encouraged a productive integration of technology and the specific work practices in question.

Brought together, these empirical cases illustrate the process and outcome of a kind of configuration development process that is not rationally controlled or centrally managed by a defined group of managers or designers. The cases fill in empirically what IS research terms such as emergence, drift, and context-dependency imply and suggest that this line of research is relevant in relation to the contemporary situation of increased deployment of standardized software packages. This case material thus confirms the need for further research on information systems development in organizations as an emergent and drifting process, and addresses the difficulties involved in balancing the projection of organizing visions with unpredictability. Furthermore, this work calls for new research on the shifting roles of professional developers, a differentiation of user types and intermediary positions, as well as a rethinking of the position of the IS professional. In closing, we want to reflect on what these cases imply for the role of the IS professionals (practitioners as well as researchers).

4 IMPLICATIONS FOR IS PROFESSIONALS

If we return to the suggestions offered by Truex et al. (1999) concerning an alternative view of ongoing development, we find little discussion of the implications of the role of the IS professional. This position is not expected to change dramatically even if the activities are more oriented toward ongoing assessment, maintenance, redevelopment, and user-led experimentation. With reference to the IS paradigms proposed by Hirschheim and Klein (1989), to Truex et al. (1999) the IS professional still seems to inhabit the role of a (technical) systems expert and/or a facilitator who may assist users to discover and make sense of the technology's potentials. We believe that this role will be different in user-initiated and user-led projects, which overflow the boundaries of a formal IS project. If at all included, the IS professional seems to be left in a more passive role. But if the IS professional is no longer the expert laying out the guidelines, plans, and models for how the process should be controlled and managed, do we have anything to offer in such user-led design situations?

We think that the answer is yes. To us, there seems to be a vacant position in these processes—that of an assessor or reflector. The changed character of the process and the impossibility of preplanning locally achieved effects makes the role of ongoing assessment crucial. The tasks of the IS professional might be to critically examine the organizing vision and evolving configurations, to subvert these if unrealistic or unjustifiable, or support these if potentially beneficial based on a here-and-now judgment. Rather than an after-the-fact generalized evaluation, we claim that ongoing and involved assessment is crucial; it should be constructive rather than just descriptive, interventional rather than detached, and objective. Again resources here are extremely scarce, exempting recent work building on a participatory design tradition (Büscher et al. 2001; Dittrich and Lindeberg 2003). These authors discuss the possibilities of shifting the locus of design into use and addressing evaluation and assessment as an open-ended and ongoing activity to be understood as part of development. Here the IS professional is uniquely equipped to contribute by playing an active part in assisting in the ongoing assessment of the outcomes and possibilities of emergence.

In all three cases the researchers, which at that time were Ph.D. students, played an observing role where participation and intervention was limited and subtle. If we had then tried to do what we now advocate, this might have consisted of more active assessment and interference, for example, by asking different questions along the way:

1. PharmaCo: The project assistants are gaining a central position as key configuration developers, their daily work practices are changing, and their influence is shaping PharmoWeb in two directions: as a broadcasting medium for entire projects and as a set of private work spaces for subgroups. Whom might benefit or loose from such an arrangement?
2. BankCo: There seems to be no overall plan for or incentive for the employment of QuickPlace and it is used very randomly and most of the time not very much. How can we reassess the initial organizing vision and bring it into line with existing practices, their differences, and evolving demands?
3. Hospital: The broadband services are not living up to the expectations but have instead generated other types of use involving a number of new issues. Is this direction of usage the most desirable for the technicians, the nurses, doctors, or students of medicine?

The objective here would not be to offer final answers, but ensure that such questions are opened up and put into focus. A new position for the IS professional might thus be thought of as exposing desirable directions to the phenomena of drift and emergence—to the processes of systems development in the wild. We call for further work that reflects upon ways in which the IS professional can participate in exploring, assessing, reworking, and transforming the organizing visions and the resulting configurations based on the practices recovered.

We conclude that user-led exploration and transformation of organizing visions are some of the urgent issues that follow from a different reality of information systems development and use. With these issues, we follow a renewed yet inescapable question of normativity which we believe is pertinent and timely for the IS community to address.

ACKNOWLEDGEMENTS

We are grateful to our informants and colleagues and the anonymous reviewers for useful comments. The Danish Research Councils (grant no. 99-00-092) and the Norwegian Research Council (grant no. 123861/320) funded this research.

REFERENCES

Aanestad, M. *Cultivating Networks: Implementing Surgical Telemedicine*, Ph.D. dissertation, Faculty of Mathematics and Natural Sciences, University of Oslo, no. 228, Oslo: Unipub, 2002.

Aanestad, M. , and Hanseth, O. "Implementing Open Network Technologies in Complex Work Practices: A Case from Tele-Medicine," in R. Baskerville, J. Stage, and J. I. DeGross (Eds.), *Organizational and Social Perspectives on Information Technology*, Boston: Kluwer, 2000, pp. 355-369.

Avison, D., and Fitzgerald, G. "Where Now for Development Methodologies?," *Communications of the ACM* (45:1), 2003, pp. 78-83.

Bansler, J., and Havn, E. "Information Systems Development with Generic Systems," in W. Baets (Ed.), *Proceedings of the Second European Conference on Information Systems*, Njienrode, The Netherlands: Njienrode University Press, 1994, pp. 707-715.

Bloomfield, B., and Vurdubakis, T. "Visions of Organization and Organizations of Vision: The Representational Practices of Information Systems Development," *Accounting, Organizations and Society* (22:7), 1997, pp. 639-668.

Büscher, M., Gill, S., Mogensen, P., and Shapiro, D. "Landscapes of Practice: Bricolage as a Method for Situated Design," *Computer Supported Cooperative Work (CSCW): The Journal of Collaborative Computing* (10:1), 2001, pp. 1-28.

Ciborra, C. U. (Ed.). *From Control to Drift: The Dynamics of Corporate Information Infrastructures*, Oxford, UK: Oxford University Press, 2000.

de Laet, M., and Mol, A. "The Zimbabwe Bush Pump: Mechanics of a Fluid Technology," *Social Studies of Science* (30:2), 2000, pp. 225-263.

Dittrich, Y., Floyd, C., and Klischewski, R. (Eds.). *Social Thinking—Software Practice*, Boston: MIT Press, 2002.

Dittrich, Y., and Lindeberg, O. "Designing for Changing Work and Business Practices," in N. Patel (Ed.), *Adaptive Evolutionary Information Systems*, Hershey, PA: Idea Group Publishing, 2003.

Fitzgerald, B. Russo, N. and Stolterman, E. *Information Systems Development: Methods in Action*, London: McGraw-Hill, 2002.

Grudin, J. "Interactive Systems: Bridging the Gaps between Developers and Users," *IEEE Computer* (24:2), 1991, pp. 59-69.

Henriksen, D. L. *ProjectWeb as Practice: On the Relevance of Radical Localism for Information Systems Development Research*, Ph.D. Dissertation, Writings on Computer Science No. 96, Roskilde University, 2003.

Hirschheim, R., and Klein, H. K. "Four Paradigms of Information Systems Development," *Communications of the ACM* (32:10), 1989, pp. 1199-1216.

Klecun-Dabrowska, E., and Cornford, T. "The Organizing Vision of Telehealth," in S. Wrycza (Ed.), *Proceedings of the European Conference on Information Systems 2002*, Gdansk, Poland, June 6-8, 2002, pp. 1206-1217.

Latour, B. *Pandora's Hope: Essays on the Reality of Science Studies*, Cambridge, MA: Harvard University Press, 1999.

Pors, J. K. *Integrating Generic Groupware and Distributed Work Practices*, Ph.D. Dissertation, Writings on Computer Science, Roskilde University, 2004.

Swanson, E. B., and Ramiller, N. C. "The Organizing Vision in Information Systems Innovation," *Organization Science* (8:5), 1997, pp. 458-474.

Truex, D., Baskerville, R., and Klein, H. "Growing Systems in Emergent Organizations," *Communications of the ACM* (42:8), 1999, pp. 117-123.

Truex, D., Baskerville, R., and Travis, J. "Amethodical Systems Development: The Deferred Meaning of Systems Development Methods," *Accounting, Management and Information Technology* (10:4), 2000, pp. 53-79.

ABOUT THE AUTHORS

Margunn Aanestad is a post-doctoral research fellow at the Institute of Informatics, University of Oslo, Norway. Telemedicine was the topic for her Ph.D. dissertation, which was completed in 2002. Her current research interests are related to large-scale information systems and infrastructures, mainly within health care. Margunn can be reached at margunn@ifi.uio.no.

Dixi Louise Henriksen is an assistant professor at the Department of Design and Use of Information Technology, IT-University of Copenhagen. Her research interests revolve around practice-based studies of systems development, design-in-use of workplace technologies, and studies of distributed work practice. Dixi can be reached at dixih@itu.dk.

Jens Kaaber Pors is a Ph.D. candidate at the Computer Science Section, Roskilde University, Denmark. His research interests are development and employment of standard-based information infrastructures for collaboration and other distributed activities, such as e-learning and e-government. Jens can be reached at pors@ruc.dk.

34 IMPROVISATION IN INFORMATION SYSTEMS DEVELOPMENT

Jørgen P. Bansler
Erling C. Havn
Technical University of Denmark

Abstract This paper discusses the role of extemporaneous action and bricolage in designing and implementing information systems in organizations. We report on a longitudinal field study of design and implementation of a Web-based groupware application in a multinational corporation. We adopt a sensemaking perspective to analyze the dynamics of this process and show that improvisational action and bricolage (making do with the materials at hand) played a vital role in the development of the application. Finally, we suggest that this case study provides an occasion to reconsider how we conceptualize information systems development (ISD).

Keywords: Web-based groupware, information systems development, sensemaking, improvisation, bricolage

1 INTRODUCTION

IS researchers have studied the process of developing and implementing information systems for a long time and a substantial body of research now offers insights into the problems and issues associated with the design and implementation of information systems in organizations. Much of this research, however, embodies assumptions about agency, knowledge, organizational change, and technological innovation that are misleading and inappropriate given the situated and emergent nature of organizational IS development (Gasson 1999; Truex et al. 1999). Information systems are no longer stable, discrete entities, but part of elaborate networks and information infrastructures that are subject to constant adjustment and adaptation (Ciborra et al. 2000).

Mainstream research is premised on the belief that information systems development (ISD) in organizations is a planned, deliberate activity—bounded in time and carried out in a systematic and orderly way.[1] ISD is commonly viewed as a rational design process, organized in formal projects with a clear purpose and a well-defined beginning and end. Another fundamental assumption is that ISD is a methodical process (Truex et al. 2000), that is, a managed, controlled activity performed in adherence to the principles and rules of a systems development method. The concept of method dominates the current discourse about ISD to such a degree and "has been so strongly impressed on our thinking about systems development, that the two concepts, *information systems development* and *information systems development method,* are completely merged in systems development literature" (Truex et al. 2000, p. 56). This privileged position of the method concept in the discourse on systems development inhibits our understanding of how information systems are developed in practice. As Truex et al. (2000, p. 74) emphasized, "When the idea of method frames all of our perceptions about systems development, then it becomes very difficult to grasp its non-methodical aspects." When adherence to methods is taken for granted, activities and situations that do not fit within a methodical frame become marginalized and practically invisible, e.g., how ISD is subject to fortuity, circumstance, human whims, talents, and the personal goals of the managers, designers, and users involved.

The objective of this paper is, then, to develop an alternative perspective on ISD—a perspective that posits improvisation and emergent change rather than methodical behavior and planned change as fundamental aspects of ISD in organizations. We recognize, of course, that ISD can be and often is performed as a deliberate, purposeful project with formal governance structures, requirement specifications, milestones, and substantial technological and organizational resources. However, we want to highlight the fact that ISD also happens in a myriad of other ways and that important activities take place outside of the formal projects—in the cracks and crevices in the official project portfolio so to speak. These development processes are more emergent, more continuous, more filled with surprise, more difficult to control, more tied to their circumstances and more affected by what people pay attention to than by intentions, plans, and methodologies (Weick 1993b). They are grounded in the situated practices of organizational actors, and emerge out of their experience with the everyday contingencies, breakdowns, exceptions, opportunities, and unintended consequences that they encounter when they appropriate, adapt, and experiment with new technologies in their work (Orlikowski 1996).

We ground our argument on the findings from a longitudinal case study of the development of a Web-based groupware system in a large multinational corporation. The findings from the case study show that improvisation, emergent change, and unanticipated outcomes may play a vital role in ISD, and that the development of IS in organizations may depend on serendipity and chance to a much higher degree than we usually want to admit.

---

[1]Alternative, less rationalistic perspectives do, of course, exist, even though they still tend to be rather marginalized in the scientific discourse on ISD. Good examples of these less rationalistic perspectives include the body of work on *reflective systems development* (Andersen et al. 1990; Mathiassen 1998), which has its root in Scandinavia, and the stream of new, *agile software development* approaches (Cockburn 2002; Highsmith 2002), which recently has attracted enormous interest from academics as well as practitioners.

2 IMPROVISATION IN ORGANIZATIONS

It is only relatively recently that organizational researchers have become interested in improvisation, but a growing number of researchers are now starting to take an interest in improvisational action and its potential value to organizations. A prominent group of scholars have used jazz and improvisational theater as a metaphor to develop their theoretical ideas about extemporaneous action in organizations (e.g., Barrett 1998; Hatch 1997, 1998, 1999; Mirvis 1998; Peplowski 1998; Weick 1993b, 1998). Others have drawn on anecdotal and empirical evidence to study organizational improvisation more directly. For instance, researchers have analyzed improvisation in new product development (Eisenhardt and Tabrizi 1995; Miner et al. 2001; Moorman and Miner 1998a, 1998b), strategy formulation and implementation (Perry 1991, 1994), implementation of new technologies (Orlikowski 1996), and during emergencies such as a failed navigational system (Hutchins 1991), an earthquake (Lanzara 1983), and a firestorm (Weick 1993a).

A working definition of improvisation can be taken from jazz music, where it connotes composing and performing contemporaneously (Barrett 1998; Weick 1998). Within organizations, it can be described as the conception of action as it unfolds, drawing on available material, cognitive, affective, and social resources (Cunha et al. 1999). It means that

(1) Improvisation is *deliberate,* meaning that it is the result of intentional efforts on the behalf of the organization and/or any of its members.
(2) Improvisation is *extemporaneous.* It deals with the unforeseen; it works without a prior plan and without blueprints and methods (Weick 1993b).
(3) Improvisation *occurs during action*, meaning that organizational members do not stop to analyze a perceived problem or an unanticipated opportunity and come up with a plan. Instead they develop their response by acting on the problem or opportunity, and can only judge its suitability by hindsight, not by foresight as in traditional planning (Cunha et al. 1999).
(4) Improvisation implies the preexistence of a set of resources, be it a plan of action, tools and technologies, knowledge, or a social structure, upon which variations can be built (Cunha et al. 1999; Weick 1998).

Most authors agree that organizational improvisation can happen in varying degrees, i.e., it occurs along a continuum ranging from spur-of-the-moment action to entirely planned action (Cunha et al. 1999; Moorman and Miner 1998a, 1998b).

Organizational improvisation is closely linked to the concept of *bricolage,* i.e., the ability to use whatever resources and repertoire one has to perform whatever task one faces (Lanzara 1999; Louridas 1999; Weick 1993b). Because improvisation means to act in an extemporaneous and spontaneous way to changing needs and conditions, improvisers cannot wait for optimal resources to be deployed and have to tackle the issues at hand with currently available resources (Cunha et al. 1999; Weick 1993b, 1998). Therefore, when improvisation happens, then necessarily, bricolage will too.

The increased interest within organizational studies in improvisation provides some conceptual grounding for the study of improvisation and bricolage in IS development.

In the research study described below, we explore the development of a Web-based groupware system intended to support globally dispersed, product development work of a large and highly successful multinational organization (BioCorp).

3 RESEARCH SETTING AND METHOD

BioCorp is a multinational biotechnological company that manufactures a range of pharmaceutical products and services. BioCorp's headquarters are in Northern Europe, but the corporation has production facilities, research centers, and sales offices in 68 countries. In 2001, BioCorp employed more than 16,000 people and the net turnover was $2.8 billion. During our field study, the corporate IT department was turned into a separate (limited liability) company. We include this information because this change had a significant impact on the relations between IT staff and users in our case.

The groupware system we studied was developed in-house, as a collaborative effort between people in the Project Management Unit within BioCorp's R&D division and the corporate IT department. The purpose of the system was (or rather turned out to be) to support communication and collaboration among participants in the company's drug development projects.

These projects are complex, large-scale, long-term endeavors. A typical project lasts 9 to 10 years and involves up to 500 people from many different areas within the company (e.g., clinical research, engineering, marketing, and regulatory affairs). Most of the activities are carried out at sites in Scandinavia and Northern Europe, but clinical trials are conducted in the United States, Singapore, Japan, and a number of other countries worldwide. The fact that a growing number of BioCorp's new drugs are developed in close collaboration with external partners in Japan, the United States, and Europe further adds to the distributed and complex nature of the projects.

The Project Management Unit (PMU), located at headquarters, is responsible for managing the development process and for ensuring efficient coordination of all the tasks and resources involved in a development project. It combines the skills of a large number of units working in matrix organization set-ups. PMU includes a number of project directors, each of whom is responsible for the management of a selected number of cross-functional drug development projects. Each project director has a personal assistant who acts as his/her "right hand." In addition, every project is headed by a group of middle managers—the so-called core group—coming from different functional areas.

Although formal as well as informal face-to-face meetings are central to communication and sensemaking within the projects, the dispersed nature of the organization means that project members must also rely heavily on a variety of communication technologies to facilitate various modes of work. At the time of our study these included familiar technologies such as mail, telephone, and fax but also more advanced technologies such as ftp, shared LAN drives, e-mail, video conferencing, and electronic calendars.

4 DATA COLLECTION AND ANALYSIS

Previous empirical studies of ISD (e.g., Bansler and Bødker 1993; Fitzgerald 1998; Gasson 1999) have focused on system development methods and how they are applied in formal projects. Rather than starting from the assumption that formal projects and methods play a key role in ISD, we wanted to observe how the development of an information system actually unfolded in a large, complex organization and track events and activities over a prolonged period of time. We wanted to focus on events in their natural setting and capture the rich array of subjective experiences of organization members during the development process.

Consistent with the focus of our research, we followed an interpretive case study approach (Myers 1997; Stake 2000, Walsham 1993). Interpretive field research is particularly appropriate for understanding human thought and action in natural organizational settings (Klein and Myers 1999). This approach allowed us to gain insights into the processes related to the development, implementation, and use of the groupware system and, in particular, to examine how different actors' technological frames and organizational priorities changed over time as they interacted with the technology. This enabled us to throw light upon the critical role of bricolage and improvisation. Moreover, this approach is also useful for discovering new insights when little is known about a phenomenon. It allows for casting a new light on complex processes whose structure, dimensions, and character are yet to be completely understood (Myers 1997).

Our field data collection lasted for more than three years and we used several data sources and modes of inquiry (for triangulation). The two primary data collection methods used were interviews and examination of archival data, but we also participated in a number of formal and informal meetings with developers and users. Finally, we examined different versions of the software under development.

Interviews. We began interviewing managers and employees of BioCorp in August 1998 and concluded the last interview three years later, in September 2001. During this period, we conducted 34 qualitative interviews of 60 to 120 minutes in length. All interviews were recorded and transcribed. Participants represented a diverse array of occupations and organizational positions, and included project directors and project assistants from PMU, members of several large development projects, as well as managers, analysts, and programmers from the corporate IT department. We interviewed developers and users throughout the research process. The goal of these ongoing interviews was to gather information about important events and actions and to track changes in the way people experienced the technology and perceived the new communicative affordances provided by it. We also wished to avoid such problems as poor recall, hindsight bias, and rationalizations.

Archival data. We reviewed public materials such as annual reports and company brochures as well as internal documents such as the company newsletter, organization charts, the corporate IT strategy, the IT project model, the project manual concerning the discovery and development of new medicinal drugs, the guidelines for organization and management of development projects, and the set of user manuals for the groupware system. This provided general information on company history, structure, core competencies, culture, IT policies, IT infrastructures, and IT expertise, as well as more specific data on the organization and management of the medicinal drug development projects (including formal planning and project management models), and the groupware system itself.

Meetings and informal conversations. We held two meetings with the director of PMU and several meetings with the IT manager responsible for the groupware system. We also participated in a one-day workshop with users and developers in spring 2001. The purpose of the workshop was to discuss user requirements for the next version of the groupware system. In addition to the formal meetings, we had many informal conversations with users and developers during our visits to the company and on the phone in connection with meetings or interviews.

Examination of the application. We had the opportunity to inspect the different versions of the groupware system on several occasions. In addition, when interviewing users, we often asked them to demonstrate how they used the system and show us the content of the document base. In this way, we gained first-hand knowledge about the system and its salient features.

We used qualitative techniques to analyze the data, informed by the overall focus on sensemaking, improvisation, and bricolage. We analyzed all data sources in a process of recursive scrutiny to get as complete a picture as possible of the design, implementation, and use of the groupware system. This process was "not unlike putting the pieces of a puzzle together, except that the pieces are not all given but have to be partially fashioned and adjusted to each other" (Klein and Myers 1999, p. 79). We endeavored to place our findings in the context of relevant literature and in interpreting our data we constantly referred to relevant bodies of research on improvisation, sense-making, information systems development, and so on. Thus, the processes of reporting the findings and conducting the analysis were highly connected and interwoven.

We shared our preliminary findings with key informants in PMU and the IT department, and they provided helpful comments that confirmed and elaborated the identified issues and conclusions drawn. By discussing our findings with the key informants, we explicitly recognize that the participants in the study—just as much as the researchers—are interpreters and analysts and that the story we tell is a result of our interaction with the participants (Klein and Myers 1999).

5 CASE STUDY

In what follows, we examine the development and use of the groupware system in PMU—from fall 1998 when the idea was first conceived by people in PMU until fall 2001 when the third version of the application had been in use for more than a year. Three versions of the application were developed during this three-year period.

The *first* version was developed by a couple of entrepreneurial people in PMU and was simply a modified piece of software that they had borrowed from BioCorp's library. It turned out to be virtually useless in practice, but it generated enough enthusiasm and inspiration to continue the development process.

The *second* version, named ProjectWeb, was developed in close collaboration with a couple of programmers from the corporate IT department. This version was well received by the intended users and the PMU management decided to make use of the system mandatory for all drug development projects.

The development of the *third* version was a very different story. By that time, the corporate IT department had been transformed to an independent company within the

BioCorp group and saw an interest in turning ProjectWeb into a generic application, which they could market broadly within—as well as outside—BioCorp. Thus, the primary driving force behind the third version was not PMU, but the IT department.

We now discuss the development and use of each version in detail, focusing on the fragmented, ambivalent, and capricious nature of the process.

5.1 Version 1

A small group of entrepreneurial people at PMU—in particular a visionary project director (Carl), his enterprising assistant (Stella) and the so-called IT supporter (Jean, a self-taught IT specialist who assisted PMU's computer users in countless ways)—were the prime movers in the development of the first version of ProjectWeb. It is difficult to pinpoint exactly when the idea of building a Web-based groupware system came into existence and who the originator was but, according to Stella, the first time the idea of using Web-technology to improve project communication was discussed was in 1997.

> It started at the annual PMU departmental seminar, where Carl [project director] called for a PMU-Web. Confidentially, the former manager was no technical wizard, and all the others hadn't given it much consideration. So, it was a discussion between the two: Carl, who was all for it, and the former manager who was against [it].

It was not until a year later, in 1998 (and after the appointment of a new PMU manager), that they first attempted to exploit the new technological opportunities. The project directors (and the new PMU manager) agreed that all projects should create a Web site to facilitate internal project communication. The task of creating and maintaining these new Web sites was assigned to the project assistants and Microsoft FrontPage was chosen as the common tool. However, it soon turned out that FrontPage was totally inadequate for the job at hand.

Jean, the IT supporter, concluded that the project assistants needed a better tool—a tool that was more advanced than FrontPage in terms of facilities, but at the same time easy to use. She contacted a Web-savvy person she knew in the corporate IT department who told her that the corporate library had similar ideas and that they had created a piece of software, which might be useful. Jean acquired the program from the library and made a few modifications to it before she let two project assistants, including Stella, test it. The result was, however, disappointing, the program was too primitive and much too difficult to use.

They had tasted blood, however, and decided that the next step would be to develop a better tool themselves.

5.2 Version 2

Jean realized that she did not possess the necessary programming skills to develop a better tool on her own and that the corporate IT department had to be involved in the process. Together with Stella and the project director Carl, she persuaded the new PMU manager to fund the project.

Development. The development of version 2 took place in close collaboration between Jean and Stella from PMU and two people from the IT department, a programmer (David) and a graphics designer (Hal). It was, however, according to David, the users who had the initiative and set the course.

> Version 2 grew out of what they [the users] came with as input. Stella has been the one giving the most input. She was also one of the main users, a very active user. That's how there was a lot of input to what new things could be done.

They took version 1 as their starting point and added new ideas along the way. It was not a formal or systematic process, but rather a process relying on informal conversations, free exchange of ideas, and extensive use of prototypes.

> I am not sure that there is any formalized [documentation]. I don't think so. Most of it was working papers. It was enough with my prioritized listings, when there is only one guy developing it. We haven't worked according to our quality department's goals and specifications. It's been based on prototyping and on close cooperation with the customer.

It is important to understand that in the beginning neither Jean nor Stella had a clear idea of what kind of system or tool they were about to design. Jean, for instance, explained that her initial idea had been that ProjectWeb should be a tool to facilitate file sharing in distributed groups. She saw ProjectWeb as an alternative to the common LAN drives, which some of the projects at that time used to share documents. In her opinion, LAN drives were not a good solution. They were difficult to manage and required too much insight into technical matters (e.g., security measures).

Stella, on the other hand, saw ProjectWeb as an alternative to e-mail. She was enthusiastic about what Web technology could offer in presenting information in a much more interesting, clear, and visually attractive way—e.g., by using different colors, fonts, graphic effects, links, and pictures. "ProjectWeb is a graphic version of e-mail," she said. She continued, "ProjectWeb is the greatest revolution since the e-mail. You can communicate in a much better way. It's more graphical, and you have a kind of library that includes all the information."

In summary, Jean and Stella had some vague ideas of how Web technology could be leveraged to improve project communication, but they did not know exactly what they wanted at the outset of the design process. It was only *during* the design process, through their interactions with David and Hal and through their joint exploration of different solutions (prototypes) that it gradually became clear what the tool should do and how it should be designed.

Implementation and use. Version 2 was finished and ready for launch in April 1999. PMU management decided that all development projects should establish an internal web site using ProjectWeb, but refrained from laying down guidelines for the design, content, or use of these Web sites. As a consequence, each project director and his or her assistant had to figure out for themselves how to use the application and for what purposes. They were, of course, uncertain about what to do, and the result was that ProjectWeb use varied significantly from one project to another. One of the directors explained that they did not really know what they were able to do with the new technology, and that they had to "experiment a little with the medium" to explore its advantages and limitations.

> The problem is, I'm not an IT nerd, and I'm not familiar with what possibilities the media has. And that's why it is about feeling your way as you go. I don't have any clear IT strategy for my project, and it may suffer as a result, if you look at it as a specialist. But we're in a process where we have to figure out what the media has to offer, and then we'll be able to get the best out of it.

At this point in time, the technology did not have a fixed or common meaning. Some directors, for instance, were mostly occupied with the problems of managing the huge amount of reports, letters, clinical data, and other records, which are produced during the lifetime of a project and have to be kept for future reference. From their perspective, ProjectWeb was an electronic document management system. Others were more focused on effective communication and team building and primarily considered ProjectWeb as a new medium of communication.

> It [ProjectWeb] is a fantastic communication tool for a multinational project group, where we don't know much about, how shall I say it, our customers and colleagues in the project. It offers the possibility of conveying information immediately. It offers a place where people can see everything that's relevant to the project, who's involved in the project, who's responsible, etc.

After having used and experimented with ProjectWeb for some time, the most active users started to come up with ideas and suggestions for improvements and additions to the system. The three most important suggestions were

(1) Many project assistants pointed out that needs and requirements varied from one project to another and they consequently wanted the system to be more flexible and easier to customize.
(2) Some of the project directors wanted the ability to restrict access to confidential or sensitive information. For instance, they would like the core group to have their own private space on the project's Web site.
(3) The directors of joint projects with external partners wanted to be able to open up their Web sites to allow external collaborators access to (at least some of) the material on the project's Web site.

5.3 Version 3

The development of version 3 started up shortly after the release of version 2. It was released in spring 2000 and was still in use when we finished our field study in the fall of 2001.

Development. The development of version 3 differed from the development of the previous versions in one important way. The IT department had taken over the initiative and managed to position itself as the leader and driving force. They considered ProjectWeb as a potential "cash cow" and believed that if they could turn it into a generic software product, marketable to other business units in BioCorp and even to other companies, then they would be able to make a substantial profit. As one of the project manager in the IT department explained,

> we thought about it. The [application] was probably something that many... could use. So, we thought that if we could make a small generic [software] package, then we could probably earn a lot of money. And that's what we want. So, we continued to make it as a small package, just like when you buy a small software package, just like when you buy [Microsoft] Word.

One consequence of this shift in direction and leadership was that PMU's influence was reduced and, more generally, that the communication between the users and the developers became less spontaneous and direct.

Furthermore, the development of version 3 entailed a shift to a new technical platform. Whereas versions 1 and 2 had been implemented by means of ASP (active server pages) technology, version 3 was constructed as an assembly of objects programmed in Visual Basic. The purpose was to obtain a more robust and reliable solution, which would scale up and accommodate large numbers of users without problems. This objective was obviously bound up with the wish to create a marketable, generic product.

Version 3 also included a number of new facilities.

(1) Facilities to change the look and feel of the user interface (e.g., the color scheme) and customize menus and toolbars
(2) Facilities to manage authorization and access control to confidential or sensitive material
(3) Facilities to create and manage project Web sites with external partners (e.g., facilities to handle security and privacy issues)

Judging from this, it is our impression that the wishes of Stella, Jean, and the other users from PMU were accommodated even though they felt they had been shunted out of the development process.

Implementation and use. In general, the project directors and assistants in PMU were excited about the many new features and began immediately to explore the new facilities in various ways. Some projects continued to treat ProjectWeb as a broadcast medium and focused on how to exploit the new facilities to create more interesting content, for instance by adding more pictures and better graphics.

Other projects were more interested in the new facilities to restrict access to sensitive information and began to explore how ProjectWeb could be used to create private work spaces and in that way support knowledge sharing and collaboration in smaller work groups. As an alternative to distributing work group documents and drafts as e-mail attachments, they started to store them in private workspaces on the project's Web site. In other words, they began to use Project Web to support small-group collaboration—something for which the application was not initially designed. The ability to restrict access to documents was essential to this new practice because work groups did not want to share private documents and unfinished drafts with outsiders.

As a final remark, we would like to add that, within BioCorp, version 3 of ProjectWeb is regarded as a highly successful Web application. More than 70 large projects with hundreds of participants each currently use the system and development of the next version (number 4) has already begun.

6 DISCUSSION

The case of ProjectWeb shows that important, innovative IS development activities may take place outside of formal ISD projects—as creative organizational actors learn about new technological opportunities, modify systems to match emergent organizational requirements, experiment with local innovations, respond to unanticipated problems, or adapt existent technologies to new uses. These activities (which may be labeled innovation "in the small") are less manifest and more difficult to grasp than formal development projects with steering committees, project managers, milestones and fixed budgets (innovation "in the large") but, we would argue, equally important.

The case also demonstrates that users may be an important source of innovation. Unlike most managers (and IT specialists), they have detailed, first-hand knowledge of how work is actually done and they personally experience the frustrations, troubles, and breakdowns caused by inexpedient work procedures, rigid rules, and inadequate or outdated technologies. As a consequence, they are often both motivated and able to come up with creative solutions to recurring problems and discover opportunities to improve their work practices by exploiting new technologies.

Three aspects of the ProjectWeb development process are thrown into relief when we contrast it with the way ISD is usually portrayed in IS research.

Sensemaking. It is striking how little the key designers of ProjectWeb (Stella and Jean) knew about what they were doing at the beginning of the development process. They did not have a clear goal or a precise idea about what they were designing or how it would fit into people's work practices. On the contrary, it was difficult for them to make sense of the Web technology and their first attempts to conceptualize the emerging system relied on comparisons with older, more familiar technologies (LAN drives and e-mail). It was not until they started experimenting with the design that they began to develop more detailed and sophisticated technological frames (Orlikowski and Gash 1994) and mental models of the technology. It was then they discovered what they wanted to do with it. Their understanding of the technology and their design goals evolved gradually and interactively as they developed new versions of the system in close collaboration with David and Hal (from the IT department) and tested it in practice together with their fellow users in PMU.

The point we want to make is that the development of ProjectWeb was fundamentally a *sensemaking process* (Weick 1995). Sensemaking is a process where people strive to convert a world of experience into an intelligible and meaningful world. It "is about sizing up a situation, about trying to discover what you have while you simultaneously act and have some effect on what you discover" (Weick 1999). It is an attempt to grasp a developing situation—in this case, the design and implementation of an innovative information system—in which the observer affects the trajectory of that development. Because new technologies are equivocal and thus lend themselves to multiple, conflicting interpretations, all of which are plausible, the development and use of technical systems require ongoing sensemaking (Weick 1990).

At the heart of sensemaking is the idea that understanding lies in the path of action. Action precedes understanding and focuses interpretation. It was by developing prototypes and trying out different versions of ProjectWeb in practice that Stella and Jean (and David and Hal) began to discover what their emergent system design meant and where they were heading. This is an example of "sense-making as manipulation" (Weick 1995). Sensemaking by means of manipulation involves acting in ways that create something (e.g., a new technology in use) that people can then comprehend and manage.

> Manipulation generates clearer outcomes in a puzzling world, and these outcomes make it easier to grasp what might be going on. Manipulation is an operationalization of the advice, "leap before you look" or the advice, "ready, fire, aim." Manipulation is about making things happen, so that a person can then pounce on those created things and try to explain them as a way to get a better sense of what is happening (Weick 1995 p. 168).

The key point is that sensemaking is an active process and that action is a precondition for sensemaking: "Action *is* intelligence, and until it is deployed, meaning and sense will be underdeveloped" (Weick 1993c).

Improvisation. The development of ProjectWeb was not guided by a preconceived plan or a systematic method. On the contrary, it was informed by hunches rather than well-developed knowledge, it relied on *ad hoc* solutions, and it had a strong core of experimentation and unjustified trial and error. In other words, the actors (users as well as developers) depended on improvisation and extemporaneous action in order to cope with unexpected problems, unanticipated opportunities, multiple meanings, and transient organizational requirements.

Improvisation deals with the unforeseen; it works without a prior plan and without blueprints and methods (Weick 1993b). "Improvisation is the deliberate and substantive fusion of the design and execution of a novel production" or performance (Miner et al. 2001, p. 314). It can be conceived of as a form of short-term learning where real-time experience informs novel action at the same time that the action is being taken. Much research on improvisation has focused on individuals, but improvisational action can occur at any level (individual, team, organization) and is often a collective process (Miner et al. 2001), as when Jean, Stella, David, and Hal together designed the second version of ProjectWeb.

The notion of improvisation implies that attention and interpretation rather than intention and decision making drives the process of designing. From this perspective, ISD is more an act of interpretation rather than an act of decision-making. The people involved improvise to make sense of unexpected possibilities and constraints that emerge. They are never in full control of the development process, but continuously challenged by having to address the unintended effects that are so commonplace in development projects. As a consequence, people are forced to revise their sense of what is happening and what can be accomplished. These revised interpretations are what guide action, not the initial decisions (Weick 1993b). Since the only things we can sense are enacted events that have already taken place, attention rather than intention becomes central to the design process.

Bricolage. The development of ProjectWeb was clearly a process that made do with whatever materials were at hand. Version 1, for instance, was a modified version of a program that Jean borrowed from the corporate library and important important elements of version 1 were again reused in version 2. Thus, ProjectWeb is a good example of the general phenomenon that "new systems are built, sometimes literally, on the ruins and with the ruins of old systems" (Lanzara 1999, p. 346). Pieces of past code become building materials and are used, together with available commercial software components (in this case a Web server and a DBMS from Microsoft), to construct new systems, which then become more or less coherent assemblies of mixed components.

In other words, the development of ProjectWeb can best be described as an instance of bricolage, i.e., a constructive activity based on transforming and reshaping what is already in use, or creatively rearranging components to fulfill new purposes (Lanzara 1999). The French word bricolage means "to use whatever resources and repertoire one has to perform whatever task one faces" (Weick 1993b). Invariably the resources are heterogeneous and less well-suited to the exact project than one would prefer but they are all there is. The materials are not project-specific, but, instead, they represent the contingent result of all of the previous uses to which those items have been put. The key to understanding the nature of bricolage as an innovative activity is Levi-Strauss' statement that materials "are not known as a result of their usefulness; they are deemed to be useful or interesting because they are first of all known" (Levi-Strauss 1966, cited in Weick 1993b).

Bricolage is closely associated with improvisation. Improvisation increases the chances that bricolage will occur because there is less time to obtain the necessary materials and resources in advance. Bricolage and improvisation are not synonymous, however, as bricolage (at least in theory) can occur in nonimprovisational contexts (Miner et al. 2001).

In summary, the development process we have described and analyzed here differs remarkably from the orderly, structured paths that most current IS theories and methods tend to assume. Instead, we have observed a more emergent, more spontaneous, more open-ended, and more continuous process involving bricolage, unjustified trial and error, small-scale practical experiments, local readjustments, and improvisations. The process has been shaped more by action than by plans, and more by attention than by intention.

As already mentioned in the introduction, we realize that this study is exploratory, that ProjectWeb belongs to a special class of information systems, and that this fact may limit the generality of our findings. Consequently, future research should engage in careful testing of our concepts and relationships in other ISD contexts.

7 CONCLUSION

The case of ProjectWeb provides an occasion to think more carefully about the way we conceptualize ISD. We believe it is about time to dismiss the tradition of viewing

ISD as an inherently rational, methodical, and orderly process. System development processes, in our experience, are never very tidy, neat, or sensible, and the importance and value of system development methods in practice are vastly overestimated (Bansler and Bødker 1993; Ciborra 1998; Fitzgerald 1998; Gasson 1999; Truex et al. 2000). System developers and users find themselves in a more complex, less stable, and less well-understood world than assumed by most IS researchers. They are placed in a world that does not always make sense and over which they often have only modest control.

Thus, there is a pressing need to develop an alternative theoretical perspective on ISD—a perspective that takes the messy reality of systems development practice seriously and makes it possible to grasp its non-methodical, unplanned, and fortuitous aspects. We believe that the concepts of sensemaking, improvisation, and bricolage proposed here constitute a useful starting point for developing such a perspective. They offer a theoretical lens for examining how people cope with ISD in practice and explaining why methods and plans have limited value in most real-life situations.

ACKNOWLEDGMENTS

We are grateful to the members of BioCorp who participated in this research. We also thank Hanne W. Nicolajsen, Dixi L. Henriksen, and Jens K. Pors for their assistance in the fieldwork. This study was supported in part by a grant from the Danish Research Councils (grant no. 99-00-092).

REFERENCES

Andersen, N. E.; Kensing, F.; Lassen, M.; Lundin, J.; Mathiassen, L.; Munk-Madsen, A.; and Sørgaard, P. *Professional System Development*, Englewood Cliffs, NJ: Prentice-Hall, 1990.

Bansler, J. P., and Bødker, K. "A Reappraisal of Structured Analysis: Design in an Organizational Context," *ACM Transactions on Information Systems* (11:2), 1993, pp. 165-193.

Barrett, F. J. "Coda: Creativity and Improvisation in Organizations: Implications for Organizational Learning," *Organization Science* (9:5), 1998, pp. 558-560.

Ciborra, C. U. "Crisis and Foundations: An Inquiry into the Nature and Limits of Models and Methods in the Information Systems Discipline," *Journal of Strategic Information Systems* (7), 1998, pp. 5-16.

Ciborra, C. U.; Braa, K.; Cordella, A.; Dahlbom, B.; Failla, A.; Hanseth, O.; Hepsø, V.; Ljungberg, J.; Monteiro, E.; and Simon, K. A. *From Control to Drift.*, Oxford: Oxford University Press, 2000.

Cockburn, A. *Agile Software Development*, Boston: Addison-Wesley, 2002.

Cunha, M. P.; Cunha, J. V.; and Kamoche, K. "Organizational Improvisation: What, When, How and Why," *International Journal of Management Reviews* (1:3), 1999, pp. 299-341.

Eisenhardt, K. M., and Tabrizi, B. N. "Accelerating Adaptative Processes: Product Innovation in the Global Computer Industry," *Administrative Science Quarterly* (40), 1995, pp. 84-110.

Fitzgerald, B. "An Empirical Investigation into the Adoption of Systems Development Methodologies," *Information & Management* (34), 1998, pp. 317-328.

Gasson, S. "A Social Action Model of Situated Information Systems Design," *The DATA BASE for Advances in Information Systems* (30:2), 1999, pp. 82-97.

Hatch, M. J. "Exploring the Empty Spaces of Organizing: How Improvisational Jazz Helps Redescribe Organizational Structure," *Organization Studies* (20:1), 1999, pp. 75-100.

Hatch, M. J. "Jazzing Up the Theory of Organizational Improvisation," *Advances in Strategic Management* (14), 1997, pp. 181-191.

Hatch, M. J. "The Vancouver Academy of Management Jazz Symposium: Jazz as a Metaphor for Organizing in the 21st Century," *Organization Science* (9:5), 1998, pp. 556-568.

Highsmith, J. *Agile Software Development Ecosystems*, Boston: Addison-Wesley, 2002.

Hutchins, E. "Organizing Work by Adaptation," *Organization Science* (2:1), 1991, pp. 14-39.

Klein, H. K., and Myers, M. D. "A Set of Principles for Conducting and Evaluating Interpretive Field Studies in Information Systems," *MIS Quarterly* (23:1), 1999, pp. 67-94.

Lanzara, G. F. "Between Transient Constructs and Persistent Structures: Designing Systems in Action," *Journal of Strategic Information Systems* (8), 1999, pp. 331-349.

Lanzara, G. F. "Ephemeral Organizations in Extreme Environments: Emergence, Strategy, Extinction," *Journal of Management Studies* (20), 1983, pp. 71-95.

Louridas, P. "Design as Bricolage: Anthropology Meets Design Thinking," *Design Studies* (20:6), 1999, pp. 517-535.

Mathiassen, L. "Reflective Systems Development," *Scandinavian Journal of Information Systems* (10:1/2), 1998, pp. 67-118.

Miner, A. S.; Bassoff, P.; and Moorman, C. "Organizational Improvisation and Learning: A Field Study," *Administrative Science Quarterly* (46), 2001, pp. 304-337.

Mirvis, P. H. "Variations on a Theme: Practice Improvisation," *Organization Science* (9:5), 1998, pp. 586-592.

Moorman, C., and Miner, A. "The Convergence Between Planning and Execution: Improvisation in New Product Development," *Journal of Marketing* (62), 1998a, pp. 1-20.

Moorman, C., and Miner, A. "Organizational Improvisation and Organizational Memory," *Academy of Management Review* (23:4), 1998b, pp. 698-723.

Myers, M. D. "Qualitative Research in Information Systems," *MISQ Discovery*, 1997.

Orlikowski, W.J. "Improvising Organizational Transformation over Time: A Situated Change Perspective," *Information Systems Research* (7:1), 1996, pp. 63-92.

Orlikowski, W. J., and Gash, D. C. "Technological Frames: Making Sense of Information Technology in Organizations," *ACM Transactions on Information Systems* (12:2), 1994, pp. 174-207.

Peplowski, K. "The Process of Improvisation," *Organization Science* (9:5), 1998, pp. 560-561.

Perry, L. T. "Real Time Strategy: Improvising Team Based Planning for a Fast Changing World," *Organizational Dynamics* (22), 1994, pp. 76-77.

Perry, L.T. "Strategic Improvising: How to Formulate and Implement Competitive Sstrategies in Concert," *Organizational Dynamics* (19:4), 1991, pp. 51-64.

Stake, R. E. "Case Studies," in N. K. Denzin and Y. S. Lincoln (Eds.), *Handbook of Qualitative Research* (2nd ed.), London: Sage Publications, 2000.

Truex, D. P.; Baskerville, R.; and Klein, H. "Growing Systems in Emergent Organizations," *Communications of the ACM* (42:8), 1999, pp. 117-123.

Truex, D.; Baskerville, R.; and Travis, J. "Amethodical Systems Development: The Deferred Meaning of Systems Development Methods," *Accounting, Management & Information Technology* (10), 2000, pp. 53-79.

Walsham, G. "Interpretive Case Sstudies in IS Research: Nature and Method," *European Journal of Information Systems* (4), 1993, pp. 74-81.

Weick, K. "The Collapse of Sensemaking in Organizations: The Man Gulch Disaster," *Administrative Science Quarterly* (38), 1993a, pp. 628-652.

Weick, K. "Improvisation as a Mindset for Organizational Analysis," *Organization Science* (9:5), 1998, pp. 543-555.

Weick, K. "Organizational Redesign as Improvisation," in G. P. Huber and H. W. Glick (Eds.), *Organizational Change and Redesign*, Oxford: Oxford University Press, 1993b, pp. 346-379.

Weick, K. "Sensemaking as an Organizational Dimension of Gobal Change," in J. Dutton and D. Cooperrider (Eds.), *The Human Dimensions of Global Change*, Thousand Oaks, CA: Sage Publications, 1999.

Weick, K. *Sensemaking in Organizations*, Thousand Oaks, CA: Sage Publications, 1995.

Weick, K. "Sensemaking in Organizations: Small Structures with Large Consequences," in J. K. Murnigham (Ed.), *Social Psychology in Organizations: Advances in Theory and Research*, Englewood Cliffs, NJ: Prentice-Hall, 1993c.

Weick, K. "Technology as Equivoque: Sensemaking in New Technologies," in P. S. Goodman and L. Sproull (Eds.), *Technology and Organizations*, San Francisco: Jossey Bass, 1990.

ABOUT THE AUTHORS

Jørgen P. Bansler is an associate professor at the Technical University of Denmark. His research interests include intranet development, computer-mediated communication, and new organizational forms. Jørgen has a Ph..D. in computer science from the University of Copenhagen, Denmark. He has published in such journals as *ACM Transactions on Information Systems*, *Industrial Relations*, *Information Technology & People*, *Knowledge and Process Management*, and *Journal of the Association of Information Systems*. Jørgen can be reached at bansler@cti.dtu.dk.

Erling C. Havn is an associate professor at the Technical University of Denmark. His research focuses on the ongoing relationship between the development of information technologies and organizing structures, work practices, communication, and control mechanisms. Erling holds a Ph.D. in social sciences from the University of Copenhagen, Denmark. He has published in such journals as *Computer Integrated Manufacturing Systems*, *AI & Society*, *International Journal in Human Factors in Manufacturing*, *Journal of the Association of Information Systems*, and *Knowledge and Process Management*. Erling can be reached at havn@cti.dtu.dk.

Part 7:

Panels and Position Papers

35 TWENTY YEARS OF APPLYING GROUNDED THEORY IN INFORMATION SYSTEMS: A Coding Method, Useful Theory Generation Method, or an Orthodox Positivist Method of Data Analysis?

Tony Bryant
Leeds Metropolitan University

Jim Hughes
Salford University

Michael D. Myers
University of Auckland

Eileen Trauth
The Pennsylvania State University

Cathy Urquhart
University of Auckland

Grounded theory has been gaining ground in Information Systems as a research method in recent years. Grounded theory has been increasingly used and discussed in IS literature spanning the past decade.

This development mirrors the establishment and wide adoption of qualitative research in information systems which has led to a diversity of approaches in qualitative analysis, among them grounded theory.

The panel examines different applications of grounded theory in information systems, and addresses a number of issues that stem from the use of grounded theory in information systems research.

- How many grounded theory studies in information systems are pure grounded theory studies that generate a whole theory about a phenomenon? Or is grounded theory used in our field as a blanket term for coding data? What is the difference, and does it matter?
- What is the epistemological position implied by using grounded theory? Does the use of grounded theory imply a positivist stance?
- Does it matter that information systems researchers often use grounded theory coding techniques without understanding the 34-year history and genesis of the methodology?
- What are the alternatives to using grounded theory in information systems? Are there other, less complicated and equally effective approaches to coding and analyzing data?

Eileen Trauth, as chair of the session, will introduce the panel and the main issues. This will be followed by a 10 minute presentation by each panelist. Each panelist will raise particular issues engendered by their own experience and use of grounded theory. After the presentations, the chair will summarize the key points of the debate. Then the audience will have an opportunity to discuss these points and others, and engage in a debate about grounded theory that we hope will both be interesting and wide ranging.

Jim Hughes will discuss the intuitive appeal of grounded theory procedures to novice or beginning researchers in qualitative research, and how IS methodologies may usefully be applied to grounded theory.

Eileen Trauth will discuss epistemological implications of different uses of grounded theory, and the epistemological challenges and benefits of coding for pure grounded theory.

Tony Bryant will discuss the need to rescue grounded theory method from some of its key protagonists. His recent work, including an exchange of ideas with Cathy Urquhart, has sought to demonstrate the philosophical inadequacies of many justifications for using the method, based on Charmaz (2000).

Michael Myers will discuss the issue of qualitative data analysis more broadly, by offering some alternatives to grounded theory analysis.

36 BUILDING CAPACITY FOR E-GOVERNMENT: Contradictions and Synergies in the Dialectics of Action Research

David Wastell
University of Manchester

Peter Kawalek
University of Manchester

Mike Newman
University of Manchester

Mike Willetts
Salford City Council

Peter Langmead-Jones
Lancashire Constabulary

Action research has been widely espoused within IS as a methodology for achieving relevant research, simultaneously addressing problems pertinent to practice as well as generating valuable IS theory. Debate, however, continues to revolve around the standing of action research. The need to address an applied problem as well as the imperative to deliver substantive research findings builds a degree of conflict into the process of action research (McKay and Marshall 2001) which has led some commentators to doubt whether action research is viable. In contrast, we believe that action research is not only feasible but an essential tool for developing and evaluating social theory. However, the need to serve the two masters of practice and research, at the heart of the action research dialectic, inevitably constrains the research process. The exigencies of practical problem solving and the need to deliver solutions limit the time and resources available for rigorous data collection and validation, and constrain the research agenda. Nonetheless the theory generated by action research reflects the dynamics and complexity of the real world milieu in which it was developed. Rather than a weakness, we see this as source of strength. The theory is richer, more relevant and holistic than its conventional counterpart. Moreover, it has been evaluated in the crucible of real-world intervention.

In this panel we explore some of the inherent conflicts of action research, seeking to show how the interaction of the practice and research can act as fertile source of new academic theory and a test-bed for established concepts. Electronic government will be the main application domain addressed. This is an important area for collaborative research between academic institutions and the practitioner community. E-Government imperatives present public sector agencies with formidable challenges, and joint work with research organizations can help build the new capacities that will be required to successfully embrace the modernization agenda. Two projects will be highlighted. First, the CRM Academy. Here the aim is to establish a research center providing tools, methods and mentoring to support the implementation of CRM within the local government sector. Manchester Business School is the lead research organization; Salford City Council the main practitioner partner. The work is based on earlier action research project which developed a BPR toolkit for implementing e-Government (Wastell et al. 2001). The second project (MADE) was aimed at the design and implementation of a multiagency decision support environment to support partnership work in the area of community safety. The project involved collaboration with Lancashire Constabulary and Lancashire County Council. The facility was successfully implemented in 2001 (Wastell et al. 2004) and has been used by all Districts across Lancashire in their crime auditing and performance monitoring activities.

REFERENCES

McKay, J., and Marshall, P. "The Dual Imperatives of Action Research," *Information Technology and People* (14), 2001, pp. 46-59.

Wastell, D. G.; Kawalek, P.; Langmead-Jones, P.; and Ormerod, R. "Information Systems and Partnership in Multi-agency Networks: an Action Research Project in Crime Reduction," *Information and Organisation*, 2004 (in press).

Wastell, D. G.; Kawalek, P.; and Willetts, M. "Designing Alignment and Improvising Change: Experiences in the Public Sector Using the SPRINT Methodology," in S. Smithson, J. Gricar, M. Podlogar, and S. Avgerinou (Eds.) *Proceedings of the 9th European Conference on Information Systems*, Bled, Slovenia, 2001.

37 NEW INSIGHTS INTO STUDYING AGENCY AND INFORMATION TECHNOLOGY

Tony Salvador
Intel Corporation

Jeremy Rose
University of Aalborg

Edgar A. Whitley
London School of Economics and Political Science

Melanie Wilson
Manchester School of Management

The question of how human action and information technology interact has been a recurrent theme in much information systems research, especially within the IFIP 8.2 community.

One of the key elements to understanding agency (Barnes 2000) is to question where the agency lies; a question that has both ontological and epistemological implications. Is the issue of agency one of human capabilities and human institutions? Is it a feature of technology or is it intimately implicated in the relationship between the two? In studying information systems, researchers have used many theoretical models to address this issue, with some of the most popular approaches being the application of critical theory (see, for example, Bloomfield and McLean 1996), structuration theory, and actor-network theory (see, for example, Latour 1996; Rose and Truex 2000).

This panel will present three proposals for how to study the question of agency and information technology and the audience will be asked to respond to each of the proposals, both in terms of their likely contributions to our understanding of information systems and their practical applicability.

In order to ground the various proposals, the panel will begin with a short presentation by Tony Salvador (panel chair) who will describe some recent research on retail environments. One of the findings of this research appeared to suggest that there was limited agency in these environments and that information and communications technologies played a role in reducing human agency.

He will then ask each of the other panelists to outline their approach to tackling the question of human agency and information technology, asking them to relate their view of how technology and agency interrelate to the case.

Melanie Wilson will draw on her previous research to argue that agency is a human attribute that is constrained and enabled by broader social factors. To that extent, information technology is only indirectly implicated in questions of agency through its role in supporting the dominant economic rationality of the day. Edgar Whitley, in contrast, will argue that questions of agency are a reflection of what we understand to be the capabilities of technology rather than humans. Questions of agency, then, depend on what the technology can't do, rather than what humans can do. Finally, Jeremy Rose will present the view that it is neither humans nor information technology in isolation that provide agency, but rather it is the imbroglios of the two that are "agentic" and that any attempts to isolate the two are artificial and hence likely to be problematic.

During the presentation audience members are encouraged to indicate their support or otherwise for each proposal. They are also encouraged to question the proposers about the likely contribution their proposal will make and the practical application of the approach proposed.

REFERENCES

Barnes, B. *Understanding Agency: Social Theory and Responsible Action*, London: Sage Publications, 2000.

Bloomfield, B. P., and McLean, C. "Madness and Organization: Informed Management and Empowerment," in W. J. Orlikowski, G. Walsham, M. R. Jones, and J. I. DeGross (Eds.), *Information Technology and Changes in Organizational Work*, London: Chapman & Hall, 1996, pp. 371-396.

Latour, B. "Social Theory and the Study of Computerized Work Sites," in W. J. Orlikowski, G. Walsham, M. R. Jones, and J. I. DeGross (Eds.), *Information Technology and Changes in Organizational Work*, London: Chapman & Hall, 1996, pp. 295-307.

Rose, J., and Truex III, D. "Machine Agency as Perceived Autonomy: An Action Perspective," in R. Baskerville, J. Stage, and J. I. DeGross (Eds.), *Organizational and Social Perspectives on Information Technology*, Boston: Kluwer Academic Publishers, 2000, pp. 371-390.

38 RESEARCHING AND DEVELOPING WORK ACTIVITIES IN INFORMATION SYSTEMS: Experiences and the Way Forward

Mikko Korpela
University of Kuopio

Jonathan P. Allen
University of San Francisco

Olav Bertelsen
Aarhus University

Yvonne Dittrich
IT-University Copenhagen and Blekinge Institute of Technology

Kari Kuutti
University of Oulu

Kristina Lauche
University of Aberdeen

Anja Mursu
University of Kuopio

The objective of this panel is to present and discuss different approaches to understanding and developing *work practice* that is facilitated by information systems. Such approaches are identified in research and practice in human-computer interaction, computer-supported cooperative work, human factors/ergonomics, developmental work research, and information systems, among others.

However, some of the approaches have remained purely academic while the more pragmatic approaches often lack a theoretical basis. The goal of this panel is to debate the potential and obstacles of various work-oriented approaches as practicable research methodologies as well as practicable requirements analysis and design methodologies in IS.

In trying to understand the role of theory in critical IS research, the concept of design artefacts has been valuable because it conveys a focus on how theories, methods, tools, etc., mediate design practice (Bertelsen 1998). A method in use is not the same as the method book; the fact that designers don't follow methods does not mean that the introduction of a method has no effect. A theory is a design artefact; misinterpreting a theory can be a step in achieving a good result. If designers don't have their own theory, theories imbedded in their tools may take over the designers' minds.

Activity theory (Engeström 1999), among other theories, has been a promising framework in IS for more than a decade; it is time to consider the practical packaging of the theory for design. Would stories, anecdotes, fairy tales, and some good examples be better ways to package the insights from activity theory informed workplace studies?

In German speaking countries, a different tradition of work-oriented research and practice has inspired systems developers (Frese and Zapf 1993). Is the Anglophone research community ready to learn from it?

Can pragmatic, design-oriented approaches like work system method (Alter 2001) and participatory design (Kensing and Blomberg 1998) be married with theoretically based approaches like activity theory? Can work-oriented IS approaches and work-oriented software engineering approaches (Robertson and Robertson 1999) be linked together?

REFERENCES

Alter, S. "Which Life Cycle: Work System, Information System, or Software?," *Communications of the AIS* (7:17), 2001, pp. 1-53.

Bertelsen, O. W. *Elements to a Theory of Design Artefacts: A Contribution to Critical Systems Development Research*, Unpublished Ph.D. Thesis, Aarhus University, 1998.

Engeström, Y. "Activity Theory and Individual and Social Transformation," in Y. Engström, R. Miettinen, and R. Punamäki (Eds.), *Perspectives on Activity Theory*, Cambridge, UK: Cambridge University Press, 1999, pp. 19-38.

Frese, M., and Zapf, D. "Action as the Core of Work Psychology: A German Approach," in H. C. Triandis, M. D. Dunnette, and L. M. Hough (Eds.), *Handbook of Industrial and Organizational Psychology* (2nd ed., Vol. 4), Palo Alto, CA: Consulting Psychologists Press, 1993, pp. 271-340.

Kensing, F., and Blomberg, J. "Participatory Design: Issues and Concerns," *Computer Supported Cooperative Work* (7:3-4), 1998, pp. 167–185.

Robertson, S., and Robertson, R. *Mastering the Requirements Process*, Reading, MA: Addison-Wesley, 1999.

39 CROSSING DISCIPLINARY BOUNDARIES: Reflections on Information Systems Research in Health Care and the State of Information Systems

Nicholas Barber
School of Pharmacy
University of London

Patricia Flatley Brennan
University of Wisconsin

Mike Chiasson
University of Calgary

Tony Cornford
London School of Economics and Political Science

Elizabeth Davidson
University of Hawaii, Manoa

Bonnie Kaplan
Yale University

Ela Klecuń
London School of Economics and Political Science

The Information Systems field has a long-standing research interest in the economic, social, and organizational implications of information technology— issues considered critically important in health care. Both IS and medical informatics are based in professional schools and draw heavily from studies of practice. To some researchers, it seems natural that the two fields would work closely together, drawing on and contributing to each other. Some think that IS research might benefit from testing, extending, or developing new theory in such a distinct organizational and institutional context. Others boldly advocate IS as a reference discipline for studying information systems in health care and call for stronger cross-referencing of IS research in the medical literature. If the IS field is ready to diffuse its knowledge into health care, what would be the terms and priority, given its usual business focus and business school base?

While IS research might be expected to flourish in health care, influencing policy and shaping practice, the reality is more sanguine. Each panel participant has experience crossing disciplinary boundaries to research information systems in health care. By examining similarities and differences between IS and medical informatics, we will offer some insight into the current state of IS as we assess methodological, institutional, and practical issues we face in our research. Comparing the two disciplines provides a provocative focus for surfacing assumptions about how IS research should be done and disseminated, for judging how well the objectives of the 1984 IFIP 8.2 conference call have been met, and for reflecting on the status of the so-called methods wars.

We aim to stimulate discussion by contrasting viewpoints on the following issues.

- How well has IS met the goals of the 1984 IFIP 8.2 conference, particularly in the field of health care?
- Can IS serve as a reference discipline for medical informatics? Should it?
- What can methodological differences in IS and medical informatics tell us about the IS methods wars?
- Medical researchers pay careful attention to ethical issues in constructing and undertaking research. What ethical considerations might be incorporated into IS research based on experience in clinical settings?
- What is generalizable or transferrable knowledge in IS? What might it look like? Have we achieved it? Should we try harder?
- Is it possible to provide theoretical bridges between some generalized IS theory and context-specific knowledge in health care so as to build and act upon multidisciplinary knowledge?
- What recommendations for cross-disciplinary work in health care (and other domains) can we provide?
- Do publication patterns and practices in each discipline reflect or drive significant or subtle differences in approaches to research and publication?

40 THE GREAT QUANTITATIVE/ QUALITATIVE DEBATE: The Past, Present, and Future of Positivism and Post-Positivism in Information Systems

Michael D. Myers
University of Auckland

Detmar Straub
Georgia State University

John Mingers
University of Kent

Geoff Walsham
University of Cambridge

Over the past 20 years there has been significant progress in quantitative and qualitative research in information systems (Lee et al. 1997; Myers 1997; Myers and Avison 2002). There has also been much progress in our understanding of the contributions that can be made by both positivist and post-positivist (i.e., interpretive and critical) perspectives.

The purpose of this panel is to review the progress that has been made and to suggest the way forward for the future. The members of the panel will discuss the proposition that there has been progress in IS (measured by the effectiveness, explanatory power, and professionalism) in quantitative/qualitative and positivist/post-positivist research methods. As part of this discussion they will debate whether some of the progress has come from a greater openness to pluralistic research methods with the field of IS. They will also debate how and to what extent such pluralism is possible.

Michael Myers will begin by outlining the motivation for the debate. He will then introduce the panel members.

Detmar Straub will review the progress that has been made in quantitative/positivist research methods in IS. He will argue that what might be viewed as progress (i.e., toward more rigor in measurement and broader acceptance and practice of new methodological tools) might not be viewed as progress by non-positivists. Nevertheless, given that a case can be made for the philosophical position of positivism (which he will **not** undertake), he will next highlight areas where there has been progress and areas where there has been little to no progress. Slow progress has been made in measurement validation, for example, but experimentation has declined as a preferred approach, which raises issues about proving causation.

Michael Myers will review the progress that has been made in qualitative/ post-positivist research methods. He will argue that, while there has been a general acceptance of qualitative and more specifically interpretive research within IS, critical IS research has remained somewhat marginal.

John Mingers will argue that the best way forward for the field is for both qualitative and quantitative researchers to collaborate and for different research methods to be combined in the one study (Mingers 2001). He will suggest existing paradigms such as positivism and interpretivism are inherently limited in their different ways so that research conducted wholly from within a single paradigm must itself be limited and one-sided. Research will be richer and more valid if it combines insights and methods from different paradigms. This can be under-written by a critical realist philosophy (Mingers, 2003).

Geoff Walsham will argue against collaboration and combining of research methodologies, broadly defined as sets of philosophical underpinnings of research activities. He will suggest that it is possible to collaborate and combine methods when there is a common philosophical base, but not when these philosophical bases are different.

Both Michael Myers and Detmar Straub will briefly comment on the various presentations, highlighting the main areas of debate.

REFERENCES

Lee, A. S.; Liebenau, J.; and DeGross, J. I. (Eds.). *Information Systems and Qualitative Research*, London: Chapman and Hall, 1997.

Mingers, J. "Combining IS Research Methods: Towards a Pluralist Methodology," *Information Systems Research* (12:3), 2001, pp. 240-259.

Mingers, J. "Real-izing Information Systems: Critical Realism as an Underpinning Philosophy for Information Systems," in L. Applegate, R. Galliers, and J. I. DeGross (Eds.), *Proceedings of the 23rd International Conference for Information Systems*, Barcelona, 2003, pp. 295-303.

Myers, M. D. "Qualitative Research in Information Systems," *MIS Quarterly* (21:2), 1997, pp. 241-242.

Myers, M. D., and Avison, D. E. (Eds.). *Qualitative Research in Information Systems: A Reader*, London: Sage Publications, 2002.

41 CHALLENGES FOR PARTICIPATORY ACTION RESEARCH IN INDUSTRY-FUNDED INFORMATION SYSTEMS PROJECTS

Karin Breu
Cranfield School of Management

Christopher J. Hemingway
Cranfield School of Management

Joe Peppard
Loughborough University

Abstract The purpose of this position paper is to open a discussion about the practicability of participatory action research (PAR) within industry-funded information systems (IS) research. We reflect on a project in which the undue exercise of power by the practitioners on the research team compromised the methodological rigor of the inquiry. Theories of power are used to articulate our reflections and develop suggestions for mitigating power imbalances on PAR research teams, although we conclude that PAR cannot be followed faithfully to its principles in industry-funded engagements.

Keywords: Industry-funded research, participatory action research, power

1 INTRODUCTION

Knowledge is power and the ability to create knowledge is power (Reason 1993). As a knowledge-creating process, social scientific research inevitably involves power relations. Participatory action research (PAR) was conceived as a way of overcoming power imbalances in social scientific research (Freire 1970). Industry-funded PAR involves collaboration with powerful professionals typically occupying executive positions in large corporations. Information systems research often takes place with such stakeholders, yet existing accounts of PAR provide neither analyses of power dynamics on researcher-practitioner teams nor guidance on how to address power imbalances.

This position paper reflects upon power dynamics in a PAR team within an industry-funded research project. The paper draws on power theory to elicit sources of power and their uses by members of the PAR team. The paper, a reflection by the researchers, develops suggestions for attaining power balance within PAR teams, although we conclude that PAR cannot be followed faithfully to its principles in industry-funded engagements.

2 PARTICIPATORY ACTION RESEARCH

The participatory paradigm upholds the epistemological stance of social reality being cocreated by all humans, rather than deterministically predefined (Skolimowski 1994). The notion of a cocreated reality implicates participatory methodologies as the only legitimate means for producing knowledge or, in other words, defining reality (Fals-Borda and Rahman 1991). Originating from research on oppressed peoples and disadvantaged minorities in developing countries, PAR recognizes knowledge as an instrument of power, domination, and control (Freire 1970). A participatory worldview posits a research practice that insists on the full participation of researchers and practitioners in the dual roles of coresearchers and copractitioners in the inquiry process (Heron and Reason 1997). The fundamental difference between PAR and action research emerges from the level of involvement of the practitioner in the research process. Action research is a strategy for doing research *on* people, whereas PAR is committed to research *with* people by inviting practitioners to participate in the analysis of their own reality (Heron 1981). PAR is also distinct from participatory research, where "members of the organization studied become active participants in the research process—but where that process itself is not linked directly to action" (Whyte 1989, p. 506). PAR, in contrast, builds action objectives into the research design from the outset (Park 1999).

3 CASE STUDY: IMPLEMENTING COMMUNITIES OF PRACTICE

This section recounts the PAR stage of a longitudinal study of communities of practice (COPs), carried out in two phases over a period of 32 months (see Figure 1). The researchers became involved with the organization through a consortium-funded project, comprising case studies of COPs in five large organizations (phase 1). Phase 2 continued the project with one sponsor, ServiceCo (a pseudonym), a public services organization. The aim of the engagement was to foster COPs as a strategy for improving knowledge sharing among its 2,500 staff who mostly work at client sites.

The project sponsor (ProSpo) worked at head office and reported directly to the board member responsible for knowledge management (KMDir). ProSpo invited five peers onto the research team. The academic researchers reported to ProSpo and the COP Steering Group. The research team agreed on PAR as the method for collaboration

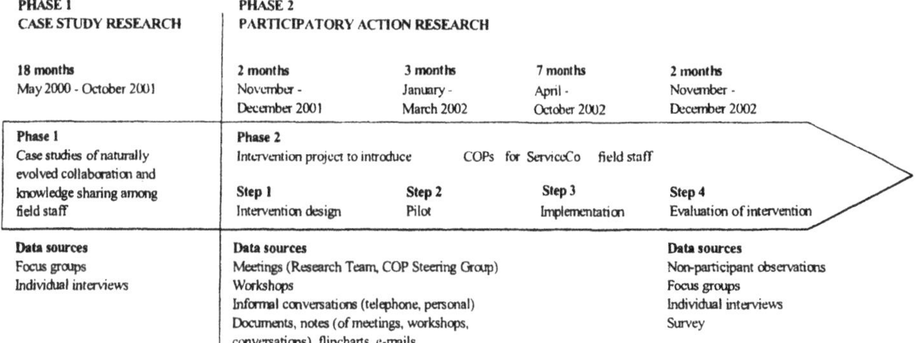

Figure 1. Research Process in ServiceCo

and jointly developed the intervention plan. To ensure that the practitioners understood the plan, the researchers coauthored the intervention plan with ProSpo and obtained its approval by KMDir.

The research team tested and subsequently revised the implementation plan through a pilot with three trial COPs and proceeded to implementation with another six COPs. Aiming to create COPs as espoused in the literature, which recommends fostering but not interfering with community development, the research team withdrew from the field after facilitating one-day start-up workshops with each COP. Given that COPs would meet bimonthly, we returned to the field after seven months, allowing each COP to meet three times. The researchers then evaluated the impact of the intervention through nonparticipant observations, focus groups, individual interviews, and a structured survey. The results revealed fundamental digressions from the intervention plan.

4 REFLECTIONS ON THE ROLE OF POWER IN THE PAR TEAM

PAR defines new roles for the actors in the research process. Researchers on PAR teams are empowered to act as change-effecting practitioners and practitioners as knowledge creators. Both parties must exchange some professional powers, an arrangement that potentially intensifies the team's power dynamics.

Power theories regard power either as an object that people possess, lose, share, fight for and win, or as force relations that people exercise in strategies and tactics. The *power as an object* view is concerned with sources and outcomes of power (Lukes 1986), whereas the *power as force relations* view draws attention to the exercise of power (Foucault 1979).

4.1 Sources of Power

The most evident sources of power in our PAR engagement concerned *knowledge* and *credibility*, on the part of the researchers, and provision of *funding* and *access* to the field on the part of the practitioners (see Figure 2).

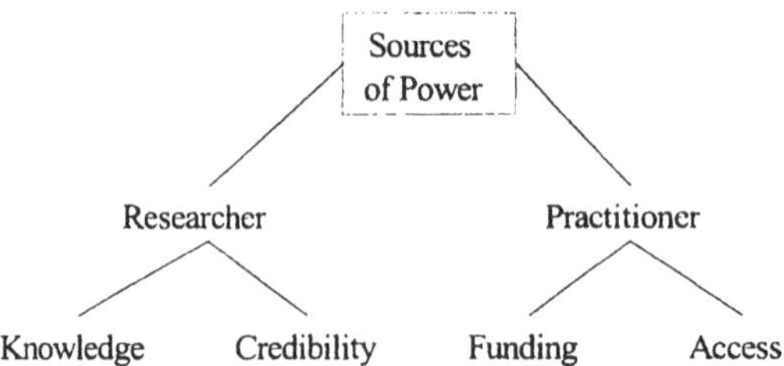

Figure 2. Sources of Power on the PAR Team

The practitioners explained that they preferred academic partners to commercial researchers and consultants. Given our commitment to theoretical and methodological rigor, they perceived our academic knowledge as generally superior and particularly relevant to the problems they were seeking to address. Our involvement in the engagement, furthermore, was seen as an endorsement of the intervention by an impartial and credible third party. We, in contrast, have obligations to raise funding for our research activities and gain access to organizations to carry out empirical research. The desire to obtain each others' resources creates dependencies in the research engagement that furnished those who hold the resources with an opportunity for exercising power.

4.2 Uses of Power

ProSpo contributed to the demise of the academic goals of the project by using his power to control access to the COPs as a means for digressing from the agreed intervention plan in several respects. (1) Rather than publicizing the COP initiative organization-wide, ProSpo discarded the jointly developed communications plan and invited COP participation selectively. (2) Rather than allowing COPs to select their leaders themselves, ProSpo recruited COP leaders from his personal network. (3) Rather than creating COPs for field staff, ProSpo allowed four out of nine COPs to convene senior decision makers who then exercised a closed access policy for peers of equal seniority. (4) Although requesting a list of the entire COP population, by the end of the evaluation it became apparent that ProSpo had provided us with a contrived sample. In this way, he silenced critical voices and neutralized negative evidence, creating a positive impression of the intervention outcomes for ServiceCo's decision makers and budget holders. Once the practical interests of the research were satisfied, ProSpo ended the engagement, although the theoretical aims of the inquiry had not yet been fulfilled. The consequence for us was that the requirements of publishable research, particularly data quality, theoretical saturation, and methodological rigor, were not met.

The condition of funding constituted a major dependency for us. The relationship was defined, from a financial point of view, as an exchange of money for research results because we used a contract format that is typical for commissioned research. It specifies no obligation for practitioners to coproduce knowledge and honor the agreed research methodology. As a result, ProSpo could insist on obtaining the research results that he had specified, whereas we had no power to enforce his compliance with the agreed methodology in the absence of such contractual obligations.

We, as the academics, share responsibility for the decline of PAR. First, we were too trusting of ProSpo to follow the plans as agreed. We assumed that our trust was well placed, as we had a history of successful work with ProSpo albeit using a different research strategy (phase 1), and also because the jointly developed intervention plan was signed-off by KMDir. Second, we were unaware of the implications of the chosen contractual terms for our ability to influence events. Once we had seen the full extent of ProSpo's deviation from the agreed plans, we had no basis for withdrawing from the engagement.

5 CONCLUSION

The application of PAR in industry-funded IS research, our experience would suggest, challenges fundamental paradigmatic assumptions, both explicit and implicit, of the participatory ideology. An explicit assumption of PAR is that knowledge is power and the ability to create knowledge is power (Freire 1970; Reason 1993). We experienced that knowledge was a weak source of power for us, as ProSpo's senior managerial authority allowed him to override intervention plans cocreated by the research team and alter previously agreed courses of action.

PAR implicitly assumes that credibility of the researchers does not influence the power dynamic within a research team and makes no mention of money as a potential source of power. In our case, credibility proved to be a weak source of power for us. Credibility helped us win the contract, and was used by ProSpo to create favorable perceptions with the stakeholders affected by the implementation, but did not help us in redressing the balance on the PAR team and rectifying the course of the implementation.

Lack of financial independence of academic institutions is a general condition that weakens academic researchers' power base. The dependency of academic institutions on funding from external sources brought about the use of contracts for research that typically reflects a transactional model of research, whereby money is exchanged for research results solely produced by the researchers. This, however, is inappropriate for participatory research, which requires the full contribution of the practitioners to the inquiry process in their capacity as coresearchers.

To address the challenges of using PAR in industry-funded research, we make the following suggestions. From a collaboration point of view, research teams should negotiate a broader collaboration base within the sponsoring organization. This would limit opportunities for individual practitioners unduly influencing the course of the engagement. From a contractual point of view, we suggest to devise contracts that reflect the principles of PAR by specifying the duty for the practitioners to fulfil the principles of *authentic collaboration* by researching, designing, and actioning all aspects of an intervention jointly (Reason 1999). While we recognize that contractual terms cannot engender genuine collaboration, they, at least, legitimize and thereby empower the researcher to withdraw from a failing PAR engagement.

Although we made suggestions for addressing the challenges of PAR, we do not believe that the PAR ideology can be followed truthfully to its principles in industry-funded IS research. The practitioner members on a research team possess authority in their organizations which exceeds, in our view, the power that academics can draw from their knowledge. This condition inevitably creates a power imbalance in a PAR team that undermines the democratic ideal of PAR, and we do not see how that can be overcome.

REFERENCES

Fals-Borda, O., and Rahman, M. A. *Action and Knowledge: Breaking the Monopoly with Participatory Action-Research*, New Delhi: Indian Social Institute, 1991.

Freire, P. *Pedagogy of the Oppressed*, New York: Herder and Herder, 1970.

Foucault, M. *Discipline and Punish*, New York: Vintage Books, 1979.

Heron, J. "Philosophical Basis for a New Paradigm," in P. Reason and J. Rowan (Eds.), *Human Inquiry: A Sourcebook of New Paradigm Research*, Chichester, England: John Wiley, 1981, pp. 19-35.

Heron, J., and Reason, P. "A Participatory Inquiry Paradigm," *Qualitative Inquiry* (3:3), 1997, pp. 274-294.

Lukes, S. *Power*, Oxford: Basil Blackwell, 1986.

Park, P. "People, Knowledge, and Change in Participatory Research," *Management Learning* (30:2), 1999, pp. 141-157.

Reason, P. "Integrating Action and Reflection through Co-operative Inquiry," *Management Learning* (30:2), 1999, pp. 207-226.

Reason, P. "Sacred Experience and Sacred Science," *Journal of Management Inquiry* (2:3), 1993, pp. 10-27.

Skolimowski, H. *The Participatory Mind: A New Theory of Knowledge and of the Universe*, London: Arkana Books, 1994.

Whyte, W. F. "Introduction," *American Behavioral Scientist, Special Issue: Action Research for the Twenty-First Century: Participation, Reflection, and Practice* (32:5), 1989, pp. 502-512.

ABOUT THE AUTHORS

Karin Breu is a senior research fellow in Information Systems at Cranfield School of Management. Her research focuses on the application of information technology to the workplace, including empirical studies of collaborative, virtual, and mobile working. Karin can be reached at k.breu@cranfield.ac.uk.

Christopher J. Hemingway is a research fellow in Information Systems at Cranfield School of Management. His research focuses on performance and management information, including studies of analytics and business intelligence. Chris can be reached at c.j.hemingway@cranfield.ac.uk.

Joe Peppard is a professor of Information Systems at The Business School, Loughborough University where he researches and teaches in the area of information systems and technology, strategy and management. Joe can be reached at joe.peppard@lboro.ac.uk.

42 THEORY AND ACTION FOR EMANCIPATION: Elements of a Critical Realist Approach

Melanie Wilson
Anita Greenhill
Manchester School of Management, UMIST

Abstract Adopting a non-Habermasian critical realist position, this paper seeks to outline some key elements of a realist ontology, on the one hand, and a radical critical stance on the other. The relationship of critical realism to positivism and interpretivism is described, and the case for methodological pluralism made. The elements of realism described are connection of the particular with general, contribution to knowledge, and construction of alternatives. The critical aims entail a commitment to emancipation, a focus on issues of equality and inequality, a questioning of the status quo, and a challenging of ideology. One key conclusion for practice concerns the alliances for researchers with less conservative members of organizations in order to avoid compromise on emancipatory aims. The contribution of the paper is summarized in a concluding table.

Keywords: Critical realism, methodological pluralism, emancipation

1 INTRODUCTION

This short paper delineates some key elements that constitute a critical realist approach to Information Systems research. Contending that there are options beyond the interpretive-positivist dichotomy and that methodological pluralism is feasible, we supplement the extant IS critical literature. Generally speaking, *critical* tends to imply critical social theory. Thus, we refer to the body of writing in this more established IS trend. Our contribution is provided by drawing on research in the social sciences concerning the validity of a critical realist approach to understanding phenomena within organizations. Within IS research, Dobson (2001) has indicated the potential of critical realism, Mingers (2001) has made the case for strong pluralism based on a realist ontology, and a guide to the use of critical realism has been well described by Carlsson (2004). Further to this, the paper is distinguished by a preference for materialist, radical writing on organizations, such as Braverman (1974) based on a Marxian approach (as opposed to the more popular Habermasian writings, based on a more idealist, Hegelian approach).

The choice of realist elements included is guided by an implicit polemic against a flat treatment of the agency structure dimension (Carlsson 2004). One of the main advantages of the adoption of a realist ontology is that it allows for the feasibility of methodological pluralism (Mingers 2001), guided by the principle that the choice of methods employed should be decided by the nature of the object of study. Underpinning the critical standpoint is a neo-humanism (Hirschheim and Klein 1989) with a radical twist. The critical elements selected serve to underline the exigency for emancipatory action. In conclusion, the implications of the approach for IS research are sketched in a summary table. Next, we pave the way for later sections by situating critical realism in relation to the positivist-interpretivist debate.

2 CRITICAL REALISM, POSITIVISM, AND INTERPRETIVISM

Critical realism is an attempt to transcend the bi-polar opposition between positivism and interpretivism (Outhwaite 1983). Evidently, research needs an amount of rigor to be of value and realism appeals to notions of scientific method and objectivity. The term critical denotes a challenge to the orthodoxy of accounts which reinforce the status quo of (unequal) power relations in organizations. While sharing many of the fundamental assumptions of the Habermasian-inspired critical social theory (Ngwenyama 1991), we favor a Marxian, non-Hegelain understanding of how radical emancipatory change in the status quo can occur. The implications of this dissimilarity must be debated elsewhere. Suffice to mention here that our approach is skeptical of the possibility of overcoming conflict in organizations, and reluctant to extend the acceptance of the socially constructed nature of knowledge to the point of extreme relativism.

The increased popularity in using critical and social constructivist approaches to study technology in IS helps in "demystifying technological imperatives and managerial rationalism justifying a particular IS design" (Cecez-Kecmanovic 2001, p. 141). Such perspectives have raised issues concerning the ontological and epistemological assumptions implied by an acknowledgment of the role of culture and language in framing our knowledge of the world. One problem faced by researchers stems from the commonly held assumption that there are two basic perspectives on offer,

> either the world is objectively and unproblematically available...or it is not knowable objectively at all; and, in the place of claims to objectivity, we find only the idea that what is known is merely the product of discourse (Ackroyd and Fleetwood 2000, p. 4).

In response, realists have argued that a dualistic or analytic ontology is necessary for studying organization and management, which asserts the reality of belief systems and cultures as elements that should feature in realist explanations (Reed 2000). Thus, conceptualizing the attitudes, values, and cultures of people will be central. However, realists do not accept that the world is *entirely* constituted by the discursive acts of people: unobservable social structures (mechanisms, relations, powers, rules, resources, institutions, technology) must also be taken into account. Hence, in critical realism, the view of social reality is materialist and structuralist in nature, with reality existing independently from the observer. Further, reality is ordered and thus open to explanation. With this realist ontology it follows that a critical realist approach, *to an extent*, accepts the applicability of scientific principles unearthed in the natural science of the social subject, while simultaneously adopting the position that there are fundamental differences between natural and social phenomena (Ngwenyama 1991).

3 WHAT CONSTITUTES A REALIST APPROACH?

In this section we detail elements that form part of the realist side of the critical realist approach.

3.1 Connection of Particular with General

Critical realism has a predilection to connect the particular with the general. This entails an examination of structures and mechanisms, such as stratification of work. An important assumption here is that IS are part of social phenomena and therefore must be viewed as open systems (Mingers 2000). The focus is on the mechanisms operating in society that are required to be in place in order for a certain set of circumstances to exist. Laws expressing the tendencies of things are a credible subject of enquiry in critical realism (Blaikie 1986, p. 8). hus, explanatory *mechanisms* are postulated and the task of research is to try to demonstrate their existence. However, for realists, causation is not understood on the model of regular successions of events (Sayer 2000). While regularity is expected, *consistent* regularities are unlikely as they occur only in closed systems. In addition, we may want to explain *why* a certain mechanism exists and so we will engage in exploring the nature of the structure or object which possesses that mechanism or power.

3.2 Contribution to Knowledge: Ontology and Epistemology

In terms of ontology, and the contribution of research to knowledge, realism assumes that what exists can be discovered (Ackroyd and Fleetwood 2000). An associated assertion is that the nature of the world being investigated dictates the appropriate methods for that investigation. In common with others (Orlikowski and Baroudi 1991), a plurality of research perspectives is recommended. However, while researchers have advocated the integration of positivist and interpretivist approaches (Lee 1991; Trauth and Jessup 2000) and the integration of case study with survey research methods (Gable 1994), or combining qualitative and quantitative methods (Kaplan and Douchon 1988), the underlying philosophy and issues of ontology are not addressed (Carlsson 2004). One exception is Mingers (2001) and his advocacy for strong pluralism.

3.2 Construction of Alternatives

The promotion of alternatives is highly dependent on the critical points made below. The argument is that, if by our research and writing we make apparent how our own acts and those of others are implicated in the reproduction of social structures and relations that stand in the way of emancipation, then we are potentially in a position to consider alternative structures and solutions (Collier 1998).

4 WHAT IS CRITICAL?

Having outlined the realist characteristics of a critical realist approach, we now construct the critical aspect.

4.1 Question Status Quo

In questioning the assumed status quo in social interaction, critical realism assumes that everyday life has superficial and often conflictual aspects in operation. In relation to IS in organizations, we would be aware and critical of capitalism and the labor process. Marx drew a central distinction between essential relations and surface appearance (Pratten 2000). In building on Marx's insights, the tradition initiated by Braverman (1974) inaugurated a radical critique of the use of technology in organizations because of the potential of increased exploitation of workers. For the IS researcher, this implies an examination of the construction and use of technology to further increase the process of exploitation on the part of management and at the expense of employees.[1]

4.2 Challenge to Ideology

Braverman's (1974) radical approach implicitly challenges views of organizations which do not seek to explore the contradictions emanating from their conflictual nature. A critical realist approach challenges *accounts* of the status quo and aims to deconstruct dominant ideology. Two notions to be criticized here are managerialism (for increasing productivity and curtailing worker resistance) and technological determinism (for excluding human agency and not examining social and organizational contexts [Lopez and Potter 2001]).

While interpretivism offers us the possibility for capturing such complex, dynamic, social phenomena, it will be vulnerable to those criticisms aimed at Habermasian approaches, namely that it does not examine the (material) conditions which give rise

---

[1]Maru and Woodford (2001) argue that in some critical approaches, the focus on emancipation has been relegated due to a concentration on pluralism. This, claims Dobson (2003), is a logical outcome of the epistemological focus of the underlying critical theory of Habermas.

to certain meanings and experiences, that it neglects to clarify historical change; unintended consequences of actions go unexplained, and that it ignores the structural conflicts and contradictions endemic to social systems (Hollis 1994). Further, interpretivism does not explicitly contain the notion of "non-corrigible accounts," constraining our ability to criticize actors' views of the world (Blaikie 1986). Critical realism, by comparison, presents the possibility of ideological error (Lopez and Potter 2001) since it allows for the notion of false beliefs and consciousness.

4.3 Focus on Equality and Inequality

Critical approaches maintain a focus on issues of equality and inequality. Critical realist researchers have recourse to the dialectic of equity: "the principle of sufficient practical reason states that there must be ground for difference. If there is no such ground then we are rationally impelled to remove them" (Bhaskar 1998, p. 676). In IS research we would be looking at patterns and conditions of employment to highlight areas of inequality, since the labor market is not only a reflection of inequalities in broader society, but rather plays a part in the generation of inequality (Peck 2000).

4.4 Commitment to Emancipation

Finally, and perhaps most importantly, a critical approach implies a commitment to contribute to changing the world for the better and for the creation of structures that are wanted, needed, or generally emancipatory (Hollis 1994; Orlikowski and Baroudi 1991). Revealing the way things are is a necessary step to demonstrate the place of human acts in the "reproduction of social structures and relations that stand in the way of emancipation" (Ackroyd and Fleetwood 2000, p. 23), thereby enabling the suggestion of alternative structures where genuine human flourishing can develop. Critical realism seeks ways of transforming asymmetric relations and calls for action to mobilize research: to act to prevent degeneration of equity, and to improve the situation. This would imply we do not support IS that lead to increased exploitation and a worsening of extant inequalities. In this regard, Mingers (2001) identifies the restrictive nature of alliances with managers who can determine what the researcher does, setting the scope and boundaries of investigation. One solution to this quandary of compromised research is the creation of alternative alliances with agencies concerned with employee welfare, such as trades unions.

5 CONCLUSION

This paper contributes to existing critical research in IS by describing elements of a critical realist approach for exploring an IS phenomenon from a more radical tendency as represented in labor process writings. The significance of this for IS is summarized in Table 1.

Table 1. Summary of Critical Realist Elements in Relation to IS Research

	Area of Interest	Example Target Areas of Concern
Realism Element		
Connecting particular to general	Existing social structures and mechanisms.	Stratification of work, impact of technology on work practices, impact of IT/ IS on workers.
Contribution to knowledge	Nature of phenomenon and appropriate method for investigation.	Quantitative analysis for examination of wages, conditions, patterns of employment. Qualitative investigation of meaning of work.
Construction of alternatives	Alternative structures and solutions. Alternative research questions, priorities and aims.	Social measures to alleviate worst of alienation. New perspectives and research foci generated via alliances with employee-centered agencies.
Critical Element		
Questioning status quo	Development and use of IS to increase exploitation of employees. Partnership with nonmanagerial sponsors in defense of employee conditions.	Extension of working day; deterioration in working conditions; routinization of tasks. Possibilities for collective representation.
Deconstruct dominant ideology	Social and organizational context of IS examined to challenge managerialism and technological determinism. Equating technical with social progress in relation to IS phenomenon.	(Non-)valuing of employees by managers. Increased/decreased conflict between work and home. Examination of talk of flexibility against lower job security.
Equality and Inequality	Areas of inequality of treatment and conditions. Patterns of work in relation to use of IS.	Access and experience of IS by employees different from managers. Propensity to part-time, degraded work.
Emancipation	Ways to transform asymmetric relations. Mobilization of research to prevent degeneration of equity.	Examination of chances for development or promotion, increased division of labor. Assemble policy making groups for dissemination of research

REFERENCES

Ackroyd, S., and Fleetwood, S. "Realism in Contemporary Organization and Management Studies," in S. Ackroyd and S. Fleetwood (Eds.), *Realist Perspectives on Management and Organizations,* London: Routledge, 2000, pp. 1-25.

Bhaskar, R. "Dialectical Critical Realism and Ethics," in M. Archer, R. Bhaskar, A. Collier, T. Lawson and A. Norrie (Eds.), *Critical Realism: Essential Readings*, London: Routledge, 1998, pp. 641-687.

Blaikie, N. *Approaches to Social Enquiry*, Cambridge: Polity Press, 1986.

Braverman, H. *Labor and Monopoly Capital: The Degradation of Work in the Twentieth Century*, New York: Monthly Press Review, 1974.

Carlsson, S. A. "Using Critical Realism in IS Research," in M. E. Whitman and A. B. Woszcynski (Eds.), *The Handbook of Information Systems Research*, Hershey, PA: Idea Group Publishing, 2004, pp. 323-338.

Cecez-Kecmanovic, D. "Doing Critical IS Research: The Question of Methodology," in E. M. Trauth (Ed.), *Qualitative Research in IS: Issues and Trends*, Hershey, PA: Idea Group Publishing, 2001, pp. 141-162.

Collier, A. "Explanation and Emancipation," in M. Archer, R. Bhaskar, A. Collier, T. Lawson and A. Norrie (Eds.), *Critical Realism: Essential Readings*, London: Routledge, 1998, pp. 444-472.

Dobson, P. J. "The Philosophy of Critical Realism: An Opportunity for Information Systems Research," *Information Systems Frontier* (3:2), 2001, pp. 199-201.

Dobson, P. J. "The SoSM Revisited: A Critical Realist Perspective," in J. J. Cano (Ed.), *Critical Reflections on Information Systems: A Systemic Approach*, Hershey PA: Idea Group Publishing, 2003, pp. 122-135.

Gable, G. "Integrating Case Study and Survey Research Methods: An Example in Information Systems," *European Journal of Information Systems* (3:2), 1994, pp. 112-126.

Hirschheim, R., and Klein, H. K. "Four Paradigms of Information Systems Development," *Communications of the ACM* (32:10), 1989, pp. 1199-1216.

Hollis, M. *The Philosophy of Social Science: An Introduction*, Cambridge: Cambridge University Press, 1994.

Kaplan, B., and Douchon, D. "Combining Qualitative and Quantitative Methods in Information Systems Research: A Case Study," *MIS Quarterly* (12:4), December 1988, pp. 571-586.

Lee, A. S. "Integrating Positivist and Interpretive Approaches to Organizational Research," *Organization Science* (2:4), 1991, pp. 342-365.

Lopez, J., and Potter, G. (Eds.). *After Postmodernism: An Introduction to Critical Realism*, London: The Athlone Press, 2001.

Maru, Y. T., and Woodford, K. "Enhancing Emancipatory Systems Methodologies for Sustainable Development," *Systemic Practice and Action Research* (14:1), 2001, pp. 61-77.

Mingers, J. "Combining IS Research Methods: Towards a Pluralist Methodology," *Information Systems Research* (12:3), 2001, pp. 240-259.

Mingers, J. "An Idea Ahead of its Time: The History and Development of Soft Systems Methodology," *Systemic Practice and Action Research* (13:6), 2000, pp. 733-755.

Ngwenyama, O. K. "The Critical Social Theory Approach to Information Systems: Problems and Challenges," in H.-E. Nissen, H. K. Klein, and R. A. Hirschheim (Eds.), *Information Systems Research: Contemporary Approaches and Emergent Traditions*, Amsterdam: North-Holland, 1991, pp. 276-280.

Orlikowski, W. J., and Baroudi, J. J. "Studying IT in Organizations: Research Approaches and Assumptions," *Information Systems Research* (2:1), 1991, pp. 1-28.

Outhwaite, W. "Towards a Realist Perspective," in G. Morgan (Ed.), *Beyond Method: Strategies for Social Research*, London: Sage Publications, 1983, pp. 321-330.

Peck, J. "Structuring the :abor Market: A Segmentation Approach," in S. Ackroyd and S. Fleetwood (Eds.), *Realist Perspectives on Management and Organizations*, London: Routledge, 2000, pp. 220-244.

Pratten, S. "Structure, Agency and Marx's Analysis of the Labor Process," in S. Ackroyd and S. Fleetwood (Eds.), *Realist Perspectives on Management and Organizations*, London: Routledge, 2000, pp. 109-137.

Reed, M. I. "In Praise of Duality and Dualism: Rethinking Agency and Structure in Organizational Analysis," *Organizations Studies* (18:1), 2000, pp. 21-42.

Sayer, A. *Realism and Social Science*, London: Sage Publications, 2000.

Trauth, E., and Jessup, L. "Understanding Computer-Mediated Discussions: Positivist and Interpretive Analyses of Group Support System Use," *MIS Quarterly* (24:1), 2000, pp. 43-79.

ABOUT THE AUTHORS

Melanie Wilson is a lecturer in Information Systems and Technology Management at Manchester School of Management, UMIST. Generally her research interests lie in the area of social and organizational aspects of Information Systems. Adopting social shaping and critical approaches to IS research, specific topics include gender success/failure, ICT enabled work practices. Melanie can be contacted Melanie.wilson@umist.ac.uk.

Anita Greenhill is a lecturer in Information Systems and Technology Management at Manchester School of Management, UMIST. Anita's research interests include social, cultural, and organizational aspects of Information Systems. Adopting social shaping and critical approaches to IS research, she researches a diversity of topics including ICT enabled work practices, space, virtuality, Web information systems development, and gender. Anita can be contacted at A.Greenhill@umist.ac.uk.

43 NON-DUALISM AND INFORMATION SYSTEMS RESEARCH

Abhijit Jain
Temple University

Abstract This paper makes a case for the grounding of information systems research within theoretical frameworks that reject the idea of subject-object dualism. In support, two rationales are offered.

(1) Research in information systems exhibits an overwhelming dominance of positivistic research methodologies. Such positivistic research approaches have their roots in the scientific method, and in the attempt to transplant the scientific method from the context of the natural sciences to the context of the social sciences. Further, according to various theorists, the scientific method has its roots in the concept of subject-object dualism put forward by Descartes and others. Thus, this paper argues that IS researchers who wish to resist the current orthodoxy, and who seek to advance a non-positivistic research program, may find it useful to anchor their research within paradigmatic and theoretical frameworks that reject the concept of subject-object dualism.

(2) Research into differences in cognitive processes among distinct cultures suggests that there are differences between Western and Eastern ways of thinking. The Eastern mindset is typically more sympathetic to the notion of subject-object non-dualism than the Western mindset. This difference may have implications for the design of IS that rely heavily on modes of human cognition (e.g., knowledge management or decision support systems). This suggests that when considering certain cultural or cross-cultural contexts, IS researchers may benefit from grounding their assumptions within non-dualistic paradigmatic and theoretical frameworks.

1 INTRODUCTION

The notion of subject-object dualism deeply permeates the existential foundations of much human activity, at least in the West. From the way communication is structured (i.e., in the rules of grammar) to the way in which epistemological and ontological notions are constructed (cf. Burrell and Morgan 1979), the impress of subject-object dualism is ineluctable. Consequently, instead of being understood as an artifact of a particular mode of thinking, dualism has been reified to the status of fact.

A fallout of this situation is that the scientific method, which the research community relies on heavily, is thoroughly immersed in assumptions emanating from implicit acquiescence of subject-object dualism (Berman 1989; Capra 1975).

However, this dualistic mode is not the only available mode of thinking. There also exist non-dualistic frameworks within which understanding can be established. In order to counter the hegemony of dualism, this paper argues for grounding information systems research within non-dualistic paradigmatic frameworks. In support of this contention, the paper discusses the potential utility of framing IS research in non-dualistic paradigmatic frameworks when conducting (1) non-positivistic IS research and (2) IS research in certain cultural and cross-cultural contexts.

The rest of this paper is organized as follows: Section 2 explicates the utility of grounding non-positivist research in non-dualistic paradigmatic frameworks, and includes a discussion of how subject-object dualism is considered to have contributed to the development of the scientific method. Section 3 elucidates the utility of non-dualistic paradigmatic frameworks to IS research in certain cultural and cross-cultural contexts. Section 4 offers concluding remarks.

2 GROUNDING NON-POSITIVIST IS RESEARCH IN NON-DUALISM

2.1 The Dominance of Positivism in the Current Orthodoxy of IS Research

In the prevailing body of information systems (IS) research, positivistic research methodologies and philosophical assumptions overwhelmingly dominate the available range of methodologies and assumptions (cf. Mingers 2001; Nandhakumar and Jones 1997; Orlikowski and Baroudi 1991). This has led to criticism from researchers who seek an alternative, more diverse order, in which multiple research approaches—that subscribe to different epistemological, ontological, and methodological assumptions—an coexist and attain significance (cf. Klein and Lyytinen 1985, Mingers 2001; Orlikowski and Baroudi 1991).

Researchers have criticized positivistic methodology for subscribing to a dogma of *scientism*, which emphasizes objectivity and methodological rigor over contextual relevance (cf. Klein and Lyytinen 1985). Consequently, researchers have questioned whether such *scientistic* methods are appropriate for social science research (ibid) and have called for increased consideration of alternative approaches, such as the interpretivist and the critical, in order to facilitate "exploration of phenomena from diverse frames of reference" (Orlikowski and Baroudi 1991, p.2).

2.2 Dualism: A Root of Positivism and the Scientific Method

According to a number of modern Western scientists, philosophers, and historians of science, modern Western scientific philosophy has been profoundly influenced by the concept of subject-object dualism (cf. Barrett 1986; Berman 1989; Capra 1975; Clapp et al. 1962; Wallace 1989, 2000). Dualistic themes have been espoused by various philosophers, going all the way back to, and including, Plato. However, the most prominent and influential proponent of dualism is generally regarded to be Descartes, who proposed that there was a dichotomy between the human mind and material things (such as the human body). According to this theory of subject-object dualism, *objective* reality consisted of the physical, material world, whereas everything that was not physical or material, e.g., feeling, belonged to a *subjective* realm. Further, according to Descartes, (1) a clean separation between subject and object was possible; (2) the objective realm was the only legitimate domain of enquiry; and (3) all of nature could be understood by studying it objectively (cf. Berman 1989; Capra 2000; Clapp et al. 1962). Knowledge obtained in this way was presumed to be better because it was objective and did away with subjective elements that were not considered particularly relevant. Such thinking gave rise to the concept of the scientific method, according to which scientists were expected to impersonally observe and measure phenomena in order to understand them. Empiricism, it was thought, transcended the limits of human subjectivity and gave an accurate account of the world. Since such inquiry was objective it was considered value-neutral and apolitical. Thus science was considered the unbiased way to solve problems, and scientists *discovered the truth* by unsympathetically observing and measuring *reality*.

As it happened, the scientific method proved remarkably successful in conducting inquiry into natural sciences such as physics, chemistry, and astronomy. This eventuality lent credibility to Descartes' metaphysical speculations, and they gradually took on the status of being fact. Over time, dualistic thinking led to objective reality acquiring primacy over subjective experience. Supposedly, reality was objectively given, and considered always true. Human contextual experience, on the other hand, was subjective, fickle, subject to continual change, and not considered worth studying. Human beings were thus regarded as spectators, irrelevant bystanders to a grand mathematical, mechanistic system that was the essence of reality. The elevation of the objective over the subjective became complete.

Subsequently, scientific thinking morphed into a dogma of scientism (Klein and Lyytinen 1985), and it gradually came to be believed that the scientific method was the only appropriate and legitimate path to knowledge in every context and domain of inquiry, prompting theorists such as Hobbes and Comte to advocate the use of scientistic, positivistic methods for the investigation of social phenomena (cf. Hirschheim 1985; Klein and Lyytinen 1985).

Thus, the dominance of positivism in the prevailing order can be traced to the successes of the scientific method in conducting inquiry into the natural sciences, and to subsequent attempts to apply the scientific method to research in the domain of the social sciences (cf. Klein and Lyytinen 1985). Further, the scientific method is said to have arisen out of the concept of subject-object dualism. Therefore, according to this chain of reasoning, subject-object dualism may be considered to be a root of positivism (or the scientific method).

2.3 The Utility of Non-Dualism to Non-Positivist IS Research

If subject-object dualism is a root of the scientific method, and by extension a root of positivism; if positivism dominates the current milieu of IS research; and if a change in this status quo is desirable; then a good starting place for a research program that seeks to effect a change in this status-quo should be from within philosophical frameworks that reject the concept of subject-object dualism and belie the suggestion that a clean separation between subject and object is possible. Thus researchers who seek to advance non-positivistic research programs should find it valuable to anchor their research within paradigmatic and theoretical frameworks that *reject* the concept of subject-object dualism.

There are several available, implicitly non-dualistic, philosophical frameworks that could prove useful in this regard. For instance, Kant (1977) espoused a theory according to which cognition was antecedent to experience, and the reception of sensory experience was necessarily determined according to preexisting concepts that existed in the mind. Thus, objective observation was unfeasible, and the subject was inextricably linked to the experience. Hegel's (1967) philosophy of absolute idealism refuted the traditional epistemological distinction between object and subject, and posited that material things existed only according how they were perceived. Heidegger (cf. Dreyfus 1991) questioned the idea that experience could be explained in terms of relationships between independent subjects and objects. He posited a "more fundamental way of being-in-the-world that cannot be understood in subject/object terms" (Dreyfus 1991, p. 5), and sought to emphasize the role of "social context as the ultimate foundation of intelligibility" (ibid, p. 7). In the case of Zen philosophy, the key aim is "the overcoming of all dualistic discrimination" (DeMartino 1981, p. 80). According to a Zen *koan*, a customer asked a butcher which piece of meat was best. The butcher replied, "Each of our pieces of meat is the best." Which piece is best would depend on a myriad of variables, such as what the customer prefers, and how one chose to define the word *best*; thus implying that everything exists only in relation to the observer.

IS researchers seeking to employ non-positivistic methodologies should benefit from anchoring their research in non-dualistic philosophical frameworks because by doing so they can avoid having to subordinate their research assumptions to the assumptions of a dualistic epistemology or ontology; and they can thus steer clear of those influences inherent in the concept of subject-object dualism that have led to the emergence of positivism.

3 GROUNDING IS RESEARCH IN NON-DUALISM IN CERTAIN CULTURAL CONTEXTS

Research into differences in cognitive processes among distinct cultures suggests that there is significant divergence between Western and Eastern ways of thinking (Capra 2000; Nisbett 2003). According to a major study devoted to this subject (Nisbett, 2003), the Western mindset is more reductionist (i.e., is more comfortable dealing with parts than with wholes), places greater value on consistency, and is more inclined toward consideration of objective attributes. In contrast, the Eastern is mindset is more holistic, more willing to accommodate contradiction and more predisposed towards consideration of subjective context. Additionally, the Western mind generally values individualism and distinctiveness, while, conversely, the Eastern mind usually places value on consensus and harmony. According to researchers, when such cognitive and attitudinal differences are aggregated, they indicate that the Western mindset is normally more predisposed toward the concept of subject-object dualism, whereas the Eastern mindset is typically more sympathetic to the notion of subject-object non-dualism (Capra 2000; Nisbett 2003).

These propositions have significant implications for the design of information systems that deeply engage human cognitive processes, and they imply that it may be naïve to attempt to transplant systems, such as knowledge management or decision support systems (among other kinds of IS), that have been developed in one cultural context (say the West) onto the other cultural context (say the East). Further, it may not be appropriate for researchers to approach IS research in certain cultural contexts (i.e., where mindsets are more inclined to non-dualistic thinking) from within dualistic paradigmatic frameworks. As explained in section 2, positivism and the scientific method have their roots in such a dualistic framework. Thus, when considering certain cultural or cross-cultural contexts where non-dualistic thinking is involved, IS researchers may benefit from framing research and development activities within non-dualistic paradigmatic and theoretical frameworks that correspond better to the cultural context at hand.

4 CONCLUDING REMARKS

This paper has sought to describe the potential utility of non-dualistic philosophical frameworks to IS research. In doing so, this paper discusses why paradigmatic and theoretical frameworks that reject or belie *a priori* assumptions about subject-object dualism may be useful for the purpose of advancing non-positivistic research, and for conducting IS research and development activities in certain cultural or cross-cultural contexts.

REFERENCES

Barrett, W. *Death of the Soul: From Descartes to the Computer*, Garden City, NY: Anchor Press, 1986.

Berman, M. *Coming to our Senses: Body and Spirit in the Hidden History of the West*, New York: Simon and Schuster, 1989.

Burrell, L., and Morgan, G. *Sociological Paradigms and Organizational Analysis*, London: Heinemann Educational Books, 1979.

Capra, F. *The Tao of Physics*, Boston: Shambhala, 1975.

Clapp, J. G.; Philipson, M.; and Rosenthal, H. M. (Eds.). *Foundations of Western Thought: Six Major Philosophers*, New York: Knopf, 1962.

DeMartino, R. J. "The Zen Understanding of the Initial Nature of Man," in N. Katz (Ed.), *Buddhist and Western Philosophy*, Atlantic Highlands, NJ: Humanities Press, 1981, pp. 8-120.

Dreyfus, H. *Being-in-the-World: A Commentary on Heidegger's Being and Time, Division I*, Cambridge, MA: MIT Press, 1991.

Hegel, G. W. F. *The Phenomenology of Mind*, J. B. Baille (Trans.), New York: Harper & Row, 1967.

Hirschheim, R. "Information Systems Epistemology: An Historical Perspective," in E. Mumford, R. Hirschheim, G. Fitzgerald, and A. T. Wood-Harper (Eds.), *Research Methods in Information Systems*, Amsterdam: North-Holland, 1985, pp. 9-33.

Kant, I. *Prolegomena to Any Future Metaphysics That Will Be Able to Come Forward as Science: A Revision of the Paul Carus Translation by James Ellington*, Indianapolis: Hackett Publishing Company, 1977.

Klein, H. K., and Lyytinen, K. "The Poverty of Scientism in Information Systems," in E. Mumford, R. Hirschheim, G. Fitzgerald, and A. T. Wood-Harper (Eds.), *Research Methods in Information Systems*, Amsterdam: North-Holland, 1985, pp. 123-151.

Mingers, J. "Combining IS Research Methods: Towards a Pluralist Methodology," *Information Systems Research* (12:3), 2001, pp. 240-259.

Nandhakumar, J., and Jones, M. "Too Close for Comfort? Distance and Engagement in Interpretive Information Systems Research," *Information Systems Journal* (7:2), 1977, pp. 109-131.

Nisbett, R. E. *The Geography of Thought: How Asians and Westerners Think Differently—and Why*, London: Free Press, 2003.

Orlikowski, W. J., and Baroudi, J. J. "Studying Information Technology in Organizations: Research Approaches and Assumptions," *Information Systems Research* (2:1), 1991, pp. 1-28.

Wallace, B. A. *Choosing Reality: A Contemplative View of Physics and the Mind*, Boston: Shambhala, 1989.

Wallace, B. A. *The Taboo of Subjectivity: Toward a New Science of Consciousness*, Oxford: Oxford University Press, 2000.

ABOUT THE AUTHOR

Abhijit Jain is a doctoral candidate in management information systems at Temple University, Philadelphia. His research interests include e-government, and defense and medical information systems. He has had papers accepted at conferences such as the International Conference on Information Systems, the Hawaii International Conference on System Sciences, and the Americas Conference on Information Systems. He can be reached at jain@temple.edu.

44 CONTEXTUAL DEPENDENCIES AND GENDER STRATEGY

Peter M. Bednar
University of Portsmouth

Abstract Analysts are often asked to help deliver systems that have a great mix of performance and features. Unfortunately, problematic organizational gender related issues have sometimes been degraded and treated as unrelated to technological issues. While information systems development in general is trying to ensure support for businesses, we might in the future expect more than just a gender-ignorant quality measure in the way the IS works to meet organizational demands. In this paper it is proposed that an interpretative and contextual analysis would support ISD in the creation of a necessary level of understanding of each specific business. The intention with an inquiry into contextual dependencies is that it helps to identify some methodological limitations which result in traps unconsciously biasing analysis of investigated problem spaces. This paper introduces a contextual analysis highlighting some contextual dependencies that are typically ignored in existing works or analysis. An initial framework is proposed using gender as an example of an inquiry into some existing contextual dependencies.

Keywords: Information systems analysis, contextual dependency, contextual analysis, gender strategy

1. INTRODUCTION

Those problematic issues which are focused upon within this paper have to do with basic underlying theoretical assumptions in which most of our work (as systems analysts) is steeped. Both unconsciously and consciously, these issues can be assumed to have a huge impact on our everyday work. Hence they create blind spots for us and support us in being unconscious in a continuous belief of gender-neutrality in our obscured state of self-denial.

One problem is that some of the central underlying and influential issues in information systems development are dependent upon cybernetics as a base in systems thinking together with psychological and sociological reductionism. Another problem is that sometimes hidden political issues are related to unresolved conflicts between assumptions of observation, interaction, and intervention. The proposed strategy in this paper to breakout of this cage of self-denial is a stronger relationship (in systems analysis) to the consideration of contextual dependencies and a higher degree of reflection. One example of a problem space where inquiry into contextual dependencies become relevant is related to managerial attitude and organizational practice of promoting ideas of decentralized power in organizations.

2 DECENTRALIZED POWER

The presupposition that extreme economic logic has to be kept in check with a pronounced distribution of power is supported in the discussion that Galbraith, Lawler and Associates (1993) put forward about how successful new forms of organizational design and organizational structures ought to be aligned with practices that distribute information, power, knowledge, and rewards throughout the organization.

Argyris and Schön's (1978, 1996) discussion on learning organizations has put a focus on empowerment from a competence raising point of view. Argyris (1990) uses the notion of "organizational defense" to explain differences between descriptions and activities. This is a notion that might be of significant importance when efforts are made to explain why a great equal opportunity policy might not work even if management is supportive. Weick's (1995) organizational sense making can be viewed as an effort that tries to bridge the gap between strategies for

1. Analyzing—quality of interpretation and understanding
2. Designing—strategies for structuring and problem solving

Contextualism and interpretative IS research are important issues in the IS field due to the way in which several new political dimensions are put on the IS research agenda (refer to Walsham 1993; Walsham and Sahay 1999). Whatever system and ICT is to be developed is supposedly to be relevant for some particular organization and some particular organizational members that are to benefit from its implementation. This is to say that systems are possible to view as being intrinsically contextually dependent. Systems analysis in practice and theory would then have to further develop an understanding of limitations of models based on assumptions of generalizability as opposed to contextual dependency.

3 CONTEXTUAL DEPENDENCY

One of the major features of the combination of contextual dependency and double loop learning is the reevaluation (reflection) of a sense making process itself. Such a reflection can be seen as a metaphysical part of an enhanced version of double loop learning and is also of major importance in the sense making process as presented in the strategic systemic thinking (SST) framework proposed by Bednar (2000). This framework is proposed as a means of accessing individual and team competencies for improved systems analysis work and the intention is to provide a systemic arena for developing a learning organization inclusive of having a constructive dialogue mechanism.

The sense making processes that often might be seen as prominent in organizations are ambiguity and uncertainty. The main purpose with the SST framework is to support an *ad hoc* creation of a systems thinking process with the aim to transfer uncertainty into structured uncertainty (this can, in some circumstances, be seen as a transformation of uncertainty into ambiguity). The impact in organizations of the two sense-making processes is somewhat different.

> In the case of ambiguity, people engage in sense making because they are confused by too many interpretations, whereas in the case of uncertainty, they do so because they are ignorant of any interpretations (Weick 1995, page 91).

One of the reasons that a problematized understanding of contextual dependency is important is that some relativism is meant to be supported, but not total relativism (e.g., extreme skepticism). The intention in this paper is to avoid an attitude of anything goes without disqualifying relativistic ideas. One example where contextual dependency may be seen as related to some kind of relativistic experience is in efforts to develop and implement gender strategies.

4 GENDER STRATEGY

The problems with gender being a significant but understudied theme are generally widespread in British industrial relations (Wajcman 2000). Wajcman concludes that

> analyses of this phenomenon have tended to be descriptive rather than explanatory. There has been increasing consciousness of the tendency to treat workers as male, but without deeper understanding of the way power inequalities between men and women shape employment relations (p. 195).

Giddens (1984, 1991) points out a necessity to analyze the dialectic between resistance (of socially constructed behavior) and power. There is always some (individual) possibility to influence social activities, and even if such a possibility would be viewed as minimal, it is still existent and has an impact.

Habermas (1984) suggests that individuals have ability for communicative actions and actively pursue resistance against those powers, which try to pacify the individual ability to take initiative and ability to act.

The basic perspective in this paper is that individuals as reflective subjects have an ability to break out of inhibiting social and cultural (control) systems. Argyris (1990) concludes in his discussion of organizational defenses that awareness of ability does not mean that this ability is easy to achieve in practical organizational settings.

A gender strategy in IS, related to a contextual dependency at micro and macro level analysis, could be interpreted as an effort to combine the experiences from the aforementioned contextually dependent combination of intra- and interindividual open systems thinking.

An inquiry into contextual dependencies at a micro level can be seen as analysis of

- The specific institutional changes that might lead to a weakening of diverse intra-individual understanding of social distinctions. This as an inquiry into individual understanding of changes in relations between different types of organizational culture, which can be explored, for example, between what can be observed as male and female bias.
- How individuals might cope with experiences in their personal construction and reconstruction of biographic storytelling about their own individual gendered selves. Such a combination of construction and deconstruction assumes a reflective approach to individual sense-making and intra-individual analysis.

An inquiry into contextual dependencies at a macro level can be seen as analysis of

- The interindividual gendered tendencies toward and relations with organizational and interorganizational, technological, cultural, and social integration, not only to analyze and inquire into such processes which could then be described in terms of fragmentation and disintegration.
- How gender-related knowledge and behavior might be viewed when related to examples of specific (organizational) contexts and specific historical situations in time. It does not necessarily exclude the possibility to create more or less temporary assumptions of *ad hoc* generalizations about global issues. It might be assumed as possible to reach systemic knowledge about diverse phenomena.

The strategy for IS proposed here is to be seen as a part of an effort to create an organizational sense-making bridge and oscillating relationship between analysis of a *past* and design of a *future* (e.g., Weick 1995).

Rationalization and resistance leads to a possibility not only for crisis but also for a richer communication and change. It is precisely through discussions about communicative action and system that Habermas (1984) tries to create a bridge between the intra-individual system and the interindividual system without giving up the idea of the active and rational individual. The social, cultural, and technical world is populated, created, and recreated by people who in different ways and aspects deal with their relation to the abstract inter- and intra-individual systems in their experienced environment (for related discussions, see Bateson 1972; Berger and Luckman, 1966; Latour 1999).

It is (in this paper) assumed, as Giddens (1984, 1991) suggests, possible and meaningful to create and design a future life space with the purpose and efforts to solve different social and cultural problems. It should also be seen as meaningful to create a gender strategy as an integral part of an IS analysis strategy. The main dimension focused on in this paper has to do with *uniqueness ignorant* theories and practices. A naive understanding of IS, organizational theories, and practices might inhibit a more adequate and progressive understanding of an unconsciously regenerative blinding behavior.

5 CONCLUSION

The IS discussion needs to take into consideration and support a greater awareness of contextual dependencies and their potential organizational and social implications. When engaging in efforts to help individuals to combine their (work and family) roles to advance organizational change and benefits, enquiries into contextual dependencies become increasingly important. Also, given that IS research practice has often treated even basic individual and organizational issues as being either nonexistent or nonrelevant in the IS arena, there is a need to expand our field to include a contextually relevant strategy.

The whole point with this discussion on contextual dependencies is to intervene in a (biased) IS community and to support an active reevaluation and reconsideration of our practical and theoretical environment. IS research can be seen as in serious need to avoid being weak on the self-critical motivational issues, and the issues of ethics and culture. An enhanced version of contextual analysis might be seen as a possible way forward.

REFERENCES

Argyris C. *Overcoming Organizational Defenses: Facilitating Organizational Learning*, Englewood Cliffs, NJ: PrenticeHall, 1990.

Argyris, C., and Schön, D. A. *Organizational Learning*, Reading, MA: Addison Wesley, 1978.

Argyris, C., and Schön, D. A. *Organizational Learning II Theory, Method and Practice*, Reading, MA: Addison Wesley, 1996.

Bateson G. *Steps to an Ecology of Mind*, New York: Ballantine, 1972.

Bednar P. M. "A Contextual Integration of Individual and Organizational Learning Perspectives as Part of IS Analysis," *Informing Science Journal* (3:3), 2000, pp. 145-156.

Berger, P. L., and Luckmann, T. *The Social Construction of Reality. A Treatise in the Sociology of Knowledge*, New York: Doubleday, 1966.

Galbraith, J. R.; Lawler, E. E.; and Associates. *Organizing for the Future: The New Logic for Managing Complex Organizations*, San Francisco: Jossey Bass Publishers, 1993.

Giddens, A. *The Constitution of Society*, Cambridge, MA: Polity Press, 1984.

Giddens, A. *Modernity and Self Identity: Self and Society in the Late Modern Age*, Stanford, CA: Stanford University Press, 1991.

Habermas, J. *The Theory of Communicative Action*, Boston: Beacon Press, 1984.

Latour, B. *Pandora's Hope: Essays on the Reality of Science Studies*, Cambridge, MA: Harvard University Press, 1999.

Walsham, G. *Interpreting Information Systems in Organizations*, Chichester, England: Wiley, 1993.

Walsham, G., and Sahay, S. "GIS for District Level Administration in India: Problems and Opportunities," *MIS Quarterly* (23:1), 1999, pp. 39-66.

Wajcman, J. "Feminism Facing Industrial Relations in Britain," *British Journal of Industrial Relations* (38:2), 2000, pp.183-201.

Weick, K. *Sense Making in Organizations*, Thousand Oaks, CA: Sage Publications, 1995.

ABOUT THE AUTHOR

Peter M. Bednar is a senior lecturer at Portsmouth University, UK, and affiliated with the department of Informatics at Lund University, Sweden. His research interest covers areas such as contextual analysis, systems thinking, and information systems development. Apart from working at teaching and research institutions, he has several years of industrial experience. He can be reached at peter.bednar@ics.lu.se.

45 INFORMATION TECHNOLOGY AND THE GOOD LIFE

Erik Stolterman
Anna Croon Fors
Umeå University

Abstract The ongoing development of information technology creates new and immensely complex environments. Our lifeworld is drastically influenced by these developments. The way information technology is intertwined in our daily life raises new issues concerning the possibility of understanding these new configurations. This paper is about the ways in which IS research can contribute to a deeper understanding of technology and the ongoing transformations of our lifeworld. As such, the paper is a conceptual exploration driven by a sincere and authentic desire to make a real difference in the way research on how technology influences our society is carried out. The article is based on the assumption that there are some foundational decisions forming research: *the question of methodology*, the question of *object of study*, and, most importantly, the question of *being in service*. In the paper we explore and propose a *research position* by taking a critical stance against unreflective acceptance of information technology and instead acknowledge people's life-world as a core focus of inquiry. The position is also framed around an empirical and theoretical understanding of the evolving technology that we label the *digital transformation* in which an appreciation of *aesthetic experience* is regarded to be a focal methodological concept.

Keywords: Information systems research, critical theory, aesthetic experiences, digital transformation, device paradigm, information technology

1 IN SERVICE OF THE GOOD LIFE

The ultimate concern for most people is to have the opportunity and capacity to live a "good life." What might constitute a good life is, of course, as difficult to define as it is to characterize basic human needs and desires. Nevertheless, in this paper we argue that information systems research should, as at least one of its intentions, create and formulate knowledge that can help people understand and reflect on their place and situation in the midst of an ongoing technological revolution. We argue that one purpose of IS research should be to explore, experiment, test, analyze, examine, explain, and reflect on how information technology can be in service of the good life. Such a purpose, even if vague, would strongly influence the way research is carried out.

An information systems researcher is always *in service* to someone or something. Ideally, as a researcher, you should be in service to the truth, and you should do this by producing true knowledge. Our contemporary research environment is, however, more complicated due to a long and intricate questioning of truth as the only objective and final goal. So, to have truth as the client has, over time, been complemented with other potential clients, leading to objectives such as organizational and/or personal efficiency and improvement, or detailed technological solutions to more specific and narrow, real or imagined problems.

We argue that a neglect of the "big" issues leads to a situation where people cannot get enough help in their everyday struggle to understand and make meaning of their rapidly changing *lifeworld*. Also, it seems as though people assume information technology to be the solution for prosperity and continuous development, while at the same time they hold technological artifacts to be a bearer of something that contradicts what they see as the core of a good life. So, while there is a strong general acceptance of information technology, there is also a fear that it will force us into a way of living that we cannot handle or do not really want.

This is a real challenge for IS research. It is a challenge that demands a creative design of the very foundation for information systems research. Such a design can be understood as a *research position*. In this paper we propose such a research position based on a critical stance against unreflective acceptance of information technology. We also propose, based on that position, the notion of the *digital transformation* as a way of framing a suitable object of study, and the idea of *aesthetic experience* as a base for a methodological approach.

2 ESTABLISHING A RESEARCH POSITION

Recently there has been an intensified debate on the status and future of information systems as an academic research discipline (Benbasat and Weber 1996; Benbasat and Zmud 2003; Holmström and Truex 2003; Orlikowski and Iacono 2001; Walsham 1993; Weber 2003).

In this debate, two of the most discussed issues are *the question of methodology* and the question of what constitutes *the object of study*. In this paper we also address a third assumption—*the question of service*—as mentioned in the introduction.

It has lately been advocated that *being in service* constitutes a distinct kind of relationship (Nelson and Stolterman 2003). If such a relationship is taken seriously, any decision of who is the major client establishes a clear *position* for IS research, a position that makes it possible to see what the purpose of the research is as well as governs what should be studied, why it should be studied, and, perhaps most important, brings a value system from which the research outcome is measured and judged as valid or not.

When a researcher decides on how to relate to the three foundations—methodology, object of study, and service—a unique *research position* is created. Even though there are several kinds of positions in IS research today, we claim that the possibilities of positions are far from being fully explored.

The basic idea in our proposal is that the most crucial challenge for IS research today is the study of the overall effects of the ongoing digital transformation of society. The digital transformation can be understood as the changes that the digital technology causes or influences in all aspects of human life. This research challenge has to be accepted on behalf of humans, not in their role as users, customers, leaders, or any other role, but as *humans living a life*. In this respect we argue that IS research must accept the challenge to overcome the predominating "one-dimensionality" in the understanding of information technology (Marcuse 1964). The position we argue for is based on the assumption that the digital transformation is the core object of study for IS research. As such, IS researchers should, instead of examining information technologies as separable and as defined along one single dimension, consider them as being a part of a greater whole. This is also expressed in some recent theories framing technological development as information ecologies, collective intelligence, and actor networks, which are more sensitive to the various ways in which information technology is entering our lives (Feenberg 1999; Latour 1993; Levy 1997; Nardi and O'Day 1999).

The suggested position is also based on the assumption that the way to study the digital transformation demands a methodology capable of reflecting the relatedness of information technology to such a larger whole. We propose an approach influenced by *critical theory* with the notions of the *device paradigm* and *aesthetic experience* as focal concepts.

3 THE DIGITAL TRANSFORMATION AND ITS CRITICAL BASE

A central aspect of information systems research is the underlying technology providing the basis and ground for any information system. We all experience in our everyday lives that information technology becomes more common and present in almost every part of our doings. We find ourselves using IT artifacts at work, in our homes, and when we exercise our hobbies. The technology is not only manifesting itself through individual IT artifacts (such as computers, software applications, PDAs, mobile phones, etc.); it also blends itself into most other artifacts. As such, information technologies are increasingly becoming embedded in all other objects.

This leads to a world that is increasingly *experienced with, through, and by information technology*. What we are witnessing is an ongoing radical digital transformation.

One of the most important changes that come with the digital transformation is that our reality by and through information technologies slowly becomes more blended and tied together. Designed objects will be parts of systems and networks where they will, or at least can, be in constant communication with all other parts and objects. These new realities, new systems, are of course designed but, at another level, they can be seen as evolving entities, where local designs contribute to systemic changes in a larger network. The notion that every design adds a new part to our reality will have a new and truer meaning.

New artifacts are not just adding to what already exists; *they are also becoming indistinguishable from the whole*. An increasing problem is knowing where one context and/or design begins and another ends. The digital transformation leads in that sense to a world where everything is connected, almost in a way that is common in many spiritual understandings of our reality.

Yet another important aspect of the transformation appears as digital objects become the basic materials in our physical reality. When this is the case the physical reality will to some degree become intelligent. Designed objects will have the power to inform themselves and the network they belong to about changes and the status of their environment and actions taken upon them by humans and other objects. This adds a new dimension to the notion of the reflexivity of information technologies. The way humans experience their lifeworld, largely influenced by digital technology, is not as separate entities that might be user-friendly or not, but as a lifeworld, as a whole. To understand this aspect of information technologies and information systems will become ever more difficult. To do it by analyzing them individually and/or by using reductionistic methods will be impossible.

At the same time, the *device paradigm,* portrayed by Borgmann (1984, 1999), pushes us to an understanding of technology as providers of commodities, designed to grant our wishes without demanding any patience, skill, or effort. Rather, the world is taken up in an instrumental and effective fashion by technological artifacts and systems that are not designed to be experienced in an active and signifying way. The device paradigm leaves us focused on the outcomes that technology provides rather than make us concerned with the way we experience reality as a whole. Information technology and the digital transformation seem to be the perfect tools for the device paradigm to be manifested.

In such a paradigm, according to Borgmann, there are important concerns and values that are being threatened, concerns that are necessary in order for people to live a good life. Experiences of what it means to live a good life are, of course, both infinite and complex. They are also experiences that are analog to their character in that the world is experienced as one and in a continuum. In such experiences, information technologies are not separated from anything else but seamlessly interwoven in a complex and complete web of meaning. To researchers with the ambition to understand the ways in which people create meaning of their realities and how information technologies transforms this process, this creates several challenges and opportunities.

One challenge inherent in the digital transformation is that information systems researchers today need to develop approaches, methods, and techniques to the study of information technology that are not based on an analytic and reductionistic stance. Another challenge, as a consequence of the device paradigm, is to take an *active stance* against a development leading to an everyday reality dominated by commodities, i.e., technological artifacts as described in the device paradigm.

In the attempt to take up this challenge we advocate the notion of *aesthetic experiences* as one possible conceptual candidate to further advance. While traditional approaches are suffering from the fact that the more complex reality becomes, the more time is needed for analysis, an approach based on the notion of the aesthetics experience makes it possible to take the whole and the immediate into account and to deal with complexity and meaning-making at another level. Since information technology is part of people's experienced life, their lifeworld, the aesthetic experience becomes a way to measure their understanding of their life in relation to the good life.

The critical stance expressed in the use of the notion of aesthetic experience is a creative and radical approach, aiming for the inherent *potentiality* of information technology (Marcuse 1964). We believe that a focus on the aesthetic experience is one way to find and explore creative abstractions that reveal reality in new ways. As Marcuse writes,

> Such abstraction which refuses to accept the given universe of facts as the final context of validation, such " transcending" analysis of the facts in the light of their arrested and denied possibilities, pertains to the very structure of social theory.

If we accept the challenge that we as information systems researchers have to grasp the way information technology changes people's lifeworld, we need conceptual tools that have the necessary scope and strength. We believe that the concept of aesthetic experience, as developed within the philosophical traditions of critical theory and pragmatism among others, is a suitable candidate. Of course, it has to be further developed as a theoretical tool, but this may be more as a fundamental methodological approach. This work is not done and will be a major task in the development of the research position we are suggesting.

4 TOWARD A RESEARCH POSITION

One of the assumptions underlying our argumentation has been that there is a need for a critical stance, a research approach that advances the idea that technology can be critically examined in the search for the good life. We have defined such a research position as being manifested by the intentional choice and definition of *methodology*, *object of study*, and *service*. Without neglecting other commonly held positions, we have proposed one research position as especially needed today when digital and device transformations are changing the preconditions for our possibilities to live a good life.

We believe that information systems research is better suited than most other academic disciplines to take on this position. However, as long as research in our field is not taking this as a serious challenge, the outcomes continue to be an efficient support of the ongoing device transformation leading us to a place were we might not want to go.

Our work has been guided by a desire to take on the big issue by taking seriously the question of *being in service* of people trying to live a good life. It is, of course, too grand a project for a single researcher or research group. At the same time, as researchers, we cannot yield to the important issues because we believe they are not researchable. The overall issue on how information technology, on a fundamental level, influences our lives is maybe the most crucial issue today. This paper is an attempt to establish a research position suitable as a starting point for such studies.

REFERENCES

Benbasat, I., and Weber, R. "Research Commentary: Rethinking 'Diversity' in Information Systems Research," in *Information Systems Research* (7:4), December 1996, pp. 389-399.

Benbasat, I., and Zmud, R. "The Identity Crisis Within the IS Discipline: Defining and Communicating the Discipline's Core Properties," *MIS Quarterly* (27:2), 2003, pp. 183-194.

Borgmann, A. *Holding on to Reality: The Nature of Information at the Turn of the Millenium*, Chicago: The University of Chicago Press, 1999.

Borgmann, A. *Technology and the Character of Contemporary Life: A Philosophical Inquiry*, Chicago: The University of Chicago Press, 1984.

Feenberg, A. *Questioning Technology*, London: Routledge, 1999.

Holmström, J., and Truex, D. "Social Theory in IS Research: Some Recommendations for Informed Adaption of Social Theories in IS Research," in J. Ross and D. Galletta (Eds.), *Proceedings of the 9th Americas Conference on Information Systems*, Tampa, 2003, pp. 2850-2856.

Latour, B. *We Have Never Been Modern*, Cambridge, MA: Harvard University Press, 1993.

Levy, P. *Collective Intelligence: Mankind's Emerging World in Cyberspace*, New York: Plenum Trade, 1997.

Marcuse, H. *One Dimensional Man: Studies in the Ideology of Advanced Industrial Society*, Boston: Beacon Press, 1964.

Nardi, B. A., and O'Day, V. L. *Information Technologies: Using Technology with Heart*, Cambridge, MA: MIT Press, 1999.

Nelson, H., and Stolterman, E. *The Design Way—Intentional Change in an Unpredictable World*, Englewood Cliffs, NJ: Educational Technology Publishing, 2003.

Orlikowski, W., and Iacono, C. S. "Research Commentary: Desperately Seeking the 'IT' in IT Research—A Call to Theorizing the IT Artifact," In *Information Systems Research* (12:2), 2001, pp. 121-134.

Walsham, G. *Interpreting Informations Systems in Organizations*, Chichester, England: Wiley, 1993.

Weber, R. "Editor's Comment: Still Desperately Seeking the IT Artifact," *MIS Quarterly* (27:2), 2003, pp. iii-xi.

ABOUT THE AUTHORS

Erik Stolterman is a member of the Department of Informatics, Umeå University, Sweden. His main work is within information technology and society, information systems design, philosophy of design, and philosophy of technology. Erik is also one of the founders of The Advanced Design Institute. Apart from the academic scholarly work, Eric is engaged in consulting, seminars, and workshops with organizations and companies. He can be reached at erik@informatik.umu.se or http://www.informatik.umu.se/~erik/.

Anna Croon Fors is a Ph.D. student and instructor at the Department of Informatics, Umeå University, Sweden. In her forthcoming dissertation, Anna analyses the meaning and consequences of information technology in people's everyday life through the phenomenological notion of "being." Her research interest covers IT use in various social settings in search of a foundation for critically oriented research beyond the common notions of either use and design. She can be reached at acroon@informatik.umu.se.

46 EMBRACING INFORMATION AS CONCEPT AND PRACTICE

Robert Stephens
University of the West of England

Abstract In this position paper it is argued that generic theories of the information concept will be an obstacle to the Information Systems discipline assuming intellectual leadership of the information portfolio associated with the growth of computing technologies beyond the organization.

Keywords: Information, information systems

The information systems discipline embraces information both as concept and as practice. A convention has been to place the concept *information* on a continuum between data and knowledge or wisdom (e.g., Avison and Fitzgerald 2003, p. 17; Walsham 2001, p. 36). The practice has been defined by the expansion of computing technologies within the business and corporate sectors. *Information systems* (IS) were variously categorized as transaction systems, management information systems, and intelligent support systems, and serviced business or organizational needs. Although problematic and contestable, the notion of information and systems could be referenced against a framework of organizational requirements giving the scope of IS practice a relatively stable focus. The contemporary growth of inter-organizational systems and the penetration of digital computing devices throughout society compromises this reference background, as does the promise of an information society. Because of this, what can be considered information and what belongs within the IS practitioners' remit is no longer defined solely by organizational requirements.

In this position paper, it will be argued that the emergence of genuinely plural horizons for IS presents both an opportunity and a threat to an IS discipline. Because the IS research community is uniquely positioned between technical and social worlds and their reciprocal interactions, it should be providing intellectual leadership on an *inclusive* discourse about the functions of information within the society at large, including business. By embracing this opportunity, an IS discipline can prevent a fragmentation of information studies into *exclusive* fields, and distinguish itself from cognate fields, such as management and computing. The challenge of an inclusive grasp of the information portfolio may be accompanied by a threat to conceptual coherence, which will be felt particularly keenly in education, where disciplinary authorities are expected to supply theoretical guidance on core concepts. However, providing a theoretical platform for IS education and practice in the shape of generic theories of information is likely to compromise the inclusive mandate to take seriously the nature of information as a practical achievement.

The contemporary form of information is highly heterogeneous. It is said to reside in molecules, cells, tissues, genes, minds, libraries, the environment, the economy, organizations, societies, all sorts of systems, and so on. We can be information rich, empowered, poor, or saturated, experience information revolutions and ages, be information analysts, scientists, workers, and managers, who have information requirements and needs. Information itself can be auditory, visual, tactile, and electronic, which may be precise, poor, candid, secure, public or private, dishonest or mischievous. In fact the limits to information appear to reside less in the environment itself and more in digital technologies and techniques themselves, and in the ingenuity of information architects. Novel technologies such as RFID (radio frequency identification) will no doubt yield novel information categories, as will the convergence of better established fields such as biometric and actuarial information. In short, the notion of information and of systems will not longer be contained by the narrow work and decision processes of organizations, but will relate to wider social and economic concerns.

Among those who work in the information industries, it is generally held to be an open scandal that there is no theory, or even definition, of *information* that is both broad enough and precise enough to accommodate the above uses of the word in a meaningful way. With respect to technology, many also regret the contribution such a theory could make to the design and construction of *information systems*, which, despite a seeming ubiquity, have made a distinctly uneven contribution to either business or society (e.g., Seely Brown and Duguid 2000; Willcocks and Lester 1999). Further, a requirement for clarity over information is likely to be more keenly felt as IS moves its concerns from information structures to information content (Willcocks and Lester 1999), interoperability, and convergence, and from the storage and manipulation of information to the qualities and characteristics of that information itself. These trends are already apparent with the arrival of ontologies (Castel 2002; Guarino 1995), the semantic Web (Berners-Lee 1999), and supporting technologies such as XML, and with the shift in concern from the medium (machines, networks, etc.) to the message (software, decoded DNA sequences, digital identity, etc.).

In an academic context, the theoretical deficit may be inhibiting an understanding of the role such systems play in a wider context, the development of a core IS curriculum in education, and a closer collaboration of researchers and practitioners. Further, if information systems as an academic discipline had its own distinct subject matter which included a robust definition of information, then it would be better equipped to defend itself from cognate, but predatory, emerging fields such as bioinformatics or new media, as well as the more established reference fields of management, computing and information science.

However, the information concept has been notoriously resistant to analytical resolution. Everything can be made to seem obvious, but very little is. Dictionaries assume it rather than explain it (Hobart and Schiffman 1998), and etymologies fail to be more precise. Within the information systems field, many have sought to approach analytical clarification, but explanations are notoriously circular and self-referential. Elsewhere I have used the philosophical argument supplied by Wittgenstein (1953) to demonstrate such a theoretical attitude should be resisted because information appears first as a conceptual term (Stephens 2001). Further, such a theory would form a barrier rather than a contribution to understanding and leave IS as a discipline with a highly stylized but inappropriately idealized and impractical subject matter.

Information Systems students are taught to distinguish information from data and knowledge, and sometimes wisdom, but considering our ever-expanding expectations from the term, we are not used to thinking any more critically about information. Agre (1995) notes that "the term 'information' rarely evokes the deep and troubling questions of epistemology that are usually associated with terms like 'knowledge' and 'belief.'" One can be a skeptic about knowledge but not about information (Nunberg 1996, p. 107). Indeed, the term appears epistemologically normative, and we are encouraged to take information for granted. Thus IS has hitherto concerned itself with information *form*, leaving *content* to other professionals, or merely taking it as given, unproblematically read off the world.

Where the IS remit is extended to working across organizational boundaries, or where formal organizations don't exist at all, the IS practitioner will need to assume responsibility for information content. Information systems work will increasingly be concerned with the assimilation and synthesis of diverse sources of information and knowledge as work and other activities become informated along heterogeneous networks (cf. Castells 1996) rather than managed hierarchical paths. Moreover, the scope of the information concept will need to be revised as a complex phenomena embracing such issues as propriety, regulation, ethics, accessibility, and even aesthetics.

As a consequence, the disciplinary substrate of IS will need to meet this diversity, or risk a subject fragmentation that would leave much of the world of digital applications beyond IS theorizing and IS academic and professional authority. Such an expansion will also demand a new type of competence from the IS practitioner. Mastery of theory and of technical and methodological *means* will be subsumed by (not replaced by) the ability to engage in rule-making and identifying *goals* appropriate to particular information and knowledge practices. This entails a shift from a bureaucratic to a charismatic personality (Gray 2001) and in many ways the IS professionals' job description will have significant parallels with the architect or designer. These tasks involve reworking the situated materials of everyday life, reconnecting the local obligations and commitments of the everyday world to the more abstract processes of public life. The new orientation will, therefore, entail the discipline and flair of information design on the one hand to a renaissance attitude to professional conduct on the other.

In this paper, I have tried to suggest that the rapid diffusion of information and information technologies through diverse areas of human endeavor will require information systems practitioners, academics, and students to consider information content closely. A generic theory of information will at best offer disengagement from the unknown and obstruct the learning processes that are essential features of IS development, much in the way methodologies can (Wastell 1996). Many IS topics will lack clear authorities, and contextual ambiguity and open-endedness relativizes theoretical studies, so the need is to review problems from a number of perspectives. To engage actively with rich information content and knowledge requires of the IS professional, first, a versatility of character and, second, a literacy, or mastery of competing information genres.

Further, academics need to recognize the historically contingent nature of what is the product of a complex of social and technological processes. Recent scholarship on the codex book (Johns 1998) and cartography (Wood 1992) demonstrates that the actually contested nature of artifacts is easily concealed by a totalizing discourse when persuasive institutions conspire to reify and naturalize their products. Both of these examples are particularly instructive for information systems, for the establishment of such intellective artifacts provides the starting point of another process whereby these very products have the appearance of a window through which the world is seen. To reveal the nature of the phenomenon, it is necessary to examine the practices of those who use information.

Ignoring the processes of production would be an intellectual denial and leave the field open to sociologists, science studies scholars, etc., who have a limited technical understanding of the information field and are inclined to pursue their own academic agendas. It would also cripple the IS discipline pedagogically because it would direct intellectual scrutiny away from the essentially practical tasks facing students and professionals—i.e., the labor and craft of information design. Moreover, unless IS researchers become an active voice within the discourse addressing information content, the discipline will be increasingly challenged for its relevance by other academic fields that are rapidly embracing digital technology. Because IS research community is uniquely positioned between technical and social worlds and their reciprocal interactions, it should be able to provide intellectual leadership by locating itself at the center of this discourse.

The information systems discipline must articulate information both as concept and practice, but this cannot be achieved exclusively. The terms are mutually referential, reflexive, and thereby unresolvable by a universal or generic theory. In fact, the problems of sense and reference surrounding conceptual terms such as information are now integral to working practice because, as Zuboff (1989) and others have argued, digital technology creates a context where work becomes an ensemble of readings, inferences, and interpretations. Conceptual issues are not detachable from empirical ones. They are there whenever a question arises about what counts as information, about how to interpret information, about what theoretical or practical inferences to draw, and how to proceed. To delegate responsibility for conceptual issues to a given theory of information would be to deny what much information work is really about: the intellectual, social, and tactile *spadework* or *articulation work* (Suchman 1996) that crafts ambiguity and openness into interpretively tractable working situations. For IS professionals, who are *ipso facto* mandated to intervene in others' lives, such circumscribed theorizing can only be an obstacle to the authentic insights required for ethical and practical action (Stephens and Probert 2000).

REFERENCES

Agre, P. E. "Institutional Circuitry: Thinking About the Forms and Uses of Information," *Information Technology and Libraries* (14:4), 1995, pp. 225-230.

Avison, D., and Fitzgerald, G. *Information Systems Development: Methodologies, Techniques and Tools*, Maidenhead, UK: McGraw-Hill, 2003.

Berners-Lee, T., and Fischetti, M. *Weaving the Web: The Original Design and Ultimate Destiny of the World Wide Web by Its Inventor*, San Francisco: Harper, 2000.

Castel, F. "Ontological Computing," *Communications of the ACM* (45:2), 2002, pp. 29-30.

Castells, M. *The Information Age: Economy, Society and Culture, Volume 1: The Rise of the Network Society,* Malden, MA: Blackwell, 1996.

Gray, J. "The End of Career," *Communications of the ACM* (44:11), 2001, pp. 65-69.

Guarino, N. "Formal Ontology, Conceptual Analysis and Knowledge Representation," *International Journal of Human-Computer Studies* (42), 1995, pp. 625-640.

Hobart, M. E., and Schiffman, Z. S. *Information Ages: Literacy, Numeracy and the Computer Revolution*, Baltimore, MD: John Hopkins University Press, 1998.

Johns, A. *The Nature of the Book: Print and Knowledge in the Making*, London: University of Chicago Press, 1998

Nunberg, G. *The Future of the Book*, Berkeley: University of California Press, 1996.

Seely Brown, J., and Duguid, P. *The Social Life of Information*, Boston: Harvard Business School Press, 2000.

Stephens, R. A. "Problems with a Theoretical Attitude to the Information Concept,"in D. Strong and D. Straub (Eds.), *Proceedings of the 7th Americas Conference on Information Systems*, Boston, 2001, pp. 1963-1967.

Stephens, R. A. and Probert, S. K. "Authentic Intervention in Information Systems Practice," in H. R. Hansen, M. Bichler, and H. Mahrer (Eds.), *Proceedings of the European Conference on Information Systems*, Vienna, July 2000, pp. 64-70.

Suchman, L. A. "Supporting Articulation Work," in R Kling (Ed.), *Computerization and Controversy: Value Conflicts and Social Choices* (2nd ed.), San Diego: Academic Press, 1996, pp. 407-423.

Walsham, G. *Making a World of Difference: IT in a Global Context,* Chichester, UK: Wiley, 2001.

Wastell, D. G. "The Fetish of Technique: Methodology as a Social Defence," *Information Systems Journal* (6), 1995, pp. 25-40.

Willcocks , L. P., and Lester, S. (Eds.). *Beyond the IT Productivity Paradox*, Chichester, UK: Wiley, 1996.

Wittgenstein, L. *Philosophical Investigations*, trans. G. E. M. Anscombe, Oxford: Blackwell, 1953.

Wood, D. *The Power of Maps*, London: Routledge, 1992.

Zuboff, S. *In the Age of the Smart Machine: The Future of Work and Power*, Oxford: Heinemann, 1989.

ABOUT THE AUTHOR

Robert Stephens is senior lecturer in information systems in the School of Information Systems, University of the West of England. He holds a Ph.D. in information systems and researches the evolution of computing and information systems in society, particularly with respect to the dialogic theory of M. M. Bakhtin. He can be reached at Robert.Stephens@uwe.ac.uk.

47 TRUTH TO TELL? Some Observations on the Application of Truth Tests in Published Information Systems Research

Brian Webb
Queen's University of Belfast

Abstract A motivation for the 1984 Manchester conference was to question the applicability of scientific truth tests to the study of socio-technical systems. While most IS researchers now agree that such tests are not appropriate, or at least are not always appropriate, the debate on the use of alternatives continues. This paper examines several truth tests applied to two truth statements in one piece of published research. Since the paper was published in a mainstream IS journal, it is argued that the standard of truth tests applied to this paper is indicative of the standard of truth tests acceptable within the IS community.

It is observed that different standards of truth test are applied, for different purposes, at different stages of the review process, reflecting the different purposes and standards of the truth statements made. Whereas the truth tests applied to the first truth statement (an inductive statement reporting the findings of the research) can be read through the text, those applied to the second truth statement (a deductive statement seeking to generalize these research findings) cannot. The observed differences in the application of internal and external validity tests point to the need for greater transparency in the application of this (second) type of truth test to better inform authors, reviewers, and readers alike; thereby improving the quality of truth statements made and of resultant publications.

Keywords: Habermas; truth; justification; fallibility, generalizability

1 INTRODUCTION

Published research includes multiple truth statements—statements which purport to represent things as they are in the real world. In accepting a paper for publication, reviewers and editors also accept that the truth statements it makes are true (since we would not knowingly accept any statement that we believed to be false). But how may we know the truth?

Habermas (2003) argues that we cannot, or at least we cannot know a real, absolute, and objective truth. Such truth exists but we do not have unmediated access to it. Rather, our concept of truth is filtered through experience and language. Even when we feel a truth, intuitively, in our everyday coping with the world—"we do not walk onto any bridge whose stability we doubt"—we cannot presuppose this truth to be unconditionally valid, since its justification is rooted in language and discourse. The objects we refer to may fail to meet the descriptions we associate with them. Even where there is a correspondence between our statements and the state of affairs that actually obtains, our descriptions can only be established by reference to other statements, which themselves are inconclusive and problematical. Our statements are linguistically determined and therefore fallible. Truth and fallibility are two sides of the same coin.

However, fallibility does not preclude knowledge, but only moderates it. Habermas distinguishes between the objective validity of a statement (*Gültigkeit*) and its *de facto* "validity for us" (*Geltung*). While the former remains elusive, the latter is accessible and assertible through our (shared) sense of normative rightness. What is true is distinct from what we hold to be true but objective validity and normative validity are interconnected. Although Habermas has now rejected his earlier concept of a consensus theory of truth— a proposition is true because it is true in the real world and not because we agree that it is true—argumentation (and consensus) remains the only valid means of justifying a truth—"truth statements that have been problematicized cannot be tested in any other way."

For all practicable purposes, then, the truth statements we make, justify, and test are fallible and, because truth (or normative rightness) is established through discourse, there are degrees of fallibility. All truth statements will not be made to the same standard, all justifications will not be equally convincing, and all tests will not be universally applied. But what is the truth for us? Which truth test should be applied, and when? What is the relationship between the acknowledged fallibility of a truth statement and the justification of it?

These questions are important in academic publishing because their resolution determines not only the standing of a particular journal but, in aggregate, the standing of the entire subject discipline. At a time when the core values of IS are once again being questioned, consensus on what constitutes legitimate truth statements has never been more important, nor elusive. Precisely because academic publishing is so obviously removed from Habermasian conditions of ideal discourse, there is a need to address such questions, and thereby make the reviewing process more explicit and transparent. This paper is presented as a modest contribution to the necessary debate.

2 BACKGROUND

In 1975, Fred Brooks asked, "Who shall be the Aristocracy?" Brooks was interested inwhether or not architects or engineers would be the aristocracy of systems development. In 1997, we posed the same question in relation to multimedia systems development. Would software engineering or graphic design emerge as the dominant paradigm? In order to answer this question, we undertook a study of multimedia development projects, interviewing 26 designers from 15 companies, over a 2 year period. Based upon a small pilot study and our knowledge of the literature, as well as personal experiences, we initially identified two dominant paradigms in multimedia systems development and then sought to prove that these paradigms did in fact exist. Data was collected mostly through interviews but also through observations and documentary sources. Eight graphic design and four software engineering texts were also examined. Data were analyzed using a variety of content analysis techniques drawn from Miles and Huberman (1994). The objective of the research was to identify distinct communities within multimedia systems development by identifying elements of Kuhn's disciplinary matrix.

We reported early outcomes of this research at the European Conference on Information Systems in Cork, Ireland (Gallagher and Webb 1997) and were sufficiently encouraged to develop a paper for journal publication in *European Journal of Information Systems* (Gallagher and Webb 2000). The major truth statements made in both papers are conveniently set out in the abstract to the second paper. For the purposes of this analysis, and in the interests of parsimony, these have been collapsed from four to two, one of each type of truth statement identified.

3 OBSERVATIONS

3.1 Truth Statement 1 (TS1)

> *We have identified two [Kuhnian] paradigms in Multimedia systems development based on the software engineering and graphic design approaches.*

This statement is immediately but only referentially substantiated as previous research is cited (Gallagher and Webb 1997). In that paper, the identification of paradigms is based on the analysis of a small number of interviews (six) with multimedia designers. There is no convincing evidence that such paradigms exist. Rather, the identification is based on the collection and categorization of quotations which are themselves informed by prior assumptions and prejudices that a dichotomy exists between graphic design and software engineering approaches. In fairness, the 1997 paper is clearly a report of some exploratory work, a pilot study, written in the context of ongoing research into the phenomenon, which was then being developed to provide the very kind of validation missing. Thus we can say that TS1 holds up because (1) it is established both referentially (in the literature) and anecdotally (six interviews) that there is some basis for making the statement in the first place and (2) the statement is established empirically in the 2000 paper, as follows.

1. Figure 1 in the 2000 paper (p. 62) sets out the number of distinct (common) and non-distinct (or non-common) elements found in the analysis of 16 interview transcripts. Although 67 elements are common across both disciplines, suggesting

non-distinct communities, 58 elements for software engineering and 54 elements for graphic design are identified as non-core elements (NCEs). Of these, 15 for each community achieved a consensus of 50 percent or greater (the element was identified in at least seven of the 14 sources used).

2. There are four pages of appendices listing each type of element—symbolic generalization, beliefs, values, exemplars—for each discipline, giving a measure of the confidence one can have in the element (the type and number of sources overall and, for each element, the number of sources in which it was found). This is immediately helpful to the reader as it provides further evidence that two paradigms do exist, and importantly provides some transparency for the research process, increasing the reader's confidence in the results.

So far so good. The first truth statement is clearly set out and is evidently supported by the data. The remainder of the paper seeks to establish a practical application for the findings substantiating and leading to Truth Statement 1, that two distinct communities of software engineering and graphic design exist within multimedia systems development. This is not so successful.

3.2 Truth Statement 2 (TS2)

> *The resulting paradigms provide a useful framework from which to inform methodological development within the multimedia field.*

The paper purports to demonstrate this through the introduction and discussion of two concepts from the literature: method evaluation and method integration. While a prima facie case is made for the use of these two concepts and the their application to the findings of this research, no test of applicability or value is offered or given.

Table 4 in the 2000 paper (p. 65) purports to show how this research maps onto the concepts. It is argued that the "common elements can act as the basis of common criteria by which to judge methods" (p. 67) —as in feature analysis. In fact, it is here (and only here) that the word *framework* in Truth Statement 2 is elaborated in any way.

> Feature analysis, therefore, provides the mechanism to determine which method offers greatest support for those mandatory features ...common elements are features (or method requirements) that are acceptable to members of both communities...while NCEs [non-common elements] can be used to derive features/method requirements that are specific to each community (p. 65).

TS2 sets up three further truth statements: (1) feature analysis is a useful way of selecting/evaluating methods, (2) common elements are acceptable to both communities, (3) non-common elements mark out those features that should be specific (within the method) to each community.

Method integration is by means of a meta model, "a framework within each individual method can be integrated in a coherent manner," or a method base from which "method fragments are combined to produce a single method for a particular context." Storyboarding is given as one example of a common denominator for integration (but this is neither developed nor tested). Assuming that methods can be successfully developed from such a methods base, what evidence is there that paradigmatic analysis can improve the process?

This is suggested but not demonstrated. Since method integration involves defining multiple viewpoints and then combining several viewpoints in order to construct a composite method for a particular situation, "the paradigms identified as part of this research programme can assist in the process of integrating several methods fragments because they effectively provide [such] viewpoints" (p. 66).

4 CONCLUSIONS

Habermas has recognized that "the truth may outrun justified belief"—truth is not defined or limited by our abilities to describe, justify, and test it, its meaning is not exhausted by consensus or normative rightness. Here we observed another kind of runaway truth—that of the truth statements made. As the paper progressed, these became less convincing, less well argued in reason or in data (or in both). The gap between the truth statement made and its justification is greatest when that statement attempts to induct or generalize the findings of the research. Here we have nested truth statements and recursive truth statements, truth statements based solely on other truth statements that are themselves problematical and inconclusive.

The justification of TS2 is not based on empirical data (as in the case of TS1) but rather on a series of assertions derived from the literature. While the interested reader can establish the *bone fides* of the referenced research, and make his or her decision as to its relevance in this context, in this paper, TS2—in the absence of evidential qualification—simply presumes the stated outcome. Given the space limitations of any published paper, this is perhaps understandable but such statements are inherently dangerous. Through repetition, under conditions of fallible discourse, they may become accepted truths that (erroneously) reflect upon the status of the original research and that (inappropriately) influence future research directions.

One interpretation of the observations made in this paper is that the reviewers and editors of *European Journal of Information Systems* are appropriately applying truth tests to truth statements, since they would not otherwise accept TS2 under the same conditions of justification applied to TS1. This interpretation is based upon two presumptions: (1) that TS2 is different from TS1 in ways that warrant a different justification and (2) the testing of that justification is qualitatively different from the testing of TS1, i.e., the standards of proof—or acceptability of the truth for us—in this instance, are different from the standards of truth applied to TS1. Acceptance of this interpretation is good news for the IS research community, or the European IS research community to be exact, since Lee and Baskerville (2003) suggest that in many top U.S. journals, tests of acceptability are inappropriately applied and "generalizations are sometimes taken to be proven statements rather than taken as well founded but as yet untested hypotheses" (p. 224).

Yet if we accept different types of truth statements and different types of truth tests, then we need also to know exactly what those tests are, as well as when and how they should be applied. While in the matter of TS1 these are accepted and asserted (through canons of positivist or interpretivist research), in the matter of TS2 they are not. What do we mean, for example, by *well founded*? It is a truism (but not necessarily true) that although not all truth statements are generalizations, all generalizations are truth statements and it is these that continue to be the least understood and most contentious.

REFERENCES

Brooks, F. P. *The Mythical Man-Month*, Reading, PA: Addison-Wesley, 1975.

Gallagher, S., and Webb, B. "Competing Paradigms in Multimedia Systems Development: Who Shall be the Aristocracy," in R. Galliers, S. Carlson, C. Loebbecke, C. Murphy, H, Hansen, and R. O'Callaghan, (Eds.), *Proceedings of the Fifth European Conference on Information Systems*, Cork, Ireland: Cork Publishing, 1997, pp 1113-1120.

Gallagher, S., and Webb, B. "Paradigmatic Analysis as a Means of Eliciting Knowledge to Assist Multimedia Methodological Development," *European Journal of Information Systems* (9), 2000, pp. 60-71.

Habermas, J. *Truth and Justification*, translated by Barbara Fultner, Cambridge, UK: Polity Press/Blackwell Publishing Ltd., 2003.

Lee, A. S., and Baskerville, R. L. "Generalizing Generalizability in Information Systems Research," *Information Systems Research* (14:3), September 2003, pp. 221-243

Miles, M. B., and Huberman, A. M. *Qualitative Data Analysis: An Expanded Sourcebook*, London: Sage Publications, 1994.

ABOUT THE AUTHOR

Brian Webb is a senior lecturer in Information Systems, School of Management and Economics, Queen's University of Belfast, N. Ireland. From January to June 2003 he was Distinguished Erskine Fellow in the Department of Accounting, Finance, and Information Systems, Faculty of Commerce, University of Canterbury, New Zealand. He holds a Bachelor's degree from Queen's, an MBA from the University of Ulster and a Ph.D. from University College London. In 1999 he was a visiting scholar in the Department of Computer Science, University of British Columbia, Canada. Prior to becoming an academic, Brian worked as a systems analyst in both the UK and the United States. His current research interests are in the areas of e-business performance measurement and multimedia systems development. Brian has been involved in a number of technology transfer programs in Ireland and the UK, and has served as an external assessor for the Engineering and Physical Sciences Research Council (UK) and the Research Council for the Humanities and Social Sciences (Ireland). Brian can be contacted by email at b.webb@qub.ac.uk.

48 HOW STAKEHOLDER ANALYSIS CAN BE MOBILIZED WITH ACTOR-NETWORK THEORY TO IDENTIFY ACTORS

A. Pouloudi
Athens University of Economics and Business

R. Gandecha
C. Atkinson
A. Papazafeiropoulou
Brunel University

Abstract Actor-network theory studies provide detailed accounts of how human and nonhuman actors gradually form stable actor networks. However, due to their focus on a particular context, there is little generic guidance on how such relevant actors can be identified when a different research context is under study. The principles of (human) stakeholder behavior presented in this paper guide the identification of human stakeholders through an iterative, interpretive, dynamic and context-contingent process. We show how they can be adopted and extended to include the identification of nonhuman actants as well. Thus, we argue that they can be instrumental in providing a generic, context-free guidance to stakeholder identification that is currently missing from ANT studies.

Keywords: Stakeholder analysis, actor-network theory (ANT), actors, stakeholders, implementation

1 ACTORS AND STAKEHOLDERS

Actor-network theory (ANT) has its origins within the work of Callon (1986, 1991) and Latour (1987, 1992). According to ANT, humans and machines interact in a multiplicity of roles, together constituting networks that act as independent autonomous actors, the *actor networks*. An information system with its information technologies and its human users may be viewed as an actor network. In actor networks, actors collectively act; actor networks are by nature heterogeneous and information-based. Networks within ANT are also perceived as actors (Latour 1999; Suchman 2002). A key proposition of ANT is to treat human and nonhuman actors as well as networks symmetrically. However, there is little guidance on how these actors are identified. According to ANT, the researcher can trace the actor network by following the actors and observing what they do. The ANT literature provides several examples of how this can be applied in a specific context. However, the idea of following the actors cannot easily be translated to practical guidance for the researcher engaging with complex socio-technical phenomena. Stakeholder analysis is an approach much more explicit in this respect (Pouloudi 1999). However, the application of stakeholder analysis (unlike ANT) in the IS context has been predominantly restricted to the study of human stakeholders.

We argue that mobilizing stakeholder analysis with ANT, within research and practice, offers a means of identifying ANT actors systematically, thus enhancing ANT analysis. Although the notions of stakeholders (in stakeholder analysis) and actors (in ANT) are implicitly related, there is little research that attempts to bring the respective approaches together. This paper makes a contribution in this area by extending the stakeholder identification process to nonhuman actors and networks, with a view to support ANT analysis.

2 STAKEHOLDER IDENTIFICATION IN ANT

In order to offer generic guidance for stakeholder identification within ANT, we need a broad definition of stakeholders. Thus, we adapt Freeman's (1984) definition of organizational stakeholders for information systems: "A stakeholder of an information system is any individual, group, organization or institution who can affect or be affected by the information system under study."

Previous research in IS proposes a number of principles of stakeholder behavior that can be used to guide an interpretive identification and analysis of stakeholders (Table 1). This approach puts forward a case for dynamic and iterative stakeholder identification. Such a dynamic and iterative process relates closely to some of the fundamental premises of actor-network theory, which "concentrates attention on a movement" (Latour 1999 p. 17).

The term *nonhuman stakeholders* has been used in information systems research already (e.g., Pouloudi and Whitley 2000; Vidgen and McMaster 1996). Here, we explore whether the stakeholder identification process cited above can include them. For this, we refer to the principles of stakeholder behavior and their implications as depicted in Table 1.

- The first principle emphasizes the importance of context and the specific time frame. It applies to human stakeholders as it does to nonhumans and networks. Indeed, ANT also emphasizes the importance of local context, as the stories told in ANT terms are largely defined by this.
- The second principle, prompting the researcher to explore links among stakeholders, is central to the notion of the actor network in ANT. Clearly, in ANT,

Table 1. Propositions for Stakeholder Identification and Analysis (Pouloudi 1999)

Principles of Stakeholder Behavior	**Implications for Stakeholder Identification and Analysis**
1. The set and number of stakeholders are context and time dependent	• stakeholder map should reflect the context • stakeholder map should be reviewed over time
2. Stakeholders cannot be viewed in isolation	• consider how stakeholders are linked
3. A stakeholder's role may change over time	• adopt a long-term perspective
4. Stakeholders may have multiple roles	• study how perceptions change
5. Different stakeholders may have different perspectives and wishes	• there are different versions of the stakeholder map to be drawn
6. The viewpoints and wishes of stakeholders may change over time	• these different versions of the stakeholder map should be reviewed over time
7. Stakeholders may be unable to serve their interests or realize their wishes	• need to consider political issues (as well as technical, economic or other)

these links or interactions occur among nonhuman and human entities. Bringing the two principles together, Latour (1999, p. 18) notes that "the network pole of actor-network...refers to...the *summing up* of interactions through various kids of devices, inscriptions, forms and formulae, into a very local, very practical, very tiny locus."

- The third and fourth principles can also be interpreted broadly to include nonhuman stakeholders. For example, technological artifacts, such as information systems, can be put to different uses at different points in time or depending on the human agents that interact with them.

The implications of these principles for identifying human *and* nonhuman stakeholders in an actor network, as well as the networks themselves, is that stakeholders should reflect the dynamics of the local context and not be treated as static elements but, rather, revisited over time for new entries. Pouloudi (1998) suggests that each stakeholder can lead to the identification of further stakeholders; for example, stakeholders refer implicitly or explicitly to other stakeholders when interviewed. While a nonhuman stakeholder cannot be interviewed, other (multiple) actors may speak on its behalf. An IT director, an upgrade specification, or a user speaking on behalf of the technology are examples. Some nonhuman stakeholders (e.g., texts) include reference to other human or nonhuman stakeholders, therefore providing cues for further stakeholder identification.

The last three principles (5, 6, and 7) of Table 1 may be seen to concern human stakeholders exclusively, as they refer to perspectives, viewpoints, interests, and wishes. However, organizational networks, through successive translations, are made up of humans and nonhumans with aligned interests. These principles imply there can be different versions of who the stakeholders are, depending on the time frame and the perspective adopted. ANT does not explicitly support this process, even though it strives to allow new actors "to define the world in their own terms, using their own dimensions and touchstones" (Latour 1999, p. 20).

The final principle draws particular attention to the political nature of stakeholder identification. While ANT acknowledges the importance of politics (Latour [1987] eloquently presents the politics of research), it does not do so explicitly in the process of *describing* the actor network. Instead, the ANT researcher is typically perceived as a reliable spokesperson for the network's establishment and/or the translation(s) of the actor network. It could be seen, though, that the researcher is acting on behalf of another (university or professional) network, one that is seeking to translate the network under investigation in line with its own (research) interests; indeed mobilizing stakeholder analysis is itself an analytical artifact, a machination, to achieve this.

3 CONCLUSIONS AND FURTHER RESEARCH

Atkinson and Brooks (2003) argue that despite ANT's call for a symmetric treatment of humans and nonhumans, the human/machine network duality "exists at the heart of the IS practice and research." Aiming to support ANT's more holistic approach, this paper has made the case for using stakeholder analysis together with ANT. The actor network binds the stakeholders together; what influences one stakeholder can ripple through the network. This idea of strongly interlinked stakeholders, that are contingent on the context, has been put forward for studying stakeholders in information systems research (Pouloudi 1998). Taking these research proposals on board, and exploring their application to nonhuman stakeholders, we have argued that stakeholder analysis can provide guidance for the systematic identification of the multiple, interdependent human and nonhuman actors and networks. In turn, the powerful notion of translations in ANT, coupled with stakeholder analysis, can contribute to a richer understanding of complex phenomena. Our point is that both stakeholder and actor network analyses have been influenced by the element of alternative interpretations of the network under study.

Starting with a set of guidelines for stakeholder identification suggested in earlier information systems research, the paper explored their applicability alongside ANT studies. We have argued that, while the value of ANT is to a great extent due to its attention to the local context, and the recommendation to follow the actors in it, it does not help the researcher recognize who the actors worth following are. The approach suggested in this paper recognizes and addresses this fundamental methodological weakness, without overlooking the importance of the local context that is key to ANT. Specifically, it guides the researcher to consider who would be the relevant stakeholders for each further actor or actor network identified. It argues, in line with ANT, that actors should reflect the local context and are interlinked, but are also likely to change over time. Importantly, it makes explicit the political nature of ANT itself: the actors and actor networks identified are contingent on the researcher; thus there are different actor networks to be identified, different explanations to be given for their formation, and different stories to be told. This argument provides an opportunity to revisit ANT within an interpretive research framework.

Thus, stakeholder analysis enhances ANT methodologically because it acknowledges explicitly the multiple stakeholder agendas, interests, and values. This prompts the researcher to recognize that there are multiple versions (or stories) of translation in each actor network, depending on the perspective adopted and the values and interests that characterize (or are inscribed in, in the case of nonhumans) stakeholder views. Research is, it may be argued, itself a translational struggle between the researchers and their university network and the subject network(s) under review.

Further research is called to explore the value of the guidelines presented in this paper in practice. We intend to look at one of the recent initiatives introduced by the National Heath Service (NHS) in the UK, the Integrated Care Record Service (ICRS). It is a "broad, continuously expanding and maturing portfolio of services covering the generation, movement and access to health records," which includes electronic prescribing in hospitals and workflow capacities to manage patient care pathways through the NHS (NHS 2002). The scale and ambitious outcomes of the ICRS project make it an interesting, complex context of information systems implementation for study. The interplay and critical role of both human stakeholders, such as patients and healthcare professionals, and nonhuman actors, such as care records, in the implementation of the ICRS have already prompted the use of ANT for its analysis (Gandecha et al. 2003). We believe that the line of research suggested in this paper can further help in the study of complex information systems implementations, such as the ICRS.

REFERENCES

Atkinson, C., and Brooks, L. "StructurANTion: A Theoretical Framework for Integrating Human and IS Research and Development," in J. Ross and D. Galletta, *Proceedings of the Ninth Americas Conference on Information Systems*, Tampa, 2003, pp. 2895-2902.

Callon M. "Some Elements of a Sociological Translation: Domestication of the Scallops and Fishermen of St Brieuc Bay," in J. Law (Ed), *Power Action and Belief: A New Sociology of Knowledge*, London: Routledge and Kegan Paul, 1986, pp. 196-233.

Callon M. "Techno-Economic Networks and Irreversibility," in J. Law (Ed.), *A Sociology of Monsters? Essays on Power, Technology and Domination*, London: Routledge, 1991, pp. 132-161.

Freeman, R. E. *Strategic Management: A Stakeholder Approach*, Cambridge, MA: Ballinger Publishing Co., 1984.

Gandecha, R.; Atkinson, C.; and Papazafeiropoulou, N. "Machinations and Machinations: The Integrated Care Records Service in the UK NHS," in J. Ross and D. Galletta (Eds.), *Proceedings of the Ninth Americas Conference on Information Systems*, Tampa, 2003, pp. 875-880.

Latour, B. "On Recalling ANT," in J. Law and J. Hassard (Eds.), *Actor Network Theory and After*, Oxford: Blackwell Publishers, 1999, pp. 15-25.

Latour, B *Science in Action: How to Follow Scientists and Engineers Through Society*, Cambridge, MA: Harvard University Press, 1987.

Latour, B. "Where Are the Missing Masses? Sociology of a Few Mundane Artefacts," in W. Bijker and J. Law (Eds.), *Shaping Technology, Building Society: Studies in Sociotechnical Change*, Cambridge, MA: MIT Press, 1992.

NHS. "The NHS Explained," National Health Service, 2002 (available online at http://www.nhs.uk/thenhsexplained/how_the_nhs_works.asp).

Pouloudi, A. "Aspects of the Stakeholder Concept and Their Implications for Information Systems Development," in R. H. Sprague (Ed.), *Proceedings of the 32nd Hawaii International Conference on Systems Sciences*, Los Alamitos, CA: IEEE Computer Society Press, 1999.

Pouloudi, A. *Stakeholder Analysis for Interorganisational Information Systems in Healthcare*, Unpublished PhD Thesis, London School of Economics and Political Science, 1998.

Pouloudi, A., and Whitley, E. A. "Representing Human and Nonhuman Stakeholders: On Speaking with Authority," in R. Baskerville, J. Stage, and J. I. DeGross (Eds.), *Organizational and Social Perspectives on Information Technology*, Boston: Kluwer Academic Publishers, 2000, pp. 339-354.

Suchman, L. A. "Human/Machine Reconsidered," Unpublished paper, Department of Sociology, Lancaster University, 2000 (available online at http://www.comp.lancs.ac.uk/sociology/soc040ls.html).

Vidgen, R., and McMaster, T. "Black Boxes, Nonhuman Stakeholders and the Translation of IT Through Mediation," in W. J. Orlikowski, G. Walsham, M. R. Jones, and J. I. DeGross (Eds.), *Information Technology and Changes in Organizational Work*, London: Chapman & Hall, 1996, pp. 250-271.

ABOUT THE AUTHORS

Athanasia (Nancy) Pouloudi is an assistant professor in the Department of Management Science and Technology at the Athens University of Economics and Business (AUEB), Greece. She holds a first degree in Informatics (AUEB, Greece), and an M.Sc. and Ph.D. degree in Information Systems (London School of Economics, UK). Her research focuses on strategic and social issues in information systems, specializing in electronic commerce, knowledge management and stakeholder issues with more than 70 publications in these areas. She is an associate editor for the *European Journal of Information Systems*, member of the Editorial Board of the *Journal of Electronic Commerce in Organizations* and the *International Journal of Society, Information, Communication and Ethics*, and has served as a Program Committee member for several international conferences. She has taught information systems at Brunel University (as a lecturer) and the London School of Economics (as a teaching assistant) and held visiting positions at Erasmus University (The Netherlands), the University of Hawaii (USA), and the Athens Laboratory of Business Administration (Greece). She has acted as scientific coordinator for AUEB on a number European projects. She is codirector of the Seventh ETHICOMP Conference and track chair for e-work and virtual organizations for the European Conference on Information Systems 2004. Nancy can be reached at pouloudi@aueb.gr or through her home page, http://istlab.dmst.aueb.gr/~pouloudi/.

Reshma Gandecha is a researcher at the Centre of Healthcare Informatics and Computing at Brunel University. Originally a pharmacist, she undertook a BPharm and MrPharmS at UWCC at Cardiff School of Pharmacy in Wales. She has since worked in multiple environments as a practitioner in industry, hospital, and community settings. Through her work experience she has been interested in developing her management and information skills. Reshma further undertook MBA specializing in Information Management at Brunel University. She then pursued a career in medicines information especially managing the new entry of drugs at Health Policy Strategy level. Her current field of study is centred around multi-professional teams, including clinicians, managers, and information systems practitioners in affecting integrated organizational development especially in the UK IT/IS healthcare arena. Her current passion is Healthcare Information Systems Management in the UK National Health Service. Her research investigates the process of the National Programme for IT from strategy through to realization particularly the effective delivery of the integrated NHS Care records for real time patient care. She can be contacted via email at Reshma.Gandecha@brunel.ac.uk.

Chris Atkinson is a senior lecturer in Information Systems in the UMIST Department of Computation. Until recently he was with Brunel University's Department of Information Systems and Computing Science. Originally a civil engineer, he undertook an M.Sc. and Ph.D. at Lancaster University in soft systems with a particular focus on systemic metaphor and its role in organizational problem solving. He has worked as both an academic and practitioner, focusing on how to integrate information systems development and organizational change, especially within healthcare settings. To that end, he has evolved and extensively deployed the Soft Information Systems and Technologies Methodology (SISTeM). Actor-network theory has recently emerged as an important framework for research, practice and methodological development. Its integration with structuration theory has proved fruitful as a further area for research and development. His field of study and practice has centred on working with multi-professional teams, clinicians, managers and information systems practitioners in affecting integrated organizational development. Chris may be contacted via christopher.atkinson@brunel.ac.uk.

Anastasia Papazafeiropoulou is a lecturer at the Information Systems and Computing department at Brunel University, where she received her Ph.D. She has worked as a technical trainee in the European Union (EU) in Brussels and as a research associate with expertise in electronic commerce at the Athens University of Economics and Business and Brunel University. She holds a first degree in Informatics (1994, Athens University of Economics and Business) and a M.Sc. in Information Systems (1997, Athens University of Economics and Business). Abastasia's research interests fall within social aspects and policy issues of electronic commerce, more specifically focusing on awareness creation and knowledge diffusion mechanisms available for electronic commerce adoption from Small and Medium Size enterprises.

49 SYMBOLIC PROCESSES IN ERP VERSUS LEGACY SYSTEM USAGE

Martin M. T. Ng
Michael T. K. Tan
National University of Singapore

Abstract Being hailed as possessing the ability to drive effective business reengineering and management of core and support processes, it is not surprising that enterprise resource planning (ERP) systems have been adopted by more than 60 percent of Fortune 500 companies at the turn of the century. In contrast, legacy systems have frequently been attached with negative connotations. Yet at the same time, it is common knowledge that some legacy systems are not replaced when companies adopt ERP solutions, while new in-house systems still continue to be developed. While risks and time involved have been highlighted as possible reasons for the non-replacement of legacy systems, little attention has been paid to process issues as well as the symbolic meanings attached to the ERP vis-à-vis the other coexisting information systems.

This research employs symbolic interactionism as the informing theoretical perspective in an ethnographic study of a large government authority in Singapore. Our findings surprisingly indicate that contrary to popular belief, the end-users in that organization tend to attach rather favorable symbols to their legacy and new in-house developed systems, while displaying relatively negative sentiments towards their ERP package. In this paper, we first discuss the different symbolism attached over the years to the coexisting systems. Next, we highlight how certain symbols gradually got sedimented over time. Finally, we demonstrate how the consequent manifestations of these symbolic realities influenced certain organizational actions that impacted the usage and perpetuation of the coexisting systems.

Keywords: Symbolic interactionism, symbolism, enterprise resource planning (ERP), legacy system

1 INTRODUCTION

By 2000, more than 60 percent of Fortune 500 companies had adopted ERP packages and this is a trend that is increasingly being followed by small- and medium-sized enterprises (SMEs) as they realize the cost effectiveness and competitive necessity to follow suit (Klaus et al. 2000). In contrast, legacy systems have been described as having a "consequentially negative impact on competitiveness" in that they tend to "resist modification and evolution to meet business requirements" (Brodie and Stonebraker 1995) while being "non-maintainable and inflexible" (O'Callaghan 1999). In fact, during the implementation process of ERP packages, legacy systems have sometimes been constructed by the organization to assume the identity of a "dying system" in order to facilitate the transition (Alvarez 2000). However, it is a well-known fact that some legacy systems are not replaced when companies adopt ERP solutions (Themistocleous and Irani 2001), while new in-house systems still continue to be developed.

In this study, we adopt symbolic interactionism as the informing theoretical perspective as it may help to shed new light on the difference in attitudes toward such coexisting systems. Besides being underutilized in IS research, this perspective is particularly appropriate in this study as little attention has been paid in extant ERP literature to process issues (Markus and Tanis 2000) as well as the symbolic meanings attached to the ERP vis-à-vis the legacy and new in-house developed systems.

2 ERP VERSUS LEGACY SYSTEMS REVISITED

Over the years, firms have been trying to replace legacy systems built on outdated technologies, based on the potential benefits and promises of standardization offered by the more powerful and comprehensive ERP package (Ross and Vitale 2000).

However, Themistocleous and Irani (2001) recently argued that ERP packages have in fact failed to achieve application integration and 38 percent of companies who adopt these ERP solutions do not replace their legacy systems. Specifically, ERP packages do not seem to be able to "cover all the business processes of an enterprise" and, as such, organizations typically do not "abandon all their existing applications when adopting ERP solutions" (Schönefeld and Vering 2000). Indeed, there is an increasing recognition of the need for legacy systems to persist in the organization in varying degrees and, toward that end, generic IS design strategies have been offered for effective coexistence of ERP and legacy systems within organizations (Holland and Light 1999).

2.1 Symbolism at Work?

In information systems development work, it has been noted that "myth, magic and metaphor" are often employed (Hirschheim and Newman 1991). Similarly, many ERP implementations are often associated with a "mythmaking" process whereby the incoming ERP package is usually slated to be the "ideal system" while the outgoing legacy systems are usually attached with the title of a "dying system" (Alvarez 2000). Clearly, there may be much symbolism at work in the implementation and use of ERP systems within organizations. Gaining an appreciation of such symbolism may, therefore, yield new and interesting insights in this ERP arena.

2.2 Symbolic Interactionism as a Theoretical Perspective

Together with phenomenology and hermeneutics, symbolic interactionism is one of several prominent interpretive approaches to social science research. The central idea of symbolic interactionism is that humans attach meanings to objects during the course of everyday social interaction with others. In this regard, there has been increasing interest over the years in the role of symbolism within organizations in general (Turner 1990) and of the symbolic nature of computers and IT in particular (Prasad 1993). In IS literature, prominent researchers have also similarly recognized the importance of symbolism when organizational and technological contexts intersect (e.g., Hirschheim and Newman 1991). However, there have been few noteworthy studies in IS literature that have explicitly used symbolic interactionism as a theoretical perspective.

In this regard, symbolic interactionism is particularly appropriate in this study because it "simultaneously emphasizes both process issues and the roles of meaning and symbols" (Prasad 1993). Consideration of process issues is important to facilitate understanding of how the symbols and meanings attached by end-users to the coexisting information systems (i.e., ERP vis-à-vis legacy and new in-house developed systems) come to be *sedimented* over time.

3 CASE DESCRIPTION

This research takes place in a large government authority (the Authority) in Singapore. Having begun operations in the early 1970s, the Authority has since grown to a strength of 10 divisions, employing approximately 2,000 people and establishing itself as a major global hub in the transportation industry. The sequence of events and the corresponding implications impacting the coexisting information systems are best represented on a timeline, elaborated in Figure 1.

4 RESEARCH METHODS

This research uses symbolic interactionism as the theoretical perspective to inform an ethnography study in which the first author was immersed in day-to-day activities at the Authority for almost three months in early 2003. During this period, he worked in the Information Systems (IS) department, whose function is to oversee project implementation of back-end systems (including the ERP package) and to ensure the smooth daily operations of these systems. Being attached to the IS department allowed the author to interact with end-users and management by following the support staff as they went on-site to solve day-to-day problems. Data collection consisted of observation, participant-observation, and interviews (Prus 1996) including informal chats and document reviews. Ethnographic notes were recorded and reflected upon contemporaneously (at the end of each day) as much as possible. The subsequent few months were spent off-site but regular contact was maintained in order to clarify and verify the findings, as the written ethnographic account took shape.

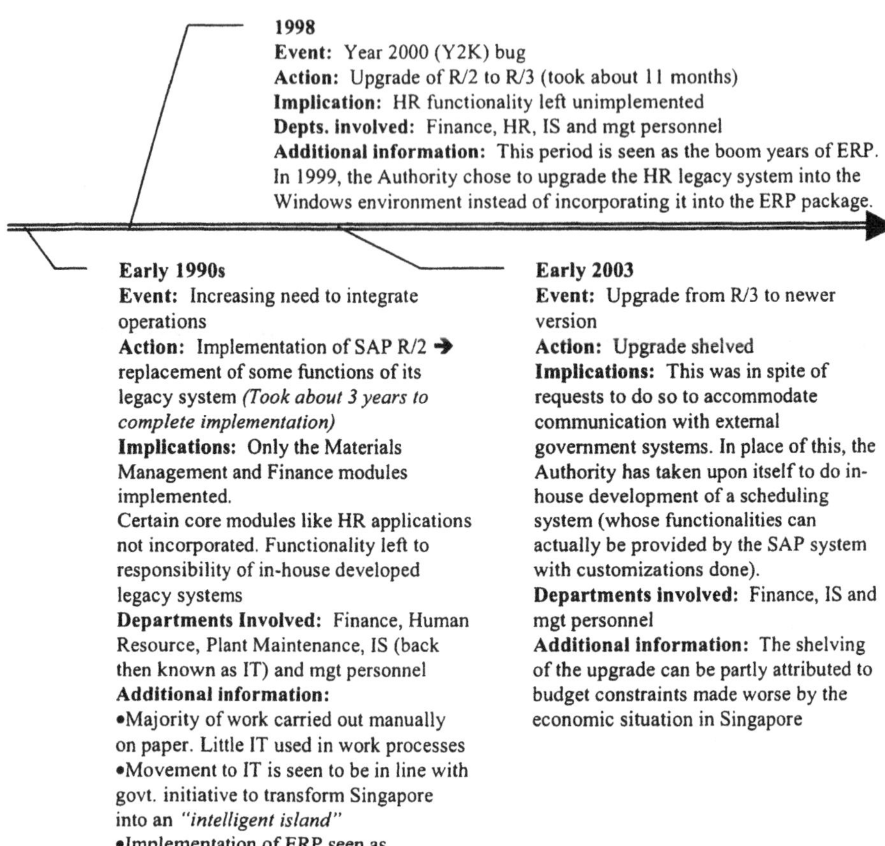

Figure 1. Timeline of Events

5 RESEARCH FINDINGS

5.1 Multiple Symbolic Representations

To begin, it is important to note that the multiple symbols attached to the ERP package vis-a-vis the legacy and new in-house developed systems (as tabulated in Table 1) are the more prominent ones to emerge from this study.

Together, these seven different symbolic realities serve to paint a clearer picture of the end-users' perspectives of the coexisting systems. Clearly, the end-users do not seem to attach any negative symbols to their legacy and new in-house systems while doing so for the ERP package. Another point worth noting is that in spite of the transition from the mainframe-based R/2 system to the client-server R/3 system, there seems to be an improvement in the status of the legacy system relative to the ERP package: some of the positive symbolism attached to the latter gradually eroded while much of negative symbolism gradually sedimented over time.

5.2 Temporal and Local Meanings of Symbolic Representations

As per Table 1, this study shows that not only do the end-users attach different symbols to the coexisting information systems but that they also interpret the various symbols in different ways (e.g., *efficiency* means better use of resources to some, while others perceive it as integration). As noted by Prasad (1993), a focus on local meanings is important to symbolic interactionists.

We also observe that beyond local meanings, some symbols signify different meanings at different times (e.g., *potential*). Close examination of these temporal and local meanings reveals that the end-users have, over time, developed an increasingly negative sentiment toward the ERP package.

5.3 Sedimentation of Symbols

The study of the sedimentation process of the symbols is as important as, if not more than, the identification of the symbols themselves and their local/ temporal meanings. This focus is supported by theorists like Fine (1992) who argue that attention should be directed to the forces behind the sedimentation.

At any one time, there are multiple symbols attached to the systems but only those that exhibit a degree of persistence and presence will develop into organizational realities. Certain symbols exhibit a strong presence within a given time frame, but fail to persist as the system undergoes a transition.

In the case of the Authority, we argue that there were three main forces that proved to be instrumental in determining the persistence and presence of the various symbols (as tabulated in Table 1). They are

- Management influence: Impact on *misalignment* and *uncontrollable*
- Innovation fit: Impact on *efficiency, potential,* and *exclusivity*
- Interaction with system: Impact on *uncontrollable, efficiency, potential, exclusivity* and *intuitive*

5.4 Symbolic Manifestation and Impact

As noted by Prasad (1993), the "process of enactment, whereby symbolic realities mediate meaningful action" is a central concern of any research project for symbolic interactionists. It was apparent that the evolving symbolism resulted in

- *Erosion of support* for the proposed upgrade of the ERP package in early 2003
- *Resistance* to the incoming human resources component of the ERP package during the initial R/2 implementation in the early 1990s
- *Inadequate use* of the ERP package over time

Table 1. Multiple Symbols of the ERP Package vis-à-vis Legacy and In-House Developed Systems

Symbols	ERP Package (Meanings of Symbols)		Legacy and In-House Systems (Meanings of Symbols)	Sedimentation Forces (Symbolism is Sustained from R/2 to R/3 Period)
	R/2	**R/3**		
Uncontrollable	Inability to use the ERP package effectively*			• Mgt Influence (*lack of effective feedback system to mgt resulting in inability to improve on package*); • Interaction with system (*bad experience attempting to customize package)*
	Inability to harness the full potential of the package's capability (*BUT more prominent in the post-R/3 period when users were more IT-savvy*) *			
Added responsibility	Use of the system resulting in tasks that were *"not their job"**			**Note:** *This symbol did not sediment (it is not prominent during the R/3 period)*
	Effort by the finance dept. to increase their workload*			
Misalignment	Not cost-saving or profit-generating*			• Mgt influence (*Mgt view that the package is not customer-facing, while favoring other information systems; this consequentially influenced the end-users' perspectives*)
	Not customer-facing* (hence seen as not vital to fulfilling the corporate vision)			

Table 1. Multiple Symbols of the ERP Package vis-à-vis Legacy and In-House Developed Systems

<table>
<tr><th rowspan="2">Symbols</th><th colspan="2">ERP Package (Meanings of Symbols)</th><th rowspan="2">Legacy and In-House Systems (Meanings of Symbols)</th><th rowspan="2">Sedimentation Forces
(Symbolism is Sustained from R/2 to R/3 Period)</th></tr>
<tr><th>R/2</th><th>R/3</th></tr>
<tr><td rowspan="4">Efficiency</td><td colspan="2">Speeding up of work proceses*</td><td>Speeding up of work processes</td><td rowspan="4">• Interaction with system (First-hand experience shaped the impressions of end-users);
• Innovation fit (Information Systems were well suited for the assigned job)</td></tr>
<tr><td colspan="2">Integration*</td><td rowspan="3">Better use of resources</td></tr>
<tr><td colspan="2">Better use of resources*</td></tr>
<tr><td colspan="2">Data collection on-site*</td></tr>
<tr><td rowspan="2">Potential</td><td>Possessing capability to improve current work processes*</td><td>Untapped capabilities "like an ocean"[#]</td><td></td><td rowspan="2">• Interaction with system (First-hand experience shaped the impressions of end-users; during the R/3 period, end-users who had experience with other information systems were well aware of the untapped capabilities of the package);
• Innovation fit (Package well suited for the assigned jobs)</td></tr>
<tr><td>Stepping stone for future organization improvements*</td><td></td><td></td></tr>
<tr><td>Exclusivity</td><td></td><td></td><td>Possessing functionalities that cannot be performed by other commercial packages</td><td>• Interaction with system (First-hand experience with system);
• Innovation fit (Suitability of system to work process)</td></tr>
<tr><td>Intuitive</td><td></td><td></td><td>User-friendliness and ease of use the result of being incorporated into intranet system</td><td>• Interaction with system (First-hand experience; system is part of the ubiquitous Windows-based environment)</td></tr>
</table>

NB: The asterisk (*) signifies that the symbolism in question has multiple local meanings. The hex (#) signifies that the symbolism has undergone a temporal change in its meaning.

Together, these manifestations give an idea as to why certain core modules like the human resources application were left out during the initial R/2 implementation and subsequent R/3 upgrade. They also shed light on why during early 2003, end-users readily accepted management's decision to shelve the SAP upgrade without much opposition, in spite of the fact that it resulted in extra in-house development work (to accommodate electronic communication with external government systems). As noted by Lindesmith (1981), "a cause must be thought of as a process—not as a condition...or event" and in the same vein, one must recognize that there are many other events working with the identified manifestations (which only constitute part of the cause of the actions taken by the Authority). At the IFIP Working Conference in Manchester, we will elaborate on this important issue.

6 DISCUSSION

As Prasad (1993) recognized in his employment of symbolic interactionism, theoretical insights offered are more like guiding propositions than testable hypotheses. As such, the findings emerging from this study may not be universally applicable statements, but they offer several empirically supported perspectives that aid in the understanding of the coexistence of the ERP package vis-à-vis the legacy and new in-house developed systems.

6.1 Relative Position of Systems over Time

Although IS literature has recognized the negative connotations usually attached to the outgoing legacy systems, this study finds that this representation may not be always accurate. In the case of the Authority, end-users instead view their legacy and new in-house systems more favorably than the commercially developed ERP package.

Specifically, this study suggests that the relative positions of coexisting systems may be the result of a combination of processes resulting in the sedimentation of various symbolic realities within the organization. For example, sedimentation forces such as management influence and technological fit have been frequently highlighted as important implementation success factors (e.g., Meyers et al. 1999). This study demonstrates that beyond the implementation phase, these same forces may also be important factors to consider in the maintenance and upgrade phases of information systems.

6.2 Perpetuation and Usage of Systems over Time

In this regard, ERP maintenance and upgrade decisions are frequently attributed in the literature to such fundamental factors as maintenance costs, availability of new versions, and benefit-realization (e.g., See 2001). However, this study suggests that the symbolism attached to the ERP system by the end-users may also be critical. This is in line with Feldman (1989), who notes that all too often, the technical aspects are focused on when considering technological change processes, while the equally important symbolic aspects are neglected. In fact, Feldman and March (1981) suggest that "information technologies are used and introduced primarily for their symbolic value."

Similarly for legacy systems, we suggest that their varying degrees of persistence may not always be a result of the time needed and the risks involved, but could be due to the organizational impact of the symbolic representations attached to them.

7 CONCLUSION

In this study, the combination of leveraging the theoretical strengths of symbolic interactionism (as a cognitive lens) and the empirical strengths of ethnography (as the strategy of inquiry) has allowed for a unique comparison of the relative positions of the ERP package vis-à-vis legacy and new in-house developed systems within a particular organization.

Our key message is that the symbolic representations attached to these systems may thus have strong implications on their perpetuation and usage. Another important point highlighted by this study is the influence of various forces on the sedimentation of such symbolic realities within the organization. The identification of such forces may serve as a valuable guideline to IS practitioners in their attempts to influence the perpetuation and use of various information systems within the organization.

In closing, through its choice of symbolic interactionism as the cognitive lens, this study points to the importance of employing different theoretical perspectives (such as critical social theory, actor network theory, and structuration theory) to examine the ERP phenomenon. Indeed, we suggest that when such complementary perspectives are purposefully employed in a portfolio of separate studies over time, they may collectively help to shed new light on the complexities of ERP implementations in organizations.

REFERENCES

Alvarez, R. "Examining an ERP Implementation through Myths: A Case Study of a Large Public Organization," in H. M. Chung (Ed.), *Proceedings of the Sixth Americas Conference of Information Systems 2000*, Long Beach, CA, 2000, pp. 1655-1661.

Brodie, M., and Stonebraker, M. *Migrating Legacy System,* San Francisco: Morgan Kaufmann Publishers, 1995.

Feldman, S. P. "The Idealization of Technology: Power Relations in an Engineering Department," *Human Relations* (42), 1989, pp. 575-592.

Feldman, M. B., and March, J. G. "Information in Organizations as Signal and Symbol," *Administrative Science Quarterly* (26:2), 1981, pp. 171-186.

Fine, G. A. "Agency, Structure and Comparative Contexts: Towards a Synthetic Interactionism," *Symbolic Interaction* (15:1), 1992, pp. 87-107.

Hirschheim, R., and Newman, M. "Symbolism and Information Systems Development: Myth, Metaphor and Magic," *Information Systems Research* (2:1), 1991, pp. 29-62.

Holland, C. P., and Light, B. "Generic Information Systems Design Strategies," in W. D. Haseman and D. L. Nazareth, *Proceedings of the Fifth Americas Conference on Information Systems,* Milwaukee, WI, 1999, pp. 396-398.

Klaus, H., Rosemann, M., and Gable, G. G. "What is ERP?," *Information Systems Frontiers* (2:2), 2000, pp. 141-162.

Lindesmith, A. R. "Symbolic Interactionism and Causality," *Symbolic Interaction* (4:1), 1981, pp. 87-96.

Markus, M. L., and Tanis, C. "The Enterprise System Experience—From Adoption to Success," in R. W. Zmud (Ed.), *Framing the Domains of IT Management: Projecting the Future through the Past*, Cincinnati, OH: PinnaFlex Educational Resources, Inc., 2000, pp. 173-207.

Meyers, P. W., Sivakumar, K., and Nakata, C. "Implementation of Industrial Process Innovations: Factors, Effects and Marketing," *Journal of Product Innovation Management* (16:3), 1999, pp. 295-311.

O' Callaghan, A. "Migrating Large Scale Legacy Systems to Component-Based and Object Technology," *Communications of the Association for Information Systems* (2:3), 1999.

Prasad, P. "Symbolic Processes in the Implementation of Technological Change: A Symbolic Interactionist Study of Work Computerization," *Academy of Management Journal* (36:6), 1993, pp. 1400-1429.

Prus, R. *Symbolic Interaction and Ethnographic Research,* Albany, NY: State University of New York Press, 1996.

Ross, J. W., and Vitale, M. R. "The ERP Revolution: Surviving vs. Thriving," *Information Systems Frontier* (2:2), 2000, pp. 233-241.

Schönefeld, M. , and Vering, O. "Enhancing ERP-Efficiency through Workflow-Services," in H. M. Chung (Ed.), *Proceedings of Sixth Americas Conference on Information Systems,* Long Beach, CA, 2000, pp. 640-645.

See, C. P. N. "A Framework for Enterprise Resource Planning Maintenance and Upgrade Decisions," in D. Strong and D. Straub (Eds.), *Proceedings of Seventh Americas Conference on Information Systems*, Boston, MA, 2001, pp. 1026-1029

Themistocleous, M., and Irani, Z. "Benchmarking the Benefits and Barriers of Application Integration," *Benchmarking* (8:4), 2001, 317-331.

Turner, B. A. (Ed.). *Organizational Symbolism*, Berlin: Walter de Gruyter, 1990.

ABOUT THE AUTHORS

Martin M. T. Ng is an honors year undergraduate in the School of Computing at the National University of Singapore (NUS). Currently on a scholarship from a government authority in Singapore, he will join the industry in mid-2004. Martin can be reached at ngmongth@yahoo.com.sg.

Michael T. K. Tan is a doctoral candidate in the School of Computing at the National University of Singapore (NUS). He received his B.Sc. and M.Sc. degrees in computer science from this same university. Prior to commencing his doctoral studies, Michael accumulated practical experience in industry including stints as Information Systems Manager and Regional Industry Consultant (ERP). His research focuses on interorganizational systems, enterprise systems, strategic use of IT, management of IS and qualitative research methodologies. Michael can be reached at mtan@comp.nus.edu.sg.

50 DYNAMICS OF USE AND SUPPLY: An Analytic Lens for Information Systems Research

Jennifer Whyte
Imperial College London

Abstract This paper describes a dual focus on the dynamics of use and supply as an analytic lens for deconstructing and examining the IT artifact. It then gives an example of the use of this approach in empirical work. Over the last 20 years, most research has focused either on organizational use of information systems or on their development and supply. There is a need to bring these two pictures into combined focus to examine creation, definition, and modification of emergent systems. Building on work in the innovation studies tradition, an analytic lens is developed to examine the evolving IT artifact across the boundaries of use and supply. Its application is demonstrated in a study of virtual reality in the construction sector. Sensemaking in the software organizations that supply virtual reality applications to the sector and in the engineering, design, and construction organizations that use them is explored. The comparison exposed a mismatch of priorities, which poses a challenge to the establishment and continued validity of these applications. The paper concludes by reflecting on how this approach builds on, and may help us extend, existing theoretical understandings of information systems and how research that uses the approach may inform practice.

Keywords: Analytic lens, IT artifact, use, supply, virtual reality

1 INTRODUCTION

"IS is characterized by the problems we study rather than the theories we use," argues Paul (2002, p. 175). Others agree. There is current concern over the lack of interest in, or sophistication of analysis of, technologies within information systems in use (Cornfield 2003; Orlikowski and Iacono 2001). Within the North American research community this has led to renewed calls to focus information systems research on the IT artifact and its immediate antecedents and consequences (Benbasat and Zmud 2003). Yet this artifact has remained impervious to our attempts to scrutinize it.

When I reread papers from the 1984 IFIP conference (Mumford et al. 1985), I am amazed at the absence in the 1980s of technological context as a motivator for theorizing or as an actor in the debate about research method. Yet it is hard to deny the extent to which information systems and the structures of their supply and use have changed since that conference. There has been a decline in the use of the mainframe and a diffusion of personal computers, Windows operating systems and related word processing, spreadsheet, and Internet applications (Campbell-Kelly 2001; Mowery 1996). The personal computer application software industry has grown and matured to serve the needs of organizational users (Campbell-Kelly 2001) and there has been a shift away from application development as a local (in-house) team activity toward organizations purchasing expertise, consultancy, software, and services through the market (Cornfield 2003). Our theoretical understandings of information systems have remained largely blind to such changing technological, historical, and industrial contexts.

This theoretical blindness may be in part due to the focus of much empirical research and theorizing over the last 20 years. The majority of work in the IS literature has focused exclusively on organizational *use* of information technologies and information systems. From this body of work, we begin to develop a rich picture of the role of the information system in the modern enterprise. We benefit from theoretical developments such as structuration theory (Brooks 1997; DeSanctis and Poole 1994; Orlikowski 1992), and from detailed empirical studies that test and extend our theoretical understandings, often through a focus on a particular application (e.g. on computer aided design: Brooks 1997; Sohal 1997; Yetton et al. 1994). However, in my opinion, this work is stunted by a "black-box" view of the technology itself and an assumption that it is possible to disregard structures of supply. These are fair boundaries and assumptions when technology is mature, but huge failings when we are studying emergent technologies that are created, defined, and modified at the same time as being implemented and used.

A quite separate picture of information systems emerges through the academic work on the *development and supply* of information technologies and information systems. Sociologists of technology, such as Latour (1987) and Callon (1987) have directed our attention to the role of the research laboratory and the political and contested nature of technologies in the making. Their approaches have gained supporters within the IS community and have been applied to information systems (e.g., Bloomfield 1995; Monteiro and Hanseth 1996). Scholars within the innovation studies tradition have also typically concentrated on supply, but they (and their supporters within the IS community) direct attention to the evolving structure and dynamics of computer hardware and software industries (Hung 2003; Mowery and Langlois 1996; Steinmueller 1995) or conduct critical studies of individual software firms and their strategies (Lee 2000). While the sociologists of science and innovation studies scholars (and their respective supporters) have markedly different epistemological positions, what they share here is a concern with the dynamic processes through which technologies come into being.

Hence we have two separate approaches to information systems, with studies of use focusing on how new information technologies *come to be used* and studies of supply focusing on how new information technologies *come to be* (Orlikowski and Iacono 2001). Information systems researchers have not brought these separate pictures into combined focus to examine the way IT artifacts are created, defined, and modified in use over time. Yet, features of relevance to our core interest in the IT artifact lie across the interface between use and supply. As long as our research community focuses exclusively on use or exclusively on supply, then the true nature of the evolving IT artifact will slip through our grasp.

In this paper I describe a dual focus on the dynamics of use and supply as an analytic lens for deconstructing and examining the IT artifact. My interest is in emerging technology that is still technology in the making but is beginning to be tentatively implemented, played with, and used within organizations. I illustrate the methodological approach that I outline through empirical work conducted as an inquiry into emerging virtual reality (VR) use in the construction sector. I then draw conclusions and outline directions for future work.

2 BRINGING THE EVOLVING IT ARTIFACT INTO FOCUS

There is growing recognition that a sense of technology as infinitely malleable has led to a failure to recognize the specificities of particular applications and the constraints they might impose (Mutch 2002). Recent work has sought to shift the attention of the IS research community back to the IT artifact itself. This paper follows in this tradition, taking as its starting point the conceptualization given by Benbasat and Zmud (2003), who show the IT artifact at the core, surrounded by the task(s), then structure(s), and then context(s), as shown in Figure 1. I try to articulate components of the IT artifact, rather than treating it as a black-box. I discuss how a focus on the dynamics of use and supply could provide a means of keeping the nature of the IT artifact in focus in empirical work.

Although the conceptualization of Benbasat and Zmud is used as a spur to further theorizing, it is problematic. One issue is that it captures only a snapshot in time, presenting a static interpretation of the IT artifact and associated task(s), structure(s), and context(s). Another is that the definition has, I contend, an unsatisfactorily recursive nature. Benbasat and Zmud conceptualize the IT artifact as

> the application of IT to enable or support some task(s) embedded within a structure(s) that itself is embedded within a context(s). Here, the hardware/oftware design of the IT artifact encapsulates the structures, routines, norms and values implicit in the rich contexts *within which the artifact is embedded* (2003, p. 186).

By attempting to evade a narrow technology focus, I feel that Benbasat and Zmud overburden the idea of the IT artifact to the extent that technological aspects simply slip from view. The aim here is to develop and extend this line of theorizing, addressing some of the weaknesses. In the following subsection, I articulate different components of a modern information system; I then discuss the dynamics of use and supply as an analytic lens for information systems research.

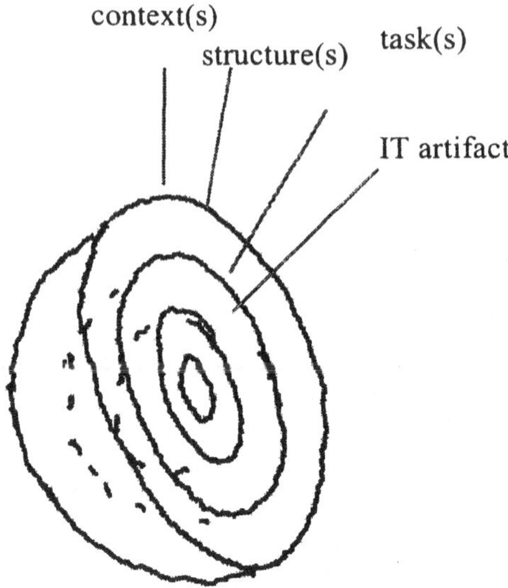

Figure 1. Context(s); Structure(s), Tasks(s) and the IT Artifact in Use, Described as Concentric Rings by Benbasat and Zmud (2003)

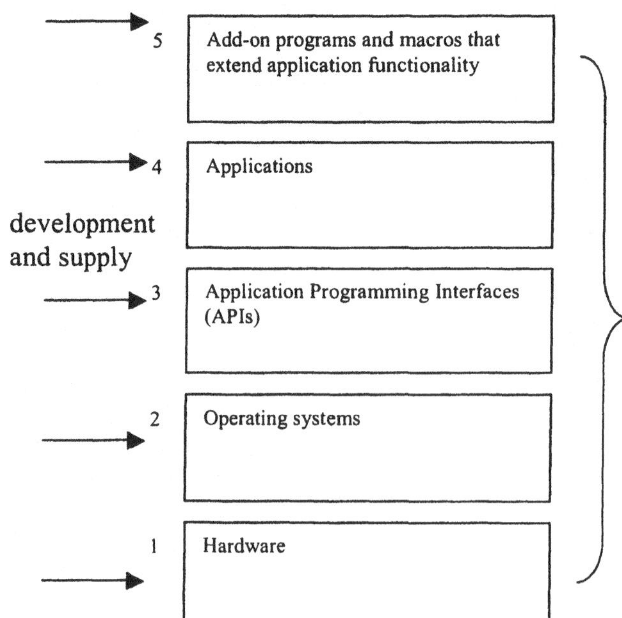

Figure 2. Articulation of Different Technological Levels Within Complex Information Systems

2.1 Articulating the Structure of the IT Artifact

Information systems are complex products. Many firms are involved in the production of systems and subsystems. Figure 2 describes a five level model of the generic (contemporary) information system as composed of (1) hardware; (2) operating system; (3) application programming interfaces; (4) applications; and (5) add-on programs, plug-ins, and macros that extend functionality. Such a model is historically and culturally situated. It not untypical of basic computing texts, but it is useful to IS researchers as it has a number of implications.

First, it allows us to be clearer about our technological unit of analysis. Much IS research is actually mainly concerned with subsets of the overall information system, such as, for example, work that focuses at level 4 on the organizational use of applications (e.g., studies of CAD implementation given earlier), with only marginal interest in the use of their add-on programs, plug-ins, and macros; or the underlying application programming interfaces, operating systems, or hardware.

Second it allows us to explain how developments at the different technological levels may be occurring at different speeds, and are driven by different sets of users and suppliers. Suppliers of a particular solution at the application level, for example, may treat the available application programming interfaces as a given. They may be surprised when competing application programming interfaces emerge that serve the needs of other application suppliers.

2.2 Dynamics of Use and Supply

Researchers in the innovation studies field have looked at the introduction of new technologies from a systemic perspective. I build on a tradition of work in this field which looks at both production and use (Lundvall 1992, 2003) to suggest that a focus on the dynamics of use and supply may be helpful as an analytic lens for IS research. Here supply is looked at from the perspective of use and use is looked at from the perspective of supply. To return for a moment to Benbasat and Zmud's suggestive diagram, this could be seen to be expanded as shown in Figure 3.

My concern here is to focus on the evolving IT artifact through a comparison and contrast of the contexts of use and supply. Like Caelli et al. (2003), I use the term *analytic lens* to refer to the methodological and interpretive presuppositions that a researcher brings to bear on his or her data. In describing a dual focus on the dynamics of use and supply as an analytic lens, I am using dualism as an analytic device to compare and contrast characteristics and explore interrelationships between use and supply. In the next section I provide an example of empirical work in which this approach has been used.

3 VIRTUAL REALITY IN CONSTRUCTION

This section describes the use of a dual focus on use and supply to interpret a study of VR in the construction sector. The study was conducted using an embedded multiple case study method, coupled with historical study of the development and status of VR technologies.

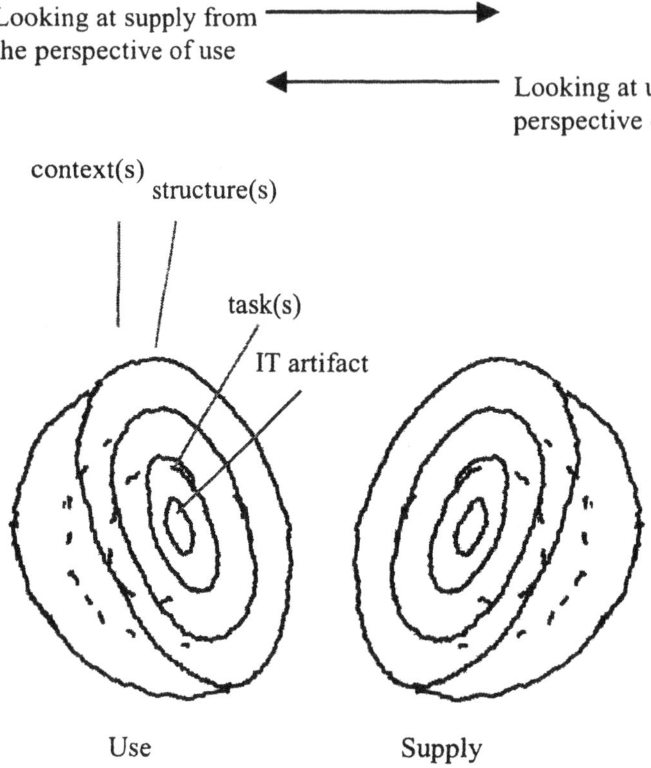

Figure 3. Context(s), Structure(s), Tasks(s), and the IT Artifact in Both Use and Supply

I was motivated to undertake this work as, in the academic literature of the late 1990s, VR was being portrayed as an important new technology awaiting a dominant design (Swann and Watts 2000; Watts et al. 1998). A government initiative was aimed at promoting its use within UK industry (DTI 2000) and this highlighted construction as a key sector for use. However, I felt that there was a thin evidential basis for such conclusions and policies. In my research I wanted to collect data on, analyze, and assess the success or failure of organizational uses of VR applications within engineering, design, and construction firms and also the potential of VR suppliers to address their needs.

The construction sector is highly fragmented with a large number of firms involved in the design, production, and maintenance of the built environment. Characteristics of the construction process which affect the nature of innovation include its project-based nature (Gann and Salter 2000), the unusual extent to which it is regulated (Pries and Janszen 1995), and the collaborative and competitive relationships between different players in supplier networks—manufacturers, designers, specialists, contractors, subcontractors, developers, users, professional industry associations, and regulators—within which changes occur (Gann 2000).

The technological context was explored through participation in the activities of the VR and visualization community in the UK and internationally (from 1997 through 2003) and desk-based research to study the historical development of graphics technologies, including the array of technologies that underlie CAD and VR applications. Case-study research was conducted in 11 user organizations and 6 suppliers in 2000-01.

Further details of the industrial context, data-collection methods, and validation have been reported elsewhere (Whyte 2003), where there is also a fuller description of findings. The focus here is on the analysis and interpretation of the findings using dynamics of use and supply as an analytic lens for decomposing and studying the IT artifact. In the following sections I describe the technological context, and then summarize the salient features of use and supply before comparing, contrasting, and exploring the role of interrelationships between use and supply in establishing legitimacy of an emerging technology.

3.1 Emergent Uses in the Construction Sector

The user organizations studied ranged from an architectural practice with low use of VR at the customer interface, to a major real estate owner that had developed a real-time interactive 3D application for viewing construction scheduling and CAD data, to a consultant engineering and project management practice that had established visualization groups in both the United States and the United Kingdom. The organizations with the largest investment in virtual reality were consultant engineers, contractors, and real estate owners. These organizations use virtual reality for systems integration, simulating dynamic operation, and coordination of detail design within the project team and supply chain. They are concerned with reducing risk, increasing technological innovation and improving business processes. Some, like the consultant engineering and project management practice mentioned above, employed specialist modelers to create and maintain virtual reality models over the life of a project, particularly on large and complex building and infrastructure projects such as railways, airports, shopping malls, and theme parks.

A wider group of organizations use virtual reality for interactions with end-users, clients, managers, funding institutions, and planners. The models used for these tasks may be quite different in nature from those used by construction professionals. Here VR was being used to offer the customer a greater understanding of the design and a limited design choice from a palate of options.

All of the user organizations using VR within construction-related production processes were found to have a pattern of requirements that included the ability to transfer data to VR packages from CAD packages, in which the majority of their design work took place. Virtual reality is also being used by architects to develop new markets outside the construction sector. This brings architects (reconceived as cyberspace architects or architects of the physical and virtual realm) into competition with Web designers, human-computer interaction experts, and programmers.

5	Add-on programs and macros that extend application functionality	**Plug-in visualization modules in CAD, GIS, and engineering analysis tools**	
4	Applications—VR packages	**New applications by specialist suppliers** ←	VR applications by established suppliers
3	Application Programming Interfaces (APIs)—3D Graphic Engines	<- VRML, Open GL, Open Inventor -> Open 3D	Iris Performer
2	Operating systems	Windows	UNIX
	Hardware	Personal Computers	Workstations and bespoke hardware solutions

Figure 4. Different Technological Components in VR Systems

3.2 Dynamics and Structure of Supply

The companies studied included three types of suppliers: (1) established VR suppliers, (2) specialist VR suppliers that have knowledge of construction practices, and (3) resellers and model builders. In this section I will concentrate on the first two types.

Both the users and suppliers struggled to make sense of competing technologies at different technological levels and Figure 4 illustrates this. The established suppliers are shown as providers of applications for UNIX-based workstations and bespoke hardware, which up until the late 1990s was the sole focus of their activity. At this time, developments in the PC market led a number of specialist suppliers to recognize the market opportunity and to deliver interactive real-time 3D applications for specific construction process; and also led established VR suppliers to make software available on the PC platform. The established VR suppliers have then subsequently developed collaborations with leading CAD, graphics information systems (GIS), and analysis packages to deliver their solutions as visualization plug-ins to standard packages.

One of the case studies was of a major VR supplier that had about 15years experience developing modeling and visualization software. Up until four years previously, everything this company had produced had been designed to run on SGI hardware as that was the only thing that served the needs of the visual simulation community.

> *Now today we can take the same techniques that we do on the high end and deliver them on common PC hardware which is directly attributable to all of the wonderful advances in processing speed but more importantly graphic*

card architecture and that is being driven...by kids! The gaming industry, I mean it's wonderful!

For the generic supplier, the construction users were a small minority group, particularly in terms of their buying capacity. They had a set of demanding requirements that did not lie comfortably within the strategic direction of these supplier companies. The suppliers were looking to diversify as the market for applications on bespoke hardware was now in decline even for the demanding real-time 3D applications that these suppliers delivered. They perceived a need to port their existing applications to new hardware platforms and operating solutions and to widen their customer base.

The specialist suppliers studied, while using broadly the same set of technologies, do not use the term *virtual reality* in their own marketing literature, preferring terms such as 4D-CAD and interactive design review. Generic suppliers and the specialist new entrants also had a very different understanding of the relationship between CAD and VR. The company described above had a background in military training applications, particularly flight simulation, and found the engineering, design, and construction firms' need for data exchange with CAD particularly problematic. The interviewee described their experience of the practices of construction users,

When they turn on the computer they are turning on their CAD program...and CAD programs and virtual reality sometimes don't mix well, at least from our perspective they don't because we are into real-time visualization and that's a whole very focused discipline in 3D visualization.

In contrast, a specialist supplier argued, "*We are not replacing the CAD engineers. We need all that source data.*"

3.3 Interrelationships between Use and Supply

It is only through an analysis of both use and supply that a mismatch in priorities becomes apparent between construction sector users, for whom data exchange with CAD is of critical importance, and the generic VR supply industry, for which such data exchange is peripheral to use. This mismatch of priorities poses a challenge to the establishment and continued validity of these applications within construction. The construction sector is seen as a difficult sector by generic VR suppliers, and with a more recent increase in United States funding for military urban simulation applications, they may have become even less relevant to these suppliers, even in their urban simulation divisions.

The analysis has implications for the central concerns of WG 8.2 in research on organizational uses of information systems. For example Orlikowski (2000) describes users' interaction with a technology as recursive, arguing that in recurrent practice, users shape the technology structures that shape their use. This work presents a challenge to such a view, arguing that it shows only part of the picture. The ability of user organizations to influence adaptation of technologies is constrained by the priorities of other users at the level of the multi-technology product and at the level of underlying technologies.

Virtual reality applications were not easily established within the construction sector. The vision of the future held by policy makers and academics in the late 1990s did not come to pass. There are questions to be asked about why it didn't, and about the future of interactive real-time 3D applications in construction. Generic VR suppliers and resellers conceived of virtual reality as an entirely separate application creating a sense of presence, but as we have seen for construction industry users, the access to engineering data and connectivity of CAD and VR was a major issues. One possible future would be for the technologies underlying VR to be used in plug-ins to major CAD packages. Indeed I see evidence that CAD packages are both developing advanced 3D graphics capabilities themselves and partnering with suppliers of VR software. These findings raise questions about how policy can or whether it should try to help foster particular visions of the future.

4 CONCLUSIONS

This paper is part of a shift in the attention of the IS research community back to the IT artifact itself. It argues that organizational use of technology is shaped and constrained by structures and dynamics that exist across the boundaries of use and supply. The analytic lens proposed examines interrelationships at this interface as a means of deconstructing and examining the evolving IT artifact as it begins to be implemented and used. I argue that this allows us to better understand the IT artifact at a range of different technological levels.

This has implications for the type of research methods used in empirical work. Using this approach, it is important that the researcher explores both structures of dynamics of use and supply. However, information systems are complex, and there are multiple boundaries between users and producers. Suppliers of multi-technology applications act as users of application programming interfaces as well as suppliers of applications to a range of industrial sectors and processes (where their users may have conflicting requirements). End-user organizations struggle to make sense of the different but often overlapping configurations of underlying technologies supplied to them in multi-technology packages. When engaged in such research, it is important to be clear about the unit of analysis and to this end I articulate different parts of the information system. In the study of virtual reality in the construction sector, I look primarily at the users and producers of virtual reality *applications*, but am able to articulate the impact of technological change at a range of different levels.

The analytic approach proposed would benefit from further theoretical and empirical work. It builds on work in the innovation studies tradition that has looked at complex product systems, and there would be merit in comparing and contrasting emerging patterns of development of information systems with historical patterns observed for other technological systems.

REFERENCES

Benbasat, I., and Zmud, R. W. "The Identity Crisis Within the IS Discipline: Defining and Communicating the Discipline's Core Properties," *MIS Quarterly* (27:2), 2003, pp. 183-194.

Bloomfield, B. "Power, Machines and Social Relations: Delegating to Information Technology in the National Health Service," *Organization* (2:3/4), 1995, pp. 489-518.

Brooks, L. "Structuration Theory and New Technology: Analyzing Organizationally Situated Computer-Aided Design," *Information Systems Journal* (7), 1997, pp. 133-151.

Caelli, K.; Ray, L.; and Mill, J. "'Clear as Mud': Toward Greater Clarity in Generic Qualitative Research," *International Journal of Qualitative Methods* (2:2), 2003.

Callon, M. "Society in the Making: The Study of Technology as a Tool for Sociological Analysis," in T. P. Pinch (Ed.), *The Social Construction of Technological Systems*, Cambridge, MA: The MIT Press, 1987, pp. 85-103.

Campbell-Kelly, M. "Not Only Microsoft: The Maturing of the Personal Computer Software Industry, 1982-1995," *Business History Review* (75), 2001, pp. 103-145.

Cornfield, K. G. "Information Systems and New Technologies: Taking Shape in Use," Working Paper Series 122, Department of Information Systems, London School of Economics and Political Science, 2003.

DeSanctis, G., and Poole, M. S. "Capturing the Complexity in Advanced Technology Use: Adaptive Structuration Theory," *Organization Science* (5:2), 1994, pp. 121-147.

DTI. *UK Business Potential for Virtual Reality*, London: Department of Trade and Industry, 2000.

Gann, D. *Building Innovation: Complex Constructs in a Changing World*, London: Thomas Telford, 2000.

Gann, D. M., and Salter, A. J. "Innovation in Project-Based, Service-Enhanced Firms: The Construction of Complex Products and Systems," *Research Policy* (29:7/8), 2000, pp. 955-972.

Hung, S.-C. "The Taiwanese System of Innovation in the Information Industry," *International Journal of Technology Management* (26:7), 2003, pp. 788-800.

Latour, B. *Science in Action: How to Follow Scientists and Engineers Through Society*, Boston: Harvard University Press, 1987.

Lee, S. Y. T. "Bundling Strategy in Base-Supplemental Goods Markets: The Case of Microsoft," *European Journal of Information Systems* (9:4), 2000, pp. 217-225.

Lundvall, B.-Å. (Ed.). *National Systems of Innovation*, London: Pinter, 1992.

Lundvall, B.-Å. *Product Innovation, Markets and Hierarchies*, Copenhagen: DRUID, 2003.

Monteiro, L., and Hanseth, O. "Social Shaping of Information Infrastructure: On Being Specific About the Technology," in W. J. Orlikowski, G. Walsham, M. R. Jones, and J. I. DeGross (Eds.), *Information Technology and Changes in Organizational Work*, London: Chapman & Hall, 1996, pp. 325-348.

Mowery, D. C. (Ed.). *The International Computer Software Industry: A Comparative Study of Industry Evolution and Structure*, Oxford: Oxford University Press, 1996.

Mowery, D. C., and Langlois, R. N. "Spinning Off and Spinning On: The Federal Government Role in the Development of the US Computer Software Industry," *Research Policy* (25:6), 1996, pp. 947-966.

Mumford, E.; Hirschheim, R.; Fitzgerald, G.; and Wood-Harper, A. T. (Eds.). *Research Methods in Information Systems*. Amsterdam: North-Holland, 1985.

Mutch, A. "Actors and Networks or Agents and Structures: Towards a Realist View of Information Systems," *Organization* (9:3), 2002, pp. 477-496.

Orlikowski, W. J. "The Duality of Technology: Rethinking the Concept of Technology in Organizations," *Organization Science* (3:3), 1992, pp. 398-427.

Orlikowski, W. J. "Using Technology and Constituting Structures," *Organization Science* (11:4), 2000, pp. 404-428.

Orlikowski, W. J., and Iacono, C. S. "Desperately Seeking the 'IT' in IT Research—A Call to Theorizing the IT Artifact," *Information Systems Research* (12:2), 2001, pp. 121-134.

Paul, R. J. "$(IS)^3$: Is Information Systems an Intellectual Subject?," *European Journal of Information Systems* (11:2), 2002, pp. 174-177.

Pries, F., and Janszen, F. "Innovation in the Construction Industry, the Dominant Role of the Environment," *Construction Management and Economics* (13:1), 1995, pp. 43-51.

Sohal, A. "A Longitudinal Study of Planning and Implementation of Advanced Manufacturing Technologies," *International Journal of Computer Integrated Manufacturing* (10:1), 1997, pp. 281-295.

Steinmueller, W. E. "The U.S. Software Industry: An Analysis and Interpretive History," in D. C. Mowery (Ed.), *The International Computer Software Industry*, Oxford: Oxford University Press, 1995.

Swann, G. M. P., and Watts, T. P. "Visualization Needs Vision: The Pre-Paradigmatic Character of Virtual Reality," in S. Woolgar (Ed.), *Virtual Society? Technology, Cyberbole, Reality*, Oxford: Oxford University Press, 2000.

Watts, T.; Swann, G. M. P.; and Pandit, N. R. "Virtual Reality and Innovation Potential," *Business Strategy Review* (9:3), 1998, pp. 45-54.

Whyte, J. K. "Innovation and Users: Virtual Reality in the Construction Sector," *Construction Management and Economics Special Issue on Innovation in Construction* (21:6), 2003, pp. 565-572.

Yetton, P. W.; Johnston, K. D.; and Craig., J. F. "Computer-Aided Architects: A Case Study of IT and Strategic Change," *Sloan Management Review* (35:4), 1994, pp. 57-67.

ABOUT THE AUTHOR

Jennifer Whyte is a research fellow at the Tanaka Business School, Imperial College London. She works within the Innovation Studies Centre and has particular interest in the development of digital tools and technologies for design and their uses within organizations. Her research is conducted in collaboration with leading firms in the UK and USA. She has published widely in journals and a book, with ongoing work looking at virtual prototyping in engineering practice. Jennifer can be reached at j.whyte@imperial.ac.uk.

51 APPLYING ADAPTIVE STRUCTURATION THEORY TO THE STUDY OF CONTEXT-AWARE APPLICATIONS

Carl Magnus Olsson
Viktoria Institute

Nancy L. Russo
Northern Illinois University

Abstract Adaptive structuration theory (AST) has been used for a number of years in the information systems discipline to study the use of new technologies in organizations. In this paper it is applied to a relatively new technology, context-aware applications. AST provides a useful lens for examining the impact of a particular context-aware application, CABdriver. Used in conjunction with the repertory grid technique and lead users in an action research study, a research approach for exploring the ways in which the technology impacts individuals within small groups is presented.

Keywords: Action research, adaptive structuration theory, context-aware applications, lead users, repertory grid technique

1 INTRODUCTION

Whereas some strides, particularly in the diversity of research approaches, have been made since the 1984 Manchester Colloquium (Mumford et al. 1985), some things have not changed. It is still true that "The rate at which information technology is being introduced in our institutions exceeds our capacity to generate knowledge about its effects and meanings" (Espejo 1985, p. 269). For example, the contexts for which we develop information systems have changed considerably over the past 20 years. Mobile and ubiquitous information environments place the interaction between the user and the technology into a much broader environment, requiring us to address systems that are integrated into everyday activities through mobile phones, PDAs, and other handheld devices, as well as wearable computing devices (Lyytinen and Yoo 2002).

This paper addresses the evaluation of a particular aspect of ubiquitous information environments. Context-aware applications (CAAs) typically use location-based data and other contextual information to trigger predefined behavior (Schmidt et al. 1998). First generation CAAs included the notification agent, the meeting reminder agent, and call-forwarding applications (Dey et al. 2001). Active badge systems (Weiser 1991) which can track wearers through particular locations, and global positioning systems that use databases of local information to provide details such as restaurant locations are other examples of CAAs.

In the same sense that DeSanctis and Poole (1994, p. 126) argue that "advanced information technologies have greater potential than traditional business computer systems to influence the social aspects of work," advanced CAAs have an even greater potential to influence the social aspects of everyday life. As Lyytinen and Yoo (2002) suggest, there are unique challenges posed by this type of system, in terms of design, use, and impact. In this environment, the technical and social aspects cannot be separated easily. There is a need for new ways of thinking about design, use, and impact, and new research approaches to study this type of system. As Galliers (1985, p. 281) reflected after the 1984 Colloquium, and is just as true today, we are faced with "an urgent need to develop our understanding of appropriate approaches to research in the IS field in a variety of different circumstances and with a variety of different objectives in mind."

This paper is part of a larger action research study exploring impact and methods for studying CAAs. To allow for a more targeted exploration of a proposed research approach for CAAs, a particular CAA is used. CABdriver (Context-Aware Backseat Driver) is a handheld interactive in-car concept currently in its first implemented version. The research approach we are taking is one that integrates the technical and social aspects of evaluating the use of a CAA. Although there are a number of approaches that fit this criterion, the approach selected for this analysis is adaptive structuration theory (AST). Lead users and repertory grids are integrated with the approach to provide early assessment of participants' perceptions of the evolving application concept.

2 CABdriver

Development of the CABdriver concept was initiated in January 2003, as a joint project with Saab, Mecel, and Vodafone. This first implementation of the CABdriver concept is a handheld game which uses contextual information to influence the game, thereby spawning social interaction between the player and the driver directly related to the changing context. The game is affected by points of interest (i.e., gas stations, tourist attractions, parking lots), traffic messages (i.e., accidents, road conditions, animal warnings), speed limit, actual car speed, fuel consumption, and driver work-load (a combination of anti-spin control and electronic stability program activity, braking, hard acceleration, navigation system indicating a turn is close-by, and recent use of the turn signal).

In this paper, as well as in the broader research project, CABdriver is used as a tangible example of a context aware application which can be used to explore the applicability of a particular evaluation approach. This approach is described in the following section.

3 RESEARCH APPROACH

Traditionally, IS research has focused on assessing the impact of deployed technology, primarily within organizational settings. However, the contexts for which we develop information systems today increasingly consist of mobile and ubiquitous information environments, putting the interaction between the user and the technology into a much broader perspective. Rather than waiting for this area to develop independently from IS, IS researchers should establish an active role in identifying the methods and approaches used to study the technology itself and its impact (Lyytinen and Yoo 2002; Orlikowski and Iacono 2001).

We are responding to this by exploring how an adaptation of action research (Baskerville and Wood-Harper 1996; Susman and Evered 1978), AST (DeSanctis and Poole 1994), repertory grids (Kelly 1955; Tan and Hunter 2002), and lead users (Von Hippel 1988) can help assess the impact of a CAA. Table 1 contains a summary of the characteristics of the research approach elements.

The representative features of action research that have so far come into play are (1) the collaborative initiative CABdriver, (2) highly involved researchers, and (3) the use of theory as a guide. As the larger study progresses, more detailed comments on the action research approach can be provided.

Out of several possible theoretical lenses, we are using AST (DeSanctis and Poole 1994) as it provides us with a rich description of the appropriation of technology and the impact this has. AST focuses not only on the social impact, but also on how technology itself affects this.

Table 1. Summary Research Approach Elements

	Characteristics	Selected References
Action research	• Involved researchers • Collaborative • Using theory as a guide • Continuous reevaluation of theory • Continuous reevaluation of application	Susman and Evered (1978) Baskerville and Wood-Harper (1996)
AST	• Social and technical aspects • Appropriation process • Impact of technology • Predictive and explanatory	DeSanctis and Poole (1994)
Repertory grids	• Qualitative and quantitative aspects • Capturing cognition	Kelly (1955) Tan and Hunter (2002)
Lead users	• Early assessment of application	Von Hippel (1988)

Our use of AST differs somewhat from the traditional use of AST. First, our two-month evaluation of the everyday use of CABdriver is not as long as traditional AST studies. Because we are evaluating a technology at an early stage, the amount of time spent on evaluation must be controlled. Second, this first implementation of the CABdriver concept is designed to directly influence one user, rather than the common case of multiple users of groupware systems. Still, the game is not a solitary game as it has been designed to be virtually impossible to play without interacting with and influencing the driver. Analysis of appropriation at this microlevel is something DeSanctis and Poole (1994, p. 133) identify as a logical starting-point "since it is in specific instances of discourse that the formation of new social structures begins." Third, AST has previously been used primarily in organizational settings, rather than in an everyday situation as is the case with CABdriver. Fourth, we are bringing the technology to the user and group, rather than studying an already existing technology. This means that we, as researchers and participants in the development, have strong structural features we hope to see, which thus forces us to take extra care to not overlook results that do not conform to our expectations.

In an effort to minimize researcher influence, we have incorporated the repertory grid technique (Kelly 1955; Tan and Hunter 2002) from personal construct theory (Kelly 1955) to assist in eliciting participants' personal constructs (also known as cognitive maps, technological frames, or mental models) related to AST. In addition, lead users (Von Hippel 1988) are included in the approach as they enable the CAA to be tested for design flaws prior to the main evaluation, something which is particularly vulnerable to neglect when researchers are involved in the development of the application.

4 THE CABdriver SETTING

Our initial strategy for evaluating CABdriver included three data collection points. At each point, we intended to individually interview each participant using repertory grids complemented by neutral probing questions (Reynolds and Gutman 1988). However, when applying this strategy to the lead user evaluation of CABdriver, it became evident that the strategy was too demanding for the participants, as each grid corresponding to particular constructs of the AST framework required close to one hour to complete. Based on this experience, the main evaluation of CABdriver instead uses two data collection points—one midway through and one after the evaluation period. Particular emphasis is put on the midway evaluation as the participants at that time have used the application long enough to have in-depth questions and experiences they can discuss with the researchers in order to better understand and further evaluate the application. Individual repertory grids capture how the task and environment are perceived by the participants while the appropriation part of AST is captured using qualitative (group) interviews (Patton 2002; Wolcott 2001). Post-evaluation, participants are asked to answer an in-depth open-ended questionnaire based on the results from the midway interview sessions. Through this two step process, initial results outlining the appropriation process and impact of CABdriver from a two month evaluation can be analyzed. Furthermore, participants are queried about their experiences from being subjected to the research approach. Together with the researchers' experiences, this enables the approach to be reevaluated for future improvement.

Table 2. Lead User Assessments Using Repertory Grids

	Core Assumptions	Insights from the CABdriver Case
Action research	• IS researchers need to take an active stance in the emerging area of context-aware applications.	• Too early in the study to discuss.
AST	• AST can provide a rich description of important aspects to evaluate for emerging technologies.	• The rich description of AST provides explanatory help rather than predictive.
Repertory grids	• The repertory grid technique can reduce the influence of preconceptions from strongly involved researchers.	• Repertory grids present participants with a valuable record of the interview session, enabling reconsideration and further explanation on issues often taken for granted. • Repertory grids present researchers with an effective lens for approaching recorded interview sessions.
Lead users	• Lead users can effectively test an application concept for design oversights prior to the main evaluation.	• Applying new or adapted research approaches to lead users prior to the main evaluation is an effective way to test the intended evaluation strategy.

As this paper focuses on the time *prior* to the main evaluation, this work does not capture the experiences from the two month evaluation of CABdriver, only the assessments made from applying the approach to lead users of the application. The contribution of action research is, therefore, difficult to discuss at this stage, and we remain neutral to this particular method. The structured approach provided by AST appears to be useful as a guideline, but will be difficult to follow strictly. This implies that the explanatory powers of AST are likely more useful than any predictive powers. Unfortunately, it will be impossible to analyze conversations between participants (as is common in AST) due to the difficulties of continuously recording families for two months without imposing too much on their privacy and the everyday situation. Instead, repertory grids in combination with open-ended, semi-structured interviews and questionnaires based on the constructs of AST appear to be a useful substitute.

As indicated by Table 2, lead user assessments of CABdriver using repertory grids provided two unexpected benefits aside from acting as a tool to capture cognition: (1) during the interviews, participants found creating grids a great support for reflecting, comparing, and discussing issues they otherwise seldom considered, and (2) using the grids as an initial filter for how to approach the recorded sessions was highly effective and likely reduced the amount of researcher bias.

Finally, although the role of lead users was originally defined as testing the application for conceptual design flaws, the lead users also provided a valuable assessment of the data collection strategy, in our case resulting in a redesign.

5 CONCLUSION

In this paper we have presented and briefly discussed the results from applying an adaptation of AST to an action research study, using the repertory grid technique and lead users. At this point (i.e., prior to the main evaluation of our CAA), we remain neutral to action research and AST as more data is needed to make a qualified discussion of their impact on the study. However, we strongly recommend using repertory grids and lead users to simultaneously assess evolving applications and research methods.

The overall project of studying CABdriver is still in the early stages. In order to continue the task of establishing an active role for IS researchers in the emerging mobile and ubiquitous information environments, additional implementations, theories, and evaluation approaches will be examined for suitability of use in this area.

REFERENCES

Baskerville, R. L., and Wood-Harper, A. T. "A Critical Perspective on Action Research as a Method for Information Systems Research," *Journal of Information Technology* (11), 1996, pp. 235-246.

DeSanctis, G., and Poole, M. S. "Capturing the Complexity in Advanced Technology Use: Adaptive Structuration Theory," *Organization Science* (5:2), 1994, pp. 121-147.

Dey, A. K.; Abowd, G. D.; and Salber, D. "A Conceptual Framework and a Toolkit for Supporting the Rapid Prototyping of Context-Aware Applications," *Human-Computer Interaction* (15), 2001, pp. 97-166.

Espejo, P. "Comment," in E. Mumford. R. Hirschheim, G. Fitzgerald, and A. T. Wood-Harper (Eds.), *Research Methods in Information Systems*, Amsterdam: North Holland, 1985, pp. 267-269.

Galliers, R. D. "In Search of a Paradigm for Information Systems Research," in E. Mumford. R. Hirschheim, G. Fitzgerald, and A. T. Wood-Harper (Eds.), *Research Methods in Information Systems*, Amsterdam: North Holland, 1985, pp. 281-297.

Kelly, G. *The Psychology of Personal Constructs*, Volumes 1 and 2, London: Routledge, 1955.

Lyytinen, K., and Yoo, Y. "Research Commentary: The Next Wave of *Nomadic* Computing," *Information Systems Research* (13:4), 2002, pp. 377-388.

Mumford, E., Hirschheim, R., Fitzgerald, G., and Wood-Harper, A. T. *Research Methods in Information Systems*, Amsterdam: North-Holland, 1985.

Orlikowski, W. J., and Iacono, C. S. "Research Commentary: Desperately Seeking the 'IT' in IT Research: A Call to Theorizing the IT Artifact," *Information Systems Research* (12:2), 2001, pp. 121-134.

Patton, M. Q. *Qualitative Research and Evaluation Methods* (3rd ed.), London: Sage Publications, 2002.

Reynolds, T. J., and Gutman, J. "Laddering Theory, Method, Analysis, and Interpretation," *Journal of Advertising Research*, February-March 1988, pp. 11-31.

Schmidt, A.; Beigl, M.; and Gellersen, H-W. "There Is More to Context Than Location," *Computers and Graphics* (23), 1999, pp. 893-901.

Susman, G., and Evered, R. "An Assessment of the Scientific Merits of Action Research," *Administrative Science Quarterly* (23), December 1978, pp. 582-603.

Tan, F. B., and Hunter, M. G. "The Repertory Grid Technique: A Method For The Study of Cognition In Information Systems," *MIS Quarterly* (26:1), 2002, pp. 39-57.
Von Hippel, E. *The Sources of Innovation*, New York: Oxford University Press, 1988.
Weiser, M. "The Computer for the 21st Century," *Scientific American*, September 1991, pp. 94-104.
Wolcott, H. F. *Writing Up Qualitative Research*, London: Sage Publications, 2001.

ABOUT THE AUTHORS

Carl Magnus Olsson is a Ph.D. student at the Department of Informatics at Gothenburg University, Sweden. He conducts his research in the Telematics Group of the Viktoria Institute in Sweden and spends part of his time as a guest researcher at the Operations Management and Information Systems Department at the College of Business of Northern Illinois University, USA. His research is situated within ubiquitous information environments and investigates how context-aware applications may be evaluated and the impact these applications have on an individual and small group level. Prior to taking his position at the Viktoria Institute in September 2002, he worked at a computer consultancy company focusing on governmental agencies in Sweden and multinational engineering and telecommunications companies. He can be reached at cmo@viktoria.se.

Nancy L. Russo received her Ph.D. in Management Information Systems from Georgia State University in 1993. Since 1991, she has been a member of the Operations Management and Information Systems Department at Northern Illinois University. During 1998, she was a visiting professor in the Department of Accounting, Finance and Information Systems at University College Cork, Ireland. Dr. Russo is currently the Chair of the OMIS Department. In addition to on-going studies of the use and customization of system development methods in evolving contexts, her research has addressed web application development, the impact of enterprise-wide software adoption on the IS function, IT innovation, research methods, and IS education issues. Her work has appeared in *Information Systems Journal*, *Journal of Information Technology*, *Information Technology & People*, *Communications of the ACM*, and other publications. Dr. Russo serves as secretary of the International Federation of Information Processing Working Group 8.2 on Information Systems and Organizations. She can be reached at nrusso@niu.edu.

INDEX OF CONTRIBUTORS